T0141799

Advances in Intelligent and Soft Computing 135

Editor-in-Chief: J. Kacprzyk

Advances in Intelligent and Soft Computing

Editor-in-Chief

Prof. Janusz Kacprzyk
Systems Research Institute
Polish Academy of Sciences
ul. Newelska 6
01-447 Warsaw
Poland
E-mail: kacprzyk@ibspan.waw.pl

Further volumes of this series can be found on our homepage: springer.com

Vol. 119. Tianbiao Zhang (Ed.)
Future Computer, Communication, Control and Automation, 2011
ISBN 978-3-642-25537-3

Vol. 120. Nicolas Loménie, Daniel Racoceanu, and Alexandre Gouaillard (Eds.)
Advances in Bio-Imaging: From Physics to Signal Understanding Issues, 2011
ISBN 978-3-642-25546-5

Vol. 121. Tomasz Traczyk and Mariusz Kaleta (Eds.)
Modeling Multi-commodity Trade: Information Exchange Methods, 2011
ISBN 978-3-642-25648-6

Vol. 122. Yinglin Wang and Tianrui Li (Eds.)
Foundations of Intelligent Systems, 2011
ISBN 978-3-642-25663-9

Vol. 123. Yinglin Wang and Tianrui Li (Eds.)
Knowledge Engineering and Management, 2011
ISBN 978-3-642-25660-8

Vol. 124. Yinglin Wang and Tianrui Li (Eds.)
Practical Applications of Intelligent Systems, 2011
ISBN 978-3-642-25657-8

Vol. 125. Tianbiao Zhang (Ed.)
Mechanical Engineering and Technology, 2012
ISBN 978-3-642-27328-5

Vol. 126. Khine Soe Thaung (Ed.)
Advanced Information Technology in Education, 2012
ISBN 978-3-642-25907-4

Vol. 127. Tianbiao Zhang (Ed.)
Instrumentation, Measurement, Circuits and Systems, 2012
ISBN 978-3-642-27333-9

Vol. 128. David Jin and Sally Lin (Eds.)
Advances in Multimedia, Software Engineering and Computing Vol.1, 2011
ISBN 978-3-642-25988-3

Vol. 129. David Jin and Sally Lin (Eds.)
Advances in Multimedia, Software Engineering and Computing Vol.2, 2011
ISBN 978-3-642-25985-2

Vol. 130. Kusum Deep, Atulya Nagar, Millie Pant, and Jagdish Chand Bansal (Eds.)
Proceedings of the International Conference on Soft Computing for Problem Solving (SOCPROS 2011) December 20–22, 2011, 2012
ISBN 978-81-322-0486-2

Vol. 131. Kusum Deep, Atulya Nagar, Millie Pant, and Jagdish Chand Bansal (Eds.)
Proceedings of the International Conference on Soft Computing for Problem Solving (SocProS 2011) December 20–22, 2011, 2012
ISBN 978-81-322-0490-9

Vol. 132. Suresh Chandra Satapathy, P.S. Avadhani, and Ajith Abraham (Eds.)
Proceedings of the International Conference on Information Systems Design and Intelligent Applications 2012 (India 2012) held in Visakhapatnam, India, January 2012, 2012
ISBN 978-3-642-27442-8

Vol. 133. Sabo Sambath and Egui Zhu (Eds.)
Frontiers in Computer Education, 2012
ISBN 978-3-642-27551-7

Vol. 134. Egui Zhu and Sabo Sambath (Eds.)
Information Technology and Agricultural Engineering, 2012
ISBN 978-3-642-27536-4

Vol. 135. Honghua Tan (Ed.)
Knowledge Discovery and Data Mining, 2012
ISBN 978-3-642-27707-8

Honghua Tan (Ed.)

Knowledge Discovery and Data Mining

Editor
Honghua Tan
Wuhan Institute of Technology
School of Electronic Engineering
Wuhan
China, People's Republic

ISSN 1867-5662 e-ISSN 1867-5670
ISBN 978-3-642-27707-8 e-ISBN 978-3-642-27708-5
DOI 10.1007/978-3-642-27708-5
Springer Heidelberg New York Dordrecht London

Library of Congress Control Number: 2011946230

Printed on acid-free paper

Springer is part of Springer Science+Business Media (www.springer.com)

Preface

Publisher's Note: This text has been removed because of overlap with another source. We have not been able to contact the Editor for comment.

The original version of the book was revised: Preface content has been removed from the book. The correction to the book is available at http://doi.org 10.1007/978-3-642-27708-5_109

Organization

Honorary Chair

ChinChen Chang	National Chung Hsing University, Taiwan
Patrick S.P. Wang	Fellow, IAPR, ISIBM and WASE Northeastern University Boston, USA

Program Co-chairs

Yuntao Wu	Wuhan University of Technology, China
Weitao Zheng	Wuhan University of Technology, China

Publication Chairs

Honghua Tan	Wuhan Insitute of Technology, China
Dehuai Yang	Huazhong Normal Universiy, China

International Committee

Wei Li	Asia Pacific Human-Computer Interaction Research Center, Hong Kong
Xiaoming Liu	Nankai University, China
Xiaoxiao Yu	Wuhan University, China
Chi Zhang	Nanchang University, China
Bo Zhang	Beijing University, China
Lei Zhang	Tianjin Institute of Urban Construction, China
Ping He	Liaoning Police Academy, China
Alireza Yazdizadeh	International Science and Engineering Center, Hong Kong
Wenjin Hu	Jiangxi Computer Study Institute, China
Qun Zeng	Nanchang University, China

Organization

Contents

Study on Detection Technology of Load Waveform Based on Fuzzy Pattern Recognition

Xiao Li Wang, Wei Wu, and Lei Yu

School of Electricity and Electronic Information Engineering,
Jilin Institute of Architecture and Civil Engineering, Changchun, Jilin, China
tieshuxiaowei2007@126.com

Abstract. In the paper a detecting system of load current waveform based on fuzzy pattern recognition is designed in order to solve a large number of existing problem of office electrical equipment safety and wasted energy. The system can real-time detect the load of office. The type, power and running state of power device can be on-line analyzed and diagnosed by the method of fuzzy pattern recognition to attain purposes of safety and energy conservation. Experimental results show that method of fuzzy pattern recognition can better reflect the effect of complex and uncertain factors. Application of this technology plays a positive role on spread of energy- conservation products and safe use of electricity.

Keywords: detection, electricity load, current waveform, fuzzy pattern recognition.

1 Introduction

Energy saving and emission reduction has become one of the main topics of modern social research, conservation of electricity is the most important. The family, office, and other public occasions electricity load and load of different types, which makes the electric energy saving becomes very complex. At present our country is in economic development period, and civilian office electrical equipment increases very fast, and the user does not have the electrical knowledge. Large electrical equipment use is bound to bring safe hidden trouble and electric energy waste problem, development of high intelligence, high efficiency monitoring equipment, implementation of safe use of electricity, energy saving and environmental protection be imperative. In 2004 we have developed with the load differential function of the electric power monitoring system, on this foundation, we present further in-depth study of various types of electric circuit load running at the same time loop current waveform, using the fuzzy pattern recognition[1] method on different load analysis and judgment, early warning. Divided into two main parts: load current waveform data acquisition and remote on-line analysis and diagnosis. This paper mainly introduces the load current waveform data acquisition and analysis of single load, load to office as the research object (including computer, printer, copier, water machine, air conditioning), on the acquisition of signal using the fuzzy pattern recognition method for waveform analysis, draw the conclusion, put forward the corresponding assumption of energy saving and energy saving measures.

H. Tan (Ed.): Knowledge Discovery and Data Mining, AISC 135, pp. 1–6.
springerlink.com

2 Current Waveform Detection

A. detecting load current waveform process

Electric load current waveform detection devices are: load, testing equipment, oscilloscope, memory. The device is connected with a load, the use of oscilloscope waveform detection, detection process is shown in figure 1:

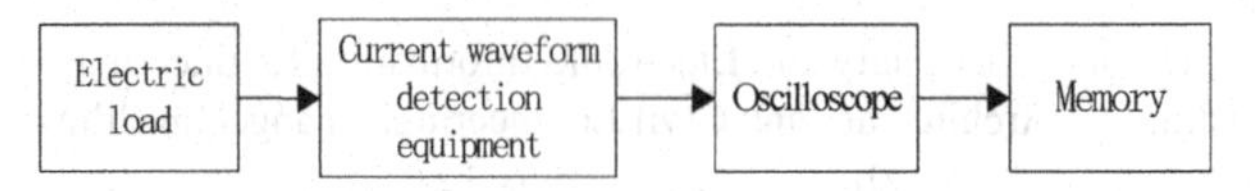

Fig. 1. Diagram of current waveform detection

B. Waveform detection instrument

Making current detection equipment and debugging. Current test instrument is mainly used for current transformer converts the current into a voltage. Select the DS1002C models of digital oscilloscope detection voltage. The oscilloscope storage is divided into internal memory and external memory, has RS232 interface with the computer direct communication to transmit data, waveform tracking, real-time data acquisition and other functions.

C. Load current waveform detection principle

Each electrical loads are different in nature, so the flow through the load current waveform characteristics of different. The current transformer is connected in series with the electrical load circuit through a resistor, the current mutual inductor current is converted to a voltage, the voltage across the sense resistor, oscilloscope waveform display and record data, including: phase, cycle, wave peak, trough, peak value, frequency difference.

The read voltage waveform of the voltage value $U_{(t)}$, get the electric load current waveform:

$$I_{(t)} = U_{(t)}/R \quad [2] \tag{1}$$

Through the analysis of the voltage waveform to get electricity load current waveform characteristics.

D. Current waveform detection process

Adjustment of oscilloscope their access to both ends of resistance, choose 1 passage port, set magnification, each grid of the horizontal axis represents the time value, the vertical axis represents the voltage value, convenient recording and computing. The apparatus is connected on the oscilloscope, waveform observation, regulation period, amplitude knob to be stable waveform, with storage devices will waveform images and data export. The completion of each kind of load testing, several different loads can be simultaneously access circuit, detection of mixed load current waveform.

3 Current Waveform Data Acquisition and Analysis

A. Waveform data acquisition

On office electricity load testing, the electricity load and the waveform detection equipment in the series circuit, testing data and storage oscilloscope. Records of the various periods of the cycle, frequency, amplitude, wave peak, trough values and phase changes are analysed. Collection of objects which are respectively the computer, copier, water dispenser, air conditioning. The dispenser and air conditioning, refrigeration and heating the two state, should be separately detected. Observe the waveform phase changes, analysis of waveform and data characteristics. The following is the detection of waveform, wherein the typical analysis were as follows:

(A) the computer working current waveforms are shown in figure 2:

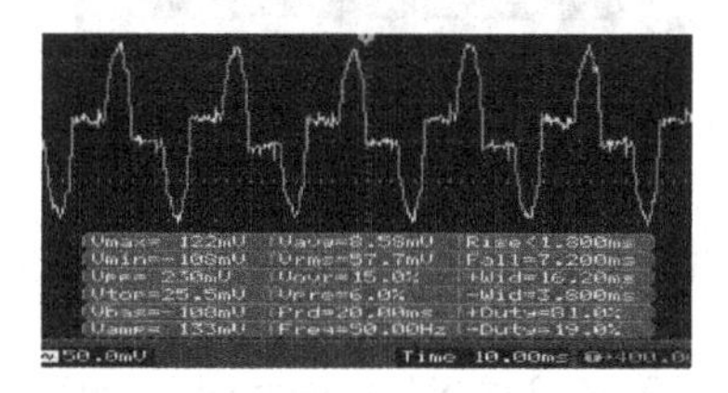

Fig. 2. Computer operating current waveform

The waveform is the computer working current waveform, observe the waveform peaks appear similar to the peak current waveform, current amplitude changes. Cycle 20ms, frequency 50HZ, amplitude of 111mV wave peak, trough, 122mV value of - 100 mV, amplitude difference is 222mV. Waveform similar to sine wave, the overall computer load characteristic of capacitive.

(B) the photocopier working current waveform as shown in Figure 3 below:

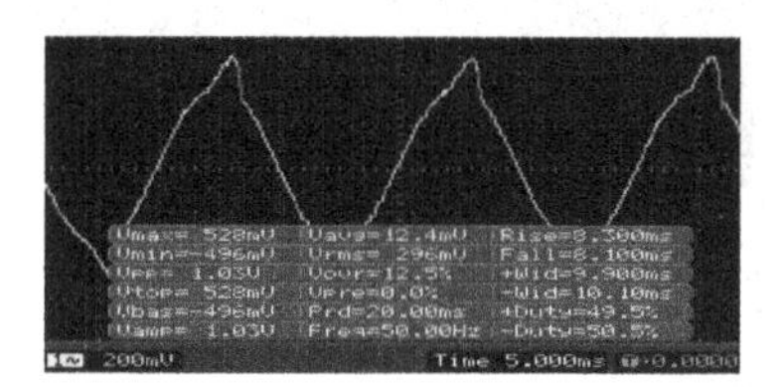

Fig. 3. The photocopier working current waveform

The wave peak value, 496mV, 520mV trough level amplitude difference 1.03V, the mean amplitude of the 48mV, frequency 50HZ. In the copying process, current is unstable, change quickly.

(C) the dispenser refrigeration current waveforms are shown in figure 4:

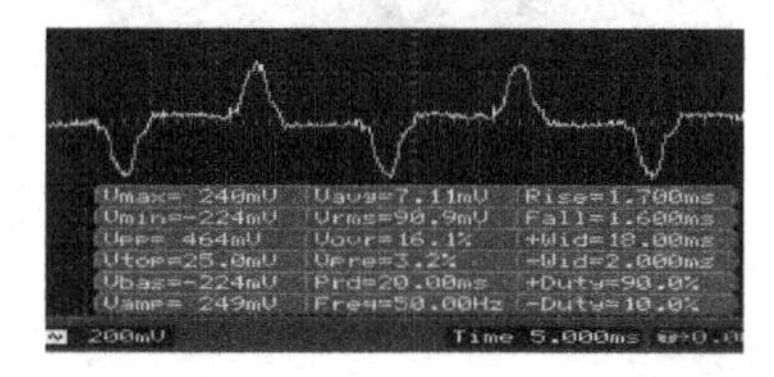

Fig. 4. Dispenser refrigeration current waveform

Cycle 20ms, frequency 50HZ, 240mV wave peak, trough value - 224mV, 464mV amplitude difference, the mean amplitude of the 232mV. The dispenser refrigeration load characteristic of sensibility, the current waveform overshoot current.

(D) drinking water mechanism when the heat current waveforms are shown in figure 5:

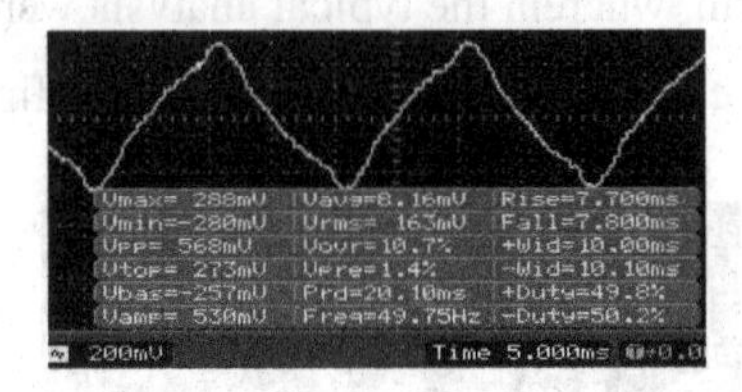

Fig. 5. Drinking machine heating current waveform

Cycle 20ms, frequency 50HZ, 288mV wave peak, trough value - 288mV, 568mV amplitude difference, the mean amplitude of the 288mV. Drinking fountains in the heating process to consume a large amount of power, the current through the load increase. From the waveform analysis, when the load is the nature of resistance.

(E) Chunlan air-conditioning refrigeration (8 degrees C) current waveforms are shown in Figure 6: (purchased in 2000, using the power of 2200W)

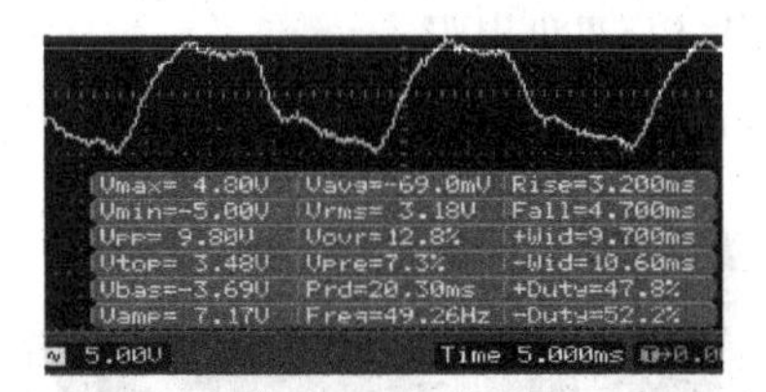

Fig. 6. Chunlan air-conditioning refrigeration current waveform

Cycle 20ms, frequency 50HZ, 4.8V wave peak, trough value - 5.0V, 9.8V amplitude difference, Chunlan air-conditioning refrigeration process waveform transformation is similar to the square wave, its characteristic of sensibility.

(F) Chunlan air-conditioning heating (25 degrees C) when the current waveforms are shown in figure 7:

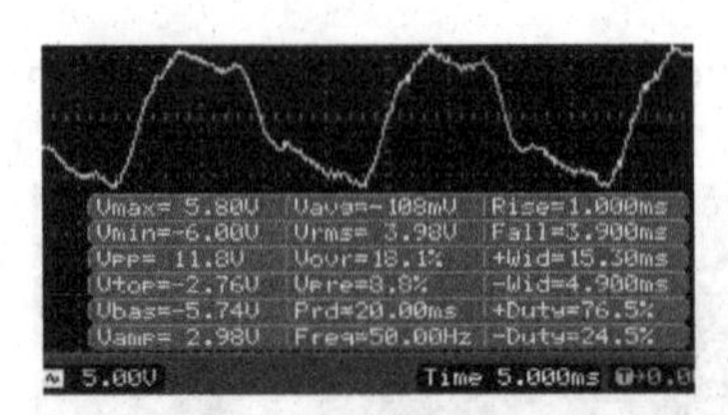

Fig. 7. Chunlan air-conditioning heating current waveform

Cycle 20ms, frequency 50HZ, 5.8V wave peak, trough value - 6.0V, 11.8V amplitude difference, Chunlan air-conditioning in the heating process of waveform transformation is similar to the square wave, properties close to resistance. But the heating process of the current than the refrigeration process, waveform change drastically jump.

B. Waveform data analysis

Through the testing, access to the data and waveform characteristics by use of fuzzy pattern recognition method to carry on the analysis, of which the most important is the phase, amplitude and wave shape. For an object recognition it belongs to what kind of problem is called pattern recognition[3]. Study on the pattern recognition method has three major categories, namely statistical pattern recognition, syntactic pattern recognition and fuzzy pattern recognition. Fuzzy pattern recognition pattern recognition is one of the most new branch. For the specific problem is easy to have a rational analysis, but the family, office, and other public occasions electricity load and load of different types, need recognition and classification of the information, and the information the boundary was not clear, has the feature of fuzziness, the need for accurate identification. So the simulation of thinking method in the identification of fuzzy objects than other method is more superior. Fuzzy pattern recognition fuzzy mathematics[4]need to use some method. Use fuzzy mathematical method to deal with this kind of problem is closer to the actual situation, and in many cases can make the complicated calculation is relatively simple. The results of experiments show that fuzzy identification method can better reflect the uncertainty of fuzzy data. Fuzzy pattern recognition in solving pattern recognition problems when introducing the idea of fuzzy logic[5]. Waveform analysis steps as shown in figure 8:

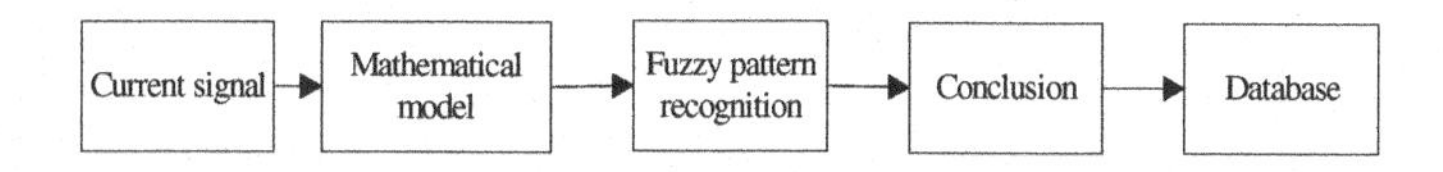

Fig. 8. Waveform analysis steps

This paper mainly describes the current signal acquisition, and then the waveform using the fuzzy pattern recognition to do thorough research. Waveform data acquisition is an important part of the whole project, it is the establishment of the database, data analysis, remote on-line analysis and diagnosis based on.

According to the office of electric load as an example, on the computer, copier, water dispenser, air conditioner respectively collecting work of the current waveform and data, the waveform and data analysis the conclusion, if aging of electric equipment or work abnormally, waveform and data change, alarm system. Through the data acquisition and data analysis, can make the following assumption of energy saving:

(A) no work can be cut off automatically.
(B) the main electrical equipment may limit the use of.
(C) uses electric equipment ageing or abnormally early warning system.

4 Conclusion

In this paper the office electricity load as an example, introduced one kind based on the fuzzy pattern recognition of the waveform of load current detection system, introduces the load current detection process, using the fuzzy pattern recognition method of current waveform analysis, draw the conclusion. This technology application to the electricity office, household electricity, schools and other public occasions, to eliminate hidden trouble of safety, electricity saving effect, to promote the use of energy-efficient products, will play a positive role in promoting.

References

1. Geng, Y., Ma, Y.: Research of Vehicle Classification Based on Fuzzy Pattern Recognition. Computer Engineering 28(1), 133–135 (2002)
2. Qiu, G.: Circuit. Higher Education press, Beijing (2006)
3. Xu, Z.: On Compatibility of Interval Fuzzy Preference Relations. Fuzzy Optimization and Decision Making (2004)
4. Li, X., Hu, Z.: Engineering Machine Fault Diagnosis Expert system Based on Fuzzy Inference and Self-learning. Computer Engineering and Apply 15, 200–202 (2005)
5. Saaty, T.L.: Making andvalidating Complex decisions with the AHP/ANP. Journal of Systems Science and Systems Engineering (2005)

On the Analysis of Speciality Construction and Development in Independent Institute

Zhuoya Wang

Xuhai College, China University of Mining and Technology, Xuzhou, China
wzycumtxh@163.com

Abstract. According to speciality construction and development status in most Independent Institutes, this paper provided the ideas and suggestions from the teaching staff construction, major settings, education model of speciality, teaching management, Characteristic speciality construction and other aspects, in order to promote speciality development and connotation construction of Independent Institute.

Keywords: Independent Institute, speciality construction and development, Characteristic.

1 Introduction

Independent Institute is an important measure of higher education in terms of reform and innovation in our country. It has played positive role in expanding our country's higher education resources and the university scale and is an important aspect of our country's higher education development in current and a certain period of future. The Independent Institute adopts the private running mechanism, highlights the running pattern of "being independent" and uses the new management system. Its high starting point and quick development and the realization of combining public university brand and community resources promote the advancement of popularizing the higher education in our country. Independent Institute has been an important part in our country's higher education system[1].

The running nature, education objects and service clients of Independent Institute are remarkably different from its parent school. With the development of Independent Institute, it can not imitate the parent school completely in terms of training orientation, specialty establishment and orientation, teaching and management system, etc. On the basis of making full use of the better resources of the parent school, Independent Institute should develop in a dislocate way and form its own characteristics so as to meet the demand of market.

2 Background of Independent Institute' Speciality Characteristic Construction Background

For basing in our country education system and maintaining the healthy development, Independent Institute must start from establishing own reasonable training objectives

H. Tan (Ed.): Knowledge Discovery and Data Mining, AISC 135, pp. 7–14.
springerlink.com

and determining service face, focus on the speciality construction, form own speciality characteristics, adapt to market demand, highlight own characteristics of training, continuously push forward the speciality construction and teaching reform work, only then can deal with the development change of our country higher education fully, compete in an invincible position. Currently, Independent Institute should pay full attention the following changing situation, and carry out the speciality construction and teaching reform research unceasingly to meet the situation development needs.

2.1 The Student Source Question

With Chinese higher education school-age population gradually decreased, and by the tide of internationalization of higher education, the student source reducing question already gradually to highlight, the source competition will become focus which each kind of university in our country work from now on. Our country ordinary College after experiencing 1999 the large-scale increased enrollment, its educational resources expanded greatly, the resources superiority was obvious, but Independent Institute, because the short running school time, fully relied on the parent school in the initial period, and lack of investment, educational resources superiority obvious insufficient, will be in the inferiority obviously in present's fresh source competition.

2.2 Social Expectations of Higher Education to Improve

As Chinese development of higher education popularization, the training of personnel increased, along with the emergence of university employment pressure, competition intensely and some other columns, so the students, parents and society set a higher request at the same time to our country higher education, on the other hand, Chinese current teaching quality was questioned. But, Independent Institute uses the private mechanisms, does not have the national investment and allocation, mainly depends on tuition income running, in the case of paying high tuition fees, the students, parents and society provides the quality education product and the high-quality service request to the Independent Institute relatively is higher.

So, Independent Institute in order to obtain long-term development in the competition and base in numerous all sorts of Colleges, must comply with the human resources market demand in the speciality development down enough effort, and take road of the characteristic speciality construction to improve quality of personnel training, improve personnel culture characteristics, the only way to continue to attract students, stable employment market, and seek long-term development.

3 Independent Institute Speciality Construction and Development Status

At present the majority Independent Institute in the speciality settings are completely consistent with the parent university, in the speciality development are too dependent on the mother, lack own characteristics, popular speciality herd and go hand in hand, speciality construction and teaching reform are not primary and secondary and key, and lack the reasonable medium and long-term speciality construction and development project, the main reason for these problems are the following:

3.1 Unreasonable Faculty Structure

Independent Institute running time is short, the faculty structure is unreasonable, the own faculty teach the public basic course primarily, the specialized teachers are relying on the parent school mainly, the own specialized teachers are scarce, and thus they are powerless for speciality interlaced development, characteristics construction and reform.

3.2 The Speciality Construction Is Similar to the Parent School

Since the majority Independent Institute fully relied on the parent university in the initial period, lack of own specialized faculty and speciality support background, depended on the parent school fully in speciality construction, is identical with the parent school in the speciality setting, although this offered advantages to attract students for Independent Institute in initial period, but also this has created the present majority Independent Institute became the parent school "the replica". Along with the continued expansion of Chinese higher education resources, and the gradually decline of students quantity, if Independent Institute imitates the parent school in the speciality setting completely, the speciality development go hand in hand, the speciality construction lacks own characteristics, not only the trained students lack the competitive power in the employment market, but gradually will compete with the parent school, which affect Independent Institute sustained and healthy development.

3.3 Speciality Setting Constitutive Question

Speciality is the important factor of forming the intersected market and non-competitive groups[2]. Independent Institute speciality setting is in favor of the popular speciality of the marked demand, although this is helping graduates to successfully enter smoothly the intersected and non- competitive market, but Independent Institute set the low-cost, fast track approach, popular speciality too many, which has tended to be saturated in the entire College's specialized layout, this kind of single structure, no characteristics speciality setting will stay hidden for employment of students in future, likely to cause structural unemployment[3].

4 The Suggestions and Ideas of Independent Institute Speciality Construction and Development

In view of Independent Institute speciality development present situation in our country, and higher education development situation in future, this article provided the ideas and suggestions from the teaching staff construction, speciality settings, education model of speciality, teaching management, Characteristic speciality construction and other aspects.

4.1 The Teaching Staff Construction

Independent Institute teaching staff construction is the major question for sustainable and stable development is also the bottleneck question which restricts the majority Independent Institute speciality development. From the long-term development, the own full-time teaching staff must be established in Independent Institute to run the school stably and maintain its vitality in the speciality construction and development. However, Independent Institute in short term to construct a rational structure , quality and efficient full-time teaching staff is very difficult considering training professional teacher 'time, the funds and the policy and difficulty of introducing talented persons, therefore, in the present stage of Independent Institute development, should make a point of the building of own full-time teaching staff, at the same time, fully rely on the parent school teachers, in the speciality development and construction take the parent school's speciality situation as the main starting point ,and on the basis of it, seek a reasonable localization and the breakthrough and foothold in Independent Institute speciality development.

Independent Institute in building own teaching staff can not absorb and introduce blindly, should work step by step on the basis of reasonable speciality construction and development medium and long-term plan. Although Independent Institute in early time has established public basic course teacher troop which has laid certain foundation for each speciality development, and the construction cost is lower, but Independent Institute to a certain stage of development should focus on own professional teacher troop and proper planning, the existing case of shortage of professional teachers, helps teachers determine a reasonable personal and professional development orientation which is consistent with the overall speciality construction plan in Independent Institute, focus on training the professional basic courses teachers, while pays great attention the “double” teachers' training. Because professional basic courses for teaching students to master a solid foundation of professional theoretical knowledge play a significant role, at the same time, their teaching quantity is big, repeatability is strong, masters the specialized front knowledge and the teacher knowledge refresh rate demanding to the teacher is less, which is advantageous to the teachers' personal growth and rapidly improve the quality of teaching, because Independent Institute is now large scale introduction of no on-site work experience, graduate teachers. Independent Colleges is currently located in the "applied" training, which will undoubtedly demand experience of teachers higher, Independent Institute in the "double" teacher training should focus on cooperation with industry and enterprises , rely on the field technicians to drive own speciality teachers training.

4.2 The Speciality Setting

The speciality setting is source of the speciality construction and teaching reform in Independent Institute. The scientific and reasonable speciality setting plays a decisive role for the quality of school education. Independent Institute in the speciality setting, first should compartmentalize and analyze the speciality of the parent school, the speciality characteristics, service face, the teachers situation and the employment market of the parent school and so on should be carried out comprehensive and in-depth research, and then according to the different speciality circumstance, combine with the above aspects to consider whether there is the possibility of dislocation and the entry point which will play an important role to the speciality construction and development of Independent Institute.

4.2.1 The Speciality Characteristic

From the applied personnel training localization of Independent Institute, and the students' knowledge foundation, Independent Institute is best not to set the speciality which requests higher theoretic foundation. Such as the "financial management speciality" mainly cultivates high-end talent for enterprise, the "Statistics speciality" requests the students' higher mathematics foundation, and the speciality is established in research and creation, and so on, Independent Institute should carefully set.

4.2.2 The Specialty Service Face

Independent Institute should be clear about own service face, generally independent college should be based on the regional economy and local development needs, and can optimize and adjust the speciality setting structure according to the local industrial upgrade, especially the specialty that the local economy development need urgently in the human resources market, Independent Institute should focus on capturing, and create the condition to speed up the development.

4.2.3 The Teachers' Situation

Some leading and characteristic speciality of the parent school, the teachers staff is abundant and competent, and have the social fine reputation, provide a lot of social services for the industry and enterprise, teacher's scientific research work load is big, therefore, the teachers devote to the scientific research and the social services more into teaching energy. For this type of speciality, Independent Institute will possibly meet many questions which can not ensure the speciality construction and development, should also be careful to set.

4.2.4 The Graduates' Employment Market

The quality of university training in the job market can be effectively tested, Independent Institute should pay full attention to the employment situation and market demand, and can effectively establish a set of good unimpeded employment feedback information system effectively to discover promptly their own lack of education and teaching, and realize the real employment-oriented, adjust flexibly the speciality development direction, and set the speciality orientation module, adjust the teaching plan, take the path of development of industry ,teaching and research to the combination, the only way to continue to provide the outstanding professionals for the society.

4.2.5 Promote the Speciality Integration, Hatch the New Speciality

Independent Institute as an emerging school entity in Chinese higher education system, its development time is shorter, at present most Independent Institute have finished the large-scale development phase, and start to march into the connotation development phase. In the scale development phase, the speciality of Independent Institute set up basically in accordance with the discipline advantages of the parent school, but in the connotation development phase, Independent Institute should optimize and adjust gradually the speciality structure, compared to other types of colleges and universities, Independent Institute also has the obvious advantage, its administration setup is relatively reduced, the system is flexible, the management is efficient, These create the positive condition for the speciality structure optimization, the speciality integration, hatching and cultivating the new speciality, and so on. Independent Institute should hold these advantages fully, and continue to study the specialized characteristic, find the common character of the speciality, achieve penetration and complementarities among the speciality, tries hard to train the applied innovative talents.

In addition, on the basis of considering fully Independent Institute adapts the local economic development and adjusts flexibly the speciality development direction, Independent Institute should also take full account of the school cost, the speciality orientation module setting both must be flexible, but can not be too many, otherwise, it will not only greatly increase the teaching cost, but also cause the speciality construction energy to be scattered, lacks the key point, difficult to form the characteristic. Independent Institute in the speciality orientation setting should be focused, and make focus clear, relatively stable, like this only then has the possibility of constructing with proper priority, makes the speciality characteristic.

4.3 The Speciality Training Model

Independent Institute in the speciality training model should not engage in "indiscriminate", if adopt the "intensive professional" mode, or the "wide professional" mode, should consider fully the each speciality characteristics, and the students characteristics to determine the different training mode of the each speciality.

The "intensive professional" training mode requires the students study and research in-depth and thoroughly in the professional field some direction, and master the professional knowledge and skills in the certain direction field, the speciality direction is clear, the curriculum is fewer and well-chosen. For the students in Independent Institute, the third group matriculate, because their enrollment base is relatively poor, it is difficult to master the professional knowledge of the certain field in-depth and thoroughly, therefore when Independent Institute determines the "intensive professional" training mode, should have the inquire deeply and close relation with the correlative enterprise, and on the basis of achieving cooperation intention of the "order form training", target-oriented advance the "intensive professional" mode training, so as to achieve the specialized characteristic training, enable the students' study can combine with the future work in the enterprise, and achieve Independent Institute application-oriented training objectives.

The "wide professional" training mode requires the students master the professional knowledge across-the-board, the curriculum is full, the teaching content is rich, but the speciality direction is unknown, which is favorable to expand the employment way for the students. Nevertheless the learning time in college is limited, in the case of more curriculum and wider teaching content, students learn various aspects of professional knowledge can only be limited to master basic concepts, and the lack of in-depth study and research, this kind of speciality training model for Independent Institute is reasonable in terms of certain speciality, such as

"International Economics and Trade Speciality", because Independent Institute mainly serves the small and medium-sized enterprise, in which the business volume is relatively small, and does not need the too many staff, but the operation processes are similar with the large enterprises, each link flow is complete, "although the house sparrow slightly be full equipped", which demand the students master the comprehensive professional knowledge and to be familiar with all aspects of each business flow, but do not involved in too deep, this kind of speciality in Independent Institute need to carry on the "wide professional" training mode.

Therefore, the specific choice of the Independent Institute, "intensive professional" of "wide professional" training patterns, we should take full account of the speciality characteristics and the students characteristics, and investigate the industry, enterprise, and employment market thoroughly.

4.4 The Teaching Management

To continue the healthy development, Independent Institute must focus on the speciality characteristic construction, and effectively take the path of combining production, study and research, increase contacts and cooperation with industry and enterprise, and enhance the students practical ability training, improve the practical teaching effect, which need the teaching management of Independent Institute to follow the education rule, at the same time must emancipate the mind fully, increase the flexibility and degree-of-freedom in the teaching arrangement, adapt the actual situation of the industry fully, and adjust the teaching arrangement nimbly, carry on seriously inviting the scene enterprise technical personnel to teach for the students, and the students' practice of entering the enterprise, thus effectively enhance the practical ability of Independent Institute students.

4.5 The Speciality Characteristic Construction

In the aspect of speciality and discipline construction, Independent Institute should take account of the situation of regional economic development fully, the education resources of the parent school, and the own actual situation. Independent Institute should follow market-oriented, practical principle, employment goal, and concentrate the limited resources, close to the community, close to the demand, close to the jobs, and eliminate the idea of doing something large and completely, should work selectively, and take the road of own disciplines characteristics construction[4]. Independent Institute should be fully emancipate the mind, carry bold attempt, through in-depth study to each speciality situation, carries out the speciality characteristic construction with emphasis. For example some speciality may take the discipline competition as an development opportunity, we should proceed from the recognized discipline competition widely by the community and the employers, carefully research the knowledge and ability which the competition requires, and combine with the speciality teaching plan, adjust the curriculum, in order to train this aspect knowledge and ability of the students, which is no doubt to provide the more advantage for the students employment, and also to provide the breakthrough in the speciality characteristics construction of Independent Institute.

The speciality construction and development in Independent Institute are the core work to improve the students training quality, Independent Institute should take full account of the overall development situation of Chinese higher education, search carefully the questions of the own speciality construction, and carry out the dislocation development with the parent school, take the characteristic development path, only then can improve competitiveness of Independent Institute, and maintain stable and healthy development.

References

1. Zhou, J.: Promoting Sustainable healthy Development of Universities Independent Institute. Exploring Education Development (8), 1–4 (2003)
2. Lou, X., Chen, L., Chen, Y.: On the employment competition advantages and disadvantages of Independent Institute graduates. Social Science Front (2), 323–325 (2006)
3. Jiang, T., Zhang, H., Zhong, X.: Problems of Characteristic Cultivation of Universities Independent Institute. Journal of Yibin University (1), 102–105 (2006)
4. He, Q., Que, W.: On the Sustainable Development of Independent Institute. Jiangxi Educational Research (4), 8–9 (2006)

EVA and DEA, Which Is Better in Reflecting the Capital Efficiency?

Yadong Shi

Institute of Defense Economics and Management
Central University of Finance and Economics
Beijing 100081, P.R. China
shiyadong@126.com

Abstract. Traditionally, people generally believe Economic Value Added (EVA) is one of the best indicators to evaluate capital efficiency of companies. This paper attempts to use Data Envelopment Analysis (DEA) in evaluating the efficiency of capital, and by comparing the samples and conclusions of DEA with those of EVA, it is found that DEA is a better way to reflect the capital efficiency. The two-stage analysis—principal component analysis and Data Envelopment Analysis in this paper provide new ideas for studying capital efficiency of companies.

Keywords: EVA, DEA, Capital efficiency.

Efficiency is a very widely used concept in economics. In general, the basic concept of efficiency in economics is referring to the relationship between input and output or costs and benefits. When the concept of efficiency applied to a single enterprise, efficiency is mainly pointing at whether the enterprise makes the maximum output of its resources, or whether a certain amount of output achieves a minimum cost. Capital efficiency can be defined as the enterprise output from a unit of capital input. Traditionally, people generally believe EVA is one of the best indicators to evaluate capital efficiency, is that so? Can capital efficiency be better evaluated by means of Data Envelopment Analysis? This paper gives the corresponding discussion through five steps. Firstly, the calculation of the EVA of sample companies is given. Secondly, the basic principle on how DEA reflects capital efficiency is illustrated. Thirdly, the input and output variables are determined in the calculation of DEA by the PCA approach. Fourthly, capital efficiency values of sample companies obtained from the EVA and the DEA respectively are compared. Finally, the main conclusion attached to the outlook of this research is drawn.

1 Calculation of the EVA of Sample Companies

EVA, as an indicator for evaluating capital efficiency developed in 1990s in the western counties, is getting more and more attention and favor in business. It is Stern Stewart(1991) who made pioneering research on EVA, they discussed the relationship between EVA and MVA by using the U.S. listed companies from 1984 to 1988 as samples. Their research demonstrated EVA and MVA were significant correlated. Their research also showed that EVA was the indicator to test whether the company substantially made profit. If the difference between net profit after tax for the year and the total cost of capital was positive, then the economic added value happened, the company created added value; Otherwise, the value of the company shrink down. Tully, Shawn's study (1993) showed that EVA and

H. Tan (Ed.): Knowledge Discovery and Data Mining, AISC 135, pp. 15–23.
springerlink.com

stock price were closely related; Uyemura, Kantor, Pettit (1996) used 100 banks between 1986-1995 as samples, MVA as explained variables, ROA, ROE, EPS and NI calculated according to financial statements and EVA as explanatory variables, to do regression analysis with MVA, found that EPS had the lowest explanatory power was: 6%; and EVA had the highest: 40%. O 'Byrne (1996) used 6551 U.S. companies during 1985-1993 as study samples, the "market value / total invested capital" as explained variable, the EVA, NOPAT and free cash flow FCF as explanatory variables, to do regression analysis with the beginning capital standards and the "market value / invested capital", found the explanatory power of each variable was: EVA: 31%, NOPAT: 33%, FCF: 0%. After eliminating scale effect and adding industrial-related indicators, the explanatory capability of EVA increased to 56%, thus he believed EVA was better than NOPAT and FCF in explaining company's performance. Lehn and Makhija (1997) took 452 U.S. companies as study samples, and empirical results showed that EVA and MVA were more associated with stock returns than other traditional accounting indexes. West and Worthington (2001) used a sample of 110 Australian companies between 1992 to 1998 to do regression analysis to see if EVA is more relevant with the stock return than traditional accounting indicators (such as Earning, Net cash flow, RI). And the results showed although earning had the highest explanatory power, EVA was able to provide more additional information than earning.

The formula for calculating EVA is: EVA is equal to net operating profit after tax minus invested capital multiplied by the weighted average capital cost rate. This paper selects 20 power listed companies of China to calculate EVA. Specific adjusted items during the calculation process are showed in the table below:

Table 1. Adjustments when calculating EVA

Net operating profit after tax adjustments		Invested capital adjustments	
Start from:	Net profit after tax	Start from:	Common equity
	Financial cost		Minority interests
	minority shareholders		Depreciation preparation and allowance for bad debts
Add:	Current increase in depreciation preparation and allowance for bad debts	Add:	Short-term borrowing
	Nonoperating expenditure		Long-term borrowing
			Long-term debt due within one year
Minus:	Nonoperating revenuce	Minus:	Construction in process
Equal to:	Adjusted net operating profit	Equal to:	Adjusted total capital

This paper selects twenty outstanding listed companies in the power plate of China as samples. In order to enlarge the comparability, this paper fully takes account of the similarity of the scale of net assets among them when choosing samples. The calculation results of EVA are shown in Figure 1. As can be seen from the EVA evaluation results, there are 14 companies whose EVA is negative, and the others' EVA is positive. The result shows that 70% of listed power companies have very low capital efficiency and lack the ability to bring companies the value continuously.

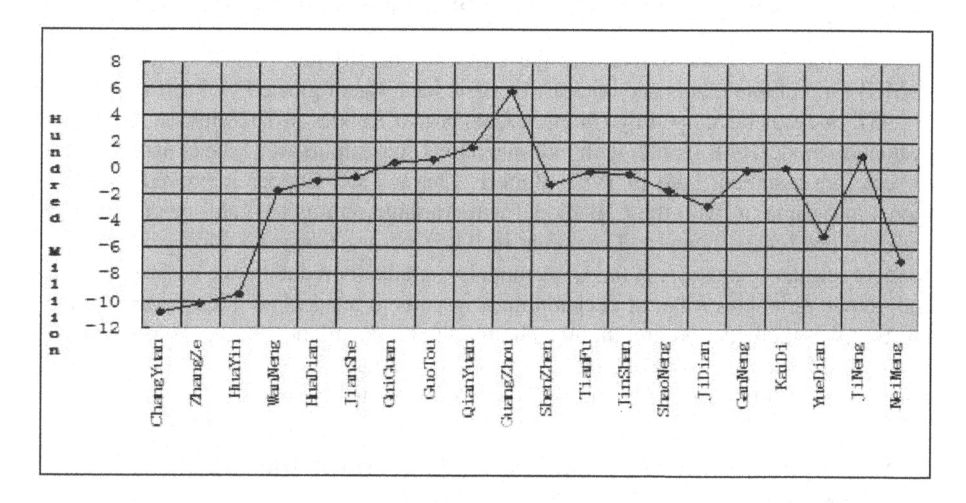

Fig. 1. Chart of EVA Comparison in 20 Samples

Among them, the lowest three include Changyuan Electric Power, Huayin Electric Power and Zhangze Electric Power, and this ranking mainly due to the low net operating profit after tax of them, and it can be further explained by their too low net profit. In 2008, the net profit of Changyuan Electric Power, Zhangze Electric Power and Huayin Electric Power is -0.977 billion, -0.959 billion and -1.052 billion respectively. Therefore, EVA is related to net profit to a large extent, but not absolutely positive correlated. It can be viewed from the figure that although the net profit of Huayin Electric Power is lower than that of Zhangze Electric Power and Changyuan Electric Power, its EVA still higher than the later two. AnHuiWenergy, Huadian Energy, Construction Energy, Shenzhen Energy, Tianfu Thermoelectric, Jinshan Share, Shaoneng Share, Jidian Share, Ganneng Share, Yue Electric Power A and Huadian Inner Mongolia belong to the middle while the EVA of them is still negative, which showed that enterprises' net operating profit after tax can not make up the cost of capital invested, and it is a waste of corporate assets, destructing the original enterprise value. For example, the net operating profit after tax of AnHui Wenergy, Huadian Energy and Construction Energy is ¥93 million, ¥659 million and ¥413 million respectively, and this is the reason why their EVA is higher than the first three companies. However, due to the high capital cost, their EVA is still negative. The EVA of Guiguan Electric Power, SDIC Huajing Power Holdings, Qianyuan Power, Guangzhou Development Industry (Holdings) , Kaidi Electric Power and Beijing Jingneng Thermal Power is positive and the net operating profit after tax of them is

¥671 million, ¥860 million, ¥498 million, ¥989 million, ¥286 million, ¥208 million respectively. Although Qianyuan Power's net profit after tax is lower, but due to low capital cost, its EVA is still higher than other enterprises. Guangzhou Development Industry (Holdings)'s EVA is the highest, and it created the highest net operating profit after tax with the lowest cost of capital. Thus, from the perspective of EVA, Guangzhou Development Industry (Holdings) made the best performance.

2 DEA's Basic Principle in Reflecting the Capital Efficiency

DEA are mostly applied to evaluate the efficiency of different decision Making Unit (DMU) for multiple inputs and outputs. Its basic idea was originated by Professor Farrell in 1957, and was developed by Charnes, Cooper and Rhodes in 1978 into an efficiency evaluation model (C^2B model) with multiple inputs and outputs while the constant returns to scale are assumed. Late in 1984, Banker, Charnes and Cooper improved the DEA model and came up with the C^2B model with variable returns to scale integrating both technical and scale efficiency. Thereafter, in 1985 Charnes, Cooper & Morey proposed to conduct sensitivity analysis in the DEA model by gradually reducing the number of input and output variables or that of decision making units to recalculate the efficiency value. DEA applies the linear programming techniques to reflect all the input and output data of decision-making units on the efficiency spaces, so as to get the efficiency frontier, which refers to a trail combined by the most efficient decision-making units.

3 Principal Component Analysis to Determine the Input and Output Variables of DEA Calculation

The first step of the DEA approach is to determine input and output variables. In order to avoid the problem that the selection of valuables is arbitrary and misleading, the PCA approach is adopted and repeated filtering among the numerous input and output variables is conducted. The following table is mainly used to test whether the data is suitable for factor analysis. Table 2 shows KMO and Bartlett's Test of original variables, and it demonstrates that KMO value of input variable is 0.794, which is larger than 0.7, and it is suitable to carry on the factor analysis; and BTS values' sig. is 0.00, which is less than 0.05, thus it is not a unit matrix and is also suitable for factor analysis. The KMO value of output variable is 0.835 and BTS values' sig. is 0.00, in the same way, they are also suitable for factor analysis.

Table 2. KMO and Bartlett's test

		Input variable	Output variable
Kaiser-Meyer-Olkin Measure of Sampling Adequacy		0.794	0.835
Bartlett's Test of Sphericity.	Approx. Chi-Square	153.157	147.263
	df	21	21
	Sig.	0.000	0.000

This paper adopts the principal component analysis as the analytical method. Principal component analysis has a common assumption: the estimated value of communalities is the square of partial correlation coefficient between each variable and other variables. That means the square of partial correlation coefficient between each variable and other variables is often took as the communalities of a set of variable. As showed in Table 3:

Table 3. Communalities

Input variables	Initial	Extraction
Equity	1.000	0.828
Operating expenses	1.000	0.964
Management cost	1.000	0.472
Financial expenses	1.000	0.406
Cash outflows from operating activities	1.000	0.943
Cash outflows from investing activities	1.000	0.598
Cash outflows from financing activities	1.000	0.737
Output variables	**Initial**	**Extraction**
Operating income	1.000	.989
Net profit	1.000	.908
Total operating revenue	1.000	.989
Operating profit	1.000	.935
Total profit	1.000	.972
Cash flows from operating activities	1.000	.979
Cash flows from investing activities	1.000	.688
Cash flows from financing activities	1.000	.738

Factor extraction threshold are set at 1, that means Eigenvalue should be larger than 1. Thus only one principal component of the input variables can be extracted, while two principal components of the output variables extracted. The total variance explained table (Table 4) shows that the Eigenvalue of the principal component extracted from the input variables is 4.948 and the extracted component can explain 70.686% of the original variables; the Eigenvalue of the two principal components extracted from the output variables is 3.937 and 3.224 respectively and they can explain 89.97% of the original variables.

Table 4. Total variance explained

Variable Type	Component	Initial Eigenvalues			Extraction Sums of Squared Loadings		
		Total	% of Variance	Cumulative %	Total	% of Variance	Cumulative %
Input variables	1	4.948	70.686	70.686	4.948	70.686	70.686
	2	0.938	13.393	84.079			
	3	0.584	8.35	92.429			
	4	0.235	3.364	95.792			
	5	0.192	2.737	98.529			
	6	0.098	1.395	99.925			
	7	0.005	0.075	100			
Output variables	1	4.556	56.955	56.955	3.973	49.665	49.665
	2	2.641	33.016	89.97	3.224	40.306	89.97
	3	0.563	7.033	97.003			
	4	0.132	1.652	98.655			
	5	0.089	1.114	99.769			
	6	0.017	0.213	99.982			
	7	0.001	0.018	100			
	8	-0.001	-0.001	100			

Table 5 shows the rotated component matrix of the input variables and output variables. As only one principal component is extracted from input variables, varimax orthogonal rotation method can not be used, thus input variables can be represented by operating expenses. And after the orthogonal rotation, two components are extracted from output variables: component 1, which mainly explains operating income, total operating revenue, cash flows from operating activities, cash flows from financing activities and cash flows from investing activities, can be comprehensively considered as the income-based indicator; component 2, which mainly explains net profit, operating profit and total profit, can be viewed as the profit-based indicator. Therefore, component 1 should be represented by operating income, component 2 by total profit. Seen from Table 5, by factor analysis, one input type indicator and two out put indicators are finally obtained to further carry out DEA.

Table 5. Rotated Component Matrix

Input variables	Component 1	Output variables	Component 1	Component 2
Equity	0.91	Operating income	0.982	0.158
Operating expenses	0.982	Net profit	0.131	0.944
Management cost	0.687	Total operating revenue	0.982	0.158
Financial expenses	0.637	Operating profit	-0.056	0.965
Cash outflows from operating activities	0.971	Total profit	0.121	0.978
Cash outflows from investing activities	0.773	Cash flows from operating activities	0.974	0.174
Cash outflows from financing activities	0.859	Cash flows from investing activities	0.589	0.584
		Cash flows from financing activities	0.845	-0.153

4 Flagrant Contrast of DEA and EVA Analysis of the Capital Efficiency of the Samples

On the bases of the obtained one input type indicators and two output type indicators through factor analysis as stated above, DEA is to be further carried out. This paper apply Data Envelopment Analysis mathematical model to calculate the sample companies' capital efficiency. By means of the software "DEAP Version 2.1" developed by Professor Coelli, University of New England in Australia, the calculation results of the sample data are shown in Figure 2.

The black line represents EVA value, and the green line DEA value. As can be seen from the figure, DEA efficiency value of Wanneng Energy is 1, and its slack variables and remnant variables are 0, thus Wanneng Energy's DEA can be effective. Huayin Electric Power has a efficiency value greater than 0.6, the rest companies' are very low, therefore, these companies are all invalid on DEA. In the previous study, it is known that Wanneng Energy's EVA is negative and its performance status is poor. It is Guiguan Electric Power, SDIC Huajing Power Holdings, Qianyuan Power, Guangzhou Development Industry (Holdings) s, Kaidi Power and Beijing Jingneng Thermal Power whose EVA is positive, while have a DEA efficiency value of 0.077, 0.033, 0.026, 0.055, 0.028 and 0.077 respectively, all is quite low. Obviously, the results of EVA

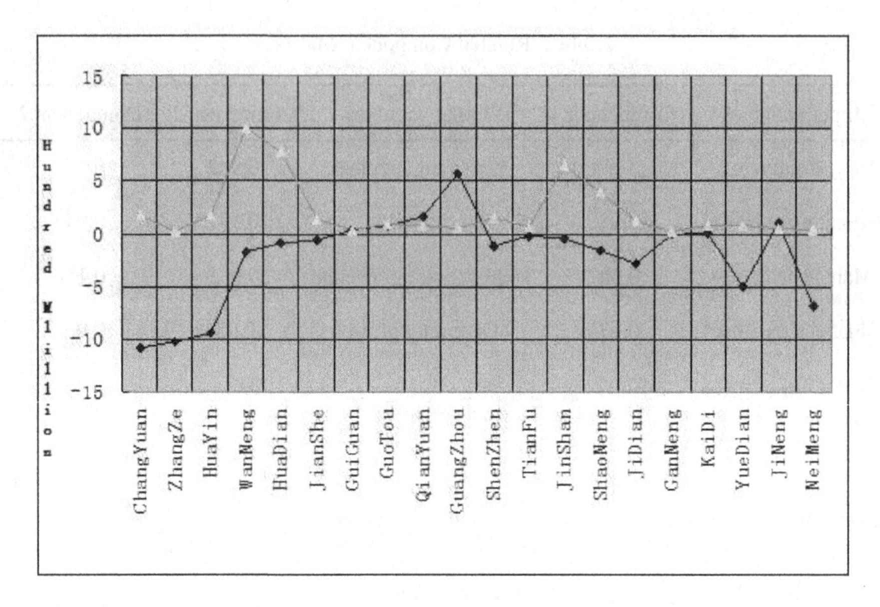

Fig. 2. Chart of EVA、 DEA Comparison in 20 Samples

analysis are inconsistent with DEA analysis. This demonstrates that those who creates a mythical EVA are not necessarily the companies that have high capital efficiency. Essentially, efficiency reflects the relativity between input and output. However, the fundamental flaw of EVA consists in the fact that it is an absolute indicator. The above empirical result reveals that it is the DEA approach that makes us avoid the confusion from the EVA value, thus evaluating capital efficiency of a company correctly.

5 Conclusions and Future Research Implications

Traditionally, EVA is generally considered as one of the best indicators to evaluate capital efficiency. This paper attempts to apply DEA method in evaluating capital efficiency, and by comparing the samples and conclusions of DEA with those of EVA, it demonstrates that DEA does a better job in reflecting capital efficiency. The EVA value always weakens our insight and misleads us to make false investment decisions. By the comparison between EVA and DEA, the phenomenon is disclosed that those listed companies that apparently create high EVA actually have low capital efficiency such as Guangzhou Development Industry (Holdings) and Qianyuan Power. Those who have apparently created poor EVA are actually outstanding from the point of capital efficiency, such as AnHuiWenergy, Huadian Energy and Jinshan Share. In order to overcome the defects of EVA evaluation, this paper innovatively proposed a two-stage capital efficiency evaluation method based on principal component analysis and Data Envelopment Analysis to assess enterprises' capital efficiency. In the future if the regression analysis, principal component analysis and DEA analysis method are combined, the input variables and output variables will be selected more properly. How to build a trinity capital efficiency model based on regression analysis, principal component analysis and DEA analysis becomes the challenging direction for further research.

References

1. Banker, R.D., Charnes, A., Cooper, W.W.: Some models for estimating technical and scale inefficiencies in Data Envelopment Analysis. Management Science 30(9), 1078–1092 (1984)
2. Charnes, A., Cooper, W.W., Rhodes, E.: Measuring the efficiency of decision making units. European Journal of Operational Research 12(6), 429–444 (1978)
3. Charnes, A., Cooper, W.W., Morey, Rousseau, J.J.: Sensitivity and stability analysis in DEA. Annals of Operations Research, 139–156 (1985)
4. Farrell, M.J.: The measurement of productive efficiency. Journal of the Royal Statistical Society, Series A, General 120,part3, 253–281 (1957)
5. Jian, F.: Decision analysis and management: a unison framework for total decision quality enhancement. Yehyeh Book Gallery (2005)
6. Lehn, K., Makhija, A.K.: EVA, Accounting Profits, and CEO Turnover: An Empirical Examination, 1985-1994. Bank of America Journal of Applied Corporate Finance 10(2), 90–97 (1997)
7. O'Byrne, S.: EVA and Shareholder Return. Financial Practice and Education, 50–54 (1997)
8. Stewart III, G.B.: The Quest for Value: The EVA TM Management Guide. Harper Business (1991)
9. Tully, S.: The Real Key to Creating Wealth. Fortune 20, 34–50 (1993)
10. Uyemura, D.G., Kantor, C.C., Pettit, J.M.: EVA for Banks: Value Creation, Risk management, and Profitability Measurement. Journal of Applied Corporate Finance 9(2), 94–113 (1996)
11. Tracey, W., Worthington, A.: The usefulness of the EVA and its components in Australian. Journal of Accounting 7(1) (2001)

Research on the Impact of VAT Transformation on Investment in Fixed Assets and the Performance of Listed Companies in Northeast China

Lan Fengyun and Wang Chunhua

School of Business, Beijing Wuzi University Beijing, 101149, P.R. China
lanfy@sina.com, Wangchunhua888@sina.com

Abstract. China has been using production-based VAT for a long time, while most countries in the world using consumption-based VAT. Compared with production-type VAT, consumption-based VAT has many advantages, such as stimulating investment, reducing the burden of enterprises. In addition, production-type VAT has not suited to China's economic development. In September 2004, the State Council required the implementation of VAT reform in Northeast China, allowing companies in some sectors offset contained in the device class of fixed assets input tax. This paper analyzes the impact of VAT transformation reform of listed companies. In theory, the transformation of VAT directly affects the company's recorded value of fixed assets and depreciation. And it can increase corporate profits, net profits. In order to test the impact of VAT transformation, this paper put forward two hypotheses: first, VAT transformation will stimulate investment; Second, value-added tax transformation will increase profit of listed company. This paper selects 2001-2003 and 2005-2008 financial data of 56 listed companies in Northeast China as the sample data, through the establishment of models, multiple linear regressions to test the value-added tax and profits, the empirical results show that the value-added tax has a positive role in the enterprise, improve the business performance.

Keywords: VAT, fixed assets, investment, performance, empirical analysis.

1 Introduction

Value-added-tax (VAT) first introduced in France in 1954. Since then, many scholars in different countries have studied VAT from different aspects. However, the VAT transformation has typical Chinese characteristics, In foreign countries, especially developed countries, VAT is a consumption-type VAT when it was first introduced in, so there is few foreign literature or monographs which is directly on the VAT. Even though this paper studies the impact of VAT transformation on companies, in fact the analysis focus on how transformation affect cash flow by reducing the tax burden of new fixed assets investment, so a large number of foreign researches involving the impact of VAT on income tax burden and the impact on fixed assets investment are still valuable references. Clement Carbonnier (2007) studied two French VAT reforms provide visual evidence of tax shifting, found consumer share of the sales tax burden is 57% in the new car sales market and 77% in the housing repair services market. In

H. Tan (Ed.): Knowledge Discovery and Data Mining, AISC 135, pp. 25–30.
springerlink.com

twelve EU countries, the lower VAT-tariff is applied to flowers and plants in order to promote the production and employment in floriculture.Bunte, F.H.J, Kuiper, W.E (2008) assesses whether the VAT-regulation for flowers and plants achieves the goals set - promoting consumer demand and production and employment in the ornamental supply chain - (effectiveness) and at what cost (efficiency). The empirical results show that the VAT-regulation for floriculture is effective, but not very efficient. As to the impact of tax on fixed assets investment,, foreign scholars have conducted empirical research. Cummins, Hassett and Hubbard (1994) found capital investment in fixed assets using the cost elasticity of 0.5 to 1.0. Blundell, R. (2009) study the likely impact of a temporary VAT cut stimulus policy on consumer demand in the UK. It suggests that around 75 per cent of the VAT reduction will be passed on to consumers and that consumers will react by maintaining their expenditure levels and therefore increasing their demand for consumption goods. The uncertainty caused by the downturn makes this a more muted impact than we might have hoped, especially on the demand for durable goods. Increases in German core inflation following the 2007 VAT hike were smaller than expected, leading to speculation about delayed inflationary effects. Carare, Alina and Danninger, Stephan (2008) find that core inflation rose by 0.36 percentage point in the run up and by a further 0.40 percentage point at the time of the VAT hike. Cumulatively, the tax hike contributed to two thirds of the increase in core inflation in 2006-07 at an estimated pass-through of 73 percent. Most of the increase in 2006 was of general nature, while about one sixth can be attributed to durable goods and items with low degree of competition.

2 Main Text

2.1 The Research Hypothesis

Business investment cycle was significantly reduced, thus the ranges of available investment projects were increased, and the transformation helped to stimulate enterprises to increase investment, improving the business of organic composition. From the actual situation of the pilot regions, the VAT transformation and other policies in Northeast region strengthened the technical transformation and equipment investment of enterprises. The eight sector investment accounted for the proportion of the total investment significantly increased.

Compared with the consumption-type VAT, production-type VAT has the disadvantage of double taxation. In theory, no double taxation is the essential characteristics of VAT. VAT Transformation allows companies to deduct the value added tax during the purchase of production equipment and other fixed assets. Thus, the transformation reduced the burden of companies. Based on above analyses, we develop the hypothesis. H1: VAT transformation will stimulate investment.

Listed companies prefer consumption-type VAT, because consumption-type VAT will directly affect the cost of purchase of fixed assets and cash flows, thereby affecting the company's total assets and annual depreciation expense. Finally, the transformation will affect net income, operating cash flow. The implementation of the VAT transformation promoted the corporate performance. Based on above discussion, we develop the hypothesis H2: VAT transformation will increase the profit of listed company.

2.2 Research Methods and Results

There are 105 listed companies in Northeast region. Getting rid of companies whose data are not complete, 72 listed companies were left and 58 out of the 72 listed companies could benefit from VAT transformation. So we use the 58 listed companies' data 2005-2008 to test our hypothesis.

Dependent variable: I_{it} means growth rate in fixed assets (such as equipment).

Explanatory variables: ANI_{it} equals new added fixed assets*17%/total asset of the year.

Control variables: $I_{i(t-1)}$ the growth rate of fixed assets of last period;

R_{it} equals net profit / average amount of fixed assets (equipment),

D_{it}: equals current liabilities / free cash flow

Regression model 1:

$$I_{it} = a_0 + \beta_1 ANI_{it} + \beta_2 I_{i(t-1)} + \beta_3 R_{it} + \beta_4 D_{it} + \varepsilon_{it} \quad (1)$$

The regression results of model 1 are presented in table 1.

Table 1. Regression results of models

	Coefficient	Standard deviation	Sig..
ANI_{it}	0.06 (0.350)*	0.004	0.082
$I_{i(t-1)}$	-1.03 (-0.404) *	0.159	0.057
R_{it}	7.42E-04 (1.07) **	0.077	0.002
D_{it}	-6.70 (-3.914)	0.010	0.36
R^2	0.204		
F	9.810		

The results presented in table 1 indicate that β_1 is positive, so the result supports the hypothesis.

Now we test hypothesis 2.

Dependent variable: EPS. There are many indicators reflect business performance, such as return on assets, return on equity and earnings per share, etc., but there is disagreement in the empirical study to choose appropriate indicators reflect the company's performance among researchers. Majority of researchers agree EPS is a relative good indicator.

Explanatory variable: ANI: New fixed assets*17%/total asset end

EPS_{it} = Net profit / number of ordinary shares, earnings per share.

ROE_{it} =Net profit/average net assets. Means net return on equity.

EM_{it} = Equity / debt

$LNTA_{it}$ = Natural logarithm of total assets,

Regression model 2:

$$EPS_{it} = \alpha_0 + \beta_1 ANI_{it} + \beta_2 ROE_{it} + \beta_3 EM_{it} + \beta_4 LNTA_{it} + \varepsilon_{it} \quad (2)$$

Table 2. EPS descriptive statistics

	Sample	Min	Max	Average
2001	52	-1.30	0.67	0.0954
2002	52	-0.57	0.67	0.1052
2003	52	-1.27	0.84	-0.0053
2005	51	-1.02	0.62	0.0089
2006	52	-0.99	1.39	0.1177
2007	51	-1.29	1.49	0.2850
2008	51	-1.18	1.53	0.2685

2、Paired samples t test

Table 3. Paired sample mean difference test

	Difference					t	df	Sig.
	Average	Standard deviation	S.E. Mean	The 95% confidence interval				
				Min	Max			
EPS before transformation - EPS after transformation	-0.24985	0.7453243	******	-0.46627	-0.033430	-2.332	47	0.025

Since the transformation was conducted in 2004, the data is removed, in order to avoid VAT transformation influence.

EPS in the year of 2001, 2002 and 2003 are the group before transformation, while EPS in the year of 2005, 2006, 2007 and 2008 are the group after transformation. Paired samples t test method is used to test if EPS increased after VAT transformation.

Table 4. Paired samples statistics analysis

	Average	Sample	Standard deviation	S.E. Mean
EPS before transformation	0.082762	48	0.2282703	0.0329480
EPS after transformation	0.332612	48	0.7859346	0.1134399

Table 4 indicates that average EPS before transformation is less than the average EPS after transformation. Table 3 shows that the result is significant. It demonstrates the difference between EPS before and after transformation is significant.

Table 5. Regression results of models

	Coefficient	Standard deviation	Sig..
ANI_{it}	0.09692 （1.064） **	1.407	0.023
ROE_{it}	0.312 （2.899） ***	0.042	0.006
EM_{it}	-0.0925 （-2.721）	0.034	0.189
$LNTA_{it}$	0.270 （0.203）	0.084	0.302
R^2	0.323		
F	6.005		

The results presented in table 5 indicate that, β_1 is positive, so support the hypothesis, but the coefficient is little. The results show that the VAT transformation improve the profits of listed companies, but not very efficient.

3 Conclusions and Future Research Direction

Empirical results suggest the value-added tax transformation significantly promote listed company investment on fixed asset, and can improve the profits but not very efficient.

Based on the conclusion, this paper suggests the following policy: the appropriate lower rate of small-scale VAT taxpayers, in the value-added tax transformation process, the biggest benefit is the large enterprises, although the small-scale VAT taxpayers to reduce the tax rate, but not enough. Further expand the scope of VAT deduction, such as houses, buildings and other real estate not included in the scope of VAT deduction, because the VAT transformation is not enough.

Acknowledgements. This work is partially supported by base of Business Administration innovation, Beijing Wuzi University (project number: WYJD200904).

References

1. Carbonnier, C.: Who pays sales taxes- Evidence from French VAT reforms. 1987–1999 Journal of Public Economics 91(5-6), 1219–1229 (2007)
2. Bunte, F.H.J., Kuiper, W.E.: Promoting floriculture using VAT regulation. European Journal of Horticultural Science 73(6), 248–253 (2008)

3. Cummins, J.G., Hassett, K.A., Hubbard, R.G.: A Reconsideration of Investment Behaviour Using Tax Reforms as Natural Experiments. Brookings Papers on Economc Activity 2, 1–60 (1994)
4. Blundell, R.: Assessing the Temporary VAT Cut Policy in the UK. Fiscal Studies 30, 31–38 (2009), doi:10.1111/j.1475-5890.2009.00088.x
5. Carare, A., Danninger, S.: Inflation Smoothing and the Modest Effect of VAT in Germany (2008), IMF Working Papers, Available at SSRN, http://ssrn.com/abstract=1266508

Design and Implementation of Tracking System for Dish Solar Thermal Energy Based on Embedded System

Jian Kuang and Wei Zhang

Beijing Key Laboratory of Intelligent Telecommunications Software and Multimedia, Beijing University of Posts and Telecommunications, Beijing, China
jkuang@bupt.edu.cn, zwcihcn@gmail.com

Abstract. Solar thermal energy has lots of advantages compare with photovoltage (PV). Most of previous study of sun tracking system looked at PV, whose accuracy and stability can't satisfy the requirements of thermal energy system. This paper gives a design and implementation of tracking system for dish solar thermal energy based on embedded system that mixes active and passive tracking, uses more stability sensors and adopts exception handling algorithm to improve both accuracy and stability. Experiments show that the design is effective.

Keywords: Embedded system, sun tracking, solar thermal energy.

1 Introduction

Solar thermal energy (STE) is a technology for harnessing solar energy for thermal energy (heat). A dish solar thermal system uses a large, reflective, parabolic dish to focus all the sunlight that strikes the dish up onto to a single point above the dish, where a receiver captures the heat and transforms it into a useful form. The advantage of a dish system is that it can achieve much higher temperatures due to the higher concentration of light. Higher temperatures lead to better conversion to electricity and the dish system is very efficient on this point. By using a Stirling engine the theory highest efficient may reach 29.4% [1], much higher than other solar power generating method such as solar photovoltage. Based on these features of dish solar thermal system, tracking system should provide higher accuracy than before. What's more, because of the high temperature is dangerous, the stability should be improved and the exception handling is necessary.

1.1 Related Works

Automatic tracking system in solar power system is deeply discussed in recent years. Wu Chunsheng and Wang Yibo [2] make some conclusion about it. According to the ways to get the position signal of the sun, the sun-tracking methods can be classified as active one and passive one. The passive control method converts the incident

H. Tan (Ed.): Knowledge Discovery and Data Mining, AISC 135, pp. 31–38.
springerlink.com

sunlight to current signal by means of the solar sensors to get sun's position, while the active one calculates the location of the sun in the sky by taking use of solar orbit function and a real-time clock to acquire precise time signal at any time. In their design, they adopt active control technology based on micro-controller, but in practical production, parameters used in calculates the location of the sun in the sky is hard to set with great precision so the errors can't be controlled in acceptable level. In this way, the design should be improved in order to satisfy the need of dish solar thermal energy system.

Xie Yuan and Guo Wencheng [3] proposed a design based on ARM microcontroller. They mixed active and passive tracking to improve accuracy of tracking system. The main question in their design is that optical camera has been used for passive tracking. Because of poor environment of solar power applications, cameras are easy to dirty and then their images are not suitable for tracking. In this reason, extra clean and maintain work should be done, the cost is increased. On the other hand, real time operation system (RTOS) is not used in their design makes tasks are high coupling and the system is hard to expand. At last, the design thinks little about exception handling and network.

1.2 Main Research Work and Content Arrangement

In this paper, we propose a new design for tracking system for dish solar thermal energy and bears the following five advantages compare with other designs:

- Active and passive tracking methods are mixed and more suitable sensors are adopted in poor environment to improve tracking accuracy;
- RTOS are used in system to improve scalability and real time;
- More sensors such as wind sensor, rain sensor and total radiometer are used so the tracking system can run smartly in conjunction with the program;
- Exception handling are considered seriously for security reasons;
- Tracking systems can work in network so that administrators can monitor and control them remotely;

We have conducted extensive experiments to test the tracking accuracy and stability of our design in various situations. Experiments show that our design is improve not only tracking accuracy but also stability greatly, and works more smartly than former designs.

The rest of this paper is organized as follows: Section 2 describes the hardware design. In Section 3, we give the software designs. We present our experiments results in Section 4. Finally, Section 5 offers the conclusion.

2 Hardware Design

The system hardware modules, including MCU, sensor subsystem, execute subsystem, storage subsystem and network subsystem five parts, as shown in figure 1.

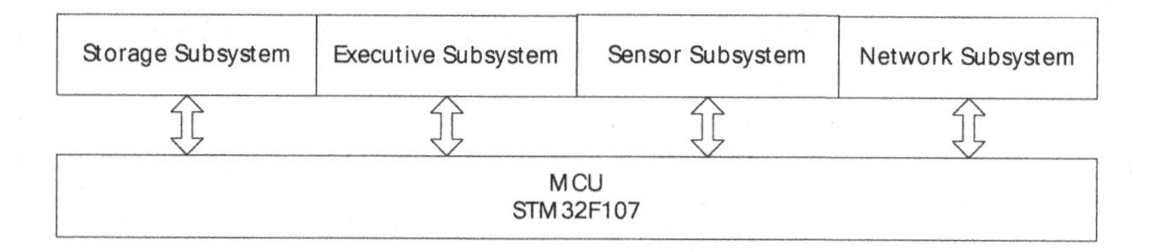

Fig. 1. System hardware modules

2.1 MCU

The system's MCU is used in ST microelectronics' STM32 family Connectivity Line STM32F107, which is based on ARM Cortex™-M3 core with 256 KB Flash, USB OTG, Ethernet, 10 timers, 2 CANs, 2 ADCs, 14 communication interfaces, fully meet system's requirements. STM32F107 is adopted because of not only its rich peripherals but also featuring architectural enhancements with the Thumb-2 instruction set to deliver improved performance with better code density, significantly faster response to interrupts, all combined with industry-leading power savings.

2.2 Sensor Subsystem

Rich sensors are used in our system. We organize sensors as a subsystem, and use a bus to connect it with MCU in order to reduce the complexity of MCU connection. A uniform sensor protocol is adopted to access data provided by sensors including absolute encoder, wind sensor, rain sensor, total radiometer, GPS, magnetic compass, digital input and thermocouple array, as shown in figure 2.

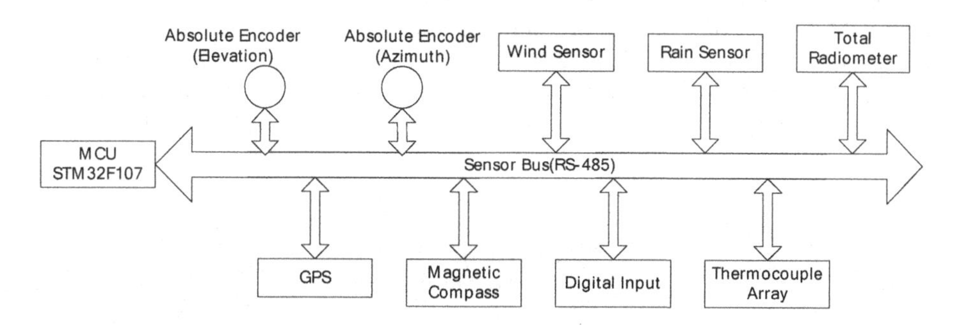

Fig. 2. Sensor subsystem

The function of each sensor is list below:

- Absolute Encoder: obtain current angles for two axis(elevation and azimuth);
- Wind sensor, rain sensor and total radiometer: obtain whether information in work environment;

- GPS and magnetic compass: obtain geographic information to provide enough parameters for active tracking;
- Digital input: obtain binary signal provide by other systems such as Stirling engine;
- Thermocouple array: Get the heat distribution on the receiver so we can use the information for passive tracking;

2.3 Execute Subsystem

Execute subsystem include motor controller and digital output. Pulse signal is used to control motor controller and binary signal is used to control or inform other system such as Stirling engine so we connect execute subsystem with MCU directly, as shown in figure 3.

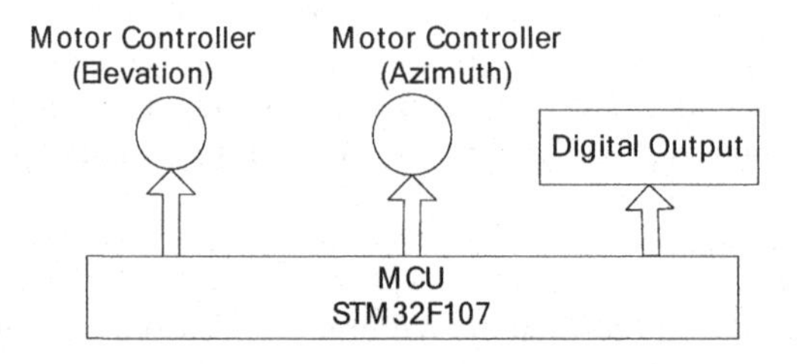

Fig. 3. Execute subsystem

2.4 Storage Subsystem

We use two types of different storage medium to store different information. EEPROM (AT24C02) is used to store important parameters needed by system operations, and flash (SST25VF032B) is used to store lots of logs. The advantage of this design is separating storages of different importance, operation frequency and capacity, improved storage stability. MCU uses I2C interface to connect EEPROM, and SPI interface to connect flash, as shown in figure 4.

Fig. 4. Storage subsystem

2.5 Network Subsystem

STM32F107 has integrated an Ethernet MAC controller on the chip. We adopt DP83848CVV as PHY controller, in this way, our system obtains the ability of connecting to network. System could upload log and accept monitor and control remotely by using a certain protocol.

3 Software Design

Our design is very complex. Lots of functions are considered so we decide to use a RTOS (RT-Thread) to reduce the coupling of functions and provide better real time performance. The tasks in software design are shown in figure 5.

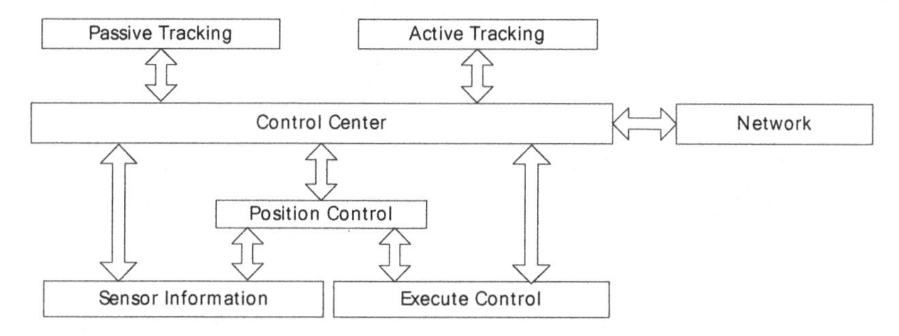

Fig. 5. Tasks in software design

3.1 Control Center

Control center task provide time-driving scheduling, intelligent operation such as deal with wind and rain, exception handling and so on. There is a finite-state machine (FSM) in the task to finish the function, as shown in figure 6.

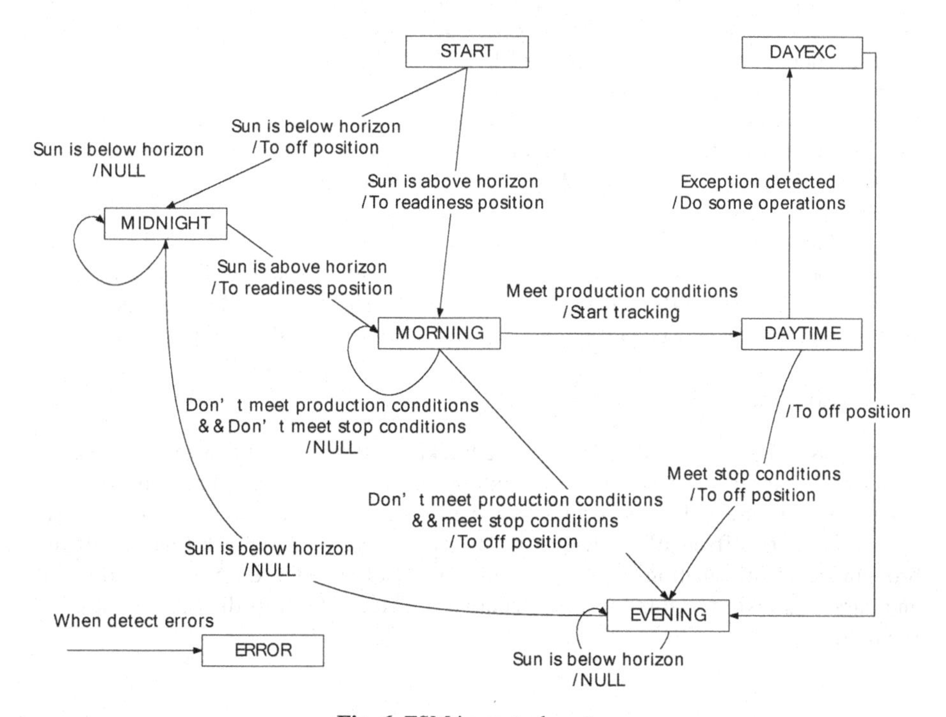

Fig. 6. FSM in control center

3.2 Position Control

Position control provides the function of moving dish to a certain position defined by elevation angle and azimuth angle. The task operates two encoders and two motors to make up a closed loop feedback system. PID is adopted to improve the movement response, as shown in figure 7.

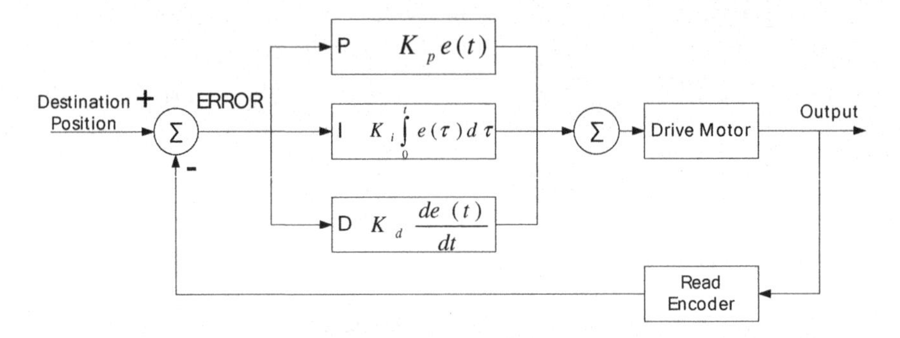

Fig. 7. Position control feedback

3.3 Active Tracking

Active tracking use lots of parameters to calculate sun's position on the sky. Some of parameters are provided by peripherals on the MCU, such as time and data; some other parameters are measured by sensors such as latitude, longitude and elevation; and the rest of parameters are stored in EEPROM, such as time zone.

We measures accuracy and computing resource consuming of NOAA and SPA, find SPA is more suitable for our system. SPA calculate the solar elevation and azimuth angles in the period from the year -2000 to 6000, with uncertainties of ±0.0003°,whose accuracy is much higher than other algorithm. The algorithm is described by Jean Meeus [4]. We also use the calculation of incidence angle for a surface that is tilted to any horizontal and vertical angle, as described by Iqbal [5].

An experiment has been done to verify the feasibility of running SPA algorithm on our system. We run SPA calculation on STM32F107 with 36MHz frequency for 10000 times and that costs 78 seconds. The result is far beyond our expectation.

3.4 Passive Tracking

We proposed a new method called passive tracking based on heat distribution. We use a thermocouple array to get heat distribution on the receiver and calculate the area whose temperature is highest on the receiver. After that, find the center of the area and estimate the offsets of the receiver. At last, passive tracking task send a message brought offset information to control center. Control center use both active tracking and passive tracking results for more accurate tracking. The flow diagram is shown in figure 8.

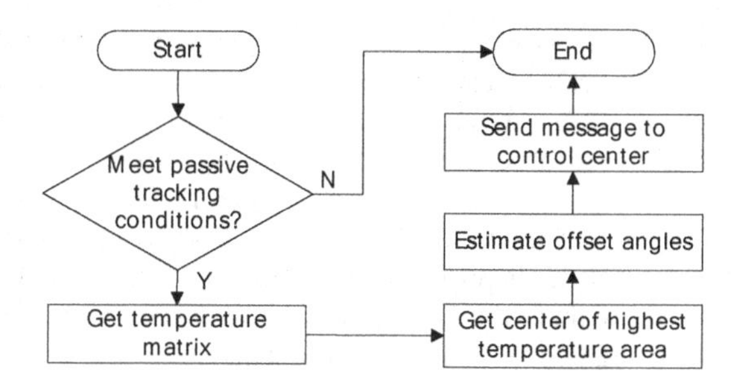

Fig. 8. Flow diagram for passive tracking

3.5 Network

We design a protocol called Modbus over UDP to make tracking system can communicate with central controller. Modbus is a widely used industrial field bus usually carried by RS-485, but we use UDP to transmit Modbus packets to improve the performance of traditional Modbus.

4 Experiments and Analysis

We have two former prototypes. One is only run active tracking (NOAA); the other one is only use passive tracking. Both of them choose DSP as controller. We use the same Stirling engine and motor to compare the power output. We do experiments in Ningxia, China. 10KW Stirling engines and dishes of 100 M^2 are adopted.

We run three prototypes at the same time for a week, and calculate the average daily power output curves shown as figure 9.

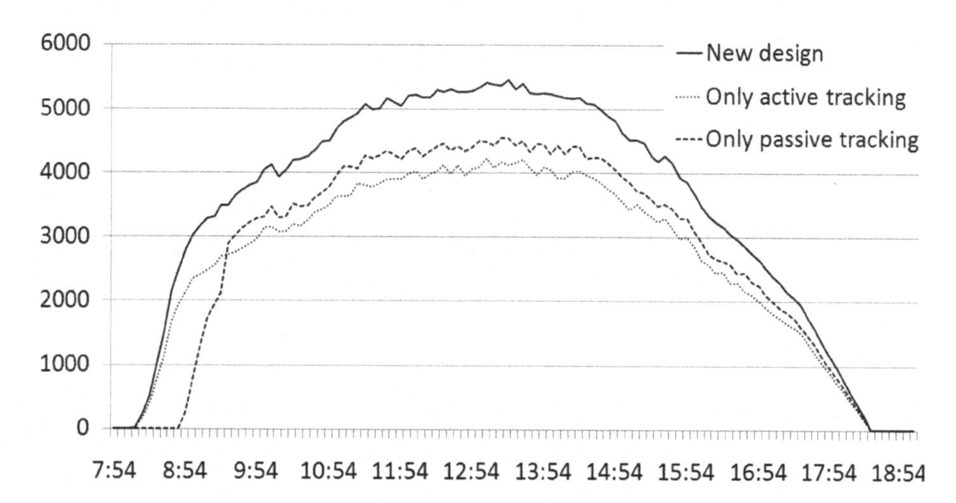

Fig. 9. Experiment result

The result shows that new design is produce more power output than the design which only uses active tracking. The reason is we use more accurate SPA algorithm instead of NOAA, and we use passive tracking result to fix errors made by active tracking. The new design also performs better than the only passive tracking design because we use more stable sensors to provide passive tracking information. It's easy to find that the only passive tracking design works bad on mornings because extra time must be used to search sun's position when sunrise without active tracking.

Experiments also show that when exceptions happen such as hardware errors and whether changes our design could handle them properly.

5 Conclusion

In this paper, we propose a new design to improve tracking accuracy by mixing active and passive tracking, using more stability sensors to make system works more smarter, and adopting RTOS to make system not only easy to expand but also provide better real-time performance. Exception handling makes system can operation reliably in poor environment for a long time. Experiments show our design is effective and reliable, has instructive significance for practice dish solar thermal energy production.

References

1. Patel, M.R.: Wind and Solar Power System, 2nd edn. CRC Press (2005)
2. Wu, C.-S., Wang, Y.-B., Liu, S.-Y., Peng, Y.-C., Xu, H.-H.: Study on Automatic Sun-tracking Technology in PV Generation. In: Third International Conference on Electric Utility Deregulation and Restructuring and Power Technologies (2008)
3. Xie, Y., Guo, W.: The Design of ARM-Based Automatic Sun Tracking System. In: International Conference on Intelligent System Design and Engineering Application (2010)
4. Meeus, J.: Astronomical Algorithms, 2nd edn. Willmann-Bell, Inc., Richmond (1998)
5. Iqbal, M.: An Introduction to Solar Radiation, New York (1983)

An Alternating Simultaneously Minimizing Diagonal Matrix Error and Covariant Matrix Error Trilinear Decomposition Algorithm for Second-Order Calibration

Jianqi Sun

College of Chemistry and Envioromental Engineering, Jiujiang University, Jiujiang, P.R. China,
sunjianqi01@sina.com

Abstract. An alternating trilinear decomposition algorithm based on simultaneously minimizing diagonal matrix error and covariant matrix error (ADCE) is developed for three-way data analysis. By alternatively optimizing three objective functions with intrinsic relationships, ADCE algorithm keeps the 'second-order advantage' of second-order calibration methods and provides a natural way to avoid the two-factor degeneracies, which is intrinsic in the traditional PARAFAC algorithm. The simulated results and real experimental results show ADCE algorithm has the features of fast convergence rate as well as insensitivity to the overestimated factor number, in other words, it not only avoids the dilemma of identifying the actual component number accurately in practical problem but also converges fast, which is rather difficult to handle for the traditional PARAFAC algorithm.

Keywords: Trilinear decomposition, ADCE algorithm, PARAFAC algorithm, Second-order calibration.

1 Introduction

With the development of hyphenated techniques, such as GC-MS, HPLC-DAD, CE-DAD, etc.[1], the second-order calibration methods used to analyze three-way data arrays have been increasingly recognized. These methods produce the so-called 'second-order advantage'[1,2].

Currently, the most commonly used trilinear decomposition algorithms for second-order calibration are iterative algorithms[3-6]. Among of them, the parallel factor analysis (PARAFAC) algorithm[3] proposed by Harshman must be the most famous and widely accepted one. PARAFAC theory has been investigated thoroughly and successfully applied to solve many chemical problems[6,7], but the requirements of accurate estimation of the number of factors and slow convergence rate may be the obstacle to its wider applications[8,9]. Thus, it is very important to develop the algorithms that converge fast and need not accurately estimate the number of factors, such as alternating trilinear decomposition (ATLD) algorithm[5].

In this paper, an alternating simultaneously minimizing diagonal matrix error and covariant matrix error (ADCE) algorithm is developed for three-way data arrays

H. Tan (Ed.): Knowledge Discovery and Data Mining, AISC 135, pp. 39–46.
springerlink.com

analysis. Results of simulated data and real data are presented to demonstrate the performance of ADCE algorithm.

2 Theory

All the following discussion is based on the famous trilinear model proposed by Harshman[3], and more detailed descriptions of model can be found in reference 6.

$$x_{ijk} = \sum_{n=1}^{N} a_{in} b_{jn} c_{kn} + e_{ijk} \quad i = 1,2,...,I; j = 1,2,...J; k = 1,2,...K \tag{1}$$

Here, N denotes the number of factors, x_{ijk} is the element of the three-way data array $\underline{\mathbf{X}}$ $(I\times J\times K)$, a_{in}, b_{jn} and c_{kn} is the corresponding element of three loading matrices **A**, **B** and **C** dimensioned by $I\times N$, $J\times N$ and $K\times N$, respectively, and e_{ijk} is the element of the corresponding three-way residual array $\underline{\mathbf{E}}$ $(I\times J\times K)$. Since trilinear model treats three loading matrices in a symmetric way, the following matrix forms may be more convenient and easier to understand:

$$\mathbf{X}_{i..} = \mathbf{B}diag(\mathbf{a}_i)\mathbf{C}^{\mathrm{T}} + \mathbf{E}_{i..}, \quad i = 1,2,...,I \tag{2}$$

$$\mathbf{X}_{.j.} = \mathbf{C}diag(\mathbf{b}_j)\mathbf{A}^{\mathrm{T}} + \mathbf{E}_{.j.}, \quad j = 1,2,...,J \tag{3}$$

$$\mathbf{X}_{..k} = \mathbf{A}diag(\mathbf{c}_k)\mathbf{B}^{\mathrm{T}} + \mathbf{E}_{..k}, \quad k = 1,2,...,K \tag{4}$$

$\mathbf{X}_{i..}$, $\mathbf{X}_{.j.}$ and $\mathbf{X}_{..k}$ is the *ith* horizontal slice, the *jth* lateral slice and the *k*th frontal slice of the three-way array $\underline{\mathbf{X}}$, respectively, and $\mathbf{E}_{i..}$, $\mathbf{E}_{.j.}$ and $\mathbf{E}_{..k}$ is separately the corresponding slice of the three-way residue array $\underline{\mathbf{E}}$. For an HPLC-DAD three-way data set, $\mathbf{a}_i$, $\mathbf{b}_j$ and $\mathbf{c}_k$ corresponds to the *i*th row of relative chromatogram matrix **A**, the *j*th row of relative spectrum matrix **B** and the *k*th row of relative concentration matrix **C**, respectively. Diag(·) denotes the diagonal matrix of order $N\times N$ in which the diagonal elements are the corresponding elements of a vector, the superscript "T" denotes the transpose of a matrix.

The following equations hold for a trilinear model.

$$\begin{aligned} &\mathbf{B}^{+}\mathbf{X}_{i..}(\mathbf{C}^{\mathrm{T}})^{+} = diag\ (\mathbf{a}_i) + \mathbf{B}^{+}\mathbf{E}_{i..}(\mathbf{C}^{\mathrm{T}})^{+} \\ &\mathbf{X}_{i..}(\mathbf{C}^{\mathrm{T}})^{+} = \mathbf{B}diag\ (\mathbf{a}_i) + \mathbf{E}_{i..}(\mathbf{C}^{\mathrm{T}})^{+} \\ &i = 1,2,...,I \end{aligned} \tag{5}$$

$$\begin{aligned} &\mathbf{C}^{+}\mathbf{X}_{.j.}(\mathbf{A}^{\mathrm{T}})^{+} = diag\ (\mathbf{b}_j) + \mathbf{C}^{+}\mathbf{E}_{.j.}(\mathbf{A}^{\mathrm{T}})^{+} \\ &\mathbf{X}_{.j.}(\mathbf{A}^{\mathrm{T}})^{+} = \mathbf{C}diag\ (\mathbf{b}_j) + \mathbf{E}_{.j.}(\mathbf{A}^{\mathrm{T}})^{+} \\ &j = 1,2,...,J \end{aligned} \tag{6}$$

$$\begin{aligned} &\mathbf{A}^{+}\mathbf{X}_{..k}(\mathbf{B}^{\mathrm{T}})^{+} = diag\ (\mathbf{c}_k) + \mathbf{A}^{+}\mathbf{E}_{..k}(\mathbf{B}^{\mathrm{T}})^{+} \\ &\mathbf{X}_{..k}(\mathbf{B}^{\mathrm{T}})^{+} = \mathbf{A}diag\ (\mathbf{c}_k) + \mathbf{E}_{..k}(\mathbf{B}^{\mathrm{T}})^{+} \\ &k = 1,2,...,K \end{aligned} \tag{7}$$

Where superscript "+" symbolizes the Moore-Penrose generalized inverse, $\mathbf{B}^{+}\mathbf{E}_{i..}(\mathbf{C}^{\mathrm{T}})^{+}$, $\mathbf{C}^{+}\mathbf{E}_{.j.}(\mathbf{A}^{\mathrm{T}})^{+}$ and $\mathbf{A}^{+}\mathbf{E}_{..k}(\mathbf{B}^{\mathrm{T}})^{+}$ can be separately called the corresponding diagonal matrix error, and $\mathbf{E}_{i..}(\mathbf{C}^{\mathrm{T}})^{+}$, $\mathbf{E}_{.j.}(\mathbf{A}^{\mathrm{T}})^{+}$ and $\mathbf{E}_{..k}(\mathbf{B}^{\mathrm{T}})^{+}$ is defined as the corresponding covariant matrix error, respectively.

To establish an algorithm with high convergence rate and insensitivity to overestimated number of factors, diagonal matrix error and covariant matrix error were simultaneously minimized in the following alternative way:

$$F(\mathbf{C})=\sum_{k=1}^{K}(\|\mathbf{A}^{+}\mathbf{X}_{..k}(\mathbf{B}^{T})^{+}-diag(\mathbf{c}_k)\|_F^2+\|\mathbf{X}_{..k}(\mathbf{B}^{T})^{+}-\mathbf{A}diag(\mathbf{c}_k)\|_F^2)$$
$$\text{(with } \mathbf{A} \text{ and } \mathbf{B} \text{ fixed, minimizing } F(\mathbf{C}) \text{ to obtain } \mathbf{C}) \tag{8}$$

$$F(\mathbf{B})=\sum_{j=1}^{J}(\|\mathbf{C}^{+}\mathbf{X}_{.j.}(\mathbf{A}^{T})^{+}-diag(\mathbf{b}_j)\|_F^2+\|\mathbf{X}_{.j.}(\mathbf{A}^{T})^{+}-\mathbf{C}diag(\mathbf{b}_j)\|_F^2)$$
$$\text{(with } \mathbf{A} \text{ and } \mathbf{C} \text{ fixed, minimizing } F(\mathbf{B}) \text{ to obtain } \mathbf{B}) \tag{9}$$

$$F(\mathbf{A})=\sum_{i=1}^{I}(\|\mathbf{B}^{+}\mathbf{X}_{i..}(\mathbf{C}^{T})^{+}-diag(\mathbf{a}_i)\|_F^2+\|\mathbf{X}_{i..}(\mathbf{C}^{T})^{+}-\mathbf{B}diag(\mathbf{a}_i)\|_F^2)$$
$$\text{(with } \mathbf{B} \text{ and } \mathbf{C} \text{ fixed, minimizing } F(\mathbf{A}) \text{ to obtain } \mathbf{A}) \tag{10}$$

Here, $\|\cdot\|_F$ denotes the Frobenius matrix norm.

Solving Eq. (8), (9) and (10) gives the iterative updates of $\mathbf{c}_k$, $\mathbf{b}_j$ and $\mathbf{a}_i$ as follows:

$$\mathbf{c}_k=(diag[(\mathbf{A}^{+}\mathbf{X}_{..k}(\mathbf{B}^{T})^{+}+\mathbf{A}^{T}\mathbf{X}_{..k}(\mathbf{B}^{T})^{+})]./diag(\mathbf{A}^{T}\mathbf{A}+\mathbf{I}_N))^{T}$$
$$(k=1,2,...K) \tag{11}$$

$$\mathbf{b}_j=(diag[(\mathbf{C}^{+}\mathbf{X}_{.j.}(\mathbf{A}^{T})^{+}+\mathbf{C}^{T}\mathbf{X}_{.j.}(\mathbf{A}^{T})^{+})]./diag(\mathbf{C}^{T}\mathbf{C}+\mathbf{I}_N))^{T}$$
$$(j=1,2,...J) \tag{12}$$

$$\mathbf{a}_i=(diag[(\mathbf{B}^{+}\mathbf{X}_{i..}(\mathbf{C}^{T})^{+}+\mathbf{B}^{T}\mathbf{X}_{i..}(\mathbf{C}^{T})^{+})]./diag(\mathbf{B}^{T}\mathbf{B}+\mathbf{I}_N))^{T}$$
$$(i=1,2,...I) \tag{13}$$

Where "./" denotes the corresponding elementwise division, $\mathbf{I}_N$ is an identity matrix dimensioned by $N \times N$.

Based on the above descriptions, ADCE algorithm can be designed as follows:

1. Initialize loading matrix **A** and **B** randomly,
2. Computer loading matrix **C** using Eq. (11),
3. Computer loading matrix **B** using Eq. (12) and normalize **B** column-wisely,
4. Computer loading matrix **A** using Eq. (13) and normalize **A** column-wisely,
5. Repeat steps 2-4 until a stopping criterion is satisfied.

3 Experimental Procedures

3.1 Simulated HPLC-DAD Data Array

Simulated HPLC-DAD three-way data set was produced by six simulated samples containing four simulated chemical species. The chromatographic profiles of these species, $\boldsymbol{a}_1$, $\boldsymbol{a}_2$, $\boldsymbol{a}_3$ and $\boldsymbol{a}_4$ were generated by $\boldsymbol{a}_{1,i}$=gs(i,35,10), $\boldsymbol{a}_{2,i}$=gs (i,25,9), $\boldsymbol{a}_{3,i}$= gs (i,15,5), and $\boldsymbol{a}_{4,i}$=gs(i,35,8) (i=1,2,…,50) and the spectral profiles $\boldsymbol{b}_1$, $\boldsymbol{b}_2$, $\boldsymbol{b}_3$, and $\boldsymbol{b}_4$ were simulated by $\boldsymbol{b}_{1,j}$=0.7gs($2j$-1,30,20), $\boldsymbol{b}_{2,j}$=0.8gs($2j$-1,20,10)+0.2gs($2j$-1,60,70), $\boldsymbol{b}_{3,j}$=0.6gs($2j$-1,50,30)+0.3gs($2j$-1,20,5), and $\boldsymbol{b}_{4,j}$=0.8gs($2j$-1,60,20)(j=1,2,…,50).

Here, gs(x, a, b) is Gaussian fucnction defined as gs (x, a, b) = exp $[-(x-a)^2/2b^2]$.

The first four samples each containing only three components were treated as the calibration samples, and the others were used for the prediction samples each containing all four components. Concentrations of all components were uniformly distributed in the range of 0-1, as shown in Table 1. The three-way responses were constructed according to Eq. (4), in which the random noise was normally distributed with central value zero and standard deviation 0.2%.

3.2 Real HPLC-DAD Data Array

Real HPLC-DAD data were collected on an Agilent-1100 liquid chromatograph with a diode array detector (Agilent Corporation, America). Separations were performed on a Hypersil-ODS column (125mm×4mm i.d., particle size 5μm). The data set was constructed by measureing six standards containing catechol, resorcinol and hydroquinone. All solvents and reagents were analytical grade unless otherwise indicated. Stock solutions were prepared by distilled water at a concentration level of 0.6000 $g \cdot L^{-1}$. Concentrations of all components in real samples were shown in Table 2. The column temperature was 25°C. The flow rate was set to 1.0 $mL \cdot min^{-1}$ with a water (1% Acetic acid)-methanol (40:60, v/v) mobile phase. Spectra were collected at 1 nm interval from 268 nm to 298 nm. The elution time was set from 1.086 min to 1.399 min, with an interval of 1/150 min. The effects of solvent were eliminated as possible by blank corrections before data treatment. The first four samples were considered as the calibration samples, and the rest were the prediction samples.

3.3 Implementation of ADCE Algorithm and PARAFAC Algorithm

All computer programs were written in Matlab and carried out on a personal computer Pentium IV processor with 256 MB RAM under the Windows XP operating system.

For all three-way data arrays, random initialization was implemented to start the iterative optimization procedures of ADCE algorithm and PARAFAC algorithm. The optimization procedures of ADCE and PARAFAC are terminated when the following criterion reaches a certain threshold ε ($\varepsilon \leq 1\times10^{-6}$ in this paper).

$$\sigma^{(m)} = \sum_{k=1}^{K} \left\| \mathbf{X}_{..k} - \mathbf{A}^{(m)} diag(\mathbf{c}_k)^{(m)} (\mathbf{B}^{\mathrm{T}})^{(m)} \right\|_F^2, \quad \left| \frac{\sigma^{(m)} - \sigma^{(m-1)}}{\sigma^{(m-1)}} \right| \leq \varepsilon \tag{14}$$

Where σ is the sum of squares of residual error between the experimental value and the predicted value, and m is the current iteration number. A maximal iteration number of 3000 is adopted to avoid possible unduly slow convergence.

4 Results and Discussion

4.1 Simulated HPLC-DAD Data Array

When N=4, namely, the actual number of factors is chosen, the perfect agreement between the resolved chromatographic profiles or spectral profiles and the corresponding simulated ones, as is shown by Fig. 1, indicates both ADCE algorithm and PARAFAC algorithm work well. If the larger number of factors is chosen, for example, N=5, it is easy to see from Fig. 2a and 2b and Table 1 that ADCE algorithm is almost not subject to the overestimated number of factors. However, the results resolved by PARAFAC algorithm are poor. There is a significant difference between the resolved profiles and simulated ones, as shown by Fig. 2c and 2d. For the results resolved by ADCE algorithm, note that the excess column representing noise and background can be easily discriminated from the desired ones (Fig. 2a and 2b), taking a zigzag shape, and the resolved concentrations of noise component are all nearly zero (Table 1). It implies that only four components are present in this chemical system. The information drawn from the Fig. 2a and 2b and Table 1 strongly suggest that ADCE algorithm is insensitive to the overestimated number of factors in mixtures.

Table 1. True concentrations of components in the simulated samples and the corresponding resolved concentrations using ADCE algorithm with N=5

Sample	True value				Resolved value using ADCE algorithm				
	1	2	3	4[b]	1	2	3	4[b]	5
#1	0.9501	0.4565	0.4565	0.0000	0.9501	0.4570	0.4558	-	4.6×10^{-5}
#2	0.2311	0.0185	0.7382	0.0000	0.2308	0.0186	0.7377	-	-1.7×10^{-5}
#3	0.6068	0.8214	0.1763	0.0000	0.6063	0.8215	0.1755	-	-1.16×10^{-4}
#4	0.4860	0.4447	0.4057	0.0000	0.4868	0.4440	0.4077	-	-2.4×10^{-5}
#5	0.8913	0.6154	0.9355	0.8132	0.8881	0.6156	0.9310	-	-9.1×10^{-5}
#6	0.7621	0.7919	0.9169	0.4565	0.7580	0.7944	0.9132	-	2.42×10^{-4}
R^a					1.0000	1.0000	1.0000	-	-

a. Correlative coefficient between the resolved value and the true value.
b. Components treated as unknown, uncarlibrated interferents.

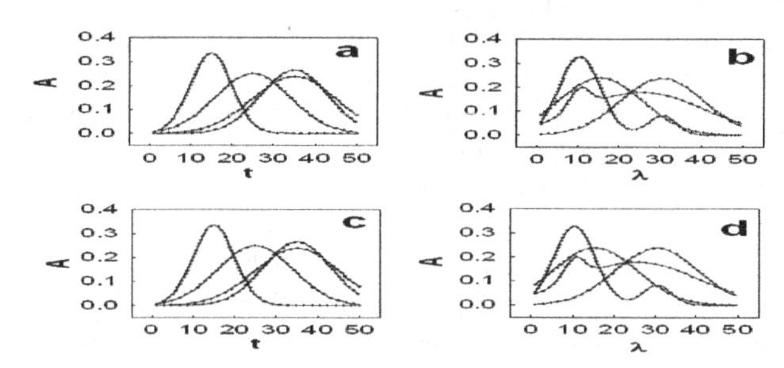

Fig. 1. Comparison of resolved chromatographic profiles and spectral profiles to simulated ones when N=4. The solid lines are resolved chromatographic profiles and spectral profiles; the dotted lines are the simulated ones. **a** and **b** are resolved by ADCE algorithm; **c** and **d** are resolved by PARAFAC algorithm.

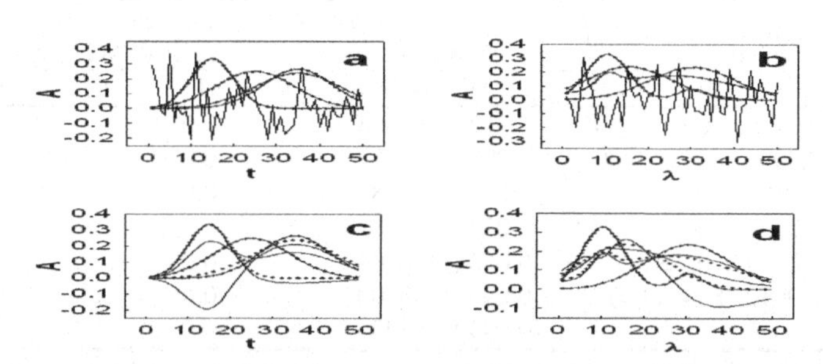

Fig. 2. Comparison of resolved chromatographic profiles and spectral profiles to simulated ones when N=5. The solid lines are the resolved chromatographic profiles and spectral profiles; the dotted lines are the simulated ones. **a** and **b** are resolved by ADCE algorithm; **c** and **d** are resolved by PARAFAC algorithm.

Another salient virtue of ADCE algorithm is its fast convergence rate. On the condition that the ture number of factors is chosen(N=4), the correct results can be obtained by both algorithms, however, ADCE algorithm always converges much faster than PARAFAC algorithm, as shown in Fig. 3.

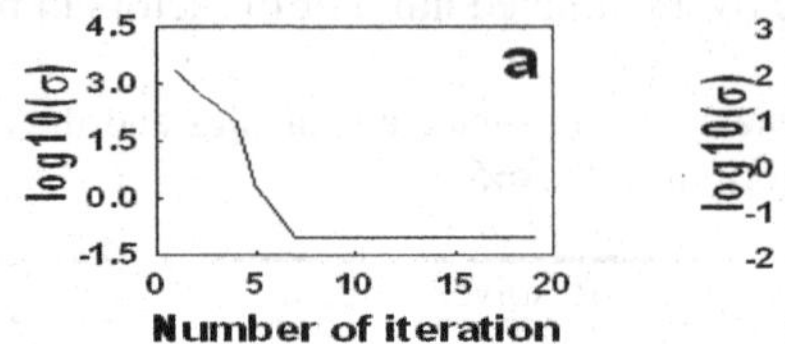

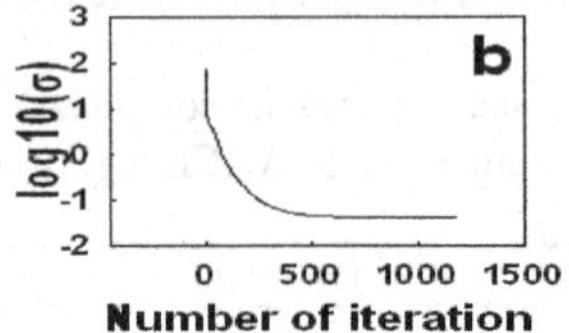

Fig. 3. Comparison of the convergency of ADCE algorithm to that of PARAFAC algorithm when N=4. **a** is obtained by ADCE algorithm and **b** is obtained by PARAFAC algorithm.

4.2 Real HPLC-DAD Data Array

When N=3, the solutions to both algorithms show small discrepancies compared with those experimentally measured for real HPLC-DAD data set, as is shown in Table 2 and Fig. 4. When N=4, the profiles resolved by ADCE algorithm still exhibit satisfying results, as shown in Fig. 5a and 5b and Table 3. It indicates ADCE algorithm is insensitive to overestimated number of factors once again. But PARAFAC algorithm gives a poor result (Fig. 5c and 5d), and it is difficult to distinguish profile representing noise and background from actual ones. In addition, the convergent features of ADCE algorithm and PARAFAC algorithm are compared in Fig. 6a and 6b, respectively. Obviously, when the correct number is chosen(N=3), ADCE algotithm has the higher convergence rate than PARAFAC algorithm. Only ADCE algorithm can give the correct solution when overestimated number is chosen(N=4). In this case, the PARAFAC algorithm may have been trapped in computing "swamp" due to two-factor degeneracies[8,9].

Table 2. Comparison of concentrations resolved by ADCE algorithm and PARAFAC algorithm with N=3 to added concentrations of all components in real samples ($g \cdot L^{-1}$)

Sample	Added concentration			Concentration resolved by ADCE algorithm			Concentration resolved by PAFARAC algorithm		
	HYD[b]	RES[b]	CAT[b]	HYD	RES	CAT[c]	HYD	RES	CAT[c]
#1	0.0768	0.0000	0.0000	0.0771	0.0003	-	0.0770	0.0004	-
#2	0.0000	0.0624	0.0000	0.0001	0.0627	-	0.0001	0.0621	-
#3	0.0576	0.0288	0.0000	0.0571	0.0282	-	0.0573	0.0295	-
#4	0.0336	0.0672	0.0000	0.0349	0.0682	-	0.0353	0.0672	-
#5	0.0504	0.0288	0.0768	0.0504	0.0301	-	0.0502	0.0312	-
#6	0.0672	0.0504	0.0384	0.0671	0.0500	-	0.0672	0.0497	-
R[a]				0.9999	0.9999	-	1.0000	0.9998	-

a. Correlative coefficient between the resolved concentration and the added concentration.
b. HYD: hydroquinone; RES: resorcinol; CAT: catechol.
c. Components treated as unknown, uncarlibrated interferents.

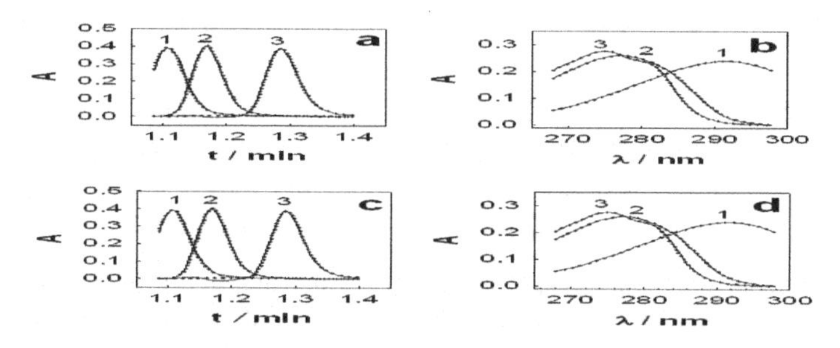

Fig. 4. Comparison of resolved chromatographic and spectral profiles to actual ones when N=3. The solid lines are the resolved chromatographic profiles and spectral profiles; the dotted lines are the actual ones. **a** and **b** are resolved by ADCE algorithm; **c** and **d** are resolved by PARAFAC algorithm. 1, hydroquinone; 2, resorcinol; 3, catechol.

Table 3. Added concentrations of all components in the real samples and the corresponding resolved concentrations using ADCE algorithm with N=4

Sample	Added concentration			Resolved concentration using ADCE algorithm			
	HYD[b]	RES[b]	CAT[b]	HYD	RES	CAT[c]	Noise
#1	0.0768	0.0000	0.0000	0.0768	0.0003	-	1.03×10^{-3}
#2	0.0000	0.0624	0.0000	0.0000	0.0627	-	2.8×10^{-4}
#3	0.0576	0.0288	0.0000	0.0576	0.0282	-	7.32×10^{-4}
#4	0.0336	0.0672	0.0000	0.0351	0.0698	-	5.82×10^{-4}
#5	0.0504	0.0288	0.0768	0.0506	0.0307	-	8.69×10^{-4}
#6	0.0672	0.0504	0.0384	0.0676	0.0510	-	6.51×10^{-4}
R^a				1.0000	0.9999	-	-

a. Correlative coefficient between the resolved concentration and the added concentration.
b. HYD: hydroquinone; RES: resorcinol; CAT: catechol.
c. Components treated as unknown, uncarlibrated interferents.

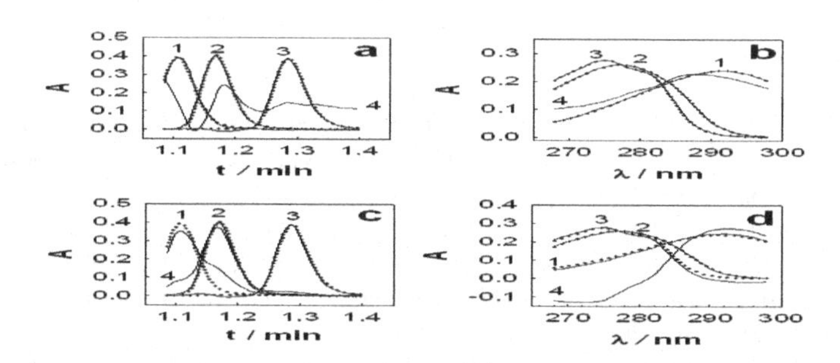

Fig. 5. Comparison of resolved chromatographic profiles and spectral profiles to actual ones when N=4. The solid lines are resolved chromatographic profiles and spectral profiles; the dotted lines are the actual ones. **a** and **b** are resolved by ADCE algorithm; **c** and **d** are resolved by PARAFAC algorithm. 1, hydroquinone; 2, resorcinol; 3, catechol; 4, noise and background.

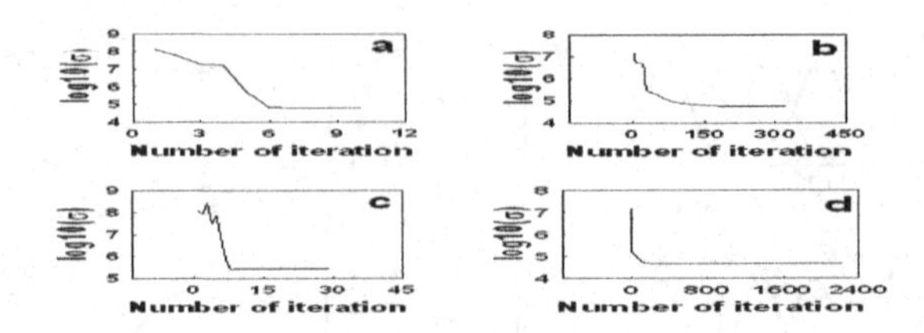

Fig. 6. Comparison of the convergency of ADCE algorithm to that of PARAFAC algorithm. **a** and **b** is separately obtained by ADCE algorithm and PARAFAC algorithm when N=3; **c** and **d** is individually obtained by ADCE algorithm and PARAFAC algorithm when N=4.

5 Conclusions

An alternating simultaneously minimizing diagonal matrix error and covariant matrix error (ADCE) algorithm, is proposed for three-way data analysis. Simulated data and real data have been carried to investigate the performance of ADCE algorithm. The presented results demonstrate ADCE algorithm has the properties of fast convergence rate and insensitivity to overestimation of the component number compared with PARAFAC algorithm. This method can be used for second-order calibration and study of chemical system or procedure characterized by three-way data.

References

1. Sanchez, E., Kowalski, B.R.: Tensorial calibration: II. Second-order calibration. J. Chemom. 2(4), 265–280 (1988)
2. Olivieri, A.C.: On a versatile second-order multivariate calibration method based on partial least-squares and residual bilinearization: Second-order advantage and precision properties. J. Chemom. 19(7), 253–265 (2005)
3. Harshman, R.A.: Foundations of the PARAFAC procedure: models and conditions for an exploratory multimode factor analysis. UCLA Working Papers Phonetics 16, 1–84 (1970)
4. Bro, R.: Tutorial and applications. Chemom. Intell. Lab. Syst. 38(2), 149–171 (1997)
5. Wu, H.L., Shibukawa, M., Oguma, K.: An alternating trilinear decomposition algorithm with application to calibration of HPLC-DAD for simultaneous determination of overlapped chlorinated aromatic hydrocarbons. J. Chemom. 12(1), 1–26 (1998)
6. Xia, A.L., Wu, H.L., Fang, D.M., Ding, Y.J., Hu, L.Q., Yu, R.Q.: Alternating penalty trilinear decomposition algorithm for second-order calibration with application to interference-free analysis of excitation–emission matrix fluorescence data. J. Chemom. 19(2), 65–76 (2005)
7. Bosco, M.V., Garrido, M., Larrechi, M.S.: Determination of phenol in the presence of its principal degradation products in water during a TiO2-photocatalytic degradation process by three dimensional excitation-emission matrix fluorescence and parallel factor analysis. Anal. Chim. Acta. 559(2), 240–247 (2006)
8. Mitchell, B.C., Burdick, D.S.: Slowly converging PARAFAC sequences: swamps and two-factor degeneracies. J. Chemom. 8(2), 155–168 (1994)
9. Rayens, W.S., Mitchell, B.C.: Two-factor degeneracies and a stabilization of PARAFAC. Chemom. Intell. Lab. Syst. 38(2), 173–181 (1997)

Indoor Comfortable Environment Self-adjust System Based on Microchip EM78F668N

Jingjing Li, Renjie Zhang, and Jun Fang

Department of Optical-Electrical and Computer
University of Shanghai for Science and Technology, Shanghai, China
shjingjing1011@163.com

Abstract. The main purpose of this thesis is to introduce an intelligent home control system which is based on MCU(EM78F668N), the PC(personal computer), wireless transceiver module FHL0611 as the core, to control the external temperature sensors, humidity sensors, light sensor and solar heating system. Thus the system achieves the automatic control of indoor home environment and use a wireless terminal on the indoor home of the remote control. This article also gives the device hardware and software design and implementation.

Keywords: EM78F668N, Intelligent control, Indoor environment, Energy conservation.

1 Introduction

Enters for the 21st century, the human society has already entered brand-new information age. Along with the social economic development, people living standard improving, the people to their own living environment requirement becomes higher and higher. Meanwhile, because of the community's progress and future development of human civilization and the general concern, People's energy-saving, green, environmental protection has become a subject of the pursuit of a more conscious action.

In modern society, people pursue the life that with perfect quality, personalized and automated. The humane and intelligent in intelligent control technology of morden home required to make smart home electronics products have been widely used. It not only optimizes the people's lifestyle and living environment, but also convenient people to effectively arrange a time and save a variety of energy. It achieve the appliances, lighting, curtain control and burglar alarm, remote control. Smart home manufacturers are mainly composed of traditional residential quarters smart system manufactures, for now mostly smart home systems are sub-systems of residential quarters smart systems, especially turning from security system and visual intercom system. Those systems are not fit for the single individuals, so it is a valuable subject that develops a low-cost temperature controlling device. we design a morden home which is based on solar heating and intelligent home control system as the core. And the morden home can meet the modern pursuit of comfort, energy saving, environmental protection, and convenient

H. Tan (Ed.): Knowledge Discovery and Data Mining, AISC 135, pp. 47–52.
springerlink.com

concept. Adopting EM78F668N chip is an attempt for us to realize controlling of comfortable living environment.

2 Design of Hardware System

This system is based on micro controller EM78F668N to control modules, as shown in figure 1: sensor module, LED module, the power modules, all kinds of indoor appliances, electric curtains module, solar energy heating module, wireless communication module, etc. EM78F668N is the core of intelligent household system.

The system mainly includes the following functions:

(1) When nobody in the home cut off all the home appliance 220 V power source, which not only eliminates all kinds of electric standby power consumption, but also avoid danger that caused by abnormal circuit or leaking indoor and other unpredictable events.
(2) According to the predetermined temperature and humidity do intelligent control to solar heating system, air conditioning and humidifier, in order to make the best comfort level indoor.
(3) Light controlled curtain according to outside light intensity to realize the curtain automatically close and open.
(4) Indoor temperature can be shown through the LED digital display tube.
(5) Adopt Labview interface design, through the wireless communication mode, the real time indoor temperature, humidity and the electric operation can be observation on the computer. The computer send command to MCU via wireless way to make control of indoor appliances, electrical power and humidity.

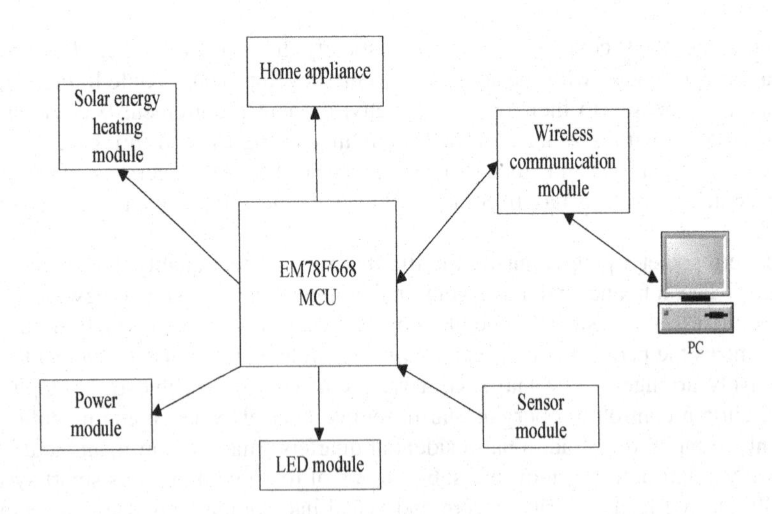

Fig. 1. Diagram of the System Hardware

2.1 Induction of Micro Controller

Compared with 8051 single chip computer EM78F668N single-chip microcomputer have many advantages: (1)There is an oscillation link inside, which can offer 8 MHz, 16 MHz etc crystal frequency, and on the condition of meet the requirements of the circuit, don't require additional crystals, therefore reduced difficulty of the circuit design and devices consumption. (2)The flexible instruction set. Through the system programming not only can use internal crystal frequency, but also can flexibly choose its crystal requency, which can bring great convenience for programming. (3)The rich I/O interface and digital communication interface. Bring convenience for communication of external modules digital signal and analog signal. (4)The optional precision AD transform module inside the chip, can do AD transform for outside input analog signal, and can directly use internal AD conversion module in precision allow.

2.2 Introduction of Wireless Communication Module

The sensors in this system includes temperature sensor and lightsensitive senser etc. Among them, the temperature sensor use the widely applied widely applied DS18B20, it only has three cables, they are digital cable, GND and VCC. Therefore, for the temperature sensor, wiring is very simple, and more important is it output digital signals, don't need AD transform. The single chip just need transform the receive digital signal through the proper way then will get the real temperature, also the operation is very simple. Light sensor adopts simple electric switch module, when there is light in the sensor, it will output a high level, when there is no light, the output is zero. Light sensor as a light-controlled switch to control of the curtain on or off in this way.

2.3 Introduction of Solar Energy Heating Module

Considering energy conservation and environmental protection requirement, design a solar energy water heater system to heating indoor area. The principle is simple, that is using solar energy water heater to make the hot water circulate indoor, the heat energy send out from the heat exchanger will applied warm for indoor environment, and the cooling water circulate to the pitcher, which can reuse through the solar panels heating.

2.4 Introduction of Wireless Communication Module

Wireless module is using FHL-0611 wireless communication module, it has multiple optional transmission frequency and communication interface ways as RS232TTL, USB, which is flexibly for use. The system uses two wireless access modules for communication, one is TTL wireless transceiver module, connect UART connected with MCU interface, the other is to use the USB and superordination machine is linked together, to realize the remote control. PC interface of software used the labview design company NI interface, through the PC, can real-time monitoring of each module indoor operation, through the wireless module send command transmitted to the microcontroller, on indoor each module control, so as to achieve the purpose of remote monitoring.

3 Design of Software System

Software system is divided into single-chip microcomputer software system and PC software system. Single-chip software design using the eUIDE simulation environment applied by ELAN electronic for C programming, as shown in figure 2 is the single-chip microcomputer software flow chart.

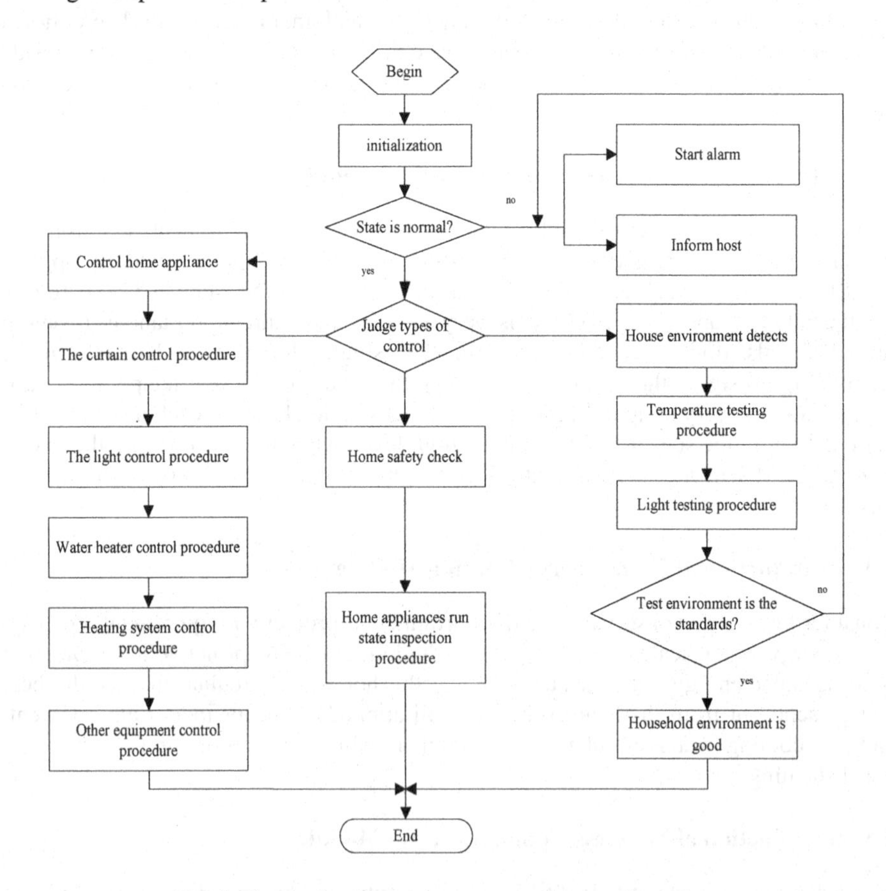

Fig. 2. The single-chip microcomputer software flow chart

PC software uses labview development environment of NI company for programming. Labview have special serial communication module visa, which bring convenience for program design. As shown in figure 3 the PC program, which realized wireless transmission baud rate digits, parity, port of the parameters such as setting, it not only can monitor the operation of each module within the household all the time , also can convenient to control each module in the household.

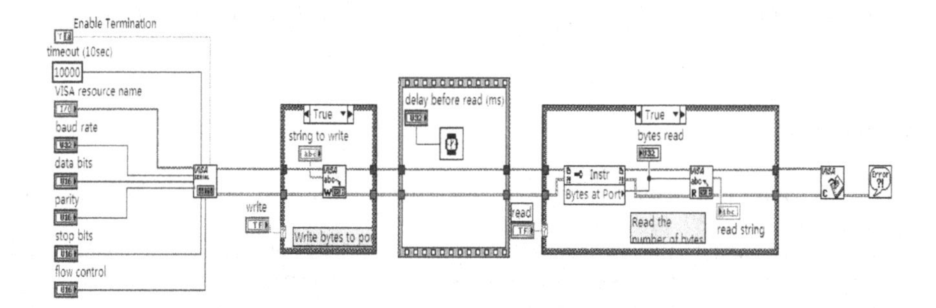

Fig. 3. The PC program

4 Conclusion

Indoor test including the temperature test, the light test. Temperature test means when we intentionally change indoor temperature, it is clearly observed that the temperature of indoor Led digital pipe will change. At the same time, the solar energy heating system would also change their working state run or stop to temperature change. The light test is defined when there exist light on light sensor, the curtain controlled by light can automatically open, instead it will automatically shut off.

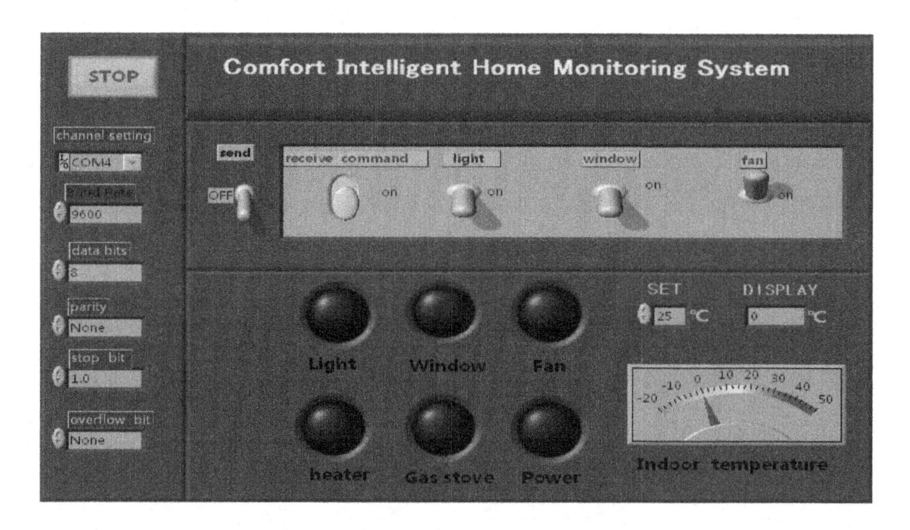

Fig. 4. The wireless monitoring interface

Outdoor test refers to using wireless module to realize remote monitoring indoor household modules through the PC. As shown in figure 4 is wireless monitoring interface. First of all, according to the EM78F668N microcontroller programming to set all parameters of the serial communication. When connect to single chip, click read command button then can read the operation of each indoor module. If required to control the indoor module, also can via send command of the object to single-chip, let MCU control.

References

1. Lee, J.B.: Smart home-digitally engineered domestic life. Lecture Personal and Ubiquitious Computing 7(20), 189–196 (2003)
2. Li, J., et al.: Mart home research. In: Proceedings of 2004 Intemational Conference on Machine Leaming and Cybemetics, 2nd edn., pp. 659–663 (2004)
3. Nemeroff, J., Garcia, L., Hampel, D.: Application of sensor network communications. In: Militan Communications Conference 2001(MILCOM 2001). Communications for Network-Centric Operations: Creating the Information Force, pp. 336–341. IEEE (2001)
4. Enrique, J., Le-Melo, D., Mingyan, L.: Dala-gathering wireless sensor networks: organization and capacity. Computer and Telecommunications Networking 43(04), 202–211 (2003)
5. EM78F568N/F668N 8Bit Microcontroller Product Specification. Elan Microelectronics corp (May 2010)
6. Orpwood, R., Gibbs, C., Adlam, T., et al.: The design of Smart Homes for People With Dementia-user-interface Aspects. Universal Access in the Information Society 4(2), 156–164 (2005)

High-Gain Stacked Minkowski Fractal Patch Antenna with Superstrate for 60GHz Communications

Yabing Shi and Wenjun Zhang

Institute of Microelectronics of Tsinghua University,
10084 Beijing, China
shiyabingthu@gmail.com, zwj@tsinghua.edu.cn

Abstract. A high-gain stacked Minkowski fractal patch antenna with superstrate and H-shaped aperture coupling for 60GHz communication is proposed.A superstrate is added above the ground plane in order to increase the gain.The maximum simulated gain of the antenna with superstrate reaches 14.1dBi at 60GHz and is more than 10dBi all over the 60GHz band(from 57GHz to 64GHz). It is noted that by stacking the antenna and using H-shaped aperture coupling the bandwidth can be increased up to 9.7% (56.4GHz—62.1GHz) which of the conventional Minkowski fractal patch antenna is only 3%.

Keywords: 60GHz, high-gain, Minkowski, fractal, stacked, superstrate.

1 Introduction

The application of 60GHz band technique has been studied recently for the spectrum is declared to be unlicensed all over the world[1].IEEE 802.15.3c supports more than 7Gbps in 60GHz communication compared with USB2.0(480Mbps) and IEEE 802.11n(300Mbps). The mmWave technique provides a solution for ultra-high-speed wireless transmission between electronic products in a short range, which is quite suitable for WPAN applications.

Moreover antenna is a key part of the 60GHz communication system. High gain, high efficiency and broad bandwidth are required characteristic of the antenna. Patch antenna is a candidate of solution for its compact size, light weight and mature process of manufacture. Because of the self similarity and space filling property the fractal antenna[3] has got a characteristic of size reduction and multi-band[4]. So a lot of research on the fractal antenna has been carried out, e.g Kock fractal slot antenna[5] and Sierpinski fractal antenna[6]. However the fractal-shaped patch antenna has always exhibited narrow bandwidth performance which is around 3%.In paper[7], it is reported that to use stacked structure and H-shaped aperture coupling can improve the bandwidth performance of fractal patch antenna. Besides, it is noted that adding a superstrate above the ground plane can increase almost 9dBi in gain at 60GHz[8][9].

In this paper, a hybrid stacked Minkowski fractal patch antenna with superstrate for 60GHz communication is designed. The simulated results show that the antenna has got a bandwidth of 9.7% and a maximum gain of 14.1dBi at 60GHz. Besides, in order to show the effect of stacked structure and superstrate, a single layer Minkowski fractal patch

H. Tan (Ed.): Knowledge Discovery and Data Mining, AISC 135, pp. 53–59.
springerlink.com

antenna with superstrate and stacked Minkowski patch antenna without superstrate are also simulated. The manufacture of the proposed antenna is now in progress, more measurement results and analysis will be provided in the future.

2 Antenna Structure

The size of the stacked Minkowski patch antenna with superstrate is 30mm*30mm. The side view of the proposed antenna is shown in Figure 1.

The 50 Ω tapered microstrip feedline with a U-shaped tuning stub[10] is placed under the F4BM220 substrate for the purpose of bandwidth improvement. Figure 2 shows the shape and related dimension of the feedline. The angle θ plays an important role in the upper edge of -10dB bandwidth and is optimized to 15° ultimately. In order to reduce surface waves the substrate is supposed to be low in thickness and permittivity. So the F4BM220 material of εr=2.2,tanδ=0.003 with h1=0.127mm is applied for substrate on which sits the 0.07mm ground plane with an H-shaped aperture. Novel aperture shape has been researched all over the world, such as e-shaped[11] and V-shaped[12].The shape of slot in this paper is selected to be H-shaped finally for bandwidth enhancement[13], the size of which is shown in Figure3.Besides two pieces of Minkowski fractal patch antenna of the same dimension are stacked above the ground plane so as to improve the bandwidth performance. Figure 4 shows the shape of the Minkowski fractal patch at 1st iteration. A Rogers RO3006 superstrate with εr=6.15 and h3=0.64mm is added above the ground plane so as to achieve gain improvement. The size of the superstrate is optimized to 10mm*10mm for high-gain and consistent radiation patterns all over the 60GHz band. The distance between the ground plane and the superstrate should be λ0/2 in the light of theory[8].A PMI foam layer of permittivity 1.125 is placed in the middle of the superstrate and the stacked antenna.

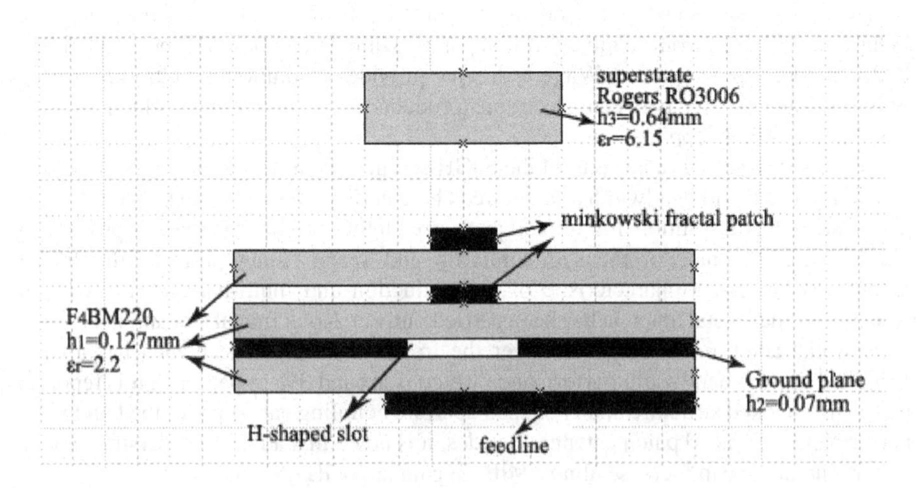

Fig. 1. Side view of the stacked Minkowski fractal patch antenna with superstrate

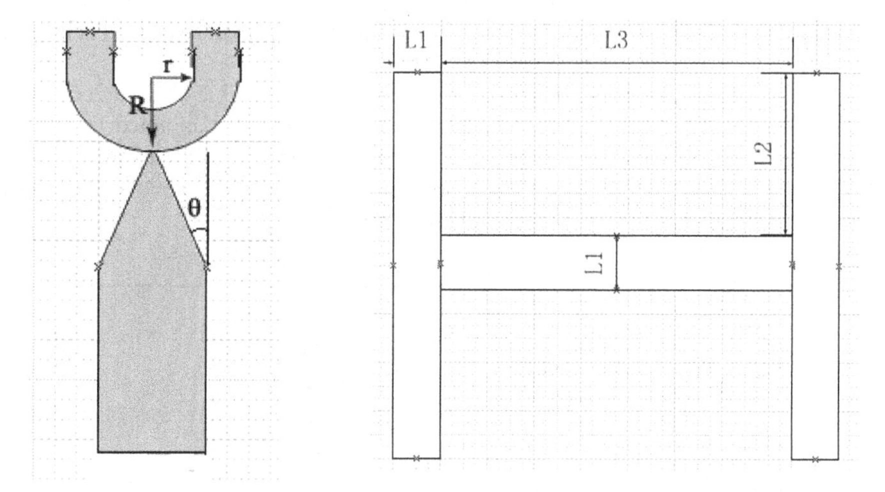

Fig. 2. Top view of the feedline, R=0.4mm, r=0.2mm,θ=15°

Fig. 3. The H-shaped aperture, L1=0.02mm, L2=0.31mm, L3=0.62mm

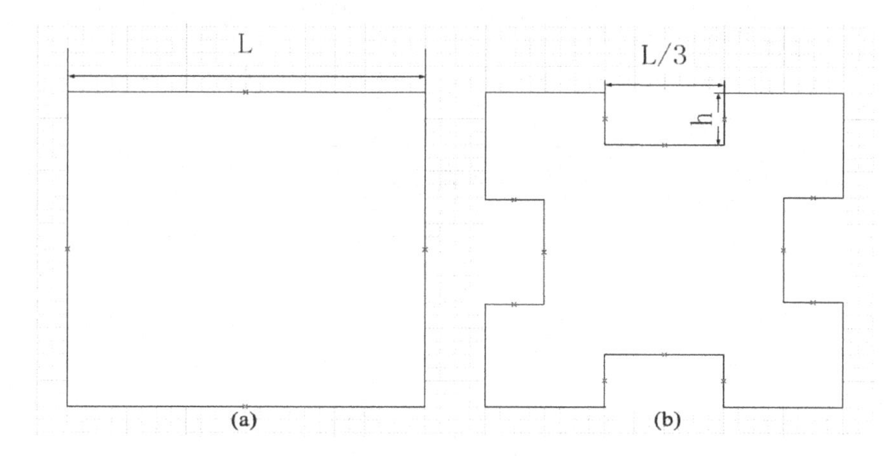

Fig. 4. (a) Initial square patch, L=1.2mm; (b) Minkowski fractal patch at 1st iteration, α=3h/L=1/2

3 Simulation Results

For comparison, the stacked Minkowski fractal patch antenna with superstrate(Antenna I),the single layer Minkowski fractal patch antenna with superstrate(Antenna II) and the stacked Minkowski fractal patch antenna without superstrate(Antenna III) are all simulated in HFSS12.

Figure 5 shows the respective return loss for Antenna I with 9.7% bandwidth and Antenna II with 2.7% bandwidth, which confirms that to use the stacked structure can improve the impedance bandwidth performance.

Figure 6 shows the respective variation of maximum gain of Antenna I and Antenna III with frequency. It is observed that adding a superstrate above the ground plane can increase the antenna gain prominently and the maximum gain of Antenna I is more than 10dBi all over the 60GHz band(from 57GHz to 64GHz).

The simulated plane and 3D radiation patterns of Antenna I at 60GHz frequency are shown in Figure 7 and Figure 8 respectively. In Figure 7 it is noted that the 3dB beamwidth of Antenna I is about 26°at 60GHz and the cross-polarization is below -37dBi in E-plane radiation patterns at 60GHz. Figure 8 shows that the maximum total gain of Antenna I at 60GHz is approximately 14.1dBi.

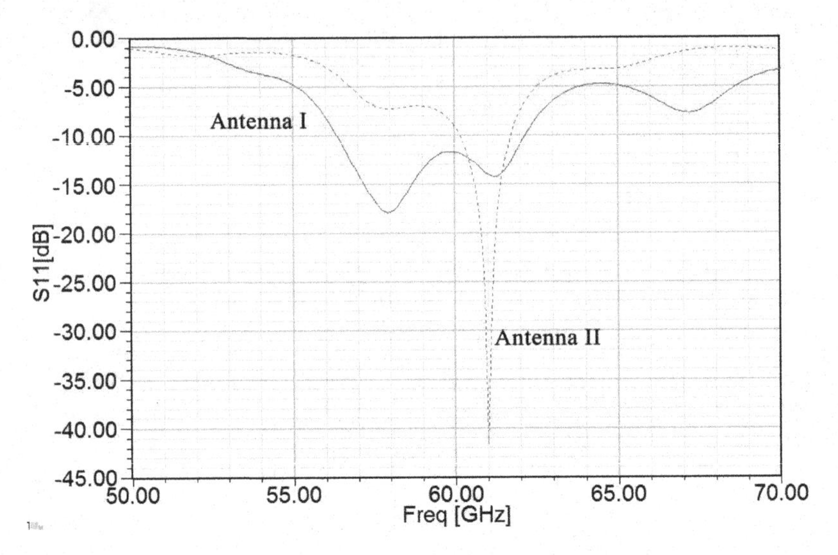

Fig. 5. The simulated S11 variation of Antenna I and Antenna II with frequency

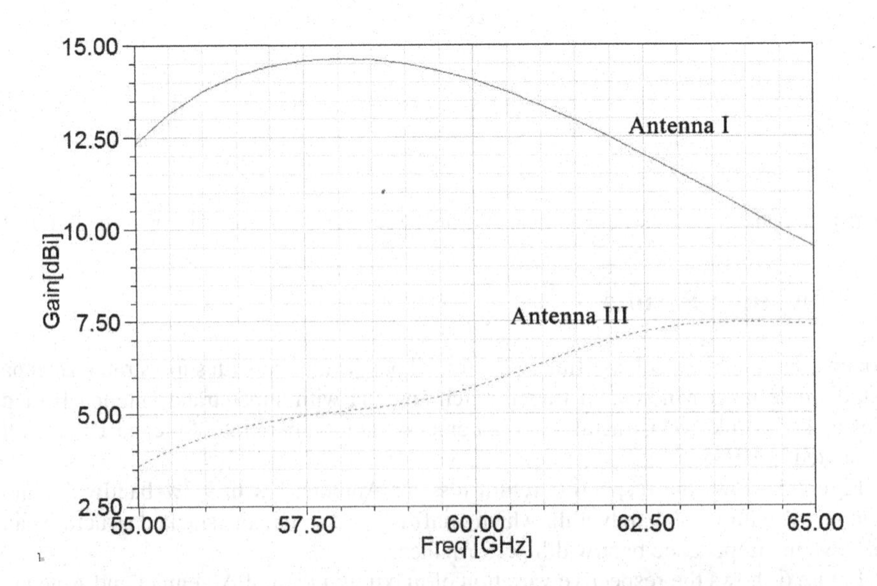

Fig. 6. The simulated maximum gain variation of Antenna I and Antenna III with frequency

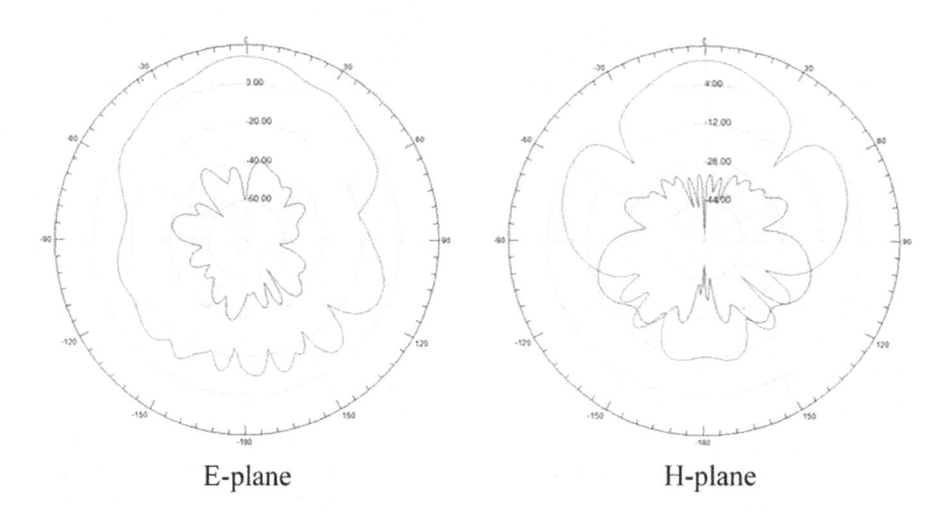

Fig. 7. Plane radiation patterns of Antenna I, the blue dash line for co-polarization, the red solid line for cross-polarization

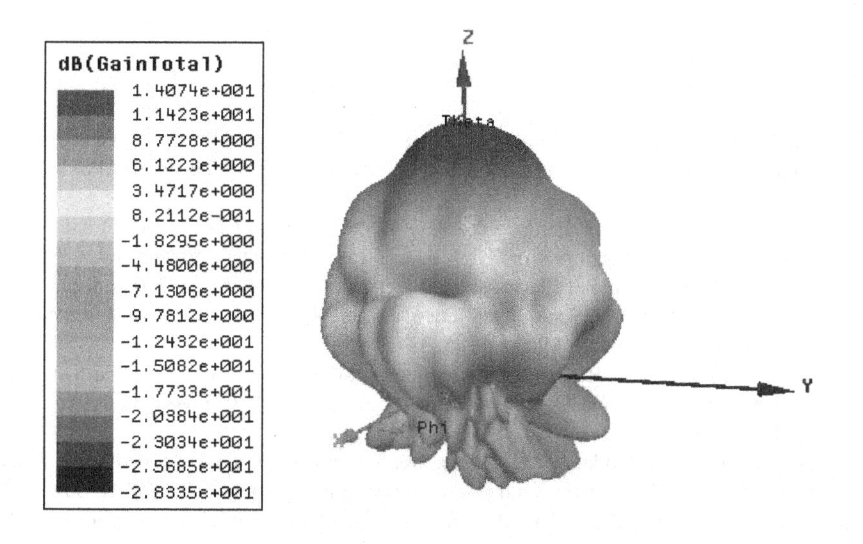

Fig. 8. 3D radiation pattern of Antenna I

From the simulation results it is noted that the proposed hybrid antenna has got 9.7% bandwidth and a maximum gain of 14.1dBi at 60GHz. Table 1 shows the main performance parameters in comparison with the previous related research on 60GHz antennas.

Table 1. Comparison in main performance parameters of 60GHz antennas

Antenna	f_L(GHz)	f_H(GHz)	BW(%)	$Gain_{max}$(dBi)
Antenna I	56.4	62.1	9.7	14.1
Paper[1]	54	64	16.7	6.4
Paper[8]	57	61.1	6.8	14.6
Paper[9]	55	64	15	13.1

4 Conclusion

In this paper, a new design of hybrid Minkowski fractal patch antenna for 60GHz communication is proposed. The gain is increased up to 14.1dBi by adding a superstrate above the ground plane .A bandwidth of 9.7% is obtained by the stacked structure, H-shaped aperture coupling and tapered microstrip line with a U-shaped tuning stub. Over the whole 60GHz spectrum the maximum gain of the antenna is stably superior to 10dBi.The manufacture is in progress and more measurement results and analysis will come out during the conference.

References

1. Mark Tan, Y.C.: Computational Modeling and Simulation to design 60GHz mmWave Antenna, Antennas and Propagation Society International Symposium (APSURSI), pp. 1–4. IEEE (2010)
2. IEEE Std.802.15.3-2003, Part5.3: Wireless Medium Access Control(MAC) and Physical Layer(PHY) Specifications for High Data Rate Wireless Personal Area Networks (WPANs), The Institute of Electrical and Electronics Engineers, Inc. (September 2003)
3. Werner, D.H., Ganguly, S.: An overview of Fractal Antenna Engineering Research. IEEE Antennas and Propagation Magazine 45, 38–57 (2003)
4. Yu, Y.-H., Ji, C.-P.: Research of fractal technology in the design of multi-frequency antenna. In: Microwave Conference Proceedings(CJMW), China-Japan Joint, pp. 1–4 (2011)
5. Sundaram, A., Maddela, M., Ramadoss, R.: Koch-Fractal Folded-Slot Antenna Characteristic. IEEE Antenna and Wireless Propagation Letters 6, 219–222 (2007)
6. Hwang, K.C.: Modified Sierpinski Fractal Antenna for Multiband Application. IEEE Antennas and Wireless Propagation Letters 6, 357–360 (2007)
7. Raje, S., Kazemi, S., Hassani, H.R.: Wideband Stacked Koch Fractal Antenna With H-shaped Aperture Coupled Feed. In: Asia-Pacific Microwave Conference, APMC 2007, pp. 1–4 (2007)
8. Vettikalladi, H., Lafond, O., Himdi, M.: High-Efficient and High-Gain Superstrate Antenna for 60-GHz Indoor Communication. IEEE Antenna and Wireless Propagation Letters 8, 1422–1425 (2009)
9. Vettikalladi, H., Le. Cop, L., Lafond, O., Himdi, M.: Broadband Superstrate Aperture Antenna for 60 GHz Application. In: 2010 European Microwave Conference (EuMC), pp. 687–690 (2010)

10. Lin, P., Liang, J., Chen, X.: Study of Printed Elliptical/Circular Slot Antennas for Ultrawideband Applications. IEEE Transaction Antennas and Propagation 54, 1670–1675 (2006)
11. Hoseini Izadi, O.: A Compact Microstrip Slot Antenna With Novel E-shaped Coupling Aperture. In: 5th International Symposium Telecommunications (IST), pp. 110–114 (2010)
12. Pantui, T., Suppitaksakul, C., Rakluea, P.: Dual V shaped slot microstrip antenna for a bi-and uni-directional radiation pattern. In: 6th International Conference on Electrical Engineering/Electronics, Computer, Telecommunications and Information Technology 2009, ECTI-CON 2009, pp. 790–793 (2009)
13. Subbarao, A., Raghavan, S.: Conductor Backed H shaped Antenna fed by CPW for Wide band Applications. In: Adeances in Recent Technologies in Communication and Computing, pp. 495–497 (2009)

Embedded Control System for Atomic Clock

Wei Deng[1], Peter Yun[1], Yi Zhang[2], Jiehua Chen[1], and Sihong Gu[1,2,*]

[1] Key Laboratory of Atomic Frequency Standards, Chinese Academy of Sciences, 430071, Wuhan, China
[2] School of Physics, Huazhong University of Science and Technology, 430074, Wuhan, China
shgu@wipm.ac.cn

Abstract. We have designed an embedded control system based on digital signal processor (DSP) for multi-negative-feedback-loop control, and applied it in a CPT atomic clock. The size of the circuit is 10 cm^3, and the power consumption is 200 mW. A CPT atomic clock is built and close loop locked by connecting the embedded control system and a physics package and the expected performances are achieved.

Keywords: Embedded System, Digital Signal Processing, Negative Feedback Loop, CPT Atomic Clock.

1 Introduction

A microwave atomic clock is a clock which uses a microwave electronic transition frequency of the electromagnetic spectrum of atoms as a frequency standard for its timekeeping element. Atomic frequency standards are the most accurate time and frequency standards known. Besides the research for high accuracy atomic clocks, most researches focus on the goals of making the clocks smaller, cheaper, and energy efficient. The passive Coherent Population Trapping (CPT) atomic clock [1][2] uses a microwave frequency-modulated Vertical Cavity Surface Emitting Laser (VCSEL) to provide a coherent bi-chromatic light field, and uses the CPT resonance signal, generated from bi-chromatic light field and atoms interaction, as the microwave frequency discriminating signal. Because the microwave cavity is eliminated by CPT technique, the CPT atomic clock is especially suitable for realizing the small size and energy efficient clock, and chip scale CPT atomic clocks have been realized [3][4]. In principle, it is the only known atomic clock which can be realized at chip scale.

The basic function of a CPT atomic clock control system is to stabilize microwave frequency by CPT resonance signal through several control loops. In our designed CPT atomic clock, five loops are arranged to control frequencies, temperatures and magnetic field separately. Analog circuits are commonly used in controlling systems [5]. However, the more control loops are involved, the more resources are consumed, because it is relatively complicated to implement several control tasks with one analog circuit control loop. Moreover, it is also difficult to adjust or extend the functions of an analog circuit.

* Corresponding author.

H. Tan (Ed.): Knowledge Discovery and Data Mining, AISC 135, pp. 61–67.
springerlink.com

Digital signal processing [6] technology is concerned with the processing of discrete time signals. Digital signal processing algorithms usually run on specialized processors called digital signal processor (DSP). By converting the signal from analog to digital using analog-to-digital converters (ADC), the DSP processes signal in digital format, and the processed signal is then converted into analog signal by digital-to-analog converter (DAC) to output. With the digital signal processing by a DSP, multi-channel signals can be processed simultaneously in a more efficient and flexible way, which makes it convenient to implement multi-loop control. The more complex the control loops are, the more resources will be saved with the DSP based circuit compared to the analog circuit. Moreover, it is easy to adjust or extend the functions of the DSP based circuit by changing algorithms and software. There have been purpose-built microprocessors that have on-chip control peripherals used for digital signal processing. For example, TMS320 series [7] from Texas Instruments Company have built-in PWM, ADC and other peripherals, which are more suitable for feedback control system.

This paper presents our work of realizing an embedded control system for a CPT atomic clock by means of constructing a DSP based controller, programming special control algorithm in the DSP, and designing peripheral analog circuits. Through connecting ADC and DAC to sensors, actuators and other external components, the DSP controller simultaneously processes five channel signals and successfully stabilizes laser frequency, microwave frequency, temperature of VCSEL, temperature of atom vapor cell, and the static magnetic field in atom-laser interaction area through five negative feedback control loops. The experimental results reveal that the DSP based control circuit scheme is promising for applying in a miniature CPT Atomic clock.

2 Control System

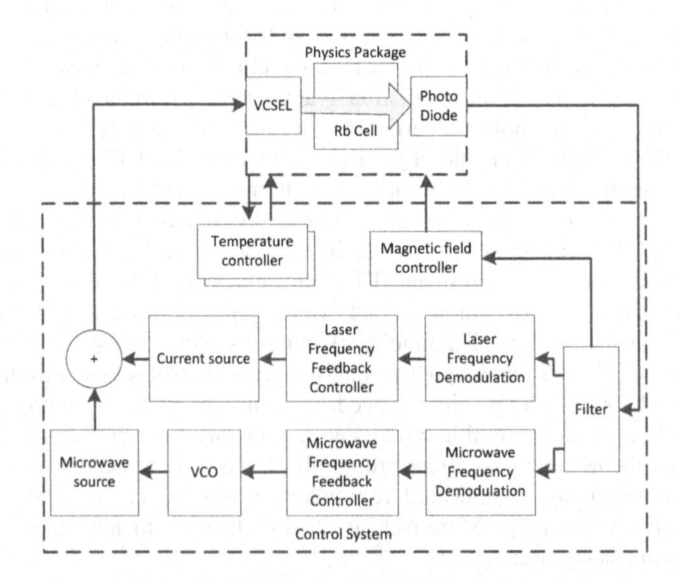

Fig. 1. Block diagram of the CPT atomic clock

A CPT atomic clock consists of a physics package and a control system is shown in figure 1. The functions of the studied CPT atomic clock control system include generating DC current to drive the VCSEL, generating microwave to modulate VCSEL, and controlling five physical quantities through five negative feedback control loops.

In laser frequency loop the atomic absorption spectral line is obtained from the photo diode (PD) output photocurrent as discriminating signal, converted into correction signal through phase-sensitive demodulation, and fed into VCSEL bias current to stabilize laser frequency through current negative feedback. Similar to laser loop, in microwave loop CPT resonance spectral line is also obtained, converted and fed to the voltage controlled crystal oscillator to stabilize microwave frequency through voltage negative feedback. The error signal of static magnetic field is obtained from the photocurrent, and through negative feedback current added into the solenoid driving current to stabilize the C-field. Two temperature control loops detect the temperature fluctuations, generate error signals, convert correction currents separately. The correction currents are added into heating (cooling) currents to stabilize VCSEL and atom vapor cell temperatures.

3 Design and Realization of Embedded Control System

3.1 Hardware

The structure of the whole control system is shown in figure 2. The hardware includes DSP, ADC, DACs and peripheral analog circuits. The DSP is connected to a high speed ADC by SPI interface, used to collect photocurrent signal from the PD output, and the multi-channel ADC built-in DSP is used to collect temperature signals. Two high performance DACs are used to control the bias current for VCSEL and the voltage of VCO, and the other Multi-Channel DAC provides three controllable voltages applied for temperature and magnetic field control, respectively.

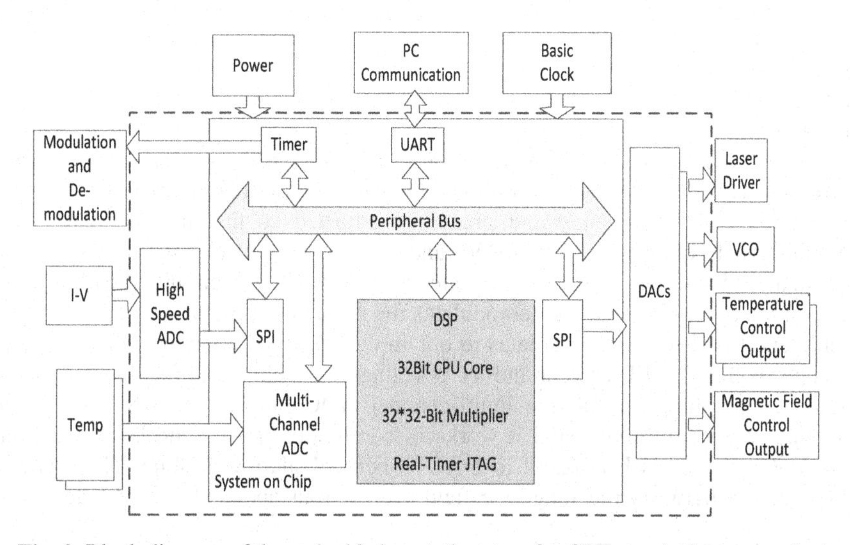

Fig. 2. Block diagram of the embedded control system for CPT atomic frequency clock

TMS320 C2000 series of Texas Instruments is chosen as the DSP of the CPT atomic clock control system. C2000 devices are 32 bit microcontrollers with high performance integrated peripherals designed for real-time control applications. Its math-optimized core and powerful integrated peripherals make the devices the ideal single-chip control solution, and the DSP is more appropriate for realizing the embedded control system for our usage. The SPI and PIO enable the MCU to communicate with the ADC and DAC, respectively. The UART is used to send temperature data to the computer through RS-232 Serial Port.

3.2 Control Algorithm

As the core component of the entire control system, the DSP is used to implement all of these control algorithms. To laser frequency loop and microwave frequency loop, the frequency division multiplexing (FDM) method is used for modulation and demodulation. The bias current for VCSEL and voltage for VCO are modulated with different modulation frequencies, thus corresponding signals are separated through arranged filters for phase-sensitive demodulations. There is a built-in multiplier in the chosen DSP which is utilized to achieve highly efficient filtering, demodulation, and real-time feedback functions for two frequency loops. In the two temperature control loops, the temperature negative feedback signals are obtained by the calculation through proportional–integral–derivative (PID) control method [8]. The intensity of the magnetic field is relevant to photocurrent data.

In addition to the functions of control and calculation, DSP communicates with the host computer, sending it the data of frequencies, temperatures other parameters for real-time monitoring and analysis, and accepting incoming commands from the host computer. Besides, the DSP controls the various peripheral devices and the control system state.

3.3 Software

The main features of DSP software includes 1) modulation and demodulation, 2) feedback of two frequency loops, 3) feedback of temperature control loops, 4) feedback of magnetic field control loop, 5)PC communication and workflow controlling.

After initialization, the Timer module of the DSP generates Pulse-width modulation (PWM) waveform for modulation and demodulation without the CPU core working. The frequency and phase are adjusted so that the modulation and demodulation frequencies match FDM, and the phases between the signals of two loops match each other. DSP communicates with ADC for PD circuit by SPI interface to obtain the photocurrent data, demodulates the photocurrent data with the Multiplier module and FDM method, calculates to obtain the feedback for two frequency loops, and controls the VCSEL current and VCO voltage by sending the feedback to DACs. For the temperature control, the multi-channel ADC built-in the DSP is used for collecting temperature data, and it works in continuous mode. The DSP reads data from ADC and writes PID result to DACs when calculation is done. There is non-linear relation between the magnetic field loop output and photocurrent data. DSP calculates the magnetic field output by a preset table obtained from experiments. Photocurrent data, temperature data and other data from DSP are sent to the host PC through UART protocol. The host PC sends commands including parameter adjustment, starting and stopping commands and so on. Communication between DSP and the host PC works in the interval time of feedback controlling to avoid affecting the real-time feedback.

Interruption is used to realize real-time controlling, which is necessary for the atomic clock control system. ADC will generate an interrupt in every sample time. Some tasks for feedback control loops need to be carried out real-time, while the less time-sensitive tasks such as calculations can be interrupted.

The flow diagram of the software is shown in figure 3. Immediately after system boot, the software configures the directions and initial value of the PIO interfaces. Then the command registers of the ADC and DAC are written to configure their work modes and execute self-calibration. The ADC starts to transform the temperature data after calibration. The DSP reads the data and starts temperature control. At the beginning the temperature is close to the environment temperature and far from the reference temperature. In this case the control value will be set as a large constant thus the system will work in an open loop state, and the temperature will be quickly driven to the set point. Then the software begins to do PID feedback control while the temperature is quite near the set point.

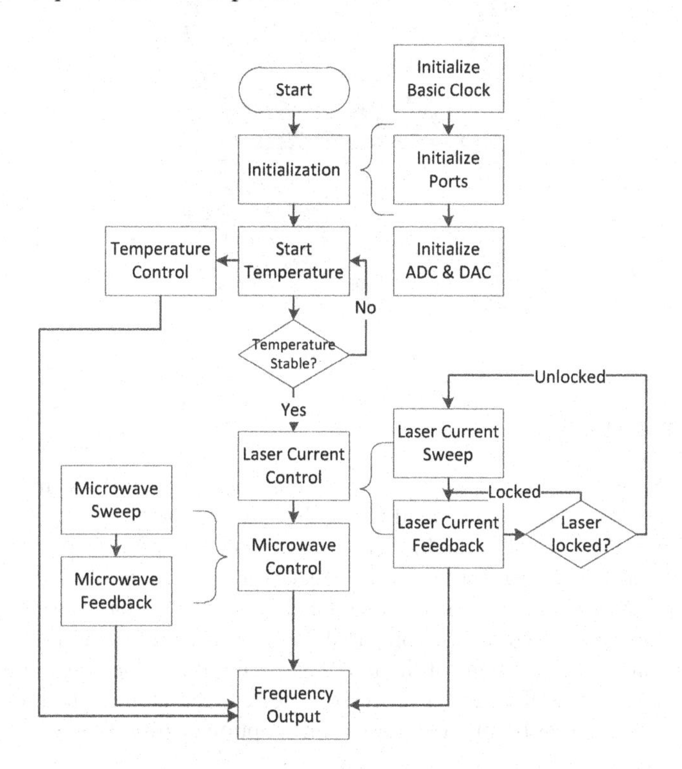

Fig. 3. Flow diagram of the software

When the DSP detects that temperatures of both VCSEL and vapor cell are stable enough, the software goes to the next stage of VCSEL laser frequency sweeping and locking. The DSP makes a current sweeping to obtain the atomic absorption spectrum, identifies the current value corresponding to frequency discriminating spectral line, sets current at the set point, and then begins the laser frequency feedback process. Microwave control loop begins to work when the laser frequency is locked. With the similar process, it sweeps the microwave frequency by changing the voltage of VCO, identifies the voltage value corresponding to CPT resonance line, and begins the feedback process.

4 Application

The realized atomic clock circuit according to the scheme is shown in figure 4, the size of which is $10cm^3$, and power consumption is 200mW. An atomic clock is formed by connecting a physics package and the circuit. The output frequency of the atomic clock is successfully stabilized and all expected functions are realized.

Fig. 4. Picture of the realized circuit

5 Conclusion

We have designed an embedded control system based on DSP for multi negative feedback loop controlling, and applied it in a CPT atomic clock.

With DSP as the controller, multi-channel signals can be processed simultaneously in a more efficient and flexible way which makes it convenient to implement multi-loop controlling. The more control loops are involved, the more resources may be saved using the DSP-based circuit compared with analog circuit. Moreover, it is easy to adjust or extend the functions of the DSP based circuit by changing algorithms and software. With DSP as the central processor, the realized atomic clock circuit costs less resources. Its size is $10cm^3$, and power consumption is 200mW.

Acknowledgments. This work has been supported by the National Natural Science Foundation of China (NSFC) under Grant No. 10927403.

References

1. Arimondo, E.: Coherent population trapping in laser spectroscopy. Progress in Optics. 35, 257–354 (1996)
2. Stähler, M., Wynands, R., Knappe, S., Kitching, J., Hollberg, L., Taichenachev, A., Yudin, V.: Coherent population trapping resonances in thermal Rb-85 vapor: D-1 versus D-2 line excitation. Opt. Lett. 27, 1472–1474 (2002)
3. Knappe, S., Shah, V., Schwindt, P., Hollberg, L., Kitching, J., Liew, L., Moreland, J.: A microfabricated atomic clock. Appl. Phys. Lett. 85, 1460–1462 (2004)
4. 2008 Compact, Low-Power Chip-Scale Atomic Clock.pdf (2008)
5. Dorf, R.C., Bishop, R.H.: Modern Control Systems, 8th edn. Addison Wesley (1998)
6. Stranneby, D., Walker, W.: Digital Signal Processing and Applications, 2nd edn. Elsevier (2004)
7. 32 bit Real-time C2000™ Microcontrollers, http://www.ti.com/lsds/ti/microcontroller/32-bit_c2000/overview.page
8. Carnegie Mellon - PID Tutorial, http://www.engin.umich.edu/group/ctm/PID/PID.html
9. TMS320C28x User's Guide (Rev. D), http://www.ti.com/litv/pdf/spru514d

Green Supply Chain: Comparative Research on the Waste Printing Plate of the Computer

Jing Zhang and Lei Su

School of Economics & Management, Beijing Jiaotong University, Beijing, 100044, China
jzhang@bjtu.edu.cn

Abstract. With the fast development in the computer and related technology, the quantity of the computers has become more and more, which has caused many environmental and resource problems. In this paper, it makes a research on the treatment technologies in the computer printing plate. Based on the supply chain management theory and the current process in dismantling different techniques, the paper does the comparative analysis, in order to identify the reasonably safe handling techniques can reduce environmental impacts.

Keywords: waste computer, green supply chain, printing plates, processing, regeneration.

1 Introduction

There are many computers and related technologies in the 20th century, which has greatly promoted the progress of society. People can enjoy the convenient taken by the updated information technology and tools, however, the waste computers are increasing quickly .It not only takes more and more social resources, but also increases the serious pollution of the environment. In a word, it has become a burden to society.

Waste computer is a special kind of the renewable resources. It has 2 special meanings both in the resource and the environment. On the one hand, it has the toxic and hazardous substances. On the other hand, if you do not separate it correctly, it will cause seriously secondary pollution. The current waste dismantling computers in China's is still done by hand. According to the investigation, it can be found both in the standard processing factory and small private workshop with no standard dismantling.

Changing our existing manual processing conditions can make a benefit to recovering the computer resources and ensuring that their resource. By the way, it will minimize the harm to the environment as well. This paper focuses on Waste printing plate in the computer, according to the dismantling process in the several processing techniques, the paper makes the comparative analysis to identify the reasonable and safe technique can reduce environmental impact.

H. Tan (Ed.): Knowledge Discovery and Data Mining, AISC 135, pp. 69–75.
springerlink.com

2 The Current Treatment Technology on the Waste Printing Plate in the Computer

Printed circuit boards are widely used in the computers, information appliances and the other communications equipment, which is made of many parts, such as polymer (resin), glass fiber, papers, high purity copper foil, printed components and so on. Printed circuit board contains copper, trace gold, silver and other precious metals. There are many methods to handle the printed circuit boards, for example burning, direct smelting (Refined by fire) and mechanical.

In the past time, the self-employed ones has adopted the non-standard pickling methods, and then dumped waste acid to the river directly, which causes the serious environmental pollution. The open burning can cause serious pollution to the atmosphere. Our regulate handling in business is the mechanical treatment with printed circuit boards. Mechanical method is divided into the dry way and the wet way. The dry is a combination in crushing and sorting classic; wet is broken the waste with a combination of hydraulic shaker sorting.

2.1 Pickling Method

Pickling method makes the use of concentrated nitric acid, sulfuric acid or strong oxidizing agents to dissolute the circuit metal board. It can easily obtain precious metals with spin-off sediment, and then reduce the gold, silver, palladium and other metal products. This method can recycle the high concentrations from the acid copper. The recycling process is shown in Fig. 1.

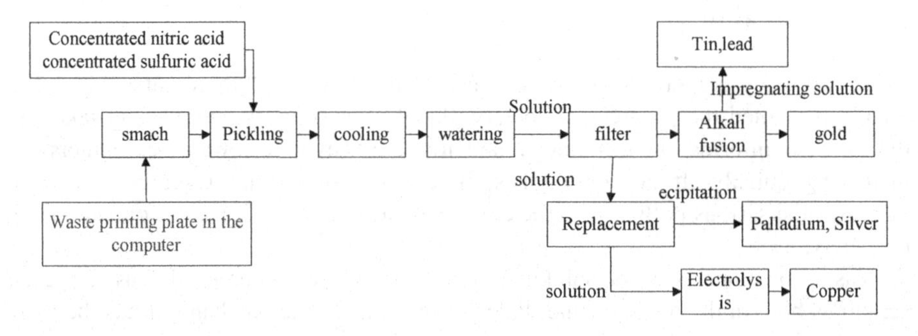

Fig. 1. Pickling method in recycling process of the pickling circuit board

2.2 Burning Method

Burning method in the waste printed circuit boards make use of the metal recovery method shown in Fig. 2. With the help of the machine, the Waste circuit boards are broken.

It was burned into the incinerator, which will be contained in the decomposition, and the remaining residues are the bare metal and fiberglass. They will be sent immediately after crushing metal smelter for the metal recovery.

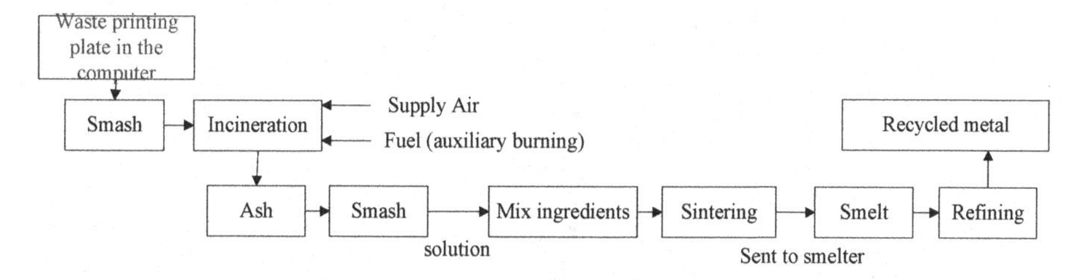

Fig. 2. Burning method in the recycling process of the circuit board scrap metal

2.3 Direct Smelting Method

The direct smelting method in the waste printed circuit boards are shown in Fig. 3. This method is widely used in developed countries.

Due to high labor costs and less demand for secondary products, developed countries prefer to use the direct smelting in the waste printed circuit boards.

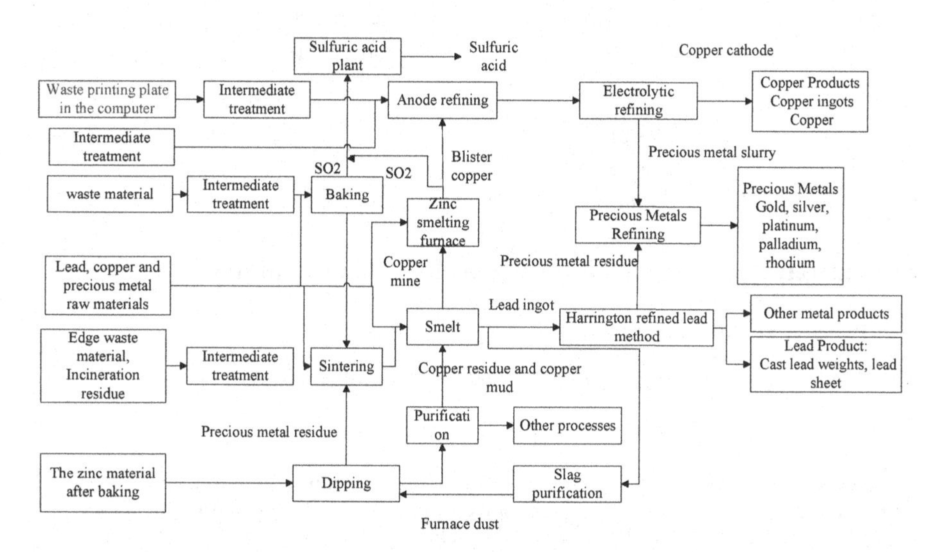

Fig. 3. Direct smelting method in recycling metals from waste circuit board

2.4 Physical Method

Physical method needs to deal with the pre-printed circuit board (automatic, semi-automatic or manual methods) with removing large pieces of pure substances (containing hazardous components) which can be used directly.

By means of broking or crushing, we can achieve the metal and nonmetal dissociation recording to the actual circumstances in case. The purpose of sorting is to achieve metal dissociation through following-up or away the medical. The overall process of physical method is shown in Fig.4.

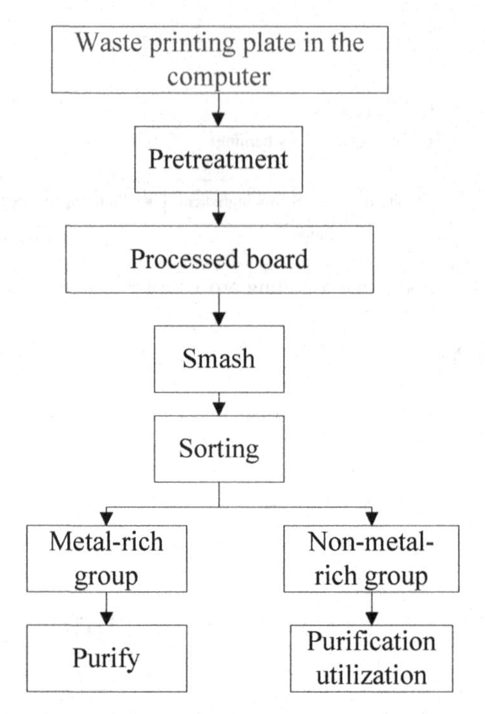

Fig. 4. Physical method in recycling metals from waste circuit board

3 Comparative Research between the 4 Technologies

3.1 Process Technology

The Comparison between the four technologies is shown in Table 1.

Table 1. Comparison of different technologies

Process technology	Fundamental	Initial investment	Running costs	Resource influence	Secondary pollution
Pickling method	Dissolve the metal board with strong acids or strong oxidizing agents, then strip the precious metals and restore it into gold, silver or palladium.	Medium	Medium	Little high	Very high
Burning method	Burn the resin in bare metal in the incinerator and send bare metal to the smelter for recovery	Very high	Very high	Low	Little high
Direct smelting method	smelt the board in the incinerator for recycling metals.	A little high	Little high	Medium	Little high
Physical method	Pre-crush the metal or non-metal waste in circuit board which can be finished in a same processing line.	Medium	Medium	Very high	Low

According to the environment, the physical methods make the minimum pollution to the environment and pickling method and the burning method makes the greater impact to the environment. According to the running costs, the pickling method is more useful than the burning technology. However, the rare metal in the waste circuit board is in a sharp drop. According to the information, only 150 ~ 400g gold can be got in the 400 ~ 1500g waste circuit boards now. In the printed circuit board process, recycling the metals is not the main purpose; On the other hand, the purpose of protecting environment with sustainable development is increasing, we want less hydrometallurgical wastewater and exhaust gas heat. Based on the different treatment results in comprehensive evaluation in the table 4.1, we can find the physical method is more favorable.

3.2 Renewable Technology

After using the physical method, it is conducive in the next regeneration step. In China, the popular recycling method in printed circuit board processing techniques are shown.

(1) Wet hydraulic shaker and sorting method

It can avoid the irritating gases and dust. The separation efficiency in the hydraulic shaker separation can reach 95%. It has the small investment with the low running costs and it is a simple and practical technology. Process is shown in Fig.5.

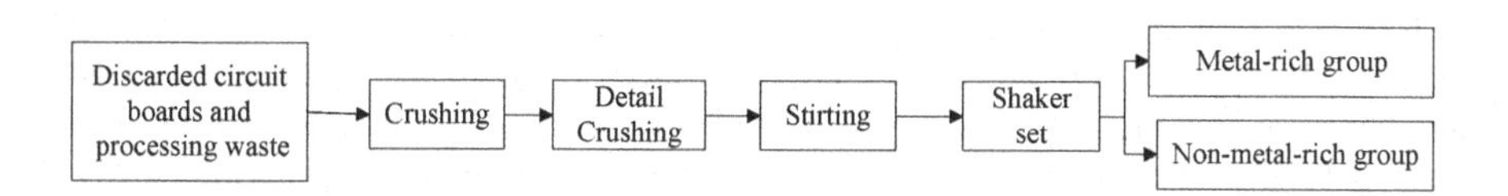

Fig. 5. Wet hydraulic shaker and sorting method in separation the circuit board

Waste printed circuit can be crushed by two grades or more. The Wet hydraulic shaker method can separate the circuit board in metal or nonmetal dissociation. After the sorting with hydraulic shaker, the metal-rich group can be sent to smelters and the non-metallic (fiberglass and epoxy resin, etc.) can be regarded as a filling material. This method generated in the process waste water (which may contain heavy metals) need to use the appropriate treatment to reduce environmental hazards.

(2) Dry electrostatic separation method

The process for dealing with waste and the side foot board material sorting is Efficient. The recycling metal grades can reach 95%.What's more, it is a small investment with low running costs. Process is shown in Fig.6.

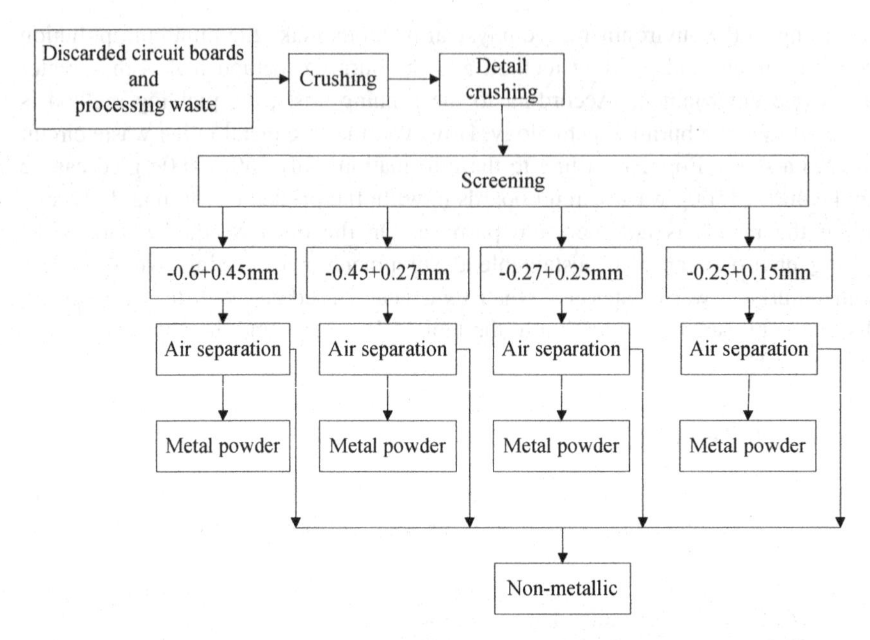

Fig. 6. Dry electrostatic separation method in the printed circuit boards

Diagram Source: All the picture information is from research on the computer dismantling environmental in China(R), Chinese Academy of Household Appliances.

The separation of metal and nonmetal in the waste circuit board can be got through the Dry electrostatic separation method. This method may cause separation of dust and debris when broken in the dry and electrostatic separation. There are some hazards to the environment. Therefore, we should choose the appropriate method according to the actual geographical and technical conditions.

4 Summary

Analyzing the above printing processing technologies, the manual dismantling should be used in the pre-stage. The recycled parts in the computer can be in high quality. When recycling materials, the refractory material can reduce the load. The removal of harmful substances can prevent proliferation. Firstly, this method can improve the e-waste re-use and improve the quality in recycling materials. Secondly, reducing waste can prevent the spread of harmful substances.

Therefore, the priority is to close a large number of personal waste computer workshops in order to stop the extensive processing. It can establish a reasonable standard in computer processing system. In China, with the constant improvement in the waste electrical and recycling legislation, the technical standards can be got in the industrial chain. The dismantling method of computer will be highly standardized and the environment impact will also get smaller and smaller.

References

1. Zhou, Q.-F., Shang, T.-M.: Parts and materials recycling in the waste computers, pp. 5–15. Chemical Industry Press (2003)
2. Li, J.-H., Wen, X.-F.: E-waste processing technology, pp. 21–27. China Environmental Press (2006)
3. Research on the computer dismantling environmental in China, pp. 9–26. Chinese Academy of household appliances (2009)
4. Sun, J., Zhao, H.: Integrated waste treatment and disposal in the computer. Renewable Resources Research (3), 23–26 (2009)
5. Xu, X.-G., Deng, X.-H., Pan, L.-T.: The waste computer status in recycling and utilization. Non-Ferrous Metals 60(4), 15–18 (2008)
6. Bowersox, D.J., Closs, D.J., Bixby Cooper, M.: Supply Chain Logistics Management. Chain Machine Press (2005)

Tree-Mesh Based P2P Streaming Data Distribution Scheme

Jianming Zhao[1], Nianmin Yao[2], Shaobin Cai[2], and Xiang Li[2]

[1] Dept. of Mathematics and Computer Science, Fuqing Branch of Fujian Normal University, Fuqing, FuJian 350300, China
[2] College of Computer Science and Technology, Harbin Engineering University, Harbin, 150001, China
yaonianmin@hrbeu.edu.cn

Abstract. This paper presents the design of a novel P2P live streaming policy, tree-mesh based P2P Streaming Data Distribution Scheme(TMDS). Both the tree-based and the mesh-based structures have respective flaws while distributing the media data. The tree-based structure is highly vulnerable, so it is very suitable for the stable environment in small scale. And the mesh-based structure is quite stable, and can guarantee certain grade of service under the large-scale environment, but its costs of control are bigger, and also creates heavy load to the network. However TMDS can effectively solve these problems, and take advantage of their merits. Confirmed in the simulated environment, TMDS can: improve the quality of service; decrease the costs of system control, lighten the burden on the network; relieve the load on the transmitting end, enhance the stability of the system.

Keywords: P2P, streaming, tree-based, mesh-based.

1 Introduction

These years, with the rapid developing of the internet, more and more streaming media applications have been deployed on the internet. Several years ago, these applications are based on proxy server[1] or content distribution system[2]; and now almost all are based on P2P scheme. The data distribution schemes based on P2P can be divided into two categories[3]: tree or multi-tree based[4] [5] [6] [7] and mesh based[7] [8] [9].

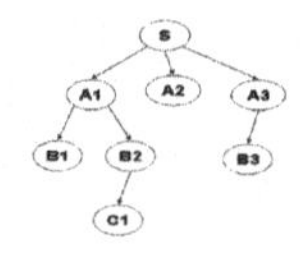

Fig. 1. Tree based data distribution scheme

H. Tan (Ed.): Knowledge Discovery and Data Mining, AISC 135, pp. 77–83.
springerlink.com

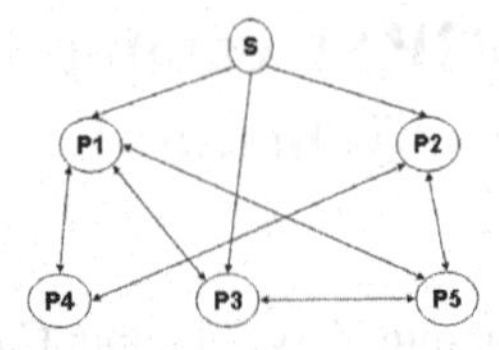

Fig. 2. Mesh based data distribution scheme

In tree based data distribution scheme as shown in fig.1, the data source S organizes a tree structure logic overlay network and data flow is from up to below. Peercast[7] is a typical example of such design. In Peercast, when a new node A wants to join the system, S first judges if its children node's number has reached the limit, if not, add A as its child; if yes, randomly selects one of its child and send its information to A, and A will again send join request to this node and so on. When a node wants to leave, its father will delete it from its children. If one node leaves accidently, the father can detect this situation in the next PING cycle and then do the same thing. This scheme has a simple topology which can be easily maintained. In the stable network environment, it has low network traffic. But its scalability is not very good especially when the network is not stable. Multi-tree structure can only solve this problem partly.

In the mesh based data distribution scheme, all the nodes are organized as a buddy relationship network as shown in fig.2. DONet[8] is a typical example of this kind. In DONet, when a node A want to join the system, S will send it some nodes information, then A will send request to these nodes to create member relationship with these nodes according to Gossip protocol. So A will join the membership and maintain a list of members. Through these members, A can create buddy relation with some nodes and distribute data through these buddies. When a node leaves, its members and buddies will update their lists. Now, the mesh based scheme is very popular. It can keep stability and provide high QoS in dynamic network situation and has high scalability. But its maintain cost and network traffic is high.

To compromise the pros and cons of the two schemes, a tree-mesh based streaming media data distribution scheme(TMDS) is presented in this paper. Now there has been some works try to combine the tree and mesh structure[10,11]. But they rely too much on the tree structure to transmit the control message which decreases the stability of the whole system. TMDS can effectively combine the good parts of tree and mesh based scheme. It uses tree structure in a neighbor domain and use mesh structure to organize nodes in different domains.

This paper is organized as follows: in section 2, TMDS is detailed described; in section 3, experiments are given; section 4 is the conclusions.

2 The TMDS Scheme

2.1 The Topology

In TMDS, the structure is shown as fig. 3. The unidirectional or bidirectional arrows in the figure represent the direction of data flow. In fig.3, node A, B, C, D, E and

F form a mesh structure. In domain 1, A is the root of the domain(Sub_Root) and A1, A2,...,A6 are its children which form a tree structure. So, in TMDS, all the nodes are organized in a mesh structure, but in some local networks, some nodes are organized as a tree structure.

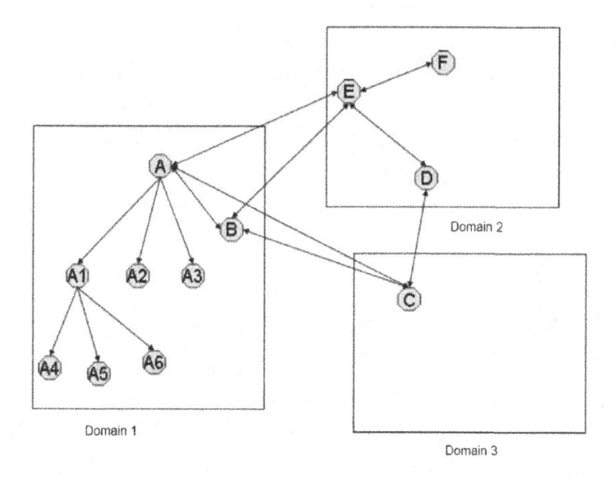

Fig. 3. TDMS data distribution scheme

2.2 Method to Divide the Domain

In a real system, nodes which are physically near to each other usually have the same behavior. Let x, y, i, and j denote the nodes, A(x, y) denote the event of x and y do the same action, and Distance(i, j) denote the physical distance of node i and j, then we have:

$$E[\ P(\ A(i, j)\ )\ ] \geq E[\ P(\ A(x, y)\ )\ ]\ ,\ \text{while distance}(i, j) \leq \text{distance}(x, y)$$

So, to group nodes which are near to each other into a domain can increase the stability of the domain and is good for tree based maintance.

The size of the domain can also affect the performance. If the size is too large, it will increase the latency of the data transmission and if it is too small, it will change into mesh based scheme and cannot utilize the advantages of the tree based structure.

2.3 Joining Protocol of the Node

First, we define the data structure of the root as follows:

```
typedef struct Sub_Root
{
UINT  Root_ID  ; //id of the node
int  num ; //number of its children nodes
Boolean  full ; //indicate if the number of children nodes has reached the limit
}Sub_Root ;
Sub_Root  Rootlist[Size]  ; //list of all the sub_root
int Tree_Size ; //the size of one tree
```

When a node N want to join the system, the steps are as follows:

1) N broadcasts its request to join a domain;
2) If N receives responses in a predefined duration, it will select the first node which sends response to it as the Sub_Root and join this domain following the Peercast rules.
3) If no answers are received in a predefined duration, it will join in the mesh organized by S following the DONet rules.

When a Sub_Root receive the request from node N, it will

1) Judge if N belongs to its domain, if not, it will do nothing.
2) Judge if its domain has been full (full==TRUE), if yes, it will do nothing.
3) It will add N to its domain following peercast rules.
4) num++; if num > Tree_Size, full = TRUE.

2.4 Leaving the System

In TMDS, when node leave the system normally,

1) When some node in the mesh leaves, it will follow DONet's rules;
2) When some node which is a child of a tree, it will follow Peercast's rules;
3) If a Sub_Root leaves, it will inform S and let S delete its information from the list. And all its children join the system again following the Peercast rules.

If a node leaves the system without informing others, its leaving can be detected by others in the next PING cycle and others can also do almost the same actions to reorganize the system.

2.5 Selection of Sub_Root

In TMDS, it is very important for the selection of the Sub_Root, since it can affect the performance of the system greatly. Two factors must be carefully considered while selecting Sub_Root: one is the performance of the node, the other is the stability.

The performance of the node is decided by its physical configuration and location in the network. Because the nodes' physical configurations have no much difference now, the location of the node plays the main role in the performance of the node. The stability of the node also is very important for the selection of the Sub_Root. If the Sub_Root frequently leave the system, then it will bring lots of costs and greatly affect its children.

In TMDS, let R(kb) denote the sum of data one node receives in a duration T(seconds) and Tlive denote the time the node has been alive. Then one node must conform the following conditions before it send request to S to become a Sub_Root:

1) R/T > V0
2) Tlive > T0

V0 and T0 are predefined threshold. When S receives the request, it checks if the Rootlist[Size] has got the limit, if not, it accepts the request.

3 Experiments

We use OPNET to simulate the P2P applications. The entire network is divided into 10 domains each of which can hold 100 nodes. 1000 nodes are added into the system

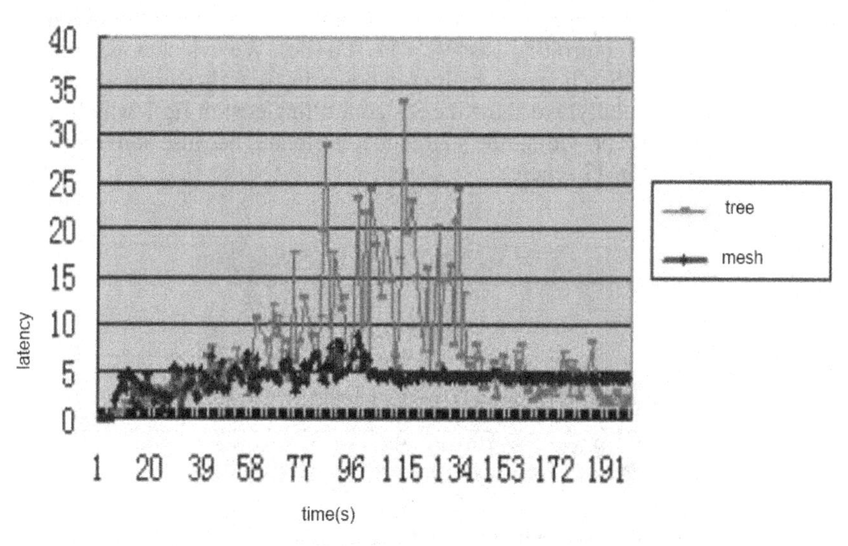

Fig. 4. Latency of tree and mesh based scheme

dynamically whose online and offline time are exponential distributed whose average values are 100s and 10s respectively. In the simulation, we set upstream bandwidth as 500kbps, the media stream is 310kbps, To=10s, Vo=300kbps and Tree_Size=10. The protocols of tree and mesh follow Peercast and DONet respectively.

Data acquisition latency means the time from node sending request to receiving the data. In fig.4, we compare the latency of tree and mesh based scheme. We can see that with the time goes on and more and more nodes join the system, tree based scheme's latency is larger, so it is not fit for large or dynamic system. In fig.5, we compare mesh and TDMS scheme. On the whole, the TDMS performs better than mesh based scheme though in some time the TDMS experiences more latency. This is because some Sub_Root leaves system which causes its children reconnect to the source. .

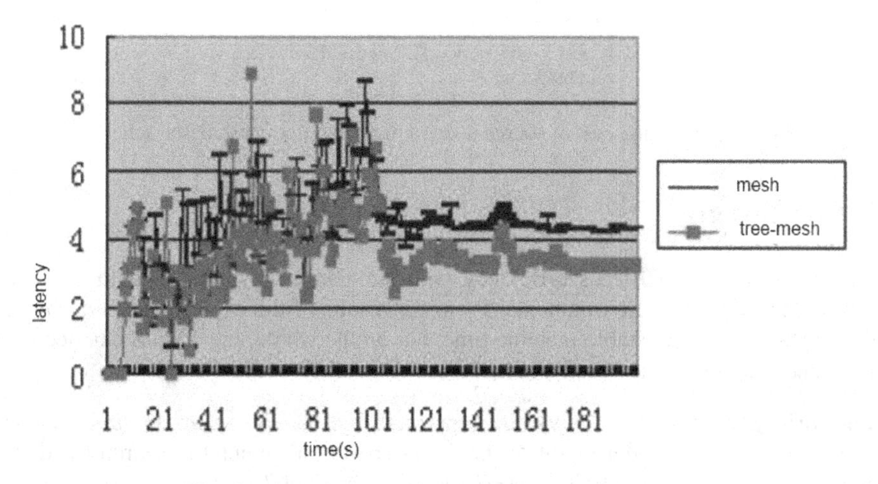

Fig. 5. Latency of mesh and tree-mesh based scheme

We define Fc as the throughput of the control message and Fd as the throughput of data message. Then the controlling cost Vc = Fc/(Fd+Fc). We evaluate the controlling cost of mesh and TDMS schemes which are shown in fig.6. It can be seen that the TDMS costs less. Especially, we show the S's controlling cost in fig.7 which shows S can have less load while providing the source media stream, because some Sub_Roots can share the load in TDMS scheme.

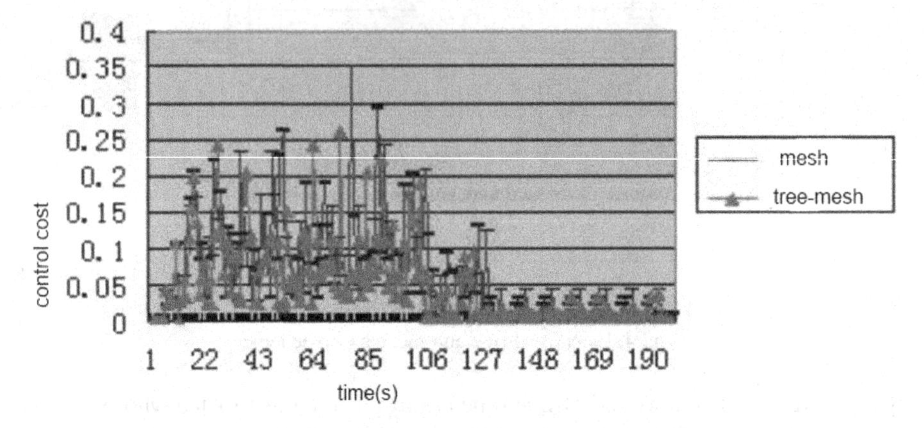

Fig. 6. Controlling cost of mesh and tree-mesh based scheme

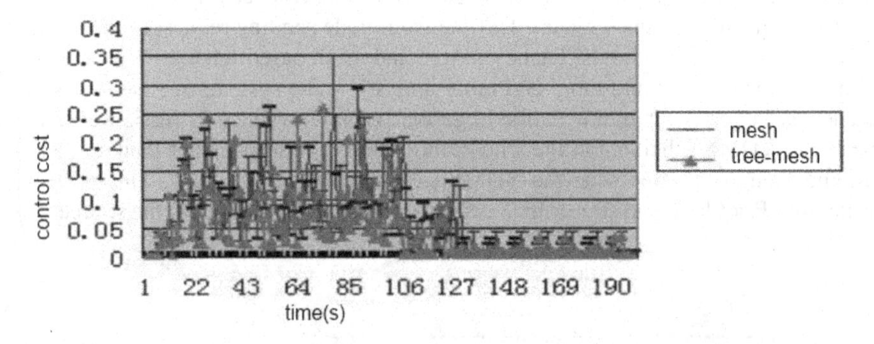

Fig. 7. Controlling cost of source node in mesh and tree-mesh based scheme

4 Conclusions

We only prove the TMDS's efficiency with simulate experiments which could be different from the real environments. From the results of the experiments, we can see that TMDS maybe not stable in some time, but on the whole, the TMDS can decrease the latency and maintance cost, and increase the stability of the system.

Acknowledgements. This work is supported by the National Natural Science Foundation of China under Grant No.61073047 and Fundamental Research Funds for the Central Universities HEUCFT1007 and HEUCF100607.

References

1. Wang, Y.W., Zhang, Z.L., Du, D.H.C., Su, D.L.: A network conscious approach to end-to-end video delivery over wide area networks using proxy servers. In: Guerin, R. (ed.) Proc. of the IEEE INFOCOM, pp. 660–667. IEEE Press, San Francisco (1998)
2. Vakali, A., Pallis, G.: Content delivery networks: Status and trends. IEEE Internet Computing 7(6), 68–74 (2003)
3. Liu, J., Rao, S.G., Li, B., Zhang, H.: Opportunities and challenges of peer-to-peer Internet video broadcast. Proceedings of the IEEE (2007)
4. Banerjee, S., Bhattacharjee, B., Kommareddy, C.: Scalable application layer multicast. In: Steenkiste, P. (ed.) Proc. of the SIGCOMM, pp. 205–217. ACM Press, Pittsburgh (2002)
5. Tran, D.A., Hua, K.A., Do, T.: ZIGZAG: An efficient peer-to-peer scheme for media streaming. In: Bauer, F. (ed.) Proc. of the IEEE INFOCOM, pp. 1283–1292. IEEE Press, San Francisco (2003)
6. Deshpande, H., Bawa, M., Garcia-Molina, H.: Streaming live media over a peer-to-peer network, Technical Report of Database Group, CS-2001-31. Stanford University (2001)
7. Peer Cast, `http://www.peercast.org`
8. Zhang, X., Liu, J., Li, B., Yum, T.-S.P.: DONet/Cool Streaming: A Data-driven Overlay Network for Peer-to-Peer Live Media Streaming. In: IEEE INFOCOM, vol. 3, pp. 2102–2111 (March 2005)
9. Liao, X., Jin, H., Liu, Y., Ni, L.M., Deng, D.: AnySee: Peer-to-Peer Live Streaming. In: Proceedings of IEEE INFOCOM, Barcelona, Spain (April 2006)
10. Wang, F., Xiong, Y., Liu, J.: mTreebone: A hybrid tree/mesh overlay for application-layer live video multicast. In: IEEE ICDCS (2007)
11. Huang, Q., Jin, H., Liu, K., Liao, X., Tu, X.: Anysee2: An Auto Load Balance P2P Live Streaming System with Hybrid Architecture, Work-in-Progress/Poster. In: Proc. of ACM INFOSCALE (2007)
12. Magharei, N., Rejaie, R., Guo, Y.: Mesh or multiple-tree: A compar-ative study of live P2P streaming approaches. In: IEEE INFOCOMM (2007)
13. Wang, F., Xiong, Y., Liu, J.: mTreebone: A hybrid tree/mesh overlay for application-layer live video multicast. In: IEEE ICDCS (2007)

The Transform of the National Geographic Grids for China Based on the Axial Minimum Enclosing Rectangle

Quanfu Bao[1,2], Zhengguo Li[1,3], Xiaochong Tong[1], and Caijiao Jin[1,4]

[1] Institute of Surverying and Mapping, Information Engineering University, Zhengzhou 450052, China
[2] 95806 Troops, Beijing 100076, China
[3] 96351 Troops, Lanzhou 730100, China
[4] 75719 Troops, Wuhan, 430074, China
baoqunfu20051@163.com

Abstract. Put forward a method of the transform between the different style of geographic grids, define the axial minimum enclosing rectangle, give the model for the code transform between the theodolite coordinate grid and the gauss coordinate grid, realize the transform of the national geographic grids based on the axial minimum enclosing rectangle, and then put forward the method of the grid transform for irregular region.

Keywords: The Axial Minimum Enclosing Rectangle, The Coordinate Grid, Gauss Coordinate Grid, Grid Code, Coordinate Transform.

1 Introduction

The concept of grid originated in the" well-field system" of the agricultural society, and it has been three thousand years history so far, as the progress of the science and technology, the grids has been widely used in various disciplines and industries, especially entering into the information ages. The paper introduces the grids in order to realize the storage, organization and rapid indexing of the plenty of geography information. There are many subdivision styles of the global sphere grids, and the theodolite coordinate grid and the gauss coordinate grid have been into practical stage, such as the standard map farming belongs to the theodolite coordinate grid and the march map belongs to the gauss coordinate grid, and there are some subdivision styles which is in the study stage, such as eight surface or twenty surface body.

There are some differences in the style of the spatial data organization, because of the different subdivision styles, so there are some problems in the transform between the different subdivision grid systems. It how to realize the conversion and transfer for the spatial data of the same region must be an important problem which the integration and use for the spatial data faced. The paper give a relating model for the transform between the theodolite coordinate grid and the gauss coordinate grid, and realize the conversion and the query effectively.

H. Tan (Ed.): Knowledge Discovery and Data Mining, AISC 135, pp. 85–91.
springerlink.com

2 The Idea of the Transform for the Geographic Grids Based on the Axial Minimum Enclosing Rectangle

The definition of the axial minimum enclosing rectangle: the axial minimum enclosing rectangle is the minimum enclosing rectangle which is parallel with the coordinate axis.

In the practical application, the grids transform displayed on the screen or a rectangular area, and the transformation area has no change on direction with before. Therefore, we can let the first grid be projected to the second grid, and then achieve its axial minimum enclosing rectangle, and figure out the grid code of the rectangle area. And we can get the code extension only based on the grid codes of the four corner points.

Based on the above ideas, the conversion process can be summed up as shown below:

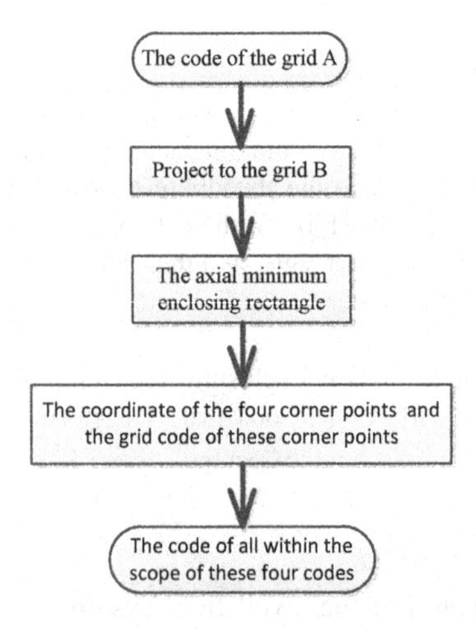

Fig. 1. The flow chart of the grid transform based on the axial minimum enclosing rectangle

3 The Acquisition of the Axial Minimum Enclosing Rectangle

According to the transform idea and flow, the key of grid transform based on the axial minimum enclosing rectangle is how to calculate the axial minimum enclosing rectangle.

First, project the first kind of grid to the second. Because the two grids may use different coordinate systems or projection methods, so it needs coordinate transform when they are different.

Second, obtain the maximum and the minimum of the coordinate of the area.

And last, take the four points from the maximum and the minimum of the coordinate as the four corner points of the rectangle, and the rectangle is the axial minimum enclosing rectangle.

Take the national standard geographic grid as an example, the paper analysis the transform between the theodolite coordinate grid and the gauss coordinate grid, and need coordinate transform because they are based on the different coordinate system. The formula (1) is the conversion formula of the transform from the theodolite coordinate grid to the gauss coordinate grid:

$$\begin{cases} x = X + \frac{1}{2}N * t * cos^2B * \ell^2 + \frac{1}{24}N * t * (5 - t^2 + 9\eta^2 + 4\eta^4) * cos^4B * \ell^4 \\ \quad + \frac{1}{720}N * t * (61 - 58t^2 + t^4) * cos^6B * \ell^6 \\ y = N * cosB * \ell + \frac{1}{6}N * (1 - t^2 + \eta^2) * cos^3B * \ell^3 + \frac{1}{120}N * (5 - 18t^2 + t^4 \\ \quad + 14\eta^2 - 58\eta^2t^2) * cos^5B * \ell^5 \end{cases} \tag{1}$$

In the formula, $\ell = L - L_0$, L is the longitude of the point, L_0 is the longitude of the centralmeridian, and :

$$X = a(1 - e^2)(A_0B + A_2\sin 2B + A_4\sin 4B + A_6\sin 6B + A_8\sin 8B)$$

$$\begin{aligned} A_0 &= 1 + \frac{3}{4}e^2 + \frac{45}{64}e^4 + \frac{350}{512}e^6 + \frac{11025}{16384}e^8 \\ A_2 &= -\frac{1}{2}(\frac{3}{4}e^2 + \frac{60}{64}e^4 + \frac{525}{512}e^6 + \frac{17640}{16384}e^8) \\ A_4 &= \frac{1}{4}(\frac{15}{64}e^4 + \frac{210}{512}e^6 + \frac{8820}{16384}e^8) \\ A_6 &= -\frac{1}{4}(\frac{35}{512}e^6 + \frac{2520}{16384}e^8) \\ A_8 &= \frac{1}{8}(\frac{315}{16384}e^8) \end{aligned}$$

$$t = \tan B \ ; \ \eta^2 = e'cos^2B \ ; \ e' = \sqrt{\frac{a^2 - b^2}{b^2}} \ ;$$

$$N = \frac{a}{W} \ ; \ W = \sqrt{1 - e^2sin^2B} \ ; \ e = \sqrt{\frac{a^2 - b^2}{a^2}} \ ;$$

Derivate the x andy to ℓ and B:

$$\begin{cases} x'_\ell = N * t * cos^2B * \ell + \frac{1}{6}N * t * (5 - t^2 + 9\eta^2 + 4\eta^4)cos^4B * \ell^3 \\ \quad + \frac{1}{120}N * t * (61 - 58t^2 + t^4) * cos^6B * \ell^5 \\ y'_\ell = N * cosB + \frac{1}{2}N * (1 - t^2 + \eta^2) * cos^3B * \ell^2 + \frac{1}{24}N * (5 - 18t^2 \\ \quad + t^4 + 14\eta^2 - 58\eta^2t^2) * cos^5B * \ell^4 \end{cases} \tag{2}$$

$$\begin{cases} x'_B = -N * t * \cos B * sinB * \ell^2 - \frac{1}{6}N * t * (5 - t^2 + 9\eta^2 + 4\eta^4)cos^3B * sinB * \ell^4 \\ \quad - \frac{1}{120}N * t * (61 - 58t^2 + t^4) * cos^5B * sinB * \ell^6 \\ y'_B = -N * sinB * \ell - \frac{1}{2}N * (1 - t^2 + \eta^2) * cos^2B * sinB * \ell^3 - \frac{1}{24}N * (5 - 18t^2 \\ \quad + t^4 + 14\eta^2 - 58\eta^2t^2) * cos^4B * sinB * \ell^5 \end{cases} \tag{3}$$

After the analysis of the formula (2) and (3),$y'_\ell>0$, and ①When$\ell > 0$, $x'_\ell>0$; ② When $\ell=0$, $x'_\ell=0$; ③When $\ell<0$时, $x'_\ell<0$. So, there are some rule as follow when transform theodolite coordinate grid to the Gauss grid coordinate:

(1) If the longitude is constant: ①and if the longitude is less than the longitude of the central meridian, x, y increases with latitude increasing; ②If the longitude equal to the longitude of the central meridian, as the latitude increasing, y is unchanged, x increases; ③If the longitude is bigger than the longitude of the central meridian, as the latitude increasing, y decreases, x increases.

(2)If the latitude is constant, as the increasing of the longitude, y increases, x reaches the minimum in the central meridian.

There are also the similar rules when transform the gauss coordinate grid to the theodolite coordinate grid:

(1) If x is constant, longitude increases with the increasing of Y, the latitude reaches the maximum in the central meridian.

(2) If x is constant: ①and ify<50000, that is located on the leftof the central meridian, as the increasing of x, latitude increase and longitude decreases;②ify=50000, that is located on the central meridian, as the increasing of x, latitude increase and longitude is constant;③ify>50000, that is located on the right of the central meridian, as the increasing of x, latitude increase and longitudealso increase.

Though the analysis we can know that there are some monotonicity in a certain range after the transform, so the ubiety of theodolite coordinate grid in the 3°or 6°projection zone can be divided into three types (Fig.2):

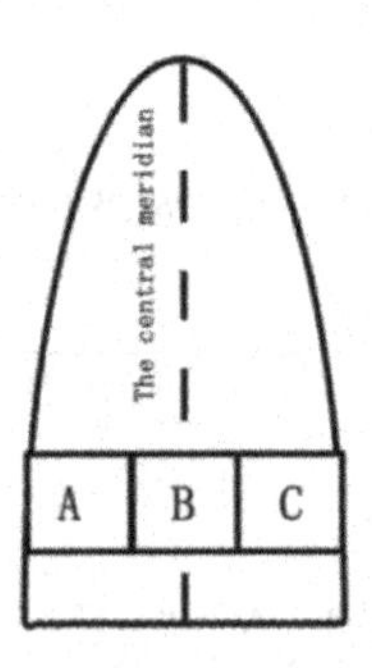

Fig. 2. The schematic diagram of the ubiety between the grid and the projection zone

According to its monotonicity in a certain range, the maximum or the minimum value of the coordinate appear at the corners and the intersection points of the central meridian with the frame, corresponding with the three ubieties; its area in the gauss grid is as shown in Fig. 3:

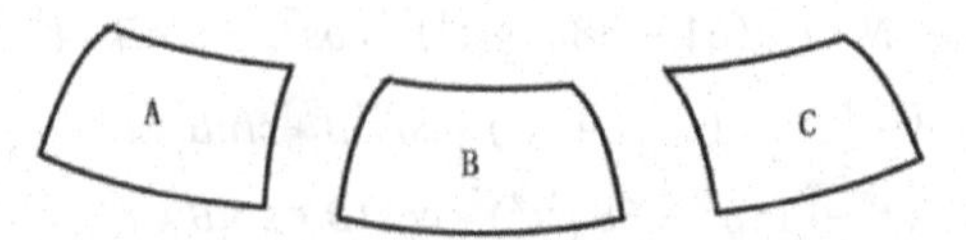

Fig. 3. The figure of the transform from the theodolite coordinate grid to the gauss coordinate grid

The ubiety of the axial minimum enclosing rectangle and central meridian is as shown in Fig. 4:

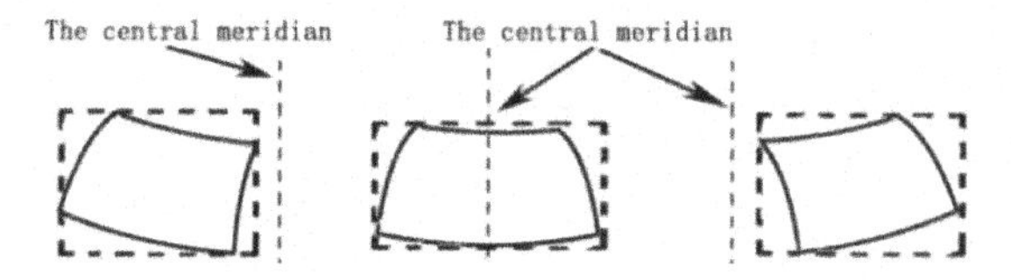

Fig. 4. The axial minimum enclosing rectangle and the positional relationship with the central meridian

Therefore, we can insure the rectangle by the four coordinates of corners and the intersection points of the central meridian with the frame, and the value of y in the central meridian is 500000. The determination method and steps are as following:

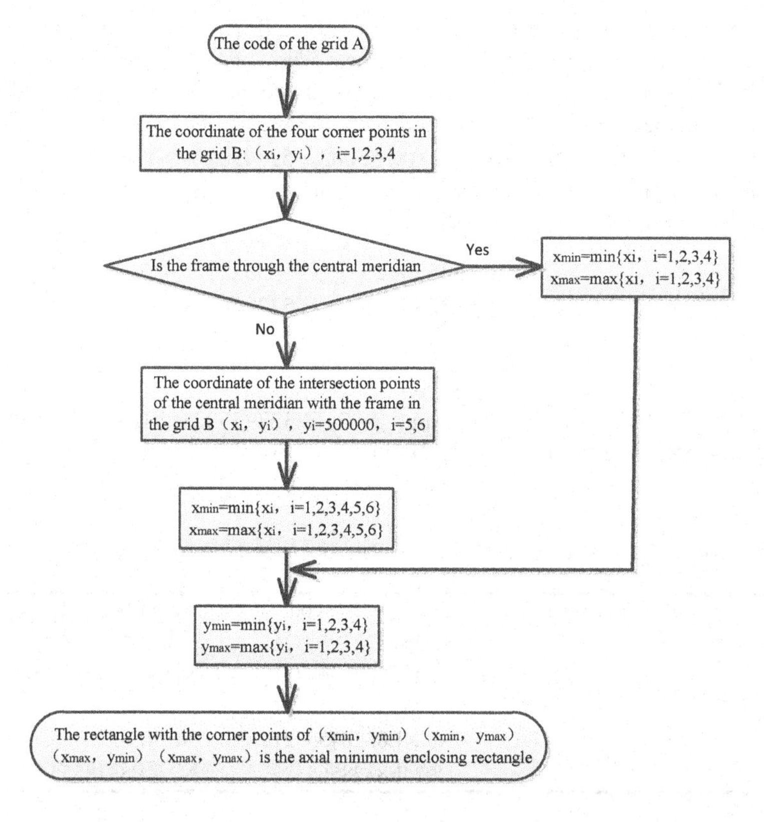

Fig. 5. The calculation process of the axial minimum enclosing rectangle

```
program Inflation
  Step1:y_max=max(y_1,y_2,y_3,y_4);
    y_min=min(y_1,y_2,y_3,y_4);
  Step2:When ((y_1-500000)(y_4- 500000)>0)
       {
           xmax=max(x_1,x_2,x_3,x_4);
           xmin=min(x_1,x_2,x_3,x_4);
       }
       else
       {
           x_max=max(x_1,x_2,x_3,x_4,x_5,x_6);
    x_min=min(x_1,x_2,x_3,x_4,x_5,x_6);
       }
   Step3:Give the rectangle from the four points of (x_max,
  y_max)、(x_max, y_min)、(x_min, y_max)、(x_min, y_min), and the rectangle
  is the axial minimum enclosing rectangle
```

4 Experiments and Analysis

Transform the theodolite coordinate 1 minute Grid (NE01M751171459) to the gauss coordinate 10 meter Grid on 6°projection zone. The code describe the extension is 75°14′-75°15′north latitude, 117°59′- 118°east longitude, and the coordinate of the four corner points in the gauss coordinate system are (8353214, 527987) 、 (8355074, 527956) 、 (8353222, 528461) 、 (8355082, 528430) .

The intersection points of the central meridian with the frame are also the corner points according the method of the paper, so:

$$x_{max}=8355082,\ x_{min}=8353214$$
$$y_{max}=528461,\ y_{min}=527956$$

Therefore, the four corner points coordinate of the axial minimum enclosing rectangle are (8353214,527956), (8355082,527956), (8355082,528461), (8353214,528461), the grid code of these points are N024E8327955321, N024E8327955508, N024E8328465508, N024E8328465321 in turn. So the conversion result is as following:

Table 1. The transform result based on the axial minimum enclosing rectangle

The code of the area in the grid A	The code of the area in the grid B
	N024E8327955321、...、N024E8327955508
	N024E8327965321、...、N024E8327965508
NE01M751171459	......
	N024E8328455321、...、N024E8328455508
	N024E8328565321、...、N024E8328555508

The method for the transform based on the axial minimum enclosing rectangle realized the conversion between the different types or scale geographic grids.

Also, this method can be used in the grids transform for irregular region, such as administrative zoning map, that is first to get the axial minimum enclosing rectangle of the irregular region, and then project the region, and get the axial minimum enclosing rectangle of the last region; Last, obtain the grid codes of the coverage area of the axial minimum enclosing rectangle according to the coding rules (Fig. 6.).

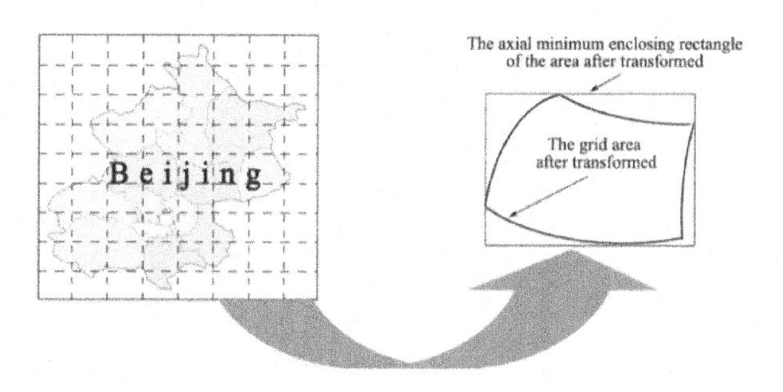

Fig. 6. The schematic diagram of the grids transform for the irregular region

Acknowledgements. National 863 program(2009AA12Z218); The National Natural Science Foundation of China(40671163).

References

1. GB/T 12409—2009. The National Standard of China on Geographic Grid. The Administration of Quality Supervision Inspection and quarantine of China, National Standardization Management Committee of China (2009)
2. Zhang, Y.-S., Ben, J., Tong, X.-C.: Discrete Global Grids for Geospatial Information: Principles, Methods and its Application. Science Press, Beijing (2007)
3. Goodchild, M., Yang, S.: A hierarchical spatial data structure for global geographic information systems. Graphical Models and Image Processing 54(1), 31–44 (1992)
4. Sahr, K., White, D., Kimerling, A.J.: Geodesic Discrete Global Grid Systems. Cartography and Geographic Information Science 30(2), 121–134 (2003)
5. Zhai, Y., Zhao, F.-L.: The Technology of Modern Surveying. PLA Book Concern, Beijing (2005)

A Dual-Decomposition-Based Resource Allocation for the Data Transmission in the Internet

Youmao Bai

School of Mechanical Electronic & Information Engineering,
China University of Mining & Technology, Beijing
Beijing, P.R. China
aazzhang.123@gmail.com

Abstract. The paper investigates the problem of allocating transmission data rates to users in the Internet. The paper provides a general problem of power and rate allocation that the utility functions can be nonconcave and nondifferetiable, which turns utility maximization into nonconvex, constrained optimization problem that is well-known to be difficult. To solve this problem, the paper derives an important property for the dual decomposition of the original optimization problem, and then proposes a simple iterative bisection algorithm for the adaptive multi-user resource allocation for the data transmission in the Internet.

Keywords: dual optimization, data transmission, Internet, adaptive resource allocation.

1 Introduction

Over the last decades, researches have a significant amount of interest in the area of Internet transmit data rate control, which aims to alleviate the congestion and provide a better quality of service (QoS) in the Internet. Currently, most services in the Internet are elastic to some degree, i.e., the network sources can be adaptively assigned to adjust their transmission data rates according to congestion levels within the network. Hence, by appropriately exploiting the elasticity through rate control, one can provide high network efficiency while at the same time alleviating network congestion. To the resource optimization problem in the Internet, the well-known concept of the utility function is typically considered to measure the level of user satisfaction or QoS at the allocated rate.

The services in the Internet can be classified into two classes based on the shape of the utility function. One corresponds to traditional data services, such as file transfer and e-mail. The QoS of these services can be greatly degraded by the network congestion, and adaptive resource allocation is a useful way to reduce the network congestion to improve the system performance. Usually, the elasticity of these services can be modeled by a concave utility function [1]. The other corresponds to delay and rate sensitive services, such as audio services and streaming video. These

H. Tan (Ed.): Knowledge Discovery and Data Mining, AISC 135, pp. 93–100.
springerlink.com

services are less elastic than traditional data services. They can maintain their transmission data rates up to a certain level with a corresponding graceful degradation in the QoS.

In the past few years, utility-based rate control problems have begun to be addressed by articles [2]–[4]. They have almost exclusively dealt with the situation where the utilities are concave. Extensive theories and algorithms are proposed to address these problems. Recently, the application of nonconvexity of the system has begun to be presented in the literature [5]–[8]. In [5][6], the authors investigate the problem of allocating transmission data rates in a distributed fashion with a sigmoid utility function. Ref. [7] develops a simple downlink ARA algorithm with three types of utility functions, such as concave, convex, and sigmoid-like utility functions. The sigmoid utility function with pricing is considered in [8], and the papers show that the pricing introduces a large flexibility for the resource optimization problem.

However, some unrealistic assumptions, such as the continue transmission data rate and the infinite modulation constellation size, significantly limit the application of above approximate utility functions. Thus, it should investigate a more general type of utility function. In this paper, in order to extend the result of the recent articles, a more general utility function, which can be nonconcave and nondifferentiable, is investigated. And the utility functions in the above references are just the special case of the extended utility function in this paper. Then, a simple iterative bisection algorithm is proposed to deal with multi-user transmission data rate control in the Internet.

The remainder of this paper is organized as follows. The system model and problem framework are given in section 2. We then present the simple iterative bisection algorithm in Section 3, and numerical results and conclusions are given in Section 4 and Section 5, respectively.

2 System Description and Problem Formulation

We consider a system that consists of N links and K users. The network has a total network resource constraint (ex. Transmit power) C_{total}, and each user has a utility function $U_{k,n}$ and resource constraint $C_{\max}$. We assume that $U_{k,n}$ has the following properties.

1) $U_{k,n}$ an increasing function of the allocated network resource $x_{k,n}$ for user k in the *n-th* link.

2) $U_{k,n}$ is a continue function of $x_{k,n}$.

3) $U_{k,n}$ can be a nonconvex and nondifferentiable function of $x_{k,n}$.

The goal of adaptive resource allocation in the network is to find the optimal network resource assignment in order to maximize the total system utility in the Internet, which can be expressed as [5].

$$\begin{aligned} &\max_{x_{k,n}} \sum_{k=1}^{K}\sum_{n=1}^{N} U_{k,n}\left(x_{k,n}\right) \\ &s.t. \sum_{k=1}^{K}\sum_{n=1}^{N} x_{k,n} \le C_{total} \\ &0 \le x_{k,n} \le C_{\max}, \qquad \forall\, k,n \end{aligned} \tag{1}$$

where C_{total} and $C_{\max}$ are the network resource constraints for users.

3 Resource Allocation Framework

Since the objective function $U_{k,n}\left(x_{k,n}\right)$ is noncovex and nondifferentiable, it is difficult to deal with the optimization problem in (1). Therefore, we first obtain some properties for the dual optimization of the optimization problem of (1).

Suppose that x has m components $x_1, x_2, \cdots, x_m$ of dimensions $n_1, n_2, \cdots, n_m$, respectively, and the problem has the form

$$\begin{aligned} &\min \sum_{i=1}^{m} f_i\left(x_i\right) \\ &s.t. \sum_{i=1}^{m} g_i\left(x_i\right) \le 0 \\ &x_i \in X_i, \quad i = 1, 2, \cdots, m \end{aligned} \tag{2}$$

where $f_i : \mathbb{R}^{n_i} \mapsto \mathbb{R}$ and $g_i : \mathbb{R}^{n_i} \mapsto \mathbb{R}$ are given functions, and X_i are given subsets of $\mathbb{R}^{n_i}$.

Define the Lagrangian function $L(\vec{x}, \mu)$ as follows.

$$\begin{aligned} &\vec{x} = \{x_1, x_2, \cdots, x_m\} \\ &L(\vec{x}, \mu) = \sum_{i=1}^{m} f_i\left(x_i\right) - \mu \sum_{i=1}^{m} g_i\left(x_i\right) \\ &\mu \ge 0 \end{aligned} \tag{3}$$

Based on Proposition 3.3.4 in [1], the optimal solution of optimization problem (1) $\left(\vec{x}^*, \mu^*\right)$ can be expressed as

$$\vec{x}^* = \arg \min_{x_i \in X_i,\, i=1,2,\cdots,m} L\left(\vec{x}, \mu^*\right) \tag{4}$$

where $\vec{x}^*$ is the global minimum of the problem.

Lemma 1: For all value μ in the feasible region, we have $L\left(\vec{x}^*, \mu\right) \le L\left(\vec{x}^*, \mu^*\right)$, which means $L\left(\vec{x}^*, \mu^*\right) = \max_{\mu}\left\{L\left(\vec{x}^*, \mu\right)\right\}$.

Proof: Based on the Proposition 3.3.1 in [9], we can see that for the optimal solution $\left(\vec{x}^*, \mu^*\right)$, we have

$$\mu^* \sum_{i=1}^{m} g_i\left(x_i^*\right) = 0 \tag{5}$$

Besides, since $\sum_{i=1}^{m} g_i\left(x_i\right) \le 0$ and $\mu \ge 0$,

$$\mu \sum_{i=1}^{m} g_i\left(x_i^*\right) \ge 0 \tag{6}$$

So, we have

$$\mu^* \sum_{i=1}^{m} g_i\left(x_i^*\right) \le \mu \sum_{i=1}^{m} g_i\left(x_i^*\right) \tag{7}$$

Based on (3) and (7), we get

$$\begin{aligned} L\left(\vec{x}^*, \mu\right) &= \sum_{i=1}^{m} f_i\left(x_i^*\right) - \mu \sum_{i=1}^{m} g_i\left(x_i^*\right) \\ &\ge \sum_{i=1}^{m} f_i\left(x_i^*\right) - \mu^* \sum_{i=1}^{m} g_i\left(x_i^*\right) \\ &= L\left(\vec{x}^*, \mu^*\right) \end{aligned} \tag{8}$$

Combining (4) and (8), we can get

$$\begin{aligned} L\left(\vec{x}^*, \mu^*\right) &= \max_{\mu} \left\{ L\left(\vec{x}^*, \mu\right) \right\} \\ &= \max_{\mu} \left\{ \min_{x_i \in X_i,\, i=1,2,\cdots,m} L\left(\vec{x}, \mu\right) \right\} \end{aligned} \tag{9}$$

Further, we can deduce

$$\begin{aligned} L\left(\vec{x}^*, \mu^*\right) &= \max_{\mu} \left\{ \min_{x_i \in X_i,\, i=1,2,\cdots,m} L\left(\vec{x}, \mu\right) \right\} \\ &= \max_{\mu} \left\{ \min_{x_i \in X_i,\, i=1,2,\cdots,m} \left\{ \sum_{i=1}^{m} f_i\left(x_i\right) - \mu \sum_{i=1}^{m} g_i\left(x_i\right) \right\} \right\} \\ &= \max_{\mu} \left\{ \min_{x_i \in X_i,\, i=1,2,\cdots,m} \left\{ \sum_{i=1}^{m} \left[f_i\left(x_i\right) - \mu g_i\left(x_i\right) \right] \right\} \right\} \\ &= \max_{\mu} \left\{ \min_{x_i \in X_i} f_i\left(x_i\right) - \mu g_i\left(x_i\right),\ i = 1, 2, \cdots, m \right\} \end{aligned} \tag{10}$$

which means that we can convert the optimization problem to be a separable optimization problem as

$$\max_{\mu} \left\{ \min_{x_i \in X_i} f_i\left(x_i\right) - \mu g_i\left(x_i\right),\ i = 1, 2, \cdots, m \right\} \tag{11}$$

Note that the minimization involved in the calculation of the dual function has been decomposed into m simpler minimizations in (11). These minimizations are often conveniently done either analytically or computationally, in which case the dual function can be easily evaluated. This is the key advantageous structure of separate problem. In (11), for a given μ, the optimal solution x_i^* should satisfy the constraint $\sum_{i=1}^{m} g_i(x_i^*) \le 0$ from (2), where $x_i^* = \arg \min_{x_i \in X_i} \{f_i(x_i) - \mu g_i(x_i)\}$.

Thus, combining the eq. (11) and the the constraint $\sum_{i=1}^{m} g_i(x_i) \le 0$, the dual optimization problem can be expressed as

$$\begin{aligned} g^* &= \max_{\mu \ge 0} \; q(\mu) \\ s.t. \; &\sum_{i=1}^{m} g_i(x_i^*) \le 0 \end{aligned} \tag{12}$$

where

$$\begin{aligned} q(\mu) &= \min_{x_i \in X_i} \; \{f_i(x_i) - \mu g_i(x_i)\} \\ x_i^* &= \arg \min_{x_i \in X_i} \; \{f_i(x_i) - \mu g_i(x_i)\} \end{aligned} \tag{13}$$

Based on (2) and (13), we can formulate its dual optimization problem as [9]

$$\begin{aligned} g^* &= \max_{\mu \ge 0} \; q(\mu) \\ s.t. \; &\sum_{k=1}^{K} \sum_{n=1}^{N} \beta_{k,n}(\mu) \le C_{total} \end{aligned} \tag{14}$$

where

$$q(\mu) = -\mu C_{total} + \sum_{k=1}^{K} \sum_{n=1}^{N} \min_{0 \le x_{k,n} \le C_{\max}} L_{k,n}(x_{k,n}, \mu) \tag{15}$$

$$\beta_{k,n}(\mu) = \arg \min_{0 \le x_{k,n} \le C_{\max}} L_{k,n}(x_{k,n}, \mu), \quad \forall k,n \tag{16}$$

and

$$L_{k,n}(x_{k,n}, \mu) = -U_{k,n}(x_{k,n}) + \mu x_{k,n} \tag{17}$$

The dual optimization problem is much simpler than the original optimization problem in (1), since the dual function $q(\mu)$ is a concave function of μ and the separable structure of the dual function decomposes the problem into KN simpler problems from (15). Thus, the dual optimization problem can iteratively search the optimal value of μ, denoted by μ^*, in the dual domain to get the optimal resource allocation $\mathbf{x}^* = \{\beta_{k,n}(\mu^*), \; \forall k,n\}$. However, this dual optimization problem is still difficult. Below, we achieve some properties of functions $q(\mu)$ and $\beta_{k,n}(\mu)$, respectively. And propose a simple bisection algorithm to iteratively search the optimal solution $(\mu^*, \mathbf{x}^*)$.

***Proposition 1*:** $q(\mu)$ is a decreasing concave function of μ in the dual domain.

***Proposition 2*:** $\beta_{k,n}(\mu)$, as in (16), is a non-increasing function of the variable μ. The objective of the optimization in (14) is to get the maximum of $q(\mu)$. Based on *Proposition 1*, this is equivalent to finding the smallest positive value of μ, i.e., μ^*, under the constraint that $\sum_{k=1}^{K}\sum_{n=1}^{N}\beta_{k,n}(\mu)\le C_{total}$. When μ decreases, from *Proposition* 2, $\sum_{k=1}^{K}\sum_{n=1}^{N}\beta_{k,n}(\mu)$ increases. So we get μ^* when $\sum_{k=1}^{K}\sum_{n=1}^{N}\beta_{k,n}(\mu)=C_{total}$. Since $\sum_{k=1}^{K}\sum_{n=1}^{N}\beta_{k,n}(\mu)$ is a non-increasing function of μ, we can use the bisection algorithm to find μ^* [10], and then get the optimal resource allocation $x_{k,n}^*=\beta_{k,n}(\mu^*)$.

Thus, the iterative bisection algorithm of adaptive resource allocation can be described as follows:

1. Initialization: Set the resource assignments of all uses to be zero.
2. Iteration: Using bisection method to iteratively search μ^* subject to the constraint $\sum_{k=1}^{K}\sum_{n=1}^{N}\beta_{k,n}(\mu)\le C_{total}$.
3. Calculation: Using μ^* to determine the assigned resource allocation $x_{k,n}^*=\beta_{k,n}(\mu^*)$. Then, users start the data transmission in the Internet.

4 Numerical Results

In this section, we provide numerical results to illustrate the performance of the proposed iterative bisection algorithm. In the simulation, we consider a system with a single bottleneck link in Fig. 1. In this figure, we provide the capacity and the propagation delay of each link. For a given link, user k transmits packets from source node S_k to destination node D_k with utility function U_k. For simple, we choose the sigmoidal utility function, as in Fig. 2, to simulate the performance.

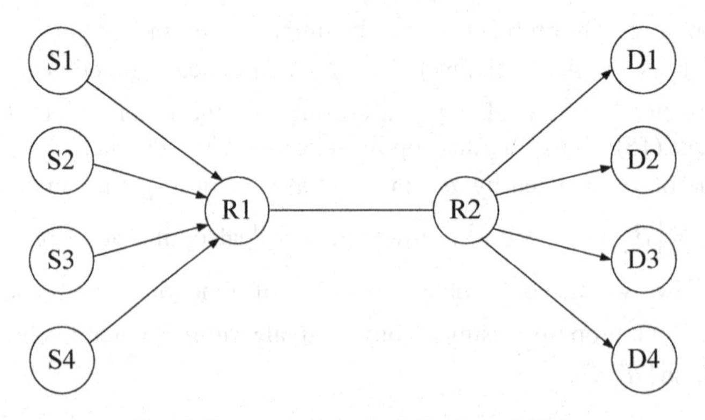

Fig. 1. System with a single bottleneck link

Fig. 3 compares the proposed iterative bisection algorithm with the equal power allocation. It is seen that the performance of the proposed algorithm is almost the same as that of the global optimization, which indicates that the duality gap of the dual optimization problem is very small. We know that the duality gap of the dual optimization problem decreases when either K or N increases [9]. So, it is reasonable for us to use the proposed iterative bisection algorithm to solve the adaptive resource allocation for the data transmission in the Internet.

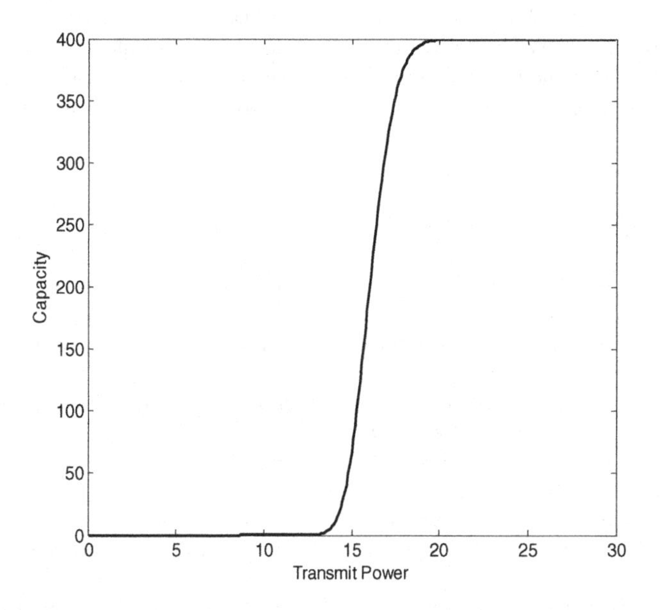

Fig. 2. Sigmoidal-like function

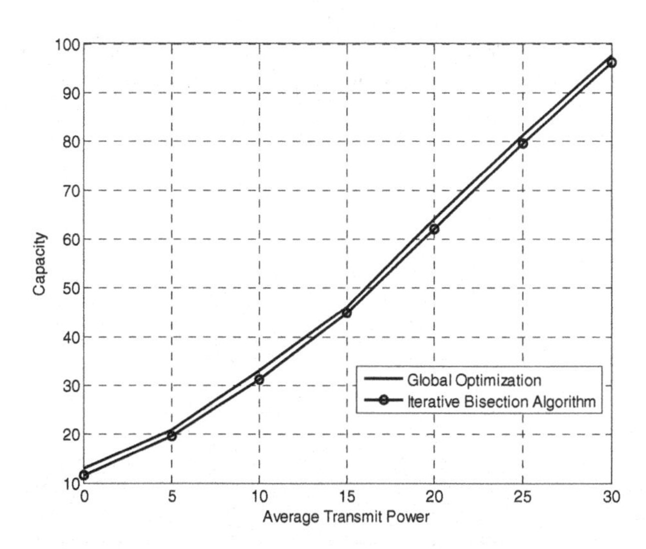

Fig. 3. Comparison of proposed algorithm and Equal power allocation

5 Conclusions

Considering a general utility function, we studied an adaptive resource allocation for the data transmission in the Internet. Since the original optimization problem is difficult, we investigated the dual-decomposition-based optimization problem, and proposed a simple iterative bisection algorithm adaptively allocate the transmit power and available subchannels by maximizing the total network capacity. The paper shows that the performance of the proposed algorithm is almost the same as the global optimization.

References

1. Shenker, S.: Fundamental design issues for the future Internet. IEEE J. Sel. Areas Commun. 13(7), 1176–1188 (1995)
2. Kelly, F.P.: Charging and rate control for elastic traffic. Eur. Trans. Telecommun. 8(1), 33–37 (1997)
3. Low, S.H., Lapsley, D.E.: Optimization flow control, I: Basic algorithm and convergence. IEEE/ACM Trans. Netw. 7(6), 861–874 (1999)
4. La, R.J., Anantharam, V.: Utility-based rate control in the Internet for elastic traffic. IEEE/ACM Trans. Netw. 10(2), 272–286 (2002)
5. Wang, Z., Peng, Q., Milstein, L.B.: Multiuser Resource Allocation for Downlink Multi-cluster Multicarrier DS CDMA System. IEEE Trans. Wire. Commun. 10(8), 2534–2542 (2011)
6. Lee, J.-W., Mazumdar, R.R., Shroff, N.B.: Non-Convex Optimization and Rate Control for Multi-Class Services in the Internet. IEEE/ACM Trans. Netw. 13(4), 827–840 (2005)
7. Lee, J.-W., Mazumdar, R.R., Shroff, N.B.: Downlink power allocation for multi-class CDMA wireless networks. In: Proc. IEEE INFOCOM, vol. 3, pp. 1480–1489 (2002)
8. Xiao, M., Shroff, N.B., Chong, E.K.P.: A utility-based power-control scheme in wireless cellular systems. IEEE/ACM Trans. Netw. 11(2), 210–221 (2003)
9. Bertsekas, D.P.: Nonlinear Programming. Athena Scientific (1999)
10. Press, W.H.: Numerical Recipes in C. Cambridge University Press (1992)

Distributed Electronic Evidence Forensics Laboratory Design and Practice That Based on High-Speed 3G and Cloud Computing Model

Qingyi Tian, Peiyun Luo, and Qingjun Liu

Cyber Security Department of Chongqing Public Security Bureau,
401121 Chongqing, China
knightcn@foxmail.com

Abstract. The ability of domestic electronic evidence forensics laboratory has been limited by the laboratory location, computing resources and information resources, which can't be obtained and shared for each-other. The authors explore the high-speed 3G technology and cloud computing model, using in the construction of the laboratory of electronic evidence, to achieve the movement of computing resources and schedule the storage resource sharing, real-time information and evidence gathered can be achieved in real, then, effectively improve the electronic forensics lab in the rapid response capability.

Keywords: High-speed 3G, Cloud computing, Electronic forensics laboratory, Construction mode.

1 Introduction

As broadband networks and high speed wireless 3G[1], cloud computing and other technologies in social business, finance, life, are closely integrated, the law enforcement agencies, in practice, investigators need to address the sharp rise in the amount of electronic evidence, importance of the information electronic evidence contained has been significantly improved. August 30, 2011, the NPC announced the "Criminal Procedure Law Amendment (Draft)", electronic evidence became one of the official categories of evidence[2]. Government agencies around the country, such as public security, judicial, taxation and other departments and sectors of society are gradually set up forensic electronic evidence forensics laboratory.

However, under construction mode there are about three defects of traditional electronic forensics lab[3-7]:

1) The evidence is difficult to achieve the movement ,which on environmental design in architecture, only pay attention to the region in the laboratory to optimize the capacity of evidence, breaking the laboratory area of the "geographic" limits. Especially as the popularity of smart phones, tablet computers and ubiquitous computing, the ability of mobile electronic evidence demand increasing. The existing electronic evidence forensics laboratory can't be able to achieve the need for inspection before Samples been sent to inspect, it is difficult to meet the actual demand;

H. Tan (Ed.): Knowledge Discovery and Data Mining, AISC 135, pp. 101–107.
springerlink.com

2) The rapid increase in hard drive capacity resulted in the traditional stand-alone analysis of the hard drive becomes a long time, difficult to adapt to actual work for quick analysis of physical evidence obtained and analyzing requirements. According to the latest report, 4T as a single hard drive has for the market. According to the U.S. ACCESSDATA's evaluation data, in the Windows7 64-bit operating system, 8GB RAM memory, AMD Phenom ™ II X4 955 Processor 3.20 GHz quad-core CPU are now mainstream advanced configuration. 100GB of image processing time is 9.08 hours, and 4T drives reach 360 hours' processing time of 15 days to complete data analysis, that can't meet the actual requirements. Meanwhile, the laboratory between the various sets of computers is difficult to share computing and storage resources, and inefficient use of resources;

3) Evidence information between the laboratory and in-spot is difficult to share among vehicles, making the information in the laboratory or between is difficult to obtain or real-time convergence. In domestic laboratory, evidence of information exchange often requires manual transmission, making the data convergence takes a long time.

Laboratory construction mode of domestic electronic evidence is difficult to meet the actual requirements. The author based on actual law, explored the establishment of building distributed e-card forensics lab in high-speed 3G and cloud-based computing which can address the shortcomings. This model can break through the geographical laboratory "wall", and make distributed mobile features come true. Laboratory evidence of the computing resources within the platform and storage resources can be shared and dynamic scheduling.

2 Key Technologies of Distributed Electronic Evidence Forensics Laboratory Construction Based on High-Speed 3G and Cloud Computing Model

The construction of distributed electronic forensics lab's key technology will rely on 3G, cloud computing and application-level distributed forensics, which will effectively avoid the traditional model of electronic forensics laboratory's construction defects.

"3G" or "the third generation" is short for the third generation mobile communication technology, which can support high-speed data transmission cellular mobile communications. 3G services can simultaneously transmit voice and data (e-mail, instant messaging, etc.). Currently there are four 3G standards[8-10]: CDMA2000, WCDMA, TD-SCDMA and WiMAX and the main use is the first three criteria. Field tests in optimal signal environment, the main three standard test date as follows:

Table 1. The speed of domestic 3G standard test

3G standard	Uplink speed	Downlink speed
WCDMA	700KB/s	1.8MB/s
CDMA2000	225KB/S	390kb/s
TD-CDMA	200kb/s	350kb/s

In test, China Unicom's WCDMA mode in vehicle moves gave us excellent performance, the network speed stability, fluctuations between the upstream and down is little, and the line of peaks and valleys of the composite is more good. Therefore, the model described in this article uses WCDMA as the standard net work to support mobile technologies. Connect the equipments and the central laboratory in order to achieve resource sharing and exchange information.

Cloud computing is a virtualized resource that can be changed dynamically, which is characterized by both hardware and software can be dynamically configured according to needs and expansion. But it is logically presented in the form of single whole through with the distributed physically sharing cloud computing infrastructure The mainstream cloud technologies are including VMware, Xen and Microsoft. According to the specific network and hardware in laboratory, evaluating robustness, ease management and other aspects, the author select VMware Vsphere cloud computing as the basis of the laboratory. By building basis cloud computing environment to effectively improve the utilization of lab resources, can ensure the laboratory's computer forensics analysis form a logical whole state, and providing a distributed architecture in application layer to handle large capacity hard disk basis.

Our forensic technology of application layer mainly relay on FTK software. FTK has a strong automatic document analysis, filtering and searching capabilities, automatically classify all documents, locating the suspect file, finding the necessary evidence quickly; FTK is recognized as the leading forensic tool for analyzing e-mail, and its sale volume is the first in world's electronic evidence analysis software. In application layer the software can be realized in distributed computing, and the topology is as follows [11-12]:

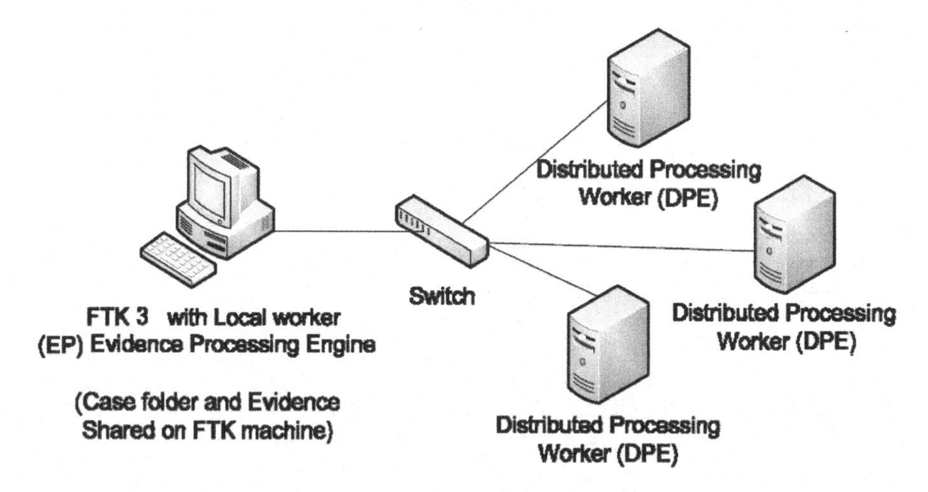

Fig. 1. The distributed computing topologies of FTK

The topology consists of a main engine and three distributed composition one. According to the author, the actual processing speed of the mirror 100G can be reduced from 9 to 2 hours. Laboratory using this topological structure of the samples described in this article analyzes the same amount of data for only 20% of original, greatly improved the speed in response to TB high-capacity hard drive.

According to the author's actual practice, FTK software's distributed computing topologies can be constructed in cloud computing, that the hardware and software in cloud environment are fully applied. Meanwhile, the evidence used in 3G mobile link can also be seamlessly integrated with the architecture.

3 The Designing of Distributed Electronic Evidence Forensics Laboratory That Based on High-Speed 3G and Cloud Computing Model

The distributed laboratory is composed by the laboratory in Urban Center, in district or lower level place, and mobile forensic vehicle. It can be distinguished into the core network of the cloud and the outside one. Among them, the Urban Lab is the main center of computing, storaging, gathering resources, and responsible for the entire distributed architecture, which provide for the core of forensic computing, the storage capacity, and the Urban Council, the county lab, the mobile survey vehicle information collected real-time aggregation and analysis. County forensics lab main responsible for the access to samples of the hardware interface, while have certain analysis capacity in the field. Mobile forensic vehicle will make evidence obtain moving, primarily as the role of information gathering, the weak capability of analysis and processing mainly dependent on the supporting of central laboratory.

The overall network topology of distributed laboratory is as shown below.

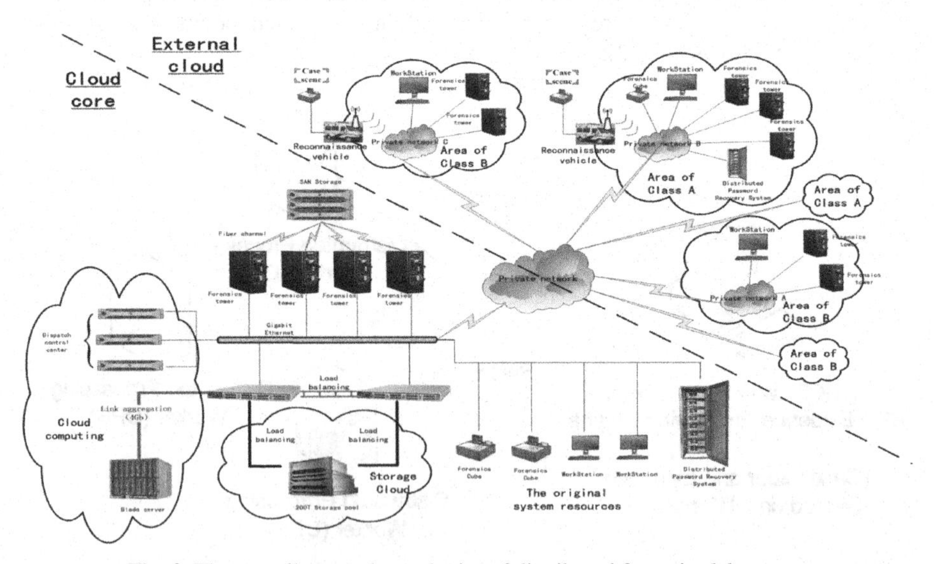

Fig. 2. The overall network topologies of distributed forensics laboratory

The advantages of the topology above are as follows:

1) By using secure remote connectivity to provide remote technical support and analysis for district laboratory. We can remotely do analysis operation, when county agencies access the hard disk to the evidence tower. Remote command center can show evidence of a interface, and remote communication and guidance comes true.

2) Any analyst of the distributed laboratory can start analysis working by accessing to the terminal, which can be easily operated. With the construction of the proprietary line network, enabling cross-regional analysis and remote guidance comes true, meanwhile a solution to the staff shortage and special operations can be carried out in many areas.

4 The Functionality of Distributed e-Platform in Real

The modules of distributed electronic forensics laboratory are as shown below.

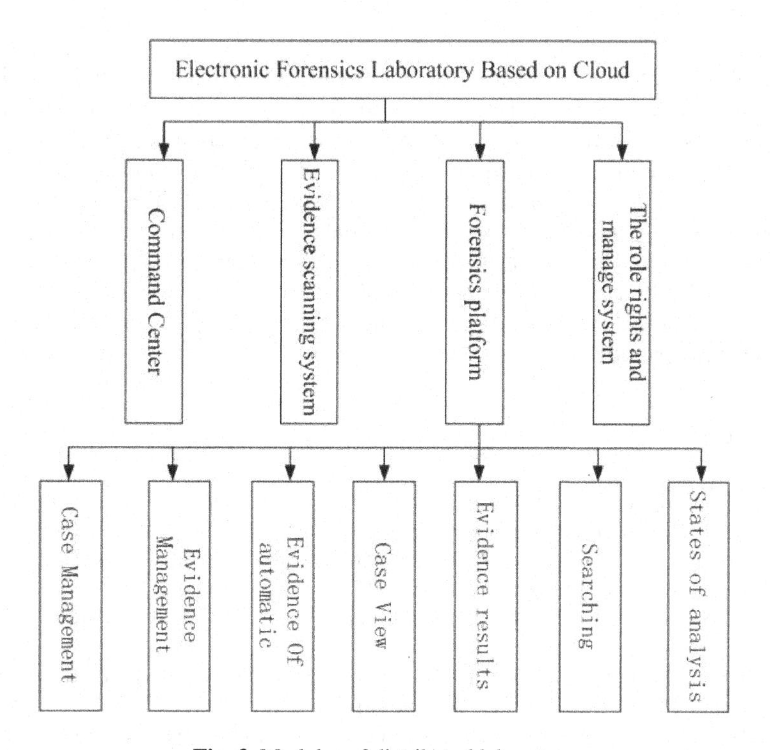

Fig. 3. Modules of distributed laboratory

Command Center: Summary displays the work of resources in the cloud, scheduling resources on-demand;

Evidence scanning system: Automatically get the date of the machines access to the media, or mirrored center data, entry the media information and image data into cloud platform;

The role rights and manage system: Personnel management, assign roles, including administrators, analysts, inspectors, etc.;

Forensics platform: Management the mirror data and media within the platform, distribution analysts, do forensic analysis.

Through this network topology and system, achieved the following functions:

1) The bulk of evidence analysis. Regardless anywhere the data in the network located, you can send commands for a unified analysis;

2) Remote analysis. In command center, you can analysis data on a remote machine; remote node can also get center resources for forensic analysis;

3) Collaboration. Many persons can log a case at the same time to resolve, discussion, learning, guidance, etc.; achieve real-time information gathering, sharing and exchange.

5 Conclusions

The distributed electronic evidence lab has been built and put into practical in the author's workplace. The laboratory that based on 3G high-speed networks and cloud computing has not been public reported by domestic literature. The lab has been actually used in multiple locations and in a series of joint action, and the design goals has been basically achieved from the actual use of effect. But because of the environment of 3G, the upload bandwidth of data has been limited, requiring a long time for large amounts of data in mobile point. It might be the bottleneck in real performance. In the next step, the author will further optimize the application of the platform, integrating more modules to improve the model's response speed and communicate capability.

References

1. Vaquero, L.M., Rodero-Merino, L., Caceres, J., Lindner, M.: A Break in the Clouds: Towards a Cloud Definition. Computer Communication Review 39, 50–55 (2009)
2. Mell, P., Grance, T.: The NIST Definition of Cloud Computing. National Institute of Standards and Technology 53, 50 (2009)
3. Armbrust, M., Fox, A., Griffith, R., Joseph, A.D.: Above the Clouds: A Berkeley View of Cloud Computing. Science 53, 7–13 (2009)
4. Armbrust, M., Fox, A., Griffith, R., Joseph, A.D.: A view of cloud computing. Communications of the ACM 53, 50–58 (2010)
5. Hensbergen, E.V., Evans, N.P., Stanley-Marbell, P.: A unified execution model for cloud computing. ACM SIGOPS Operating Systems Review 44, 12 (2010)
6. Zhang, Q., Cheng, L., Boutaba, R.: Cloud computing: state-of-the-art and research challenges. Journal of Internet Services and Applications 1, 7–18 (2010)
7. Langmead, B., Schatz, M.C., Lin, J., Pop, M., Salzberg, S.L.: Searching for SNPs with cloud computing. Genome Biology 10, R134 (2009)
8. Honkasalo, H., Pehkonen, K., Niemi, M.T., Leino, A.T.: WCDMA and WLAN for 3G and beyond. IEEE Wireless Communications 9, 14–18 (2002)
9. Lozano, A., Farrokhi, F.R., Valenzuela, R.A.: Lifting the Limits on High Speed Wireless Data Access using Antenna Arrays. IEEE Communications Magazine 39, 156–162 (2001)

10. Ekstrom, H., Furuskar, A., Karlsson, J., Meyer, M., Parkvall, S., Torsner, J., Wahlqvist, M.: Technical solutions for the 3G long-term evolution. IEEE Communications Magazine 44, 38–45 (2006)
11. Configuring Distributed Processing with FTK 3, Access Data Group, LLC, http://accessdata.com/downloads/media/Configuring%20Distributed%20Processing%20with%20FTK%203.pdf
12. Kolding, T.E., Frederiksen, F., Mogensen, P.E.: Performance Aspects of WCDMA Systems with High Speed Downlink Packet Access (HSDPA). Vehicular Technology Conference 1, 477–481 (2002)

Research on Indoor Positioning Technology Based on RFID

Huaichang Du

School of Information Engineering, Communication University of China
Beijing, China
hc_du@yahoo.com.cn

Abstract. The global positioning system (GPS) in the indoor is difficult to use, ultrasound, infrared and other indoor positioning technology in positioning accuracy and the accuracy has evident defects. RFID based indoor positioning technology is a hot topic at present. The paper introduces the indoor location algorithm based on RFID, especially LANDMARC indoor positioning theory. And on this basis, it improves traditional method by introducing cluster analysis. The simulation results show that the improved algorithm is better than the original algorithm in positioning accuracy.

Keywords: Radio frequency identification, Indoor localization, positioning algorithm, clustering algorithm.

1 Introduction

Radio Frequency Identification (RFID) is widely used for electronic identification and tracking. RFID offers substantial advantages for businesses allowing automatic inventory and tracking on the supply chain. In the indoor positioning application, the no contact, low cost, high precision and non-line-of-sight nature of this technology are significant advantages common among all types of RFID systems. At present, for all kinds of positioning technology and localization algorithm research and improvement become the focus of research [1]. Especially all kinds of the researches and improvement on positioning technology and localization algorithm have become hot point. Therefore, research on indoor positioning technology based on RFID not only has the theory significance, but also practical significance. In this paper, we first discuss the background knowledge of RFID technology, and then discuss the relevant position technology used in recent years. We will further discuss LANDMRC algorithm. Finally, the improved algorithm and the simulation result are given.

2 The Related Theory and Technology

2.1 Radio Frequency Identification

Radio frequency identification (RFID) is a technology that uses radio waves to transfer data from an electronic tag, called RFID tag or label, attached to an object,

H. Tan (Ed.): Knowledge Discovery and Data Mining, AISC 135, pp. 109–115.
springerlink.com

through a reader for the purpose of identifying and tracking the object. The tag's information is stored electronically. The RFID tag includes a small RF transmitter and receiver. An RFID reader transmits an encoded radio signal to interrogate the tag. The tag receives the message and responds with its identification information. All RFID tags use radio frequency energy to communicate with the readers. However, the method of powering the tags varies. An active tag embeds an internal battery which continuously powers it and its RF communication circuitry. Readers can thus transmit very low-level signals, and the tag can reply with high-level signals. An active tag can also have additional functionalities such as memory, and a sensor, or a cryptography module. On the other hand, a passive tag has no internal power supply. Generally, it backscatters the carrier signal received from a reader. Passive tags have a smaller size and are cheaper than active tags, but have very limited functionalities. The last type of RFID tags is semi passive tags.

This technology plays a key role in pervasive networks and services [2]. Indeed, data can be stored and remotely retrieved on RFID tags enabling real-time identification of devices and users. However, the usage of RFID could be hugely optimized if identification information was linked to location [3].

RFID readers have two interfaces. The first one is a RF interface that communicates with the tags in their read range in order to retrieve tags' identities. The second one is a communication interface, generally IEEE 802.11 or 802.3, for communicating with the servers. Finally, one or several servers constitute the third part of an RFID system. They collect tags' identities sent by the reader and perform calculation such as applying a localization method. They also embed the major part of the middleware system and can be interconnected between each other's [2].

2.2 Indoor Positioning

At present, there are several types of location-sensing systems, each having their own advantages as well as limitations. Infrared, 802.11, ultrasonic, and RFID are some examples of these systems. The main positioning algorithms include Cell of Origin (CoO), Time of Arrival (ToA), Time Difference of Arrival (TDoA), Received Signal Strength (RSS), Angle of Arrival (AoA).

Ultrasonic. The Cricket Location Support System and Active Bat location system are two primary examples that use the ultrasonic technology. Normally, these systems use an ultrasound time-of-flight measurement technique to provide location information. Its accuracy can reach cm level. However, the use of ultrasonic this way requires a great deal of infrastructure in order to be highly effective and accurate, and the cost is so exorbitant that it is inaccessible to most users.

Infrared. Active Badge, developed at Olivetti Research Laboratory (now AT&T Cambridge), used diffuse infrared technology to realize indoor location positioning. The line-of-sight requirement and short-range signal transmission are two major limitations that suggest it to be less than effective in practice for indoor location sensing.

IEEE 802.11. RADAR is an RF based system for locating and tracking users inside buildings, using a standard 802.11 network adapter to measure signal strengths at multiple base stations positioned to provide overlapping coverage in a given area.

The disadvantages of "RADAR" position location system are (1) the track object has to worn related department; (2) the communication technology is apt to receive the interference of other communication apparatuses; (3) the system mean error about 3~4meter, therefore the system accuracy doesn't accord with expecting. The above are popular technologies for indoor location sensing. Some other technologies, such as ultra-wideband, are also being investigated. It has advantage of high transmission rate, low power; strong penetration ability, high accuracy, but high cost limits its application.

RFID networks are composed of three different entities, RFID tags, readers, and servers, as shown in Fig. 1. One well-known location sensing systems using the RFID technology is SpotON, I will introduce it in the follow text.

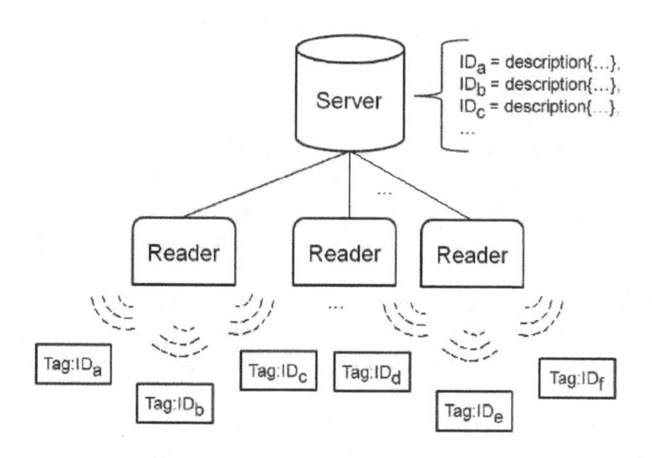

Fig. 1. Architecture of a classic RFID system

2.3 Research on Indoor Positioning Technology Based on RFID

The mature indoor positioning scheme based on RFID includes the following [4]:

1. SpotON. SpotON is one well-known location sensing systems using the RFID technology. It uses an aggregation algorithm for three dimensional location sensing based on radio signal strength analysis. SpotON researchers have designed and built hardware that will serve as object location tags. In the SpotON approach, objects are located by homogenous sensor nodes without central control. SpotOn tags use received radio signal strength information as a sensor measurement for estimating inter-tag distance. However, a complete system has not yet been made available [5].

2. Pin Point scheme. This scheme is commercialized now which uses wireless local network technology. The positioning algorithm is based on time difference, it measures the distance of different object by way of RF signal transmission time. It needs antenna array set in known place to realize multilateral measuring. Pin Point scheme uses base station and tags to get indoor object position. Its accuracy is 1 meter to 3 meters, and it needs larger investment.

3. LANDMARC scheme. LANDMRC have been developing by a research team of Michigan State University and Hong Kong University of Science and Technology. In the LANDMRC method, it uses cheaper active tags to assist RFID readers for position location. These auxiliary tags would become reference tags in the position system. These reference tags put on stationary location beforehand. These reference tags would enhance the available cover range of RFID readers and promote the accuracy of position. The displacement of expensive RFID readers is replaced by active tags. This scheme would enhance the feasibility of RFID technology for indoor position and save cost.

3 LANDMARC Scheme Analysis

The LANDMARC approach does require signal strength information from each tag to readers, if it is within the detectable range. However, the current RFID system does not provide the signal strength of tags directly to readers. Readers only report the power level (1 to 8 levels) of the tag detected. We might do a preliminary measurement to know which power level corresponds to what distance in free space. We have to develop an algorithm to reflect the relations of signal strengths by power levels. Because the power level distribution is dynamic in a complicated indoor environment, the physical distance cannot be computed accurately by using power levels directly.

The LANDMARC approach is more flexible and dynamic and can achieve much more accurate and close to real-time location sensing. Obviously, the placement of readers and reference tags is very important to the overall accuracy of the system. Reference tags which are fixed tags with known positions are deployed regularly on the covered area. Readers have eight different power levels. This approach consists in selecting the k nearest reference tags from the unknown active tag with the following indicator for each reference tag:

We locate the tag using the method of LANDMARC. Suppose we have n RF readers along with m tags as reference tags and u tracking tags as objects being tracked. We define the signal strength vector of a tracking/moving tag as $\vec{S_i} = (S_1, S_2 \cdots S_n)$, where S_i denotes the signal strength of the tracking tag perceived on reader i. For the reference tags, we denote the corresponding signal strength vector as $\vec{\theta_j} = (\theta_1, \theta_2 \cdots \theta_m)$, θ_j denotes the signal strength of the reference tag j perceived on reader. We introduce the Euclidian distance in signal strengths. For each individual tracking tag, we define

$$E_j = \sqrt{\sum_{i=1}^{n} (\theta_i - S_i)^2} \tag{1}$$

as the Euclidian distance in signal strength between a tracking tag and a reference tag j. Let E_j denotes the location relationship between the reference tag j and the tracking tag. The nearer reference tag to the tracking tag is supposed to have a smaller E_j

value. When there are m reference tags, a tracking tag has its E_j vector; $\vec{E}_j = (E_1, E_2, \cdots E_m)$. The vector $\vec{E}_j$ denotes the distance between a reference tag j and a tracking tag. We find the k unknown tracking tags' nearest neighbors by comparing different E_j values. Since these E values are only used to reflect the relations of the tags, we use the reported value of the power level to take the place of the value of signal strength in the equation. When we use k nearest reference tags' coordinates to locate one unknown tag, we call it k-nearest neighbor algorithm. The unknown tracking tag' coordinate (x, y) is obtained by:

$$(x, y) = \sum_{i=1}^{k} w_i (x_i, y_i) \tag{2}$$

Where w_i is the weighting factor to the i-th neighboring reference tag. The choice of these weighting factors is another design parameter. Giving all k nearest neighbors with the same weight would make a lot of errors. Thus, we have to determine the weights assigned to different neighbors. Intuitively, w_i should depend on the E value of each reference tag; w_i is a function of the E values of k-nearest neighbors.

Empirically, in LANDMARC, weight is given by:

$$w_l = \frac{\frac{1}{E_l^2}}{\sum_{i=1}^{k} \frac{1}{E_i^2}} \tag{3}$$

LANDMARC approach' calculation is relatively large. It needs reader to calculate signal strength difference between all reference tags in the cell and tracking tag when locating, and then compare these data. When a lot of tags to be located, computing speed is low and complicated [6][7][8].

4 The Improve to LANDMARC Based on Clustering Algorithm

Clustering analysis algorithm is in fact a classification standard, some having similar physical quantity of a collection of objects. In the LANDMARC method, the algorithm requires calculate all indoor reference tag signal strength value. In fact, some reference tags are a little bit far from the tracking tag. It needn't to calculate these data and this will save time.

The main step of this method is to select the master node and sub-nodes of the network using a timer device, and select other and clustering a group from the node. When the RFID network is established, all nodes are considered to be the sub- nodes. The message is broadcasted by the timer. Compare with the neighbor node, the message of node i first received, the node i is the master node. And node i broadcast to its neighbor node, all the nodes that can receive the message from the node i are sub-nodes to cluster.

The global clustering coordinate is transformed from similar node clustering domain according master node ID size. In the process of conversion requires the boundary nodes. Boundary node is defined as the non-master node but belongs to the two node domain; it connects two adjacent nodes domain. We take the smaller coordinate system of master node ID for the standard; transform all the original local coordinate of the larger master node ID to another node coordinate domain. Repeat these steps, until global coordinate is established.

5 Simulation and Result

According to above improved method, we simulate it with computer to verify its performance. In the simulation, we place 4 RF readers (n=4) and 16 tags (m=16) as reference tags while the other 4 tags (u=4) as objects being tracked, as illustrated in Fig. 2.

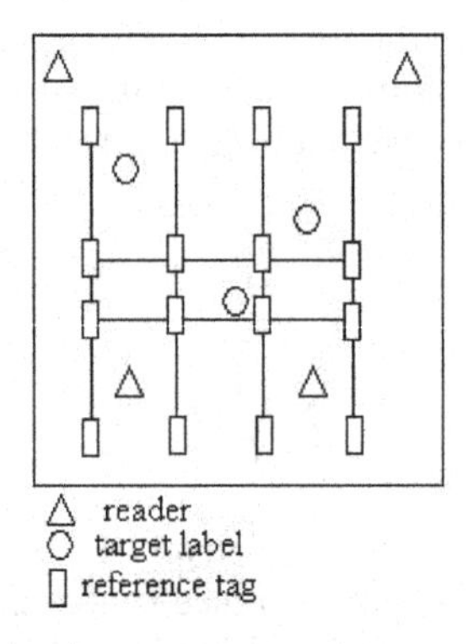

Fig. 2. Placement of RF readers and tags

To quantify how well the LANDMARC system performs, the error distance is used as the basis for the accuracy of the system. The definition of the location estimation error e is given by:

$$e = \sqrt{(x - x_0)^2 + (y - y_0)^2} \tag{4}$$

Where (x_0, y_0) is the tracking tag's real coordinates, and (x,y) is the computed coordinates. In the simulation, we have repeated the positioning the same place 100 times to avoid statistical errors.

We choose different k values and compute the coordinates of the tracking tags with two different methods, respectively. Fig. 3 shows the results.

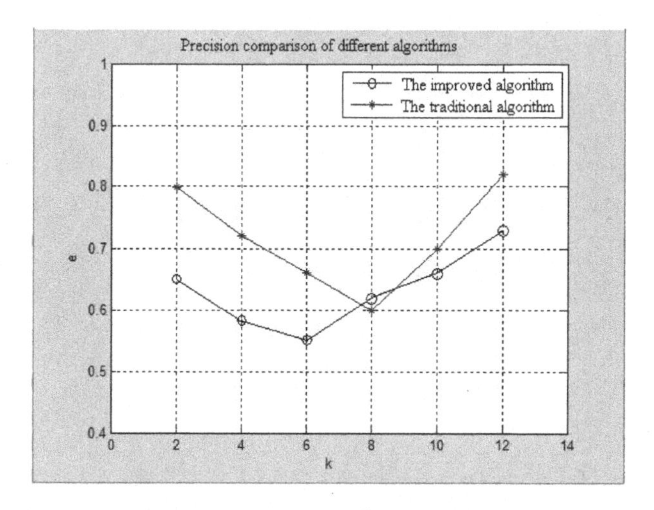

Fig. 3. Simulation results with different *k*

As shown in Fig. 3, k value has no big difference (8 and 6 respectively) in getting the smallest error using traditional LANDMARC algorithm and improved method. But in order to speed up the positioning process, generally k value is as small as possible. As shown in Fig. 3, the improved algorithm has better performance than the traditional algorithm performance when k decreases. Therefore, even with the use of fewer nearest neighbors, the improved method can also obtain high positioning accuracy.

References

1. Bolic, M., Simplot-Ryl, D., Stojmenovic, I.: RFID Systems Research Trends and Challenges, ch. 3. Wiley (2011)
2. Bouet, M., dos Santos, A.L.: RFID Tags: Positioning Principles and Localization Techniques. In: 1st IFIP Wireless Days. WD 2008 (2008)
3. Ni, L.M., Liu, Y., Lau, Y.C., Patil, A.P.: LANDMARC: Indoor Location Sensing Using Active RFID. In: Proceedings of the First IEEE International Conference on Pervasive Computing and Communications, PerCom 2003 (2003)
4. Yang, W., Zhong, W.: Based on the CC2431 wireless positioning system communication technology, pp. 9190–9192 (2009)
5. Hightower, J., Want, R., Borriello, G.: SpotON: An Indoor 3D Location Sensing Technology Based on RF Signal Strength. Technical Report, UW CSE 00-02-02. University of Washington (2000)
6. Shih, S.-T., Hsieh, K., Chen, P.-Y.: An Improvement Approach of Indoor Location Sendsing Using Active RFID. In: ICICIC, pp. 453–456 (2006)
7. Tesorieroa, R., Tebara, R., Gallud, J.A., Lozanoa, M.D., Penicheta, V.M.R.: Improving location awareness in indoor spaces using RFID technology. Expert Systems with Applications 37(1), 894–898 (2010)
8. Want, R.: An introduction to RFID technology. IEEE Pervasive Computing 5(1), 25–33 (2006)

[illegible]

References

1. [illegible]
2. [illegible]
3. [illegible] LANDMARC [illegible] RFID [illegible] (2003)
4. [illegible] Based on [illegible] positioning system [illegible]
5. [illegible] Technology Based on [illegible] Signal [illegible] University of Washington (2000)
6. [illegible] Improvement [illegible] Indoor Location Sensing [illegible] pp. 453–[illegible]
7. [illegible] Improving the accuracy of indoor [illegible] using RFID technologies. Expert Systems with Applications 37(1), [illegible]–898 (2010)
8. Want, R.: An introduction to RFID technology. IEEE Pervasive Computing [illegible], 25–[illegible] (2006)

Virtual Polarization Technology in Adaptive Polarization Radar System

Xi Su[1], Feifei Du[2], Peng Bai[1], and Yanping Feng[1]

[1] Science Institute of Air Force Engineering University
710051 Xi'an China
[2] Military Transportation University
Tianjin China
suxi60@163.com

Abstract. Adopting optimal polarization state can be highly conducive to anti-interference as well as target enhancement while virtual polarization is an effective method to realize adaptive polarization decision in radar system. It can freely vary the polarization state of received waves to match targets' polarization characteristics or be orthogonal to interference polarization state on the purpose of promoting SIR. This paper has not only made thorough research on virtual polarization's concepts and principals but also presented its specific realization methods as well as application in radar system.

Keywords: virtual polarization, orthogonal polarization, match polarization.

1 Introduction

Except amplitude, phase and Doppler frequency shift, the received radar waves' polarization state is another significant characteristic. The concept of virtual polarization was originally put forward by Britain scholar A. J. Poelman in 1981[1-2] and it is crucial for the design and research of adaptively polarized radar especially the corresponding digital signal process. The conventional method changes the polarization state of radar receiving system through antennas or feeding networks. However, virtual polarization can directly vary polarization state in the receiving channel. Actually, it does not convert the polarization state of receiving antennas and feeding networks. Virtual polarization attains same results of conventional methods by properly processing signals in orthogonal dual polarization receiving channels. Accordingly, received waves can be freely polarized through virtual polarization to match target waves or be orthogonal to noise or interference.

2 Virtual Polarization Algorithm Analysis

The received wave with some original polarization state would be first divided into two orthogonal polarization components which usually are horizontal and vertical polarized. Through frequency shifting and amplifying them in two independent receiving channels, each of the two components is divided into two orthogonal

H. Tan (Ed.): Knowledge Discovery and Data Mining, AISC 135, pp. 117–122.
springerlink.com

signals: signal I and signal Q. Accordingly, the relationship of amplitudes and phases of two components are successfully stored. And then, after sampling and quantization the four signals are converted into digital signals and sent to the virtual polarization module. Taking advantage of previously stored parameters, four digital signals will be freely weighted in amplitude as well as phase in the module and their vector sum is the target polarization state.

The wave's polarization components can be expressed as follow:

$$[E_H, E_V]^T = |E| \cdot [\cos\alpha, \sin\alpha e^{j\varphi}]^T \tag{1}$$

Polarization field is a continuous field and all polarization states can be found on Poincare sphere in Fig.1. The parameters α and φ of polarization point P are also presented on the Poincare sphere.

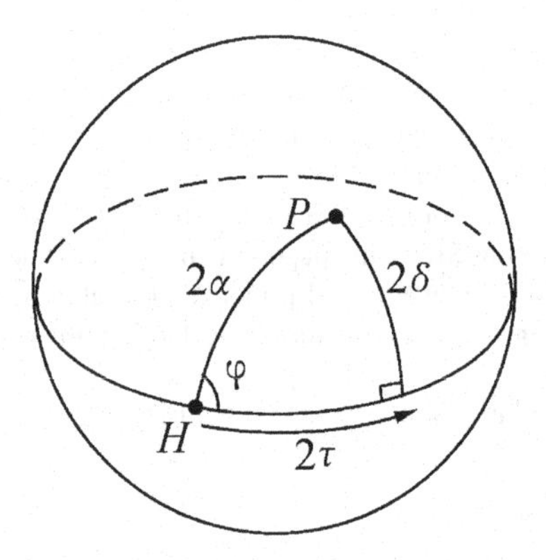

Fig. 1. Poincare Sphere

In expression (1), E_H , E_V denote horizontal and vertical components respectively. Through orthogonal division the two components are divided into four signals $[I_H, Q_H]^T, [I_V, Q_V]^T$ which will be weighted in amplitude and phase according to target polarization state (α_0 , φ_0) by virtual polarization module.

Horizontal component is only weighted in amplitude with the coefficient $\cos\alpha_0$:

$$\begin{bmatrix} I'_H \\ Q'_H \end{bmatrix} = \cos\alpha_0 \times \begin{bmatrix} I_H \\ Q_H \end{bmatrix} = |E| \cdot \begin{bmatrix} \cos\alpha_0 \cos\alpha \\ 0 \end{bmatrix} \tag{2}$$

Vertical component should be amplitude weighted with $\sin\alpha_0$ before being shifted φ_0 in phase. The weighting process is expressed as:

$$\begin{bmatrix} I'_V \\ Q'_V \end{bmatrix} = \cos\alpha_0 \times \begin{bmatrix} \cos\varphi_0 & -\sin\varphi_0 \\ \sin\varphi_0 & \cos\varphi_0 \end{bmatrix} \times \begin{bmatrix} I_V \\ Q_V \end{bmatrix} = |E| \sin\varphi_0 \cos\varphi_0 \begin{bmatrix} \cos(\varphi_0 + \varphi) \\ \sin(\varphi_0 + \varphi) \end{bmatrix} \tag{3}$$

After weighting horizontal and vertical components, the final step is to get the vector sum of the two components. The output of virtual is polarization presented as follow:

$$E_{out}^2=\left(I_H'^2+Q_H'^2\right)+\left(I_V'^2+Q_V'^2\right)+2\sqrt{\left(I_H'^2+Q_H'^2\right)}\times\sqrt{\left(I_V'^2+Q_V'^2\right)}\times\cos\left[\tan^{-1}\left(\frac{Q_V'}{I_V'}\right)-\tan^{-1}\left(\frac{Q_V'}{I_V'}\right)\right]$$

$$=E^2\cos^2\alpha_0\cos^2\alpha+E^2\sin^2\alpha_0\sin^2\alpha+2E^2\cos\alpha_0\cos\alpha\sin\alpha_0\sin\alpha(\varphi_0+\varphi) \quad (4)$$

As two components are in-phase, the output power reaches a high equal to input power:

$$E_{outmax}^2=\left[\sqrt{\left(I_H'^2+Q_H'^2\right)}+\sqrt{\left(I_V'^2+Q_V'^2\right)}\right]^2=E_{in}^2 \quad (5)$$

The output power reaches the minimum when two components are out of phase:

$$E_{outmin}^2=\left[\sqrt{\left(I_H'^2+Q_H'^2\right)}-\sqrt{\left(I_V'^2+Q_V'^2\right)}\right]^2 \quad (6)$$

3 Match Polarization and Orthogonal Polarization

Frequently, there are two purposes of virtual polarization. First is to attain polarization match with target polarization characteristics which means largest output while the other is to get polarization orthogonality which denotes minimal output to reduce noise and interference. To get the largest amplitude of vector sum, the horizontal and vertical components should be converted to in-phase state. To get the zero vector sum, the two components should be converted to out of phase state with same amplitude.

Unless it is setted to the conjugate polarization state, the receiving antenna cannot match the received waves' polarization in one coordinate system. The definition of conjugate polarization is showed in expression (7).

$$p_r=p_t^*=\left(\tan\alpha e^{j\varphi}\right)^*=\tan\alpha e^{-j\varphi} \quad (7)$$

In which p_r and p_t denote polarization ratios of receiving antenna and received waves respectively. Two conjugate polarization points are symmetric to the equator of Poincare sphere. On the other hand, the receiving antenna must be setted to the cross polarization of received waves in one coordinate system on purpose of being orthogonal to waves' polarization [3-4]. In this case, the relation between p_r and p_t is expressed as follow:

$$p_r=-\frac{1}{p_r}=-\frac{1}{\tan\alpha e^{j\varphi}}=-\cot\alpha e^{-j\varphi} \quad (8)$$

Coordinate system is very important in the analysis. If different coordinate systems are adopted, the whole expression would change.

4 Applications in Radar System

4.1 Virtual Polarization Module's Horizontal State

As the virtual polarization module is setted to horizontal state ($\alpha_0 = 0^\circ$, φ_0 is random), responses of various polarized signals are analyzed follow.

When input is horizontally polarized signal, signals in horizontal and vertical channels are $[I_H \quad Q_H]^T$ and $[0 \quad 0]^T$ respectively. Amplitudes of I_H and Q_H are both $E/\sqrt{2}$. Outputs of the two channels are showed in expression (9).

$$\begin{bmatrix} I'_H \\ Q'_H \end{bmatrix} = \cos 0^\circ \times \begin{bmatrix} I_H \\ Q_H \end{bmatrix} = \begin{bmatrix} I_H \\ Q_H \end{bmatrix} \quad \begin{bmatrix} I'_V \\ Q'_V \end{bmatrix} = \begin{bmatrix} 0 \\ 0 \end{bmatrix} \tag{9}$$

The power of vector sum is equal to input signal's power.

$$E_{\text{out}}^2 = I_H^2 + Q_H^2 = E^2 \tag{10}$$

As a vertically polarized signal is inputted, signals in horizontal and vertical channels are $[0 \quad 0]^T$ and $[I_V \quad Q_V]^T$ respectively. Amplitudes of I_V and Q_V are both $E/\sqrt{2}$. Outputs of the two channels are presented in expression (11).

$$\begin{bmatrix} I'_H \\ Q'_H \end{bmatrix} = \begin{bmatrix} 0 \\ 0 \end{bmatrix} \quad \begin{bmatrix} I'_V \\ Q'_V \end{bmatrix} = \sin 0^\circ \times \begin{bmatrix} \sin\varphi_0 & \cos\varphi_0 \\ \cos\varphi_0 & -\sin\varphi_0 \end{bmatrix} \times \begin{bmatrix} I_H \\ Q_H \end{bmatrix} = \begin{bmatrix} 0 \\ 0 \end{bmatrix} \tag{11}$$

According to expression (11), the output of virtual polarization module is zero.

When input is left circularly polarized signal, signals in horizontal and vertical channels are $[I_H \quad Q_H]^T$ and $[I_V \quad Q_V]^T$ respectively. Amplitudes of I_H , Q_H , I_V and Q_V are all equal to $E/\sqrt{2}$. Outputs of the two channels are expressed as follow:

$$\begin{bmatrix} I'_H \\ Q'_H \end{bmatrix} = \cos 0^\circ \times \begin{bmatrix} I_H \\ Q_H \end{bmatrix} = \begin{bmatrix} I_H \\ Q_H \end{bmatrix} \begin{bmatrix} I'_V \\ Q'_V \end{bmatrix} = \begin{bmatrix} 0 \\ 0 \end{bmatrix} \tag{12}$$

From expression (12) we can see that the output becomes a horizontally polarized signal and its power is half of input's power.

$$E_{\text{out}}^2 = I_H^2 + Q_H^2 = E^2/2 \tag{13}$$

Therefore, a conclusion could be drew from the analysis above: only the horizontal component of the input signal can go through the virtual polarization module setted to horizontal polarization, no matter the input has what polarization state.

4.2 Virtual Polarization Module's Right Circular Polarization State

When the virtual polarization module is setted to right circular polarization state in which $\alpha_0 = \pi/4$ and $\varphi_0 = -\pi/2$, responses of various polarized signals are analyzed follow.

As the input is in horizontal polarization state, signals in horizontal and vertical channels are $[I_H \quad Q_H]^T$ and $[0 \quad 0]^T$ respectively. Amplitudes of I_H and Q_H are both $E/\sqrt{2}$. Outputs of the two channels are showed in expression (14).

$$\begin{bmatrix} I'_H \\ Q'_H \end{bmatrix} = \cos\frac{\pi}{4} \times \begin{bmatrix} I_H \\ Q_H \end{bmatrix} = \begin{bmatrix} I_H/\sqrt{2} \\ Q_H/\sqrt{2} \end{bmatrix} \quad \begin{bmatrix} I'_V \\ Q'_V \end{bmatrix} = \begin{bmatrix} 0 \\ 0 \end{bmatrix} \tag{14}$$

Accordingly, output is still a horizontally polarized signal with a halved power compared to input signal's power.

$$E_{\text{out}}^2 = I_H^2 + Q_H^2 = E^2/2 \tag{15}$$

As a left circularly polarized signal is inputted, signals in horizontal and vertical channels are $[I_H \quad Q_H]^T$ and $[I_V \quad Q_V]^T$ respectively. All the amplitudes of I_H, Q_H, I_V and Q_V are $E/2$. Outputs of the two channels are presented in expression (16).

$$\begin{aligned} \begin{bmatrix} I'_H \\ Q'_H \end{bmatrix} &= \cos\frac{\pi}{4} \times \begin{bmatrix} I_H \\ Q_H \end{bmatrix} = \begin{bmatrix} I_H/\sqrt{2} \\ Q_H/\sqrt{2} \end{bmatrix} \\ \begin{bmatrix} I'_V \\ Q'_V \end{bmatrix} &= \sin\frac{\pi}{4} \times \begin{bmatrix} -1 & 0 \\ 0 & 1 \end{bmatrix} \times \begin{bmatrix} I_V \\ Q_V \end{bmatrix} = \begin{bmatrix} -I_V/\sqrt{2} \\ Q_V/\sqrt{2} \end{bmatrix} \end{aligned} \tag{16}$$

Obviously, they have the same amplitude and the power of vector sum can be expressed as:

$$E_{\text{out}}^2 = \left(\sqrt{\frac{1}{2}\left(I_H^2 + Q_H^2\right)} + \sqrt{\frac{1}{2}\left(I_V^2 + Q_V^2\right)} \right)^2 = E^2 \tag{17}$$

When input is right circularly polarized signal, signals in horizontal and vertical channels are $[I_H \quad Q_H]^T$ and $[I_V \quad Q_V]^T$ respectively. Amplitudes of I_H, Q_H, I_V and Q_V are all equal to $E/2$. Outputs of the two channels are expressed as follow:

$$\begin{aligned} \begin{bmatrix} I'_H \\ Q'_H \end{bmatrix} &= \cos\frac{\pi}{4} \times \begin{bmatrix} I_H \\ Q_H \end{bmatrix} = \begin{bmatrix} I_H/\sqrt{2} \\ Q_H/\sqrt{2} \end{bmatrix} \\ \begin{bmatrix} I'_V \\ Q'_V \end{bmatrix} &= \sin\frac{\pi}{4} \times \begin{bmatrix} -1 & 0 \\ 0 & 1 \end{bmatrix} \times \begin{bmatrix} I_V \\ Q_V \end{bmatrix} = \begin{bmatrix} -I_V/\sqrt{2} \\ Q_V/\sqrt{2} \end{bmatrix} \end{aligned} \tag{18}$$

From expression (18) we can see that they have the same amplitude and the output's power is zero.

5 Conclusions

In Along with the development of modern wars, radar as a main weapon has also ceaselessly evolved for generations. To adapt to various new needs of electric wars, virtual polarization increasingly plays a key role in modern radar systems. Here are two tips of virtual polarization's application. First of all, setting the virtual polarization module to state (α_0, φ_0) has two steps. One is to multiply amplitude of horizontal component and $\cos\alpha_0$. The other is to multiply amplitude of vertical component and $\sin\alpha_0$ on the base of a forward shifted phase φ_0. Secondly, as the match receiving mode is required, virtual polarization module should be set to conjugate polarization while it must be in cross polarization state when orthogonal mode is requisite.

References

1. Poelman, A.J.: Virtual polarization adaptation, a method of increasing the detection capabilities of a radar system through polarization vector processing. IEEE Proc. Communication, Radar and Signal Processing 128(10), 465–474 (1981)
2. Zhuang, Z., et al.: Polarization information processing and application in radar. National Defense Industry Press, Beijing (2005)
3. Jian, P., Xubo, C., Bo, L.: A Simulation study of radar varied polarization. Modern Radar 28(4), 53–59 (2005)
4. Li, Y., Xiao, S., Wang, X.: Anti-interference technology based on radar polarization. National Defense Industry Press, Beijing (2010)
5. Jian, P., Bo, L., Erke, M.: The Method and Application of Adaptive Receiving Polarization Process. Modern Radar 26(1), 53–55,61 (2004)
6. Zeng, Q., Li, R., Guo, H.: Effect analysis on polarization converter in electronic countermeasures. Modern Radar 23(5), 72–76 (2001)
7. Zeng, Q.: Radar Polarization Technology and Application of Polarization Information. National Defense Industry Press, Beijing (2006)

Applying Fourth-Order Partial Differential Equations and Contrast Enhancement to Fluorescence Microscopic Image Denoising

Yu Wang and Hong Xue

School of Computer Science and Information Engineering,
Beijing Technology and Business University, 100048, Beijing, China
wangyu@th.btbu.edu.cn

Abstract. This paper proposes a new denoising diffusion model for fluorescence microscopic images, in which fourth-order partial differential equations (PDEs) and contrast enhancement are utilized to overcome the blocky effect and false edges usually caused by second-order PDEs. Experimental results show that the proposed method not only makes the denoised images subjectively natural and clear, but also achieves better performance in terms of objective criterion such as peak signal to noise ratio (PSNR) compared to the second-order PDEs diffusion models.

Keywords: PDE, Contrast enhancement, Image denoising, Regularization.

1 Introduction

Fluorescence microscopy is an effective technique to study properties of organic or inorganic substances using fluorescence probe in biomedicine field [1]. Noise, however, can't be avoided because of imperfect optical imaging system, photoelectric conversion, specimen tissue structure and human errors etc. during the course of optical imaging. The obtained images usually are obscured by a lot of noise, especially in deep tissue with strong excitation laser power. Therefore noise removal is of great importance for fluorescence microscopic images.

In the field of image processing, partial differential equation (PDE) is considered as an useful tool for image denoising and enhancement [2-3]. In essence PDE belongs to physics. Witkin [4] noticed firstly that the convolution of the signal with Gaussians at each scale was equivalent to the solution of the heat equation of this signal, which makes PDEs become a bridge between image processing and physics. But This model filters the image in all directions at the same time, so it destroys the edges of the image when smoothing this image. Later many researchers improved upon this approach. P-M method [5] is a very successful example by introducing a diffusion function $g(|\nabla u|)$ which is a nonincreasing function about the image gradient $|\nabla u|$. $|\nabla u|$ is large at edges while small within regions, so the modified P-M method smoothes the edges of the image slowly and the edges are kept, while the diffusion within the regions is enhanced. Hongmin Zhang et al. [6] added the local gray-level

H. Tan (Ed.): Knowledge Discovery and Data Mining, AISC 135, pp. 123–128.
springerlink.com

variance σ^2 to the diffusion function $g(|\nabla u|)$ because the edge's σ^2 was generally smaller than noise's σ^2 in terms of magnitude. Cattéet et al. [7] replaced the function $g(|\nabla u|)$ by $g(|\nabla(G_\sigma * u)|)$, where G_σ is a Gaussian smoothing kernel and the symbol "$*$" denotes convolution operation. Alvarez et al. [8] not only used diffusion function $g(|\nabla(G_\sigma * u)|)$, but also imposed their PDEs diffusion model only along the tangent direction of the edge as filtering noise.

Although the above second order PDEs methods can achieve a good trade-off between noise removal and edge preservation, they tend to cause the processed image to look "blocky". This effect is visually unpleasant and is likely to cause false edges in computer vision. This blocky effect is, to a large extent, inherent in the nature of second-order PDEs. Yu-Li You made a detailed explanation about this in [9]. While fourth-order PDEs can conquer effectively this problem. Therefore, in this paper we design a new diffusion model based on fourth-order PDEs for fluorescence images denoising.

During the course of smoothing the images, contrast enhancement technique can improve image definition in human vision to a great extent. In addition, the regularization theory [10] has been demonstrated to be very effective in the field of image processing. Villa et al. [2] used regularized technique and second-order PDEs to denoise electronic speckle pattern interferometry (ESPI) fringes.

Illuminated by these ideas, we use regularized technique to combine fourth-order PDEs with contrast enhancement technique for obtaining natural and clear denoising images.

This paper is organized as follows. Section 2 explains detailedly the diffusion model proposed by this paper for noise removal of fluorescence images which uses fourth-order PDEs and contrast enhancement technique. Section 3 shows the experimental results of the method proposed by this paper and the other related methods via processing simulated and real fluorescence microscopic noisy images. Finally, a conclusion is drawn.

2 Methodology

Supposing $E(u)$ is an energy functional which measures the oscillations in a digital image $u(x,y)$, and a general formulation of the noise removal problem is to solve the minimization of $E(u)$ in the image support Ω. Here we propose a new energy functional $E(u)$, which combines fourth-order PDEs with image contrast enhancement to deduce diffusion model of noise removal. For obtaining a clear image $u(x,y)$ we define a functional described as,

$$E(u) = \int_\Omega \frac{1}{2}\left(\left|\nabla^2 u\right|\right)^2 dxdy + \gamma \int_\Omega \frac{1}{2}\left(\left|\nabla u - W^*\right|\right)^2 dxdy \,. \tag{1}$$

In this equation the first term imposes the smooth constraint, in which $\nabla^2(u)$ is Laplacian operator and $\nabla^2(u)=u_{xx}+u_{yy}$. u_{xx} and u_{yy} are the second-order partial derivatives of $u(x,y)$ respectively. On the other hand, the second term imposes the estimated image gradient field ∇u (i.e. contrast field) to be close to the target gradient field W^* in the sense of least squares, and $W^* = k\nabla u = (ku_x, ku_y)$. k is magnifying multiple of image contrast field and $k = 1+\lambda e^{(-|\nabla u|)}$. $\lambda > 0$ denotes the largest magnifying multiple of image contrast. $|\nabla u| = \sqrt{(u_x)^2+(u_y)^2}$, u_x and u_y are the first-order partial derivatives of $u(x,y)$ respectively. γ is the weight about regularization term and the term dependent on original data, and $\gamma > 0$.

By derivation Eq. (1) we can obtain the following functional,

$$E(u)=\int_\Omega \frac{1}{2}(u_{xx}+u_{yy})^2\,dxdy+\alpha\int_\Omega \frac{1}{2}e^{-2\sqrt{u_x^2+u_y^2}}\left(u_x^2+u_y^2\right)dxdy, \tag{2}$$

where α is a regularized parameter, and $\alpha > 0$, $\alpha = \gamma\lambda^2$. The Euler equation of Eq. (2) is shown as,

$$\frac{\partial f}{\partial u}-\frac{\partial}{\partial x}\left(\frac{\partial f}{\partial u_x}\right)-\frac{\partial}{\partial y}\left(\frac{\partial f}{\partial u_y}\right)+\frac{\partial}{\partial xx}\left(\frac{\partial f}{\partial u_{xx}}\right)+\frac{\partial}{\partial yy}\left(\frac{\partial f}{\partial u_{yy}}\right)=0, \tag{3}$$

In this equation, f is represented using the following equation,

$$f=\frac{1}{2}(u_{xx}+u_{yy})^2+\alpha\cdot\frac{1}{2}e^{-2\sqrt{u_x^2+u_y^2}}\left(u_x^2+u_y^2\right), \tag{4}$$

According to the Euler equation (3) and gradient descent procedure, we can obtain the discrete diffusion model

$$u_{i,j}^{n+1}=u_{i,j}^{n}+\Delta t\left[\left(\frac{\partial\left(e^{-2|\nabla u|}u_x(1-|\nabla u|)\right)}{\partial x}+\frac{\partial\left(e^{-2|\nabla u|}u_y(1-|\nabla u|)\right)}{\partial y}\right)-\alpha\nabla^2\left(\nabla^2 u_{i,j}^n\right)\right], \tag{5}$$

where, the subscripts (i, j) denote the pixel position. $u_{i,j}^n$ is the numerical solution expressing denoised image, n is the iteration time and Δt is the time step. For avoiding image shift, forward differences and backward differences are adopted alternately for different order partial derivatives of $u(x,y)$.

3 Experimental Results

To study the performance of our proposed method, we compare it with several methods, including P-M method[2] and Alvarez's method[4]. All of the experiments are implemented in Windows XP and Matlab environment of personal computer (Memory unit 512 M and CPU 1.60 GHz).

Figures 1(a) and 1(b) are original 'Brain' image and its simulated Gaussian noisy image ($\sigma = 0.05$) of size 193×220 respectively. Here for fair test we adjust the parameters of these methods, and use the best results to compare. In figures 1(c), 1(d) and 1(e) we show respectively the results by P-M method (n=8, $\Delta t = 0.1$, peak signal to noise ratio/PSNR=20.8406), Alvarez et al.'s method (n=7, $\Delta t = 0.01$, PSNR=20.0452) and our method (n=7, $\Delta t = 0.01$, $\alpha = 10^4$, PSNR=21.5467). Our model required 2.1 s.

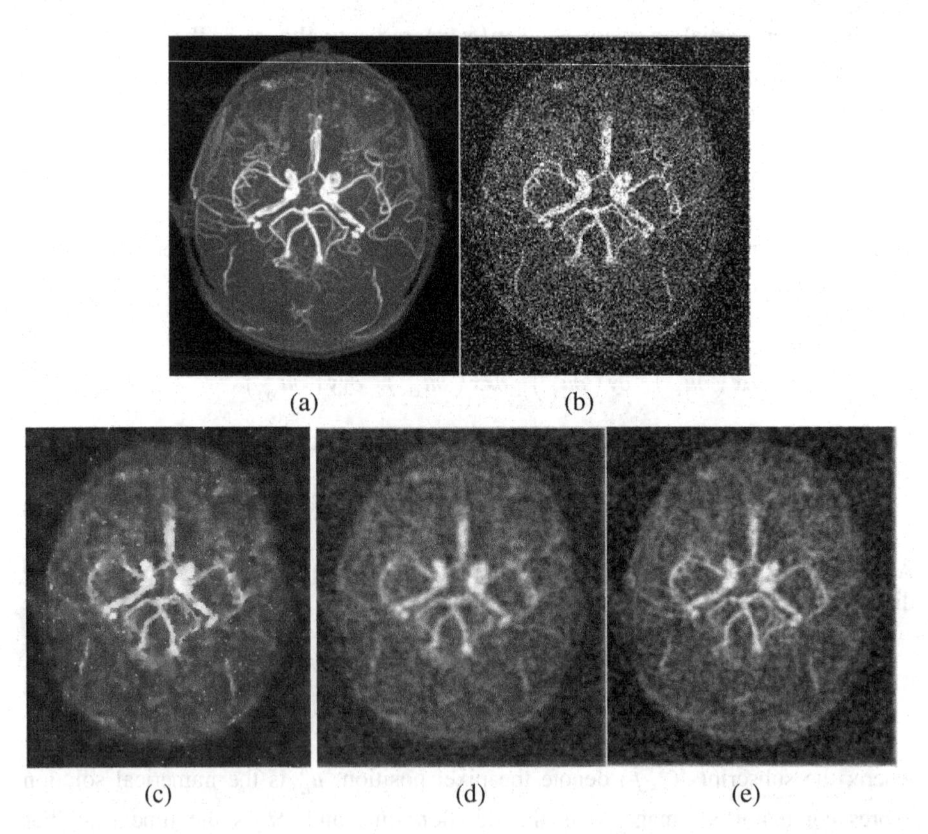

Fig. 1. Original 'brain' image, computer-simulated noisy image and its denoised images: (a) original image, (b) Gaussian noisy image, (c) the denoised image by P-M method, (d) the denoised image by Alvarez et al.'s method, and (e) the denoised image by our method

Figure 2(a) is an experimentally obtained noisy microscopic image of rat olfactory bulb neurons by two-photon laser scanning microscope and its size is 512×512 pixels. Its filtered images by the P-M method with n=7, the Alvarez et al.'s method with n=15 and our method with n=10 are given in figures 2(b), 2(c) and 2(d), respectively.

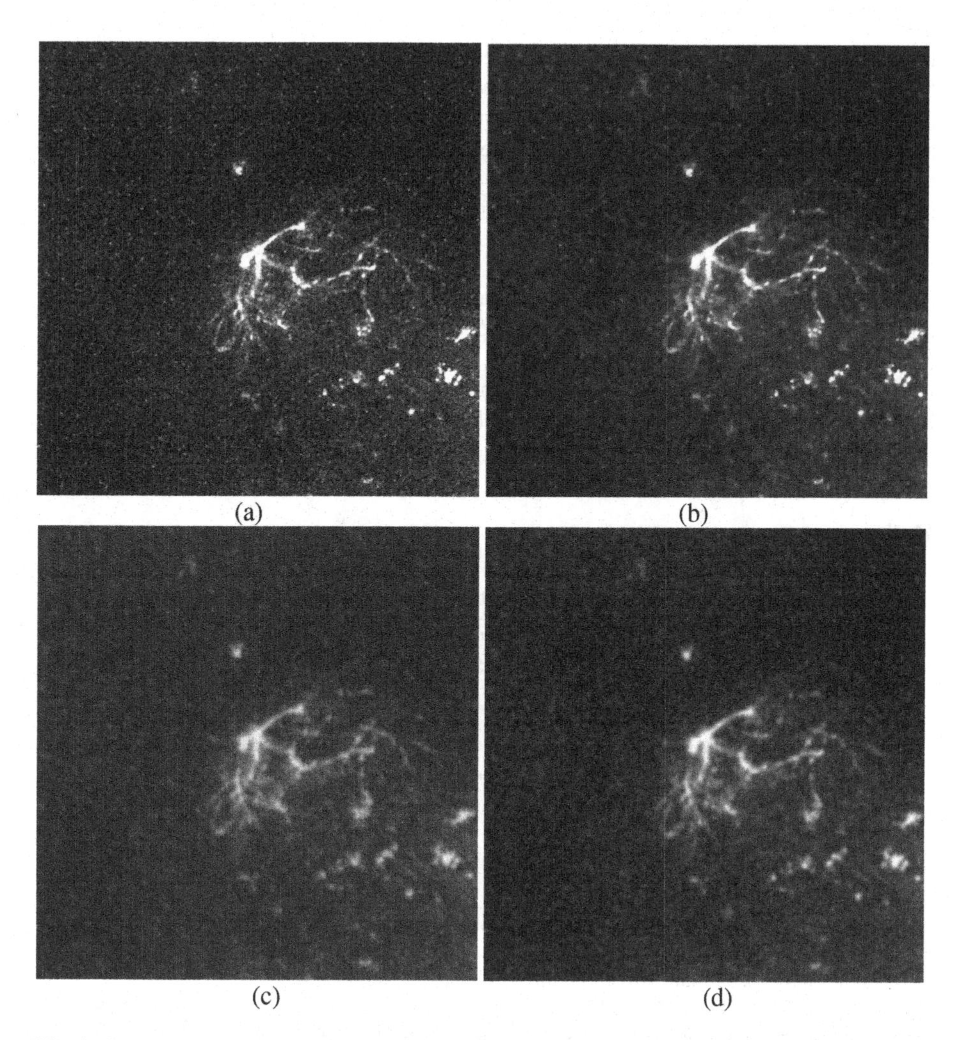

(a) (b)

(c) (d)

Fig. 2. Experimentally obtained noisy microscopic image and its denoised images: (a) the noisy microscopic image, (b) the denoised image by P-M method, (c) the denoised image by Alvarez et al.'s method, and (d) the denoised image by our method

As can be observed, the imposed images by the previous PDE model look rigid (such as the figures 1(c) and 2(b)) or blurring (such as the figures 1(d) and 2(c)). While our method not only gives superior performance in terms of PSNR, but also makes the denoised image look more natural and clearer.

4 Conclusion

A new denosing diffusion model based on fourth-order PDEs and contrast enhancement has been proposed to remove noise of the fluorescence microscopic images and to overcome the blocky effect often caused by second-order PDEs. Experimental results show that compared with the related second-order PDEs methods, our model shows superior performance in terms of both objective criteria and subjective human vision via processing simulated and real fluorescence microscopic noisy images. In addition, our fourth-order PDEs can also be applied to other texture images, which is of a great significance in microscopic image analysis and PDE-based image processing.

Acknowledgments. The authors are grateful to the professors Shaoqun Zeng (Huazhong University of Science and Technology in China) and Hongmin Zhang (Chongqing University of Technology in China) for providing rat olfactory bulb neurons images. Thanks to the anonymous reviewers for their constructive and helpful comments. This work is supported by the national natural science foundation of China (NSFC No. 61171068).

References

1. Sarder, P., Nehorai, A.: Deconvolution methods for 3-D fluorescence microscopy images. IEEE Signal Processing Magazine 23(3), 32–45 (2006)
2. Jesús, V., Juan, A.Q., Ismael, R.: Regularized quadratic cost function for oriented fringe-pattern filtering. Optics Letters 34(11), 1741–1743 (2009)
3. Tang, C., Zhang, F., Li, B.T., Yan, H.Q.: Performance evaluation of partial differential equation models in electronic speckle pattern interferometry and the mollification phase map method. Applied Optics 45(28), 7392–7400 (2006)
4. Witkin, A.P.: Scale-space filtering. In: Proc. Int. Joint Conf. on Artificial Intelligence, Karlsruhe, vol. 2, pp. 1019–1021 (1983)
5. Perona, P., Malik, J.: Scale space and edge detection using anisotropic diffusion. IEEE Trans. on PAMI 12(7), 629–639 (1990)
6. Zhang, H.M., Luo, Q.M., Zeng, S.Q.: Restoration of fluorescence images from two-photon microscopy using modified nonlinear anisotropic diffusion filter. In: Proc. Int. Conf. on SPIE, Wuhan, vol. 6534, pp. 6534H-1–6534H-8 (2006)
7. Catté, F., Lions, P.L., Morel, J.M., Coll, T.: Image selective smoothing and edge detection by nonlinear diffusion. SIAM J. Numer. Anal. 29(1), 182–193 (1992)
8. Alvarez, L., Lions, P.L., Morel, J.M.: Image selective smoothing and edge detection by nonlinear diffusion. SIAM J. Numer. Anal. 29(3), 845–866 (1992)
9. You, Y.L., Kaveh, M.: Fourth-order partial differential equations for noise removal. IEEE Trans. on Image Processing 9(10), 1723–1730 (2000)
10. Rivera, M., Marroquin, J.L.: Efficient half-quadratic regularization with granularity control. Image Vis. Comput. 21, 345–357 (2003)

Design of Programmable Laser Receiver Signal Simulator

Mingang Wang and Chuanxin Sun

College of Astronautics, Northwestern Polytechnical University, Xi'an, 710072, China

Abstract. A rapid changeable signal should be provided for Laser receiver testing. To this meet requirement, this paper raised a design of Laser receiver Signal Simulator adopting DSP, FPGA and high precision DAC. The Signal Simulator can change output amplitude and frequency once per 1ms, provide different output wave combinations and give a smooth transition when two different waves connect. The design has a simple structure, strong anti-interference ability, overcomes the shortcoming of traditional signal generator, meets the excepted performance. This design has been applied to a real laser receiver's system for simulation and test.

Keywords: Generator, Wave combinations, DSP, FPGA.

1 Introduction

Though traditional signal generator has good signal quality and high accuracy on signal frequency and amplitude, but it can't satisfy some special usage. There is one kind of laser simulator, which testing input signal requires several frequencies(Y) in one period(X) of tens of milliseconds. What's more, the value of X, Y, signal frequency, amplitude and lasting time should be programmable and controlled by PC. According to these requests, this paper proposes a design of programmable laser signal simulator based on DSP and FPGA. This simulator could output sine wave, square wave, triangular wave, which frequency is between 8K~20K, amplitude is between -5V~+5V. The PC software is programmed by LabWindowsCVI.

2 Hardware Design

The programmable simulator has two parts, one is core board and the other is DA board. They are connected by PC104 pins. This kind of design is easy to function expansion and the design of core board could be reused by other systems. The MCU, TI's Defino floating micro-controller TMS320F28335, is responsible for data calculating and communicating with PC. The FPGA, Altera's EP2C8Q208, is responsible for controlling DA Chip which outputs -5V~+5V signal using bipolar circuit. DSP exchanges data with FPGA by XINTF parallel interface. System structure block diagram shown in Figure 1.

H. Tan (Ed.): Knowledge Discovery and Data Mining, AISC 135, pp. 129–134.
springerlink.com

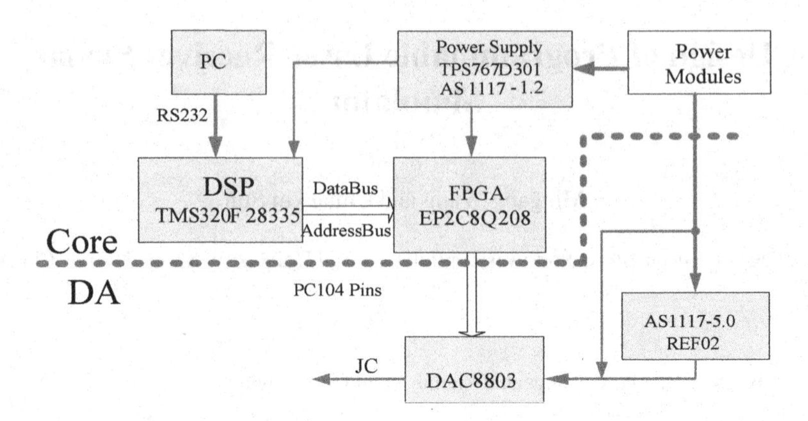

Fig. 1. System structure block diagram

TMS320F28335 is a TI's latest floating-point DSP of TMS320C28X series. It has 150MHz processing speed, 32 bits floating unit for complex algorithm, and low cost, low power consumption, high performance, high integrity of outside device and high storage capacity.

Cyclone II is the second generation of Altera low cost Cyclone series. It is mainly used for consumption electronics, telecommunication, wireless, computer devices, industry and vehicles and so on.

The core board includes power supply part, RS232 communication, the minimum system of DSP and FPGA and data exchanging part between DSP and FPGA based on XINTF. The DA board includes reference source part and bipolar changing circuit.

3 Software Design

Software includes DSP software and FPGA software. DSP software includes four parts, which are system configuration, RS232 data receptiont, wave parameter quantification and XINTF data transmission. FPGA software includes address decoding, state machine, DA control part, etc.

3.1 RS232 Data Reception

The PC data output sequence is shown in Table 1. First, The PC outputs period X, number of waves Y, the kind of waves WaveMode, and start flag 0xF0; second, it outputs the frequencies, amplitudes and lasting times of Y kinds of waves. The kind of wave includes sine wave (WaveMode=0), square wave (WaveMode=1), triangular wave (WaveMode=2).

DSP receives data from PC using FIFO interrupt mode. The FIFO Level is 5, which means DSP will activate one interrupt after receiving 5 data from PC, then it will execute interrupt service routing, read data from FIFO to array Data, set flag bit. Every data configuration will activate Y+1 interrupts.

Table 1. PC and DSP communication protocol

	0th interrupt	1st~Yth interrupts
Data[0]	X	Freq H
Data[1]	Y	Freq L
Data[2]	Wave Mode	Amp L
Data[3]	0xFF	Amp H
Data[4]	0xF0	Last

3.2 Wave Parameter Quantification

Wave parameter quantification is to change data received from PC into parameters for DA control of FPGA. The clock of FPGA's is 50MHz. The resolution of DAC8803 is 14 bits.

Wave parameter quantification is part of main circle. Figure 2 shows the flow chart of one interrupt. Data[4] =0xF0 means the first receiving interrupt, specifying X, Y and wave types. Data[4]=0xFF means receiving interrupts for 2~Y+1 times, specifying the amplitudes, frequencies and lasting times for each wave, then send data to FPGA by ports in Zone 6.

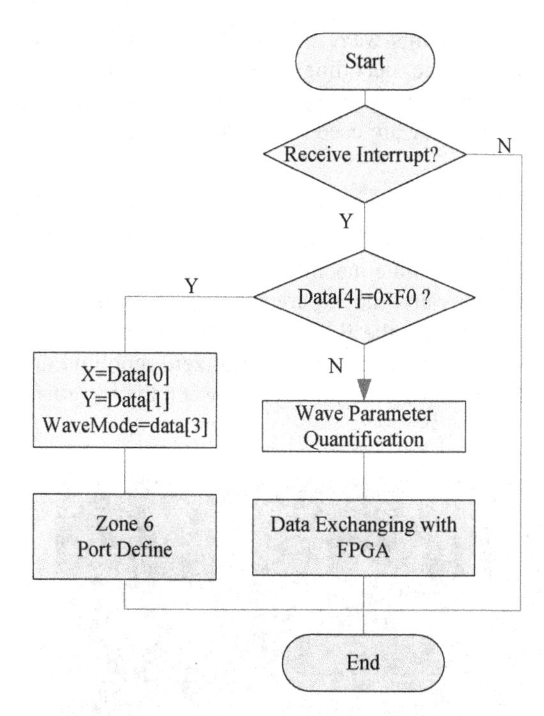

Fig. 2. Parameter quantification flow

3.3 Data Exchanging between DSP and FPGA

DSP exchanges data with FPGA using XINTF. Set the data bus of XINTF 16 bits and operation space Zone 6 , the effective address bus is 10 bits, XA[9]~XA[6] attends address decoding. FPGA receives data at XWE0's raising edge, saves it in FPGA's

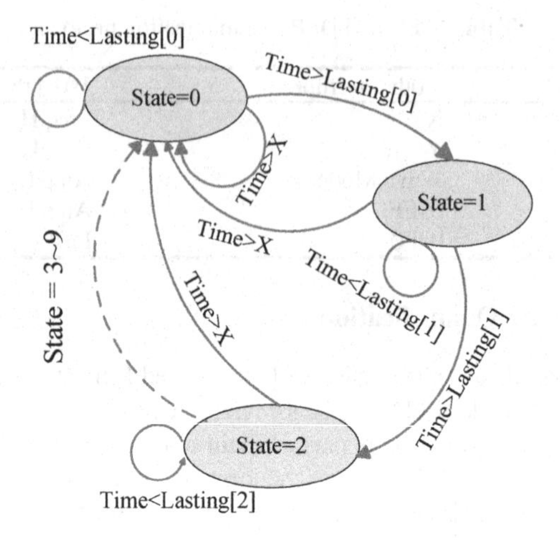

Fig. 3. Wave combinations' state changing

internal SRAM. The data contains wave 0 to wave 9's whole period data, which is used to descript amplitude, the decoding address index is 0~9. After parameter quantification, the interval between waves is slot[10], periodic compensate is Modify Front[10], Modify Back[10], they are used to descript frequency.

3.4 State Machine

Waveform changing state machine is Gray code synchronize state machine. The figure 3 shows states changing. State machine is activated each millisecond, each time when state changes, frequency and index parameters are assigned to Backup variable, after i^{th} whole periods waveform, the Backup variable value is assigned to operation variable. Then the two different waveforms joint at zero amplitude point. The actual effect is showed in figure 4, there is no jumping between the two different waveforms and the waveform is better.

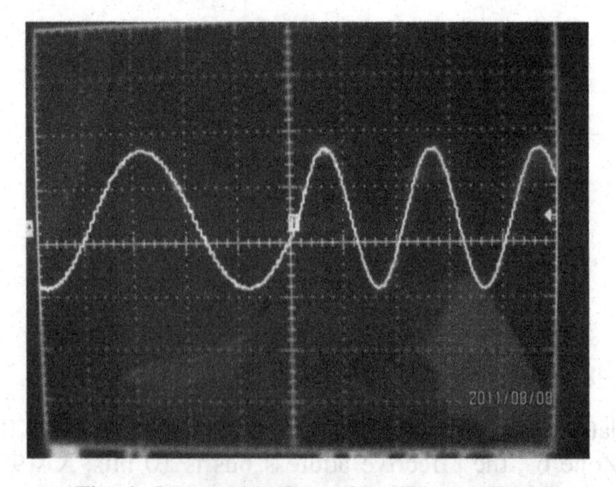

Fig. 4. Output waveform (25us/Div, 1V/Div)

3.5 Digital to Analog Control

DAC8803 is a 14 bits quad current output Digital to Analog convertor, double buffer serial interface adopts CS, SCLK and SDI to provide 3 lines, high speed SPI input. The following figure is ModelSim simulation chart of DA control sequence.

When Cs is low potential, shifting register is enabled and Sdi data is moved into it at Sclk's rising edge. After finishing 16 bits data's moving, the content in shifting register will be moved into DA register at LDAC's falling edge.

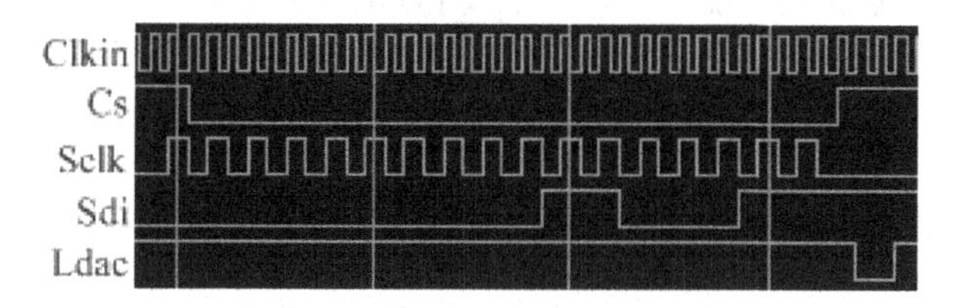

Fig. 5. DA control sequence chart

4 Experiment Results

DSP software adopts C language; FPGA software adopts Verilog language and is simulated under ModelSim6.0 environment. First, X, Y and waveform types are setup on upper level PC, then each wave's amplitude, lasting time and frequency is setup separately. All parameters are transmitted to DSP by RS232. It is proved by experiment that the signal generated has frequency accuracy higher than 0.1%, amplitude accuracy 10% (30mVpp~100mVpp), 5% (100mVpp~1Vpp), 1% (1Vpp~10Vpp). This design satisfies the project's requests.

5 Conclusion

This paper raised a design of programmable laser signal simulator based on DSP and FPGA, which also is a programmable high speed signal generator. This design makes full use of the flexibility of FPGA, high performance processing capability of DSP and plenty of peripherals on chip. This paper introduces system hardware design, system block diagram and explains specifically the communication between upper level PC and DSP, wave parameter quantification and waveform combination state machine. This signal generator has a stable performance, strong anti-interference ability, and great extendable function, low cost and good engineering application.

References

1. Wang, H.-L., Huang, Y.-W.: Design of Programmable Multi-signl Generator based on FPGA. Fire Control & Commamd Control 35(6), 97–99 (2010)
2. Liao, L., Wan, P., Zhang, B., Tang, D.: Design and Implementation of HF Signal Sources for Ionospheric Vertical Sounding Instrument. Computer Measurement & Control 18(6), 1458–1461 (2010)
3. Zhu, S., Li, F.: Design of Signal Source in Sonar Receiver's Testing. Modern Electronics Technique 312(1), 103–105 (2010)
4. Zhao, C.-K., Zhen, G.-Y., Jiao, X.-Q.: Design of High Accuracy Multi-Signal Source Based on FPGA and AD768, vol. 158(2), pp. 139–141 (2010)
5. TMS320x2823x, TMS320x2833x System Control and Interrupts Reference Guide (Rev .B). Texas Instruments (September 2007)

Fault Current Limit (FCL) Technology (Magnetic Valve Controlled Reactor-Type Fault Current Limiter Principle and Simulation)

Chunzhe Shi

Henan Polytechnic University, P.R. China, 454150
chunzhe2001@126.com

Abstract. Summarized the FCL practical research which faces to the key technical problems, briefly introduces the study of magnetic valve controllable reactor type fault current limiter principle, and the simulation results validate the superiority of the limiting current performance.

Keywords: Magnetic valve controlled reactor-type, Simulation, Superior performance-limiting.

1 Introduction

The rapid development of power system, the single and power plant, substation capacity, city and industrial center load and the load density is increased, and the power system interconnection, leading to the modern power system at all levels of the grid in the short-circuit current level of increasing [1-5]. On how to effectively restrict the short circuit current level, on the safe operation of the power system, has important significance.

This paper introduces the magnetic valve controllable reactor fault current limiter principle and simulation results.

2 The Key Technical Issues of the FCL

Ideal fault current limiter system should not adversely affect, but in practical application, due to their respective principle defects, FCL can not completely satisfy the technical requirement of current limiter. When short-circuit fault, FCL participation system transient process, and system between the various devices to generate interactive effects. In depth study of FCL and system between the interactive effects of implementation of FCL project is the prerequisite, following induction of FCL engineering applicable must be faced with in the process of key technical problems.

1) short circuit fault induced transient angle oscillation and even instability threaten the safe operation of power grids, and FCL in limiting short circuit current and also involved in the transient process, and the transient behavior of the system is worth studying [6-10].

2) FCL the dynamic characteristics, the stray capacitance will impact near the high voltage circuit breaker breaking transient recovery voltage characteristics, is not

H. Tan (Ed.): Knowledge Discovery and Data Mining, AISC 135, pp. 135–141.
springerlink.com

conducive to the circuit breaker [11-14]. Meanwhile, FCL and the circuit breaker reclosing can also apply the key to effective co-ordination [12, 16].

3) FCL string into the line, on line over current protection setting of impact, at the same time as the dynamic characteristic of FCL, this will make the line dynamic impedance characteristics become complicated, affecting distance protection equipment [15-17].

4) As everyone knows, fault current limiter is used in high voltage, large capacity, to highlight its technical economy. At present, FCL research has stuck in the middle, low pressure areas, engineering application is not much, how to design a high voltage, large capacity FCL research.

5) Study of FCL economic evaluation index, quantitative effect of technical factors, build reasonable technological economics evaluation model, the FCL application significance. In technological economics evaluation model based on regional power grid is presented, aimed at a specific FCL best configuration, it has important practical value.

FCL generally connected to the transmission line with the system normal operation, only in the event of a system failure before action. On-line monitoring of FCL long-term job characteristics can ensure the reliable action in the event of a system failure. Study of a set of comprehensive digital online monitoring, fault diagnosis, trigger control and protection strategies for engineering application of FCL research, is another key element.

3 The Principle and Simulation of the Magnetic Valve Controlled Reactor-Type Fault Current Limiter

3.1 Working Principle

Magnetic valve controlled reactor-type fault current limiter is essentially a DC controlled adjustable reactance, its basic structure shown in Figure 1.

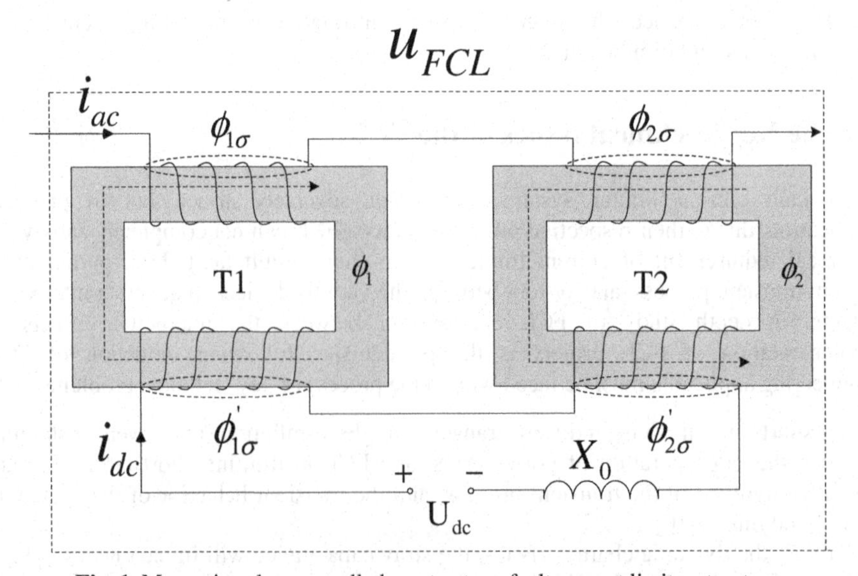

Fig. 1. Magnetic valve controlled reactor-type fault current limiter structure

Fig. 1. which consists of two cores, each core has a coil winding and the exchange a DC bias winding, one of the core on the AC and DC winding around the winding to the same, and the other core on the contrary, two AC windings in series after series in the transmission line, in order to increase the limiting effect, the DC circuit in series with a current-limiting reactor.

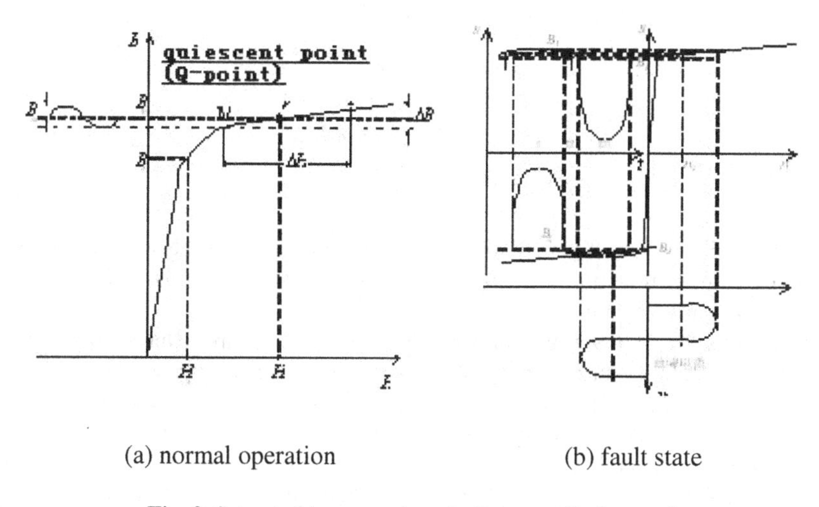

(a) normal operation (b) fault state

Fig. 2. Saturated iron core-type fault current limiter works

Fig. 2. Normal operation, the DC winding produces a strong DC magnetic field, the core depth of saturation, as shown in Figure 2 (a) shows the magnetic core work at the quiescent operating point of H0. Limiter rated AC current flowing through the small, winding through the exchange of AC magnetic field generated is not enough to core out of saturation, changes in magnetic flux density within the core is very small, almost can be considered the same, accordingly, through the exchange of winding flux is constant, so the exchange of winding ends of the induced electromotive force is almost zero. In the DC circuit, two coils wound to the contrary, the induction out of the offset voltage, DC over a constant stream of DC, the equivalent of limiting short-circuit reactance X0. In the DC circuit, two coils wound to the contrary, the induction out of the offset voltage, DC over a constant stream of DC, the equivalent of limiting short-circuit reactance X0. At this point, the fault current limiter on the voltage drop is small, its impact on the system can be ignored. Failure, a sudden increase of the fault current flows through the exchange of winding generates a magnetic field makes the positive and negative half-cycle alternating two cores out of saturation, as shown in Figure 2 (b) shows.

3.2 Simulation Analysis

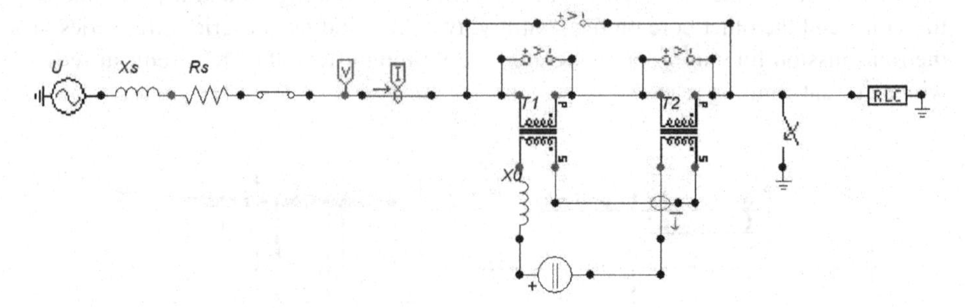

Fig. 3. FCL simulation model

Using ATP-EMTP electromagnetic transient simulation software to build a core-type FCL 6kV system saturated model, limiting reactance X0 take 10mH, voltage loss by 3% design, shown in Figure 3.

Transmission system simulation to 6kV transformer three-phase short circuit occurs, for example exported, the system parameters are as follows: The system's maximum line voltage of 6.6kV, the largest single-phase line-to-ground voltage effective value of 3.8kV, as a power supply voltage, the system equivalent inductance L_s = 0.85mH, resistance R_s = 0.05Ω, the line rated current 200A, power factor 0.98, the most serious cases to consider short-circuit fault, the fault occurred in the power potential just had zero time, at this time have the greatest short-circuit current value, such as Fig. 4 (a). Shows the steady-state value to short-circuit current 14kA (steady-state peak 20kA), the impact of short-circuit current up to 36kA.

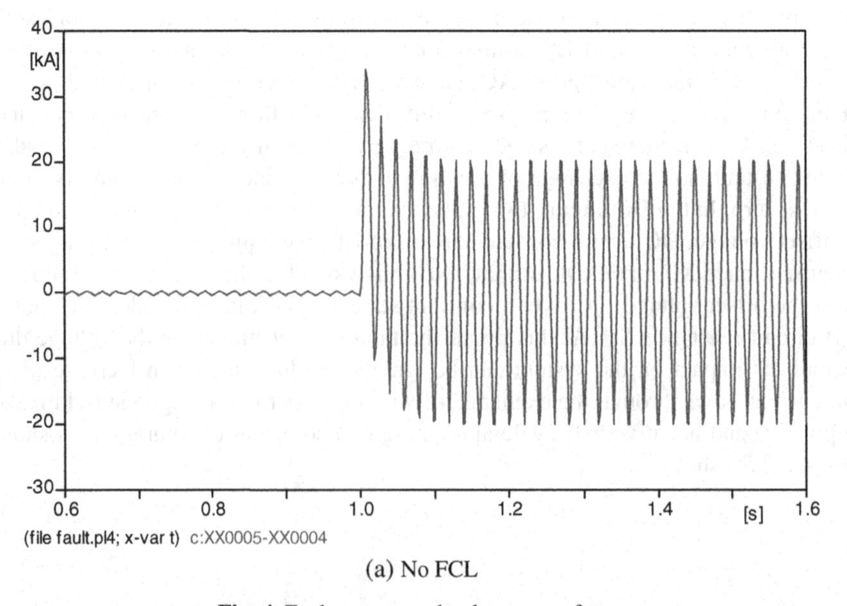

(a) No FCL

Fig. 4. Fault current and voltage waveforms

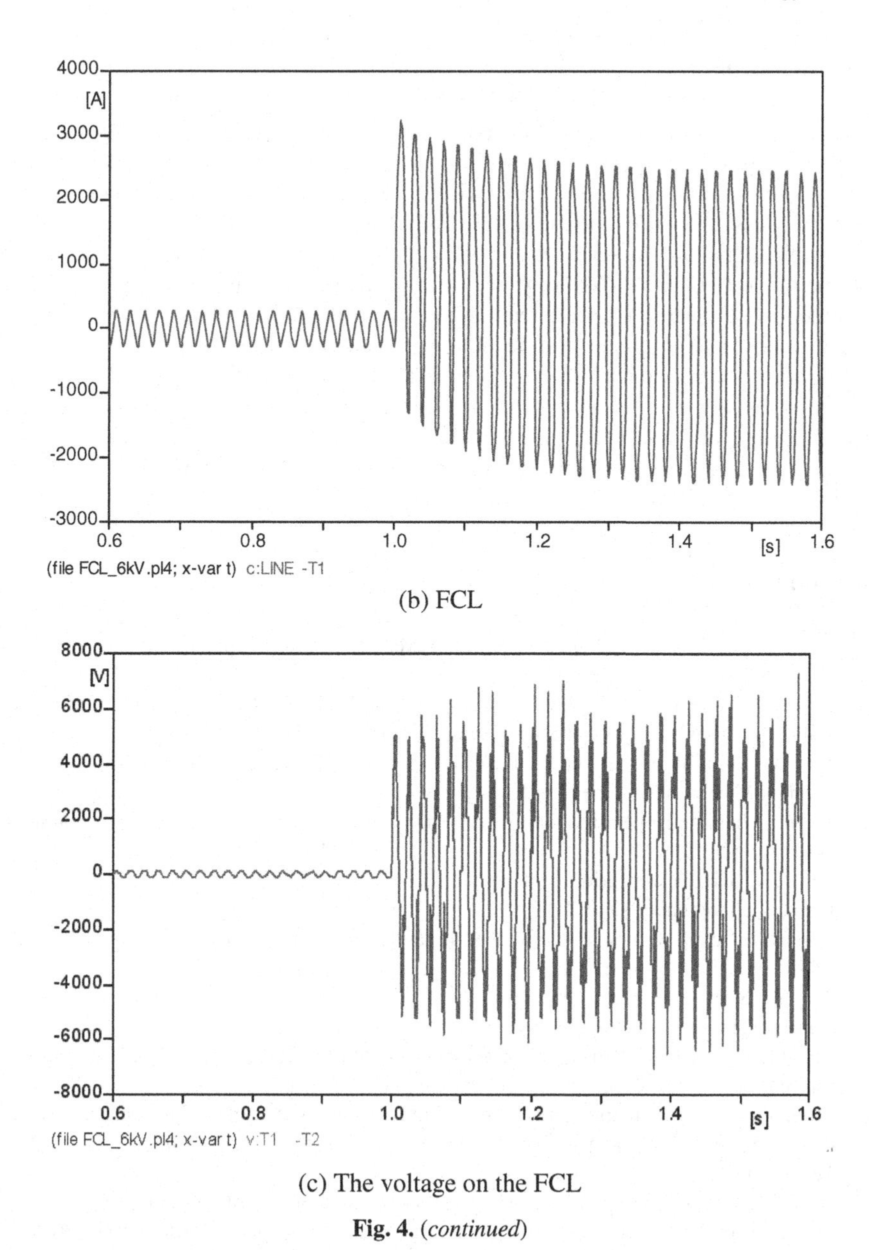

(b) FCL

(c) The voltage on the FCL

Fig. 4. *(continued)*

Fig. 4(a). Shows short circuit current steady-state value reached 14kA (20kA steady peak), short circuit impulse current can reach 36kA.

Fig. 4(b). Can be seen, with the FCL, the short circuit current is well circumscribed, peak inrush current for 2.4kA 3.2kA, steady peak. The FCL can limit the short circuit current, reduce the power equipment on the move, the heat stability.

Fig. 4(c). For the FCL on the voltage waveform, normal operation, the voltage loss is very small (3%, depending on the FCL design), when a short circuit, When there is a short circuit, FCL load almost all system voltage.

4 Conclusion

1) Install fault current limiter is to limit the power short-circuit current of the most effective measures. It can effectively limit the short circuit current, lower power system short-circuit flow of other devices, thermal stability requirements, and t Reduce the loss, and the economic benefit is improved.

2) Limiting reactor in series is the only high-pressure system can be applied to limit the maturity of the technology; the traditional current-limiting reactor to control the direction of current-limiting reactor may be out of the current transition fault current limiter technology the best development path bottleneck.

3) Magnetic valve controlled reactor-type fault current limiter structure is simple, without any power electronics control, fast response (5ms), high reliability, ideal for current-limiting device fully meet the technical requirements, is a very good fault current limiter, and easy-to high-pressure high-capacity, thus, expected to be widely used in the power system.

References

1. Yuan, J., Liu, W., Dong, Q.: Northwest power grid short-circuit current restrictions of Electrical Engineering, vol. (10) (2007)
2. Yuan, Q.: Future of Shanghai power grid short circuit current control status and countermeasures of Electrical Engineering, vol. (02) (2005)
3. Wu, D.: Limit the level of the bulk power system short-circuit current measures of research. Master thesis. Zhejiang University (2005)
4. Li, M.: Zhenjiang operation of the grid hierarchy partitions. Master thesis. Southeast University (2006)
5. Chen, Y.: Big power grid short-circuit current restrictions research. Master's degree thesis. Zhejiang university (2008)
6. Hao, Z., Jiang, D., Cai, Y.: New solid-state fault current limiter on power system transient stability. The Effect of Automation of Electric Power Systems (08) (2004)
7. Hao, Z.: Solid fault fault current limiter on power system operation effects. Master's degree thesis. Zhejiang university (2004)
8. Liu, H., Li, Q., Xu, L.: Zinc oxide arrestor type fault fault-current limiter on power system transient stability. The Influence of Electric Power Automation Equipment (08) (2007)
9. Yang, Z.: Superconducting fault current limiter on power system transient stability of analysis on the influence. Master's degree thesis. University of north China electric power (hebei) (2007)
10. Gu, X., Yang, Z.: A symmetric when short-circuit fault circuit reactance type FCL is transient stability of the influence of the journal. An Electrician Technology (09) (2009)
11. Liu, H., Li, Q., Lou, J.: Inductance type in a circuit breaker FCL resuming voltage current-rising-rate influence. Electrician Technology Journal 22(12), 84–91 (2007)
12. Liu, H., Li, Q., Zou, L.: Installation fault FCL is hidden power transmission lines and single phase reclosing arc characteristic strategy. Chinese Electrical Engineering (31) (2008)
13. Geng, S., Zhang, J., Zhang, M.: Network (FCL type of high voltage circuit breaker open circuit fault). The Influence of Power System Protection and Control (18) (2009)

14. Hong, J., Guan, Y., Xu, G.: Different fault serial resonant FCL to restore the influence of high voltage circuit breaker voltage. High Voltage Apparatus (05) (2009)
15. Zhou, X., Xu, X., Ma, Y.: Superconducting fault current limiter on the influence of the relay protection. China Electric Power (03) (2007)
16. He, Y., Chen, X., Tang, Y.: Superconducting fault current limiter for automatic reclosing and after the influence. High Voltage Technology (10) (2008)
17. Zou, L., Li, Q., Liu, H.: Plan and fault current limiter and fault of interim resistance grounding distance protection compensation algorithm. Power Automation Equipment (03) (2009)

The Scattering from an Object below 2D Rough Surface

Huang Xue-Yu and Tong Chuang-Ming

Air Force Engineering University
P.O. Box 25, Sanyuan, Shannxi 713800, China

Abstract. A rigorous fast numerical method called E-PILE+SMCG is introduced and then used in a Monte Carlo study of Scattering from a three dimensional perfectly electrical conductor (PEC) object below lossy soil rough surface. This method is the three dimensional (3D) extendability of PILE (Propagation-Inside–Layer Expansion) method which is proposed for two dimensional (2D) scattering problem. The rough surface with Gaussian profile is used to emulate the realistic situation of statistically rough surface, while the tapered incident wave is chosen to reduce the truncation error. The 3D angular correlation function (ACF) and bistatic scattering coefficient (BSC) are studied and applied to the detection of a target embedded in the clutter. The ACF is computed by using numerical method with circular azimuthal angle averaging technique. Because of its successful in suppressing the clutter, the technique appears attractive in real life implementation.

Keywords: E-PILE+SMCG, ACF, BSC, buried object.

1 Introduction

Recently, Dechamps et al. developed a fast numerical method, PILE (Propagation-Inside–Layer Expansion), devoted to the scattering by a stack of two one dimensional (1D) interfaces separating homogeneous media. Bourlier C. et al. have applied the PILE method for an object located below a 1D rough surface. The main advantage of the PILE method is that the resolution of the linear system (obtained from the method of moments) is broken up into different steps: Two steps are dedicated to solving for the local interactions, which can be done from efficient methods valid for a single rough surface, such as SMCG and FB/NSA. Two steps are dedicated to solving for the coupling interactions. The purpose of this paper is to extend the PILE method to the 3D scattering problem of a object buried under soil rough surface. In order to accelerate the extended PILE method and to treat a large problem, the local interactions of the rough surface are computed by using SMCG because of its relative simplicity of programmation and its low complexity of computation.

The ACF is the correlation function of two scattering fields in two different directions, corresponding to two incident fields. The ACF has been applied to the detection of a buried object. For the detection of a object under a rough surface, the scattering of the object is often obscured by clutter such as rough surface scattering and random medium scattering, which makes it difficult to detection the object. In this paper, we study the ACF of wave scattering by an object buried under soil rough

H. Tan (Ed.): Knowledge Discovery and Data Mining, AISC 135, pp. 143–150.
springerlink.com

surface for 3D scattering problems. For the ACF applied in detecting buried object, the object is associated with one realization of the random rough surface, therefore realization averaging is meaningless. In the 3D ACF, we make the circular azimuthal angle averaging, which is successful for detection the 3D object.

1.1 The Boundary Integral Equations

Fig.1 illustrates the basic geometry considered in this paper: a PEC object is located below a rough surface. The rough interface between free space and a dielectric medium with relative complex permittivity ε_1 is described by $z = f(x, y)$, and generated by Monte Karlo method. The θ_i and φ_i are incident angles, and the θ_s and φ_s are scattering angles.

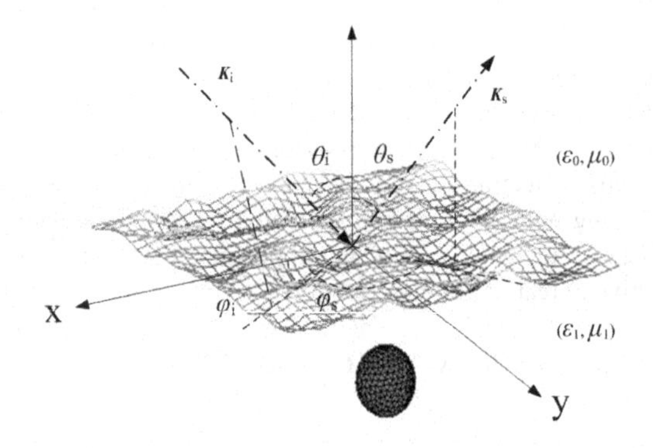

Fig. 1. The geometry of a PEC object buried under rough surface

To avoid edge limitations, the incident field is tapered so that the illuminated rough surface can be confined to the surface area $L_x \times L_y$. Consider the tapered plane wave incident on the structure along $\hat{\boldsymbol{k}}_i = \sin\theta_i \cos\varphi_i \hat{\boldsymbol{x}} + \sin\theta_i \sin\varphi_i \hat{\boldsymbol{y}} - \cos\theta_i \hat{\boldsymbol{z}}$, Then the incident fields can be expressed in terms of spectrum of the incident wave

$$\boldsymbol{H}^{inc}(x, y, z) = -\frac{1}{\eta} \int_{-\infty}^{\infty} dk_x \int_{-\infty}^{\infty} dk_y \exp(ik_x x + ik_y y - ik_z z) \cdot E_{TE}(k_x, k_y) \hat{\boldsymbol{h}}(-k_z) \tag{1}$$

$$\boldsymbol{E}^{inc}(x, y, z) = \int_{-\infty}^{\infty} dk_x \int_{-\infty}^{\infty} dk_y \exp(ik_x x + ik_y y - ik_z z) \cdot E_{TE}(k_x, k_y) \hat{\boldsymbol{e}}(-k_z) \tag{2}$$

Where

$$E_{TE}(k_x, k_y) = \frac{1}{4\pi^2} \int_{-\infty}^{\infty} dx \int_{-\infty}^{\infty} dy \exp(-ik_x x - ik_y y) \cdot \exp(i(k_{ix} x + k_{iy} y)(1+w)) \exp(-t) \tag{3}$$

$$t = t_x + t_y = (x^2 + y^2) / g^2 \tag{4}$$

$$t_x = \frac{(\cos\theta_i \cos\varphi_i x + \cos\theta_i \sin\varphi_i y)^2}{g^2 \cos^2\theta_i} \tag{5}$$

$$t_y = \frac{(-\sin\varphi_i x + \cos\varphi_i y)^2}{g^2} \tag{6}$$

$$w = \frac{1}{k_0}\left[\frac{(2t_x - 1)}{g^2\cos^2\theta_i} + \frac{(2t_y - 1)}{g^2}\right] \tag{7}$$

The parameter g controls the tapering of the incident wave.

In the above k_0 and η are the wave-number and wave impedance of free space respectively, and $\hat{e}$, $\hat{h}$ denote the polarization vectors. Let $r' = x'\hat{x} + y'\hat{y} + f(x', y')\hat{z}$ and $r = x\hat{x} + y\hat{y} + f(x, y)\hat{z}$ denote source and field point separately. The fields in region 0 and region 1 satisfy the following equations:

$$\frac{\hat{n}\cdot E_s(r)}{2} - \hat{n}\cdot\left\{\int_S J_s(r')i\omega\mu G_0 ds' + P\int_S [M_s(r')\times\nabla' G_0 + \hat{n}\cdot E_s(r')\nabla' G_0]ds'\right\} = \hat{n}\cdot(E^{inc}(r) + E_b^s(r)) \quad (r\in S) \tag{12}$$

$$\frac{J_s(r)}{2} - \hat{n}\times\left\{\int_S (-i\omega)M_s(r')\varepsilon_0 G_0 ds' + P\int_S [(J_s(r'))\times\nabla' G_0 + \hat{n}\cdot H_s(r')\nabla' G_0 ds'\right\} = \hat{n}\times(H^{inc}(r) + H_b^s(r)) \quad (r\in S) \tag{13}$$

$$-\frac{M_s(r)}{2} - \hat{n}\times\left\{\int_S J_s(r')i\omega\mu G_1 ds' + P\int_S [(M_s(r'))\times\nabla' G_1 + \hat{n}\cdot E_s(r')\nabla' G_1]ds'\right\} = -\hat{n}\times E_b^s(r) \quad (r\in S) \tag{14}$$

$$-\frac{\hat{n}\cdot H_s(r)}{2} - \hat{n}\cdot\left\{\int_S (-i\omega)M_s(r')\varepsilon_1 G_1 ds' + P\int_S [(J_s(r'))\times\nabla' G_1 + \hat{n}'\cdot H_s(r')\nabla' G_1]ds'\right\} = -\hat{n}\cdot H_b^s(r) \quad (r\in S) \tag{15}$$

$$E_S^s(r)\Big|_{\tan} = i\omega\mu_0\int_{S_b}[J_b(r') + \frac{1}{k_q^2}\nabla(\nabla'\cdot(J_b(r')))]G_1 ds' \quad (r\in S_b) \tag{16}$$

Where, $G_{0,1} = \frac{\exp(ik_{0,1}R)}{4\pi R}$ $R = \sqrt{(x-x')^2 + (y-y')^2 + (f(x,y) - f(x',y'))^2}$, k_0 and k_1 are the wave-number of the upper and lower medium, S denotes the rough surface, and S_b denotes the surface of object. $J_b(r) = \hat{n}_b\times H_b(r)$, $M_S(r) = \hat{n}\times E_S(r)$, $J_S(r) = \hat{n}\times H_S(r)$. The unit normal vector $\hat{n}$ and $\hat{n}'$ refer to primed coordinates and point away from the lower medium, $\hat{n}_b$ and $\hat{n}_b'$ point away from the object.

$$H_b^s(r) = \int_{S_b} J_b(r')\times\nabla' G_1 ds' \tag{17}$$

$$E_b^s(r) = -\frac{i}{\omega\varepsilon_1}\nabla\times\int_{S_b} J_b(r')\times\nabla' G_1(r,r')ds' \tag{18}$$

$$E_S^s(r) = -\int_S [(-i\omega\mu_1)J_S(r')G_1 + M_S(r')\times\nabla' G_1 + \nabla' G_1\hat{n}'\cdot E_S(r')]ds' \tag{19}$$

The H_b^s and E_b^s denote the scattering field from object to rough surface, and E_S^s denote the scattering field from upper surface to object.

1.2 The E-PILE Method

The integral equations of rough surface is discretized by using MoM with pulse base function, therefore the integral equations of object is discretized by using MoM with RWG base function. After discretizing equations (12)-(19), we get the following matrix equation

$$\boldsymbol{ZI} = \boldsymbol{V} \tag{20}$$

Where $\boldsymbol{Z}$ is the total impedance matrix of size $(N+M)\times(N+M)$ (N is the unknown number of rough surface, and M is the unknown number of object). $\boldsymbol{I}$ is the unknown vector.

$$\boldsymbol{I}^T = \left[\boldsymbol{I}_r^T \boldsymbol{I}_o^T\right] \tag{21}$$

$\boldsymbol{I}_r^T$ is the unknown vector of rough surface, and $\boldsymbol{I}_o^T$ is the unknown vector of object.

In order to solve efficiency the linear system (20), the impedance matrix is expressed from sub-matrices as

$$\boldsymbol{Z} = \begin{bmatrix} \boldsymbol{Z}^r & \boldsymbol{Z}^{o\to r} \\ \boldsymbol{Z}^{r\to o} & \boldsymbol{Z}^o \end{bmatrix} \tag{22}$$

$\boldsymbol{Z}^r$ and $\boldsymbol{Z}^o$ correspond exactly to the self-impedance matrices of rough surface and object. The matrices $\boldsymbol{Z}^{o\to r}$ and $\boldsymbol{Z}^{r\to o}$ can be interpreted as coupling impedance matrices between object and upper surface. The reverse of the matrix Z is:

$$\boldsymbol{Z}^{-1} = \begin{bmatrix} \boldsymbol{T} & \boldsymbol{U} \\ \boldsymbol{Q} & \boldsymbol{W} \end{bmatrix} \tag{23}$$

Where

$$\boldsymbol{T} = (\boldsymbol{Z}^r - \boldsymbol{Z}^{o\to r}(\boldsymbol{Z}^o)^{-1}\boldsymbol{Z}^{r\to o})^{-1} \tag{24a}$$

$$\boldsymbol{U} = -(\boldsymbol{Z}^r - \boldsymbol{Z}^{o\to r}(\boldsymbol{Z}^o)^{-1}\boldsymbol{Z}^{r\to o})^{-1}\boldsymbol{Z}^{o\to r}(\boldsymbol{Z}^o)^{-1} \tag{24b}$$

$$\boldsymbol{Q} = -(\boldsymbol{Z}^o)^{-1}\boldsymbol{Z}^{r\to o}(\boldsymbol{Z}^r - \boldsymbol{Z}^{o\to r}(\boldsymbol{Z}^o)^{-1}\boldsymbol{Z}^{r\to o})^{-1} \tag{24c}$$

$$\boldsymbol{W} = (\boldsymbol{Z}^o)^{-1} + (\boldsymbol{Z}^o)^{-1}\boldsymbol{Z}^{r\to o}(\boldsymbol{Z}^r - \boldsymbol{Z}^{o\to r}(\boldsymbol{Z}^o)^{-1}\boldsymbol{Z}^{r\to o})^{-1}\boldsymbol{Z}^{o\to r}(\boldsymbol{Z}^o)^{-1} \tag{24d}$$

By using equations (23) and (24), the total field on the rough surface can be expressed as

$$\boldsymbol{I}_r = (\boldsymbol{Z}^r - \boldsymbol{Z}^{o\to r}(\boldsymbol{Z}^o)^{-1}\boldsymbol{Z}^{r\to o})^{-1}(\boldsymbol{V}_r - \boldsymbol{Z}^{o\to r}(\boldsymbol{Z}^o)^{-1}\boldsymbol{V}_o) \tag{25}$$

Because the object is buried under the rough surface, so we have $\boldsymbol{V}_o^T = \boldsymbol{0}$. Thus equation (25) can be expressed as:

$$\boldsymbol{I}_r = (\boldsymbol{Z}^r - \boldsymbol{Z}^{o\to r}(\boldsymbol{Z}^o)^{-1}\boldsymbol{Z}^{r\to o})^{-1}\boldsymbol{V}_r = (\boldsymbol{I} - (\boldsymbol{Z}^r)^{-1}\boldsymbol{Z}^{o\to r}(\boldsymbol{Z}^o)^{-1}\boldsymbol{Z}^{r\to o})^{-1}(\boldsymbol{Z}^r)^{-1}\boldsymbol{V}_r \tag{26}$$

Where $\boldsymbol{I}$ is the identity matrix. Let us introduce the characteristic matrix $\boldsymbol{M}_c$ as

$$\boldsymbol{M}_c = (\boldsymbol{Z}^r)^{-1}\boldsymbol{Z}^{o\to r}(\boldsymbol{Z}^o)^{-1}\boldsymbol{Z}^{r\to o} \tag{27}$$

Then the first term in equation (25) can be expanded as an infinite series over p

$$(\boldsymbol{I}-(\boldsymbol{Z}^r)^{-1}\boldsymbol{Z}^{o\to r}(\boldsymbol{Z}^o)^{-1}\boldsymbol{Z}^{r\to o})^{-1}=\sum_{p=0}^{p=\infty}\boldsymbol{M}_c^p \tag{28}$$

For the numerical computation, the sum must be truncated at order P_{PILE} . Combing the equations (27)and (28), the total unknown vector on the rough surface is then expressed as

$$\boldsymbol{I}_r=\left[\sum_{p=0}^{p=p_{PILE}}\boldsymbol{M}_c^p\right](\boldsymbol{Z}^r)^{-1}\boldsymbol{V}_r=\sum_{p=0}^{p=p_{PILE}}\boldsymbol{Y}_r^{(p)} \tag{29}$$

Where

$$\begin{cases}\boldsymbol{Y}_r^{(0)}=(\boldsymbol{Z}^r)^{-1}\boldsymbol{V}_r & \text{for } p=0\\ \boldsymbol{Y}_r^{(p)}=\boldsymbol{M}_c\boldsymbol{Y}_r^{(p-1)} & \text{for } p>0\end{cases} \tag{30}$$

We define the norm $\|\boldsymbol{M}_c\|$ of a complex matrix by its spectral radius, i.e. the modulus of its eingenvalue which has the highest modulus. Expansion (28) is accurate if $\|\boldsymbol{M}_c\|$ is inferior to 1.

1.3 Acceleration of E-PILE by Using the SMCG

The advantage of the PILE method is that the most complex operations, which is $(\boldsymbol{Z}^r)^{-1}\boldsymbol{u}$ ($\boldsymbol{u}$ is a vector), only concerns the local interactions on rough surface, and can be calculated by fast numerical methods that already exist for single 2D rough surface, like for instance the SMCG. The SMCG will be used in this paper because of its relative simplicity of programmation and its low complexity of computation. It is based on the fact that performing $(\boldsymbol{Z}^r)^{-1}\boldsymbol{u}$ is equivalent to solve $\boldsymbol{Z}^r\boldsymbol{v}=\boldsymbol{u}$ for $\boldsymbol{v}$, this latter problem can be solved iteratively by using a conjugate gradient scheme as Pre-BICGtab. At each iteration, it is necessary to compute $\boldsymbol{Z}^r\boldsymbol{v}$, where $\boldsymbol{v}$ is an updated of $(\boldsymbol{Z}^r)^{-1}\boldsymbol{u}$. The SMCG is employed at this step to speed up the product $\boldsymbol{Z}^r\boldsymbol{v}$. We choose a neighborhood distance r_d as the distance which defines the boundary between the weak and strong element of the impedance matrix. Then the impedance matrix of rough surface is decomposed into the sun of a strong matrix $\boldsymbol{Z}^{r(s)}$ and a weak matrix $\boldsymbol{Z}^{r(w)}$ which is a Toeplita matrix. Then the $\boldsymbol{Z}^{r(s)}\boldsymbol{v}$ is computed by using traditional MoM method, and $\boldsymbol{Z}^{r(w)}\boldsymbol{v}$ can be accelerated by using 2D FFT. As a conclusion, we call E-PILE+SMCG this method, in order to differentiate it from the PILE method for 2D scattering problem, which usually using only exact matrices inversions for the local interactions. Because the unknown number of rough surface is much large than that of object, so the complexity of E-PILE+SMCG is approximately the same to that of SMCG.

2 Numerical Results

The soil rough surface profiles used in the study are realizations of a rough surface with a Gaussian spectrum

$$W(k_x,k_y)=\frac{l_x l_y h^2}{4\pi}\exp(-\frac{l_x^2k_x^2+l_y^2k_y^2}{4}) \tag{31}$$

Here, l_x and l_y are the correlation lengths in x- and y-directions, respectively, and h is the rms height of the rough surface.

In order to study the convergence of E-PILE+SMCG versus its order P_{PILE}, define the Relative Error (RE) of nth order as

$$\text{RE:}\,\tau(n)=\frac{\text{norm}\left|\boldsymbol{Z}^r(\boldsymbol{I}^{r,(n)}-\boldsymbol{I}^{r,(n-1)})\right|}{\text{norm}\left|\boldsymbol{V}^{inc}\right|} \tag{32}$$

The norm of a vector of components X_i and of length N is expressed as $\text{norm}(\boldsymbol{X})=\sum_{i=1}^{i=N}\left|X_i\right|^2$. The vector $\boldsymbol{I}^r$ represents the current on the rough surface. The $\boldsymbol{V}^{inc}$ represents the initial incident field.

We calculate the scattering amplitudes for 10 azimuthal angles at $0°$, $36°$, $72°$,…,and $324°$, respectively. There is only one realization of the random rough surface. The BSC and ACF are calculated by using azimuthal angular averaging. The results are plotted as function of the scattering angle θ_{s2}. Parameters for other angles are $\theta_{i1}=20°$, $\theta_{s1}=-20°$, and $\theta_{i2}=20°$. The fully polarimetric results of BSC and ACF are calculated and shown in Fig. 2 and Fig.3. Fig.2(a) shows the results of BSC for HH-polarization and VH-polarization component, and the results for VV-polarization and HV-polarization component are shown in Fig.2(b). Both the results with and without the object sphere are shown for comparison. As expected, there is a peak in the specular direction, which is due to the slightly rough surface. We see that the differences of BSC between with and without the object for co-polarizations are larger than those for cross-polarizations. This is because the cross-polarization components are mainly due to the rough surface scattering. Because the object is a sphere, it is only a small cross-polarization contribution in BSC. It is also found that there are larger differences for the VV component than for the HH component. Fig.3 shows the results of ACF. We can see the large difference of ACF between with and without a object in both the co-polarization and cross-polarization result. As shown in Fig.3, the difference of ACF is large even for angles closed to the nadir direction. This is because the memory effect is avoided and rough surface scattering is minimized in the ACF.

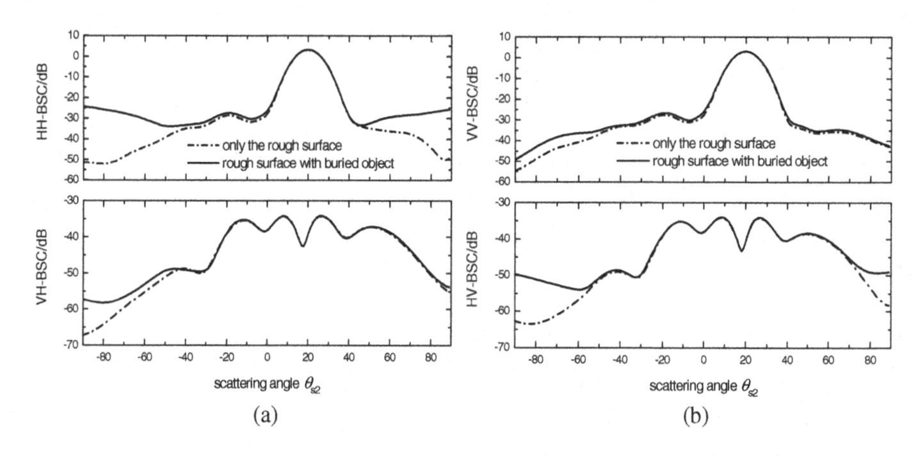

Fig. 2. The BSC of a buried sphere

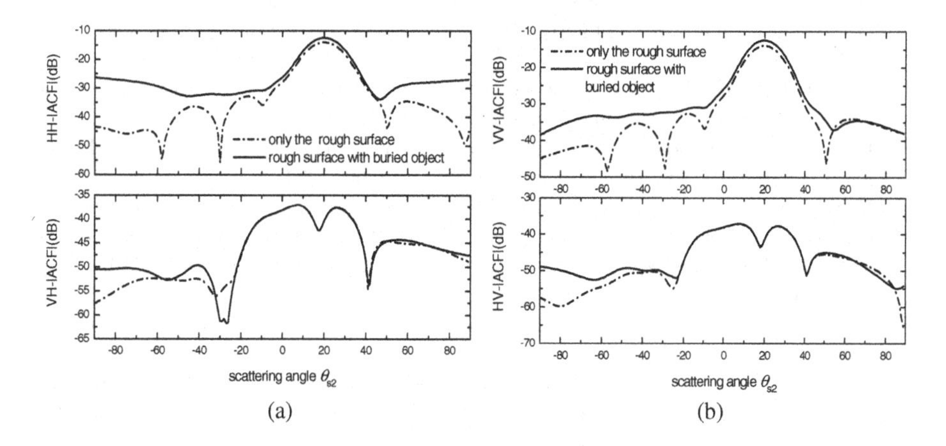

Fig. 3. The ACF of a buried sphere

2.1 Conclusions

By extending the PILE method which is introduced for 2D scattering problem, a rigorous fast numerical method E-PILE+SMCG is proposed for scattering from 2D rough surface with 3D buried object (3D problem). The SMCG method is used for accelerating the calculation the integral equations of the rough surface, and the object is computed by traditional MoM. The complexity of this method is similar to that of SMCG ($N \log N$). Then this method is used to compute the scattering from soil rough surface with buried sphere object. The results of E-PILE+SMCG are agreement with that in literature, which shows the validity of this method. At the same time, the constringency of E-PILE+SMCG is also discussed. Then, this method is used to study the scattering field and ACF of soil rough surface with buried object. The results show that the ACF has great advantage for detecting buried object over BSC. At last, the ACF is studied for different parameters of rough surface such as rms height and correction length, anf for different parameters of sphere object such as radii and depth. The simulation result shows that the ACF can be influence greatly by parameters of rough surface and object, and this character can be very useful in realistic detecting technology.

References

1. Zhang, Y., Yang, Y.E., Braunisch, H., Kong, J.A.: Electromagnetic wave interaction of conducting object with rough surface by hybrid SPM/MOM technique. PIER: Progr. Electromagn. Res. 22, 315–335 (1999)
2. Ishimaru, A., Rockway, J.D., Kuga, Y.: Rough surface Green's function based on the first-order modified perturbation and smoothed diagram methods. Waves Random Media 10, 17–31 (2000)
3. O'Neill, K., Lussky, R.F., Paulsen, K.D.: Scattering from a metallic object embedded near the randomly rough surface of a lossy dielectric. IEEE Trans. Geosci. Remote Sensing 34, 367–376 (1996)
4. Dogaru, T., Carin, L.: Time-domain sensing of targets buried under a rough air-ground interface. IEEE Trans. Antennas Propagat. 46, 360–372 (1998)
5. Johnson, J.T., Burkholder, R.J.: A Study of Scattering From an Object Below a Rough Surface. IEEE Trans. Geosci. Remote Sensing 42, 59–66 (2004)
6. El-Shenawee, M., Rappaport, C.M., Miller, E.L., Silevitch, M.B.: Three-dimensional subsurface analysis of electromagnetic scattering from penetrable/PEC objects buried under rough surfaces: Use of the steepest descent fast multipole method. IEEE Trans. Geosci. Remote Sensing 39, 1174–1182 (2001)
7. Tsang, L., Kong, J.A., Ding, K.H.: Scattering of Electromagnetic Waves: Numerical Simulations, New York (2000)
8. Tsang, L., Chang, C.H., Sangani, H.: A Banded Matrix Iterative Approach to Monte Carlo simulations of scattering of waves by large scale random rough surface problems: TM case. Electron. Lett. 29, 1666–1667 (1993)
9. Tsang, L., Chang, C.H., Sangani, H., Ishimaru, A., Phu, P.: A Banded Matrix Iterative Approach to Monte Carlo simulations of large scale random rough surface scattering: TE case. J. Electromag. Waves Appl. 29, 1185–1200 (1993)
10. Torrungrueng, D., Chou, H.T., Johnson, J.T.: A Novel Acceleration Algorithm for the Computation of Scattering From Two-Dimensional Large Scale Perfectly Conducting Random Rough Surfaces with the Forward Backward Mehod. IEEE Trans. Geosci. Remote Sensing 38, 1656–1668 (2000)

Moving Target Localization Based on Multi-sensor Distance Estimation

Zhao Wang[1,2], Chao Zhang[1,2], and Zhong Chen[1,2]

[1] School of EECS, Peking University, 100871, Beijing, China
[2] Key Laboratory of High Confidence Software Technologies, MoE, Beijing, China
wangzhao@pku.edu.cn, chao945@yahoo.com.cn, chz@pku.edu.cn

Abstract. Based on the results of a number of single-sensor distance estimates, this paper presents how to calculate the actual position of the flight target based on multi-sensor distance estimates. The time 0 unification issue of estimation model for each sensor is solved. The goal that the sensors need to use acoustic signals that target emitted in the same location but received by different sensors, this article gives the corresponding solution.

Keywords: Multi-sensor, target localization, distance estimation, instantaneous frequency.

1 Introduction

Target positioning and tracking technology has important application value both in the military and in civilian areas. Moving target tracking is one of the most popular international research directions. As wireless sensor network is self-organizing, distributed, and with nodes that can be densely deployed, the perception of moving targets at close range can be achieved. Target localization and tracking system based on wireless sensor network has significant advantages in contrast with the traditional target localization and tracking, which lies in: more precise and reliable positioning, flexible deployment means, more subtle deployment, distributed tracking, and environmental adaptability, etc [1].

Doppler Effect is the occurrence of frequency change due to the relative motion between the sound source and the receiver, which implies the motion information of the target. According to this principle, a method that obtains the motion parameters of the target noise by estimating the instantaneous value of its fundamental frequency was proposed in the literature [2]. On this basis, this paper presented a method to carry out DOA estimation of the motion target based on the ranging information of multiple single sensors.

2 Passive Acoustic Detecting Model of the Single Sensor

Aiming at the localization of battlefield helicopters, assume that the helicopter flies straight maintaining constant speed at the same altitude, with flight speed less than the speed of sound; wind component is invariant in time and space; the frequency of

H. Tan (Ed.): Knowledge Discovery and Data Mining, AISC 135, pp. 151–157.
springerlink.com

source is constant. Assume C is the velocity of sound, V is the flight speed of the helicopter, L is the height of the helicopter as well as the distance at the point of closest approach, f_0 is the fundamental frequency of the helicopter rotor, Literature [2] and [3] according to Doppler Effect give the fundamental frequency of the helicopter rotor, which the sensor receives at the time t — $f_r(t)$, :

$$f_r(t)=\frac{f_0C^2}{C^2-V^2}\left[1-\frac{V/C(t-t_c)}{\sqrt{(L/(VC))^2(C^2-V^2)+(t-t_c)^2}}\right]=A+B\cdot Z(t_c,s) \tag{1}$$

In the equation, t_c is the time when the helicopter arrives at the point of closest approach:

$$t_c=-L/C\,,\quad A=f_0C^2/(C^2-V^2)\,,\quad B=-f_0CV/(C^2-V^2)\,,$$
$$s=L\sqrt{C^2-V^2}/(CV)\,,\ Z(t_c,s)=(t-t_c)\sqrt{s^2+(t-t_c)^2}\,.$$

The time t is defined as

$$t=\mp d(t)/V+(\sqrt{L^2+d^2(t)}-L)/C \tag{2}$$

$d(t)$ is the horizontal projection distance of the distance between the helicopter and the sensor in the flight direction when the helicopter noise reaches the sensor at time t:

$$d(t)=\frac{(tC^2V+LCV)}{C^2-V^2}-\frac{CV^2\sqrt{(L/(VC))^2(C^2-V^2)+(t-t_c)^2}}{C^2-V^2} \tag{3}$$

Give the velocity of sound C, helicopter flight speed V, the distance at the point of closest approach L and the rotor fundamental frequency f_0. As the distance between the helicopter and the sensor $r(t)$ is only relevant to the instantaneous frequency $f_r(t)$, these parameters can be determined by a set of measured values. The estimation of the helicopter flight parameters V , L, f_0 and t_c can be obtained by taking the minimum value of the mean square error of the estimation frequency and the predictive value of their frequency, which also means that the value of the following function is minimum [2].

$$\sum_{j=1}^{N}\left(\hat{f}_{tj}-f_{tj}\right)^2=\sum_{j=1}^{N}[\hat{f}_{tj}-A-BZ_{tj}(t_c,s)]^2 \tag{4}$$

In the equation, $\hat{f}_{tj}$ is the estimation of the instantaneous frequency at the time t_j, which can be obtained by using the technique of instantaneous frequency estimation. f_{tj} is the frequency value obtained by the estimation of equation (1). According to the estimated flight speed of the helicopter $\hat{V}$, the distance at the point of closest approach $\widehat{L}$ and the fundamental frequency of the rotor $\widehat{f}_0$, the distance estimation between the helicopter and the sensor can be obtained.

$$r(t)=\sqrt{\widehat{L}^2+\widehat{d}^2(t)}=\sqrt{\widehat{L}^2+\left(\left(tC^2\widehat{V}+\widehat{L}C\widehat{V}-CV^2\sqrt{\left(\frac{\widehat{L}}{\widehat{V}C}\right)^2(C^2-V^2)+(t-\widehat{t}_c)^2}\right)/(C^2-V^2)\right)^2} \quad (5)$$

3 The Promotion of Passive Acoustic Detecting Model of the Single Sensor

The above algorithms indicate that the range estimation of helicopter with uniform rectilinear movements by single sensor is merely dependent on the frequency of the receiving signals with Doppler shift sent by the target in rectilinear motion at different times. As show in Figure 1, the shortest distance between the sensor and flight track $L_1 = L_2$ is 2 different flight tracks (linear track with the same speed). The instantaneous frequency of acoustic signal with Doppler shift received by sensor has the same theoretical value, the sensor cannot distinguish whether the acoustic signal received belongs to track A or track B.

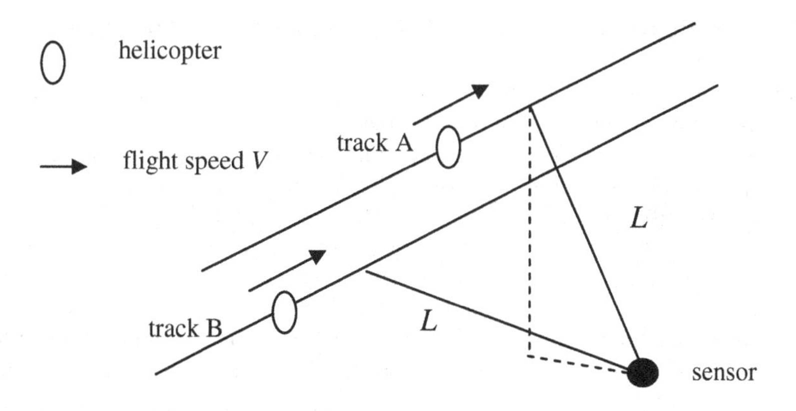

Fig. 1. Different linear flying track with the same the same flight speed

As the formula (1) and (2) are only correlated with the V and L of flight track, the situation that flight track A and B have the same signals in Figure 1 can be regarded as the overlapped track after rotating the coordinate system. Similarly, it can be applied to flight track maintaining constant speed at non-uniform altitude, if this flight track has the same distance at the point of closest approach L with an linear flight track maintaining constant speed at uniform altitude, the sensor will receive the same frequency, so the distance $r(t)$ s of the 2 corresponding points also keep the same, as shown in Figure 2. As a result, the range estimation utilizing single sensor based on the instantaneous frequency estimation can be used to observe the uniform linear flight track at various directions.

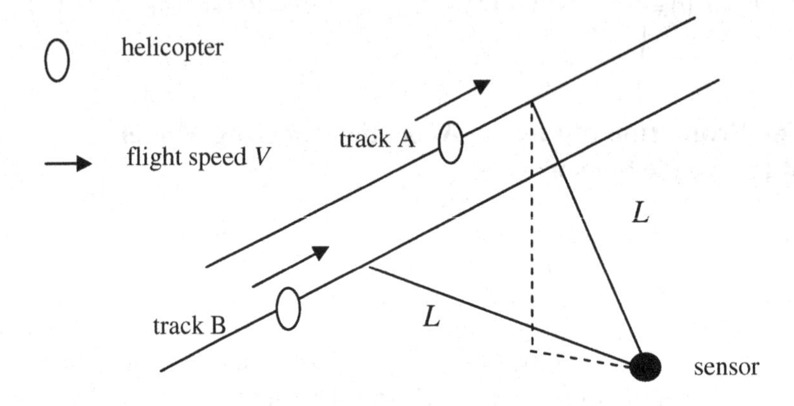

Fig. 2. Equal height and the uplifted flight track within the same vertical plane

4 The Unification Issue in Multi-sensor Distance Measurement

The distance-measurement method utilizing single-sensor regards the acoustic signal sent by the helicopter at the point of closest approach to the sensor as the observation time 0. As the sensor's location varies, the time that helicopter reaches at the point of closest approach to the sensors is also different; in addition, the distance between helicopter and different sensors also varies even though helicopter is in the same position, so the time that the acoustic signal sent by the helicopter reaches at the sensor is also different. Therefore, for the same flight, different sensors have different timer shafts, that is, they have different observation time 0. Then, the first reasonable solution is to unify the observation time of multi-sensor. For example, transform the observation time t_B of sensor B to t_A under the benchmark of observation time of sensor A. The time that the sensor can receive the rotor fundamental frequency f_0 can be obtained by observation, due to the high SNR (Signal to Noise Ratio) of rotor frequency, its estimated relative error is below 0.5% and the moment that the instantaneous frequency is equal to the rotor fundamental frequency is the observation time 0. Taking sensor A and sensor B as an example, as shown in Figure 3, due to the fact that the time the helicopter arrive at the point of closest approach of 2 sensors is different, if the observation time 0 of sensor A is set as t_{A0}, the observation time 0 of sensor B will be $t_{B0} = t_{A0} + \Delta t_{AB}$, Δt_{AB} can be obtained by the subtraction of receipt time f_0 between two sensors [4].

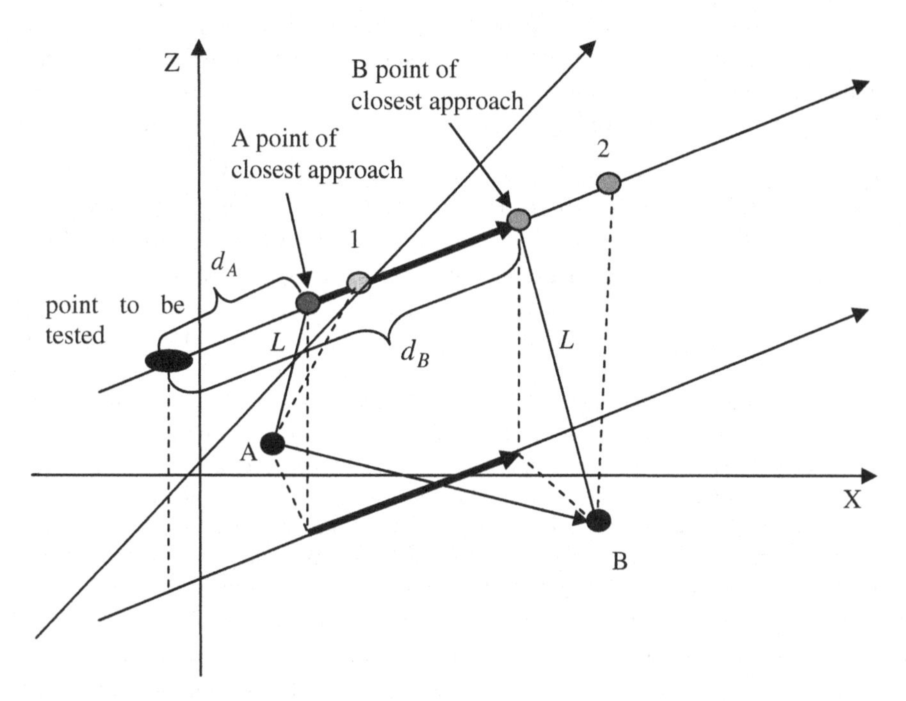

Fig. 3. Unified schematic diagram of range estimation of two points

For any $t_B = t_A + \Delta t_{AB}$, $d_A(t_A)$ and $d_B(t_B)$ are respectively on behalf of the distances between the point the helicopter at certain moment t_A and t_B and the point of closest approach in a flight direction, whether the distances $d_A(t_A)$ and $d_B(t_B)$ correspond to the same point on the flight track is an issue which requires further research.

The corolla change of observation time t is unequal corresponding to the change of d, that is, there exists a non-linear relationship between them. For the moment t_A and t_B , in addition to the time at the point of closest approach, the non-linear relationship between d and t indicates that $t_A - t_{A0} = t_B - t_{B0}$ doesn't mean $d_A(t_A) - d_A(t_{A0}) = d_B(t_B) - d_B(t_{B0})$. This shows that $d(t_A)$ and $d(t_B)$ obtained by t_A and t_B may correspond to different points on the track, so it is unreasonable to directly utilize $d_i(t)$ of different sensors i at the same time t to locate the target.

Therefore, how to solve the unification issue about the observation time of sensors is the key point of calculating the target position, only in this way, the receipt time t of each sensor corresponding to acoustic signals emitted by any point on the flight track can be found, and then the distance from this point to each sensor can be calculated according to the function $r(t)$ of estimated distance of each sensor regarding t as the variable. Analysis indicates that the relationship between the distance d from the point on the flight track to the point of closest approach to the sensor and the observation time t can be clarified by the formula (3). However, if this formula is utilized, first we should use d to solve t which will then be put into formula (5) to get distance estimation r between helicopter and the sensor. However, in this way, r should be derived by d and t; so we can elide t and directly obtain the relationship between r and d by $r = \sqrt{d^2 + L^2}$. The parameters L, V of the flight track can be estimated by the method in literature [1]. Therefore, the function of r and d can be obtained by the estimated L, V , which can avoid the unnecessary conversion from d to t to r, thus simplifying the calculation process, as shown in formula (6).

$$r(t) = \sqrt{L^2 + d^2(t)} = \sqrt{L^2 + V^2 \cdot (\tau - t_c)^2} \tag{6}$$

The current key problem lies in how to unify the distance d from a point on the flight track to various sensors. Now, the unification issue of the observation time of two sensors will be taken as an example to carry out the discussion.

When the frequency issued by the helicopter at the closest point of sensor A is received, the helicopter has moved to point 1. This time can be gained by matching the received frequency with its estimated fundamental frequency of the rotor (it can be regarded as the same with the real, when relative error below 0.5%), then obtain T_{Arec} which is the time that the reception corresponding to the clock time of the sensor A's own system produces signals at the point of closest approach. As a result, the time when the helicopter arrives at the point of closest approach to sensor A $T_{Anear} = T_{Arec} - L_A / C$ can be obtained. Similarly, T_{Brec} which is the time the signals at the point of closest approach to sensor B is received corresponding to the clock time of the sensor B's own system (at this time, the helicopter has continued to fly to point 2), and furthermore, the time when the helicopter arrives at the point of closest approach to sensor B $T_{Bnear} = T_{Brec} - L_B / C$ is calculated. In calculating the distance, for sensor A, the time difference that the helicopter flies from its position to point of closest approach to sensor A $T - T_{Arec}$ is noted as τ_A, while for sensor B, the time difference that the helicopter flies from its position to point of closest approach to sensor B $T - T_{Brec}$ is noted as τ_B, and then there is $\tau_B = \tau_A - (T_{Bnear} - T_{Anear})$. Meanwhile, $d_A(\tau_A) = v \cdot \tau_A$, $d_B(\tau_B) = v \cdot \tau_B = v \cdot (\tau_A - T_{Bnear} + T_{Anear})$, r_A and r_B corresponding to it can be solved by $r = \sqrt{d^2 + L^2}$.

Other sensor nodes C, D, and E, etc. can find the corresponding $r(\tau)$ of $r_A(\tau)$ through the above methods. Calculate the V , f_0 and L for each sensor node, then a set of r can be solved for positioning by using $r(\tau)$.

5 Conclusions

In the single-sensor ranging model based on instantaneous frequency estimation, the observation time 0 of each sensor is the moment when the acoustical wave that the helicopter emitted at the point of closest approach reaches at the sensor, therefore, when multi-sensors are utilized to conduct fusion estimation, it is necessary to solve the problem about the unification of observation time 0 for each sensor's estimated model. This paper analyzed this issue and drew the specific conclusions.

Since the distance between the target and each sensor is different, the time that the sound in the same location reaches at different sensors also varies from each other. At the same observation time, the acoustical signals received by different sensors are actually signals of target in different locations, while the estimation on the target location of multi-sensor need to use acoustic signals that target emitted in the same location but received by different sensors. In this paper, the corresponding solutions have been proposed.

At last, On the basis of the distance estimation for various single-sensors, by means of multilateral measurements [5] or other methods the real-time location of the target can be estimated.

References

1. Shen, X.: The Distributed Location Tracking System based on Wireless Sensor networks. Zhejiang University (2007) (in Chinese)
2. Ferguson, B.G., Quinn, B.G.: Application of the short-time Fourier transform and the Wigner–Ville distribution to the acoustic localization of aircraft. J. Acoust. Soc. Am. 96(2), 821–827 (1994)
3. Reid, D.C., Zoubir, A.M., Boashash, B.: Aircraft flight parameter estimation based on passive acoustic techniques using the polynomial Wigner–Ville distribution. J. Acoust. Soc. Am. 102(1), 207–223 (1997)
4. Zhang, C.: The Design and Analysis of a Kind of Moving Target Acoustic Passive Localization Method. Peking University (2009) (in Chinese)
5. Xue, W.: Measuring system of wireless sensor network. Machinery Industry Press, Beijing (2007) (in Chinese)

A Wavelet Transform Based Structural Similarity Model for Semi-structured Texts

Jie Su and Junpeng Bao

Department of Computer Science & Technology, Xi'an Jiaotong University,
Xi'an 710049, P.R. China
baojp@mail.xjtu.edu.cn

Abstract. The semi-structured texts including Xml and Html texts are a basic information format in the Internet and World Wide Web. The text content values and the tree-organized structure are two aspects of a semi-structured text. Usually, the same text contents with different structures imply different objects. So the structural similarity of semi-structured texts is an essential key point to search, index, retrieve, query, or compare information in web pages. We presents a Wavelet Transform Based Structural Similarity Model (WTBSSM) in order to fast measure the structural similarity of semi-structured texts and compress the structural information into a short vector so as to develop an efficient semi-structured text index system. This paper introduces the Binary Encoding Method to convert a semi-structured text into a {-1, 1} sequence. Then the text structure signals are decomposed by means of Discrete Wavelet Transform to get the approximation coefficients, which is only a half length of the original signals. Finally, the structure similarity is measured by the Euclidean distance of approximation coefficients. The experimental results show that the WTBSSM can keep almost the same distance distribution to the direct distance of the original signals with a half or a quarter of information. The comparisons with a method of shorten DWT coefficients suggests that WTBSSM is better than it.

Keywords: Semi-structured Text, Structural Similarity, Wavelet Transform.

1 Introduction

Both Xml (eXtensible Markup Language) text and Html (HyperText Markup Language) text are semi-structured text that contains some tags organized in a tree. Obviously, semi-structured texts are the most popular and most population information container in the Word Wide Web(WWW) because there are so many web pages in the internet as well as a variety of knowledge, configuration, ontology documents written in Xml. Every day a mountain of Html or Xml texts appends in the WWW. Indeed, many of them are very similar even identical in their contents and structure. A serious question is that how similar or identical those semi-structured texts are. It is important not only to protect intellectual property, but also to save time and promote system efficiency for Web applications such as searching engine,

H. Tan (Ed.): Knowledge Discovery and Data Mining, AISC 135, pp. 159–167.
springerlink.com

information retrieval, information extraction, web classification, and web clustering. The similarity metric model of semi-structured texts is one of key technique for them.

A semi-structured text organizes its content in a tree structure, which is labeled with some tags, while a plain text has no structure tags at all. Two Xml texts in different structure can not be considered as the same one, they are not identical no matter their text contents are identical or not. Hence, the methods designed to measure similarity of plain texts are not good at semi-structured texts because that structural similarity is ignored. The similarity or identity of two semi-structured texts means that (1) the text contents are similar or identical between them; (2) their structures are also similar or identical to each other.

This paper concentrates on the problem of pure structural similarity of semi-structured texts. We present a wavelet transform based model to compress the structural information of a semi-structured text into a short vector, and measure the structural similarity based on the short vector.

The rest of the paper is organized as the follows. The section 2 introduces some related methods on the problem. We present a novel Wavelet Transform Based Structural Similarity Model in the section 3. The experimental results are illustrated in the section 4. At last, conclusions and future work are stated in the section 5.

2 Related Work

The methods based on Edit Distance are a straight way to measure structural similarity of semi-structured texts, such as [1], [2], and [3]. The Edit Distance is the cost of operations that delete, insert or update nodes in a tree in order to change it to another tree. Unfortunately, the time complexity of this method is $O(N^2)$[7], in which N is the amount of tags in a semi-structured text. It is not fit for detecting large semi-structured texts.

Another way is to find similar paths or sub-trees in two semi-structured texts, and then calculate the portion of similar paths or sub-trees out of the whole structure tree, such as [4-6]. While someone else proposed a vector model for the Xml documents' similarity measure. They represented the structure and content information in an Xml document as a vector respectively. At last, the Xml documents' similarity was defined as the cosine measure of two structure vectors, such as [8-11]. They are all time consuming methods to find out the similarity in two semi-structured texts.

The serial sequence methods convert the tree structure of a semi-structured text into a sequence of code, and then calculate the similarity of code sequences to assess the similarity of semi-structured texts, such as [12-14]. An attracting idea is that a code sequence is considered as a sequence of time serial signals so that signal analysis methods can be applied to detect similarity of semi-structured texts. Flesca et al.[7] introduced an Xml structural similarity detection method based on Fourier Transform. It is well known that the Fourier Transform, which observes signals in a fixed scale, can not get the varying details in both time and frequency domain. That means the Fourier Transform based method can not find some structure information in other views so that its capability is limited in a fixed range.

3 Measure of Structural Similarity

3.1 The Varying Similarity

A pair of semi-structured texts can be measured in different scales to fulfill different application. The wavelet transform has a brilliant feature to analyze time serial signals. So it is a straight idea to encode a semi-structured text into a digital sequence, which can be considered as time serial signals, and then a wavelet transform is performed to extract signals' feature vectors at various scales, at last the structural similarities are represented by distances between those feature vectors. Flesca et al.[7] applied a Discrete Fourier Transform(DFT) based method to calculate the structural similarity of Xml documents. The DFT method inspired our idea with the structure encoding and the Xml analysis on the frequency spectrum.

In this paper, we present a Wavelet Transform Based Structural Similarity Model (WTBSSM) to store the structural information of a semi-structured text in a short vector, and fast measure the structural similarity based on the short vector with few errors. The WTBSSM considers only structure information of a semi-structured document and ignores all text values so that it evaluates similarity of two trees indeed. An attractive feature of WT method is that the time complexity of Discrete Wavelet Transform (DWT) is less than that of Fast Fourier Transform (FFT). Moreover, many WT methods can compress signals in a short vector. As a result, we believe that WT based method is a promising technique to promote efficiency for a semi-structure text indexing or query system.

3.2 Wavelet Transform Based Structural Similarity Metric

The following is the pseudo code of the WTBSSM.

Procedure name: WTBSSM
Input: Semi-structured text D_1, D_2, and the decomposition level M.
Output: The structural similarities of D_1 and D_2 at level M.

a) Read semi-structured text D_1.
b) Read semi-structured text D_2.
c) Get D_1's structural encoding sequence L_1,
d) Get D_2's structural encoding sequence L_2,
e) Get the Signal Length n,
f) Validate the decomposition level M,
g) Execute the Discrete Wavelet Transform(DWT) with the Haar Wavelet on L_1 at level M to get the L_1's approximation coefficients H_{1a},

$$(H_{1a}, H_{1d}) = DWT(L_1, Harr, M) \tag{1}$$

h) Execute the Discrete Wavelet Transform(DWT) with the Haar Wavelet on L_2 at level M to get the L_2's approximation coefficients H_{2a},

$$(H_{2a}, H_{2d}) = DWT(L_2, Harr, M) \tag{2}$$

i) Calculate the Euclidean distance between H_{1a} and H_{2a}, which is the structural similarity value of the semi-structured text D_1 and D_2 at level M,

$$d_{H_a}(D_1, D_2, M) = \sqrt{\sum_{i=1}^{l}(H_{1a}(i) - H_{2a}(i))^2} \quad (3)$$

j) End

where $|L_1|$ denotes the length of the signal L_1. The length of DWT approximation coefficients (H_a) shrunk to half while the level goes up a step, i.e. the length of H_a at the level j is,

$$|H_a^j| = \frac{1}{2^j}n \quad (j = 1, 2, \ldots, k) \quad (4)$$

where n is the length of the original signal, j is DWT decomposition level, k is the max level.

The DWT approximation coefficients (H_a) at level 1 contain only half information of the original signals. The other half information is contained in the detail coefficients (H_d), though H_d is discarded. At the next level, H_a^j instead of the original signal is decomposed into H_a^{j+1} and H_d^{j+1}. As a result, the higher level, the shorter length of the approximation coefficients (H_a), and the more information is discarded. According to the wavelet transform theory, H_a keeps the global basic trend of the original signal, and holds most parts of signal energy. So H_a simulates the signal with only a half length vector. It implies that the whole structure information of a semi-structured text can be stored in a very short vector with few errors.

3.3 The Binary Encoding Method

A well-formed semi-structured text is organized in a tree with nested tag pairs. A tag pair consists of a start tag denoted as "<Tag Name>" and an end tag denoted as "</Tag Name>". The tag value is enclosed in the start tag and the end tag. The attribute-value pairs in the start tag can be considered as son tags. The tag name and tag value are full of semantic information. Since we aim to measure the pure structural similarity, the semantic information is ignored. On the first encoding step, all tag names and values are discarded, only the start tag symbol "<>" and the end tag symbol "</>" are considered. Then all start tags are encoded to 1, all end tags are encoded to -1. As a result, a well-formed semi-structured text is converted to a sequence of binary code on {1, -1}, i.e. a segment of digital signal, so it is called the Binary Encoding Method (BEM).

4 Experimental Results

We downloaded the Sigmod Record Data[15] in Xml edition, and selected 40 Xml documents of the ordinary issue page to examine our method. All Xml documents are encoded by BEM into {1, -1} serials, which are original signals. The average length of those signals is 457. A pair of Xml documents produces 13 distances, an Xml document produces 13 distance curves against the rest Xml documents. In the paper, we randomly select an Xml document to illustrate the distance curve distribution.

The Fig. 1 shows the original signals' distance (labeled as "Signals") and the approximation coefficients (H_a) distances (labeled as "Ha at leveli") of the selected Xml documents to the rest Xml documents. We can note that the distance distribution of the original signal distance and that of H_a at level 1 is almost the same, though the latter's vector length is only half of the former length. On the whole, the distributions of the H_a distance at level 1 to level 3 still adhere to that of the original signal distance. While the level is up to 4 to 6, the H_a distance distributions apparently fail to keep with the original signals' distribution because that too much information is discarded.

The Fig. 2 shows the original signals' distance and the Distance of the Shorten DWT Coefficients (DSDC), which is proposed by Chan[16] et al. and Liu[17] et al.(labeled as "$K=2^m$"). The K value to the DSDC plays the same role as the decomposition level to H_a distance. In order to compare the 2 methods with the same rule, we also choose 6 values to illustrate the trend. The structure signal length of the selected Xml documents is 502, the vector length of the H_a at level 1 is 251. The first K value $K=2^8=256$ is a little more than half length of the original signal. In this case, the vector length of the DSDC is a bit longer than that of the H_a. In the Fig. 2, the DSDC has similar distributions to the H_a distance.

In order to precisely evaluate the H_a distance and the DSDC, we define the distance error as follows to measure the standard deviation of the difference between the original signal distance and another distance.

$$DE = \sqrt{\frac{1}{P}\sum_{i=1}^{P}(\Delta d_i - \overline{\Delta d})^2} \tag{5}$$

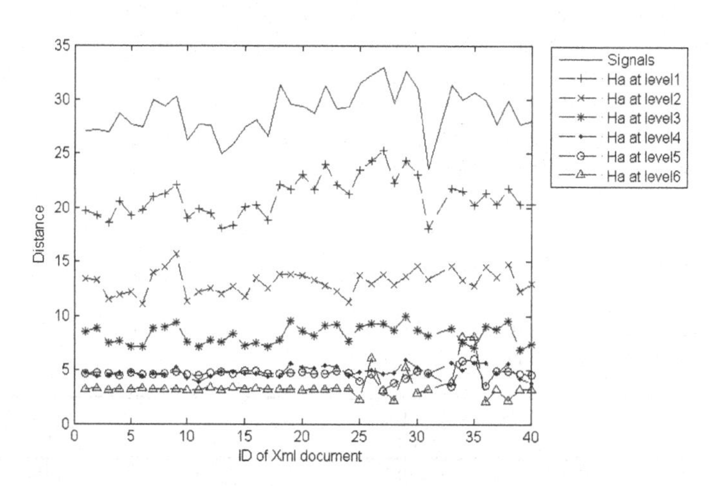

Fig. 1. The selected Xml document's (322.xml) distances of the original signals and the approximation coefficients (H_a)

Where, P is the number of Xml document pairs. $\Delta d_i = ds_i - dc_i$ denotes the difference between the original signal distance of the ith Xml document pair and the H_a distance of that pair. The $\overline{\Delta d}$ is the average of Δd_i. Chan et al.[16] proposed a

Vertical Shift Distance (VSD) to measure the difference between 2 signals. Both distance error and VSD are used to assess the DSDC and the H_a distance in the Table 1 and the Table 2.

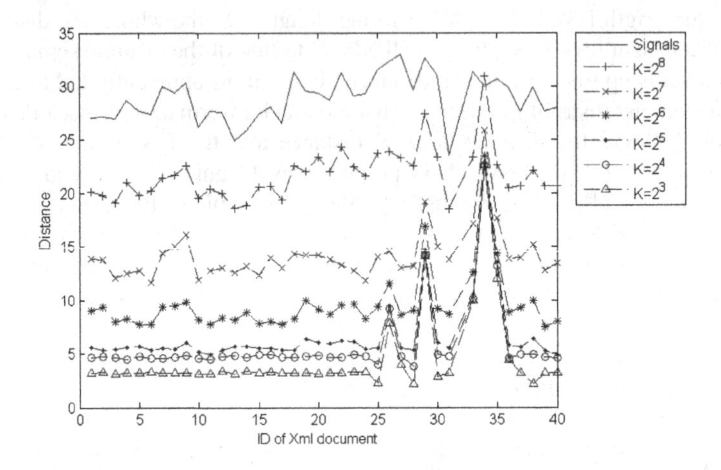

Fig. 2. The selected Xml document's (322.xml) distances of the original signals and the Shorten-DWT coefficients (DSDC)

Table 1 shows the DE and VSD of the H_a distance and DSDC on the selected Xml document at different levels (K values). The DE and VSD of the H_a distance are smaller than that of the DSDC at each DWT decomposition level.

Table 1 implies that the H_a distance outperforms the DSDC on one selected distance distribution. We repeat the experiment on all Xml documents in the data set so as to get the average capability of the H_a distance and the DSDC. In Table 2, we list the mean distance error and mean VSD of the 2 methods on the whole data set (among 40 Xml documents). It also shows both DE and VSD of the H_a distance are smaller than that of the DSDC at all levels. So we can say that the H_a distance is a little better than the DSDC, it is not an accident.

Table 1. The distance errors and vertical shift distances of the H_a distance and the DSDC on the selected Xml document (322.xml)

Level	K	Distance Error		Vertical Shift Distance	
		H_a Distance	DSDC	H_a Distance	DSDC
1	2^8	1.5025	1.9699	9.3834	12.3020
2	2^7	3.0939	3.4059	19.3213	21.2699
3	2^6	3.7125	4.1376	23.1847	25.8391
4	2^5	4.2477	4.7785	26.5269	29.8416
5	2^4	4.4710	5.0208	27.9211	31.3547
6	2^3	4.5526	5.3575	28.4311	33.4574

Moreover, the vector length of the H_a can be shorter than that of DSDC at the same distance error. In fact, the WTBSSM need not decompose the signal at the highest level to get the H_a. Whereas the DSDC method has to decompose the signal at the highest level first to get the full coefficients, and then cut the top K coefficients. Hence the decomposition steps in the WTBSSM are less than the DSDC. It implies that the WTBSSM is faster than the DSDC method, and more efficiency promotion can be expected. Another merit is that the level factor in the WTBSSM is easy to set and understand for common users because it is not variable to the text length, whereas the K value in the DSDC method has to increase along with text length in order to keep a fixed compress level.

Table 2. The mean distance errors and vertical shift distances of the H_a distance and the DSDC on the whole data set (among 40 Xml documents)

Level	K	Mean Distance Error		Mean Vertical Shift Distance	
		H_a Distance	DSDC	H_a Distance	DSDC
1	2^8	1.3799	2.1719	8.6173	13.5638
2	2^7	2.8183	3.3818	17.6001	21.1193
3	2^6	3.5322	4.0782	22.0587	25.4686
4	2^5	4.0545	4.6153	25.3203	28.8226
5	2^4	4.2881	4.9098	26.7794	30.6620
6	2^3	4.4112	5.1377	27.5482	32.0848

5 Conclusions and Future Work

We present a fast approach WTBSSM to measure the structural similarity of semi-structured texts, and compress the structural information into a short vector. The empirical results show that the WTBSSM can keep almost the same distance distribution to the direct distance of the original signals. The main contributions of the paper are:

(1) To apply the wavelet transform to the structural similarity metric of semi-structured texts. The WTBSSM can fast compare the structural similarity of two semi-structured texts, which has a great effect on the engines of the information searching, retrieval, extraction, classification, and clustering.

(2) The WTBSSM outperforms the shorten DWT coefficients method. Both of them can produce almost the same structure distance. But the former needs less DWT coefficients and decomposition steps than the latter. That does a great favor for saving time and promoting efficiency. Moreover, the WTBSSM is easy to set because the model parameter, i.e. decomposition level, is not variable to the text length.

Though the idea of WTBSSM is plain, the ongoing research focuses on 2 points to develop it. (1) The BEM is simple but may not be the best encoding method. Since Flesca et al.[7] mentioned some other encoding methods, we are going to find the optimal one. (2) The WTBSSM is a pure structural similarity metric model, we will develop another method to harmonize the WTBSSM with a pure text similarity metric in order to evaluate the whole integrated similarity of semi-structured texts in variable levels.

Acknowledgements. This research is supported by National Natural Science Foundation of China (Grant 60903123) and the Fundamental Research Funds for the Central Universities.

References

1. Zheng, S.H., Zhou, A.Y., Zhang, L.: Similarity Measure and Structural Index of XML Documents. Chinese Journal of Computers 26(9), 1116–1122 (2003)
2. Tekli, J., Chbeir, R., Yetongnon, K.: A Fine-Grained XML Structural Comparison Approach. In: Parent, C., Schewe, K.-D., Storey, V.C., Thalheim, B. (eds.) ER 2007. LNCS, vol. 4801, pp. 582–598. Springer, Heidelberg (2007)
3. Xie, T., Sha, C., Wang, X., Zhou, A.: Approximate Top-k Structural Similarity Search over XML Documents. In: Zhou, X., Li, J., Shen, H.T., Kitsuregawa, M., Zhang, Y. (eds.) APWeb 2006. LNCS, vol. 3841, pp. 319–330. Springer, Heidelberg (2006)
4. Moon, H.J., Yoo, J.W., Choi, J.: An Effective Detection Method for Clustering Similar XML DTDs Using Tag Sequences. In: Gervasi, O., Gavrilova, M.L. (eds.) ICCSA 2007, Part II. LNCS, vol. 4706, pp. 849–860. Springer, Heidelberg (2007)
5. Viyanon, W., Madria, S.K.: XML-SIM-CHANGE: Structure and Content Semantic Similarity Detection among XML Document Versions. In: Meersman, R., Dillon, T., Herrero, P. (eds.) OTM 2010. LNCS, vol. 6427, pp. 1061–1078. Springer, Heidelberg (2010)
6. Leung, H.P., Chung, F.L., Chan, S.C.: On the use of hierarchical information in sequential mining-based XML document similarity computation. Knowledge and Information Systems 7, 476–498 (2005)
7. Flesca, S., Manco, G., Masciari, E., Pontieri, L.: Fast Detection of XML Structural Similarity. IEEE Transactions on Knowledge and Data Engineering 17(2), 160–175 (2005)
8. Yang, J.W., Chen, X.O.: Similarity measures for XML documents based on kernel matrix learning. Journal of Software 17(5), 991–1000 (2006)
9. Jeong, B., Lee, D., Cho, H., Kulvatunyou, B.: A kernel method for measuring structural similarity between XML documents. In: Proceedings of the 20th International Conference on Industrial Engineering and other Applications of Applied Intelligent Systems, pp. 572–581 (2007)
10. Zhang, L.J., Li, Z.H., Chen, Q., Li, N.: Structure and Content Similarity for Clustering XML Documents. In: Shen, H.T., Pei, J., Özsu, M.T., Zou, L., Lu, J., Ling, T.-W., Yu, G., Zhuang, Y., Shao, J. (eds.) WAIM 2010. LNCS, vol. 6185, pp. 116–124. Springer, Heidelberg (2010)
11. Antonellis, P., Makris, C., Tsirakis, N.: XEdge: Clustering Homogeneous and Heterogeneous XML Documents Using Edge Summaries. In: Proceedings of the 2008 ACM Symposium on Applied Computing, pp. 1081–1088 (2008)
12. Kim, W.: XML document similarity measure in terms of the structure and contents. In: Proceedings of the 2nd WSEAS International Conference on Computer Engineering and Applications, pp. 205–212 (2008)

13. Wen, L., Amagasa, T., Kitagawa, H.: An Approach for XML Similarity Join Using Tree Serialization. In: Haritsa, J.R., Kotagiri, R., Pudi, V. (eds.) DASFAA 2008. LNCS, vol. 4947, pp. 562–570. Springer, Heidelberg (2008)
14. Bertino, E., Guerrini, G., Mesiti, M.: Measuring the structural similarity among XML documents and DTDs. Journal of Intelligent Information Systems 30(1), 55–92 (2008)
15. Sigmod Record, http://www.sigmod.org/publications/sigmod-record/Xml-edition
16. Chan, F.K.P., Fu, A.W., Yu, C.: Haar Wavelets for Efficient Similarity Search of Time-Series: With and Without Time Warping. IEEE Transactions on Knowledge and Data Engineering 15(3), 686–705 (2003)
17. Liu, B., Wang, Z., Li, J.-T., Wang, W., Shi, B.-L.: Tight Bounds on the Estimation Distance Using Wavelet. In: Yu, J.X., Kitsuregawa, M., Leong, H.-V. (eds.) WAIM 2006. LNCS, vol. 4016, pp. 460–471. Springer, Heidelberg (2006)

Performance of Adaptive Modulation with Space-Time Block Coding and Imperfect Channel State Information over Rayleigh Fading Channel

Xiangbin Yu[1,2], Yuyu Xin[1], Xiaoshuai Liu[1], and Xiaomin Chen[1]

[1] College of Electronic and Information Engineering, Nanjing University of Aeronautics and Astronautics, Nanjing, China
[2] National Mobile Communications Research Laboratory Southeast University, Nanjing, China

Abstract. The performance analysis of adaptive MQAM/MPSK with space-time block coding (STBC) and imperfect channel state information (CSI) over Rayleigh fading channels is presented. The fading gain value is partitioned into a number of regions by which the modulation is adapted according to the region the fading gain falls in. Under a target bit error rate (BER) and fixed power constraint, the fading gain switching thresholds and the probability density function of fading gain under imperfect CSI are respectively derived. Based on the above results, the closed-form expression for the SE is obtained. A tightly closed-form approximation of average BER is derived by means of the approximation of complementary error function. Computer simulations verify the effectiveness of the derived theoretical SE and BER expressions.

Keywords: Adaptive modulation, Spectrum efficiency, Bit error rate, Space-time coding, MQAM, MPSK.

1 Introduction

With the fast development of modern communication techniques, the demand for high data rate service is grown increasingly in the limited radio spectrum. For this reason, the future wireless communication system will require spectrally efficient techniques to increase the system capacity. To satisfy this requirement, adaptive modulation (AM), as a powerful technique for improving the spectrum efficiency, has obtained fast development recently [1]. It can take advantages of the time-varying nature of wireless channels to transmit higher speeds under favorable channel conditions and to reduce throughout as the channel degrades by varying the transmit power, symbol rate, code rate, and their combination. Thus it can provide much higher spectrum efficiency without sacrificing bit error rate (BER) [2-5]. Moreover, multiple antennas technique is well known to offer improvements in spectrum efficiency along with diversity and coding benefits over fading channels. Especially, space-time coding in multi-antenna system provides effective transmit diversity for combating fading effects [6-9]. Hence, the effective combination of adaptive modulation and space-time coding techniques has received much attention in the literature [7-9, 2-3]. Discrete-rate adaptive MQAM scheme in conjunction with space-time block coding (STBC)

H. Tan (Ed.): Knowledge Discovery and Data Mining, AISC 135, pp. 169–176.
springerlink.com

scheme with constant power over flat Rayleigh fading is studied in [7]. A general performance analysis of STBC with fixed and variable rate QAM is given in [8]. An adaptive coded modulation scheme with 2-antennas STBC is considered in [9] that make use of turbo code to improve the system coding gain. The performance of adaptive modulation with single antenna system is analyzed in [2]. The performance analysis of single-and multicarrier AM with constant power under perfect channel state information (CSI) is provided in [3]. Variable-power AM are investigated in Rayleigh fading channel in [4-5], where AM schemes are both designed for single antenna systems with perfect CSI.

The above AM schemes, however, basically do not consider the channel estimation error, and assuming that perfect channel estimation can be available at the receiver, thus the expected performance is obtained. In practice, the CSI is imperfect due to channel estimation errors. For this reason, in this paper, we will investigate the effect of imperfect CSI on spectrum efficiency (SE) and average BER with channel estimation errors modeled as complex Gaussian random variables. By theoretical analysis and mathematic derivation, the exact closed-form expression of SE is obtained. Besides, using the tightly approximate BER to replace the existing looser BER upper bound, the closed-form approximate expression of average BER is derived. This expression is shown to provide better agreement with simulation than existing approximate BER expression. Moreover, the derived SE expression is also in good agreement with the simulated value.

2 System Model

In this paper, a wireless communication system with N antennas at the transmitter and K antennas at the receiver are considered, and the system operates over a flat and quasi-static Rayleigh fading channel. Given that $\mathbf{H}=\{h_{kn}\}$ is $K \times N$ fading channel matrix, where h_{kn} denotes the complex channel gain from transmit antenna n to receive antenna k. Considering quasi-static channel, the corresponding channel gains are constant over a frame (P symbols) and vary from one frame to another. The channel gains are modeled as samples of independent complex Gaussian random variables with zero-mean and variance 0.5 per real dimension. A complex orthogonal STBC, which is represented by an $L \times N$ transmission matrix $\mathbf{D}$, is used to encode P input symbols into an N-dimensional vector sequence of L time slots. The matrix $\mathbf{D}$ is a linear combination of P symbols satisfying the complex orthogonality: $\mathbf{D}^H\mathbf{D}=\varepsilon(|d_1|^2+\ldots+|d_P|^2)\mathbf{I}_N$, where $\mathbf{I}_N$ is the $N \times N$ identity matrix, $\{d_p\}_{p=1,\ldots,P}$ are the P input symbols, and ε is a constant which depends on the STBC transmission matrix [6]. Hence, the transmission rate of the STBC is $R=P/L$.

Utilizing the complex orthogonality of STBC, the instantaneous SNR per symbol after space-time decoding is expressed as [6]

$$\rho = E_s \,\|\mathbf{H}\|_F^2 \,/(RN\sigma_z^2) = \alpha E_s \,/\,(RN\sigma_z^2) \tag{1}$$

where E_s is the total transmitted energy on the N transmit antennas per symbol time. σ_z^2 is the noise variance, and $\alpha = \|\mathbf{H}\|_F^2 = \sum_{n=1}^{N}\sum_{k=1}^{K} |h_{kn}|^2$. For Rayleigh fading channel, α is a central χ^2 distributed random variable with $2NK$ degrees of freedom.

According to Eq.(2-1-110) in [10] and using the transformation of random variable, the pdf of ρ can be obtained as follows

$$f(\rho)=(NR\rho/\bar{\rho})^{NK}\exp(-NR\rho/\bar{\rho})/[\rho\cdot\Gamma(NK)] \tag{2}$$

where $\bar{\rho}=E_s/\sigma_z^2$ is the average SNR per receive antenna.

When the channel estimation errors are modeled as complex Gaussian random variables [11], the pdf of ρ with estimation errors can be expressed as

$$f(\rho)=e^{-NR\rho/\bar{\rho}}\sum_{a=1}^{NK}A_{a,NK}(\mu)(NR\rho/\bar{\rho})^a/[\Gamma(a)\rho]\ ,\ \rho\geq 0 \tag{3}$$

where $\mu=|c|^2$ is the squared correlation, and $c=E\{\hat{h}_{kn}h_{kn}^*\}$ is the correlation between the actual channel gain h_{kn} and its estimate $\hat{h}_{kn}$, $A_{a,NK}(\mu)=\binom{NK-1}{a-1}(1-\mu)^{(NK-a)}\mu^{(a-1)}$ is a Bernstein polynomial with the following properties: $\sum_{a=0}^{NK}A_{a+1,NK+1}(\mu)=1$ and $A_{a,NK}(1)=\delta(a-NK)$. For the case that channel estimation has error, we refer the obtained CSI as imperfect CSI. When the channel estimation is perfect, i.e. $\hat{h}_{kn}=h_{kn}$, the squared correlation $\mu=1$, using $A_{a,NK}(1)=\delta(a-NK)$, the pdf in (3) is equal to (2). This result shows that the pdf of ρ for perfect CSI can be viewed as a special case of (3) under imperfect CSI.

3 Adaptive Modulation with STBC and Imperfect CSI

In this paper, a multi-antenna system with adaptive modulation and space-time block coding is referred as AM-STBC. The MPSK or MQAM is considered for AM. For discrete-rate MPSK or MQAM, the constellation size M_l is defined as $\{M_0=0, M_1=2,$ and $M_l=2^{2l-2}, l=2,\ldots,q-1\}$, where M_0 means no data transmission. The instantaneous SNR range is divided into q fading regions with switching thresholds $\{\rho_0, \rho_1,\ldots, \rho_{q-1}, \rho_q; \rho_0=0, \rho_q=+\infty\}$. The constellation size M_l is used for modulation when ρ falls in the l-th region $[\rho_l, \rho_{l+1})$. Consequently, the data rate is $m_l=\log_2 M_l$ bits/symbol with $m_0=0$. According to [2,5,10], the BER of MPSK or square MQAM of size M_l with Gray code and received SNR ρ over additive white Gaussian noise (AWGN) channel is approximately given by

$$BER_\rho\approx 0.2\exp\{-g_l\rho\}\quad \text{for } \rho\in[\rho_l,\rho_{l+1}) \tag{4}$$

The exact BER is tightly bounded from above by this approximated BER for $M_l\geq 4$, where $g_l=\sin^2(\pi/M_l)$ for MPSK and $g_l=1.6/(M_l-1)$ for square MQAM. Suppose we set a target BER equal to BER_0, then the fading gain switching thresholds can be set to the required SNR to achieve the target BER_0 over an AWGN channel as follows:

$$\rho_1=[erfc^{-1}(2BER_0)]^2, \tag{5}$$

$$\text{and } \rho_l=-(1/g_l)\ln(5BER_0),\ \ l=2,\ldots q-1. \tag{6}$$

where $erfc^{-1}(.)$ denotes the inverse complementary error function. When the switching thresholds are chosen according to (5) and (6), the system will operate with a BER below target BER_0, as will be confirmed in the following numerical results in section 4. We now derive closed-form expressions for the spectrum efficiency and average BER of the system, which will be employed to evaluate the performance of AM-STBC under imperfect CSI.

Based on the switching thresholds described in (5)-(6), using (3), we can calculate the probability of the effective SNR ρ falls in the lth region $[\rho_l, \rho_{l+1})$, denoted as b_l.

$$b_l = \int_{\rho_l}^{\rho_{l+1}} f(\rho)d\rho = \sum_{a=1}^{NK} A_{a,NK}(\mu)\{\Gamma(a, NR\rho_l / \bar{\rho}) - \Gamma(a, NR\rho_{l+1} / \bar{\rho})\} / \Gamma(a) \quad (7)$$

where $\Gamma(\cdot,\cdot)$ is incomplete Gamma function [12].

For discrete-rate adaptive scheme, the SE is defined as the ensemble average of effective transmission rate. So according to (7), the SE of the adaptive MPSK or MQAM scheme with space-time block coding can be given by

$$\eta = \sum_{l=1}^{q-1} Rm_l b_l = \sum_{l=1}^{q-1} Rm_l \sum_{a=1}^{NK} A_{a,NK}(\mu)\{\Gamma(a, NR\rho_l / \bar{\rho}) - \Gamma(a, NR\rho_{l+1} / \bar{\rho})\} / \Gamma(a) \quad (8)$$

We define ensemble average BER for AM-STBC system as

$$\overline{BER} = \sum_{l=1}^{q-1} m_l \int_{\rho_l}^{\rho_{l+1}} BER_l f(\rho) d\rho \Big/ \left[\sum_{l=1}^{q-1} m_l b_l\right] \quad (9)$$

where BER_l is the BER of MPSK or MQAM with constellation size M_l and Gray coding. For MQAM, its exact expression is given by [13]

$$BER_l = \sum_j \zeta_{lj} \mathrm{erf}\{\sqrt{\kappa_{lj}\rho}\} \quad (10)$$

where ζ_{lj} and κ_{lj} are constants which depend on M_l, and can be found in [13]. For MPSK, high-accuracy approximate expression of BER_l is given by [3]

$$\begin{aligned} BER_l &\cong (1/m_l)[erfc\{\sqrt{\rho \sin^2(\pi / M_l)}\} + erfc\{\sqrt{\rho \sin^2(3\pi / M_l)}\} \\ &= \sum_j \zeta_{lj} \mathrm{erf}\{\sqrt{\kappa_{lj}\rho}\} \end{aligned} \quad (11)$$

where the set of constants $\{(\zeta_{lj}, \kappa_{lj})\}$ is given by $(1/m_l, \sin^2(\pi/M_l))$, $(1/m_l, \sin^2(3\pi/M_l))$.

According to [14], the complementary error function $erfc(x)$ can be approximately expressed as $erfc(x) \cong \sum_{i=1}^{2} u_i \exp(-v_i x^2)$, where $\{u_i\}=\{1/6, 1/2\}$, $\{v_i\}=\{1, 4/3\}$. By using this approximation, (10) and (11) can be rewritten as:

$$BER_l \cong \sum_{i=1}^{2} u_i \sum_j \zeta_{lj} \exp\{-v_i \kappa_{lj} \rho\} \quad (12)$$

Substituting (12) and (3) into (9) yields

$$\overline{BER} \cong \frac{\sum_{l=1}^{q-1} m_l \sum_{i=1}^{2} u_i \sum_j \zeta_{lj} \sum_{a=1}^{NK} A_{a,NK}(\mu)[1 + v_i \kappa_{lj} \bar{\rho}']^{-a} \{\Gamma(a, s\rho_l) - \Gamma(a, s\rho_{l+1})\} / \Gamma(a)}{\left[\sum_{l=1}^{q-1} m_l b_l\right]} \quad (13)$$

where $s = \kappa_{lj} v_i + 1/\bar{\rho}', \bar{\rho}' = \bar{\rho} / (NR)$.

Substituting (7) into (13), the average BER can be obtained as:

$$\overline{BER} \cong \frac{\sum_{l=1}^{q-1} m_l \sum_{i=1}^{2} u_i \sum_{j} \zeta_{lj} \sum_{a=1}^{NK} A_{a,NK}(\mu)[1+v_i\kappa_{lj}\bar{\rho}']^{-a}\{\Gamma(a,s\rho_l)-\Gamma(a,s\rho_{l+1})\}/\Gamma(a)}{\sum_{l=1}^{q-1} m_l \sum_{a=1}^{NK} A_{a,NK}(\mu)\{\Gamma(a,\rho_l/\bar{\rho}')-\Gamma(a,\rho_{l+1}/\bar{\rho}')\}/\Gamma(a)} \quad (14)$$

Eq.(14) is a tightly closed-form approximate expression for the average BER of the AM-STBC system under imperfect CSI. When the CSI is perfectly known with $\mu=|c|^2=1$, (10) and (14) will be changed to the following (15) and (16), respectively. Specifically,

$$\eta = \sum_{l=1}^{q-1} Rm_l[\Gamma(NK, Nr\rho_l/\bar{\rho})-\Gamma(NK, Nr\rho_{l+1}/\bar{\rho})]/\Gamma(NK) \quad (15)$$

$$\overline{BER} \cong \frac{\sum_{l=1}^{q-1} m_l \sum_{i=1}^{2} u_i \sum_{j} \zeta_{lj}[1+v_i\kappa_{lj}\bar{\rho}']^{-NK}\{\Gamma(NK,s\rho_l)-\Gamma(NK,s\rho_{l+1})\}}{\sum_{l=1}^{q-1} m_l\{\Gamma(NK,\rho_l/\bar{\rho}')-\Gamma(NK,\rho_{l+1}/\bar{\rho}')\}} \quad (16)$$

The Eq.(16) is a closed-form expression of the ensemble average BER for the AM-STBC system under perfect CSI. This theoretical formula will be shown to provide better agreement with simulation results than the existing closed-form expressions in [2] and [7].

4 Simulation Results

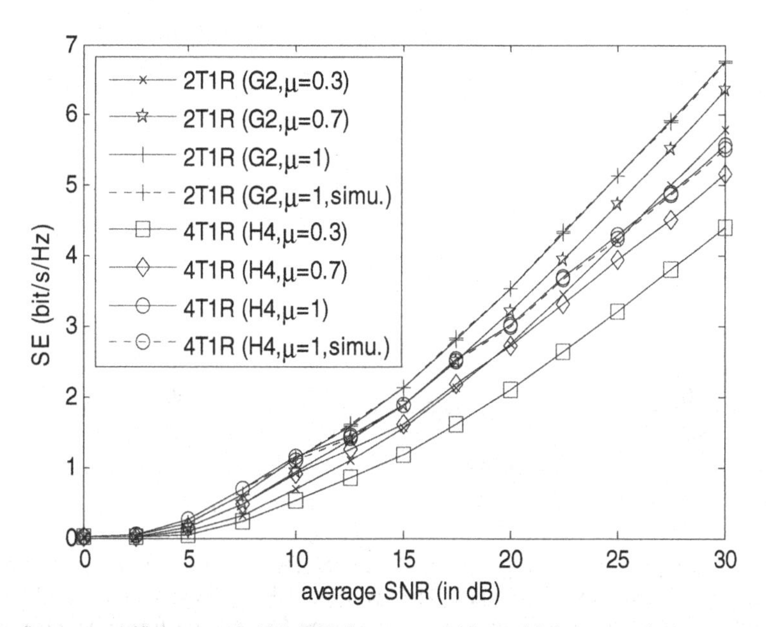

Fig. 1. Spectrum efficiency of AM-STBC system with multiple transmit antennas and one receive antenna for imperfect CSI

In this section, we use the derived theoretical performance formulae and computer simulation to evaluate the SE and average BER of the multi-antenna system with adaptive modulation and STBC over Rayleigh fading channel. In simulation, the channel is assumed to be quasi-static flat fading, square MQAM is used for modulation, and Gray code is employed to map the data bits to MQAM constellations. The set of MQAM constellations is $\{M_l\}_{l=0,1,\dots,5}=\{0, 2, 4, 16, 64, 256\}$. The target *BER* equals BER_0=0.001. Different space-time block codes, such as G_2, G_3, G_4, and H_4, are adopted for evaluation and comparison. In the following figures, *x*T*y*R denotes a multiple antenna system with *x* transmit antennas and *y* receive antennas.

In Fig.1, we plot the theoretical SE of the AM-STBC system for different transmit antennas and one receive antenna. Under imperfect CSI, the squared correlation μ=0.3, 0.7, 1 is considered for evaluation. The theoretical SE for perfect and imperfect CSI are respectively calculated by using (15) and (10) with the switching thresholds defined by (5) and (6). As shown in Fig.1, adaptive modulation are capable to increase SE with SNR. The theoretical SE is in good agreement with the corresponding simulated value. The 2T1R system of using G_2 provides larger SE than 4T1R system of using 3/4-rate H_4 because the G_2 is a full rate code. From the figures, it is observed that the larger is the squared correlation, the bigger is the SE.

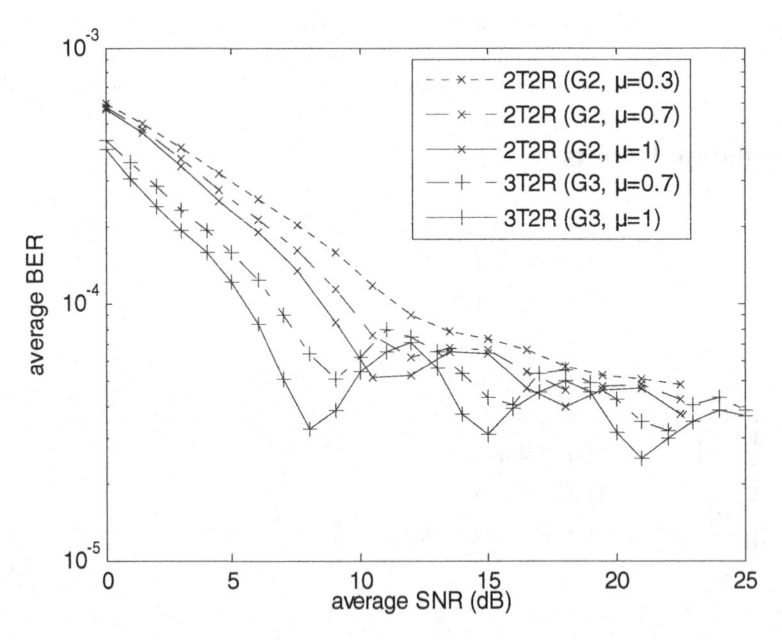

Fig. 2. Average BER of AM-STBC system with multiple transmit antennas and two receive antennas for imperfect CSI

In Fig.2, we plot the average BER of AM-STBC system for different transmit antennas and two receive antennas. The average BER is evaluated by (14). It is observed that the adaptive modulation scheme using G_3 code performs better than that of using G_2 code because the former has greater diversity than the latter. Most importantly, the average BER of the AM-STBC system is always less than the target BER_0 while the SE effectively increases with SNR. This result verifies the effectiveness of the AM-STBC system on SE. Besides, the average BER curves of the system have ripples phenomenon due to the use of the individual discrete constellations. Furthermore, those fluctuations are more obvious at medium SNR range because the distance between the SNR switching thresholds are larger in that SNR range where the fluctuations occur.

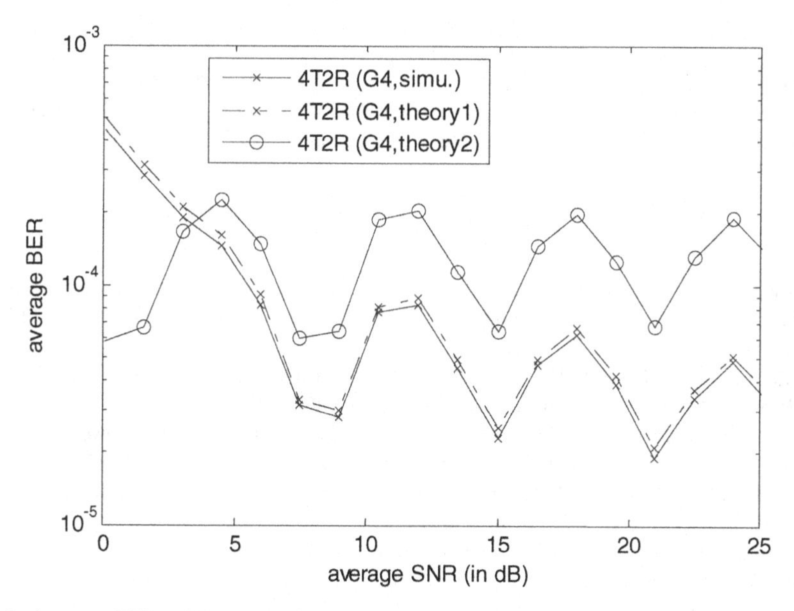

Fig. 3. Average BER of AM-STBC system with multiple transmit antennas and two receive antennas for imperfect CSI

In Fig.3, we plot the theoretical average BER and simulation of AM-STBC system with four transmit antennas and two receive antennas for perfect CSI, where G_4 code is used. Two theoretical average BERs are considered in the comparison. The average BER described by (16) is referred to as 'theory 1' and the one described by [2, Eqs.(35)-(37)] is referred to as 'theory 2'. From Fig.3, it is observed that both the theoretical evaluation and simulation results are having the BER values below the target BER. It means that the AM with perfect CSI is also successful in increasing SE while the target BER is maintained. It is found that the theoretical analysis of (16) can match the simulation results well, whereas the theoretical evaluation of using [2] is very different from the simulation results. These results show that our derived BER expression is valid and has more accuracy.

5 Conclusions

We have investigated the performance of adaptive modulation with space-time block coding and imperfect CSI over Rayleigh fading channel. Subject to the target BER and fixed power constraint, we achieve the switching thresholds of the fading gain under imperfect CSI case. Based on the switching thresholds, the average BER of the systems can be maintained less than the target BER. Besides, by using the obtained probability density function of effective SNR and switching thresholds, the closed-form expressions of the SE and average BER for AM-STBC system under imperfect CSI are derived, respectively. Numerical results show that the theoretical SE agrees with the corresponding simulation result, and the derived theoretical BER is more accurate than the existing BER expression.

Acknowledgments. This work is supported by Nanjing University of Aeronautics and Astronautics Research Funding (NS2010113, NP2011036) and Open Research Fund of National Mobile Communications Research Laboratory (N200904) of SEU.

References

1. Chae, C.-B., Forenza, A., Heath, R.W., et al.: Adaptive MIMO transmission techniques for broadband wireless communication systems. IEEE Commun. Magazine 48, 112–118 (2010)
2. Alouini, M.-S., Goldsmith, A.J.: Adaptive Modulation over Nakagami Fading Channels. Wireless Personal Communications 13, 120–143 (2000)
3. Choi, B., Hanzo, L.: Optimum mode-switching-assisted constant-power single-and multicarrier adaptive modulation. IEEE Trans. Veh. Technol. 52, 536–559 (2003)
4. Goldsmith, A.J., Chua, S.G.: Variable-rate variable-power MQAM for fading channel. IEEE Trans. Commun. 45, 1218–1230 (1997)
5. Chung, S.T., Goldsmith, A.J.: Degrees of freedom in adaptive modulation: a unified view. IEEE Trans. Commun. 49, 1561–1571 (2001)
6. Shin, H., Lee, J.H.: Performance analysis of space-time block codes over keyhole Nakagamim fading channels. IEEE Trans. Veh. Technol. 53, 351–362 (2004)
7. Maaref, A., Aissa, S.: Rate-adaptive M-QAM in MIMO diversity systems using space-time block codes. In: IEEE PIMRC 2004, pp. 2294–2298. IEEE Press, New York (2004)
8. Carraro, H.M., Fonollosa, A.J., Delgado-Penin, A.J.: Performance analysis of space-time block coding with adaptive modulation. In: IEEE PIMRC 2004, pp. 493–497. IEEE Press, New York (2004)
9. Dong, J., Zhou, Y., Li, D.: Combined adaptive modulation & coding with space-time block code for high data transmission. In: IEEE ICCT 2003, pp. 1476–1479. IEEE Press, New York (2003)
10. Proakis, J.G.: Digital communications, 5th edn. McGraw-Hill, New York (2007)
11. Tomiuk, B.R., Beaulieu, N.C., Abu-Dayya, A.A.: General forms for maximum ratio diversity with weighting error. IEEE Trans. on Commun. 47, 488–492 (1999)
12. Gradshteyn, I.S., Ryzhik, I.M.: Table of Integrals, Series, and Products, 7th edn. Academic, San Diego (2007)
13. Simon, M.K., Hinedi, S.M., Lindsey, W.C.: Digital Communication Techniques Signal Design and Detection. Prentice-Hall, Englewood Cliffs (1995)
14. Chiani, M., Dardari, D., Simon, M.K.: New exponential bounds and approximations for the computation of error probability in fading channels. IEEE Trans. Wireless Commun. 2, 840–845 (2003)

A Spectrum Allocation System Model of Internet of Things Based on Cognitive Radio

Liguo Qu, Yourui Huang, Ming Shi, and Chaoli Tang

School of Electronic and Information Engineering
Anhui University of Science and Technology
Huainan 232001, Anhui province, China
qlg77@163.com

Abstract. Aiming at the growing demand for spectrum space issues of Internet of Things, a spectrum allocation system model is structured based on the Homo Egualis(HE) social model. The model solves the problem of spectrum requirements for Internet of things, reutilizes idle spectrum and improves spectrum utilization by cognitive radio technologies, In model, the access point calculations and updates node access probability with the Homo Egualis community access schemes periodically. Because node control access time and frequency based on the access automatically, we can use idle spectrum fairly and efficiently. The results of the simulation show that the mode improves the spectrum utilization greatly.

Keywords: Internet of Things, cognitive radio, spectrum allocation, HE model.

1 Introduction

With presenting the strategy of Internet of Things, it has been paid more attention to development and application of Internet of Things. It not only applies to information sensing equipment on identification layer, such as RFID device, infrared sensors, global positioning system, laser scanner; but also uses to M2M operation with little capacity, including street lamp management, water quality monitoring; and high bandwidth business mainly with video images, including peace city and public transit[1].

In addition, owing to the large business scale of Internet of Things, nothing can give expression to the huge demand of spectrum, including 2G and 3G. So, it occupies a very important position in development and application of technology of Internet of Things for radio spectrum resources.

Aiming at the growing demand for spectrum space issues of Internet of Things, it structures a spectrum allocation system model based on the HE social model in paper. The model solves the Internet of Things spectrum needs by "cognitive radio" technologies, and reutilizes free spectrum utilization and improves spectrum utilization.

H. Tan (Ed.): Knowledge Discovery and Data Mining, AISC 135, pp. 177–184.
springerlink.com

2 System Modeling

In general, spectrum allocation scheme of Internet of Things, what we know of, is static. It has low spectrum utilization, and spectrum resources are not fully taken advantage of. In view of this situation, we put cognitive radio to use. It is a smart radio communications technology, perceives surrounding, and then obtains information by "understanding-building", adapts to the changing surroundings by changing the transmission parameters in real time, such as transmitted power, carrier frequency, modulation technique. That is, the technology of cognitive radio, can detect the idle channel at special time and space. In order to avoid the interference for other user, utilize this spectrum to transmit. It improves spectrum utilization, and realizes negotiation with other spectrum users and shares spectrum resources effectively. Meanwhile, cognitive radio technology can make each different communication system, with different spectrum and transfer mode, to join hands with each other. In brief, by detecting, cognizing, reasoning and learning for spectrum, cognitive radio technology allows secondary user, who has no permission, to use the allocated frequency range, without interference for primary user.

The architecture, of which is a spectrum allocation system model of Internet of Things based on the HE social model, uses the CRA II cognitive ring [2,3] in "cognitive radio", as showing in Fig.1. The system model is mainly made up by the follow five parts.

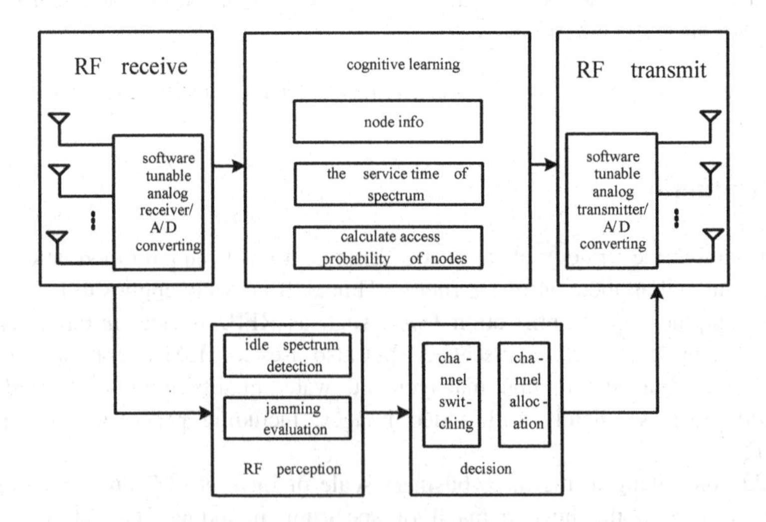

Fig. 1. Spectrum allocation system model based on the HE social model

① RF Receive: It mainly contains antenna, software tunable receiver and A/D converting. It is used to receive the access information of node and RF perception.

② Cognitive Learning: It is mainly used to store node information, records communication time of node, and calculates the access probability of each node. Then, plan the behavioral strategy of node according to the above information.

③ RF Perception: Detect the idle spectrum and jamming evaluation.

④ Decision: Allot spectrum for the access nodes effectively, mainly including channel switching and allocation. It makes sure that the access node has available channel in real time.

⑤ RF Transmit: It mainly contains antenna, software tunable transmitter and A/D converting. It is used to broadcast the action strategy and normal communication of each node.

In model, spectrum allocation makes node collaborate with AP. Which, AP is controlling user, and node is controlled user. To AP, calculate the behavioral strategy (access probability) of each node in the next cycle by a game, according to collecting node information, including channel, flow load and priority. And then, the AP transmits the access probability to each node periodically by broadcast. Each node adjusts its behavioral strategy, aiming to use idle spectrum fairly. On the other hand, when node joins in and according to the access strategy, AP makes certain channel adjusting, which is channel switching and finishes channel allocation.

As the controlled user, node adjusts its service time and access times, according to its behavioral strategy.

3 System Model Descriptions

The system model is made up of RF receiver, cognitive learning, RF perception, decision and RF transmit. Some key links are introduced in detail as following.

3.1 RF Perception

RF perception uses centralized perception [2]. That is, make a central unit to collect perception of message, coming from cognitive equipment, and then determine the usable spectrum. RF perception contains two parts: idle spectrum detection and jamming evaluation. And the math descriptions are spectrum idle matrix and spectrum interference matrix.

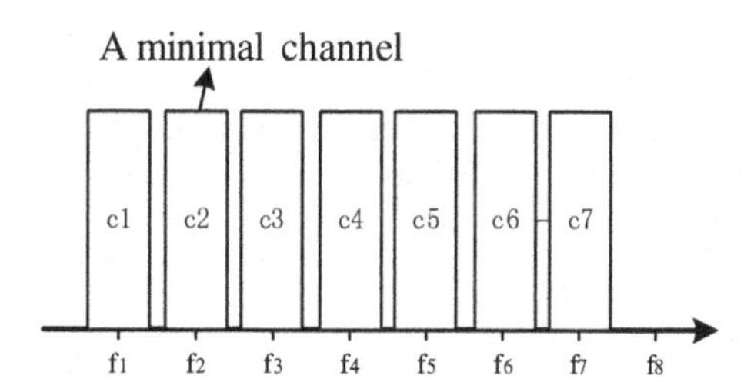

Fig. 2. Spectrum segmentation and number

(1) Spectrum idle Matrix

In conventional practice, radio communication uses frequency-division technology to subdivide spectrum. Take the carrier frequency f1 as the center and Δf as the interval. And after segmentation, each small frequency band is a channel. In a word, according to a certain standard, subdivide the spectrum to small channels, and then number them, showing in Fig.2.

The main function of RF perception is to detect the spectrum hole, and then subdivides the spectrum to different small channels by a certain standard, and establishes a spectrum idle matrix:

$$\mathbf{L} = \{l_{i,j} | l_{i,j} \in \{0,1\}\}_{I \times J} \quad (1)$$

Which, I is the user number, and J is the number of whole frequency bands. $l_{i,j} = 1$ expresses that, the frequency band j is available for user i. Conversely, $l_{i,j} = 0$ expresses that frequency band j is not available for user i.

(2) Interference Matrix [8,9,10]

Because of the different spatial location and transmitted power, to whether the same channel is idle, it may be different for different users. There may be interference when they use the same channel. And it is concerned with the overlap circumstance of protective range between cognitive users. The interferences are indicated by an interference matrix. The interference matrix C is defined as:

$$\mathbf{C} = \{c_{i,k,j} | c_{i,k,j} \in \{0,1\}\}_{I \times I \times J} \quad (2)$$

Which, $c_{i,k,j} = 1$ expresses that, user k can't use frequency band j, when user i uses. And $c_{i,k,j} = 0$ expresses user i and user k can use frequency band j at the same time.

3.2 Cognitive Learning

In the cognitive ring CRA II, cognitive learning is a function of the perception, observation, decision and action [2~4]. The initial learning is that the perception of message establishes a corresponding strategy by the function. And then compare the next strategy with the former one to produce a new strategy.

In the model, the learning is to collect the information and spectrum use time of the node, and establish the behavioral strategy by the utility function. The math descriptions are information matrix, spectrum use time matrix and formula of node access probability.

(1) Node's information

Node's information contains the number of small channels which is occupied by the access node and its available channel. The information matrix of node, U, is defined by the following:

$$\mathbf{U}' = \{k_i\}_{I \times 1} \quad (3)$$

$$\mathbf{U}'' = \{u_{i,j} | u_{i,j} \in \{(0,1)\}\}_{I \times J} \quad (4)$$

$$\mathbf{U} = \{\mathbf{U}' \quad \mathbf{U}''\} \quad (5)$$

Which, $u_{i,j}$ is the status of cognitive user j to use the channel i. when $u_{i,j} = 0$, cognitive user j don't have the channel i. Conversely when $u_{i,j} = 1$, it owns this channel. k_i is the number of channel when cognitive user i communicates.

(2) Spectrum use time of node

Spectrum use time of node is one of the important measurements in the cognitive radio technology. It is the average time of each radio system in air [2]. The time in air is corresponding with reference time, and the time in air, of which each radio system, is:

$$\text{time in air}_{\text{type=A,B}} = \frac{1}{N_{\text{type}}} \sum_{i=1}^{N_{type}} \frac{\text{allocated time of type i}}{\text{reference time}} \tag{6}$$

In order to ensure the effectiveness and fairness of spectrum interest, the model brings in weighting fairness, showing in formula (7):

$$\frac{\text{time in air}_{\text{type i}}}{L_i} = \frac{\text{time in air}_{\text{type j}}}{L_i} = K \qquad \forall i, j \tag{7}$$

Which, K is a constant, and the weight $L_i = \theta_i \lambda_i$. θ_i is a priority parameter and λ_i is the flow load of system of type i.

The time in air of system i is $Onlinetime_i$. That is the average time of access spectrum of node i. To adapt different flow load and priority, normalize $Onlinetime_i$.

$$x_i = \frac{Onlinetime_i}{L_i} \tag{8}$$

Which $L_i = \theta_i \lambda_i$.

In this model, take each node as a system. The access point will record the communication time Onlinetime of each node. The Li in formula (8) can be get in information matrix U.

(3) access probability of node — based on Homo Egualis access method [1]

The access method, Homo Egualis, is based on its Social model. And it is the utility function of the player i, when n players are gaming, showing as follows:

$$u_i = x_i - \frac{\alpha_i}{n-1} \sum_{x_j > x_i} (x_j - x_i) - \frac{\beta_i}{n-1} \sum_{x_j < x} (x_i - x_j) \tag{9}$$

Which, x=(xi,…xj) is payment of each player. And 0≤βi≤αi≤1. β≤α means that, when playing better than others, the desire to pursue imparity of Homo Egualis is weak, and when playing not better, the desire will be strong.

In this scheme, make sure the fairness of access spectrum by the antipathy characteristics of Homo Egualis to inequality. Each radio system control access behavioral strategy probability pi by access probability. The behavioral strategy probability pi is calculated and published by AP termly.

The update of probability pi is defined as following:

$$p_i = \max\left(0, \min\left(1, p_i + \frac{\alpha_i}{n-1} \sum_{x_j \ge x_i} \left(\frac{x_j - x_i}{x_j}\right) - \frac{\beta}{n-10} \sum_{x_j < x_i} \left(\frac{x_i - x_j}{x_i}\right)\right)\right) \tag{10}$$

α and β are two important control parameters in HE access scheme. When α>β, the performance of system is basic consistently, no matter what value is used. And when α<β, the performance of each system descend observably.

3.3 Decision

The decision stage is to distribute the idle channel optionally. According to its access probability pi, and when a node applies for channel to AP at appropriate time, AP will make the necessary channel switching and allocation on the basis of node information. Decision can be described as channel switching and allocation.

(1) Channel switching

Because of the different channel occupied by each node and in order to access multiband node more effectively, package the frequency band on the basis of action probability pi. The package show in Fig.3:

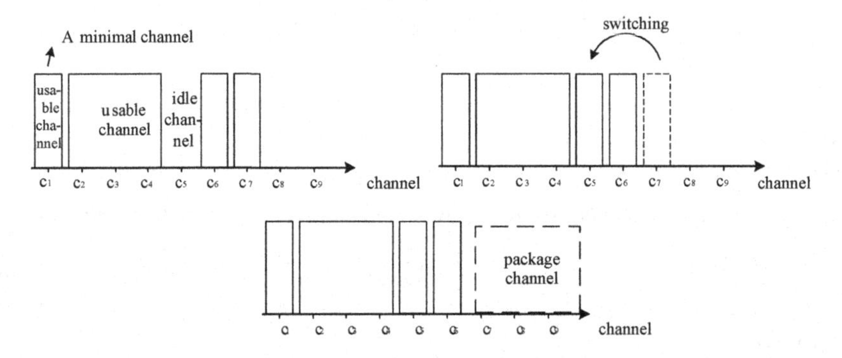

Fig. 3. Channel switching and package

The principles of channel switching and package are:

① Switch according to the value of pi. That means, pi is prior to own a channel.

② Query each user to ensure whether they have enough idle channels, of which the number of communication channels is larger than one. If not, establish package channel by switching as far as possible. But, make sure the existing communication not interrupting.

(2) Channel allocation

Channel allocation is based on action probability pi distributed greedy algorithm. Distribute only one channel (after package) to one node every time. Arrange behavioral strategy of each node by size. Distribute channel to node with big behavioral strategy first. And if there is no channel match, turn to next node until channel can't subdivide (including zero) and no node is left. It has been finished to distribute channel.

4 Simulation of System

On the basis of modeling, evaluate the fairness and effectiveness of spectrum allocation system using the average throughput and throughput variance, by simulation on MATLAB.

In the process of simulation, set:

(1) Distribute 5 APs and N nodes randomly in 30×30 area, including N1 nodes with one mid-frequency of carrier frequency and N2 nodes with three mid-frequencies of carrier frequency (N1 < N2).

(2) Set d < 10, that is, when the distance between AP and one node is less than ten, AP will disturb for this node.

(3) In probability formula pi, α=0.1 and β=0.01, and the initial value pi=1.

(4) Because of normalization, xi (i=1,2…N) is a constant.

The simulation results show in Fig.4(a) and Fig.4(b) for average throughput and throughput variance when N1/N2=3, and in Fig.5 for throughput variance when N1/N2=3 and N1/N2=4.

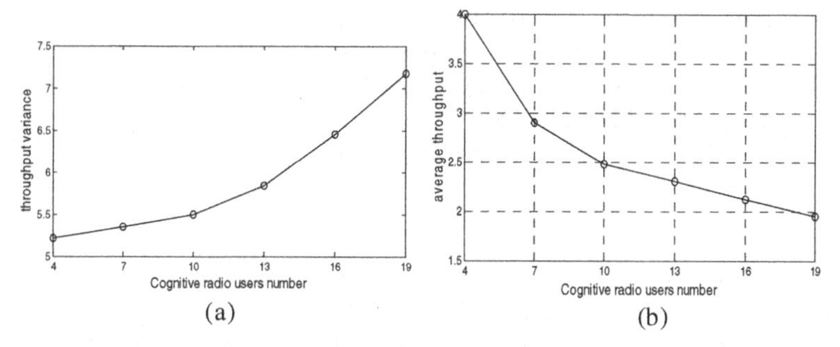

(a) (b)

Fig. 4. (a) Simulation result of average throughput, (b) Simulation result of average throughput

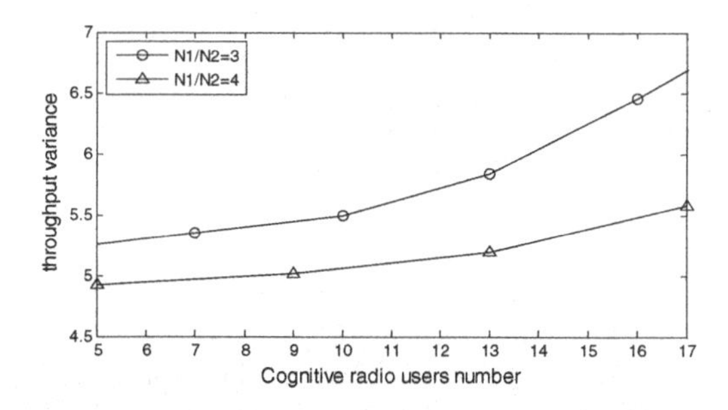

Fig. 5. Simulation result of throughput variance (N1/N2=3,4)

It turned out that, increasing single channel node and with a specified capacity of idle channel, the downswing of average throughput slow down. It also means the channel is used effectively. But, the throughput variance accelerated, which makes the channel use is unfair.

That is to say, when channel is limited and along with the more and more multichannel cognitive users, the success rate of channel package is going down. When N1/N2=4, throughput variance is improved significantly from Fig.5, and this proved well the above idea.

This model is great appropriate for the condition when N1<N2. And if there is no cross-channel between multichannel node and single channel node, it will not be falling fairly.

5 Conclusions

Utilizing "cognitive radio" technologies, it builds a spectrum allocation system model based on the HE social model for Internet of Things in paper. It reutilizes free spectrum utilization and improves spectrum utilization, by adjusting spectrum in real time. Simulation results indicate this model improves spectrum utilization, and meets the demand for spectrum space issues of Internet of Things.

Acknowledgments. This work is supported by Program for New Century Excellent Talents in University(No.NCET-10-0002),supported by Anhui Provincial Natural Science Foundation (No.1108085J03), supported by Natural Science Research Project of Anhui Province (No.KJ2010B061), supported by Huainan City Science and Technology Project (No.2011A07908).

References

1. He, Y.: The effective supply of spectrum resources is precondition of the internet of Things development. Communications World Weekly 3, 28 (2010)
2. Arslan, H.: Cognitive Radio, Software Defined Radio, and Adaptive Wireless Systems, p. 123. Springer, Heidelberg (2007)
3. Mitola III, J.: Cognitive Radio Architecture, p. 235. Wiley, NY (2006)
4. Zhou, X., Zhang, H.: Principle and Application of Cognitive Radio, p. 1. Bejing University of Posts and Telecommunications Press (2007)
5. Mitola, J.: Cognitive radio: Making sofware radios more personal. IEEE Personal Communications Magazine 6(4), 13–18 (1999)
6. Mitola, J.: Cognitive radio for flexible mobile multimedia communications. In: IEEE International Workshop Mobile Multimedia Communications 1999 (MoMuc 1999), November 15-17, pp. 2–10 (1999)
7. Mitola, J.: Cognitive radio: An integrated agent architecture of software defined radio. Doctor of Technology. Royal Inst. Technol. (KTH), Stockhokm, Sweden (2000)
8. Wang, W., Liu, X.: List Coloring based channel allocation for open spectrum wireless networks. In: 2005 IEEE 62nd Vehicular Technology Conference, VTC-2005-Fall (2005)
9. He, X.: Base of Graph Coloring Spectrum Allocation in Cognitive Radio. China New Telecommunications 4, 71 (2009)
10. Wang, W., Liu, X.: List-coloring based channel allocation for open-spectrum wireless networks. In: Vehicular Technology Conference, pp. 690–694 (2005)

Low Cost Dual-Basis Multiplier over $GF(2^m)$ Using Multiplexer Approach

Hung Wei Chang[1], Wen-Yew Liang[1], and Che Wun Chiou[2]

[1] Department of Computer Science and Information Engineering,
National Taipei University of Technology, 1, Sec. 3,
Chung-hsiao E. Rd.,Taipei City. 10608, Taiwan, R.O.C.
{t6599009,wyliang}@csie.ntut.edu.tw

[2] Department of Computer Science and Information Engineering, Ching Yun University
229, Jianxing Road, Zhongli City, Taoyuan County. 320, Taiwan, R.O.C.
cwchiou@cyu.edu.tw

Abstract. Information security is heavily dependent on public key cryptosystems such as RSA. However, RSA is not available for the resource-constrained devices like embedded systems. Therefore, the new elliptic curve cryptosystem with very low cost as compared to RSA is now available and suggested for information security. Galois/Finite field multiplication is the most important operation in elliptic curve cryptosystem. There are three popular types of bases for representing elements in finite field, termed polynomial basis (PB), normal basis (NB), and dual basis (DB). A novel low-cost bit-parallel DB multiplier which employs multiplexer approach is presented. As compared to traditional DB multiplier using XOR gates, the proposed design saves at least 40% space complexity.

Keywords: Elliptic curve cryptosystem, finite field arithmetic, dual basis multiplier, information security, cryptography.

1 Introduction

Recently, finite field arithmetic operations attracted many attentions because of their importance and practical applications in the field of error correcting code [1], cryptography [2], digital signal processing [3, 4], switching theory [5], pseudorandom number generation [6], encoding of Reed-Solomon code [7], and solving the Wiener-Hopf equation [8]. In particular, two public-key cryptography schemes, elliptic and hyper elliptic curve cryptosystems [9-11] require arithmetic operation to be performed in finite fields. The finite field arithmetic operations include addition, multiplication, division, inversion, and exponentiation. Addition is a simple bit-by-bit XOR operation. Complicated operations such as division, inversion, and exponentiation can be performed by repeated multiplications. Moreover, the elliptic scalar multiplication is based on the multiplication over $GF(2^m)$. As a result, the multiplication over $GF(2^m)$ is the most important operation in finite field arithmetic.

There are three popular types of bases over finite fields: polynomial basis (PB) [12-18], normal basis (NB) [19-23], and dual basis (DB) [24-27]. Each basis

H. Tan (Ed.): Knowledge Discovery and Data Mining, AISC 135, pp. 185–192.
springerlink.com

representation has its own distinct advantages. With the advantages of low design complexity, simplicity, regularity, and modularity in architecture, the PB is widely used for producing efficient VLSI multipliers. The major advantage of the normal basis is that the squaring of an element could be performed simply by cyclically shifting its binary form. As compared to the other two bases, the DB multiplier requires less chip area.

Both prime field *GF(p)* and extended binary field $GF(2^m)$ are now considered in elliptic curve cryptosystems. The prime field *GF(p)* contains p elements *{0, 1, ..., p-1}*. Finite field means a field that contains finite numbers of elements. Both binary field and prime field are necessary in elliptic curve cryptosystem. Extended binary field is more efficient for the hardware implementation due to its carry-free addition operation. Therefore, extended binary field becomes more and more important research area in recent years. Elements of extended binary field $GF(2^m)$ are m-bit strings. The rules for arithmetic operations in $GF(2^m)$ can be defined by either of the polynomial representations. Since $GF(2^m)$ operates on bit strings, modern computers can perform arithmetic in this field very efficiently. In recent years, unified architectures for both extended binary field $GF(2^m)$ and prime field *GF(P)* have been developed for low hardware cost purpose [28-35].

Addition operations in finite fields over $GF(2^m)$ are simpler compared with other operations, because there is no carry propagation, and addition of any two bits can be performed by a simple logical XOR operation. Among the $GF(2^m)$ operations, multiplication is the most important one, because other complex operations such as exponentiation, inversion, and division can be carried out through iterative multiplications.

In this paper, a novel DB multiplier using multiplexer approach is introduced. The proposed DB multiplier utilizes multiplexer rather than XOR gates used in the traditional DB multipliers. The hardware complexity can be drastically reduced about 40% by performing the accumulation operations through multiplexers.

The organization of this paper is as follows. Section 2 briefly reviews the mathematical background. The new novel DB multiplier is discussed in Section 3. Finally, results and comparisons are given in Section 4.

2 Preliminaries

In this section, we will briefly review conventional DB representation and multiplication. Readers can refer to [24-27] for more detail.

Let two elements *A* and *B* be in $GF(2^m)$. *A*, *B* are represented in the PB ($\phi=\{\alpha^0,\alpha^1,...,\alpha^{m-1}\}$) and its DB ($\psi=\{\psi_1,\psi_2,...,\psi_{m-1}\}$), respectively, as follows.

$$A=\sum_{i=0}^{m-1} a_i^* \psi_i, B=\sum_{i=0}^{m-1} b_i \alpha^i,$$

where a_i^* and $b_i \in GF(2)$ for $0 \leqq i \leqq m\text{-}1$, and

$$P=1+p_1\alpha^1+p_2\alpha^2+...+p_{m-1}\alpha^{m-1}+\alpha^m.$$

The product C of A and B is computed as follows: $C=AB \bmod P$, where C is represented in the DB ψ by

$C=\sum_{j=0}^{m-1} c_j^* \psi_j$, for all $0 \leqq j \leqq m-1$, $c_j^* \in GF(2^m)$. The coefficient c_j^* of C is computed by

$$c_j^* = Tr\left(\gamma\alpha^j C\right) = Tr\left(\gamma\alpha^j AB\right) = Tr\left(\gamma\alpha^j A \sum_{i=0}^{m-1} b_i \alpha^i\right) = \sum_{i=0}^{m-1}\left(b_i Tr\left(\gamma\alpha^{i+j} A\right)\right) \tag{1}$$

Based on Eq.(1), the product C can be calculated in the following form

$$\begin{bmatrix} c_0^* \\ c_1^* \\ c_2^* \\ \cdots \\ c_{m-1}^* \end{bmatrix} = \begin{bmatrix} Tr\left(\gamma\alpha^0 A\right) & Tr\left(\gamma\alpha^1 A\right) & Tr\left(\gamma\alpha^2 A\right) & \cdots & Tr\left(\gamma\alpha^{m-1} A\right) \\ Tr\left(\gamma\alpha^1 A\right) & Tr\left(\gamma\alpha^2 A\right) & Tr\left(\gamma\alpha^3 A\right) & \cdots & Tr\left(\gamma\alpha^{m} A\right) \\ Tr\left(\gamma\alpha^2 A\right) & Tr\left(\gamma\alpha^3 A\right) & Tr\left(\gamma\alpha^4 A\right) & \cdots & Tr\left(\gamma\alpha^{m+1} A\right) \\ \cdots & \cdots & \cdots & \cdots & \cdots \\ Tr\left(\gamma\alpha^{m-1} A\right) & Tr\left(\gamma\alpha^{m} A\right) & Tr\left(\gamma\alpha^{m+1} A\right) & \cdots & Tr\left(\gamma\alpha^{2m-2} A\right) \end{bmatrix} \times \begin{bmatrix} b_0 \\ b_1 \\ b_2 \\ \cdots \\ b_{m-1} \end{bmatrix} \tag{2}$$

Let $a_{m+d}^* = Tr\left(\gamma\alpha^{m+d} A\right)$, for $0 \leqq d \leqq m-2$, Eq.(3) based on Eq.(1) can be rewritten as:

$$\begin{bmatrix} c_0^* \\ c_1^* \\ c_2^* \\ c_3^* \\ \cdots \\ c_{m-2}^* \\ c_{m-1}^* \end{bmatrix} = \begin{bmatrix} a_0^* & a_1^* & a_2^* & a_3^* & \cdots & a_{m-2}^* & a_{m-1}^* \\ a_1^* & a_2^* & a_3^* & a_4^* & \cdots & a_{m-1}^* & a_m^* \\ a_2^* & a_3^* & a_4^* & a_5^* & \cdots & a_m^* & a_{m+1}^* \\ a_3^* & a_4^* & a_5^* & a_6^* & \cdots & a_{m+1}^* & a_{m+2}^* \\ \cdots & \cdots & \cdots & \cdots & \cdots & \cdots & \cdots \\ a_{m-2}^* & a_{m-1}^* & a_m^* & a_{m+1}^* & \cdots & a_{2m-4}^* & a_{2m-3}^* \\ a_{m-1}^* & a_m^* & a_{m+1}^* & a_{m+2}^* & \cdots & a_{2m-3}^* & a_{2m-2}^* \end{bmatrix} \times \begin{bmatrix} b_0 \\ b_1 \\ b_2 \\ b_3 \\ \cdots \\ b_{m-2} \\ b_{m-1} \end{bmatrix} \tag{3}$$

As stated previously, α is a root of P, thus

$$\alpha^m = 1+p_1\alpha+p_2\alpha^2+ \ldots +p_{m-1}\alpha^{m-1} \tag{4}$$

Therefore,

$$a_{m+d}^* = Tr\left(\gamma\alpha^{m+d} A\right) = \sum_{i=0}^{m-1} p_i a_{d+i}^* \quad \text{for } 0 \leqq d \leqq m-2 \tag{5}$$

Each a^*_j for $j \geqq m$ should be replaced by sum of a^*_is for $0 \leqq i \leqq m\text{-}1$ and p_i=1 based on Eq. (5). For clarity, $A=a_0+a_1\alpha+a_2\alpha^2$ and $B=b_0\beta_0+b_1\beta_1+b_2\beta_2$ over $GF(2^3)$ are used as an example to illustrate the general bit-parallel DB multiplier as shown in Fig. 1.

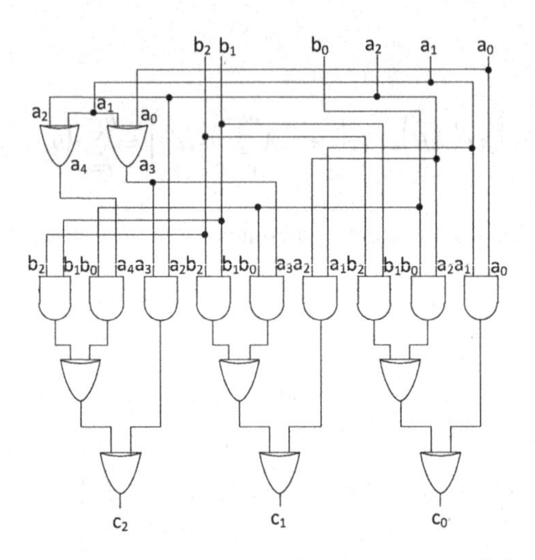

Fig. 1. Bit-parallel DB multiplier over $GF(2^3)$

3 The Proposed DB Multiplier Using the Multiplexer Approach

The CMOS circuits of the inverter, 2-input AND gate and 2-to-1 multiplexers can be constructed from the CMOS gates by 2, 6 and 6 transistors, respectively [36]. However, in standard CMOS manufacturing construction requires 12 transistors to implement a 2-input XOR gate. Thus, this paper presents a low-cost bit-parallel DB multiplier over $GF(2^m)$ using non-XOR approach. The space complexity can be reduced when using the non-XOR approach to replace the XOR-gate in a finite field multiplier.

One non-XOR approach consists of one inverter gate and one 2-to-1 multiplexer. It is composed of eight transistors. The non-XOR approach can be instead of XOR gates by the same Boolean function. The general XOR-gate Boolean function is $R=P\oplus Q$, where both P and Q are two inputs and R is its output. We propose the non-XOR approach which can implement the same Boolean function by $R=S\times P+\overline{S}\times\overline{P}$. The Boolean function comparison table between the XOR-gate and the non-XOR approach is shown in Table 1. The table indicates that both Boolean functions are equivalent. Moreover, the *U*-cell which is constructed using Non-XOR approach is shown in Fig.2.

As mentioned above, the *U*-cell can be employed to replace XOR gates and applied in conventional DB multipliers to reduce space complexity. The proposed bit-parallel DB multiplier using the non-XOR approach is shown in Fig. 3.

Table 1. Comparison of the Boolean-function between the XOR-gate and the non-XOR unit

input of XOR-gate	Output of XOR-gate	Input of Non-XOR unit		Output of Non-XOR unit
PQ	R	$P\overline{Q}$	S	R
00	0	01	0	0
01	1	01	1	1
10	1	10	0	1
11	0	10	1	0

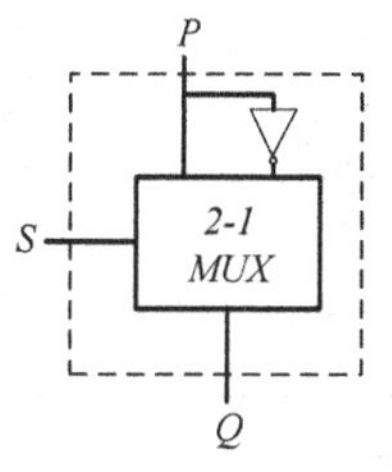

Fig. 2. Detailed *U*-cell structure using Non-XOR approach

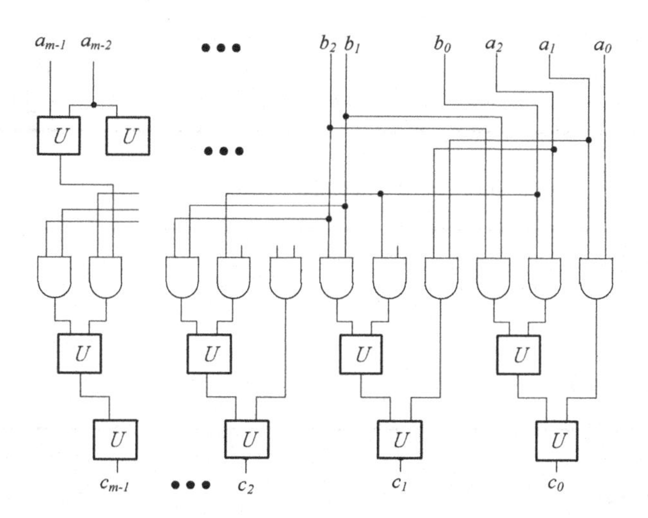

Fig. 3. Bit-parallel DB multiplier for $GF(2^m)$

4 Comparisons of Complexity

There is no DB multiplier using bit-parallel architecture in the literature. Thus, we compared the systolic array type to the proposed design. Systolic array multiplier is the typical bit-parallel multiplier using XOR gate in circuit design. We use the 2-to-1 multiplexers to replace the XOR gates in our architecture. As shown in Table 2, the proposed DB multiplier using non-XOR approach has smaller space complexity than traditional design.

We take the transistor count using a standard CMOS VLSI realization. In the CMOS VLSI technology, inverter, n-input AND, n-input OR, n-input XOR and 1-bit latch are composed of 2, 2n+2, 2n+2, 2n+2, and 8 transistors, respectively [36]. The results of comparison for multipliers are listed in Tables 2 and 3. In general CMOS manufacturing, 2-to-1 multiplexers are only using 6 transistors, and the XOR gate using 12 transistors. The results demonstrate that the proposed multiplier saves approximately 40% space complexity as compared to Singh's multiplier [37].

Table 2. Comparisons among various DB multipliers

Multipliers	Singh [37]	Lee [26]	Proposed design [Fig.3]
Basis	Dual Basis	Dual Basis	Dual Basis
Num. of cells	2	2	1
Array type	systolic	systolic	bit-parallel
		Cell complexity	
inverter	0	0	m^2
2-input XOR	$2m^2$ -m	m^2	0
2-input AND	$2m^2$	m^2	m^2
2-to-1 MUX	0	2m	m^2+m
1-bit Latch	0	$2m^2$+m	0
transistor counts	$24m^2$-6m	$28m^2$ +20m	$14m^2+8m$
Latency	2m	2m	$\lceil \log_2 m \rceil$

Table 3. Comparison of space complexity for NIST suggested m values

#Transistors Multipliers	m=163	m=233	m=283	m=409	m=571
Singh [37] (1)	636,678	1,301,538	1,920,438	4,012,290	7,821,558
Proposed design [Fig. 3] (2)	373,262	761,902	1,123,205	2,345,198	4,569,134
Comparing results (2/1)	58.6%	58.5%	58.4%	58.4%	58.4%

5 Conclusions

Until now, no existing bit-parallel DB multiplier adopts the non-XOR approach. In this paper, a novel DB multiplier using the non-XOR approach was presented. The proposed bit-parallel DB multiplier uses 2-to-1 multiplexers instead of XOR gates in traditional DB multipliers. The proposed bit-parallel DB multiplier has low space complexity so the chip area can be used more efficiently. As compared to general multiplier architecture, the proposed DB multiplier reduces approximately 40% less space complexity.

References

1. MacWilliams, F.J., Sloane, N.J.A.: The Theory of Error-Correcting Codes. North-Holland, Amsterdam (1977)
2. Lidl, R., Niederreiter, H.: Introduction to Finite Fields and Their Applications. Cambridge Univ. Press, New York (1944)
3. Blahut, R.E.: Fast algorithms for digital signal processing. Addison-Wesley, Reading (1985)
4. Reed, I.S., Truong, T.K.: The use of finite fields to compute convolutions. IEEE Trans. Information Theory, IT 21(2), 208–213 (1975)
5. Benjauthrit, B., Reed, I.S.: Galois switching functions and their applications. IEEE Trans. Compute. 25, 78–86 (1976)
6. Wang, C.C., Pei, D.: A VLSI design for computing exponentiation in GF(2^m) and its application to generate pseudorandom number sequences. IEEE Trans. Computers 39(2), 258–262 (1990)
7. Berlekamp, E.R.: Bit-serial Reed-Solomon encoder. IEEE Trans. Inform. Theory, IT 28, 869–874 (1982)
8. Morii, M., Kasahara, M., Whiting, D.L.: Efficient bit-serial multiplication and the discrete-time Wiener-Hopf equation over finite fields. IEEE Trans. Inform. Theory 35(6), 1177–1183 (1989)
9. Schroeppel, R., Orman, H., O'Malley, S., Spatscheck, O.: Fast Key Exchange with Elliptic Curve Systems. In: Coppersmith, D. (ed.) CRYPTO 1995. LNCS, vol. 963, pp. 43–56. Springer, Heidelberg (1995)
10. De Win, E., Bosselaers, A., De Gersem, P., Vandenberghe, S., Vandewalle, J.: A fast software implementation for arithmetic operations in GF(2^n). In: Kim, K.-c., Matsumoto, T. (eds.) ASIACRYPT 1996. LNCS, vol. 1163, pp. 65–76. Springer, Heidelberg (1996)
11. Menezes, A.J.: Applications of finite fields. Kluwer Academic, Boston (1993)
12. Lee, C.Y.: Low complexity bit-parallel systolic multiplier over GF(2^m) using irreducible trinomials. IEE Proc. Computer Digit. Tech. 150(1), 39–42 (2003)
13. Paar, C.: A new architecture for a parallel finite field multiplier with low complexity based on composite fields. IEEE Trans. Computers 45(7), 856–861 (1996)
14. Chiou, C.W., Lin, L.C., Chou, F.H., Shu, S.F.: Low complexity finite field multiplier using irreducible trinomials. Electronics Letter 39(24), 1709–1711 (2003)
15. Chiou, C.W., Lee, C.Y., Lin, J.M.: Efficient systolic arrays for power-sum, inversion, and division in GF(2^m). International Journal of Computer Sciences and Engineering Systems 1(1), 27–41 (2007)
16. Lee, C.Y., Chen, Y.H., Chiou, C.W., Lin, J.M.: Unified Parallel Systolic Multipliers over GF(2^m). Journal of Computer Science and Technology 22(1), 28–38 (2007)
17. Lee, C.Y., Lin, J.M., Chiou, C.W.: Scalable and systolic architecture for computing double exponentiation over GF(2^m). Acta Applicandae Mathematicae 93(1-3), 161–178 (2006)
18. Chiou, C.W., Lee, C.Y., Lin, J.M.: Concurrent Error Detection in a Polynomial Basis Multiplier over GF(2^m). Journal of Electronic Testing: Theory and Applications 22(2), 143–150 (2006)
19. Massey, J.L., Omura, J.K.: Computational method and apparatus for finite field arithmetic. U.S. Patent Number: 4,587,627 (1986)
20. Reyhani-Masoleh, A., Hasan, M.A.: A new construction of Massey-Omura parallel multiplier over GF(2^m). IEEE Trans. Computers 51(5), 511–520 (2002)

21. Lee, C.Y., Chiou, C.W.: Efficient design of low-complexity bit-parallel systolic Hankel multipliers to implement multiplication in normal and dual bases of GF(2^m). IEICE Transactions on Fundamentals of Electronics, Communications and Computer Science E88-A(11), 3169–3179 (2005)
22. Chiou, C.W., Lee, C.Y.: Multiplexer-based double-exponentiation for normal basis of GF(2^m). Computers & Security 24(1), 83–86 (2005)
23. Chiou, C.W., Chang, C.-C., Lee, C.-Y., Hou, T.-W., Lin, J.-M.: Concurrent Error Detection and Correction in Gaussian Normal Basis Multiplier over GF(2^m). IEEE Trans. Computers 58(6), 851–857 (2008)
24. Wu, H., Hasan, M.A., Blake, I.F.: New low-complexity bit-parallel finite field multipliers using weakly dual bases. IEEE Trans. Computers 47(11), 1223–1234 (1998)
25. Lee, C.Y., Chiou, C.W., Lin, J.L.: Concurrent error detection in a bit-parallel systolic multiplier for dual basis of GF(2^m). Journal of Electronic Testing: Theory and Applications 21(5), 539–549 (2005)
26. Lee, C.Y., Horng, J.S., Jou, I.C.: Low-complexity bit-parallel multiplier over GF(2^m) using dual basis. Journal of Computer Science & Technology 21(6), 887–892 (2006)
27. Chiou, C.W., Lee, C.Y., Lin, J.-M., Hou, T.-W., Chang, C.-C.: Concurrent error detection and correction in dual basis multiplier over GF(2^m). IET Circuits, Devices & Systems 3(1), 20–40 (2008)
28. Sava, E., Tenca, A.F., Koç, Ç.K.: A Scalable and Unified Multiplier Architecture for Finite Fields *GF*(*p*) and tex2html_wrap_inline111. In: Paar, C., Koç, Ç.K. (eds.) CHES 2000. LNCS, vol. 1965, pp. 277–296. Springer, Heidelberg (2000)
29. Goodman, J., Chandrakasan, A.P.: An energy-efficient reconfigurable public-key cryptography processor. IEEE Journal of Solid-State Circuits 36(11), 1808–1820 (2001)
30. Großschädl, J.: A Bit-Serial Unified Multiplier Architecture for Finite Fields GF(*p*) and GF(2). In: Koç, Ç.K., Naccache, D., Paar, C. (eds.) CHES 2001. LNCS, vol. 2162, pp. 202–219. Springer, Heidelberg (2001)
31. Wolkerstorfer, J.: Dual-Field Arithmetic Unit for *GF*(*p*) and *GF*(2^m). In: Kaliski Jr., B.S., Koç, Ç.K., Paar, C. (eds.) CHES 2002. LNCS, vol. 2523, pp. 500–514. Springer, Heidelberg (2003)
32. Savas, E., Tenca, A.F., Çiftçibasi, M.E., Koç, Ç.K.: Multiplier architectures for GF(p) and GF(2^n). In: IEE Proceedings-Computers and Digital Technology, vol. 151, pp. 147–160 (2004)
33. Satoh, A., Takano, K.: A scalable dual-field Elliptic Curve cryptographic processor. IEEE Trans. Computers 52, 449–460 (2003)
34. Chiou, C.W., Lee, C.-Y., Lin, J.-M.: Unified Dual-Field Multiplier in GF(P) and GF(2^k). IET Information Security 3(2), 45–52 (2009)
35. Lee, C.-Y., Chen, Y.-H., Chiou, C.W., Lin, J.-M.: Unified Parallel Systolic Multipliers over GF(2^m). Journal of Computer Science and Technology 22(1), 28–38 (2007)
36. Weste, N., Eshraghian, K.: Principles of CMOS VLSI Design: A System Perspective. Addison-Wesley, Reading (1985)
37. Singh, A.K., Bera, A., Rahaman, H., Mathew, J., Pradhan, D.K.: Error Detecting Dual Basis Bit Parallel Systolic Multiplication Architecture over GF(2^m). In: IEEE Circuits and Systems International Conference on Testing and Diagnosis, pp. 1–4 (2009)

Multi-precision Indoor Location System Based on Ultrasonic and ZigBee

Lei Zhang, Yang Tian, and Xiaomei Yu

Department of Engineering, Ocean University of China, Qingdao Shan Dong Province, China
zhl@ouc.edu.cn

Abstract. Propose a multi-precision indoor location system which is constituted by a new receiver network. The system integrates advantages of ZigBee and Ultrasonic location system, using TOA and RSSI algorithm to complete the location. There are multi-precision, low-cost, low-power, flexible installation and maintenance features. Experimental results show that the location precision of the location system can be changed by arranging the receiver network flexible.

Keywords: ZigBee, Ultrasonic, Indoor Location, multi-precision.

1 Introduction

With the development of location technology, there have been a variety of indoor location technologies, such as RF (Radio Frequency) location, infrared location, ultrasonic location and optical location. RF location subdivided into WLAN (Wireless Local Area Network) location, RFID (Radio Frequency Identification) location, Bluetooth location, UWB (Ultra Wide Band) location and ZigBee location. Because of the location technologies have its own shortcomings which cannot overcome, it is difficult to use in a large area independently. As shown in table 1.

In addition to the features described in Table 1, the range of precision of location technology is limited, so different location technologies suit for different circumstances. If only using the high-precision location technology, it will be wasteful for the location equipment in low-precision location area and then increase the total cost. If only using the low-precision location technology, it cannot satisfy the requirements of the location precision in high-precision location area. To develop a multi-precision location system will solve the problems. The system can satisfy the requirement of location precision by flexibly changing the layout which depended on the features of environment and the requirement of location precision.

This paper proposes a multi-precision indoor location system which is constituted by multi-precision location receiver group. This system integrates the advantages of ZigBee and Ultrasonic location system and TOA and RSSI algorithm to achieve the multi-precision, low-cost, low-power, flexible installation and maintenance indoor location function.

H. Tan (Ed.): Knowledge Discovery and Data Mining, AISC 135, pp. 193–199.
springerlink.com

Table 1. Technical Features of the Location

Technology	Feature	Example
RF	Wide scope, NLOS relation constraint; Cheap;	RADAR System[1]
Infrared	Cheap, Low power; Low location precision; Susceptible to the sun interference; NLOS relation constraint;	Active Badge[2]
Ultrasonic	High location precision; Sending and receiving angle constraint; Inconvenience to cable installation inconvenience;	Active Bat[3] Cricket System[4]
Optical	Uncertain location precision; Expensive; NLOS relation constraint;	EasyLiving[5] Trip System[6]
UWB	Expensive; High location precision; NLOS relation constraint;	Ubisense[7]

2 System Structure and Hardware

This system includes many MTs (mobile terminals), control centers and position receiver network. The MT installed on pending objects determines the position coordinates of pending objects by location system. Control centers calculate the data obtained from system to position and offers the results to users in real time.

The position receiver network consists of many networks constructed by multi-precision position receiver groups which generally installed in indoor ceiling panel. The position receiver network is a most important part of location system.

Multi-precision position receiver groups is called "receiver group" for short. It consists of a power supply module, a ZigBee wireless node and 0~10 ultrasonic receiving nodes, the latter two nodes of which are connected to communicate by RS485 bus. ZigBee wireless node is used as master node in RS485 bus. It has wireless communication function with location system. The number of ultrasonic receiving node in one receiver group is optional within the permission scope, which is completely increased or decreased by installation environment, location accuracy, etc. Ultrasonic receiving node can receive the ultrasonic signal sent by MTs, and it has timing function. The power in the group is uniformly supplied by the power supply module which is consists of high power batteries to reduce the cost of installation and maintenance.

Multi-precision position receiver groups is called "receiver group" for short. It consists of a power supply module, a ZigBee wireless node and 0~10 ultrasonic receiving nodes, the latter two nodes of which are connected to communicate by RS485 bus. ZigBee wireless node is used as master node in RS485 bus. It has wireless communication function with location system. The number of ultrasonic receiving node in one receiver group is optional within the permission scope, which is completely increased or decreased by installation environment, location accuracy, etc. Ultrasonic receiving node can receive the ultrasonic signal sent by MTs, and it has timing function. The power in the group is uniformly supplied by the power supply module which is consists of high power batteries to reduce the cost of installation and maintenance.

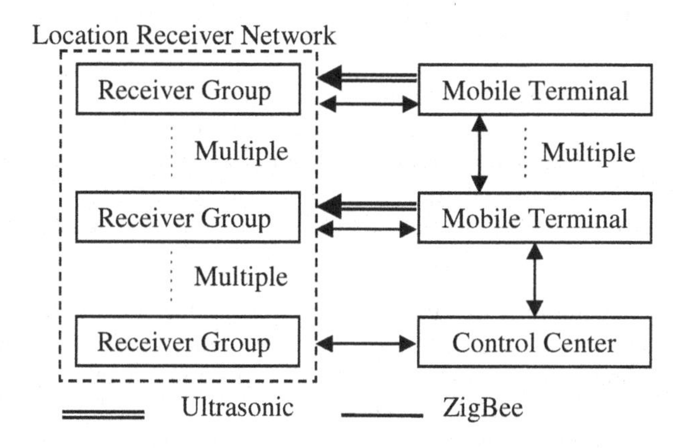

Fig. 1. Structure Diagram of Multi-precision Location System

The control centers and MTs two-way communicate with position receiver network by ZigBee. Receiver group with ultrasonic node can receive the one-way ultrasonic signal sent by MTs to position, as is6 shown in Figure 1.

MTs consist of MCU, ZigBee wireless module, ultrasonic generator, position switch as well as power supply. As shown in figure 2. And ZigBee wireless module creates communication with location system, creating corresponding RSSI value by received ZigBee signal. Ultrasonic generator can send ultrasonic signal. Position switch is used to control the location mode. Power supply is supplied by carried battery.

Control center consists of PC as a host-computer and RS232 to ZigBee wireless node, which connects to RS232 bus to realize the communication with location system.

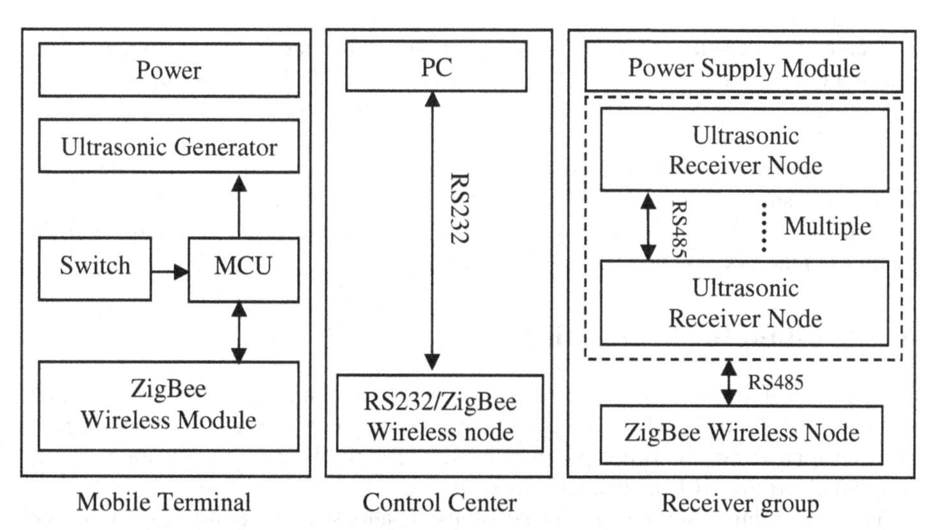

Fig. 2. Structure Diagram of Mobile Terminal, Control Center and Multi-precision Position Receiver Group

3 Software Implementation

This system fully takes the advantages of the two location technology and compensates their shortcomings. it achieved the target of multiple precision locations under low-power consumption through the software program. The target of low-precision location is achieved by ZigBee location technology, and high-precision location is achieved by ultrasonic location technique. Specific location process is as follows:

Step 1: Start position, it has two different ways. One way is designated by the control center to send the start location command to the mobile terminal, waking up the mobile terminal which is in sleep mode to start location; another way is to presses the location switch on the mobile terminal.

Step 2: The location network receives the broadcast messages transferred by the mobile terminal and start location.

Step 3: The multi-precision location receiver sends its two dimensional coordinate to the mobile terminal, and then the wireless nodes come into sleep mode.

Step 4: The mobile terminal receives the data transferred by the wireless nodes and then generates the corresponding signal values. It transfers them to the control center, the control center achieves to locate and display the result by algorithm.

Step 5: After calculating the location range, the control center send the ultrasonic position command to the nearly multi-precision receiver depended on the multi-precision receiver group distribution and their carrying case of ultrasonic receiving node. The multi-precision receiver wakes up the sleeping ultrasonic receiving nodes as soon as it receives the position command and send confirmation message to the control center, it will transfer the ultrasonic location command to the mobile terminal and the timing command to the multi-precision receiver until it receives confirmation information.

Step 6: The mobile terminal which received the ultrasonic location command send ultrasonic to the location network by their own ultrasonic transmitter modules. The ultrasonic receiving node start to timing after the location network received the timing command and end up timing till the node received the ultrasonic signal. The node transfers its two dimensional coordinate and time value to the ZigBee nodes by RS485 bus, and the ZigBee nodes transfer the data to the control center in batches.

Step 7: The control center calculates and displays the position values, and the modules of the system enter into sleep mode.

4 Algorithm Implementation

To realize multi-precision location function, system organically combines two kinds of location techniques. The ultrasonic position used TOA algorithm [8] and ZigBee position used RSSI algorithm [9] in order to measure distance, then using trilateral location algorithm [10] calculate the position.

TOA algorithm takes the advantage of the relationship of transmission time and distance. Supposed that sender and receiver both know the time when signal trans-mission begins. Receiver can use the signal arriving time to calculate spreading time of signal in the media. And then using signal transmission speed calculates spreading distance.

RSSI algorithm chooses the following mode, as formula (1):

$$RSSI = -(10n\log_{10}d + A) . \quad (1)$$

That means ignoring the influence of RSSI to blocking factor. In a realistic environment, it is the effect of non-line-of-sight that affects RSSI most. Radio frequency (RF) parameter A is defined as absolute value of average energy (received signal str-ength as well) with unit dBm, which is received from sender in 1 meter; N is for sig-nal trans-mission constant related to signal transmission environment; D is for the dis-tance from the sending node.

5 Experimental Results and Analysis

Do the location experiments in $50m^2$ office environment. Personnel carries mobile terminal. Location receiver network installs in the ceiling. Control center monitors clerk and provides navigation service.

Ultrasonic generator carried by mobile terminal vertical face the ceiling. There are 2m (-10% ~ +10%) away between them in real-time experimentation.

The ultrasonic transducer used by site, includes 90°sending and receiving angle, round projecting scope whose radius is equal to projecting distance. For seeing signal scope in ceiling panel projected by ultrasonic generator in the spot is a circle with a radius of 2 meters. So long as there are three or more ultrasonic receiving notes in this scope, it can be accomplished to ultrasonic position. Wireless signals sent from ZigBee are easy to be adsorbed by wall, so the installment of ZigBee wireless nodes should keep away from the corner. The distribution of nodes and receiver groups are given in Figure 3, which should be reduced cross wire and installed evenly in all areas.

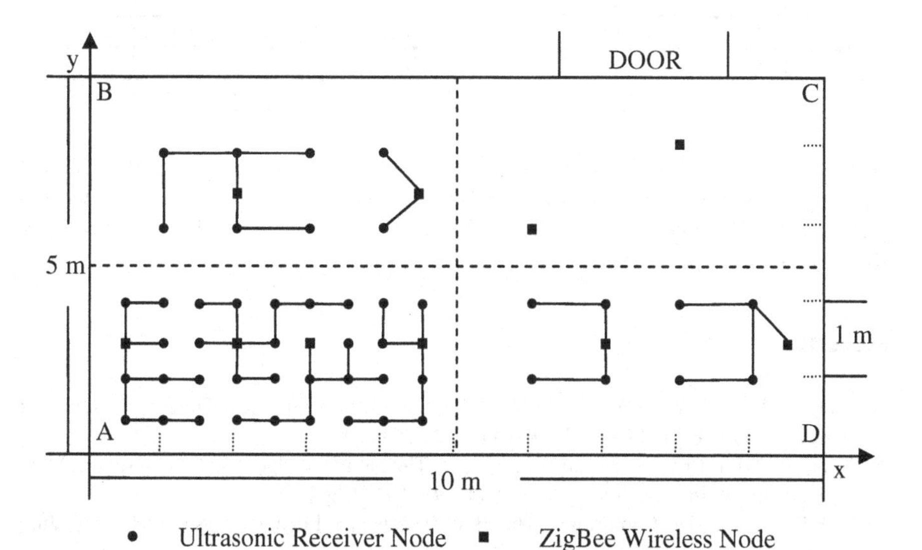

Fig. 3. Distribution Schematic Diagram for Nodes of Testing Area

Experimental environment is divided into A, B, C and D four equal areas. High-precision location area A provides the location information such as important items. Low-precision location area C is only used to check in/out. General-precision location area B and D is used to monitor staff's activities. According to the requirements of location precision, node distribution is shown in table 2.

Table 2. Table for Regional Number of Node Distribution and Requirements of Location Precision

Area	ZigBee	Ultrasonic	Request(m)
A	4	32	0.0500
B	2	10	0.1500
C	2	0	4.0000
D	2	10	0.1500

The experiment uses total 10 ZigBee wireless nodes and 52 Ultrasonic receiver nodes. They formed 10 multi-precision location receiver groups. The experimental results are shown in table 3.

Table 3. Table for Location Situation (m)

Actual coordinate /Area	ZigBee location coordinate	Ultrasonic location coordinate	Error (ZigBee/ coordinate)
(1 , 1)/A	(2.5 , 2.2)	(1.03 , 0.98)	1.9209/0.0361
(3 , 2)/A	(1.6 , 3.0)	(2.98 , 2.01)	1.7205/0.0224
(1 , 4)/B	(4.1 , 2.8)	(1.12 , 4.09)	3.3242/0.1500
(7 , 4)/C	(9.5 , 2.6)	NULL	2.8653/NULL
(9 , 4)/C	(5.8 , 2.9)	NULL	3.3838/NULL
(9 , 1)/D	(5.7 , 2.1)	(8.90 , 1.11)	3.4785/0.1487

The study of experimental process and result shows that the way presented in this paper, which location system adopts adjusting node density or exclusively uses ZigBee location technique, realizing the information processing of location accuracy in different areas.

References

1. Bahl, P., Radar, P.V.N.: An Inbuilding RF-based User Location and Tracking System. In: Proceeding of IEEE INFOCOM 2000, pp. 775–784 (2000)
2. Want, R., Hopper, A., Falcao, V., et al.: The active badge location system. ACM Transactions on Information Systems 10(1), 91–102 (1992)
3. AT&T Laboratories Cambridge. The BAT Ultrasonic Location System[EB/OL] (July 2009), http://www.cl.cam.ac.uk/research/dtg/attarchive/bat/

4. Priyantha, N.B., Chakraborty, A., Balakrishnan, H.: The cricket location-support system. In: Proceedings of the 6th Annual International Conference on Mobile Computing and Networking, pp. 32–43 (2000)
5. Krumm, J., Harris, S., Brian, M., et al.: Multi-camera Multi-person Tracking for Easy Living. In: Proceedings of the 3rd IEEE International Workshop on Visual Surveillance, pp. 3–10 (2000)
6. Diego, L.I., Mendona, P.R.S., Hopper, A.: TRIP: a Low-cost Vision-based Location System for Ubiquitous Computing. Personal and Ubiquitous Computing Journal 6(3), 206–219 (2002)
7. Steggles, P., Cadman, J.: A comparison of RF tag location products for real world applications. Ubisense limited (2004)
8. Harter, A., Hopper, A., Steggles, P., Ward, A., Webster, P.: The anatomy of a context-aware application. In: Proc. of the 5th Annual ACM/IEEE Int'l Conf. on Mobile Computing and Networking, pp. 59–68. ACM Press, Seattle (1999), `http://www-lce.eng.cam.ac.uk/lce-pub/public/files/tr.2002.2.pdf`
9. Patwari, N., Ash, J.N., et al.: Locating the nodes: Cooperative localization in wireless sensor network. IEEE Signal Processing Magazine 22(4), 54–69 (2005)
10. Savvides, A., Han, C.C., Strivastava, M.B.: Dynamic fine-grained localization in ad-hoc networks of sensors. In: Proceedings of the 7th Annual International Conference on Mobile Computing and Networking, pp. 166–179 (2001)

A Trust-Third-Party Based Key Management Protocol for Secure Mobile RFID Service Based on the Internet of Things

Tao Yan and Qiaoyan Wen

State Key Laboratory of Networking and Switching Technology, BUPT
Beijing, China
yantao1982@hotmail.com, wqy@bupt.edu.cn

Abstract. The Internet of Things as an emerging global, information service architecture facilitating the exchange of goods in global supply chain networks is developing on the technical basis of the present Domain Name System. Technological advancements lead to smart objects being capable of identifying, locating, sensing and connecting and thus leading to new forms of communication between people and things and things themselves. In this paper, we assume that all the communication channels among Tag, Reader, and Server are insecure. We implement a TTP (Trust-Third-Party) based key management protocol to construct a secure session key among the tag, reader and server to construct a security Mobile RFID (Radio Frequency Identification) service based on the IOT(Internet of Things). A smart mobile phone with a RFID reader can provide Mobile RFID (MRFID) services based on RFID tagged objects. While the mobile RFID system has many advantages, the privacy violation problems at reader side, consumer tracking with malicious purpose and the sensitive information transmission problems among Tag, Reader and Server, are deeply concerned by individuals and scholars. The proposed idea is security and efficiency and supply three advantages for a secure IOT architecture.

Keywords: The Internet of Things, Key Management Protocol, Trust Third Party, RFID, Mobile RFID, RFID Security.

1 Introduction

The Internet of Things (IOT) is a technological phenomenon originating from innovative developments and concepts in information and communication technology associated with: 1) Ubiquitous Communication/Connectivity, 2) Pervasive Computing and Ambient Intelligence.

These concepts have a strong impact on the development of the IOT [1]. Ubiquitous Communication means the general ability of objects to communicate (anywhere and anytime). Pervasive Computing means the enhancement of objects with processing power (the environment around us becomes the computer). Ambient Intelligence means the capability of objects to register changes in the physical environment and thus actively interact in a process. Typically, objects fulfilling these

H. Tan (Ed.): Knowledge Discovery and Data Mining, AISC 135, pp. 201–208.
springerlink.com

requirements are called "smart objects" [2]. Hence, the IOT is defined as the ability of smart objects to communicate among each other and building networks of things, the Internet of Things.

There are also some other explanation of the IOT is that the world's objects build a dynamic network which is connected by the Internet and sometimes designated as the network of networks.

The key players in enabling the IOT are smart objects, which are characterized by four technological attributes:

- Identification
- Location
- Sensing
- Connectivity

Furthermore, smart objects can be active (local decision making is possible) or passive (sensor data is stored and can be read out–no local decision making is possible).

The major enabling technologies for smart objects are Radio Frequency Identification (RFID), Global Positioning system (GPS), developments in sensor networks, Micro Electro-mechanical Systems (MEMS) and further developments in wireless connectivity. RFID tags allow objects to be uniquely identified, to determine the location, to sense changes in physical data and to connect and communicate with a corresponding transponder.

The IOT architecture is technically based on data communication tool, primarily Radio Frequency Identification (RFID) tagged items. A smart mobile phone with a RFID reader can provide Mobile RFID (MRFID) services [3] based on RFID tagged objects. All these IOT services need to be carefully designed so that they can be intuitively used by humans, devices and things, to mash up services, to federate new services along with others and overcome the future technological challenges.

Considering the requirements for MRFID services, there is a need for service level architectural solutions as the services themselves could be created on a MRFID reader alone or in combination with some end-to-end systems. Various issues such as enabling a mobile device to act as a MRFID service platform, managing the services and contents, MRFID service classification, recognizing the pattern of use cases, type of service solutions and factors affecting the quality attributes of services are all part of the solution design process. Moreover, when more than one solution option is available at different levels, a way to make a rational judgment is needed.

This paper introduces a mobile RFID Architectural for IOT. Chapter 2 introduces related work. Chapter 3 describes the mobile RFID network. In Chapter 4 we describe TTP (Trust-Third-Party) based secure mobile RFID system and Chapter 5 proposed our key management protocol. We analyze security and privacy problems of our scheme in efficiency and security in the chapter 6 and conclude in Chapter 7.

2 Related Work

Despite vast literatures have been reviewed on several occasions and new secure communication mechanisms have been proposed [4,5,6,7,8,9], none of which considered the security and privacy issues of M-RFID as a whole system. This scheme only considered the privacy of readers and it would greatly increase the burden on system for the practical application if key agreement occurred for each tag access.

3 MRFID Network

Besides these privacy threats, we introduce a mobile RFID network consists of a tag, a mobile RFID reader and network servers which will be more security.

3.1 RFID Tag

A tag receives a query from a reader and transmits EPC (Electronic Product Code), unique information of an object using RF signals. The tag is composed of an antenna for a RF communication and the microchip which stores information of objects and implements operation function. The tag is divided into a passive tag and an active tag in power supply method.

3.2 Mobile RFID Reader

In mobile RFID systems, the mobile device plays a role as a reader of the RFID systems. The mobile reader is the mobile including reader of the RFID systems. It transmits a query to a tag and identifies the received information from the tag. In some case, the mobile reader rewrites new information to some tag. The mobile reader has a high computation ability and a large storage capacity unlike the reader of previous RFID systems.

3.3 Network Server [10,11]

ONS(Object Name Service) server: When a base-station requests information of URL for EPCs, the server informs the URL(Uniform Resource Location) of server which manages information related to EPC to the base station (similarity DNS (Domain Name Server)).

OIS (Object Information Service) server: The server stores the information related to EPC of tags. When the base station requests the information of the tags, the server sends it to the base station.

4 Trust-Third-Party Based Secure MRFID System

The architecture for TTP (Trust-Third-Party) based secure mobile RFID system. When a mobile RFID user possesses a RFID tagged product, TTP enables the owner to control his backend information connected with the tag such as product information, distribution information, owner's personal information, and so on. TTP ensures the owner to communicate with the tag and application server in secure channel through the key agreement among the tag, reader and server.

In our system, a consumer reads the tag from the tagged product with his or her mobile RFID reader. The consumer then browses the product-related information from the application server and purchases the product using one of payment methods. Then the consumer becomes the tag owner. Next, the security and privacy protection will be provided by TTP. TTP requests the owner defined privacy profile such as privacy protection level used in [8] to access the application server. The TTP server receives and forwards the owner's privacy protection policy for his application service to the application server and distributes certificates to the owner and server, respectively. The tag, reader and server then set up the key for secure transmission. Anyone requesting information associated with this tag can browse all information provided by the application server in secure channel if the requestor is the owner, but otherwise, the requestor only has access to a limited amount or no information according to the limits established by the owner in insecure channel. Step (1) - (6) in Figure 1 depicts the process. In the following section we will elaborate on the distribution of certificates and key agreement.

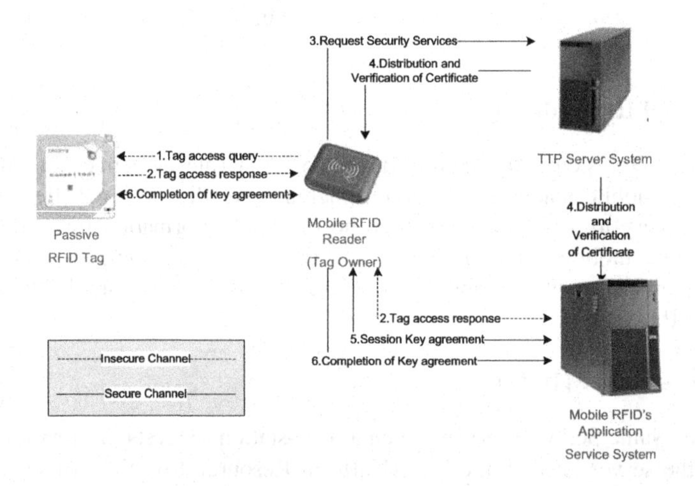

Fig. 1. Data transmission and filtering

5 Key Management Protocol

In our scheme, a trusted third party server is introduced as a key distribution authority to solve the privacy disclosure problem for mobile RFID-enabled devices.

5.1 Preliminary Notations

We use the same way in generation and distribution of certificates as [9]. We assume each tag, reader and server is hard coded a unique identity ID_T, ID_R and ID_S. An one-way hash function $H(\ )$ and a stream cipher algorithm are utilized among the tag, reader, server and TTP. We assume the tag has the computation power to do hash function and stream cipher encryption. To realize the anonymity of reader, the protocol is based on the elliptic curve cryptosystem [12]. We assume the reader has the computation power to do elliptic curve cryptography algorithm. Another two reasonable assumptions are that the application server shares symmetric key k_{ST} and k_{TPS} with the tag and TTP, respectively.

The symbol $E_k(M)$ denotes the encryption of message M using key $Cert(\ )$ denotes a certificate. The protocol is divided into three parts, key initialization, key agreement and key update. The process of proposed scheme is shown in Figure 1.

5.2 Key Initialization

Once label a RFID tag on certain product, the tag ID_T will be bundled with an application server ID_S. The tag and server will set up shared symmetric key k_{ST}. The application server will provide default privacy protection. The TTP server set up shared symmetric key k_{TPS} with all RFID application servers. To avoid tag tracking, it is required to establish an ID flag, denoted by ID_{T-FLAG}, which will be used to show whether the corresponding ID_T is possessed by fixed users. If the tagged product isn't belong to private users, ID_{T-FLAG} will be closed. That is, the identity of the tag ID_T remains unchanged, any mobile RFID user obtain the same information according to the default privacy protection policy using the fixed ID_T to access the information server. The process is:

Step 1: $reader \rightarrow tag$: Access query.

Step 2: $tag \rightarrow reader \rightarrow server: ID_T || ID_S || ID_{T-FLAG} || E_{k_{ST}}(T_n)$.

Where the symbol || denotes concatenation and the value of ID_{T-FLAG} is False. T_n is a random number generated by tag to protect its identity at the next steps.

5.3 Key Agreement

The user purchases the product using one of payment methods and requests security services to TTP server.

Step 3: $reader \rightarrow TTP$: $E_{P_{k_{TP}}}\left(ID_T || ID_S || ID_{T-FLAG} || ID_R || k_R || t_{r_1}\right)$ Reader applies elliptic curve encryption with the public key

$$P_{k_{TP}} = i_1 B_1 + i_2 B_2 \tag{1}$$

to protect its own identification ID_R, its secret key k_R, the current timestamp t_{r_1} to TTP server. Where (i_1, i_2) is the sect key $s_{k_{TP}}$ of TTP and randomly selected from the interval $[1, n-1]$.

Step 4: $TTP \rightarrow reader: \; E_{k_R}\left(R_C \| Cert(R_C) \| r\right)$

$TTP \rightarrow server: E_{k_{TPS}}\left(ID_T \| ID_S \| P_C \| Cert(P_C) \| r\right)$

Step 5: $reader \rightarrow server: \; E_{p_{k_S}}\left(P_i \| Cert(P_i) \| r \| n_1 G \| t_{r_2}\right)$

$server \rightarrow reader: \; P_i \| n_2 G \| E_{k_{auth}}\left(t_{r_2}\right) \| E_{k_{ST}}\left(k_{auth} \| Cert(P_i) \| T_n\right)$ Once reader verified the certificate of

$$R_C = (c_1, c_2), \tag{2}$$

a pseudonym

$$P_i = (p_1, p_2) \tag{3}$$

of reader is generated for the usage of future communication. Reader compute

$$Cert(P_i) = \left(e'', S_{e_1''}, S_{e_2''}\right), \tag{4}$$

where

$$e'' = H(P_i, R_S), S_{e_1''} = d_1 + c_1 e'' (\bmod n), \tag{5}$$

$$S_{e_2''} = d_2 + c_2 e'' (\bmod n), R_S = d_1 B_1 + d_2 B_2, \tag{6}$$

d_1 and d_2 is randomly selected from the interval $[1, n-1]$.

Step 6: $reader \rightarrow tag : E_{k_{auth}}\left(Cert(P_i)\right) \| E_{K_{ST}}\left(k_{auth} \| Cert(P_i) \| T_n\right)$

$tag \rightarrow reader \rightarrow server: ID_T \| ID_S \| ID_{T\text{-}FLAG} \| E_{k_{ST}}\left(T_n \| Cert(P_i)\right) \| k_{auth}\left(Cert(P_i)\right)$

5.4 Key Update

The session key k_{auth} update is very simple, repeat Step 5 and Step 6, reader, server and tag will renegotiate the session key.

6 Security Analysis

In this section, we analyze our scheme as follow:

6.1 Session Key Security

Attackers can't know the message content, because every message is enciphered by session key k_{auth}. Based on the $ECDLP$ problem [12], the session key $k_{auth} = n_1 n_2 G$ cannot be derived without knowing the value n_1 and n_2.

6.2 Impersonation Attack

Attacker can intercept the transmitted message under the wireless communication environment and attend to impersonate any communication party in a normal operation. However, unknowing the secret key k_{ST}, k_R and k_{TPS}, attacker can't obtain any information from eavesdropped messages. Furthermore, based on the value ID_{T-FLAG} stored in server, attacker can't engage the Step 3 to apply for the key pair (P_C, R_C) and corresponding certificates. Thus, the impersonation attack can be prevented.

6.3 Forward Security

The benefit of forward security is to protect the new session key even if the current session key is leaked, the backward security is just the opposite. Since the session key k_{auth} is dynamic and the key is generated by the random selection of n_1 and n_2, the attacker can't derive the previous or new session key from the current session key.

7 Conclusion

A mobile RFID system is a new service using a mobile device like RFID reader and wireless internet. In this paper, we propose a scheme that protects personal privacy. This paper introduces a mobile RFID Architectural for IOT and introduces a key management protocol for secure mobile RFID service to protect data privacy against the threats of replay attack, impersonation attack and provide excellent privacy protection features forward and backward security at the same time. In conclusion, the proposed scheme can offer data security enhancement and privacy protection capability at reader side under an insecure and wireless mobile RFID system.

Acknowledgments. This work is supported by National Natural Science Foundation of China (Grant Nos. 61170270, 61100203, 60873191, 60903152, 61003286, 60821001).

References

1. Kim, H., Kim, Y.: An Early Binding Fast Handover for High-Speed Mobile Nodes on MIPv6 over Connectionless Packet Radio Link. In: Proc. 7th ACIS International Conference on Software Engineering, Artificial Intelligence, Networking, and Parallel/Distributed Computing (SNPD 2006), pp. 237–242 (2006)

2. Information Society Technologies Advisory Group (ISTAG): Revising Europe's ICT Strategy. Report from the Information Society Technologies Advisory Group (ISTAG) (2009)
3. Seidler, C.: RFID Opportunities for mobile telecommunication services. ITU-T Lighthouse Technical Paper (2005)
4. Chang, G.C.: A Feasible security mechanism for low cost RFID tags. In: International Conference on Mobile Business, ICMB 2005, pp. 675–677 (2005)
5. Osaka, K., Takagi, T., Yamazaki, K., Takahashi, O.: An Efficient and Secure RFID Security Method with Ownership Transfer. In: Wang, Y., Cheung, Y.-m., Liu, H. (eds.) CIS 2006. LNCS (LNAI), vol. 4456, pp. 778–787. Springer, Heidelberg (2007)
6. Kim, H.W., Lim, S.Y., Lee, H.J.: Symmetric encryption in RFID authentication protocol for strong location privacy and forward-security. In: International Conference on Hybrid Information Technology, vol. 2, pp. 718–723 (2006)
7. Yang, M.H., Wu, J.S., Chen, S.J.: Protect mobile RFID location privacy using dynamic identity. In: 7th IEEE International Conference on Cognitive Informatics (ICCI 2008), pp. 366–374 (2008)
8. Park, N., Won, D.: Dynamic privacy protection for mobile RFID service. In: RFID Security: Techniques, Protocols and System-on-Chip Design, pp. 229–254. Springer, US (2009)
9. Lo, N.W., Yeh, K.H., Yeun, C.Y.: New mutual agreement protocol to secure mobile RFID-enabled devices. Information Security Technical Report (2008)
10. EPCglobal Inc.: EPCglobal Network: Overview of Design, Benefits, and Secutiry (2004)
11. EPCglobal Inc.: EPCglobal Object Name Service (ONS) 1.0 (2005)
12. Miller, V.S.: Use of Elliptic Curves in Cryptography. In: Williams, H.C. (ed.) CRYPTO 1985. LNCS, vol. 218, pp. 417–426. Springer, Heidelberg (1986)

The Smart Home System Design Based on i.MX51 Platform

Haibin Sun

Computer Department, College of Information Science and Engineering,
Shandong University of Science and Technology, 579 Qianwangang Road Economic &
Technical Development Zone, Qingdao Shandong Province, 266590 P.R. China
Hisense State Key Laboratory of Digital Multimedia Technology, 11 Jiangxi Road,
Qingdao Shandong Province, 266071, P.R. China
Offer_sun@hotmail.com

Abstract. A smart home system is designed in this paper. The hardware platform and software architecture are presented, and the key software solutions are stated. A smart home prototype system is developed to prove the adopted technologies,which can satisfy the smart home market and will lay the foundation for the future extension.

Keywords: Smart Home, Visual Intercom, Multimedia.

1 Introduction

With the research and development of sensor network in China, the smart home system has been endowed with the new meaning. What is the smart home life like? In the home equipped with the smart home system, the appliances, lighting and home theatre can be controled; the public information from the community can be received at real time; the community online mart service,remote health ware and express mail service can be provided more conveniently. If you are out of home, you can connect with the smart terminals by telephone,cell phone or internet to be aware of the using of appliances and monitor the situation of home; the visitor can leave message with voice and video; if the gas leak happens and a thief enters your home, the safety device can alarm automatically and shoot a picture of the thief,which can be sent to you instantly. When you are on the way home, you can open the air conditioner and hot water system in advance through your telephone. In a word, you can enjoy a more intelligent, convenient, and comfortable life with the smart home system[1,2,3,4,5,6].

According to the topological structure of the system, the smart home system can be divided into three types: american series, european series and korea series. The characteristics of american series products are as following: a controller is the center of the whole system, which has the strong function of assigning and controling multimedia signals and has the impressive touch screen user interface. This system can provide users with resplendent audio-visual enjoyment, which fits in with deluxe villas. But it also has shortcomings. This system relies on central controller so much that once the central controller fails, the whole system is prone to be paralyze, i.e. low liability.

H. Tan (Ed.): Knowledge Discovery and Data Mining, AISC 135, pp. 209–216.
springerlink.com

The european series products adopt the non-controller dispersed control mode. Every module has its control program, which means that the malfunction of an individual device will not induce the failure of the whole system, i.e. high liability. But the european series products don't have the ability of asigning and controling the video signals, the touch screen user interface of which is simple and the multimedia controling ability is weak. This type of products are fit for the middle and high class villa whose main function is control.

The korea series products add some control functions to the visual intercom system such as lighting and curtain controling. The advantage is that such system has the ability of video transfering. But it also relies on the central device as the american series. Furthermore, its control function is very simple and most of these kind of systems cannot play the background music. So this type of system fits with the Luxury Residence.

A complete smart home system is not simply controling of lighting and curtain or the only integration of visual intercom and home security system, but the organic integration of appliance control(including lighting and curtain), security system(visual intercom, home safety,etc.), home entertainment(including the control of background music and home theatre) and message communication, etc. and makes itself be the center of controling, security, entertainment and information, which has been agreed on by the professionals in the industry.

This paper will design a smart home system that integrates visual intercom, multimedia, security alarm and information sharing,etc.. and the related key technologies are analyzed and solved.

This paper is structured as follows. First, the system design including hardware and software is presented; Second, the architecture of visual intercom is explained, Last, the conclusion and future work are presented.

2 System Design

As shown in Fig.1, there are an indoor machine, several indoor machine extensions, and a door machine in a house. The indoor machine is connected to the security devices, lights, appliances(air conditioner, refrigerator, etc.), curtain, meters(water, electric and gas meters) by wire or wireless means; a unit building is equipped with a unit door machine and maybe several extensions; the doorway of the district is equipped with a district door machine and maybe several extensions; the management machine and management computer are located at the property management center. These devices are communicated with each other through the ethernet network to realize the audio-video stream transfer and information exchange.

2.1 Hardware Design

The i.Mx51 series chip of FreeScale is chosen as the main hardware platform, whose main characteristics are as follows:

（1） ARM Cortex A8 processor, 800MHz;

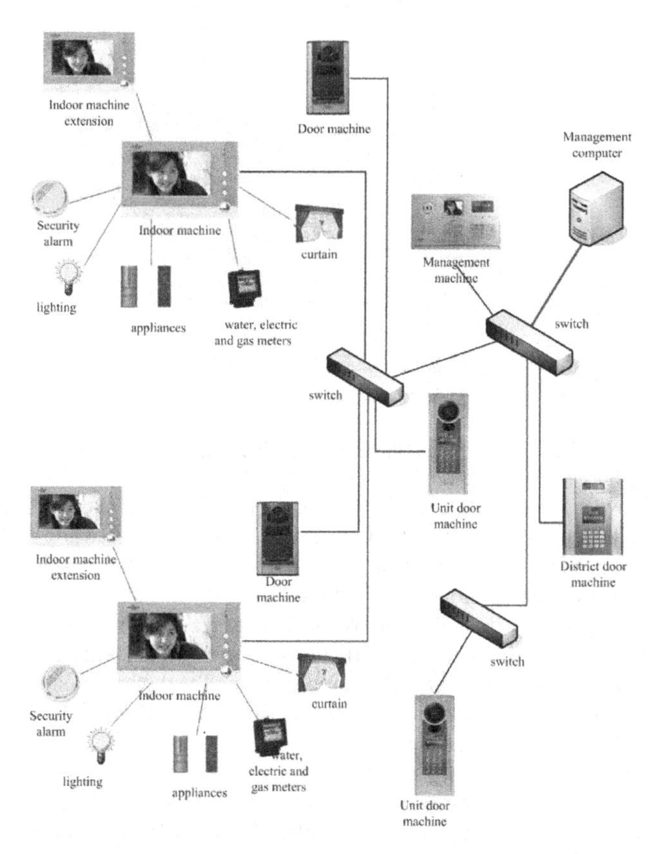

Fig. 1. The schematic diagram of smart home system

（2） Having the Video Process Unit and Image Process Unit with hardware accelerators;

（3） Supporting MPEG4-SP, H.264 BP and MJPEG Baseline encoding;

（4） Supporting MPEG-2/MPEG-4/H.264/H.263/VC-1/DivX/Real Video/MJPEG decoding;

（5） Supporting power management;

（6） Supporting TV encoding (NTSC/PAL, SD and HD,1080p).

This chip can fully satisfy the needs of our system. The block diagram is shown in Fig. 2.

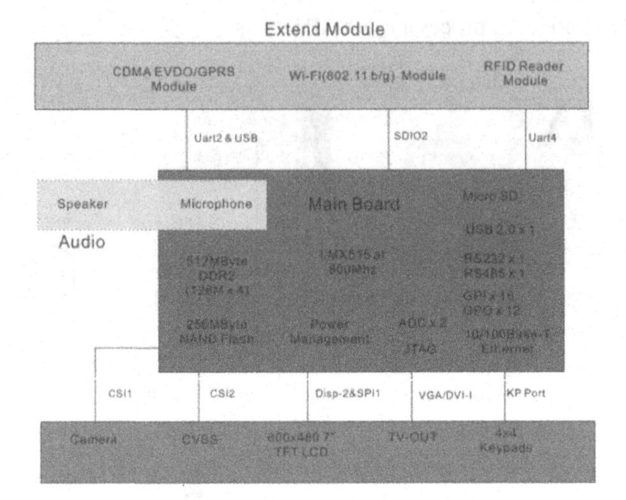

Fig. 2. The block diagram of the hardware platform

From Fig.2 we can see that the size of the RAM is 512M and the NAND Flash 256M, which can fully support the running of the system software and the video decoding. The NAND Flash is used to host the linux kernel and system programme, and the database and configuration files are located in the TF card.

As shown in Fig.2 the platform support touch screen whose resolution is 800X480. The camera interface is designed to fulfil the needs of visual talk and surveillance. The TV out interface can make the video output to the TV device. A small keypad can be set up to satisfy system input and hotkey functions. The RS485 and serial ports are used to control the peripheral devices such as 3G module, appliances,etc. The platform supports many GPIO ports that can be used to connect security devices.

Through attaching 3G module, the system can provide the function of sending alarm messages; through connected to ID/IC/RFID reader, fingerprint and palm vein identification modules, a entrance guard system can be implemented. The platform also support the WI-FI module that enables the wireless application.

The board used in this paper is shown in Fig.3.

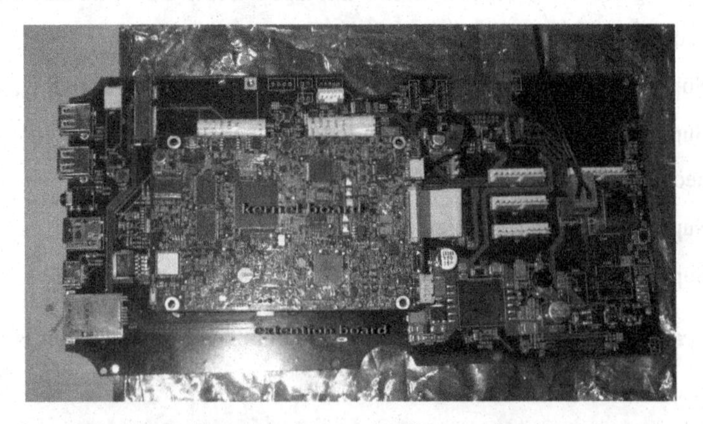

Fig. 3. The hardware platform for the smart home system

The external interfaces of the board are shown in Fig.4.

Fig. 4. The external interfaces of the platform

The USB1 and USB2 can be connected to USB flash disks; the user can use ear phone to listen music; a TF card can be plugged into the TF slot for storing programs and data; HDMI slot can be used to connect the T V HDMI input; the SIM slot can host the SIM card for the 3G or GSM functions.

2.2 Software Design

The software in the indoor machine is the most complete and complicated, from which the software in the other devices can be tailored. So this paper will take the software in the indoor machine as the example of software design. The operating system in this paper is embedded linux. QT/Embedded is used to develop GUI(Graphic User Interface), network application and database application. The programming language is C/C++. The SQLite is used as the database.

Linux is an efficient, open source and free operating system, which is extensively used in the embedded application development. Especially the linux kernel can be tailored according to the application requirement and linux supports multi-thread and device file, which perfectly fits for the smart home system development.

QT is a framework of cross-platform application and user interface development. The programme written in QT can be compiled and run under many platforms. QT includes plenty of C++ class libraries, supporting graphic,image, network and database programming APIs. QT also provides the convenient and efficient signal-slot mechanism which facilitates the development. QT/Embedded is the QT platform optimized and specially used for embedded linux.

SQLite is a lite relational database management system that complies with ACID, whose design object is for embedded system and it has been extensively applied on many kinds of embedded products. SQLite occupies very little resource and needs only several hundred kbytes of memory.

Based on the design principle of modularity, the software architecture is designed as follows(Fig.5):

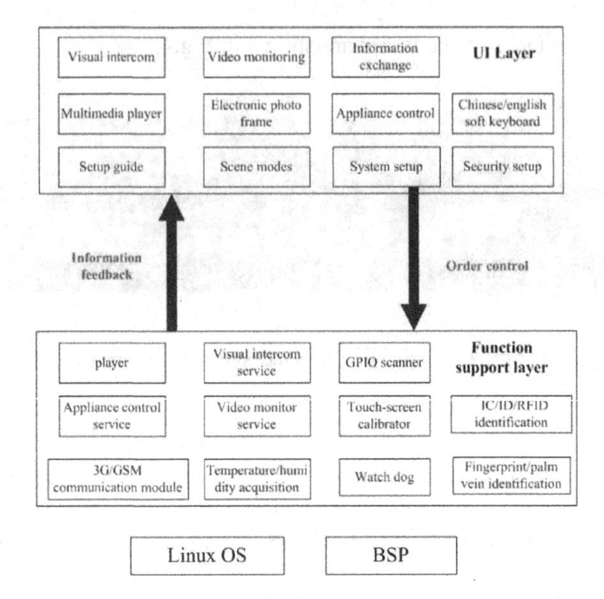

Fig. 5. The software architecture of indoor machine

The software in this paper is divide into two layers: the UI layer and function support layer. The UI layer is in charge of interaction with users, network message sending and receiving, image and video rendering, and system setup, etc. the function support layer is in charge of the functions related to the harware platform such as the player using decoder to play the media files, the visual intercom service using codec to encode and decode the video stream, the GPIO scanner scanning the GPIOs for security devices and hotkey keyboard at real time, the watch dog feeding the watchdog hardware at regular time, etc..

The function support layer transfering messages to UI layer (e.g. the player feedback of current palying progress) and the UI layer tranfering orders to the function support layer (e.g. the video on/off in visual intercom) are realized through pipeline technology. The advantages of the design are as follows:

First, if there are some problems in the function support layer, the UI layer will not be heavily affected and can recover in time; Second, the portability is very good, i.e. when the hardware platform changes, only the support function layer needs to be developed again.

3 The Design of Visual Intercom

In this paper, the video is encoded and decoded according to the H.264 standard. By scaling the parameters, stable tranfering of high quality video and low delay can be achieved. The realtime H.264 codec is supported by the i.MX51 platform. The raw audio stream is transferred, not by encoding method, which can lead to a low delay.

The procedure of initiative call in visual intercom is designed as shown in Fig.6.

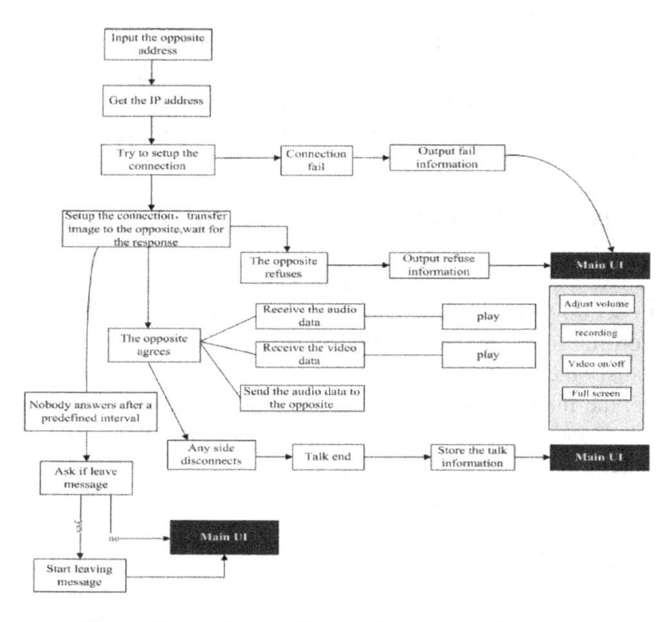

Fig. 6. The initiative call procedure in visual intercom

The procedure of passive call in visual intercom is designed as shown in Fig.7.

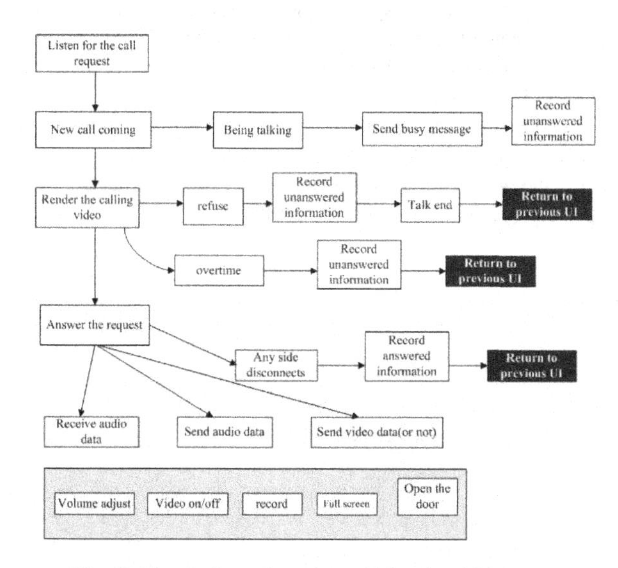

Fig. 7. The design of passive call in visual intercom

According to the above design of talk procedure, not only the audio and video data on the both sides should be transferred, but also the control orders(e.g. accept, refuse, and nobody etc.) should be transferred. In this paper, several network transfer channels are designed including data transfer channel and order transfer channel, of which there are four data transfer channels and two order transfer channels. Every channel is maintained using a thread. The data transfer channel is started or stopped according to the orders received from the order transfer channel. Furthermore the audio data channel and video data channel must be synchronized.

4 Conclusions and the Future Work

This paper designs a smart home system that integrates visual intercom, multimedia entertainment, security alarm and information service. A prototype system is developed according to the architecture designed in this paper, which functions very well and can satisfy the smart home needs.

In the future, the functions such as appliance control, remote video monitoring,etc. will be developed, and more devices such as cell phone, tablet pc will be added to the smart home system.

References

1. Zhang, Z., Sun, J.: Design about network system of intelligent family based on ethernet. Computer Engineering and Design 26(11), 3133–3135 (2005)
2. Kang, P., Song, X., Wang, L., et al.: Serial communication in the intelligent home network. Journal of Northwest University 35(4), 388–391 (2005)
3. Xiang, J., Xie, Z.: SmartHome system model based on embedded Internet/Intranet. Computer Engineering and Design 26(9), 2425–2427 (2005)
4. Liu, X., Dong, Y.: Design of Visual-speaking System. Computer Knowledge and Technology 6(8), 1951–1953 (2010)
5. The Visual Inter-conversation Door-entry Control System Based on Internet line. Microcomputer Information 23(26), 125–127 (2007)
6. Xie, Y., Liu, J.: The Application of AT89S51 in Visual Speaker-phone. Techniques of Automation and Applications 25(11), 95–97 (2006)

Study of Self-optimizing Load Balancing in LTE-Advanced Networks

Zhengrong Xiao, Yaoxu Zhang, and Zhaobiao Lv

China Unicom Research Institute, Beijing, China

Abstract. With the rapid development of mobile communication, mobile networks become more and more complex. Self-organizing network (SON) is a research focus. This article firstly gives briefly review on the performance of SONs that without physical backhaul interface, and then introduces a MLB algorithm which meets the requirements of this kind of SONs. Finally, a new scheme to improve the performances of LTE-Advanced network is proposed, and simulation shows that the scheme is of good performance.

Keywords: Self-organizing Network, LTE-Advanced, Self-optimizing Load Balancing.

1 Introduction

With the demands of people increased rapidly, the future development of network service will become more complex, efficient and convenient. So further optimized and advanced network schemes are needed [1]. LTE-Advanced standard was born in this case. As an advanced version of LTE (Long Term Evolution), LTE-Advanced was formally proposed by 3GPP partners, including ARIB, ATIS, CCSA, ETSI, TTA and TTC. It represents the standard of North American, European and Asian regions and makes 3GPP a truly global initiative [2]. Motivated by IMT (International Mobile Telecommunications)-Advanced (4G), the technology standards designed in LTE-Advanced have already exceeded 4G requirements actually [2]. Be different from previous generations' mobile systems (for example, UMTS based on WCDMA), LTE/LTE-Advanced adopts a new deployment plan of networks, which is similar to an "IP" based network. In LTE/LTE-Advanced, eNB is treated as a base composed element in telecommunication systems, which performs independently according to its collected information (Maybe from UE, maybe from neighbor NBs). By doing this, new added equipments can be easily configured, which may lead to a more efficient and intelligent network.

2 Challenges and Solutions

When it comes to the operations of the networks, it is widely accepted that the revenue of a mobile network highly depends on its operational efficiency. Consequently, advanced optimizing strategies are very important in reducing operational expenditures of networks, of course, LTE-Advanced network is also included. It is this reason that

H. Tan (Ed.): Knowledge Discovery and Data Mining, AISC 135, pp. 217–222.
springerlink.com

motivates the research on the radio resource management of LTE-Advanced networks. Actually, in LTE system, self-organized technology has already been adopted, and in LTE-Advanced system, the target of self-optimizing scheme is similar to LTE, which is to realize the efficient usage of limit wireless resource in the case of meeting the requirements of the users such as good quality of service, efficient support for higher layer and good load balance mechanism. However, due to higher system performance (for instance, higher resource utilization ratio, faster transform rate and so on) required compared to LTE, LTE-Advanced must develop new algorithms and optimizing plans. Load balance is an important part that could be improved, which is also the core technology of self-organized networks. And via new schemes designed in LTE-Advanced, the transactions of cells are allocated more reasonable so that radio resource can be used more efficiently, level of QoS can be higher and the talk drop ratio becomes little. Conventionally, load balance in communication systems is realized by handover, which hand over traffics from hot spot cells to other available cells so that transaction wouldn't congested in some hot spot cells only. So to optimize the load balance design is actually to optimize the hand over mechanism [8] [9]. However, in practical design, parameters that affect the results usually perform oppositely, which means, one parameter may increase the performance of the results but another one may decrease this effect. So, choose proper design schemes, including the parameter, sub-plan design and so on, are also very important. Greedy thought [10] is an important thought in engineering designs, which usually solve problem by four steps: use mathematic language to describe the problems, divide the domain problem into numbers of sub-problems, find best solution for each sub-problem and at last take all the effect factors into consideration and get an optimized solution for the entire design, which maybe not the best solution for each sub-problem. In wireless communication system, due to the complex conditions of the transform channels and uncertainties of users' motion, greedy thoughts seem to be one of the good choices. A MLB algorithm proposed in [3] that is derived make steps on optimizing performance of LTE-Advanced networks. It considers the effective factors that contribute to radio resource management together, and use greedy algorithm to achieve a trade-off between user backlog and times of hand over. Also, it improves the throughputs of the total system by almost 20 percents. However, as an optimized design, other performance should also be enhanced as much as possible, for example, the talk drop ratio of the user.

3 Mobility Load Balancing with Low Cost

The conventional load balance algorithm can be found in [6], to decide MRC and TP is the key problem in LB algorithm design, and reference [7] gives more information on this.

It is reasonable to assign u to a cell providing the largest [3]$W = QB(u) \times DR(u,c) \times (1-PF)$, where *QueueBacklog(u)*is the waiting time for a user. *DataRate(u,c)* stands for the MAX achievable transform rate of the available target eNB stations. *Penalty actor* means a control parameter that is meaningful to the performance of the network.

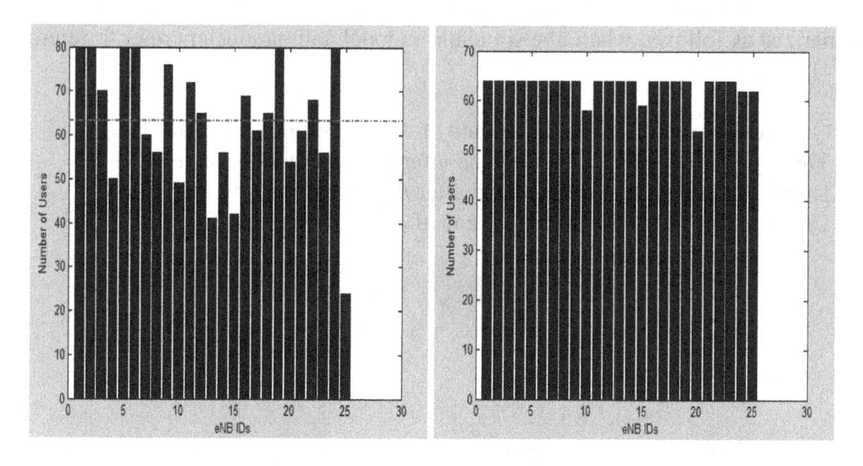

Fig. 1. Comparisons of queue traffic in each eNB before and after improvement

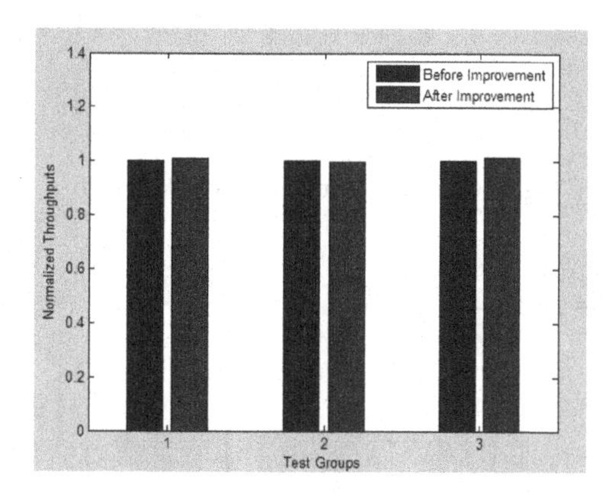

Fig. 2. Comparisons of normalized system throughputs

In the previous research, *PF* is designed only as a factor manually control the frequency of the hand over, and in my research, proper *PF* values are not only meaningful to the hand over costs, but dynamic *PF* scheme can also reflect the effects of target loads and consequently reduce hand over drop ratio in high load case. The *PF* solution for high load case can be given as follow:

$$PF = PF\^{}\ \{-1/S\times L+1\} \quad (1)$$

Where S is Unused queue space after all users been deployed uniformly to the eNBs.

Where L is load differences between actual load and average load.

By revising the MLB algorithm, load factor of the target eNB is taken into consideration and the user distribution problem occurred from this can also be solved by relating traffic conditions and algorithm parameter together. The test steps are summarized as follows, where the simulation model and parameters refer to reference [3], [4], [5]:

1) Queued user comparisons before and after improvements: figure 1
2) Throughput comparisons before and after improvements: figure 2
3) Drop ratio comparisons before and after improvements: figure 3
4) User backlog comparisons before and after improvements: figure4

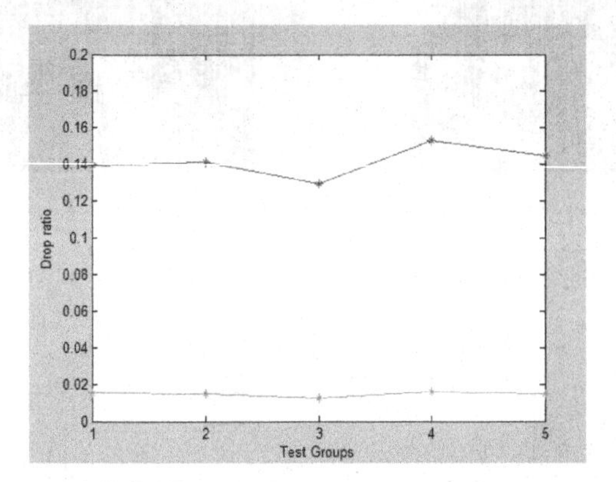

Fig. 3. Comparisons of hand over drop ratio

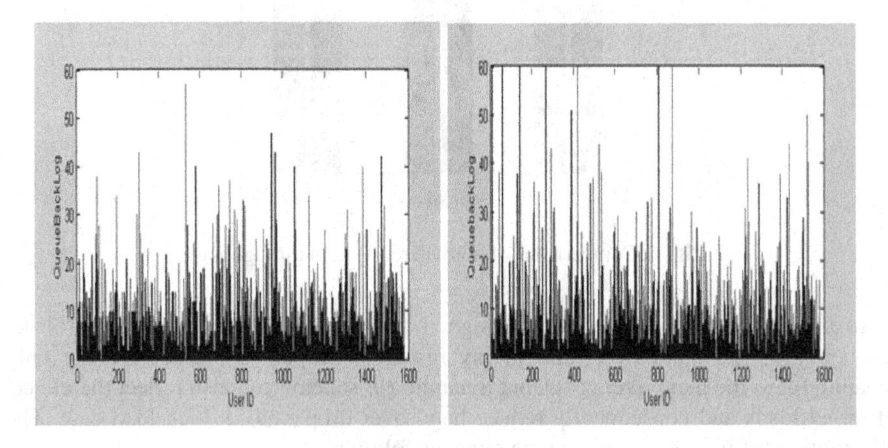

Fig. 4. Comparisons of user backlog

In Figure 4, because the extra time delay introduced is very small, the average time delay before and after are given: Left part-5.45, Right part-6.00

4 Analysis

In figure 1, although the reference MLB algorithm [3] take user backlog factor into consideration, user backlog is not equal to queue length here, which means User B can get a large QueueBacklog value due to its transaction priority, so may be HO from the shorter queue to the longer one, while User A, who is one of the user in the longer queue will not HO from the original queue due to its activities. And in most cases, the communication quality of the alternative base station is not that differential. Consequently, the hand over drop ratio would occur in hot spot cells, especially in high load case. The reason that causes this problem can be showed as figure 5.

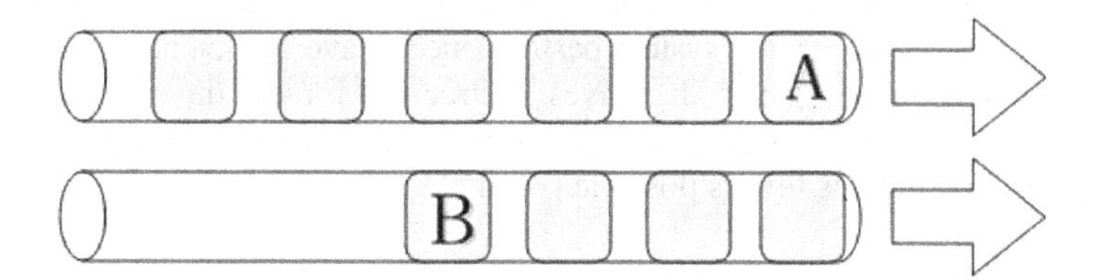

Fig. 5. Sketch map for status of queued users in two base stations

As the improved scheme only makes revise on the way selecting penalty factor, it would not affect the system throughputs, which is coincide with figure 2. Introduce of extra time delay in figure 4 also agrees with reference [3], because the relationship between user backlog and penalty factor has already been proved. And figure6 shows the sketch map.

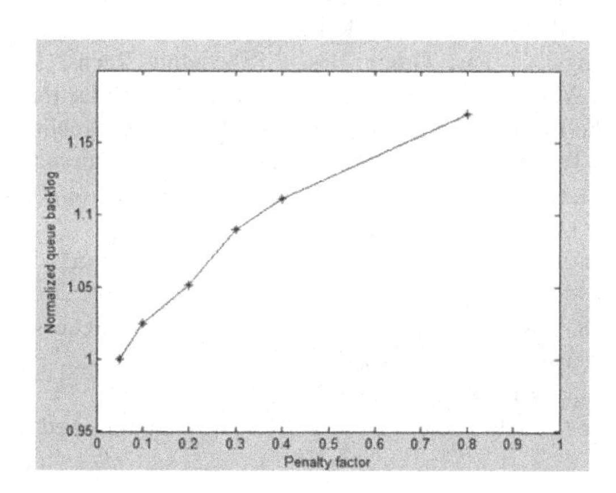

Fig. 6. Relationship between normalized queue backlog and penalty factor

Generally speaking, the improved scheme firstly takes load factor of the target eNB into consideration and fixed the problem by relating traffic conditions and algorithm parameter together. Then new improvements keep the advantages of the original algorithm in a large degree, which meets the requirements of increasing.

5 Conclusion

SON technology play an important role in LTE networks, and clearly, its advantages will be extended in LTE-Advanced standards. So it's very important to propose more efficient algorithms and network structures. The newly proposed MLB algorithm realize the addition of new eNBs without backhaul interfaces, and the improvements on the algorithm, not only keep the advantages of the original algorithm but also smooth the user accumulations in the system and as a result, decrease the hand over drop ratio of the whole system. The idea of switch penalty factor, which is an important factor to improve the system performance, leave sufficient space for further algorithm improvements and it also gives a direction for the developments of new algorithm in LTE-Advanced standard: consider as many as possible factors together and integrate them into as few as possible parameters.

References

1. Lee, S., Kim, S.-C., Shin, Y.-S.: Soft Load Balancing Over Heterogeneous Wireless Networks Hyukmin Son. IEEE Xplore (2008)
2. 3GPP. LTE-Advanced networks, http://www.3gpp.org/LTE-Advanceddvanced (retrieved June 15, 2010)
3. Honglin, H., Jian, Z., Xiaoying, Z., Yang, Y., Ping, W.: Self-Configuration and Self-Optimization for LTE Networks. IEEE Communications Magazine (2010)
4. 3GPP TR 25.996 V7.0.0(2007-06), 3rd Generation Partnership Project; Technical Specification Group Radio Access Network; Spatial channel model for Multiple Input Multiple Output (MIMO) simulations (Release 7)
5. Honglin, H., Jing, X.: Fast Group Based Scheduling Techs. In: 3GPP LTE Key Technologies on Wireless Links, ch. 11. Publishing House of Electronics Industry, China
6. Keping, Z.: Radio Resource Managements. In: LTE-B3G/4G Mobile Telecommunication System Wireless Technologies, ch. 6. Publishing House of Electronics Industry, China
7. 3GPP R3-071254, Load Balancing Self optimization Use case. Ref Type: Generic (2007)
8. Baidupedia (July 14, 2010). Ping-Pong effect, http://baike.baidu.com/view/1065268.htm?fr=ala0_1_1 (retrieved August 1, 2010)
9. 3GPP R3-071598. Information to be included at Hand Over request messages, to avoid Ping-Pong Hand Over. Ref Type: Generic (2007)
10. Baidupedia (August 2010). Greedy algorithm, http://baike.baidu.com/view/1628576.htm (retrieved August 10, 2010)

Research on the Methods of Similarity Measures between Vague Sets

Jun-min Li and Hu Li

Department of Computer Science, Xi'an University of Science and Technology, Xi'an, Shaanxi 710054, China

Abstract. Vague set is a generalization of the concept of fuzzy set , similarity measure is the key to research vague set theory. At present, many methods of similarity measure have been proposed, and even been used in several areas. Based on the analysis and studies the methods of similarity measure proposed by domestic and foreign scholars, a new kind of similarity measures of vague set or value is put forward in this article, as well as compares this method with other scholars' methods, it is verified that the new measures are more rational and practical, and shown that it can be a more useful way to measure the degree of similarity between vague set or value.

Keywords: Vague sets, similarity measure, membership function.

1 Introduction

Since Zadeh brought forth the Fuzzy Set Theory in year 1965, the Fuzzy Set Theory has developed sound and has been applied widely in actual projects. While it is found that the subordinate function value of a fuzzy set is a single value, so it cannot show the proofs of both support and opposition meanwhile. To describe the fuzzy information more entirely, in year 1993, Gau, Buehrer and others put forward an expanded form of the Fuzzy Set, that is, the Vague Set Theory. The traditional fuzzy sets cannot express and or process this kind of fuzzy information. Compared to the traditional fuzzy sets, the vague sets can express information more entirely, well and truly.

As a new theory to express the fuzzy knowledge, the vague set is used widely in the reasoning process of the intelligent control system. Currently, the vague set is successfully used in fuzzy control, decision analysis, mode recognition and other areas, where the similitude degree measurement is the core technology. While up till now, there is no reasonable criteria to define the similitude degree measurement. Now, all similitude degree measurement methods rely on two thoughts. One is based on the difference between the real membership degree and the fake membership degree of a vague set. Chen[1] proposed that the S function can be used to measure the similitude degree. The defect of this method is that when vague sets change in a very small degree, their similitude degree changes in a large degree. The other thought is based on the distance measurement. Liang[10], together with others, analyzed comprehensively the methods of similitude degree measurement of the vague set and found that the degrees and the equalizing values of the vague sets' subintervals were also important factors that could affect the similitude degree between two vague sets. Then they proposed

H. Tan (Ed.): Knowledge Discovery and Data Mining, AISC 135, pp. 223–229.
springerlink.com

improved the method of similitude degree measurement, which was based upon the distance of the Vague sets. While there are still some defects in the improved method. For example, different vague sets may have the same similitude degree. By analyzing many domestic and abroad scholars' similitude degree measurement methods[5-9], in this paper, we propose a new method of the similitude degree measurement of the vague set or value and it is proved that this method meets some rules. And this method is compared with similitude degree measurement methods through several examples.

2 The Foundation of Vague Set Theory and the Concept of Similitude Degree

Definition 1. Set U is a domain, use x to represent arbitrary element in the domain, one Vague set V in the domain U is represented with a real membership function $t_v(x)$ and a fake membership function $f_v(x)$. $t_v(x)$ and $f_v(x)$ associate the real number in the interval [0,1] with the each one point in the set U, $t_v : V \to [0,1]$ and $f_v : V \to [0,1]$,meanwhile $t_v(x) + f_v(x) \le 1$.

When U is continuous, vague set can be written as

$$V = \frac{1}{x}\int_x [t_v(x), 1 - f_v(x)] \quad x \in U ;$$

When U is discrete, vague set can be written as

$$V = \sum_{i=1}^{n} [t_v(x), 1 - f_v(x)] / x_i \quad x_i \in U .$$

Definition 2. Suppose VFS(U)said the all of the vague sets in the domain U, set A,B,C $\in$ VFS(U),if the mapping M: VFS(U)×VFS(U)→[0,1] satisfies a condition: ① M(A,B)=M(B,A); ②M(A,A)=1; ③if $A \subseteq B \subseteq C$, then $M(A,C) \le \min(M(A,B), M(B,C))$.Then M(A,B) is said the degree of similarity between vague set A and vague set B.

Definition 3. Set A and B are the two vague sets in domain U, the complementary set $\overline{A}$ of A can be defined as

$$t_{\overline{A}}(x) = f_A(x), 1 - f_{\overline{A}}(x) = 1 - t_A(x) .$$

Definition 4. Set A and B are the two vague sets in domain U,

$$A \subseteq B \Leftrightarrow \forall x \in U, t_A(x) \le t_B(x), f_A(x) \ge f_B(x) .$$

3 Vague Value Similarity Measures Method and Related Standards

3.1 Some Rules of Similarity Measure between Vague Value

According to actual background of the uncertain information processing[4], put forward some operative principle to verify whether the similarity measures is rational or not. Set vague values $x=[t_x, 1-f_x]$and $y=[t_y, 1-f_y]$,then

Rule 1 :$0 \leq M(x,y) \leq 1$;
Rule 2 :$M(x,y)=M(y,x)$;
Rule 3 :$M(x,y)=M(\bar{x}, \bar{y})$;
Rule 4 :$M(x,y)=1 \Leftrightarrow x=y$; $M(x,y)=0 \Leftrightarrow x=[0,0],y=[1,1]$,or $x=[1,1],y=[0,0]$;
Rule 5 :if $x \subseteq y \subseteq z$,then $M(x,y) \geq M(x,z)$,and$M(y,z) \geq M(x,z)$;
Rule 6 :If for any x, has $M(x,y)=M(x,z)$, then $M(y,z)=1$.

3.2 Some Methods to Measure the Degree of Similarity between Vague Value

In literature [1], Chen define the degree of similarity between vague values $x=[t_x ,1-f_x]$ and $y=[t_y ,1-f_y]$ as

$$M_C(x,y)=1-(|S(x)-S(y)|)/2, \tag{1}$$

where, $S(x)=t_x-f_x$, $S(y)=t_y-f_y$.

In literature [2], Chen define the degree of similarity between vague values $x=[t_x ,1-f_x]$ and $y=[t_y ,1-f_y]$ as

$$M_H(x,y)=1-(|t_x-t_y |+|f_x-f_y|)/2 . \tag{2}$$

In literature [3], Chen define the degree of similarity between vague values

$x=[t_x ,1-f_x]$and $y=[t_y ,1-f_y]$ as

$$M_L(x,y)=1-(|S(x)-S(y)|)/4 -(|t_x-t_y |+|f_x-f_y|)/4. \tag{3}$$

4 A New Method to Measuring the Degree of Similarity

To measure more reasonably the similitude degree between two vague sets, we do some analysis and study. According to the current vague set knowledge and given the similitude degree measuring formulas of scholars both domestic and abroad, in this paper, we propose a new method to measure the similitude degree between vague sets or values.

4.1 New Method to Measure the Similitude Degree between Vague Values

Supposing $X=[t_X ,1-f_X]$ and $Y=[t_Y ,1-f_Y]$ are two vague values in the domain U, the degree of similarity between X and Y can be calculated by the following function $M_{new}(X,Y)$ as

$$M_{new}(X,Y)=1-\frac{|S_X-S_Y|}{8}-\frac{3}{4}\cdot(\frac{|t_X-t_Y|}{1+t_X+t_Y}+\frac{|f_X-f_Y|}{1+f_X+f_Y}) \tag{4}$$

where, $S_X=t_X-f_X$ and $S_Y=t_Y-f_Y$.

Now we prove that the following function $M_{new}(X,Y)$meet the rules for measuring the similarity between vague values(Rule 1 to Rule 6).

Proof:

(1) Because there are $0 \le 2 \cdot |t_X - t_Y| \le 1 + t_X + t_Y$ and $0 \le 2 \cdot |f_X - f_Y| \le 1 + f_X + f_Y$, we have $0 \le \frac{3}{4} \cdot (\frac{|t_X - t_Y|}{1+t_X+t_Y} + \frac{|f_X - f_Y|}{1+f_X+f_Y}) \le \frac{3}{4}$

Because there are $-1 \le S_X \le 1$ and $-1 \le S_Y \le 1$, we have $0 \le (|S(x)-S(y)|)/8 \le 1/4$, then we have $0 \le M_{new}(X,Y) \le 1$.

(2) According to the symmetry property, we have $M_{new}(X,Y)= M_{new}(Y,X)$.

(3) Because there are $\overline{X}=[f_X, 1-t_X]$ and $\overline{Y}=[f_Y, 1-t_Y]$, we have $M_{new}(\overline{X},\overline{Y}) = 1 - \frac{|S_Y - S_X|}{8} - \frac{3}{4} \cdot (\frac{|f_X - f_Y|}{1+f_X+f_Y} + \frac{|t_X - t_Y|}{1+t_X+t_Y})$, therefore, we have $M_{new}(X,Y) = M_{new}(\overline{X},\overline{Y})$.

(4) $M_{new}(X,Y)=1 \Leftrightarrow 1 - \frac{|S_X - S_Y|}{8} - \frac{3}{4} \cdot (\frac{|t_X - t_Y|}{1+t_X+t_Y} + \frac{|f_X - f_Y|}{1+f_X+f_Y}) = 0 \Leftrightarrow |S_X - S_Y| = 0$, $|t_X - t_Y| = 0$ and $|f_X - f_Y| = 0 \Leftrightarrow t_X = t_Y$. So we have $f_X = f_Y \Leftrightarrow X = Y$.

Furthermore,

$$M_{new}(X,Y) = 0 \Leftrightarrow \frac{|S_X - S_Y|}{8} + \frac{3}{4} \cdot (\frac{|t_X - t_Y|}{1+t_X+t_Y} + \frac{|f_X - f_Y|}{1+f_X+f_Y}) = 1 \Leftrightarrow |S_X - S_Y| = 2$$

So we have

$\frac{|t_X - t_Y|}{1+t_X+t_Y} = \frac{|f_X - f_Y|}{1+f_X+f_Y} = \frac{1}{2} \Leftrightarrow t_X = 0, t_Y = 1, f_X = 1, f_Y = 0$, or $t_X=1$, $t_Y=0$, $f_X=0$, $f_Y=1 \Leftrightarrow X=[0,0]$, $Y=[1,1]$ or $X=[1,1]$, $Y=[0,0]$.

(5) We suppose $X=[t_X,1-f_X]$, $Y=[t_Y,1-f_Y]$ and $Z=[t_Z,1-f_Z]$.

If there is $X \subseteq Y \subseteq Z$, we have $t_X \le t_Y \le t_Z$, $f_X \ge f_Y \ge f_Z$ and $S_X \le S_Y \le S_Z$.

Now we have $M_{new}(X,Y) - M_{new}(X,Z) = 1 - \frac{|S_X - S_Y|}{8} - \frac{3}{4} \cdot (\frac{|t_X - t_Y|}{1+t_X+t_Y} + \frac{|f_X - f_Y|}{1+f_X+f_Y}) - \left[1 - \frac{|S_X - S_Z|}{8} - \frac{3}{4} \cdot (\frac{|t_X - t_Z|}{1+t_X+t_Z} + \frac{|f_X - f_Z|}{1+f_X+f_Z})\right] > 0$, and we have $M_{new}(X,Y) \ge M_{new}(X,Z)$, In a similar way, we have, $M_{new}(Y,Z) \ge M_{new}(X,Z)$.

(6) Let $X=[0,0]$. Substituting the the equation $M_{new}(X,Y)=M_{new}(X,Z)$, after necessary arrangements, we have

$$1 - \frac{1+S_Y}{8} - \frac{3}{4} \cdot (\frac{t_Y}{1+t_Y} + \frac{1-f_Y}{2+f_Y}) = 1 - \frac{1+S_Z}{8} - \frac{3}{4} \cdot (\frac{t_Z}{1+t_Z} + \frac{1-f_Z}{2+f_Z}) \quad (5)$$

Let X=[0,1], and $S_Y>0, S_Z>0$. Substituting the equation $M_{new}(X,Y)=M_{new}(X,Z)$, after necessary arrangements, we have

$$1-\frac{S_Y}{8}-\frac{3}{4}\cdot\left(\frac{t_Y}{1+t_Y}+\frac{f_Y}{1+f_Y}\right)=1-\frac{S_Z}{8}-\frac{3}{4}\cdot\left(\frac{t_Z}{1+t_Z}+\frac{f_Z}{1+f_Z}\right)\cdot \tag{6}$$

Let the formula (5) subtract the formula (6). After necessary simplification, we have $(7+5\cdot(f_Y+f_Z)+4\cdot f_Y f_Z)\cdot(f_Y-f_Z)=0$. Because there is $(7+5\cdot(f_Y+f_Z)+4\cdot f_Y f_Z)\neq 0$, then $f_Y-f_Z=0$, that is, $f_Y=f_Z$ .So we have $t_Y=t_Z$, and then we have $M_{new}(Y,Z)=1$.

4.2 The Similitude Degree Measurement of M_{new} at Vague Sets

We suppose A,B are two vague sets belonging to the set $X=\{x_1,x_2,\ldots,x_n\}$. Then the similitude degree between the two vague sets A and B can be computed by the following formula:

$$MS(A,B)=\frac{1}{n}\sum_{i=1}^{n}\left[1-\frac{|S_A(x_i)-S_B(x_i)|}{8}-\frac{3}{4}\cdot\left(\frac{|t_A(x_i)-t_B(x_i)|}{1+t_A(x_i)+t_B(x_i)}+\frac{|f_A(x_i)-f_B(x_i)|}{1-f_A(x_i)+f_B(x_i)}\right)\right] \tag{7}$$

where there are $S_A(x_i)=t_A(x_i)-f_A(x_i)$ and $S_B(x_i)=t_B(x_i)-f_B(x_i)$.

According to the proof method of formula (4), we know that MS(A,B) satisfies the Rule 1 to Rule 6 of the similitude degree measurement between vague sets.

4.3 Comparison and Analysis of Examples

Let the vague set x=[0.2,0.4] and the vague set y=[0.1,0.4]、[0.2,0.3]、[0.2,0.5]、[0.3,0.4]. The Table 1 shows the results computed by different similitude degree measurement methods.

Table 1. Comparison of Different Similitude Degree Methods

	1	2	3	4
x	[0.2,0.4]	[0.2,0.4]	[0.2,0.4]	[0.2,0.4]
y	[0.1,0.4]	[0.2,0.3]	[0.2,0.5]	[0.3,0.4]
$M_C(x,y)$	0.950	0.950	0.950	0.950
$M_H(x,y)$	0.950	0.950	0.950	0.950
$M_L(x,y)$	0.950	0.950	0.950	0.950
$M_{new}(x,y)$	0.930	0.955	0.952	0.938

We can learn from the vague values that, when measuring the similitude degree between the vague set x=[0.2,0.4] and different vague sets y, none of the measurement functions M_C, M_H and M_L can distinguish different vague sets y, the measurement function M_{new} proposed in this article is able to distinguish different vague sets y. Therefore we know that the similitude degree measure formula proposed in this article is better than the other three formulas.

Example 1. Let A、 B and C be vague sets in the universe of discourse U={x_1, x_2, x_3, x_4, x_5}, where A is a regular set for comparing the sets B and C. Then let

$$A = [0.1,0.3]/ x_1 +[0.2,0.6]/ x_2 +[0.4,0.8]/ x_3+[0.6,0.8]/ x_4+[0.8,1]/ x_5,$$
$$B= [0.2,0.5]/ x_1 +[0.3,0.7]/ x_2 +[0.5,0.8]/ x_3+[0.7,0.9]/ x_4+[0.9,1]/ x_5 , \text{ and}$$
$$C = [0.1,0.2]/ x_1 +[0.2,0.5]/ x_2 +[0.5,0.8]/ x_3+[0.7,0.8]/ x_4+[0.9,1]/ x_5.$$

According to the formula (7), then we have MS(B,A) =0.9020, and MS(B,A)= 0.9536.

The value of MS(B,A) is smaller than that of MS(C,A),which shows the similitude degree between the vague set B and the vague set A is smaller than the similitude degree between the vague set C and the vague set A, which further shows that the vague set C is more similar to the vague set A than the vague set B.

It can be seen that, for any two vague sets A and B, if the MS(A,B) is bigger, the similitude degree between the two vague sets is bigger.

Example 2. Supposing there are three known modes defined in X={x_1,x_2,x_3}, with their vague sets as follows:

$$A_1 = [0.1,0.9]/ x_1 +[0.5,0.9]/ x_2 +[0.1,0.1]/ x_3,$$
$$A_2 = [0.5,0.5]/ x_1 +[0.7,0.7]/ x_2 +[0.1,0.2]/ x_3 \text{ and}$$
$$A_3 = [0.7,0.8]/ x_1 +[0.1,0.2]/ x_2 +[0.4,0.6]/ x_3.$$

Let B be a sample to be distinguished. with its vague set characteristic as

$$B = [0.4,0.4]/ x_1 +[0.6,0.8]/ x_2 +[0.1,0.1]/ x_3.$$

Please analyze the Sample B should be classified to which mode.

According to the formula (7), we can compute the similitude degree between the vague A_i(i=1,2,3) and B and we can get the results of MS(A_1,B)=0.837, MS(A_2,B)=0.926 and MS(A_3,B)=0.548. Therefore we know that the vague set B is more similar to A_2, that is, the Sample B should be classified to the mode A_2. This result is the same as that in the document[10], and the size of the similitude degree value is reasonable and direct viewing.

5 Conclusion

Based on the analysis and research of the method to measuring the degree of similarity between vague sets(values) in the literature [1, 3], a new method to measuring the degree of similarity between vague sets(values) was put forward ,to prove its meet six rules of similarity measure between vague sets (values) , through further comparing the new method M_{new} put forward in this paper with three similar measure methods M_C、 M_H and M_L (see table 1), shows that the effectiveness and superiority of the proposed method of the similarity measure, also can see the similar measurement method presented in this paper has better resolution and differentiate.

References

1. Chen, S.-M.: Measures of similarity between vague sets. Fuzzy Sets and Systems 74(2), 217–223 (1995)
2. Hong, D.H., Kim, C.: A note on similarity measures between vague sets and between elements. Information Sciences 115(1), 83–96 (1999)
3. Li, F., Xu, Z.-Y.: Measures of Similarity between Vague Sets. Journal of Software 12(6), 922–927 (2001)
4. Zhang, C.-Y., Dang, P.-A.: On Measures of Similarity between Vague Sets. Computer Engineering and Application 39(17), 92–94 (2003)
5. Shi, Y.-Q., Wang, H.-X.: A method to measuring the similarity between Vague sets. Computer Engineering and Application 41(27), 178–180 (2005)
6. Zadeh, L.A.: Fuzzy sets. Information and Control 8(3), 338–356 (1965)
7. Fun, J.-L.: Similarity Measures on Vague Values and Vague Sets. System Engineering Theory and Practice 26(8), 95–100 (2006)
8. Zhou, X.-G., Zhang, Q.: Comparision and improvement on similarity measures between vague sets and between elements. Journal of Systems Engineering 20(6), 613–619 (2005)
9. Fan, P., Liang, J.-R., Li, T.-Z.: New Method of Similarity Measures between Vague Sets. Computer Engineering and Application 34, 70–72 (2006)
10. Liang, Z.-Z., Shi, P.-F.: Similarity measures on intuitionistic fuzzy sets. Pattern Recognition Letters 24(15), 2687–2693 (2003)

A Signal Processing Scheme for CPT Atomic Clock

Yi Zhang[1], Wei Deng[2], Yuan Tian[3], and Sihong Gu[1,2,3,*]

[1] School of Physics, Huazhong University of Science and Technology, 430074, Wuhan, China
[2] Key Laboratory of Atomic Frequency Standards, Chinese Academy of Sciences, 430071, Wuhan, China
[3] College of Optoelectronic Science and Engineering, Huazhong University of Science and Technology, 430074, Wuhan, China
shgu@wipm.ac.cn

Abstract. Coherent Population Trapping (CPT) atomic clock can be realized to be a chip scale atomic clock. This paper presents our signal processing scheme, including the equations we derived for the scheme, and the corresponding circuit based on digital signal processor. The power consumption of the circuit is 200mW, and the volume is $10cm^3$. Connected with an experimental physics package to form a CPT atomic clock, the circuit has realized all the expected functions. The atomic clock circuit based on this scheme is suitable for further reducing size and power consumption, and is promising to be applied in the chip scale atomic clock.

Keywords: Digital signal processing, CPT atomic clock, CSAC.

1 Introduction

Atomic clocks [1] have been widely used in measurement, navigation, communication etc. Coherent Population Trapping (CPT) atomic clock is especially suitable for atomic clocks of small size and low power consumption. With the use of micro electric-mechanical system (MEMS), it can be fabricated into Chip Scale Atomic Clock (CSAC). In principle, it is the only atomic clock suitable for CSAC [2][3].

A CPT atomic clock consists of a physics package and a clock circuit. In a CPT atomic clock, there are two frequencies, laser frequency and microwave frequency, which need to be locked [4][5]. When a CPT atomic clock begins to work, the locking process consists of two steps: first, the circuit modulates and scans the direct current (DC) injected to the physics package, processes the output of the physics package to find out the value of DC corresponding to the minimum of the spectrum in figure 1(a). It then drives DC to the value at the end of scanning, and mixes the DC with negative feedback current obtained with phase-sensitive detection. Thus, DC is stabilized at the value, i.e., the laser frequency is stabilized by DC servo. The locking process of microwave frequency is similar [6][7]. Second, right after laser frequency

* Corresponding author.

H. Tan (Ed.): Knowledge Discovery and Data Mining, AISC 135, pp. 231–238.
springerlink.com

is locked, the circuit modulates and scans the microwave frequency injected to the physics package, identifies the value of microwave frequency corresponding to the maximum of the spectrum in figure 1(b). It then drives microwave frequency to the value, and mixes the control voltage of Oven-Controlled Crystal Oscillator (OCXO) with negative feedback voltage obtained with phase-sensitive detection. Thus, microwave frequency is stabilized by microwave servo, i.e., the output frequency of the atomic clock is stabilized by microwave servo.

The circuit can be realized by either analog circuit or digital circuit. However, the power consumption of the digital circuit is lower and its size is smaller. In addition, it is more convenient to modify or adjust the functions of the circuit through software. All of these make digital circuit more suitable for CSAC. This paper presents our study on digital circuit scheme for applications in CSAC.

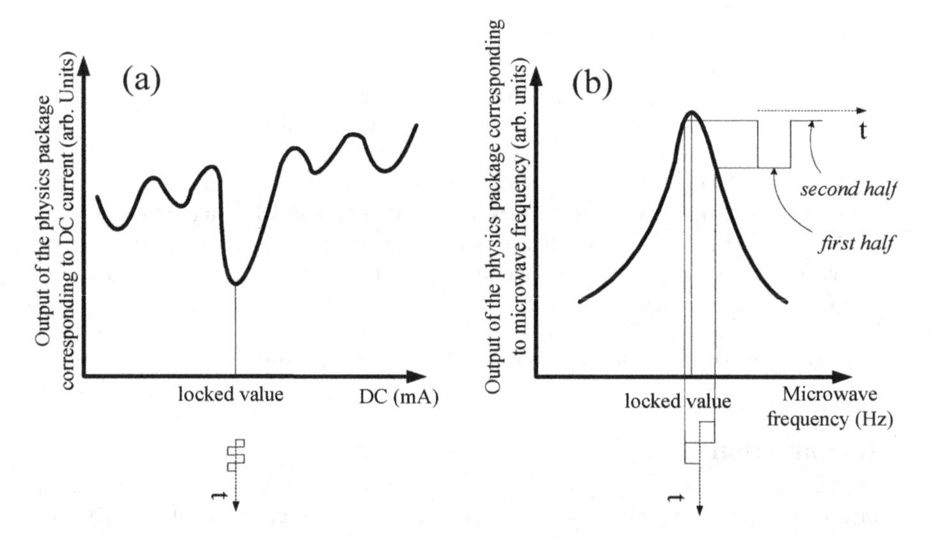

Fig. 1. CPT atomic clock working process, (a) DC servo: the *thick line* presents the response of the physics package to DC. The *thin line* presents the amplitude of DC modulated in time. (b) Microwave servo: the *thick line* presents the response of the physics package to microwave frequency. The *thin line* presents the frequency of microwave modulated in time and the signal corresponding to the modulation frequency extracted from the output of the physics package.

2 Signal Processing Scheme

The DC servo and the microwave servo are similar. As an example, we introduce our signal processing scheme of microwave servo. In order to lock the microwave frequency, we modulate the microwave with a low frequency signal, and extract signal corresponding to modulation frequency from the output of the physics package through a low pass filter. The obtained signal is shown in figure 1(b). The signal is integrated for first and second half of modulation period separately. The difference between the two integrated results, which is the differential function of the discriminating signal, is the frequency correction signal for locking microwave frequency. Hence, the difference is utilized as negative feedback error signal to lock microwave frequency. The key functions of our signal processing scheme are to integrate the difference signal using phase-sensitive detection and to achieve negative feedback.

Figure 2 presents the analysis model of microwave servo. Constant control word x_6 added by modulation signal $x_7(n)$ becomes Phase lock loop (PLL) control word $x_8(n)$. The Digital to Analog Converter (DAC) control word $x_9(q)$ determines $x_{10}(t)$, the output of DAC, which adjusts the output frequency of oven controlled crystal oscillator (OCXO) $f_{REF}(t)$. PLL synthesizes $f_{REF}(t)$ and $x_8(n)$, and gets microwave signal $f_{RF}(t)$, which is injected into the physics package with TEE. The signal $x_2(t)$ is extracted from the output of the physics package $x_1(f_{RF})$ through an analog filter, transformed to a digital signal with an Analog to Digital Converter (ADC). The demodulated signal $x_4(n)$ is obtained by multiplying $x_2(n)$ with local digital oscillator $x_3(n)$. Then, $x_4(n)$ is filtered by low pass filter and the result is $x_5(q)$, which is sent into scan&lock function in figure 2 as an error signal. In scan&lock function $x_9(q)$ and $x_5(q)$ are processed with newton iteration method to obtain the initial value of next cycle $x_9(q+1)$.

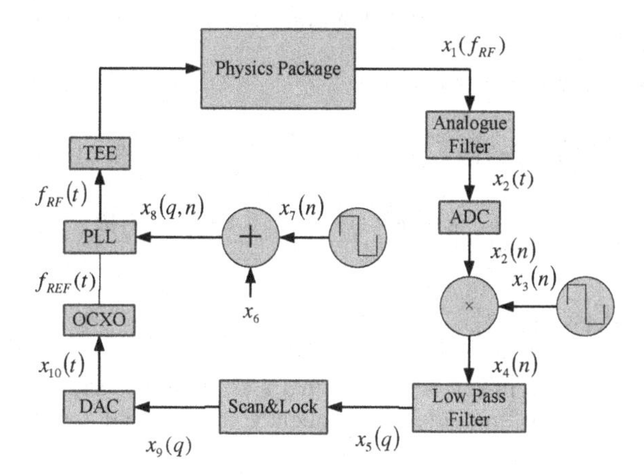

Fig. 2. Analysis model of microwave servo

To simplify the analysis, we assume the physics package transfer function $x_1(f_{RF})$ as a quadratic function in the vicinity of the locked value in figure 1(b).

$$x_1(f_{RF}) = af_{RF}(t)^2 + bf_{RF}(t) + c \tag{1}$$

in which f_{RF} satisfies $\begin{cases} f_{RF} = f_0, & x_1' = 0 \\ f_{RF} < f_0, & x_1' > 0 \\ f_{RF} > f_0, & x_1' < 0 \\ f_{RF} = f_0, & x_1 = X_0 \end{cases}$.

Then, we have

$$a < 0, \quad b = -2af_0, \quad c = X_0 + af_0^2$$

Where f_0 is the locked value in Fig. 1(b), X_0 is maximum value of the function. a, f_0 and X_0 are fixed by the physics package.

The eq. (1) becomes

$$x_1(f_{RF}) = af_{RF}^{\ 2} - 2af_0 f_{RF} + (X_0 + af_o^2) \tag{2}$$

The injected microwave frequency is given by

$$f_{RF}(t) = f_{REF}(t)\cdot\frac{M}{R} = (A + \eta x_{10}(t))\cdot\frac{M_0 + Bx_7(t)}{R} \tag{3}$$

Where A is the output frequency of OCXO with 0V of voltage controlled, η is the deviation slope of OCXO, R is the division factor of PLL, and t is time. M_0 equals to x_6, $x_8(n)$ equals to $M_0+Bx_7(n)$ which is the M division factor of PLL, where B is Frequency Shift Keying (FSK) modulation depth, which is a constant. $x_7(t)$ is the analog form of a square wave $x_7(n)$ defined as:

$$x_7(n) = \sum_{n=0}^{+\infty}\left[U(nT - k\Delta T) - 2U(nT - k\Delta T - \frac{\Delta T}{2}) + U(nT - n\Delta T - \Delta T)\right] \tag{4}$$

Where T is sampling period, U is unit step function, ΔT is total sampling time. Integer n is sample sequence time index.

Using (3) and (4), we rewrite (2) as

$$\begin{aligned} x_1(t) = &\left\{a[A + \eta x_{10}(t)]^2\cdot\frac{M_0^{\ 2}}{R^2} - 2af_0\frac{[A + \eta x_{10}(t)]\cdot M_0}{R} + \left(X_0 + af_0^{\ 2}\right)\right\} \\ &+\left\{a[A + \eta x_{10}(t)]^2\cdot\frac{2BM_0}{R^2} - 2af_0\cdot\frac{[A + \eta x_{10}(t)]B}{R}\right\}x_7(t) \\ &+ a[A + \eta x_7(t)]^2\cdot\frac{B^2}{R^2}x_7^{\ 2}(t) \end{aligned} \tag{5}$$

Notice that the first term in (5) is a DC signal and the third terms in (5) is a second order harmonic signal of modulation frequency; the second term is a first order harmonic signal of modulation frequency which is needed by the servo. We extract the second term with analog filter in figure 2 and get

$$x_2(t) = \left\{a[A + \eta x_{10}(t)]^2\cdot\frac{2BM_0}{R^2} - 2af_0\cdot\frac{[A + \eta x_{10}(t)]B}{R}\right\}\cdot\sin(2\pi ft + \theta)$$

Where f is the FSK modulation frequency.

Local digital oscillator $x_3(n)$ is a square wave, written as $x_3(n)=x_7(n-j)$, where j is the phase delay (θ) from the physics package and the printed circuits board. $x_2(t)$ is sampled and the digitized signal $x_2(n)$ multiplied by $x_3(n)$ becomes $x_4(n)$

$$x_4(n) = k\left\{a[A + \eta x_{10}(nT)]^2\cdot\frac{2BM_0}{R^2} - 2af_0\cdot\frac{[A + \eta x_{10}(nT)]B}{R}\right\}\cdot\sin(2\pi fnT + \theta)\cdot x_3(n)$$

Where $x_{10}(nT)$ is the digital form of $x_{10}(t)$, k is the linear transfer coefficient of the analog circuit and the ADC. We choose the value of j so that

$$\sin(2\pi fnT + \theta)\cdot x_3(n) = \left|\sin(2\pi fnT + \theta)\right|.$$

We accumulate $x_4(n)$ when n is n-j to n-j+Q-1, which acts as a low pass filter, and get

$$x_5(q)=k\cdot\sum_{k=n_0}^{n_0+Q-1}\left\{a[A+\eta x_{10}(n_0T)]^2\cdot\frac{2BM_0}{R^2}-2af_0\cdot\frac{[A+\eta x_{10}(n_0T)]B}{R}\right\}$$
$$=k\cdot\frac{2aB[A+\eta x_{10}(n_0T)]}{R}\sum_{k=n_0}^{n_0+Q-1}\left\{[A+\eta x_{10}(n_0T)]\cdot\frac{M_0}{R}-f_0\right\} \quad (6)$$

Where the new time index n_0 is n-j, the quantity of accumulated samples Q is KN, the quantity of samples in every modulation period N is f_s/f, the sampling frequency f_s is $1/T$, the feedback sequence time index q is n_0/Q, which is an integer.

It is seen from (6) that,

$$f_{tem}=[A+\eta x_{10}(n_0T)]\cdot\frac{M_0}{R}=f_0,\ x_5(q)=0,$$
$$f_{tem}=[A+\eta x_{10}(n_0T)]\cdot\frac{M_0}{R}>f_0,\ x_5(q)<0,$$
$$f_{tem}=[A+\eta x_{10}(n_0T)]\cdot\frac{M_0}{R}<f_0,\ x_5(q)>0.$$

Therefore, $x_{10}(n)$ could be used as a new locked quantity of the servo and $(f_0R/M_0-A)/\eta$ is the locked value, and $x_5(q)$ is the error function [8]

$$e(q)=x_5(q)=\frac{2akQM_0B}{R}[f_{RF}(q)-f_0]=C[f_{RF}(q)-f_0]$$
$$=\mathrm{C}[\frac{2\mathrm{M}_0}{\mathrm{R}}\eta x_9(\mathrm{q})+\frac{2\mathrm{M}_0}{\mathrm{R}}\mathrm{A}-\mathrm{f}_0]=\frac{2M_0C\eta}{R}x_9(q)+C[\frac{2M_0}{R}A-f_0]$$
$$=Dx_9(q)+E$$

Where $C=\frac{2akQM_0B}{R}$, $D=\frac{2M_0C\eta}{R}$, $E=C[\frac{2M_0}{R}A-f_0]$, $x_9(q)=x_{10}(n_0T)$.

$E[e^2(q)]$, the mean of $e^2(q)=D^2x_9^2(q)+2DEx_9(q)+E^2$, is a quadratic error function, which is also a concave function. Therefore, the iterative formula (7) can be obtained with the least mean square-root error criterion $E[e^2(q)]=0$ and the steepest descent method

$$x_9(q+1)=x_9(q)+\mu[-\nabla(q)] \quad (7)$$

Where μ is the convergence factor of the steepest descent method, $\nabla(q)=\partial\{E[e^2(q)]\}/\partial x_{10}(q)$ is grads of $E[e^2(q)]$. According to the least mean square-root method [9], $\nabla(q)$ could be replaced with its expectation $\hat{\nabla}(q)$

$$\hat{\nabla}(q)=\frac{\partial[e^2(q)]}{\partial x_{10}(q)}=2e(q)\frac{\partial e(q)}{\partial x_{10}(q)}=2De(q)=2Dx_5(q)$$

So, we have

$$x_9(q+1)=x_9(q)-2\mu Dx_5(q) \quad (8)$$

Meanwhile, $e^2(q)=D^2x_9^2(q)+2DEx_9(q)+E^2$

Using newton iteration method, we have

$$x_9(q+1) = x_9(q)\frac{e^2(q)}{\partial[e^2(q)]/\partial x_9(q)} = x_9(q) - \frac{x_9(q)}{2D} \tag{9}$$

Comparing (7) and (9), we get $\mu=1/(4D^2)$.

Overall, we have verified the signal processing scheme, and derived the iterative formula (9) and the convergence factor μ. Applying the scheme, we can realize a practical CPT atomic clock circuit.

3 Realization

Figure 3 presents the realization of our signal processing scheme. A Digital Signal Processor (DSP) is the key controller. The output of the DAC1 is connected to the control port of OCXO, so that, the digital signal of the processor adjusts the output frequency of OCXO. The processor changes PLL control word periodically for FSK modulation through controlling DMA. The PLL locks the frequency of the microwave generated by Voltage Controlled Oscillator (VCO) to the output frequency of OCXO and modulates it with a constant modulation depth. The DC source is controlled by the output of the DAC2 and modulated by PWM controller. Bias-TEE mixes the DC and the microwave, and sends the mixed signal to the physics package. After being amplified by AMP, the output of the physics package is filtered by BPF1 and BPF2 to get the signal respectively, corresponding to the microwave modulation frequency and the DC modulation frequency, then ADC1 and ADC2 collect sample signal and send the digitized signal to the processor. The processor processes the digital signal using the formed signal processing scheme, and controls DAC1 and DAC2 with the processing results.

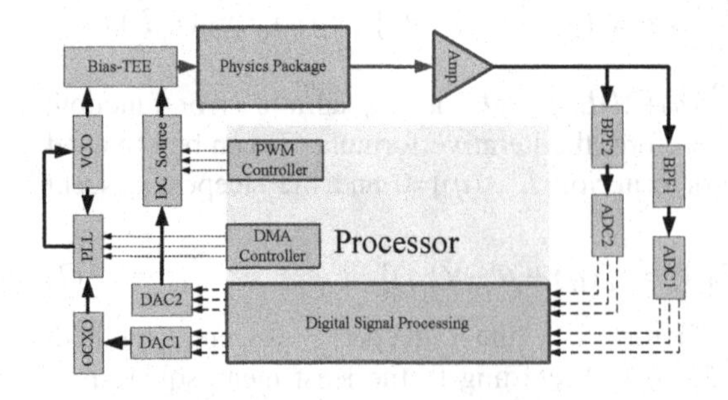

Fig. 3. Signal processing flow of the realized circuit

Fig. 4. Realized circuit

Figure 4 shows our realized circuit for CPT atomic clock, for which the power consumption of the circuit is about 200mW and the size is 10cm^3. Connected with an experimental physics package, the circuit has achieved all functions as shown in figure 5.

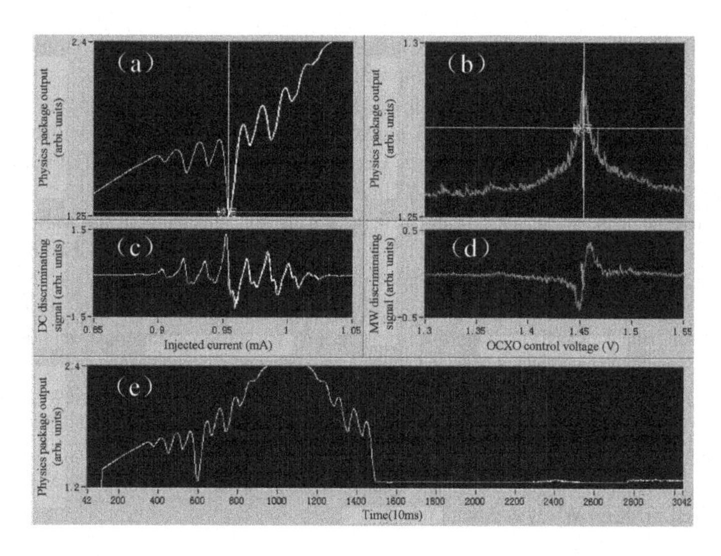

Fig. 5. The scanning and locking process of the DC and the microwave frequency (a) The output of the physics package to DC, (c) demodulation result to DC, (b) the output of the physics package to the output of the DAC1, (d) demodulation result to the output of the DAC1, (e) the output of the physics package to time

4 Conclusion

We have studied the principles of CPT atomic clock, derived the equations for our chosen signal processing scheme, and obtained the expected results. With the study results, we have realized signal processing function with our developed atomic clock circuit based on the scheme. The power consumption of the circuit is 200mW, the volume is $10cm^3$, and most of the signal processing is performed by the processor. Connected with an experimental physics package to form a CPT atomic clock, all of the expected functions of the circuit have been realized.

The digital circuit works on the chosen scheme is small in size, low in power consumption, easy to be integrated, and its functions can be conveniently modified or adjusted, which makes it a preferable one for application in the CPT atomic clock.

The atomic clock circuit based on this scheme is suitable for further reducing size and power consumption, and is promising for application in the chip scale atomic clock.

Acknowledgments. This work has been supported by the National Natural Science Foundation of China (NSFC) under Grant No. 10927403.

References

1. Risley, A.S.: The physical basis of atomic frequency standards. University of Michigan Library (1971)
2. Lutwak, R., Emmons, D., English, T., Riley, W.: The Chip-Scale Atomic Clock-Recent Development Progress. In: Proceedings of the 34th Annual Precise Time and Time Interval Systems Applications Meeting (2003)

3. DeNatale, J.F., Borwick, R.L., Tsai, C.: Compact, low-power chip-scale atomic clock. In: Position, Location and Navigation Symposium, 2008 IEEE/ION (2008)
4. Vanier, J., Godone, A., Levi, F., Micalizio, S.: Compact, Atomic clocks based on coherent population trapping: basic theoretical models and frequency stability. In: Proceedings of the 2003 IEEE International Frequency Control Symposium and PDA Exhibition Jointly with the 17th European Frequency and Time Forum (2003)
5. Ben-Aroya, I., Eisenstein, G.: Characterizing absorption spectrum of natural rubidium by using a directly modulated VCSEL. In: Proceedings of the 2005 IEEE International Frequency Control Symposium and Exposition (2005)
6. Vanier, J., Levine, M.W., Kendig, S., Janssen, D., Everson, C., Delaney, M.J.: Practical realization of a passive coherent population trapping frequency standard. IEEE Transactions on Instrumentation and Measurement (2005)
7. Vanier, J.: Atomic clocks based on coherent population trapping: a review. Appl. Phys. B 81, 421–442 (2005)
8. Wittenmark, B.: Adaptive Control, 2nd edn. Prentice Hall (1994)
9. Macchi, O.: Adaptive Processing: The Least Mean Squares Approach with Applications in Transmission. John Wiley & Sons (1995)

Mobile Application of Land Law Enforcement and Supervision

Zhirong Chen* and Tianhe Yin

Ningbo University of Technology, Fenghua Road 201,
Jiangbei District, Ningbo, Zhejiang 315211, China
chenzr29@gmail.com

Abstract. This paper discusses how to apply the mobile technologies to the land law enforcement and supervision business, to build a "sky view, ground search, and online management" inspection mechanism. After analyzing the traditional way, the new land law enforcement and supervision method is put forward to achieve the digitization, standardization, and spatial visualization of land law enforcement, and improve land resource management. The physical framework and functional design are discussed in detail. The system has been used in Bureau of Land Resources Fenghua for land resource management, and gotten good result. So the requirements of scientifically manage the land resource and establish e-government are completed met.

Keywords: Law enforcement and supervision, PDA, inspection, land resource management.

1 Introduction

Limited and scare are two natural features of Land resource. With the continuous deepening of reform, developing of economy and adjustment of industrial structure, the conflicts between land resource protection and economic development become more acute. Rational and efficient use of land resource can be directly related to the consolidation and coordination of social and economic health and sustainable development. In order to strengthen the management of land resource, firmly curb and prosecute violations of land law, ensure rational use and effective protection of China's land resource, the land law enforcement and supervision work is essential.

In practice, however, this work is difficult to carry out, mainly in five areas, difficult to find, difficult to qualitative, difficult to stop, difficult to process and difficult to implement. Timely detection of illegal land is not only the needs to strengthen management and protect arable land, but also the need for sustainable economic development of district-level. Currently, the conventional means of investigation is difficult to quickly and directly acquire the land-use change

* Corresponding author.

H. Tan (Ed.): Knowledge Discovery and Data Mining, AISC 135, pp. 239–246.
springerlink.com

information of illegal land, such as the changes in the spatial distribution and quantity. Integrating with Pocket PC (PDA), Global Positioning System (GPS) and 3G mobile communications network technology [1], the new land law enforcement and supervision method can quickly capture and transfer the illegal land's position, quantitative and changes information, and provide a basis for leadership decision making, greatly improve efficiency of dynamic land law enforcement and supervision.

2 Traditional Methods and Problems

At present, it is mainly using traditional inspection methods. Inspector needs to carry related paper maps (including remote sensing image maps, land-use maps, land use planning maps, etc.) for site inspections, and camera equipment to get the picture evidence when the illegal land use was found. It brings the following major issues:

1. Long inspection cycle and low probability of find out illegal land use

Due to lack of advanced scientific technology, the current land inspection efficiency is relatively low, the inspection cycle is long, and inspector is not convenient to obtain information of the suspected illegal land use in spot. Thus it is easily to lead missed investigation and underreporting, further to lead low probability of find out illegal land use.

2. Low efficiency of process illegal land use event

There are multi steps to qualitative illegal land use. Due to lack of basic data support, inspector can not determine whether the change of land use is illegal in spot when they found the suspected land. They need to manually delineate the red line range of illegal land use on the survey map, and then back to office for legitimacy analysis of the red line. As a filing material, the red line scope needs to be measured by a professional survey team after approval. So the efficiency of inspection is greatly reduced.

3. Hard to manage the original paper-based data

Inspector needs to carry related paper materials, and be familiar with the land surrounding terrain. On the other hand, it is difficult to divide regions and arrange lines for inspection because of the absence of the application system support. Therefore, the land can not be accepted for continuous supervision and tracking.

In order to establish an effective land enforcement mechanism, it requires the application of sophisticated modern scientific technologies, such as 3S (RS, GPS, GIS), network communications and computer technology, to monitor and take analysis to the land in real-time.

3 Mobile Methodology

The portable personal digital assistant, which not only have the computer power, but also can be combined with communication and internet technologies, has been widely used in line security monitor, cadastral survey, geological mapping, etc [2]. Based on the land "one map" database, combining PDA, GPS and 3G communication technology, the new land law enforcement and supervision method will make the cumbersome and complex work become easy, fast, efficient, and accurate. It will build a "sky view, ground search, and online management" inspection mechanism, and achieve the digitization, standardization, and spatial visualization of land law enforcement, improve land resource management.

The PDA-based management of land law enforcement and supervision has following advantages:

1. Through the land law enforcement and supervision system, the land planning, approval and status information can be obtained in real-time. The illegal land uses can be "early discovered, early reported, and early stopped." Violations are promptly nipped in the bud stage, to avoid illegal land intensify social contradictions.
2. Inspectors will be effectively regulated and scheduling, so that the work has become visible and controllable. Tasks and responsibilities can be assigned to the primary branch and specific inspectors. It has realized responsibility to the people, providing technical support for the lower assessment and higher supervision.
3. Together with satellite monitor, dynamic inspection, public report and approval supervision, it assigns inspection tasks according to the satellite image polygons, mass report and other information. The inspections can be reviewed and evaluated for quality, to improve the land management department's public image.
4. It will build a new model of land law enforcement and supervision, to meet the requirement of "process-oriented operation, systematic procedure, standardized management, quick disposal, information opening, reward and punishment Institutionalized". The land law enforcement and supervision work gradually transit from the current manual paper-based information operation to electronic operation.

4 System Architecture and Design

4.1 Framework Design

The PDA-based land law enforcement and supervision system consists of two parts, monitor control center (server side) and real-time inspection system (mobile terminal). Fig. 1 shows the system physical framework. The server side includes business acceptance, monitor and command, filing and investigation, system maintenance, data statistics, data exchange, data calls and other functions. The mobile terminal includes professional inspection navigation, creating law cases, and other functions. Inspectors transmit information between server and mobile terminal through wireless communication network. Officer grasps the inspection and law cases in real-time through the intranet.

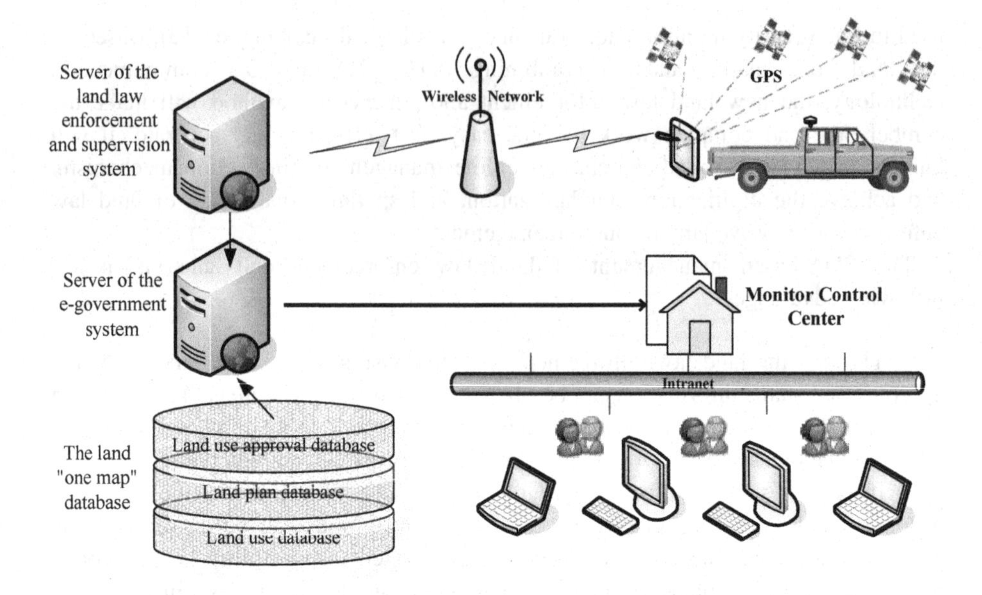

Fig. 1. The system physical framework

4.2 Functional Design

System functions can be divided into two parts:

■ **Server functions**

- map operation

Display image maps, vector maps and a variety of map annotation; pan, zoom in, zoom out; area measurement, position, distance measurement.

- Real-time track monitor

Show the inspection track in real-time; track a specific inspector; view the inspection basic information.

- Query of historical track

Provide inspection path search based on region, person, or time, and simulated playback the historical track.

- Leader supervision

Specify a supervision leader for the law case, and let him monitor the case status in real-time.

- Violations Management

Create, query, process, and update violations, to realize the electronic office of violations.

- Law enforcement and inspection by satellite images

Do operations such as additional block, legal review, filing investigation and implementation, according to the changed polygons. Finally generate the satellite images of law enforcement registration card, for the seamless integration with the satellite images of law enforcement registration system which running in the Ministry of Land.

- Scheduling management

Send tasks to the inspectors; track the inspections; achieve a focused inspection.

- Dynamic inspection

Set up and manage inspection program, including inspection area setting, responsibility and so on. Inspectors work with plan and purpose, thus ensuring a true understanding of the use of land resources.

- Red line analysis

Collect the red line, and analyze the land use information in red zone. It will determine the legality of land quickly and effectively, and improve the case processing efficiency.

- Schedule issues

Prompt the logged user which task needs to handle.

- Statistical analysis

Query, statistical, and analyze the records; print electronic documents directly for paperless office.

- System management

Manage the roles, users, equipments, and vehicles; ensure system security and flexibility.

■ **Mobile terminal functions**

- map operation

Display image maps, vector maps and a variety of map annotation; pan, zoom in, zoom out.

- Real-time position display

Acquire the GPS information, display and upload the current position and track to server.

- Red line collection and analysis

Generate the red zone by GPS information and query its category information and planning information. So that inspector can quickly access to the land information, and make the right judgments.

- Report suspected violations

According to the red line results, create suspected violations; fill out basic information and timely report to the server. So that the violations are early detected, and early stopped.

- Law enforcement and inspection by satellite images

Query the changed polygons information; add or update the suspected land; take photos and upload to the server; collect land coordinates. So inspectors can work in the field. This will not only improve efficiency, but also to avoid redundant operations.

- Schedule task feedback

  Receive the scheduling tasks; report the task status timely; feedback the task photographs; keep smooth contact between inspector and server.
- Data synchronization

  The information, including suspected violations, red lines, field photos, track lines, not only can be upload to the server, but also can be temporarily stored in the mobile terminals. User can upload manually to ensure the integrity of information.
- Support offline mode

  Both online and offline modes are compatible, to maximally ensure inspectors away from the signal or network disruption.

4.3 System Application

The system has been applied to the land law enforcement and supervision in Bureau of Land Resources Fenghua, and the expected effect is achieved. Fig. 2 and Fig. 3 show the system interface. The inspection system will strengthen the land law enforcement and supervision modernization, increase the transparency of administrative business and enhance the government's rapid response capability. Furthermore, it provides valuable experience for the monitoring mode to be more scientific and effective.

Fig. 2. The system interface of mobile terminal

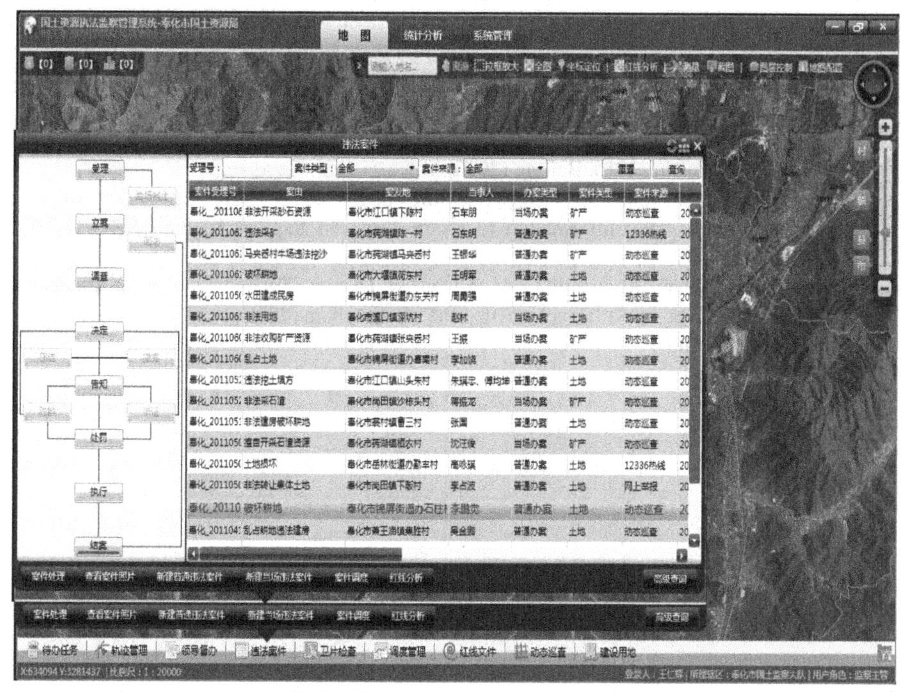

Fig. 3. The system interface of server

5 Conclusion

With the rapid development of the computer hardware, software and network technology, the mobile device has become widely used and much cheaper, and the wireless communication has become very convenient [3]. Based on PDA, the "sky view, ground search, and online management" inspection mechanism can be easily contributed. The illegal land uses can be "early discovered, early reported, and early stopped." The study has realized the promotion of land resource scientific management by information technologies.

Acknowledgments. We would like to thank the anonymous reviewers for their constructive suggestions and advice on improving the quality and presentation of this paper. This effort is sponsored by the National Natural Science Foundation of China (NSFC) No. 40901241 and Natural Science Foundation of Zhejiang province No. Y5090377.

References

1. Li, D.R., Li, Q.Q., Xie, Z.Y., Zhu, X.Y.: The Technique Integration of the Spatial Information and Mobile Communication. Geomatics and Information Science of Wuhan University 27(1), 1–6 (2002)

2. Pan, X.F., Li, J., Yue, J.W., Wang, B., Liu, Y.: Design and Implementation of PDA-based Land Law Enforcement System. Science of Surveying and Mapping 33(S1), 142–143 (2008)
3. Maguire, D.: Mobile Geographic Services Come of Age. Geoinformatics 4(2), 6–7, 9 (2001)
4. Chen, Z.R., Huang, Y.Z.: An Integrated Geographic Information Service Model Based on 3G. In: Geoinformatics 2010, Beijing (2010)
5. Meeks, W.L., Dasgupta, S.: Geospatial Information Utility: An Estimation of the Relevance of Geospatial Information to Users. Decision Support Systems 38(1), 47–63 (2004)
6. David, C., Huffman: Applications of Advanced Display Technology for Dismounted Combatants. In: Proc. of SPIE, vol. 5801. SPIE, Bellingham (2005)
7. What is 'Mobile GIS', http://www.gispark.com/html/Mobile-GIS/2007/1014/1477.html

A New Mechanism of EAB in RCS

Peng Zhao, Qun Wei, Hailun Xia, and Zhimin Zeng

Beijing University of Posts and Telecommunications, Beijing, China
zhaopengj@gmail.com

Abstract. This paper presents a new mechanism, Optimal of PRESENCE, of Enhanced Address Book (EAB) in Rich Communication Suite (RCS). Optimal of PRESENCE provides the ability of communication capability discovery for the contacts in EAB. It inherits the advantages of both PRESENCE and OPTIONS (the existing mechanisms of EAB), and largely decreases the network and terminal loads. Furthermore, we make a detailed comparison between Optimal of PRESENCE and two existing mechanisms, and give some advices for operators in the aspect of deployment.

Keywords: RCS, EAB, PRESENCE, OPTIONS, optimal.

1 Introduction

RCS, the Rich Communication Suite initiative is the first go-to-market program of IMS based services. The objective of RCS is to define an interoperable and convergent rich communication portfolio of services, and the scope of RCS is based on the following features: Enhanced Address Book (EAB), Enriched Call, Enhanced Messaging and File transfer.

EAB, as the core feature of RCS, is an evolution of the traditional address book. With EAB feature, the address book becomes more convivial and gives new reasons to communicate.

Up to August 2011, GSMA had issued 4 RCS releases and a RCS market project – RCS-e, in the meantime, different mechanisms of EAB emerged. However, both of the current mechanisms, PRESENCE and OPTIONS, cannot adapt to all the situations. PRESENCE is too complicated to deploy, and so many features, which are not much useful, make large amount of loads to the whole system. Meanwhile, OPTIONS mechanism demands additional requirements of terminal, and is difficult to update in the future. In order to overcome the weaknesses of PRESENCE and OPTIONS, a new EAB mechanism is presented.

In this paper, firstly, we analyze the characteristics of PRESENCE and OPTIONS mechanisms as well as some necessary principal introduction. Then, based on the previous analysis, we present Optimal of PRESENCE, a new mechanism of EAB and introduce this mechanism for the aspects of capability discovery process, UE behaviors and configuration. At last, we make a detailed comparison between these 3 mechanisms in both functional and requirement aspects, and give some suggestion in RCS deployment perspective.

H. Tan (Ed.): Knowledge Discovery and Data Mining, AISC 135, pp. 247–254.
springerlink.com

2 PRESENCE and OPTIONS Mechanisms

2.1 EAB Based on PRESENCE

Enhanced Address book based on PRESENCE mechanism is the mandatory feature of RCS in all the 4 RCS releases in GSMA. Besides providing a whole set of friendly UI to the end users, PRESENCE can realize real-time update of the social information for all the contacts in the enhanced address book. Presence information here includes users' social relationship, personal profiles and communication capability.

The principal of PRESENCE is based on OMA [PRESENCE]. PRESENCE includes 8 network entities: Presence Server, Presence XDMS (XML Data Management Server), RLS (Resource List Server), RLS XDMS, AP (Aggregation Proxy), Shared XDMS, Permanent XDMS, Content XDMS and 3 functional entities: Presentity, Watcher and Watcher Information Subscriber.

PRESENCE mechanism is worked in Subscribe-Notify mode: RCS users subscribe their RCS friends' Presence information, once Presence information changes, the new capability and personal profiles will be put to the network and make a notification to all his Watchers.

- **Social relationship** is one of the most significant concepts in PRESENCE mechanism. For a RCS user, his social relationships with contacts are uniquely decided by the lists kept in the Shared XDMS (RCS List, RCS Restricted List, Block List and Revoke List). User can execute different policies by adding contacts in different lists.

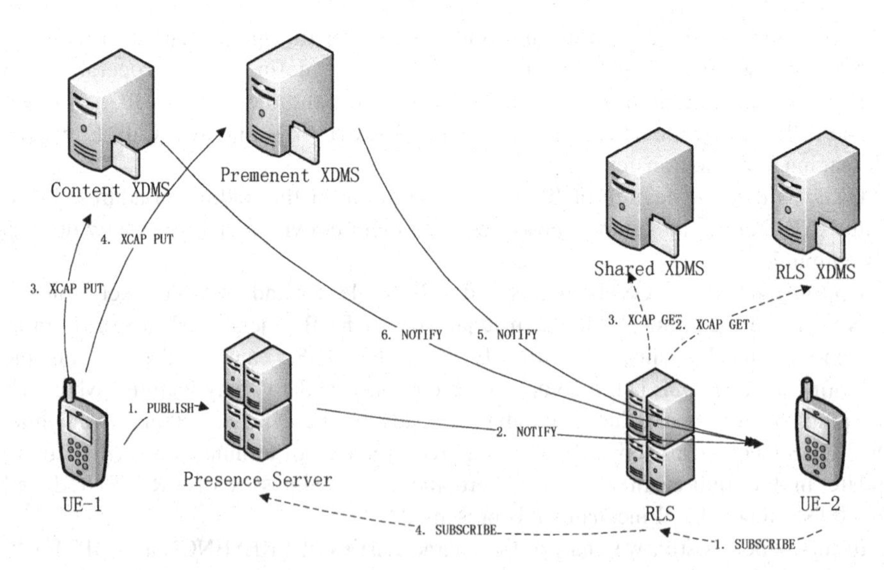

Fig. 1. Subscribe-Notify flow chart

- **Presence Information exchange** worked in Subscribe-Notify mode and it contains 2 processes (Fig. 1).

Subscribe: RCS user (e.g. UE-2) can subscribe his/her RCS friend's (UE-1) Presence Information by SIP SUBSCRIBE.

Notify: Either UE-1 first register in RCS or change UE-1's Presence Information, it will cause a publish process: UE-1 will put (XCAP PUT) the new personal profiles to the relevant XDMS and the publish (SIP PUBLISH) the new communication capability to the Presence Server.

Based on the subscribe relationship existed (UE-2 has subscribed UE-1's Presence Information), all the changes of UE-1 will cause a notification (SIP NOTIFY) to UE-2 through RLS.

2.2 EAB Based on OPTIONS

Another mechanism of EAB is OPTIONS, which is presented in RCS-e. OPTIONS mechanism abandons the complicated PRESENCE to make all the EAB system much easier. As the result, OPTIONS cannot provide the exchange of personal profiles and the concept of social relationship, it only realize the real-time query of the communication capability.

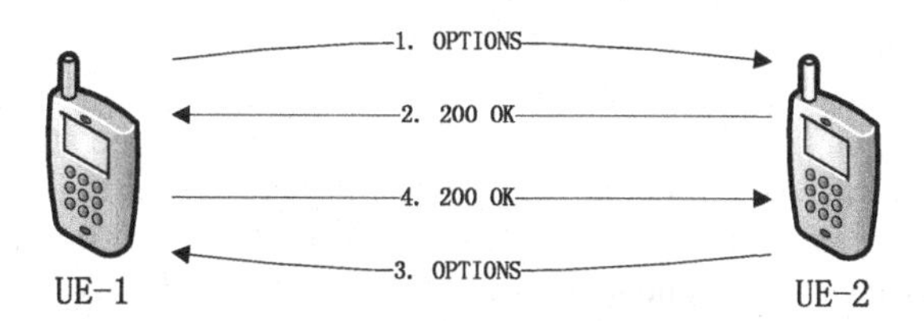

Fig. 2. SIP OPTIONS flow chart

As is shown in Fig. 2, OPTIONS is a pure UE-to-UE capability check, once receiving the check query (SIP OPTIONS), a reply (200 OK) will be sent as a respond with the communication capability in it.

Both of PRESENCE and OPTIONS mechanisms have weaknesses and cannot adapt all the situations: PRESENCE is too complicated, and it will cause a large amount of network flow; while OPTIONS's UE-to-UE mode demands particularly support of RCS client and increase the load of terminal, what's worse, OPTIONS cannot interoperate with PRESENCE in the future. Based on the above reasons, we present the Optimal of PRESENCE.

3 Optimal of PRESENCE

In this part, we present an Optimal of PRESENCE mechanism. Before of that, we point out the main disadvantages of PRESENCE and OPTIONS mechanisms, which are precisely the points need to be improved. Then, we make an introduction of capability discovery which is the core process in Optimal of PRESENCE. At last, we give some recommends on UE behaviors and configuration.

3.1 Capability Discovery

In our program, capability discovery can be split into two processes: Capability Publish and Capability Fetch:

Capability Publish process flow is extended from PRESENCE by add some new communication capability tags.

Capability Fetch is original from Anonymous Subscribe in PRESENCE. We make some changes about that process to meet our demand of Capability Fetch.

All these optimizing are based on PRESENCE to ensure a seamless interoperate with PRESENCE.

- **Capability Publish**

A RCS user should publish his communication capability on Presence Server when he/she register on RCS services at the first time. Then, once the communication capability changes, it should be re-published to insure the update. Capabilities in SIP PUBLISH message should contain the following types:

- Voice call via CS domain (always available)
- Video call via CS domain
- SMS and MMS (always available)
- Instant Message
- File Transfer
- Image share
- Video share

Communication capabilities of all the online RCS users can be storage in Presence Server. But the storage of communication capability is much different from personal profiles: the former is storage in cache which will be lost once the user gets off the line; while as permanent documents in the Server, the latter are saved in XDMS in the format of XML date.

- **Capability Fetch**

Capability Fetch is an optimizing of Anonymous Subscribe which provides capability subscribe for contacts who are not your RCS friends in PRESENCE mechanism. We just use Capability Fetch to get all the RCS contacts' communication capabilities, because, there is no concept of social relationship in Optimal of PRESENCE. And The policy of whether allow to fetch my own communication capability can be configured by the user, and the policy is storage in Presence XDMS.

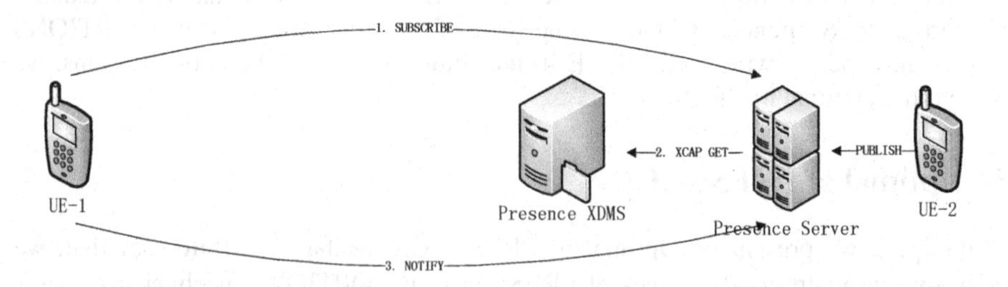

Fig. 3. Capability discovery flow chart in Optimal of PRESENCE

As is shown in Fig. 5, UE-2 is online and has published (SIP PUBLISH) its communication capability (as described in 3.3.2.1) on Presence Server, then UE-1 starts a Capability Fetch process (SIP SUBSCRIBE) towards UE-2.

For that SIP SUBSCRIBE message (Table 1), we make two revises in SIP SUBSCRIBE message:

- Revise the From Header to "Anonymous" and add a Privacy Header as a display name;
- Set the Expires time as 0.

Table 1. Revised SIP SUBSCRIBE message used in Optimal of PRESENCE

From: "Anonymous" ;tag=Je8PbbAzBA
To: <tel:+1234567890>
Expires: 0
P-Access-Network-Info: 3GPP-UTRAN-TDD
Privacy: user-id
Event: presence
P-Preferred-Identity: sip:+1234567891@ims
*(ignore some unrelated parameters)

The first revise make the SIP SUBSCRIBE as an Anonymous Subscribe to fit our non social relationship program. And the second one make this Anonymous Subscribe become invalid once received, in fact, it is not a subscribe any more but a single time query, because, we do not need the further notifications when communication capability changes. All these two revises above make the SIP SUBSCRIBE message suitable for Capability Fetch query of our program.

Once received the Capability Fetch query, Presence Server check the policy in Presence XDMS (XCAP GET) to know whether allow this query, if yes, Presence Server will send a respond (SIP NOTIFY) back to UE-1 with communication capability in it.

3.2 UE Behaviors and Configuration

The principal of UE behaviors contain Capability Publish and Capability Fetch:

Capability Publish is performed when either the first time register or communication capability changes.

Capability Fetch is in motion when a contact is chosen (willing to make a call or RCS communicate to whom) in EAB.

Capability Fetch can also be triggered in other scene (period polling) depends on the configuration as show in the following table:

Table 2. Configurable parameters in Optimal of PRESENCE

Parameters	COMMENTS
Polling Period	This is frequency to run a periodic capabilities update for all the contacts in EAB whose capabilities are not available or are expired. If set to 0, this periodic update is no longer performed.
Capability Info Expiry	When performing periodic update, capability discovery with timestamp kept together takes place only if the time since the last capability update is greater than this expiration parameter.

4 Comparison of the 3 Mechanisms

As an EAB realization mechanism, Optimal of PRESENCE makes many optimizing, and is different from PRESENCE and OPTIONS mechanism. We make a comparison between these 3 mechanisms in function and requirement aspects.

4.1 Function Aspect

In function aspect, Optimal of PRESENCE, on the one hand, inherits the most important and characteristic features of PRESENCE, on the other hand, sucks up the advantage of OPTIONS to simplify the EAB system. We can see the detailed differences between these 3 mechanisms in functional and user experience aspect in the following table.

Table 3. Comparison in function aspect

	PRESENCE	OPTIONS	Optimal of PRESENCE
Personal Profile	Support	Not support	Not support
Social relationship	RCS (Restricted) friend, Block, and Revoke.	Not support	Not support
Period update	Period subscribe to maintain subscribe relationship	Depends on the configuration	Depends on the configuration
Network address book	Support	Not yet	Not yet
First register behaviors	Consistency test; Publish communication capability to Presence Server; Put personal profiles and policy to XDMS;	None	Put policy to Presence XDMS; Publish communication capability to Presence Server.
User experience	Maintain all the RCS friends' Presence Information (including personal profiles and communication capability).	Fetch a certain RCS user's communication capability when needed.	Fetch a certain RCS user's communication capability when needed.

Presence Information: All the 3 mechanisms support communication capability exchange which is considered as the most significant feature of EAB. Personal Profile (including portrait, free text, link, nickname, timestamp, and geo-location.) and social relationship are supported in PRESENCE mechanism. Network address book storage the contact list of users, providing the synchronization between network and UE (maybe has more than one UE per user). It can be easily supported by PRESENCE, because, RCS List is storage in Shared XDMS, and it can be seen as a part of social relationship feature; for Optimal of PRESENCE and OPTIONS, it need an additional XDMS to support network address book.

First register: When a RCS user first register to RCS service, PRESENCE mechanism need a Consistency Test and publish many Presence Information to the network, while Optimal of PRESENCE only need to publish the communication capability and policy which can save more than 80% the process and traffic. OPTIONS mechanism does not need the support of network and should do nothing while register.

User experience: In the aspect of user experience, PRESENCE provides the most Presence Information and maintains them of all the RCS friends, while Optimal of PRESENCE and OPTIONS only support communication capability exchange and get them if only initiatively ask for.

4.2 Requirement Aspect

In the view of requirement, the more features we achieve, the higher requirements we demand for the network and client.

Table 4. Comparison in requirement aspect

	PRESENCE	OPTIONS	Optimal of PRESENCE
Network entity requirement	IMS core, Presence Server, Presence XDMS, etc.	IMS core	IMS core, Presence Server, Presence XDMS
Client functional requirement	PRESENCE (social relationship, personal profiles and communication capability), SIP OPTION (for content share)	Extended SIP OPTION	Simplified PRESENCE (communication capability only)
Network performance requirement	Very High	None	Low
Client & UE performance requirement	Very High	Middle	Low

Network requirement: Besides the IMS core, PRESENCE requests additional 8 network entities, and only 2 entities needed in Optimal of PRESENCE. Too many network entities make the whole EAB system much more complicated and cause a lot of network traffic.

Performance requirement: so, for PRESENCE mechanism, both the network and client need a Very High performance to handle so many signaling, it is also a big challenge for the battery of mobile device. Because of the UE-to-UE mode to get communication capability, OPTIONS mechanism do not need any support of the network, while, RCS Client should deal with capability queries (SIP OPTIONS) from other RCS users, and it increases load of RCS Client and UE.

Optimal of PRESENCE has largely simplified PRESENCE mechanism to reduce the load of both of the network, client and UE. Once publishing the communication capability for one time, Presence Server instead of Client can deal with Capability Fetch, it reduces the load of UE, as well as the traffic between UE and network.

5 Conclusion

EAB, as the core and most fundamental feature in RCS, is still disputed. Because both of the existing mechanisms, PRESENCE or OPTIONS, are not perfect in the aspects of deployment cost, load, user experience and future update.

In this paper, we present Optimal of PRESENCE which is a new mechanism of EAB. We have designed the whole capability discovery process, which is comprised by capability publish and capability fetch. Furthermore, we make a detailed description on UE behaviors and configuration to make this mechanism complete.

After that, we make a comprehensive comparison between Optimal of PRESENCE and two existing mechanisms in both function and requirement aspects. Optimal of PRESENCE, on the one hand, inherits the most significant features and advantages of PRESENCE and OPTIONS: it reserves capability discovery which is most valuable feature, and abandons others which is not much useful. On the other hand, it largely simplifies EAB system to realize a sharp load decrease of both network and terminal.

Based on the analysis in this paper, Optimal of PRESENCE is an effective and appropriate mechanism of EAB, and it is a considerable choice for major operators who do not have a rigid demand of Personal Profile exchange at the beginning. Optimal of PRESENCE has no special demand on UE, and it is much cheaper than PRESENCE in aspect of investment, as well as operation. What's more, it can seamlessly interoperate with PRESENCE which OPTIONS cannot, and smoothly updated to the "full-PRESENCE" while protect the original investment.

References

1. GSMA Rich Communication Suite (RCS) Functional Description, Release 2, version 1.1 (February 25, 2010)
2. GSMA Rich Communication Suite (RCS) Technical Realization, Release 2, Version 1.1 (February 25, 2010)
3. Arias, J.A., Telefónica, et al.: RCS-e Advanced Communications: Services and Client Specification v1.1 (April 08, 2011)
4. OMA-TS-Presence_SIMPLE_MO-V1_0-20060725-A: OMA Management Object for SIMPLE Presence, Approved Version 1.0 (July 25, 2006)
5. 3GPP TS 24.167, Technical Specification Group Core Network and Teminals, 3GPP IMS Mangment Object (MO)
6. 3GPP TS 24.229, IP multimedia call control protocol based on Session Initiation Protocol (SIP) and Session Description Protocol (SDP)
7. Dorbes, G., Hue, C., Alcatel-Lucent: Beyond RCS: Capitalizing on address book to embrace multimedia services, IEEE, 978-1-4244-4694-0/09 (2009)

Design of Security Solution to Mobile Cloud Storage

Xiaojun Yu and Qiaoyan Wen

State Key Laboratory of Networking and Switching Technology
Beijing University of Posts and Telecommunications, Beijing, China
{yuxiaojun,wqy}@bupt.edu.cn

Abstract. The cloud storage owns advantages in pay for use and elastic scalability. However, the data security risk destroys the trust relation between the cloud service provider and user. A direct method to avoid this problem is to encrypt data before data stored in the cloud. Thus, without the decryption key, the leakage data cannot be decrypted. While the encryption technology is good, it is not always suitable for the mobile user. When using the mobile device, such as smart phone, to access the data that stored in cloud storage system, the performance issue should be considered, because the encryption scheme involves high workload. This paper is focus on the design of security solution to mobile cloud storage. It detailed the design principle, security function model, and typical deploy model. It also proposed a design case based on searchable encryption to guide the further research.

Keywords: Mobile cloud storage, security solution, data security, searchable encryption.

1 Introduction

The mobile cloud storage is a typical application that people use cloud storage as their information store and communicate with the cloud storage by mobile devices.

However, the data security problem has not been solved in mobile cloud storage. On one hand, it is hesitant for enterprise to store the information related to enterprise benefit in cloud storage system, where the attacks occurred frequently, resulting the relation between user and cloud storage provider is low trust. On the other hand, people use the mobile devices to get the cloud storage service where the security level is low [1]. So the attacker can intrude into cloud storage system from mobile platform. Thus, the security problem has become a top research issue in mobile cloud storage field.

One of technologies to solve the data security problem is the searchable encryption. It achieves the data confidentiality by using the encryption and it also supports the search ability on cipher text. Song et.al [2] proposed a sequential scanning searchable encryption. This scheme is low performance as every word in all files will be compared in searching process. Goh[3]proposed the keyword-based searchable encryption. The pain file is encrypted with symmetric encryption scheme and the keyword indexes are encrypted with other encryption. The search work could be conducted on the encrypted keywords. Other related work of searchable encryption is in [4][5].

H. Tan (Ed.): Knowledge Discovery and Data Mining, AISC 135, pp. 255–263.
springerlink.com

However, the searchable encryption technology brings much work load to client, such as the encryption and decryption, resulting low performance in mobile device which is in limited computing ability.

This paper is focus on the design of security solution. The functions architecture and security deploy model are proposed according to the threat model and design principle. It also proposed a design case of security solution base on searchable encryption. In this case, given the performance problem in searchable encryption for mobile user, a user proxy is introduced to executing the encryption and decryption work for mobile devices.

The structure of paper as flowing: section two is the threat model in mobile cloud storage. Section three is the detail design of solution, including the design principle, functions architecture model, solution deploy model and a detail case to apply the designation idea. Section four is related work. The last section is conclusion.

2 Threat Model

In this paper, it assumed that there are three participators: the mobile user, attacker, cloud manager. These actors run in different place as descripted in the mobile cloud storage system threat model, see fig.1

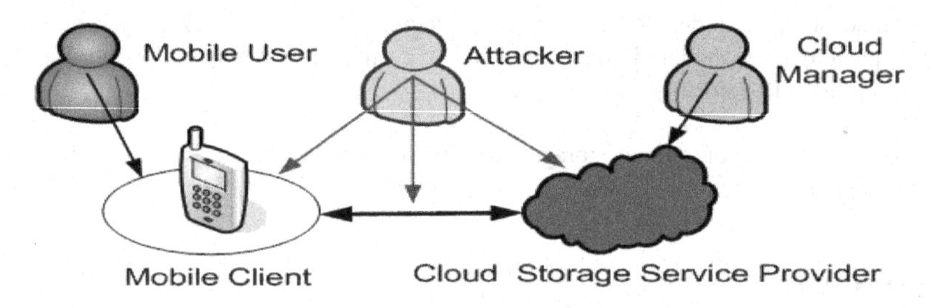

Fig. 1. Threat Model

As above shows, the mobile client act on the mobile platform and face the security risk from the misuse and the intrude actions from the attacker. The attack actions include data modified or data leakage both in mobile client platform or data transplant procedure between the mobile client and cloud storage service provider. The cloud storage service is untrusted for the data security risk cause by the unsafe cloud system and the untrusted cloud manager.

3 Design of Security Solution

3.1 Design Principle

When designing the solution for mobile cloud storage, the basic principles are two sides:

Security: the security should be considered completely, not only for cloud storage system, but also for mobile client. The security requirements including the confidentiality, integrity, access control and so on. The security technology should be proving security or computing security that means it is hard to compromise.

Performance: this means the security solution should be not too complex or resource consuming. As we know the security measure often involve the low performance problem. In the mobile cloud storage application, we can take use of the compute ability of cloud system. But this is not always the best choice because of the security worries when executing all the action in untrusted cloud. This principle is important for keep user expediencies too. Thus, the good design may consider the practical user conditions.

3.2 Functions Architecture Model

Given the data security risk in cloud storage, a simple security functions architecture model is summarized, see fig. 2:

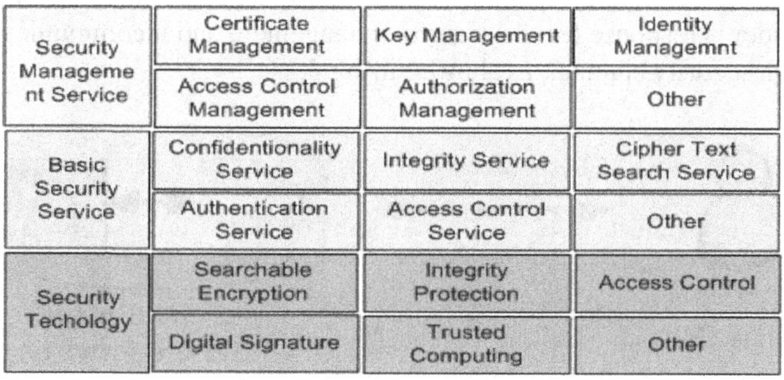

Security Manageme nt Service	Certificate Management	Key Management	Identity Managemnt
	Access Control Management	Authorization Management	Other
Basic Security Service	Confidentionality Service	Integrity Service	Cipher Text Search Service
	Authentication Service	Access Control Service	Other
Security Techology	Searchable Encryption	Integrity Protection	Access Control
	Digital Signature	Trusted Computing	Other

Fig. 2. Function Architecture Model

As above shows, the cloud storage security functions could be described three layers:

Security technology layer: this layer mainly includes the searchable encryption, integrity protection technology, access control, digital signature, trusted computing and so on. Specially, the searchable encryption can achieve the data confidentiality and search ability based on cipher text. The integrity protection technology is used to verify the data consistent or find the modification. The access control and digital signature are used in identity authentication. The trusted computing is a platform security technology based on hardware. This technology can be used to seal the encryption keys and verify the integrity of application.

Basic security service layer: this layer including the data confidentiality service, integrity service, cipher text based search service, user authentication service, access control service and so on. Above service cover the basic security requirement of cloud storage. Specially, the data confidentiality service, integrity service, cipher text based search service are three topics of current research field in cloud storage security. Other services may include the data backup, security audit as so on.

Security management service layer: This layer includes the certificate management, key management, user identity management, access control management, authorization management and so on. These management functions are also necessary for the security protection solution. These management work may conducted by different entity, such as mobile user and cloud manager. Other management work may include the history management, trust management and so on.

It should be noted that our security function architecture is proposed based on data security. Thus, these functions may be considered through the data lifetime cycle, not only in static storage stage in cloud storage system, but also the transport stage and use stage in mobile client.

3.3 Solution Deploy Model

According to the practical conditions of users and security functions architecture model of cloud storage, the security service may be deployed in cloud storage system or mobile user or trusted third partner, resulting different security deploy architectures.

Model I: the security service deployed in cloud storage service provider. Thus the cloud service provider is response for the security management and it communicates with the client through security channel, as showed in fig.3

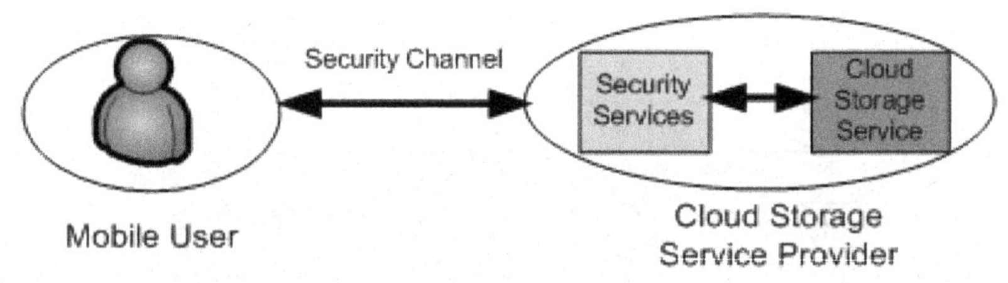

Fig. 3. Model I: In this model, the client need low security measure, thus the computing work is low. This model is popular in the current cloud service product. However, the data security is full dependent on the trust to cloud storage system. The data access is controlled by the condition that the user should gain the authorization or secret from the data owners. Though it seemed security, how to make sure the user data is not leakage during executing the data operation untrusted cloud. The trusted computing may be used to monitoring the executing progress.

Model II: Data Security Service is mainly deployed on client, and the mobile user is response for the security management. The cloud storage service provider take little care about the data protection response, as the used data is encrypted before stored in cloud storage. See fig.4

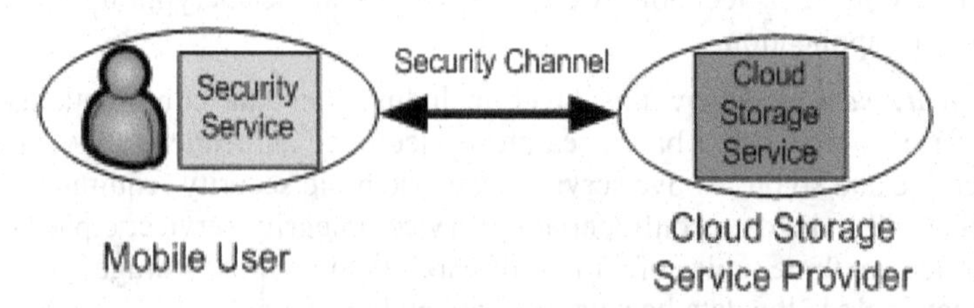

Fig. 4. Model II: In this model, the data security level is high ,but the workload is also high, thus the client need strong computing ability. For enterprise user, there are huge and important data need to encrypted and stored in cloud storage system. However, for the mobile user, as the computing ability is limited, this deploy is unsuitable. So it is necessary to segment the workload in the client. A reference design is detail in next section based on this model.

Model III: the security service is neither deployed in client nor cloud storage system, but in the trusted third partner or TTP. In such model, both the client and cloud storage service provider set the trust relation with the TTP. The TTP is a cooperator and administrator, and provide the basic security service, such as the data encryption and decryption, but not all the dataflow passed it.

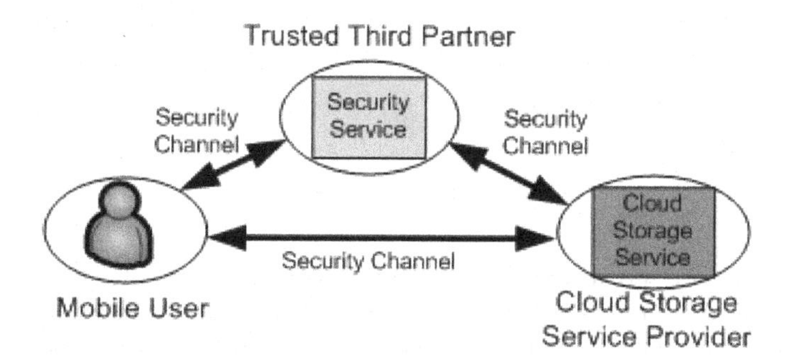

Fig. 5. Model III: In this model, the trust relation is changed. The mobile user is benefit from such model. Because the security of data protection and the advantage of cloud storage cloud be keep good compromise. However, the TTP will face security problem two or it must be enhanced in security carefully.

3.4 A Security Solution Case

1) Design Idea

In this part, we describe a solution case where the deploy model is based on the model II discussed in the above section. The security solution idea is in three parts:

First, use the searchable encryption to solve the data security problem in cloud storage system. The searchable encryption can support the data confidentiality and cipher text operation. The keyword based searchable encryption and fuzzy matching search scheme are used.

Second, use the trusted computing technology to enhance the mobile platform security. Our previous work [7] has proposed a trusted client model which combined the trusted computing and transparent encryption technology. The mobile client can prevent many intruding actions from the attacker.

Third, the user proxy is introduced to solve the performance limitation when using the searchable encryption directly on mobile platform. Thus, user proxy can mitigate burden of the mobile client. The user proxy is the service access point and security gateway for the cloud storage system.

2) Security FunctionsDeployment

The security deploy model of mobile cloud storage shows the basic work to implement the oursolution, see the flowing fig.6:

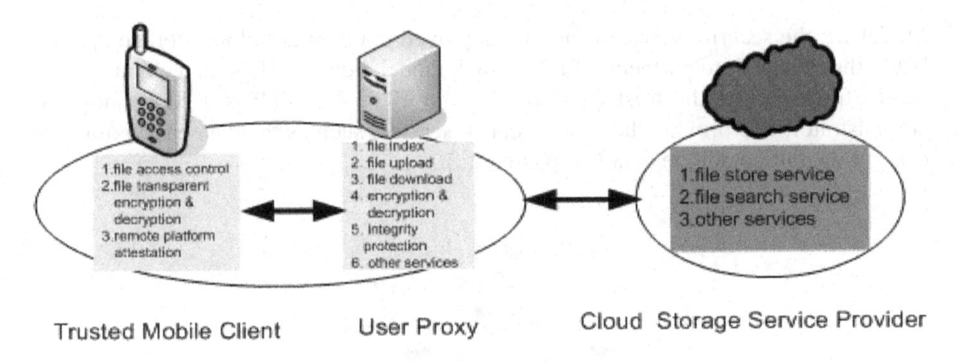

Fig. 6. Security Solution Deployment

As above shows, this model shows the division of work in protecting the mobile cloud storage. The trusted client is response for the platform security with functions, such as file access control, file transparent encryption, platform remote attestation, and so on. The user proxy is the bridge to the cloud storage system. It realizes the most security services, such as file upload and download proxy, file index and so on. The cloud storage service provider is response for the basic functions, such as file store service, file search service and so on.

➢ ***Trusted Mobile Client***

The trusted mobile client combined the trusted computing and transparent encryption to reduce the risk of file leakage. The transparent encryption means that the encryption and decryption work is conducted in kernel space, so have little influence to the application.

- ***Application integrity verification:*** the function is realized based on the MTM[6]. This function aim to check if the original application is tempered before loading or execute it, avoiding the risk from malware attack to change the application action.
- ***File access control:*** there may be different entity accessing the user files, access control is aim to protect file from illegal entity. This function is executed during all the encryption and decryption process.
- ***Platform remote attestation:*** This function is realized based on MTM too. This function aim to report the mobile platform status to user proxy which will compare with its security policy to find if the mobile client in the default security requirement. If so then a trust relation could be made and the further communication will continue.

➢ ***User Proxy***

The user proxy is the core component in the solution. It is response for most work for mobile client. These works include:

- ***File index:*** the function include the index creation and index update. In order to defend the data leakage from the index, the index file is encrypted. The encrypted index file and encrypted user file are stored in cloud. When new file upload or old file update, the index file should be update.

- ***File integrity protect:*** This function uses the integrity protection measures, such as MAC or digital signature to keep the file consistent .When file upload, a integrity proof is created and attached to the encrypted file. When file is downloaded to the user proxy, the integrity verification is executed to find the any error.
- ***File encryption and decryption:*** This function is transparent to the mobile user. This function need strong computing abliltiy and there may be different choice for encryption arithmetics, include the symetiric encryption scheme and asysmetiric encryption scheme.
- ***File upload and download:*** the file is uploaded as cipher text form or through a security channel between mobile client and user proxy. The upload file will be re-encrypted using the searchable encryption. This function also needs to call the file index service. The encrypted file and related information will be sent to the cloud storage system. When receiving the file download request, the user proxy will send download request to cloud storage which will finally send back the target file to the user proxy. After get the encrypted file from cloud storage service provider, the user proxy will call the file verification function and decryption function. This work is transparent for mobile user, but if there is any error in file verification or decryption process, the warning information will report to user.
- ***Other functions:*** include the certificate management, key management and other service.

➢ ***Cloud Storage Service Provider***

In the mobile storage service, the cloud storage service provider must to support some basic service, these include:

- ***File store service:*** the cloud storage service provider interface for the user proxy to execute the upload and download work. This service should support single file store operation and batch files operation fashion.
- ***File search service:*** this service work bases on the keyword and the authorized token provided by the user proxy. user proxy will pre-process the search request from the mobile user. The basic work is the keyword filtering and keyword encryption. The search work is executed on cipher text.
- ***Other services:*** there are other functions can deploy in cloud storage system. Such as the data backup service that keeps the use data in different places.

4 Related Work

There is lots of research work about the data security in cloud storage or mobile cloud storage. Curino et al[8]proposed the Relational Cloud system. This system consist of three core components, including the workflow balance component for multi-tenancy, the data partitioning placement and migration engine and the CryptDB[9]. The system architecture is as flowing fig.7.

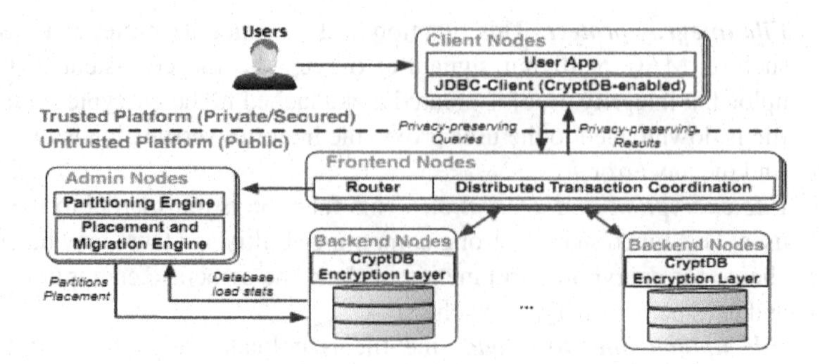

Fig. 7. Relational Cloud. In the Relational Cloud, the CryptDB supports the basic SQL operation on encrypted data. The advantage of CryptDB is that there is no any modification to database software. This component is deployed in the untrusted platform or public (cloud). While it could be efficient in executing the database operations in cloud, the client has to do the workload of encryption and decryption. Thus, for the mobile user, some modification should be taken, considering the idea of proposed solution case in this paper.

Sven et al[10] proposed a Two Cloud solution for the cloud storage security based on trusted cloud. The basic idea is that a trusted cloud response for keeping data security and data operation security. The system architecture is in fig.8

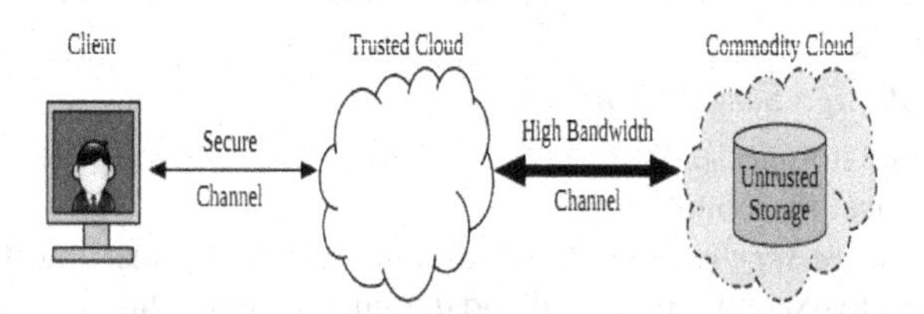

Fig. 8. Two Cloud. In this architecture, the user has to build the trust relation with trusted cloud which will be response for the key operations such as the data encryption and decryption, data integrity verification. The commodity cloud is only response for the encrypted data storage and data search. The trusted cloud could be the other cloud service provider or the private cloud that owned by user. The trusted cloud takes use of the trusted computing technology to improve the platform security and cloud trust level.

Frank[11]proposed a private virtual infrastructure(PVI)model aiming to improve the trust to cloud service. PVI is a new management and security model that shares the responsibility of security management in cloud computing between the service provider and client, decreasing the risk exposure to both. The PVI datacenter's security posture is set by the client, while the cloud's configuration is under control of the service provider. Clients can then protect their information independently of the cloud configuration.

5 Summary

It is urgent to solve to data security risk in mobile cloud storage. This paper discusses how to design the security solution to mobile cloud storage .According to the threat model and design principle, It proposes the security function model and compare different deploy model. It also proposes a case to that combined searchable encryption and trusted computing and for the mobile user, it not only could use the advantage of cloud storage, but also solve the trust risk of the cloud storage. The further work is the detail design and implement according to the different deploy model.

Acknowledgement. This work is support by National Natural Science Foundation of China (Grant Nos. 61170270, 61100203, 60873191, 60903152, 61003286, 60821001), the Fundamental Research Funds for the Central Universities(Grant Nos. BUPT2011YB01, BUPT2011RC0505).

References

1. Jansen, W., Scarfone, K.: Guidelines on Cell Phone and PDA Security. Recommendations of the National Institute of Standards and Technology NIST Special Publication 800-124 (2008)
2. Song, D., Wagner, D., Perrig, A.: Practical techniques for searches on encrypted data. In: Proceedings of IEEE Symposium on Security and Privacy, pp. 44–55 (May 2000)
3. Goh, E.-J.: Secure Indexes Cryptology ePrint Archive, Report 2003/216 (2003), `http://eprint.iacr.org/2003/216/`
4. Li, J., Wang, Q., Wang, C.: Fuzzy Keyword Search over Encrypted Data in Cloud Computing. In: IEEE Proceedings of INFOCOM 2010 (2010)
5. Hwang, Y.-H., Lee, P.J.: Public Key Encryption with Conjunctive Keyword Search and Its Extension to a Multi-user System. In: Takagi, T., Okamoto, T., Okamoto, E., Okamoto, T. (eds.) Pairing 2007. LNCS, vol. 4575, pp. 2–22. Springer, Heidelberg (2007)
6. TCG Mobile Trusted Module Specification Specification version 1.0 Revision 6 (June 26, 2008)
7. Yu, X., Wen, Q.: Trusted MobileClient for Document Security in Mobile Office AutomationI. J. Information Technology and Computer Science 1, 54–62 (2011)
8. Carlo, C., Jones, E.P.C., Popa, R.A., et al.: Relational Cloud: A Database-as-a-Service for the Cloud. In: 5th Biennial Conference on Innovative Data Systems Research, CIDR 2011, Asilomar, California, January 9-12 (2011)
9. Popa, R.A., Zeldovich, N., Balakrishnan, H.: CryptDB: A Practical Encrypted Relational DBMSMIT-CSAIL-TR-2011-005 (January 26, 2011)
10. Bugiel, S., Nuurnberger, S., Sadeghi, A.-R., et al.: Twin Clouds: An Architecture for Secure Cloud Computing (2011), `http://www.zurich.ibm.com/~cca/csc2011/submissions/bugiel.pdf`
11. Krautheim, F.J.: Building Trust Into Utility Cloud Computing. Phd Dissertation University of Maryland (2010)

A Study on Usability of Finger's Moving Direction in Direct Touch Environment

Feng Wang, Hui Deng, Kaifan Ji, and Lipeng Yang

Computer Tehcnology Application Key Lab of Yunnan Province,
Kunming University of Science and Technology, Kunming,
Yunnan, 650500, China
wangfeng@acm.org, dh@kmust.edu.cn, jkf@cnlab.net, ylp@cnlab.net

Abstract. Moving user's finger in a certain direction is a fundamental action when interacting with a touch screen. Previous literatures have discussed the techniques on how to exploit finger's directional movement in the design of user interface widgets. However, few studies have investigated the usability issues of finger's directional movement. In this paper, we investigate the performance of two algorithms to determine the slant angle of finger's moving direction. The experimental results prove that both algorithms can determine moving direction accurately. And the approximating the direction of finger movement using all samples from a trajectory generates better results than using only 2 samples.

Keywords: moving direction, usability, direct-touch.

1 Introduction

Many novel human-computer interaction techniques require using finger as a primary input device [4,5]. Input by using finger has many advantages. For instance, it is often considered to be simple, intuitive, and reliable [1]. An add-on benefit of using fingers is that it does not require additional input devices. Users can manipulate a virtual object by tapping or sliding by using their bare finger(s). The recent introduction of low-cost computer-vision based multi-touch tabletops [7,10] encourages a wide range of research in multi-touch interactions[3].

There are, however, several distinct drawbacks limit the use of multi-touch techniques on touch-displays. Prior studies have showed that input by using bare finger often suffers from occlusion and precision issues [1,13,16]. In addition, after using a tabletop for a long period of time, user can suffer from arm fatigue [1]. These issues can limit the usability of tabletops. Thus, effective solutions are needed in order to provide better user experience.

Researchers have proposed a variety of methods to address these issues [13,1]. Among these previous researches, an effective approach aims to improve the input bandwidth of finger input. This approach is based on a deep understanding of some of the fundamental properties of human finger [15]. Existing studies have investigated many finger properties, such as position [3,5,9], motion velocity and acceleration [9], the size of contact area [2], the shape of contact [15], orientation [15,14] and the tap or flick event properties [12].

H. Tan (Ed.): Knowledge Discovery and Data Mining, AISC 135, pp. 265–270.
springerlink.com

Notice that sliding a finger in a particular direction is a fundamental action that is carried out in many daily activities to fulfil many tasks, such as zooming or rotating a picture. Various interaction techniques, such as gesturing [6], rely heavily on such action. However, our survey of the existing work suggests that there is very limited work has been conducted to investigate the properties of the directionality of finger movement. In this paper, we discuss the accuracy of two direction determination approaches and further make a deep investigation on usability of the moving direction.

2 Related Work

Human fingers are the primary input device on touch displays. Directional finger movement is a fundamental action when interacting with a touch display. It is a basic action for various interaction techniques, such as gesturing. For instance, a pinch gesture [8] requires a user to slide his/her finger in a particular direction. More advanced techniques, such as marking menu rely on the direction of a hand motion to trigger a desired command. Directional finger gestures were also shown to be effective in selecting small touch icons. Escape [16] assigns a directional mark to a small target. Users can slide their finger in the corresponding direction to select a corresponding target. Due to the fact that only 8 directions can be effectively distinguished by a user, Escape can only work with a cluster of less than 8 targets.

Moscovich [11] showed how to design touch-screen widgets that respond to a directional sliding gesture. Unlike a traditional user interface (UI) widget, activating a Sliding Widget requires a user to slide the target in a predefined direction. Sliding Widgets help reduce the ambiguity of selecting small and clustered targets. It provides designers with a rich body of self-disclosing interaction mechanisms.

In summary, our literature review shows that direction finger motion has been widely used in designing interaction techniques on touch displays. However, there is no systematic investigation on the range of directional finger motion, especially in the effective algorithm of determining the moving direction.

3 Approaches for Determining Finger Moving Direction

3.1 Goal

Figure 1(a) shows a track of finger movement. Notice that a user won't be able to see the immediate update of this track due to the occlusion introduced by the finger tip. Without such important visual feedback, simple tasks such as dragging a straight line become nearly impossible by using a finger because users can not correct their movement immediately after the finger deviates from the straight line. In this study, we have to present algorithms designed to address this issue. The algorithms are capable of providing a system with an approximation of user's intent moving direction. In the rest of the paper, we describe the details of the algorithms. We also describe an empirical evaluation measuring the performance of the proposed algorithms.

We considered two algorithms to find a user's intent moving direction: 1) line fitting using all sample points of a trajectory; and 2) line fitting using start and end points of a trajectory. In the first approach, we used linear regression to fit user's moving trajectory into a straight line. We then calculated the slope of this line to determine the angle of user's moving direction. The second approach finds the slope of the line between the first and last point of the trajectory. We choose the first point as the origin point, and the last point as the destination point. Obviously, the time complexity and space complexity of using all the sample points of a trajectory is much greater than those of using only 2 points.

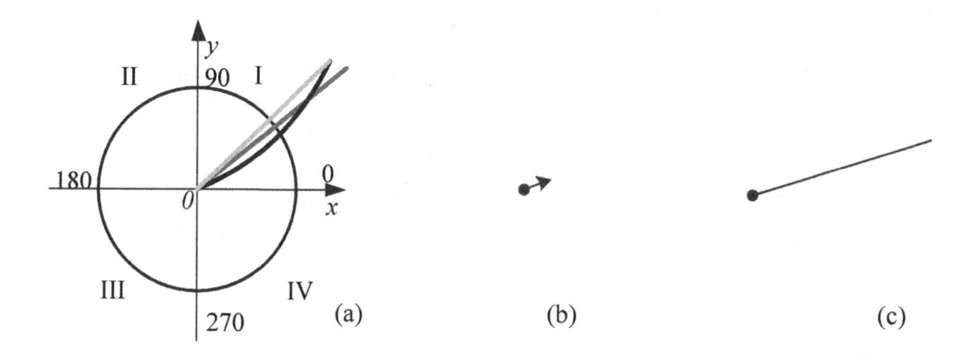

Fig. 1. The schema of finger's moving. (a) The black trajectory represents the track of finger movement. The red line represents the line generated by linear regression using all sample points from the black trajectory. The green line represents the line generated by linear regression using the start and end points of the trajectory. (b) Directional arrow mode. (c) Directional line mode.

3.2 Experimental Apparatus

We used a Lenovo ThinkPad T410s taptop computer (Intel i5 560M CPU, Nvidia NVS 3100M display card, 4GB memory, and multi-touch LCD display). The display of the laptop is 14.1" wide with a resolution of 1440 × 900 pixels. The size of each pixel is 0.21mm. Our program was written in *C#* using Windows Presentation Foundation.

3.3 Participants

Twelve volunteers (nine male and three female) between the age of 21-25, participated in the experiment. Ten participants were right-handed. Seven of them had no previous experience with touch displays.

3.4 Tasks

We used two tasks to evaluate the accuracy of the two methods.

Task 1. We showed a small circle with an arrow in a predefined direction (see Figure1(b)). Participants were asked to tap on the circle, followed by moving their finger in the direction of the arrow as accurate as they could.

Task 2. We showed a line on the screen to guide participants in a desired moving direction (see Figure1(c)). Participants were asked to move their finger by following the line as accurate as they could.

3.5 Procedure and Design

In each trial, participants were asked to perform the task in one of 4 different directions (30°,120°,210°,300°). We decided to test these directions instead of oblique directions (45°, 135°, 225° and 315°) because we wanted to make the results more pivotal role. The tested directions were presented in an increasing order. Each trial was repeated 6 times. Participants were asked to perform the tasks using any finger they liked. In summary, the experiment consisted of:

12 participants ×
2 tasks ×
4 subtasks ×
6 repetitions ×
= 576 trials.

3.6 Results and Analysis

For each trial, we recorded a user's actual direction of finger movement. We then calculate an angle deviation (Deviation) by subtracting the recorded direction from the pre-determined direction.(30°,120°,210° and 300°, each from different quadrant of a circle). We list the average of the deviation (Mean) and stand deviation (SD) in Table 1.

The results show that both approaches have a decent accuracy in approximating the direction of user's finger movement. According to the data, SD of fitting approach is only one half of SD of 2-point approach. We found that using all the samples from a trajectory gave us significantly better results than using only 2 points. This is because the former uses treats all data with the same weight, thus presenting more information of the direction of finger movement. In spite of the effect of "fat" finger, the mean values is less than 3°. This means our approaches gave sufficient accuracy for designing user interface in touch displays.

In order to verify the usability of two approaches, we further calculate the correlation of two approaches. ANOVA results show there are significant differences between two algorithms in all four quadrants ($p < .05$).

Table 1. Contrast experimental results under two different navigation modes. One is directional arrow mode and another is directional line mode.

	Fitted Angle			2-Point Angle		
	Mean	Deviation	SD	Mean	Deviation	SD
			Directional Arrow Mode			
30°	31.428	1.428	3.734	31.912	1.912	5.006
120°	121.298	1.298	3.536	122.728	2.728	5.148
210°	210.356	0.356	3.399	210.214	0.214	4.403
300°	301.819	1.819	3.725	301.191	1.191	4.557
AVG		1.225			1.511	
SD			3.623			4.853
			Directional Line Mode			
30°	30.486	0.486	1.962	30.040	0.040	2.547
120°	120.447	0.447	1.729	120.304	0.304	2.322
210°	210.643	0.643	1.638	211.459	1.459	2.315
300°	300.320	0.320	1.731	300.205	0.205	1.903
AVG		0.475			0.502	
SD			1.764			2.340

4 Conclusions

In this paper, we discuss several usability issues of finger's moving direction. We conducted a study to measure the performance of two algorithms to approximate the slant angle of the direction of finger movement. We also measured user performance with two visual feedbacks, arrow and line. Our results show that the approximating the direction of finger movement using all samples from a trajectory generates better results than using only 2 samples. We believe that the results of the empirical evaluations in this study will contribute to the HCI community with a better understanding of the fundamental issues of touch input, in particular user's perception of the direction of their finger movement. Our results can help improve the design of interaction techniques on touch displays.

Acknowledgements. This work was funded by National Science Foundation of China (61063027).

References

1. Albinsson, P., Zhai, S.: High precision touch screen interaction. In: Proceedings of the SIGCHI Conference on Human Factors in Computing Systems, pp. 105–112. ACM, Ft. Lauderdale (2003)
2. Benko, H., Wilson, A.D., Baudisch, P.: Precise selection techniques for multi-touch screens. In: Proceedings of the SIGCHI Conference on Human Factors in Computing Systems, pp. 1263–1272. ACM, Montroal (2006)

3. Buxton, B.: Multi-touch systems that i have known and loved. Microsoft Research (2007)
4. Buxton, W., Hill, R., Rowley, P.: Issues and techniques in touch-sensitive tablet input. SIGGRAPH Comput. Graph. 19(3), 215–224 (1985)
5. Buxton, W., Myers, B.A.: A study in two-handed input. In: Proceedings of the SIGCHI Conference on Human Factors in Computing Systems, pp. 321–326. ACM, Boston (1986)
6. Cao, X., Wilson, A.D., Balakrishnan, R., Hinckley, K., Hudson, S.: Shapetouch: Leveraging contact shape on interactive surfaces. In: 3rd IEEE International Workshop on Horizontal Interactive Human Computer Systems, TABLETOP 2008, pp. 129–136. IEEE, Amsterdam (2008)
7. Han, J.Y.: Low-cost multi-touch sensing through frustrated total internal reflection. In: Proceedings of the 18th Annual ACM Symposium on User Interface Software and Technology, pp. 115–118. ACM, Seattle (2005)
8. Krueger, M.W., Gionfriddo, T., Hinrichsen, K.: Videoplace—an artificial reality. In: Proceedings of the SIGCHI Conference on Human Factors in Computing Systems, pp. 35–40. ACM, San Francisco (1985)
9. Lee, S.K., Buxton, W., Smith, K.C.: A multi-touch three dimensional touch-sensitive tablet. In: Proceedings of the SIGCHI Conference on Human Factors in Computing Systems, pp. 21–25. ACM, San Francisco (1985)
10. Microsoft: Microsoft surface (2009), `http://www.surface.com`
11. Moscovich, T.: Contact area interaction with sliding widgets. In: Proceedings of the 18th Annual ACM Symposium on User Interface Software and Technology, pp. 13–22. ACM (2009)
12. Reetz, A., Gutwin, C., Stach, T., Nacenta, M., Subramanian, S.: Superflick: a natural and efficient technique for long-distance object placement on digital tables (2006)
13. Vogel, D., Baudisch, P.: Shift: a technique for operating pen-based interfaces using touch. In: Proceedings of the SIGCHI Conference on Human Factors in Computing Systems, pp. 657–666. ACM, San Jose (2007)
14. Wang, F., Cao, X., Ren, X., Irani, P.: Detecting and leveraging finger orientation for interaction with direct-touch surfaces. In: Proceedings of the 22nd Annual ACM Symposium on User Interface Software and Technology, pp. 21–30 (2009)
15. Wang, F., Ren, X.: Empirical evaluation for finger input properties in multi-touch interaction. In: Proceedings of the 27th International Conference on Human Factors in Computing Systems, pp. 1063–1072. ACM, Boston (2009)
16. Yatani, K., Partridge, K., Bern, M., Newman, M.W.: Escape: a target selection technique using visually-cued gestures. In: Proceeding of the Twenty-Sixth Annual SIGCHI Conference on Human Factors in Computing Systems, pp. 285–294. ACM, Florence (2008)

Double R-Tree and Double Indexing for Mobile Objects

Ye Liang

Department of Computer Science, Beijing Foreign Studies University, Beijing, China
liang_ye@sohu.com

Abstract. To effectively achieve various query requests oriented at mobile objects, this thesis proposes the DTDI index structure based on grid by extending GG TPR tree index structure. By adding corresponding spatial grid model to GG TPR leave node records, this structure realizes the correspondence between the mobile objects and their moving space, thus makes it possible to predict the current and future locations of mobile objects on the grid road. Meanwhile, as we improve the children node of leaves and information unit in GG TPR index, DTDI index structure is able to achieve the integrated maintenance, query and pruning of one-group mobile objects according to road curves and crossing points and thus to reduce the information updates of mobile objects, enhancing the efficiency of the index and maintenance of mass data within a limited range and the query performance of mobile objects, being available to respond to various types of query requests. Experimental results show that the performance of DTDI structure is better than the other indexing structure on indexing and querying a great capacity of moving objects within a limited area.

Keywords: Moving query, grid, GG TPR tree, space dividing tree, double R-tree and double indexing.

1 Introduction

Since the moving objects are in a state of motion, constantly changing their positions, the query technology of the moving objects is full of inherent complexities compared to that of static ones.

Theoretically, since the rest is also a special sport, the theoretical system of the moving objects query contains the management of all objects including the static ones, and the query technology of the moving objects can also be applied to that of the static ones. Summarizing the research results at home and abroad over the years, the researches on queries of the moving objects center on the simple real-time query [2], the simple continuous query [3], the current window query [4], the continuous window query [5], motional condition query, complex query containing two or more queries and so on.

In the moving objects query system, the simplest index is to query the location of the moving objects in the launching moment, called the simple real-time query. Since

H. Tan (Ed.): Knowledge Discovery and Data Mining, AISC 135, pp. 271–278.
springerlink.com

any object belongs to a coordinate point in space at a given time, the query to static objects can be seen as a subset of the simple real-time query in the objects query system. And therefore we can put the query management of the static objects, like gas stations, hotels, hospitals, into the moving objects query system. Based on the simple real-time query, if we extend the query, that is, indexing the location changes of the moving objects within a period of time, we call it the simple continuous query. If the query target is uncertain, we need to index all the current moving objects within the window, and then it is called the current window query. On a basis of the window query, if we extend the query, that is, indexing the location changes of all the moving objects within a window for a period of time, we call it the continuous window query.

The query themselves, the simple real-time query, the simple continuous query, the current window query and the continuous window query, do not change. The index requirement is satisfied only if the location information of one or some moving objects is read. For instance, the query problem, like the moving vehicles looking for the parking lot, belongs to the static query problem in the moving query system. Sometimes, the query does not only require the moving objects to be in motion, but also require the query itself changes over time. This is the dynamic conditions query which is based on a specific given moving object and indexes other objects consistent with the control condition, such as the nearest neighbor query represented by the emergency center vehicles looking for the shortest path to reach the hospital according to the dynamic changing road condition and the reverse nearest neighbor query represented by the police force deployment while chasing the fleeing vehicles. In the motional conditions query, there are two types of queries that are regarded to be seriously important by the researchers because of wide range of applications. One is the nearest neighbor query represented by the vehicles looking for the parking lot and the other is the reverse nearest neighbor query represented by the police force deployment.

2 The Existing Query Technology and Its Shortage

The moving objects query usually deals with the moving objects' data sets that their locations change all the time and the query itself is also dynamic. In order to meet the forecast query request that facing the moving objects, in recent years, researchers propose a variety of moving objects index structure and its query algorithms. The representatives are the grid dividing method and the query method based on PRA-tree index structure.

2.1 Grid Dividing Method and Its Shortage

Giving that the users concerns more about the query's response speed in some cases and allow certain error in the query results, Jimeng Sun[7] et al proposed a grid based on the two-dimensional space dividing. The moving objects update their positions in the way of data stream and the grid saves the moving objects' information collection that fall in it. They use statistical and sampling method to do the approximate query. This method can respond to the real-time query in a moment of the history, current and future and can not respond to the continuous query and the window query.

2.2 PRA-Tree Index Structure and Its Shortage

In order to meet all kinds of forecast query quest that facing the moving objects, establishing an efficient moving objects index structure is a necessity. In 2000, Simonas Saltenis[8] et al proposed TPR-tree index structure to meet the forecast query request; however, it contains the problem that the index performance declined rapidly ,with time in the process of indexing. Therefore, many domestic scholars have proposed a variety of improved methods. Among those methods, PRA-tree (predictive range aggregate) index structure and its query method is the most effective one which is proposed by Liao Wei et al in 2007, National Defense Technology University. Its idea is that while indexing the moving objects, firstly, it divides the velocity field into many velocity barrels with same speed taking the maximum and minimum values of the moving objects' speed as the extreme points in the indexing group. And then map the objects into different velocity barrels according to the size of the velocity vector in this velocity field. At last, indexing the moving objects in every velocity could barrel with the existing TPR-tree.

3 The Query Method Based on DIDI Structure

The establishment of GG TPR-tree reduces the efficiency of the index and management maintenance to the mass moving objects and improves the responding time of the simple real-time query and the short-term predicted simple continuous query. However, the simple GG-TPR-tree index can not achieve the high quality query results against the long-term predicted continuous query, window query and dynamic condition query. So we bring in the Grid-based spatial dividing technique into the process of query to the moving objects. And then we combine this technique and GG TPR-tree to complete the responding process of query requirements to the moving objects.

In most cases, the quantity of the moving objects is much larger than that of the roads, and the establishment of GG TPR-tree makes it possible to manage the moving objects overall. To complete the high efficient query to the various requirements under the guidance of the space partitioning tree, we preserve the original establishment thoughts of the GG TPR-tree and adjust the definition of GG TPR-tree's leaf node as follows:

Please note that, if your email address is given in your paper, it will also be included in the meta data of the online version.

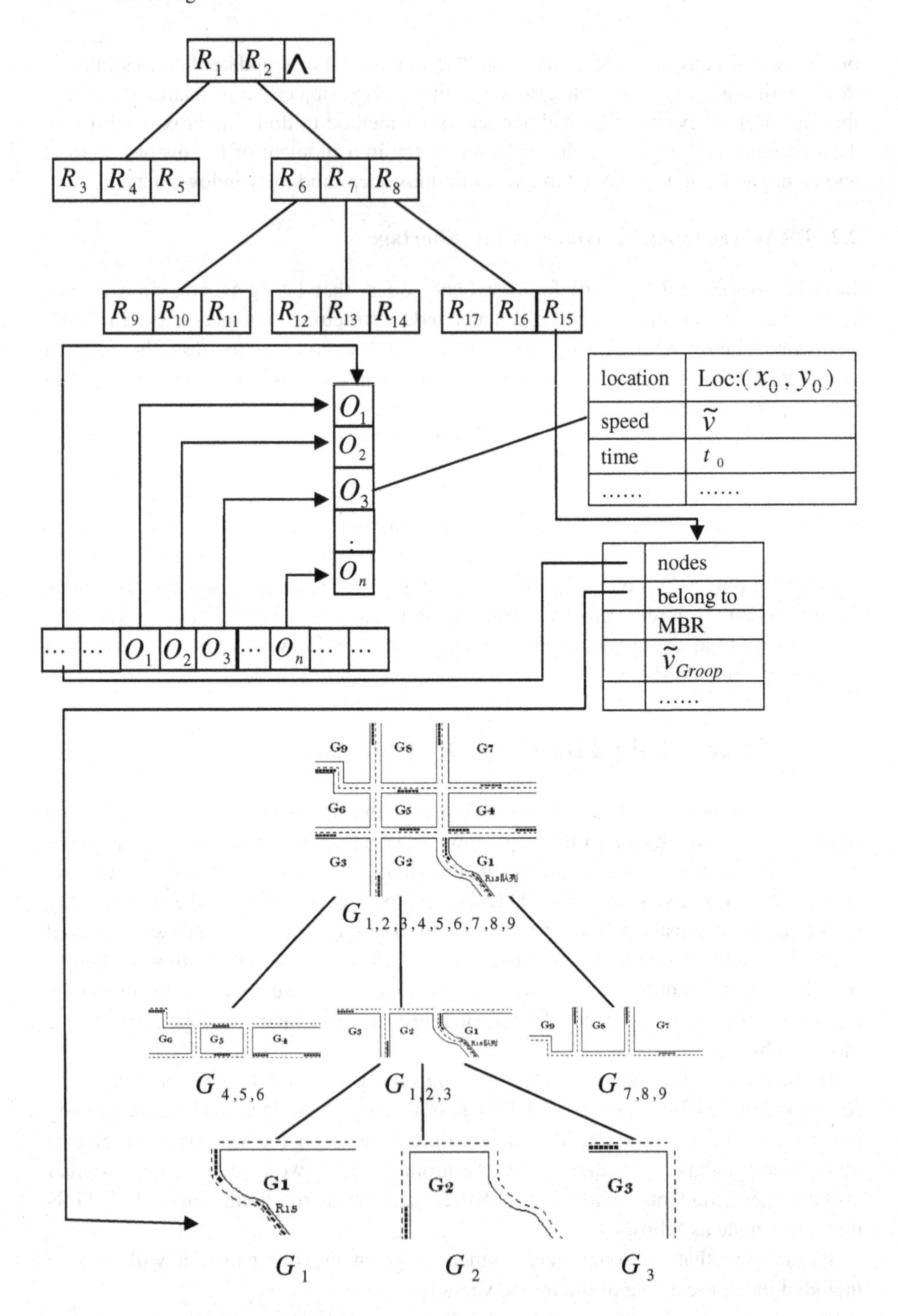

Fig. 1. Double R-tree and double indexing for moving objects

1) Each of the moving objects in the queue maintain a triple record < Loc_{ref} , $\tilde{v}$, t_{ref} >, t_{ref} , Loc_{ref} and $\tilde{v}$ separately represent the information update time of last time (reference time), the position coordinates of the moving objects in the moment Loc_{ref} :(x_0 , y_0) , and the velocity accumulation scalar value. If there is no triple record information update while indexing, which represents that the moving objects progress along the common direction of this group with the velocity accumulation value from the reference time t_{ref} and the position (x_0 , y_0).

Ideally, once the velocity and direction of the moving objects change, we should update the objects' triple information record in the GG TPR-tree index. But when the scale of the moving objects is large in the system, this method is unpractical and unnecessary. As the moving objects' moving speed and direction fluctuates inevitably, to avoid frequent updates of the record information, the moving objects adopt the non-equal time update method to maintain each record information. When the error of the position calculated with the location and velocity in the reference time and the actual position is larger than the distance error allowed by the system, it triggers information update request. Meanwhile, in order to reduce the frequency of the information updates, we need to extract the velocity value with high stability as the information record value in the index structure.

To solve the problem of the future rate prediction, document [11] proposes the Time Delay Model of the rate prediction, $\hat{v}_n = v_{n-1}$. that is, the future rate scalar value and direction both adopt those of the previous moment. This model can not obviously adapt to the section of the roads in which the moving objects always change their velocity. Document [12] proposes the Moving Average Model based on Time Delay Model, $|\hat{v}_n| = \left(\sum_{i=1}^{p} v_{n-i}\right)/ p$, that is, the future rate scalar value is the accumulation of the previous p rates and the direction is same with that of the previous moment. This model is more suitable than the Time Delay Model, but it does not take into account that the closer the velocity value is to the current time, the larger is the similarity with the current moving velocity value.

After taking the advantages and the disadvantages of two models above, we propose Rate Accumulation Model: the moving objects go into the grid when t=0, the moving object's moving rate accumulation scalar value in the road of the grid is:

$$\tilde{v} = \sum_{t=1}^{T}(t \times v_t) / \sum_{t=1}^{T}(t) \tag{1}$$

2) The Quintuples records of GG TPR-tree leaf nodes are changed into <grid, $ptr_{objects}$, MBR, $\tilde{v}_{Groop}$, ptr_{parent} >, respectively representing the grid that this group of moving objects belong to, the pointer pointed to the moving objects queue, the bounding rectangle of the nodes, the node's velocity stable value, the pointer pointed to the parent nodes, the record form of GG TPR-tree non-leaf node does not change.

Estimating equation (1) shows that supposing there are n members in a group of the moving objects, the MBR stable value is achieved by the equation as follows:

$$\tilde{v}_{Groop} = \sum_{t=1}^{n}((T - t_i) \times \tilde{v}_i) / \sum_{t=1}^{n}(T - t_i) \quad (2)$$

Under the help of GG TPR-tree and spatial partitioning tree, we can predict the approximate location moving direction according to $\tilde{v}_{Groop}$ and road network structure, achieving the exact answer of the statistical query. Meanwhile, in order to update the dynamic information of the moving objects efficiently, we establish Hash index list pointed to every moving object in the memory. This list may also achieve the exact answer of the deterministic query in collaboration with the spatial partitioning tree. Therefore, we establish the moving objects' complete query basis based on DTDI structure.

4 Experiment

In order to evaluate the query performance of mobile objects based on DTDI structure, we conduct the following experiments: the space region is a rectangular frame consisting of 800*600 pixels. There are "four horizontal, five vertical" simulate road in it. Each mobile object on the road is represented by a circle with a 40-pixel radius. The objects are distributed randomly on all roads. We adopt the proportion of 100 steps versus 100 steps to allocate the time of red lights and green lights at the crossing roads. For simplicity, we assume that all the roads are two-way and one-lane, that is to say, the mobile objects are incapable of surpassing the objects ahead. In addition, all the turnings (include the right turnings) are controlled by signals and the turning probability of each mobile object is known already.

The hardware environment of the experiment: CPU, Pentium 1.8GHz; Memory, 512MB; HD, 40G, 7200RPM.

The software environment: Microsoft Visual C++6.0.

4.1 Contents and Settings of the Experiments

The simulator program is carried out step by step. Each mobile object moves 0~200 pixels in one-step time, and moves as fast as possible when no other objects block the way. Before the movement, it is necessary to meet that there is no other objects from the current location to the destination. When the object is at the n fork, its turning direction is decided according to the turning probability of crossing points $p_1, p_2, \ldots\ldots, p_n$. If the fork road is filled with mobile objects, the turning direction will be re-decided. If all the fork roads are full of other mobile objects, the mobile object will not move this time.

Under the conditions of the experiments, we respectively adopt PRA tree and DTDI index structure to conduct continuous window queries of the 350 mobile objects on the spatial roads. Query windows are available in the randomly-selected 200*200-pixel region of the roads. The duration stars from 5 steps, adding 5 steps to each time to 50 steps altogether. The experiments are conducted 10 times. The precision of continuous window prediction query is calculated as follows:

Veracity = (the predicted quantity of moving objects which come into the query area after step n) / (the actual quantity of moving objects which come into the query area after step n) * 100%.

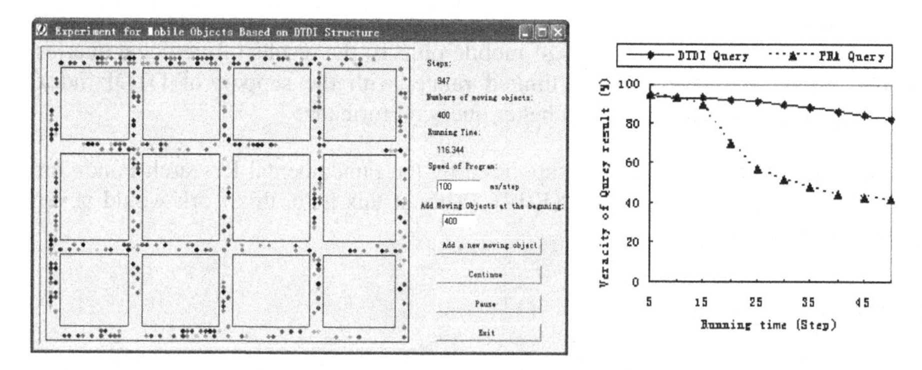

Fig. 2. The map for experiment and the chart for results

4.2 Results and Analyses of the Experiments

We can see from Figure 3 that during the beginning process, as the VBR of PRA tree differs little from the $\widetilde{V}$ of DTDI and both approximate the actual velocity of mobile objects, therefore, the predicted data of mobile objects within a recent time range are relatively accurate. With the increase of the predicted time range, under the guidance of network structure, turning probability of crossing point, the accumulated EXP velocity value $\widetilde{V}$ recorded in history record, DTDI is capable of ensuring that the predicted direction and velocity approximate the possibly actual direction and velocity of the mobile objects in the future, therefore, the predicted precision is relatively high. However, as $\widetilde{V}$ is adopted as EXP after passing the crossing point, and there exist certain errors compared with actual velocity value in the future, as a result, the precision becomes gradually low as the predicted time range increases. To PRA tree, there is no guidance of network structure in it, so it is incapable of considering turning, therefore, its predicted success rapidly decrease. Meanwhile, because the mass data can compensate for the error, accordingly, the ratio between the predicted number and the actual number will not be extremely low.

5 Conclusion

To effectively achieve various query requests oriented at mobile objects, this thesis proposes the DTDI index structure based on grid by extending GG TPR tree index structure. By adding corresponding spatial grid model to GG TPR leave node records, this structure realizes the correspondence between the mobile objects and their moving space, thus makes it possible to predict the current and future locations of mobile objects on the grid road. Meanwhile, as we improve the children node of leaves and information unit in GG TPR index, DTDI index structure is able to achieve the integrated maintenance, query and pruning of one-group mobile objects according to road curves and crossing points and thus to reduce the information updates of mobile objects, enhancing the efficiency of the index and maintenance of mass data within a limited range and the query performance of mobile objects, being available to respond to various types of query requests. The experiments prove that compared with other existing index structures of mobile objects, the results of predicted queries of mass mobile objects within a limited range, with the support of DTDI index structure, have higher precision and better query performance.

Acknowledgments. This paper is supported by the Fundamental Research Funds for the Central Universities (No.2010XJ017). Without this help, this work would never have been completed.

References

1. Chon, H.D., Agrawal, D., El Abbadi, A.: Data management for moving objects. In: Proceeding of International Telematics and LBS Workshop (2003)
2. Tao, Y., Papadias, D., Zhai, J., et al.: Venn sampling: A novel prediction technique for moving objects. In: ICDE 2005, Tokyo, Japan (2005)
3. Cheng, R., Lam, K.-Y., Sunil, P., et al.: An efficient location update mechanism for continuous queries over moving objects. Information Systems 32 (2007)
4. Xing, G., John, S., Hurson Ali, R.: Window query processing with proxy cache. In: 7th International Conference on Mobile Data Management, Nara, Japan (2006)
5. Liu, G., Liu, X., Yu, J., et al.: Continuous monitoring of skycube queries over sliding windows in data stream environment. In: 10th IASTED International Conference on Software Engineering and Applications, Dallas, TX, United States (2006)
6. Sun, J., Papadias, D., Tao, Y., et al.: Querying about the past, the present, and the future in spatio-temporal database. In: VLDB 2004, Toronto, Canada (2004)
7. Saltenis, S., Jensen, C., Jensen, S.: Indexing the positions of continuously moving objects. In: SIGMOD 2000, pp. 331–342. ACM Press, New York (2000)
8. Lee, J., Cho, W., Edgar Thomas, F.: Control system design based on a nonlinear first-order plus time delay model. Journal of Process Control 7, 9 (1997)
9. Yu, X.-L., Chen, Y., Ding, X.-C., et al.: A Moving Object Database Model Based on Road Network. Journal of Software 14(9), 8 (2003)

Activity Recognition in Ubiquitous Computing Environment

Tao Lu, Xin Wang, and Shaokun Zhang

System Engineering Institute, Dalian University of Technology
Dalian, China
lutao@dlut.edu.cn

Abstract. Activity recognition is one of research issues in ubiquitous computing. In this paper we present an approach to recognize complex activity given the formal context representation. Complex activity refers to the activity composed of simple actions to achieve a certain goal, and the formal context representation is inferred context derived from sensor data. We first propose activity model, into which the relations between activity and context patterns are integrated. The relations are expressed by a set of temporal constraints on context. Then the recognition algorithm, which is based on the activity model and activity flow, is described. We further interpret the approach by the case of single-crystal X-ray diffraction experiment.

Keywords: ubiquitous computing, activity recognition, context reason, temporal logic.

1 Introduction

In the early 90s of 20 century, Mark Weiser proposed the concept of ubiquitous computing [1], which is constituted by communication technology and computers, aimed to merge the physical space and the information space together, and set up a customer-centric environment in order to support the acquirement of individual information services transparently anytime and anywhere.

In ubiquitous computing environment, context awareness is a significant technique to develop ubiquitous computing applications which are flexible, adaptable, and capable of acting autonomously on behalf of users [2]. By context, we refer to any information that can be used to characterize the situation of an entity, where an entity can be a person, place, or physical or computational object [3]. A context-aware application is defined as one that uses the context of an entity to modify its behavior to best meet the context of the user [3]. One of the important functionalities provided by a context-aware infrastructure is context reasoning to derive high-level contexts from low-level contexts. Low-level contexts are information that is directly acquired from hardware sensors and software programs, while high-level contexts are the information that is used by application and related to user and user's demanding [4].

Activity can be regarded as one type of high-level context, based on which the application adapt its behavior. In this paper, we focus on recognizing the complex activity, usually composed of simple actions, to achieve a certain goal. For example,

H. Tan (Ed.): Knowledge Discovery and Data Mining, AISC 135, pp. 279–286.
springerlink.com

in single-crystal X-ray diffraction experiment, there are eight activities which should be constructed as predefined sequence. Every activity is composed of many simple actions.

Various activity recognition approaches have been proposed in ubiquitous computing research area. The main approaches can be divided into data-driven and knowledge-driven approaches [5]. Data-driven techniques recognize the basic physical activities through the use of machine learning methods and the acquired data from sensors. Probabilistic methods such as Bayesian networks and Markov models are popular method in activity recognition to deal with uncertainty. In [6], Markov networks are applied to recognize high-level activities such as shopping or dining out.

Knowledge-driven techniques have a long history. Among them, situation calculus and the event calculus are two well-known methods [7] [8]. They are mainly applied to model the temporal feature of activities and the relationships between activities and events. Ontology is another technique. In Ref.[9] activities has been recognized through ontological reasoning.

The rest of the paper is organized as follows. Section 2 provides the activity model. Section 3 proposes the activity recognition algorithm. Case study is described in section 4 and section 5 concludes the paper.

2 Activity Model

Activities may result in different context and the context change, by which activities are recognized. Therefore, the relations between activities and contexts should be integrated into activity models.

2.1 Context Pattern

Suppose that the application is related to a set of context $c_1, c_2, \ldots c_n$, and the domain of context c_i is D_i, i=1,2,…n. At given time t, the value of context c_i is denoted as $c_i(t)$. The values of context $c_1, c_2, \ldots c_n$ at time t is denoted as $C(t)$, $C(t)=(c_1(t),c_2(t),\ldots c_n(t))$.

Suppose $d_j^i \subseteq D_i$, (c_i, d_j^i) is called atomic context pattern which represents the condition that context satisfies. The context pattern (c_i, d_j^i) holds at time t, denoted as $(c_i, d_j^i)|t$, if and only if $c_i(t) \in d_j^i$

We use conjunction($\wedge$), disjunction($\vee$) and negative($\neg$) operation on atomic context pattern to derive composite context pattern. The meanings of conjunction, disjunction and negative are as their usual.

Let p,q be atomic context pattern, obviously we have following rules:

$$(p \wedge q)|t \equiv p|t \wedge q|t$$

$$(p \vee q)|t \equiv p|t \vee q|t$$

$$(\neg p)|t \equiv \neg(p|t)$$

Definition 1. Context pattern is defined recursively as follow:

(1) If α is a atomic context pattern, then α is a context pattern.

(2) If α, β are context patterns, then $\alpha \wedge \beta$, $\alpha \vee \beta$, $\neg \alpha$ are context patterns.

$(v_1, D_1) \wedge (v_2, D_2), \ldots \wedge (v_n, D_n)$ is a special context pattern that holds any time, denoted as p_T .

Definition 2. Let $v = (v_1, v_2, \ldots v_n), i = 1,2,..n$, p be a context pattern. v is called an instance of context pattern p if and only if $(C(t) = v) \rightarrow p|t$

We use *Instance*(p) to denote the set of instance of context pattern p.

Apparently, if p and q are context patterns, $p \rightarrow q$, Then *Instance*(p)$\subseteq$ *Instance*(q).

Any activity has consumed some time which is called duration. So do the holding context pattern.

Let A be an activity set, P be a context pattern set. *Equal*, *During*, *Start* and *Finish* are functions: $A \rightarrow P$, mapping the activity to context pattern in different phase of the whole process. The semantic meanings of these functions are as follows:

Equal(a): context patterns that holds in the time period the activity is performed.

During(a): context patterns which ever holds during the period the activity a is performed.

Start(a): context patterns whose beginning time is same with activity.

Finish(a): context patterns whose ending time is same with activity.

The semantic meaning of the functions has a few differences with their usual meaning in temporal logic representation, and not all the temporal relations are represented by the above functions, such as overlap which may exist in real world problem. This is because the context patterns are used to judge the beginning, ending or suspending of activities. Some relation like overlap has no use here.

Definition 3. Let A be an activity set, P be a context pattern set. Activity model of each activity is tuple <ActivityID, *Equal*, *During*, *Start*, *Finish*>, where ActivityID is activity identifier, *Equal*, *During*, *Start* and *Finish* are functions: $A \rightarrow P$

For simplicity, *Start*(a) and *Finish*(a) are called beginning pattern and ending pattern of activity a respectively. *Equal*(a) is called running pattern of activity a.

Definition 4. Let A be an activity set, $a \in A$. If the activity model of a satisfies the following rule, the model is consistent.

Equal(a), *During*(a), *Start*(a) and *Finish*(a) should not be the context pattern that never hold.

During(a) $\Rightarrow$ *Equal*(a)

During(a) $\Rightarrow$ *Start*(a)

During(a) $\Rightarrow$ *Finish*(a) □

That the model of activity is consistent is the basic premise of recognizing the activity.

2.2 Condition of Distinguishable Activity

In this paper, we focus on the activity which has to been conducted in predefined sequences. The activity process can be described in digraph. For example, the sequence of activities in single crystal X-Rays diffraction experiment is shown in Fig.1.

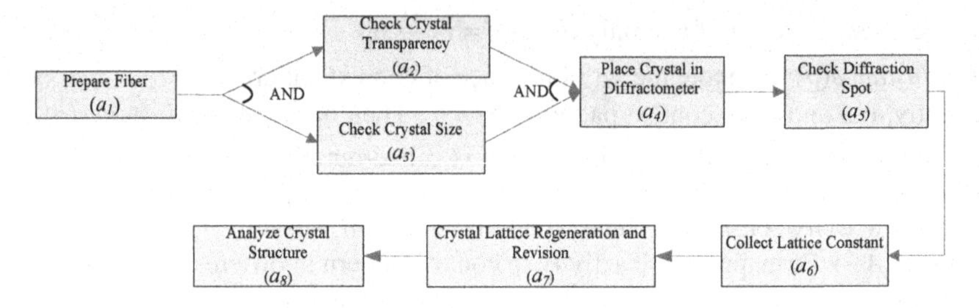

Fig. 1. Activity flow in single crystal X-Ray diffraction experiment

Definition 5. Activity process is 4-tuple <*A*, *E*, *join*, *split*>, where *A* an activity set, $E \subseteq A \times A$, *join*: $A \rightarrow \{AND, XOR, NULL\}$, *split*: $A \rightarrow \{AND, XOR, NULL\}$.

<*A*, *E*> is a directed graph which represents the dependency between activities. *AND*, *XOR* and *NULL* are join or split types of nodes. *NULL* means the corresponding activity has a single incoming or outgoing activity.

Definition 6. Let $a \in A$, if *split*(a)=*XOR*, then a is called *XOR-split* activity. *AND-split*, *XOR-join*, *AND-join* can be defined similarly.

Definition 7. Let $a,b,c \in A$, if $(a,b) \in E, (a,c) \in E, split(a) = XOR$, then b and c are called *XOR-Sibling* activities of each other.

The basic assumption of the paper is that the activity is completed by one person or one teams, so that the activity can't be performed concurrently.

The aim to define context pattern for every activity is to recognize the activity by context. This requires that the context resulted from some activities should have some differences. And the critical step is that the beginning and ending should be determinable. Therefore the beginning pattern and ending pattern of activity should not be p_T.

Since the activity is constructed as predefined sequence, and only after one has been completed, there is possibility to select next to execute, we just need to compare the beginning pattern of the activities which have possibility to be selected to execute at the same time.

Suppose $a \in A$, let *neighbor*(a) to denote the set of activities which have possibility in selecting-to-execute-next candidate set including a. Obviously the predecessor activities and postdecessor activities can't be included in *neighbor*(a). Besides, the activities which has common *XOR-split* predecessor with a but in other branches, should be excluded in *neighbor*(a) if *XOR-split* predecessor is not direct predecessor of a. Similarly, the postdecessor activities of *XOR-sibling* of a should be also excluded.

Let A be an activity set. Suppose the models of each activity are consistent. The rules listed below guarantee that the activities are distinguishable.

Rule 1: for each $a \in A$, $Start(a) \neq p_T$, $Finish(a) \neq p_T$

Rule 2: for each $a' \in neighbor(a)$, $start(a') \Rightarrow \neg start(a)$.

3 Activity Recognition Algorithm

If the activity models of every activity are consistent, and satisfy Rule 1 and Rule 2, activity can be recognized by algorithm 3.1.

For an activity, its starting can be judged by holding of its beginning pattern. When the finishing pattern starts to hold, activity ending can be determined when the finishing pattern ceases holding. Besides, the activity may be interrupted by some events. This can be recognized if the context values violate its normal pattern when the activity has been started and not finished. When one activity has been finished and the next activity has not been started, we call the state is *waiting*.

Therefore, we define four states which represent the state of execution of activity flow: *waiting*, *on*, *suspend*, *finishing*. Here *on* means that the activity is performed but ending pattern has not held, *finishing* means the activity is performed normally, but the ending pattern has held, implying that the activity end once the pattern cease holding. The activity and states express the recognized situation.

Algorithm 3.1 can be used to determine the activity and its state when the context value is input and the activity and its state of last time is known.

Algorithm 3.1

```
Input: LastAct,LastState,v
Output:CurAct, CurState
BEGIN
  CurAct=LastAct, CurState=LastState
  CASE CurState OF
  waiting:FOR each a ∈next(LastAct)
            IF v ∈Instance(start(a))
                   AND v ∈Instance(finish(a))
            THEN  CurAct=a, CurState=finishing ;
            ELSE IFv ∈Instance(start(a))
                   AND v ∉Instance(finish(a))
                  THEN CurAct=a, CurState=on
  on: IF v ∉Instance(equal(CurAct))
      THEN CurState=suspend
      ELSE IF v ∈Instance(finish(CurAct))
            THEN CurState=finishing;
```

```
Suspend : IF v ∈Instance(equal(CurAct))
              AND v ∈Instance(finish(CurAct))
          THEN CurState=finishing
          ELSE IF v ∈ Instance(equal(CurAct))
               THEN CurState=on;
finishing: IF v ∉Instance(finish(CurAct))
           THEN CurState=waiting
           ELSE IF v ∉Instance(equal(CurAct))
                THEN CurState=suspend
                ELSE CurState=finishing;
END
```

In algorithm *next*() is a function which mapping an activity to its next possible activity.

4 Case Study

As the model presented in section 2, we design activity models of activities in single-crystal X-ray diffraction experiment.

The single-crystal X-ray diffraction procedure consists of three phases, which are conducting in three smart rooms. The details of the experiment are as follows [10]:

(1) Select a crystal. The activities in the phase are conducted in smart room 101.

(2) Analyze the crystal. The activities in this phase are conducted in smart room 102.

(3) Structural determination. The activities in the phase are conducted in smart room 103.

There are 19 types of contexts related to the experiment. Table 1 shows default value of each context. Since in any time point, the values of most of contexts are default values, for simplicity, they are ignored in context pattern description.

Table 1. The default value of each context

Context	Default value	Context	Default value
User_Loc	*Room 101*	*Crystalcenter_Loc*	*Screenleft*
Copperbar_Loc	*Lab Cabinet*	*Diffractometer*	*Off*
Slicer	*Off*	*SpotResult_Bool*	*Off*
Microscope	*Off*	*Matrix*	*Off*
OcularlensI_Loc	*Ocularlensbox*	*Saint*	*Off*
OcularlensII_Loc	*Ocularlensbox*	*Shelxt*	*Off*
Adjustmentscrew	*Off*	*ConstantResult*	*False*
Temperature	20℃	*RevisedResult*	*False*
Voltage	*0KV*	*AnalyzeResult*	*False*
Controller	*Off*		

Table 2 shows the activity models of eight activities.

Table 2. Activity Models

Activity	Function	Context Pattern
Prepare Fiber	*Equal*	$(User\_Loc, Room\ 101) \wedge (Copperbar\_Loc, Cuttingmat)$
	During	$(Slicer, On) \wedge (User\_Loc, Room\ 101) \wedge (Copperbar\_Loc, Cuttingmat)$
	Start	$(User\_Loc, Room\ 101) \wedge (Copperbar\_Loc, Cuttingmat)$
	Finish	$(Copperbar\_Loc, \neg Styrofoam)$
Check Crystal Transparency	*Equal*	$(User\_Loc, Room\ 101) \wedge (Microscope, On) \wedge (OcularlensI\_Loc, Microscope)$
	During	$(User\_Loc, Room\ 101) \wedge (Microscope, On) \wedge (OcularlensI\_Loc, Microscope) \wedge (Adjustmentscrew, On)$
	Start	$(User\_Loc, Room\ 101) \wedge (Microscope, On) \wedge (OcularlensI\_Loc, Microscope)$
	Finish	$(OcularlensI\_Loc, \neg Ocularlensbox)$
Check Crystal Size	*Equal*	$(User\_Loc, Room101) \wedge (Microscope, On) \wedge (OcularlensII\_Loc, Microscope)$
	During	$(User\_Loc, Room101) \wedge (Microscope, On) \wedge (OcularlensII\_Loc, Microscope) \wedge (Adjustmentscrew, On)$
	Start	$(User\_Loc, Room101) \wedge (Microscope, On) \wedge (OcularlensII\_Loc, Microscope)$
	Finish	$(OcularlensII\_Loc, \neg Ocularlensbox)$
Place Crystal in Diffractometer	*Equal*	$(User\_Loc, Room102) \wedge (PCI, On) \wedge (Temperature\ [10,25]) \wedge (Voltage, 50KV) \wedge (Copperbar\_Loc, Diffractometer)$
	During	$(User\_Loc, Room102) \wedge (PCI, On) \wedge (Temperature\ [10,25]) \wedge (Voltage, 50KV) \wedge (Copperbar\_Loc, Diffractometer) \wedge (Controller, On)$
	Start	$(User\_Loc, Room102) \wedge (PCI, On) \wedge (Temperature\ [10,25]) \wedge (Voltage, 50KV) \wedge (Copperbar\_Loc, Diffractometer)$
	Finish	$(Crystalcenter\_Loc, \neg\ Screencenter)$
Diffractometer Spot Check	*Equal*	$(User\_Loc, Room102) \wedge (Voltage\_Val, 50KV) \wedge (Diffractometer, On) \wedge (Copperbar\_Loc, Diffactometer)$
	During	$(User\_Loc, Room102) \wedge (Voltage\_Val, 50KV) \wedge (Diffractometer, On) \wedge (Copperbar\_Loc, Diffactometer)$
	Start	$(User\_Loc, Room102) \wedge (Voltage\_Val, 50KV) \wedge (Diffractometer, On) \wedge (Copperbar\_Loc, Diffactometer)$
	Finish	$(SpotResult\_Bool, on)$
Lattice Constant Collection	*Equal*	$(User\_Loc, Room102) \wedge (Voltage, 20KV) \wedge (matrix, on)$
	During	$(User\_Loc, Room102) \wedge (Voltage, 20KV) \wedge (matrix, on)$
	Start	$(User\_Loc, Room102) \wedge (Voltage, 20KV) \wedge (matrix, on)$
	Finish	$(ConstantResult\_Bool, False)$
Crystal Lattice Regeneration and Revise	*Equal*	$(User\_Loc, Room102) \wedge (Saint, On)$
	During	$(User\_Loc, Room102) \wedge (Saint, On)$
	Start	$(User\_Loc, Room102) \wedge (Saint, On)$
	Finish	$(RevisedResult\_Bool, False)$
Crystal Structure Analyze	*Equal*	$(User\_Loc, Room103) \wedge (Shelxt, On)$
	During	$(User\_Loc, Room103) \wedge (Shelxt, On)$
	Start	$(User\_Loc, Room103) \wedge (Shelxt, On)$
	Finish	$(AnalyzeResult\_Bool, False)$

5 Conclusion

Activity recognition is one of issues in ubiquitous computing research. In this paper, we propose an approach to recognize complex activity which is usually composed of single actions. We first present the concepts of context pattern. The activity model is defined by its beginning pattern, ending pattern and normal running pattern. Considering the sequence constraint by the activity flow at the same time, we analyze the condition to distinguish the activities. We develop an activity recognition algorithm which can be used in activity recognition if the models satisfy the distinguishable condition. According to the approach, we design activity models in single-crystal X-ray diffraction experiment.

The limitation of our approach is the uncertainty has not been considered which exists in many real world problems. Besides, the activity recognition is under ideal assumption which is strict for many applications. These problems should be further taken into account in future research work.

Acknowledgments. This research work was supported by the National Natural Science Foundation of China (Grant No. 70771017).

References

1. Weiser, M.: The computer for 21st century. Scientific American 261(30), 94–104 (1991)
2. Bettini, C., Brdiczka, O., Henricksen, K., Indulska, J., Nicklas, D., Ranganathan, A., Riboni, D.: A survey of context modelling and reasoning techniques. Pervasive and Mobile Computing 6(2), 161–180 (2010)
3. Dey, A.K., Abowd, G.D., Salber, D.: A conceptual framework and a toolkit for supporting the rapid prototyping of context-aware applications. Human-Computer Interaction 16(2-4), 97–166 (2001)
4. Ranganathan, A., Al-Muhtadi, J., Campbell, R.H.: Reasoning about uncertain contexts in pervasive computing environments. IEEE Pervasive Computing 3(2), 62–70 (2004)
5. Riboni, D., Bettini, C.: OWL 2 modeling and reasoning with complex human activities. Pervasive and Mobile Computing 7(4), 379–395 (2011)
6. Liao, L., Fox, D., Kautz, H.: Location-based activity recognition using relational Markov networks. In: Proceedings of the Nineteenth International Joint Conference on Artificial Intelligence, pp. 773–778. Professional Book Center (2005)
7. Kowalski, R., Sergot, M.: A logic-based calculus of events. New Generation Computing 4(1), 67–95 (1986)
8. Brdiczka, O., Crowley, J.L., Reignier, P.: Learning Situation Models for Providing Context-Aware Services. In: Stephanidis, C. (ed.) UAHCI 2007. LNCS, vol. 4555, pp. 23–32. Springer, Heidelberg (2007)
9. Chen, L., Nugent, C.D.: Ontology-based activity recognition in intelligent pervasive environments. International Journal of Web Information Systems 5(4), 410–430 (2009)
10. Hwang, G.J., Yang, T.C., Tsa, C.C.: A context-aware ubiquitous learning environment for conducting complex science experiments. Computers and Education 53(2), 402–413 (2009)

Circuit Diagram Design for DAS Based on Single Chip Microcosm

Xiaolin Lu

College of Information and Electronic Engineering,
Zhejiang University of Science and Technology, Hangzhou, China, 310012
XiaolinLu@139.com

Abstract. The labor and social security information system needs DAS (Data Acquisition Systems) to exchange information. The DAS of the social insurance applications has a very important practical value. This paper presents the circuit diagram design for DAS based on single chip microcosm, a phone dialer, modulation devices, central processing unit, buttons, input devices and display devices. It can be used to transmit data through the public switched telephone network. It is suitable for constructing DAS in the labor and social security information systems with low cost for data acquisition through the PSTN.

Keywords: DAS, PSTN, circuit design, labor and social security.

1 Introduction

A labor and social security information system needs DAS to exchange information that is the basis for building monitoring tools that enable the supervision of local and remote systems. Information communication with distance computer or equipment is of great importance in numerous labor and social security information systems and public data services. Many kinds of communication and data-transmitting systems have used DAS to exchange information [1-3]. The special-purpose DAS has the important practical value in those systems because they are cheaper and more convenient than system built up by the computer network to form a distribute environment.

The DAS can often satisfy the need special-purpose communication system in many special fields. A number of experiments and research have been done regarding the distance data acquisition systems in recent years. Many investigators have explored the special devices that exchange data through public switched telephone network [4-5]. V. Reddy reported a device that can transmit data over telephone networks. He proposed the basic principles of data transmission over the PSTN [6]. H. Sun investigated a kind of HUST-B distance terminal device, which consists of computers, measurement devices. It can communicate with distance equipments; gather the parameter data of power, power factor, and frequency from the distance industrial equipments. It can also be used to control the remote devices [7]. S. Mao designed a real-time distance data gather device based on Neuron chip and Lon works technologies [8]. Z. Jia reported a device that utilize PC ISA slot. It realized data exchange between a power system and the distance terminal machines [9]. H. Huang

H. Tan (Ed.): Knowledge Discovery and Data Mining, AISC 135, pp. 287–293.
springerlink.com

suggested a system structure of Internet based radiotherapy diagnose system with some kind of distance device [10]. Although many researchers have investigated the distance data exchange system with the special-purpose distance terminal devices, little work has been published on the system and circuit design of the data acquisition systems that can transmit the data based on public switched telephone network.

This paper we investigated a mini-type distance data acquisition system that can be used to exchange the data with the distance host computer through PSTN. The system principle and the circuit design of the device have been studied in detail. The results of this study may help the design of the PSTN based data acquisition systems and the system construction of the general used special purposed communication network.

2 Backgrounds and Related Work

Multipoint Data Conferencing and real-time communication protocols (Multipoint Data Conferencing Protocol Suite) ITU T.120 standard by a group composed of communication and application protocols. Protocol for multi-point data conferencing and real-time communications, including multi-protocol greatly improves the multimedia, MCU and multimedia digital signal codec control. The product can establish a connection, transmit and receive data and mutual cooperation, which is through the use of compatible data conferencing features, such as application sharing, whiteboard sessions and file transfers. The main functions are as follows:

(1) public switched telephone network (PSTN);
(2) Integrated Switched Digital Network (ISDN);
(3) circuit-switched digital network (CSDN);
(4) packet-switched digital network (PSDN);
(5).Novell Netware IPX;

The Mini-type distance data acquisition system is considered in non-industrial the field. The public has great demand to exchange data and information with a distance computer and database in daily life. Such device should satisfy following requirements:

(1) The data is transmitted through PSTN.
(2) The expenses are relative low.
(3) It requires relatively small size and easy to carry with.
(4) It can be easily operated and managed without certain training.

The mini-type distance data acquisition system is used to exchange data with distance host computer's database when data users are dispersed in the different geographical positions. User can set up connection distance host computer with mini-type distance data acquisition system. The device can be used at where, such as at home or hotel and at any place where PSTN exists, to connect the host computers for exchanging through telephone wire. It would improve the working efficiency and reduce expenses.

Because digital signal is unsuitable to be transmitted in the pair line of the public switched telephone network, communication between user and server must use the modem device to modulate the digital signal. ITU-T has made the standard of many kinds of modems. V. 34 standards support the speed as 28Kpbs. The upgrading edition V. 34, V. 34 annexes 1201H supports the speeds of 33. 6Kbps, by using data compress (V. 42 bis), the supreme speed can reach 133. 8Kbps. The new standard is V. 90 can transfer data at 56 Kbps and higher rate through data compression.

The mini-type distance data acquisition system is used to small and simple terminal that can communicate with distance host computer. One datum server on the host computers and many mini-type distance data acquisition systems build up a small-scale communication network and data exchange application system. The database is in the host computer. The host computer also can communicate with other computers in Internet or intranet. The mini-type distance data acquisition systems are used to exchange the data with host computer through the public switched telephone networks. Figure 1 show the system infrastructure.

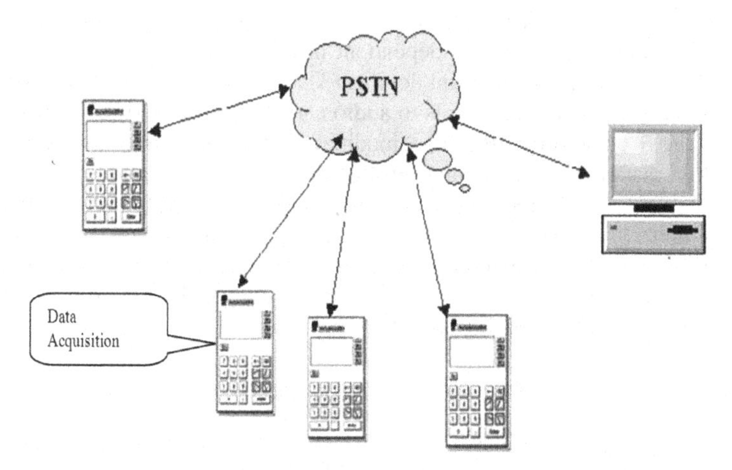

Fig. 1. A small-scale communication network and data exchange application system

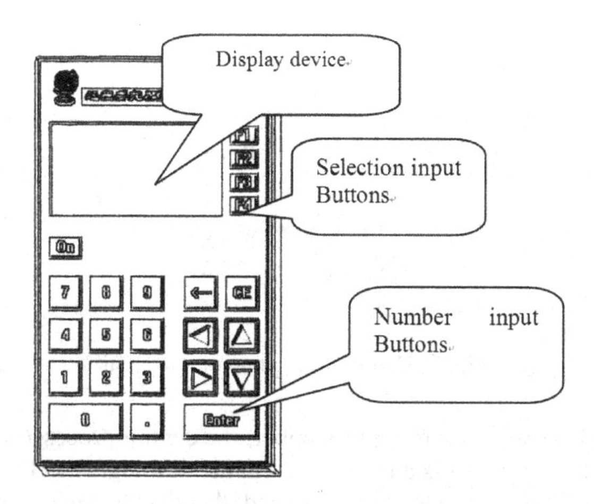

Fig. 2. The mini-type distance data acquisition system

The mini-type distance data acquisition system is consisted by numbers of integrated circuit such as telephone dial device, modulate integrated circuit, central processing unit, button input device and display device. The device has many advantages such as small, easy to carry, relatively cheap, easy to operate and low data transmission cost. Figure 1 shows the appearance of the device.

3 Circuit Design of Mini-type Distance Data Acquisition System

DAS System Infrastructure
(1). Central processing unit: Namely CPU. It is a single chip microcosm (SCM), such as INTEL 51 SCM series. It is used to control all other components and to coordinate work of all devices. (2) Telephone dial device: It is used to dial and produce Dual-tone-multi-frequency (DTMF) signal or pulse signal. It is used to choose the datum data exchange server at distance host computer before set up connection with the host computer. This device can be an independent integrated circuit or be included in the following Modulation-Demodulation device. (3) Modulation-Demodulation device: It is used to modulate the digital signals to audio carrier frequency signals for sending to host computer. At the same time it demodulate the audio carrier frequency signals received from distance host computer through the public switched telephone networks to digital signals. The Demodulated digital signal will be processed by the central processing unit. (4) The button imports the device: It consists of a small-scale keyboard and several electronic elements. User can use the device to input data and operate the mini-type distance data acquisition system. (5) Display device: It is used to show device the data in the display screen.

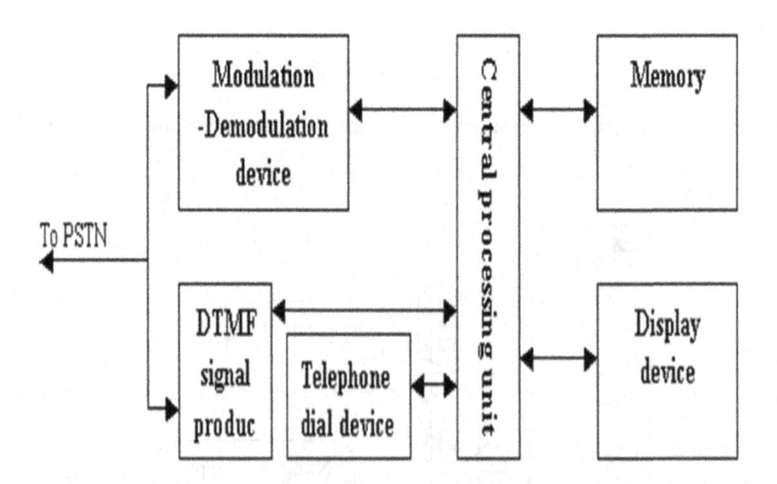

Fig. 3. The block-diagram of the circuit

Operator use button device to input number, the central processing unit process the input data and transmit to distance host computer through the PSTN. The display device shows the data the user inputted and the data received from distant host computer. User operates the mini-type distance data acquisition system by the input device. The digital signals user inputted will be moderated by modem and transmitted to the distance host computer. Audio carrier frequency signals received from host computer will be demodulated by modern and show in the display device.

The distance host computer received the requests from the mini-type distance data acquisition system and makes corresponding treatment with different application program, and sends the results back to the mini-type distance data acquisition system with same modulation method through public switched telephone network. This device used the back-echo method to confine the data inputted by user. The modulation demodulation devices receive the demodulation signal and demodulate to the digital signal. The central processing unit reads and deals with data and sends the data to display device to show the results.

DAS System Circuit Design

The circuit design is showed in figure 4.

1) Ul is an 89C5l Single Chip Microcosm.
2) U2 and U11 flip-latch.
3) U3 is a 6116 static random access memory
4) U4 and U5 is 27C512 ROM.
5) U6 is a liquid crystal display component based on T6963C TM24064B.
6) U7, U10 and U17 are data transceiver.
7) U8 is programmable array logic chip of ISPl016.
8) U9 is an inputting keyboard.
9) Ul2 is a UM91531 signal producer for Dual-tone-multi-frequency (DTMF) signal.
10) U14, U15 and U16 are linear amplifiers.
11) XU is telephone network's interface.

DAS System Data Workflow. (1) After the machine started, Ul run the initial procedure, take out corresponding word models (lattice type matrix storehouse) from U4 U5 and U6; (2)Demonstrates the control signal for connection; (3) the distance host computer; (4) User operates the device with U9, makes U8 connect with at Ul and INTO foot, produces low electronic signal at the same time. Ul responds the interrupt, shows the corresponding scan keys of U8 and U7 through U6 at low third digital of Pl. At the same time, Ul deposits Dual-tone-multi-frequency (DTMF) signal at Ull, sends the DTFM to XU through U8 and U12 to produce one operation and finishes data inputting. By repeating above procedure, we can impute all the data; (5)

After connect with distance host computer, the host computer sends back the modulated signal. It will be transmitted thorough T, C5 and C8 and linear amplifier U14 then enter RC foot Ul3. U13 will demodulate the signal and it makes its foot become high-level electricity. At the same time, this high level produce an interrupt signal at the INTl feet of Ul. Ul respond interrupt signal read and enter 1 bit of demodulated data from RD foot of U13. (6) After U1 read a bit of data, U1 transforms the bit to word code and shows signals that the device has connected to the distance host computer. (7) Then, Ul shows a menu for operator to choose on U6. The operator chooses or inputs essential data. U1 will transform the inputted data to a serious of bit code that will be modulated to DTMF signal through U10 to the TD foot of U13 and be amplified through U15 and U16, be sending to the distance host computer. (8) The TS, RDR1 and RDR2 of U13 are the work statutes of used or controlled and the transmit speed. Through the different combination of three control ends, user can choose one work status and one transmit speed. The highest data transmission rate of this chip is 1200 B/S. (9) U9 is a programmable ISP1016 array logic chip from LATTICE Company. It is mainly used to store the data out of 64 KB, and the address space of chip, the ranks decipher of the keyboard, the space for the limited address visited by U1. If TD and TD of U13 feet were connected to RXD and TXD feet of U1, it can realize the same function of data communication.

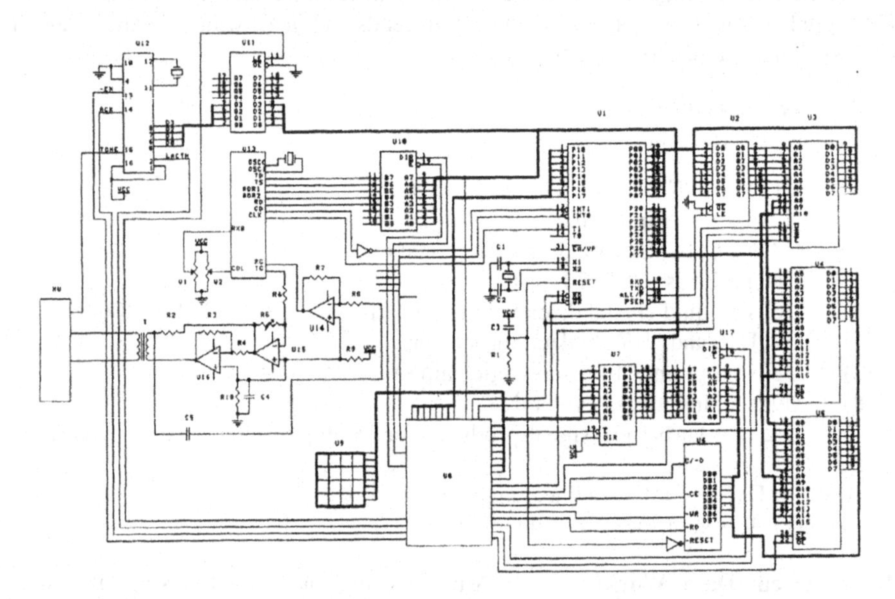

Fig. 4. Circuit diagram of the mini-type distance data acquisition system

4 Discussions and Further Work

The system and circuit design the mini-type distance data acquisition system in paper has been experimented and implemented. We have built up the device and an experimental network for data exchange through the PSTN.

The results of the experiment show that the mini-type distance data acquisition system can send and received data with a distance host computer through the PSTN. Although the device only consists of a center processor unit, a telephone dial device, modern, a button input device, memory and display device, it can perform that work of connecting to distance host computer through public switched telephone network.

The device has been used in a practical application for public to query the current stock price information. In the stock price information inquiring system, the mini-type devices are used for inputting a stock code, connecting the host computer, inquiring data from database and reporting current stock price for users.

This paper presents the system and circuit design of the mini-type distance data acquisition system. Further works should be performed to establish the general labor and social security information systems with the host computer and devices through the PSTN. Also, the communication between the devices should be considered the essential function for exchange information between the device's users. Further research could explore the possibility to design more functions of the data acquisition systems. The PSTN based mini type data acquisition systems can be used in a wide application filed, such as the stock information querying, the public database and bank account querying.

5 Conclusions

This paper proposed the circuit design of a new kind of mini-type distance data acquisition system. It can be used to transmit data through public switched telephone network. It consists of a small number of integrated circuit, telephone dial device, modulation device, central processing unit, button input device and display device. The device can be used in many applications with the advantages of small size, easy to carry, relatively cheap, easy to operate.

Acknowledgements. The funding support of this work is by National Science and Technology Support Program. The item number: 2008BAH32B00. Project Name is Key technologies and major application research of Human resources and social security public service system. Project Topic 6 Number: 2008BAH32B06. Project Issue 6 Name: The typical application development of Human resources and social security public service system. This paper is supported by Zhejiang Jandar technology Co., Ltd. (Jandar Company Information: Zhejiang Jandar technology Co., Ltd. Address: 11F,Jiangong Building,No.20, Wensan Road, Hangzhou. Tel: 86-0571-88,822,288 Zip: 310012 E-mail: djx@jandar.com.cn). Zhejiang Jandar technology Co., Ltd is greatly appreciated.

References

1. Hamdi, M., Verscheure, O., Hubaux, J., Dalgic, I., Wang, P.: IEEE Communications Magazine 37, 104–111 (1999)
2. Bellazzi, R., Montani, S., Riva, A., Stefanelli, M.: Computer Methods and Programs in Biomedicine 64, 175–187 (2001)
3. Gbaguidi, C., Hubaux, J., Pacifici, G.: IEEE Journal on Selected Areas in Communications 17, 1563–1579 (1999)
4. Luo, L., Liu, J., Shao, L., Lu, W., Ye, M.: IBM Systems Journal 45, 160 (2006)
5. Nieva, T., Wegmann, A.: Computers in Industry 47, 215–237 (2002)
6. Reddy, V.: Resonance: Journal of Science Education 6(5), 60–70 (2001)
7. Shun, H.: Information Technology 27(6) (2003)
8. Jia, Z., Li, Y.: Modern Electronic Technique (1) (2002) (in Chinese)
9. Huang, H., Cai, H.: Application Research of Computers 20(11) (2003) (in Chinese)

Network Management System for Employment Service and Social Insurance Application

Xiaolin Lu

College of Information and Electronic Engineering,
Zhejiang University of Science and Technology, Hangzhou, China, 310012
XiaolinLu@139.com

Abstract. The social insurance applications often consist of a combination of numerous computing systems. This paper presents network management architecture for employment service and social insurance application platform. The architecture is based on the J2EE and EJB technologies that allow all service and social insurance application system located on separate computers in different location. A practical employment service and social insurance application network management platform has been implemented. The experiments show that the architecture can be used to construct large-scale network consisting of numerous social insurance applications.

Keywords: Network management, employment service, social insurance, Web-NMS, J2EE.

1 Introduction

The large-scale computer and social insurance application networks often consist of a combination of numerous different operating systems spreading across various computing systems such as mainframes, workstations, PCs and some other telecom devices. The different programming languages in different programming environment are developing network management system deployed on these systems. Distributed object-computing promises a distinct software engineering paradigm for developing a kind of software infrastructure that can unify computers available in the network and facilitate communication between them. Also, in large-scale network management system, there are many problems compared to a small network management system. Network management data consist of millions up to hundreds of millions of records are commonplace. The system needs to deploy on a large and powerful mainframes with large relational database.

Traditional centralized network management systems running on UNIX environments and using the Graphical User Interface (GUI) suffer from the problems of high cost and only adapted in very small-scale networks. The management system is only available through the GUI on the NMS hardware console. The Web based NMS is the trend of next generation NMS solution for large-scale network with the highly distributed, unified NMS architecture. It offers dramatic extensible and flexible and highly unified heterogeneous NMS integration. It offers management access via any Web browser interface [1, 2]

H. Tan (Ed.): Knowledge Discovery and Data Mining, AISC 135, pp. 295–302.
springerlink.com

To be another choice, multi-tier, disturbed technology can perfectly apply to the large-scale network management system. In this paper, we present network management for employment service and social insurance application - a scalable and extensible architecture for web based NMS to construct the large-scale, heterogeneous, complex, and multi-tier network management system. By using a group of lower cost workstation servers instead of large powerful workstation, the total cost of the system can be lowered. Also, the system is highly scalable. When the network increased, the network management for employment service and social insurance application system will not change the architecture of the system but increase another workstation server.

2 Distributed Enterprise JavaBeans (EJB) Infrastructure

Distributed computing allows an application system to be more accessible. Distributed systems allow parts of the system to be located on separate computers located in different location. Enterprise JavaBeans (EJB) is a comprehensive technology that provides the infrastructure for building enterprise-level server-side distributed Java components [3, 4].

The EJB technology provides a distributed component architecture that integrates several enterprise-level requirements such as distribution, transactions, security, messaging, persistence, and connectivity to mainframes. When compared with other distributed component technologies such as Java RMI and CORBA, the EJB architecture hides most the underlying system-level semantics that are typical of distributed component applications, such as instance management, object pooling, multiple threading, and connection pooling. Secondly, unlike other component models, EJB technology provides us with different types of components for business logic, persistence, and enterprise messages.

Any distributed component technology should have the following requirements [5,6,7]. Firstly, there should be a mechanism to create the client-side and server-side proxy objects. A client-side proxy represents the server-side object on the client-side. As far as the client is concerned, the client-side proxy is equivalent to the server-side object. On the other hand, the purpose of the server-side proxy is to provide the basic infrastructure to receive client requests and delegate these request to the actual implementation object. Secondly, we need to obtain a reference to client-side proxy object. In order to communicate with the server-side object, the client needs to obtain a reference to the proxy. In the end, there should be a way to inform the distributed component system that a specific component is no longer in use by the client.

In order to meet these requirements, the EJB architecture specifies two kinds of interfaces for each bean. They are home interface and remote interface. These interfaces specify the bean contract to the clients. However, a bean developer need not provide implementation for these interfaces. The home interface will contain methods to be used for creating remote objects. The remote interface should include business methods that a bean is able to serve to clients. One can consider using the home interface to specify a remote object capable of creating objects conforming to the remote interface. That is, a home interface is analogous to a factory of remote objects.

These are regular Java interfaces extending the javax.ejb.EJBHome and javax.ejb.EJBObject interfaces respectively.

The EJB architecture specifies three types of beans - session beans, entity beans, and message-driven beans. A bean developer has to specify the home and remote interfaces and also he has to implement one of these bean interfaces depending upon the type of the bean. For instance, for session beans, he has to implement the javax.ejb.SessionBean interface. The EJB architecture expects him to implement the methods specified in the bean interface and the methods specified in the home and remote interfaces. The EJB container relies on specific method names and uses delegation for invoking methods on bean instances.

Thus regarding the first requirement, the EJB container generates the proxy objects for all beans. For the second one, the EJB container for each bean implement a proxy object to the home interface and publishes in the JNDI implementation of the J2EE platform. One can use JNDI to look for this and obtain a reference. As this object implements the home interface only, he can use one of the creation methods of the home object to get a proxy to the remote interface of the bean. When one invokes a creation method on the home proxy object, the container makes sure that a bean instance is created on the EJB container runtime and its proxy is returned to the client. Once the client gets hold of the proxy for the remote interface, it can directly access the services of the bean.

Finally, once the client decides to stop accessing the services of the bean, it can inform the EJB container by calling a remote method on the bean. This signals the EJB container to disassociate the bean instance from the proxy and that bean instance is ready to service any other clients.

3 Distributed Architectures

Distributed computing allows business logic and data to be reached from remote locations at any time from anywhere by any one. Distributed systems allow parts of the system to be located on separate computers located in different location. The different application architectures between single server architecture and distributed server architecture are explained as flowing.

3.1 Single Server Architecture of Network Management for Employment Service and Social Insurance Application

In single server architecture, all the tier of the system including the database is in a server. In single server application architecture, there is usually less than 10-20 clients will be exercising the server at the same time. The data collection, status polling and event processing requirements are modest. The following diagram gives an overview of the single server architecture. In the single server application architecture, the middle tier of the network management solution (servers) is deployed to run in one Java Virtual Machine (JVM) process. This could be deployed on a JRE or in a J2EE server environment.

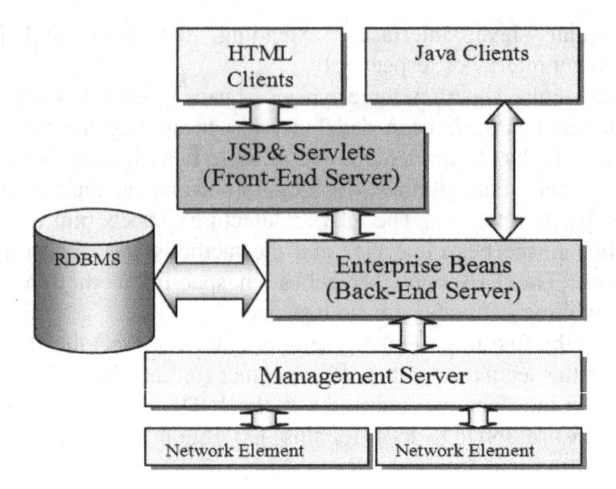

Fig. 1. Network management for employment service and social insurance application: Single-server and distributed multi-tier network management system architecture

To build a highly scalable management solution for large-scale network management system where a large number of clients will be accessing the management data, the multi-tier and multi-servers can be used. The network management system should use the distributed server architecture. The following diagram gives an overview of the distributed server architecture. In the distributed server architecture, the middle tier of the management solution is distributed to run on multiple systems on the network. There can be multiple front-end servers to scale to a large number of clients. Based on scalability requirements, the back end server can also be run in a single server or multiple servers. There can be multiple management servers, based on the data collection, status polling and event processing capabilities, required out of the management solution. The servers support fails over and load balancing capabilities. There is a choice of rich, thin clients based on Java or thin clients based on HTML. The client session management services are deployed in the front-end servers and they access the back-end business logic APIs using RMI or messaging.

3.2 Multi-server Architecture of Network Management for Employment Service and Social Insurance Application

When there exist multi-server architecture of the network management system and the different type of network management for employment service and social insurance application, the communication between the different servers and different network management for employment service and social insurance application is needed for integrated them together to a universal network management for employment service and social insurance application. In the telecom network, the network management systems become increasingly large and complex. There exist different network management systems to management deferent devices. How to integrate the deferent network management systems together are need to considered in the infrastructure of a universal network management systems. The high degree of integration characteristic of programming at the class library level makes it difficult to construct heterogeneous applications that use modules from different systems. The core technologies needed to achieve this flexibility are the message bus, which provides flexible services and methods for distributed network management systems to communicate with one another and share data.

The message bus is a facility used to bind, at runtime, distributed objects (program components) to construct network management systems. Each network management system communicates at runtime via the message bus. Systems can dynamically connect to or disconnect from the message bus at runtime. In effect the message bus framework is a virtual machine, built from highly modular, interchangeable components based on an open architecture, with applications being the software available for this virtual machine.

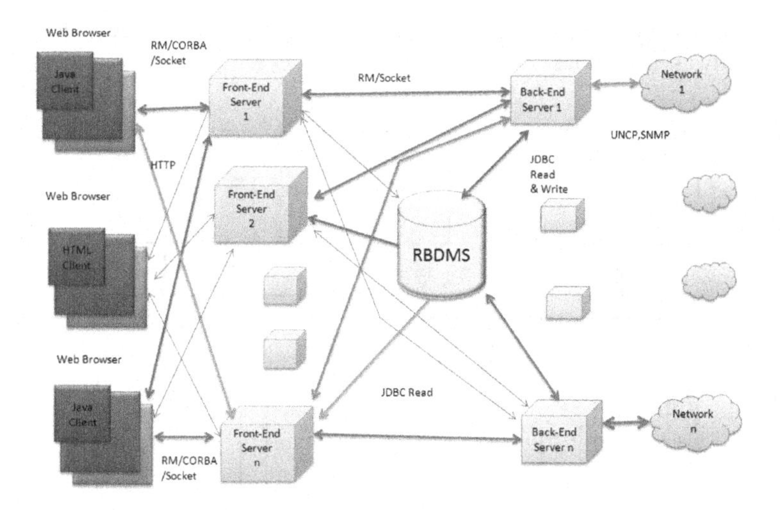

Fig. 2. Network management for employment service and social insurance application: Multi-server and multi-tier network management architecture

3.3 Communicate via Message in Network Management for Employment Service and Social Insurance Application

The message bus is also a means of structuring software, in that it defines system-level software architecture. Single network management for employment service and social insurance application with multi- tier, distribute techniques are used to construct the individual network management systems to manage special kinds of telecom devices, which can be very different from one another internally. A universal network management system is composed at a high level, relying upon network management systems for most of their functionality. The universal network management system can be large, while substantial network management for employment service and social insurance application capable of independent execution. From the point of view of the message bus architecture they are merely interchangeable systems, which share a standard interface to the message bus.

In addition to providing a range of messaging capabilities, the message bus provides facilities for service registration and lookup, data services, and distributed execution. Message bus clients (e.g., network management systems or services) can execute on any host computer connected to the message bus, allowing distributed network management systems to be easily constructed. A message bus console is used to examine and control the state of the message bus and monitor system activity.

As illustrated in Figure 3, most messages fall into one of two primary classes. Requests are messages sent to particular network management systems or to request that it perform some action. Requests may be either synchronous or asynchronous. A synchronous request is similar to a remote procedure call. Events are messages that are broadcast by a producer system, and received by zero or more consumer systems. Systems subscribe to the classes of events they which to receive. The message bus keeps track of the event subscription list for each system and uses it to construct a distribution list when broadcasting an event. In general a producer system does not know what other systems, if any, may be consuming the events it generates. Events are inherently asynchronous.

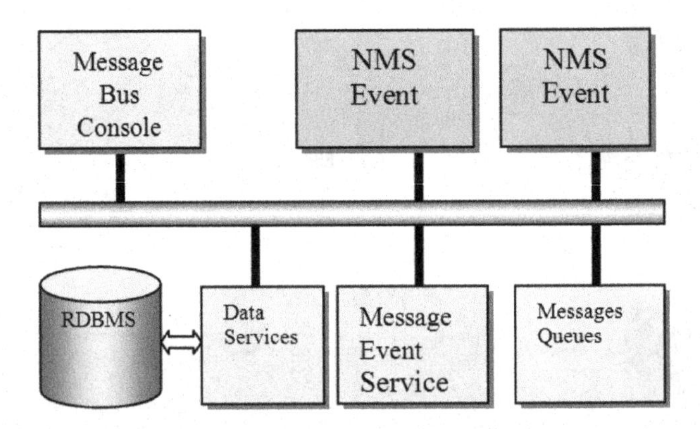

Fig. 3. Network management for employment service and social insurance application: Network management system components communicate via message bus

At the simplest level the message bus is responsible only for classifying messages by type and handling message distribution and delivery. The actual content of a message, that is the data contained in a message, is determined by the messaging protocol used by the systems. Multiple messaging protocols are possible on the same bus although not necessarily recommended.

The message bus is responsible for the reliable delivery of messages and ensures that messages are received in the order in which they were sent. There is no fixed limit on the size of a message. The bus will queue messages for delivery if needed.

The message bus provides a powerful way to structure large network management systems to control complexity. Universal network management systems can be developed at a high level, relying upon sub network management systems for much of their functionality. Universal network management systems based on the message bus architecture are inherently modular and open, allowing very different sorts of systems to be intermixed. Since network management systems can be large and complex products, with few restrictions on how they are implemented internally, it becomes easier for a large number of people to make significant contributions to the universal network management systems, reducing the dependence on key personnel.

4 Implement

We implemented the proposed network management for employment service and social insurance application to construct the large-scale, heterogeneous, complex, and multi-tier network management system. The network can be partitioned into small pieces each of which is monitored by a dedicated back-end server. All the server tasks are distributed symmetrically without any server dedicated for one specific task. In this mode, the servers including front-end server and back-end server construct a matrix of server nodes. In case of updating a managed object, front-end server will locate the back-end server where this object is managed and send updating request to this back-end server. And every back-end server will execute the core tasks on the sub-network assigned to it. This architecture is more complicated than asymmetric one. It will have the best performance, reliability and scalability.

The back end server modules are highly customizable through the configuration files. The back end server can be extended through the interfaces available for providing custom managed objects, filters, parsers, etc. The back end server can be run in a single JVM or each module of the back end server like topology, maps & FCAPS modules can be run as separate processes, in a highly scalable solution. The services of the back end server are available through Java API (when both the back end & front end servers run in the same JVM) RMI or Socket (when the back end and front end servers are distributed).

The front-end will handle most of the requests from the client and only for sensitive and critical operations like changing some configuration properties it will pass on the requests to the back end server, where they would be executed. It means front-end server can do read only operation on database, while back-end server can do read/write on database.

The front-end server has a web server to provide web-based access to clients. This web server will deploy along with a Servlets container to support dynamic servlets and JSPs.

The HTML client can be customized and extended by developing custom HTML pages, Servlets and JSP's using the Java Beans components and custom view module available in the front-end server. The figure 4 is a screenshot of the HTML clients of the distrusted multi-tier network management system. The front-end server supports the integration and deployment of device, system and application management applets. Also, front-end server supports extensive customization of the Java clients. The figure 5 is a screenshot of the java clients of the distrusted multi-tier network management system.

5 Conclusions

This paper we presented the network management for employment service and social insurance application - a scalable and extensible architecture for web based NMS to construct the large-scale, heterogeneous, complex, and multi-tier network management system. The network management for employment service and social insurance application allows all the parts of the system to be located on separate computers located in different location. EJB is a comprehensive technology that provides the infrastructure for building enterprise-level server-side distributed java components. The whole network management system can be partitioned into small pieces. We can manage a geographically distributed network or very large-scale network consisting of thousands devices in a reliable, good performance environments.

Acknowledgements. The funding support of this work is by National Science and Technology Support Program. The item number: 2008BAH32B00. Project Name is Key technologies and major application research of Human resources and social security public service system. Project Topic 6 Number: 2008BAH32B06. Project Issue 6 Name: The typical application development of Human resources and social security public service system. This paper is supported by Zhejiang Jandar technology Co., Ltd. (Jandar Company Information: Zhejiang Jandar technology Co., Ltd. Address: 11F,Jiangong Building,No.20, Wensan Road, Hangzhou. Tel: 86-0571-88,822,288 Zip: 310012 E-mail: djx@jandar.com.cn). Zhejiang Jandar technology Co., Ltd is greatly appreciated.

References

1. Spolidoro, F., Rodriguez, N.: Distributed Environment for Web-Based Network Management. In: 26th Annual IEEE International Conference on Local Computer Networks (LCN 2001), pp. 41–46 (2001)
2. Advent Network Management Inc.: AdventNet WebNMS, `http://www.adventnet.com/`
3. Kahani, M., Beadle, H.: Decentralized Approaches for Network Management. ACM Computer Communication Review 27(3), 36–47 (1997)
4. Wies, R., Mountzia, M., Steenekamp, P.: A Practical Approach Towards a Distributed and Flexible Realization of Policies Using Intelligent Agents. In: Proc. 8th IFIP/IEEE Int. Workshop on Distributed Systems: Operations & Management (DSOM 1997), Sydney, Australia, pp. 292–308 (1997)
5. Choi, J., Lee, K.: A Web-based management system for network monitoring. In: IEEE Workshop on IP Operations and Management, pp. 98–102 (2002)
6. Seo, J.-C., Kim, H.-S., Yun, D.-S., Kim, Y.-T.: WBEM-Based SLA Management Across Multi-domain Networks for QoS-Guaranteed DiffServ-over-MPLS Provisioning. In: Kim, Y.-T., Takano, M. (eds.) APNOMS 2006. LNCS, vol. 4238, pp. 312–321. Springer, Heidelberg (2006)
7. Yemini, Y.: The OSI Network Management Model. IEEE Communications Magazine 31(5), 20–29 (1993)

CSCW Based Labor and Social Security and Employment Information System

Wei Zhou and Dong Zhang

Zhejiang Jandar Technology Co., Ltd.,
Construction Building, 11F, No.20 Wen San Road, Hangzhou, China, 310012
djx@jandar.com.cn

Abstract. The information technology is imperative for managing the extensive labor and social security related information. The CSCW based labor and social security information system can provide an environmental that allows government hall located in different halls of human resources and social security to achieve a cooperative management through the cooperative tools, telemedicine, electronic and video-conference. The system would have very important application value and practical prospect in labor and social security processing and bring great benefits to small and medium labor and social security government hall.

Keywords: labor and social security, system, CSCWD, design.

1 Introduction

Labor and Social Security social insurance and employment related to two core business. In the end of 2006, national employment population reached 760 million. With the basic medical insurance for urban residents, the rural social old-age insurance and a series of new labor and social security policies, labor and social security services are further expanded the coverage of social groups. The business will gradually involve all socio-demographic the country employing about 30 million units and tens of thousands of labor and social security services. The large service groups, the labor and social security sector efficiency and service levels face enormous challenges, but also for labor security department of information technology put forward higher requirements. As one of the main target is the business managers at all levels of labor and social security department, The growing demand for social services, labor and social security needed to be established an effective public service system, especially the use of modern technology and e-government public service concept to provide efficient and convenient public services, labor and social security.

The human resources and social security organizations are facing significant challenges today. The oral government must process a great amount of information manually. This will greatly increase the chances for error and decrease efficiency. In a great number of human resources and social security organizations, the information technology is imperative for managing the extensive labor and social security related information [1]. On the other hand, a considerable portion of human resources and

H. Tan (Ed.): Knowledge Discovery and Data Mining, AISC 135, pp. 303–310.
springerlink.com

social security human resources and social securities are willing to search related human resources and social security information before seeking labor and social security advice. Most of the human resources and social security human resources and social security reported positive attitudes towards a cooperative human resources and social security clinic hear thecae information system for organizations Social security service [2].

Computer Supported Cooperative Work (CSCW) provides technologies to design and integrate the cooperative distributed labor and social security information system for government processes [3]. The CSCW based labor and social security information system is established to support the growing needs of human resources and social security government organizations to enhance their operational activities, reduce costs and significantly increase resource efficiency. It brings immediate benefits to small and medium labor and social security government organizations.

Although there is a number of labor and social security information management software in the market, but most of them are focus on the financial information management, little research has been done on the cooperative labor and social security information system.

In this paper we describe a system design and implementation of a CSCW based labor and social security information system that can provide the labor and social security service for the remote districts located by different organizations sites.

2 Background and Related Work

CSCW and groupware emerged in the 1980s from shared interests among product developers and researchers in diverse fields. CSCW is a technology to support group and organizational work practices. It aims to develop computer-controllable networks, virtual workspaces to support synchronous collaboration. CSCW systems support the work of individual and cooperative activities by a set of common support services [3].

The research field of the labor and social security information system may one of the most important application areas of the CSCW technology. There are a lot of cooperative requirements in the labor and social security information system in modern halls of human resources and social security to make communication and collaboration between health professionals and departments. Followings are some of the collaborative aspects in the labor and social security work:

The CSCW based cooperative distributed labor and social security information system can make the government or labor and social security experts and the human resources and social security to carry on the remote distance labor and social security consultation and labor and social security cooperative labor and social security management and labor and social security recordings. The human resources and social security can effectively obtain the economical labor and social security service in anywhere. Through the cooperative distributed labor and social security information system, the labor and social security data, the labor and social security picture, the labor and social security record and the human resources and social security record file can be transformed from one place to another. The government or labor and social security experts also can carry on the discussion harmonious cooperation work with the remote colleague without the region limit.

3 System Design of the CSCW Based Labor and Social Security Information System

The system arms to set up a CSCW based labor and social security environmental for social security service and dentists for providing the labor and social security service to the remote districts located in different locations. It would be used as a fast channel for government to human resources and social security, social security service-to-social security service, social security service to clinic to provide the long-distance labor and social security service, the long-distance labor and social security consultation and the electronic labor and social security records exchanging.

3.1 The Principles of System Functions Design

The overall infrastructure of the cooperative human resources and social security information system is shown in figure 1. The main functions and principles for the system design are as the following:

(1) It provides the human resources and social security's basic information to the long-distance labor and social security consultation system, the government's advice and the examination, the inspection treatment report list, the medicine image material and diagnoses and treatment record and information.

(2) Information Sharing: The social security service information system can receive and stores the opposite part government' consultation information for human resources and social security's diagnosing and treatment, returns to original state and satisfies the precision request, which the organizations management needs. The local labor and social security data, the human resources and social security picture, the labor and social security record, the human resources and social security file can be transmitted to differently place. Also it can be used for long-distance education observation.

(3) Dynamic inquiry: The system would be designed to respond immediately for the human resources and social security request of the consultation from government at long-distance places. The system can receive the government consultation result information; provide the labor and social security record materials that preserved in remote halls of human resources and social security.

(4) Social security service graph data administration. The human resources and social security graph is needed to provide the exactly information about the human resources and social security human resources and social security disease information. All human resources and social security graphs of human resources and social security's EHRSSR database can be preserved in the system. It has enormously enriched the Social security service specialization electronic labor and social security record.

(5) The video conference room: The video conference room could provides a visual, bi-directional real-time video transmission for human resources and social security graphics that support the long-distance labor and social security service and long-distance government' consultation.

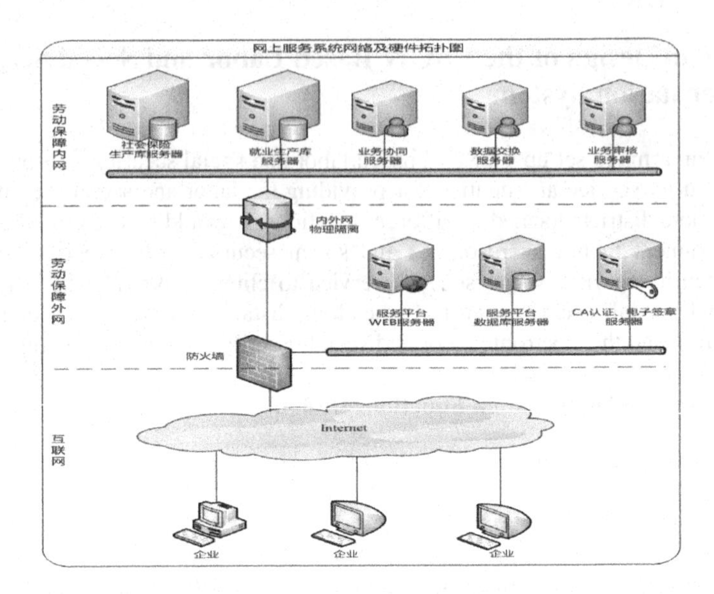

Fig. 1. System Framework of CSCW Based Labor and social security Information System

The system architecture is shown in figure 2. The architecture is composed of three layers. The first the layer of the system is the database layer. The first system layer is the database layer. It saves as the information center of the system. The database of the system included the human resources and social security and the human resources and social security electronic human resources and social security records (EHRSSR) data, human resources and social security images data, multimedia data, CSCW data. The second layer is the system transaction layer, which is consisted of the system function modules, such as the electronic labor and social security records data management, user management, human resources and social security images information collection, collaborative system functions, etc. The third layer is the user interface, which included the client side of the system, video conference, camera, and CSCW workspaces, etc.

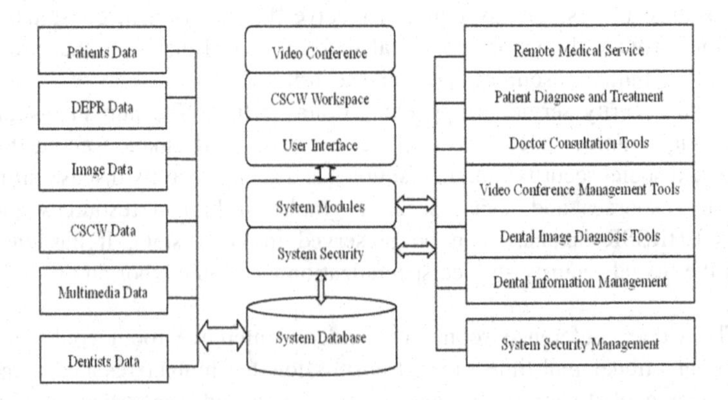

Fig. 2. System Architecture of Multi-layers CSCW based labor and social security Information System

3.2 The Video Conference System

A video conference room is designed to support the dentist to communicate with human resources and social security. The system supports the point-to-point and multi-points conference pattern. It satisfies the ITU H.323 video conference standard. Video and audio stream data can be processed synchronously. The video hardware devices include a video camera and microscope. The video and audio media data can be saved in the server according to the human resources and social security identify number as a part of the EHRSSR. It can be made as a compact disc to put on file for further information exchanging.

The video conference room is also designed as bi-directional, real-time, multi-points meeting for government carries on the discussion and the exchange through the picture and the sound with the conference room functions. Based on ITU-T H.323 standard and T.120 standard, it can be realized the point-to-point, the multi-points video conference over Internet.

3.3 System Infrastructure Based on J2EE

Because there is many computers with different kinds of operating system will work in coordination in the system, the distributed, platform independent system architecture is needed to adapt the change in future use. To enable application system be accessed by lots of different computers in network, we choice the J2EE and EJB technology that offered a framework of service system to construct enterprise system structure.

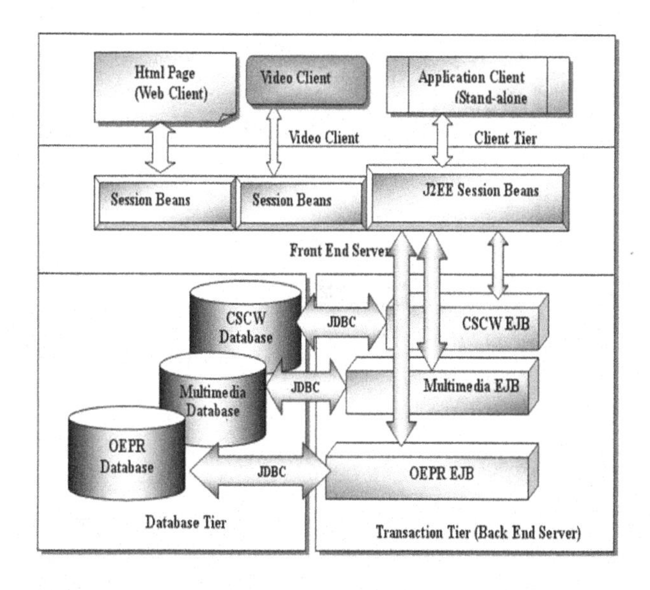

Fig. 3. J2EE based System Architecture of Multi-Layers Cooperative Human resources and social security Telemedicine Information System

The J2EE based system architecture is shown in figure 3. The middleware of CSCW, EHRSSR, human resources and social security images data, and multimedia EJB offers distributed transaction processing. Many host computers can join to offer the many services. Compared with other distributed technology, such as CORBA technology, the system structure of EJB has hidden the lower detail, such as distributed application, the events management, the target management, multi-thread management and connoting pool management etc. In addition, J2EE technology offers many kinds of different middleware to be applied to business logic. The data are stored and managed with EJB. Distributed computing enable users operate in any time, any place, and obtain business logic and data processing in remote server. The distributed systems enable the databases and services in the same or different computers. The database uses JBBC to communicate with EJB (back-end server). The system front-end uses the back-end server to provide service. The front-end client communicates with back end EJB.

4 Implementation and Further Work

According to the system framework, we have implemented a prototype of the CSCW based labor and social security information system. We have implemented the system

Fig. 4. A Screenshot of the CSCW based labor and social security information system

of the CSCW based labor and social security information system and it has been used by more than ten human resources and social security halls of human resources and social security and organizations in china. This study only focused on introducing the cooperative aspect of the system architecture design and more detail system developing work was omitted in the paper. Since the requirements of the practical labor and social security information system in different application social security service are quite different and complicated, Further work should be performed to establish the common CSCW based labor and social security software tools that could be easily used and integrated in different human resources and social security applications.

5 Conclusions

The CSCW based labor and social security information system has integrated the most advanced technologies of CSCW, multi-layers architecture, computer graph, video conference, and the information processing to provide a cooperative environment for government and human resources and social securitys to communicate for dealing the labor and social security consultations and human resources and social security heal care affairs. The system can be also be operated by number of government in different location to negotiate and diagnose the same human resources and social security labor and social security disease through the human resources and social security image tools, video conference tool and distributed software environment. The system would enhance the organizations labor and social security management level and the benefit to raises government's labor and social security work efficiency.

Acknowledgements. The funding support of this work is by National Science and Technology Support Program. The item number: 2008BAH32B00. Project Name is Key technologies and major application research of Human resources and social security public service system. Project Topic 6 Number: 2008BAH32B06. Project Issue 6 Name: The typical application development of Human resources and social security public service system. This paper is supported by Zhejiang Jandar technology Co., Ltd. (Jandar Company Information: Zhejiang Jandar technology Co., Ltd. Address: 11F,Jiangong Building,No.20, Wensan Road, Hangzhou. Tel: 86-0571-88,822,288 Zip: 310012 E-mail: djx@jandar.com.cn). Zhejiang Jandar technology Co., Ltd is greatly appreciated.

References

1. Yinhuan, C., Xiangping, Y.: The model investigation of digital human resources and social security clinic information system. Chinese Labor and social security Equipment Journal 26(1), 6–8 (2005)
2. Hu, J., Luo, E., et al.: Human resources and social securitys' attitudes towards online human resources and social security information and a web-based virtual reality program for organizations Social security service: A pilot investigation in China. International Journal of Labor and Social Security Informatics 78(3), 208–215 (2009)

3. Jonathan, G.: Computer-Supported Cooperative Work: History and Focus. Computer 27(5), 19–26 (1994)
4. Gómez, E.J., del Pozo, F., Quiles, J., et al.: A telemedicine system for remote cooperative labor and social security imaging management. Computer Methods and Programs in Biomedicine 49(1), 37–48 (1996)
5. Siriwan, S.: A collaborative intelligent tutoring system for labor and social security problem-based learning. In: Proceedings of the 9th International Conference on Intelligent User Interface, Funchal, Madeira, Portugal, pp. 14–21 (2004)
6. Istvan, B., Tamer, F., Dieter, S.: Remote collaboration using Augmented Reality Videoconferencing. In: Proceedings of the 2004 Conference on Graphics Interface, London, Ontario, Canada, pp. 89–96 (2004)
7. Lim, S., David, D., Weidong, C.: A web-based collaborative system for labor and social security image analysis and management. Selected Papers from the Pan-Sydney Workshop on Visualisation, Sydney, Australia, vol. 2, pp. 93–95 (2000)

XML Based Asynchronous Communication for Labor and Social Security Systems

Jian Yu

College of Information and Electronic Engineering,
Zhejiang University of Science and Technology
Hangzhou, China, 310012
Jianmyu@yeah.net

Abstract. Labor and social security systems of public service users are online services units and individuals. This paper proposed an embedded XML data exchange framework for asynchronous communication between different labor and social security systems. The XML data is used to exchange the labor and social security systems data between the special labor and social security systems system. The XML data is embedded into mail messages and delivered to the merchants and received from the mail server transformed to the labor and social security systems data items. The embedded XML data exchange framework can realize the asynchronous data exchange and data flow management for labor and social security systems system.

Keywords: Data exchange, labor and social security systems, XML, data flow management.

1 Introduction

Labor and social security systems of public service users are online services units and individuals. Users of the system are the unit labor and ordinary individuals, rather than have the business skills of labor support professional business managers. Therefore, typical applications in the public service development must be convenient and easy to use. System must have a strong operational.

Most labor and social security systems are used the centralized united platforms that need every participant to use the labor and social security systems and exchange synchronously. However, loosely coupled communication is often needed to exchange commerce data asynchronously between various different Labor and social security systems and participants. The data exchange process is very important to establish the B2B and E-GOVERNMENT labor and social security systems business. To setup the data exchange system, the key problem is how to transmit the datum with different data structure between data exchange systems and existing management information systems. The management information systems only recognize the special data structure. Therefore, the data in the data should be contained both the data and the data structure. Extensible Markup Language (XML) is a markup language for documents containing structured information that contains both content

H. Tan (Ed.): Knowledge Discovery and Data Mining, AISC 135, pp. 311–318.
springerlink.com

and structure of the content. Therefore, the XML technology is most suitable to design the exchange systems.

Data exchange system based on XML has been widely researched in recent years. The data exchange is the problem of transmit the data between the source and the target with given an instance of specification schema. Marcelo Arenas [1] researched the basic notions of XML data change for source to target constraints, data exchange settings, consistence and query answering problems. The researchers are more concerned the theoretical foundations of XML data exchange. The system realization the how the data transmitted are not mentioned. In the reference [2], a system for data exchange is designed between the systems of different application, XML is used as data exchange file. But the data transmit method used ftp server and lacked the flexions.

This paper proposed an embedded XML data exchange framework for asynchronous communication between different labor and social security systems. The XML data are used to exchange the labor and social security systems data between the customers and the special labor and social security systems. The framework selected the mail system to transmit the data. Compare to other data transmit protocols, the mail based XML data exchange system is inexpensive, easily and quickly to set up in the case of asynchronous information treatment. It is quite suitable for huge users to exchange the labor and social security systems data in the short time. The XML data are embedded into mail messages and delivered to the merchant by STMP protocol and received from the mail server by POP3 protocol and transformed to the labor and social security systems data items that can be recognized by the labor and social security systems.

2 Backgrounds and XML

2.1 XML Technology

XML is a markup language for documents containing structured information that contains both content and structure of the content. XML is derived from the Standard Generalized Markup Language (SGML). SGML and XML are text-based formats that provide mechanisms for describing document structures using markup tags (words surrounded by '<' and '>'). XML is able to represent both structured and semi-structured data [3,4].

XML has a number of characteristics that have caused it to be widely adopted as a data representation format. XML is extensible, platform independent, and supports internationalization by being fully Unicode compliant. It is a mechanism for describing structured and semi-structured data, which provides access to a rich family of technologies for processing such data.

XML facilitates universal data access. XML is a plain text, Unicode-based meta-language that is not tied to any programming language, operating system, or software vendor. XML provides the technologies for manipulating, structuring, transforming and querying data [5,6].

2.2 XML Based Documents for the Labor and Social Security Systems

More and more organizations have used the XML documents to distribute and transmit the data and materials now. The technologies of XML based documents are quite suitable for the electronic Labor and social security systems in order to exchange the Labor and social security systems data between the customers and the merchants. The Labor and social security systems data can be collected through the XML based documents that contain both the Labor and social security systems data and the data structure. The Labor and social security systems data based on XML can be processed conveniently with many XML processing software. Therefore, XML based electronic Labor and social security systems system is suitable for various network environments and various the operating systems. The XML based documents can be edited by a lot of XML editor software. XML documents are very suitable for transmitting on the Internet.

XML technologies could be used in the process of the electronic Labor and social security systems in the following aspects.

Distributing XML based labor and social security systems forms: The merchant could use the XML based documents to distribute the labor and social security systems forms. Users can download the XML documents, and fill in the XML based electronic form. It means that the merchant does not need to store and distribute thousands of paper labor and social security systems forms again.

Filling in the XML based forms: The customers could fill in the Labor and social security systems forms with the XML based documents and send the form back to the merchant. If the paper forms are still need, users can just print the XML forms and submit the paper form to merchant. Also, the XML based documents have the abilities of calculating, checking the mistake or leak in the forms. If there were mistakes in the XML based forms it would require users to correct them immediately. In this way, it would improve the quality of filling in forms notably and avoid the trouble of the wrong forms that need to alter afterwards. It could save much time and expenses in the Labor and social security systems greatly.

3 Embedded XML Data Exchange System

The goal of the embedded XML data exchange system based email messages is to setup up a Labor and social security systems system on the top of the existing management information system. It provides a convenient method for customers to exchange labor and social security systems data by email. The XML based Email is used to transmitted the labor and social security systems data between the customers and merchants and be transmitted on Internet with standard email proto-cols such as STMP and POP3 protocols. The XML based Labor and social security systems documents are used as the core theologies to exchange the Labor and social security systems data from the email to existing management information system in the merchant.

3.1 System Framework

The system architecture was composed of three layers as showed in the figure 1.

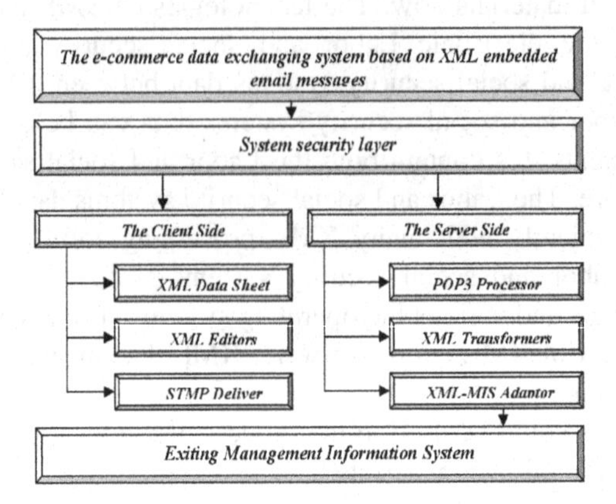

Fig. 1. System architecture for the embedded XML data exchange system based on mail messages

The lowest layer is the existing management information system. The next layer is a system security layer to deal with the system security problems. The top layer is the layer of the XML based email system. The XML based email system consist the client side and server side. The client side is consisted of the XML data sheet, XML editor, and STMP deliver. The Labor and social security systems data is inputted into the XML sheet with XML editors and delivered to mail server by STMP protocol. The server side is consisted of the POP3 processor, XML trans-former and the XML to MIS adaptors. The Labor and social security systems data sheet is received from the mail server by POP3 protocol and processed by the XML transformer. XML based Labor and social security systems data sheet in the mail is transform to the data items that can be recognized by the management information system and inputted into the database by XML to MIS adaptor.

3.2 The Business Flow for the XML Based Email Labor and Social Security Systems System

The customers download and fill in the relevant XML based labor and social security systems tables or forms, then return them to the merchant through the Email system. In the server side of the system in the merchant, the XML based Emails are be collected and dealt with in the XML transformer. After Labor and social security systems data extracted from the email by XML transformer and XML-MIS adaptor, the merchant deliver the messages of customers account number and corresponding data to the appointed bank to deduct the money from customers. After the information of money deducting confirmed from the bank, the system will send Email to the customers automatically to inform customers the labor and social security systems results.

The business flow of the XML based email Labor and social security systems system is explained in figure 2.

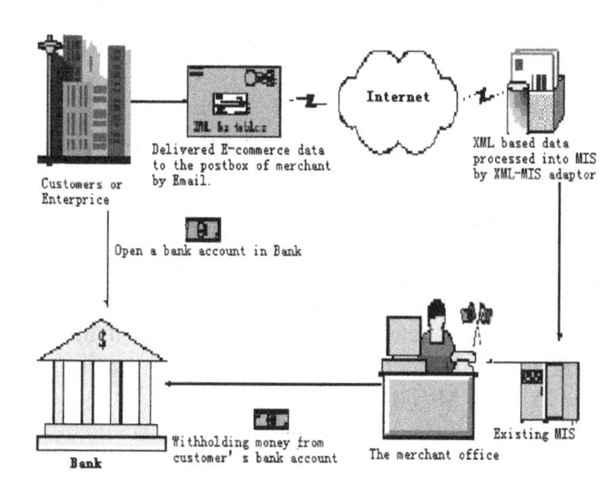

Fig. 2. The business flow for XML based data exchange

There are four kinds of system actors in the system business flow.

Merchant: The merchant uses the system to accept the XML based Labor and social security systems email from the customers, deal with the Labor and social security systems data, and withhold money from the appointed in accounts of customers in directly.

The ISP: The ISP offers online technical support for the system. He is responsible for maintaining the declare datum and dividing into groups and deposit temporarily to legal users, cooperating in corresponding tax departments for collecting the data of Labor and social security systems.

The banks: The bank offers the customers a labor and social security systems bank account and transmits the money from customers to merchant.

The administrative center of certification: The administrative center of certification issues CA digital certificate in order to offer the security of the authentication from the third part to prevent the divulging a secret or loss on the Internet.

The detail system flow of the XML based email Labor and social security systems information system is shown as in the figure 3. When customers want to exchange labor and social security systems data, he only needs to do the following steps.

Step 1: Customers download and fill in XML based labor and social security systems form from the website of the merchant.

Step 2: The filled XML based labor and social security systems sheet is delivered to the postbox of merchant by Email.

Step 3: All the filled XML based labor and social security systems sheets send by customers are collected and processed into data items that can be understood by XML-MIS.

Step 4: After the labor and social security systems data items have inputted the XML-MIS, the normal labor and social security systems process will be deal with by XML-MIS according to the data of labor and social security systems. For example, XML-MIS will deduct corresponding withholding money from customer's bank account, automatically send the Email to the relevant customers ac-cording to the confirmed information of withholding from bank, inform customers the final process result, etc.

3.3 The Original XML Based Labor and Social Security Systems Sheet Be Shared as the Initial Data and Legal Evidence

The XML document keeps the same as the paper forms of the Labor and social security systems. It keeps the consistency with the form of paper.

As the paper from of the Labor and social security systems, the XML based form can be operated as the initial data in the process of the making report, reference, distribute. The XML based electronic form is more easily used for the data exchanging with the existing MIS database.

Because the XML document can be used as the announcement and legal contracts, the XML based Labor and social security systems sheet has more advances than the online Labor and social security systems system. By utilizing the XML based Labor and social security systems sheet, the Labor and social security systems information in the original self-structured XML files could be used and attributed by the different tax department and different government organization.

The original XML document could be served as the initial data and legal evidence. XML documents realized the original Labor and social security systems documents be shared by different department. As shown in figure 4, XML based Email system keeps the original XML documents, a lot of departments in the same security system could handle the work of examining and approving through sharing the XML based Labor and social security systems data. Though the Labor and social security systems data be used and deal with in different departments, the original papers will not needed changed. Because the data exchanged based on the XML forms, it is quite easily to combine of original XML data sheet with different existing MIS system.

3.4 System Security Design

During the practical application on the Internet, the security of the system is a key problem that must be considered. The security of the large-scale Email Labor and social security systems system should be considered for the following unsafe factors.

System uses he Firewall and Virtual Private Network (VPN). The firewall was built to rises the barrier of safety for intranet. There are many computers in the merchant. Some secret information stored in these computers. To enable only the staff member of merchant could use these computers; the simplest method is that these computers would not be connected into the network. But this method would bring a lot inconvenience for the work. The firewall is good way to prevent others from invading through Internet. It isolated the computers from the public Internet of outside. Though ordinary people could visit the Web site of merchant and could send the Email to the merchant, they would generally unable to know which computers works behind the firewall and could not use them. VPN sets up special-purpose pass way in the common network. The principles of VPN are to encrypt first and then transmit the information of different firewalls. If the information is intercepted and capture by others while passing through public Internet, others are unable to decipher the captured information.

Using Public Key Infrastructure (PKI): Public-key infrastructure (PKI) is the combination of software, encryption technologies. It services that enables the system to protect the security of system communications and business transactions on the Internet.

4 Discussions

The embedded XML data exchange system based mail messages is proposed to provide a convenient method to exchanging labor and social security systems data for the customers from the Internet. Because the system has combined Email with XML technologies, it has the advantages of the XML and email, which XML documents could be transmitted by email conveniently, and be used to exchange the information with different existing MIS. The system could realize the automation of the Labor and social security systems flow possessing and controlling. The embedded XML data exchange system has following advantages compared to the online Labor and social security systems information system.

Safe: The XML based email Labor and social security systems information system just receives and deals with the emails. The users need not to log in the operating system. Therefore, it is safer than the online labor and social security systems system.

Reliable: The XML based Email would not be easily lost in the process of delivering. The reliable mailbox would support the email system. The mail-dir guarantees the system from the suddenly system collapse and destroy the whole mailbox. To solve user store problem by mail, the system would used the mail-dir store technology that each piece of mails keep as independent one a piece of files. It avoided adding the lock with other emails.

High efficiency: Run on UNIX and BSD operating system, the system can deliver 200,000 Emails easily every day.

Simple: The embedded XML data exchange system is much more simple and small than other the online Labor and social security systems system. The system dealt with and delivered XML based labor and social security systems sheets. It needs not to consider much of systematic security, stability, set up perfect user's database and offer online service and user management. Therefore, it would reduce the work of system management.

5 Conclusions

The embedded XML data exchange system based mail messages is inexpensive, easily and quickly to set up. In the case of asynchronous information treatment it is quite suitable for huge users to exchange the labor and social security systems data in the short time. The email based information exchange system is very suitable for the B2B and E-GOVERNMENT labor and social security systems business processes. It could provide a convenient way for customers to exchange labor and social security systems data to the merchant.

Acknowledgements. The funding support of this work is by National Science and Technology Support Program. The item number: 2008BAH32B00. Project Name is Key technologies and major application research of Human resources and social security public service system. Project Topic 6 Number: 2008BAH32B06. Project Issue 6 Name: The typical application development of Human resources and social security public service system. This paper is supported by Zhejiang Jandar technology Co., Ltd. (Jandar Company Information: Zhejiang Jandar technology Co., Ltd. Address: 11F,Jiangong Building,No.20, Wensan Road, Hangzhou. Tel: 86-0571-88,822,288 Zip: 310012 E-mail: djx@jandar.com.cn). Zhejiang Jandar technology Co., Ltd is greatly appreciated.

References

1. Arenas, M., Libkin, L.: XML data exchange: Consistency and query answering. Journal of the Association for Computing Machinery 55(2), 1–72 (2008)
2. Guohui, H., Yinbo, Q.: Data exchange system design based on XML. Computer Engineering and Design 28(3), 583–585 (2007)
3. Consortium: Extensible markup language (XML) 1.0. (2000), `http://www.w3.org/TR/REC-xml`
4. Dashofy, E.M., Hoek, A., Taylor, R.N.: An infrastructure for the rapid development of XML-based architecture description languages. In: Proceedings of the 24th International Conference on Software Engineering, Orlando, Florida, pp. 266–276 (2002)
5. Conrad, R., Scheffner, D., Freytag, J.-C.: XML Conceptual Modeling Using UML. In: Laender, A.H.F., Liddle, S.W., Storey, V.C. (eds.) ER 2000. LNCS, vol. 1920, pp. 558–571. Springer, Heidelberg (2000)
6. Xiaou, R., Dillon, T.S., Chang, E., Feng, L.: Modeling and Transformation of Object-Oriented Conceptual Models into XML Schema. In: Mayr, H.C., Lazanský, J., Quirchmayr, G., Vogel, P. (eds.) DEXA 2001. LNCS, vol. 2113, pp. 795–804. Springer, Heidelberg (2001)

Research on Grain Moisture Curve Fitting Based on Support Vector Machine Regression

Cheng-shi Luo

School of Physics and Electronic Engineering, Taizhou University, Taizhou 318000, China

Abstract. This paper introduces the experimental method which acquires relation curve between grain moisture gauge output and multi-parameter input by modeling with the use of existing software library directly. This method is taking support vector machine regression fitting forecast as core. The results show that speed of modeling established by support vector machine regression fitting forecast is faster than that of traditional modeling. Besides, this method has higher accuracy and is easy to use. Moreover, it can overcome drawbacks of traditional forecast. Therefore, it has higher accuracy and generalization capability, being able to realize on-spot grain moisture forecast. As a result of these, it can provide good theoretical foundation for environmental variation compensation.

Keywords: Support vector machine, radial basis function, moisture detection, environmental variation compensation.

1 Introduction

In the process of research on portable instrument of measuring grain moisture, prototype experiment equipment has to be used to conduct all trail researches with grains of different moistures. Besides, all influential factors on moisture have to be studied to gain trail data which can get relation between moisture value and other values. An effective way is to describe the fitting curve of trail data. From the fitting of curve, their rules can be known and forecast research can be conducted.

Commonly used data processing methods in scientific trail are as follows: when the rules of data are approaching linear, it is regarded as a standard and reliable method to use linear regression to summarize rules. If the rules deviate linear, artificial neural network will be used to summarize rules. Besides, adding square to linear equation or other higher order term to conduct non-linear regression. The traditional way and neural network have drawbacks shown in applications. With the defects of lack of fitting or over fitting [1-3], the large sample generalization capability of them is not strong.

This paper proposes a grain moisture regression modeling based on LIBSVM (Library for Support Vector Machines) which belongs to support vector machine modeling. This method was initially used in data mining. After that, it was gradually introduced to fault diagnose, model recognition and regression fitting forecast

H. Tan (Ed.): Knowledge Discovery and Data Mining, AISC 135, pp. 319–326.
springerlink.com

in engineering field, Different from the above methods, modeling based on LINSVM is centered on support vector machine, using directly the existing software bank to model. This method is faster, more accurate and more feasible than traditional one. The results show that with higher accuracy and more generalization capability, this method can overcome the drawback of traditional forecast method, thus, providing environmental variations compensation and fine theoretical foundation on time.

2 Measurement of Grain Moisture Data

Traditional moisture measurement is drying method. As a major trail method, drying is considered as standard measurement method. However, drying method can only be done in laboratory. Besides, it takes a lot time to be done. Therefore, it can used to do calibration. Normally, portable moisture measurement will use capacitance sensor method which utilizes measured grain as the dielectric of capacitance sensor. With various moisture contents, outputs of capacitance sensors with dielectric parameters will be different.

2.1 Capacitance Sensor Principle

Capacitance moisture gauge adopts column container as sensor. And it uses red copper as material. Compared with sensor with plate container, the influence of edge effect on measurement value is less. Furthermore, it is easier to operate. The structural principle [4] is shown in figure 1.

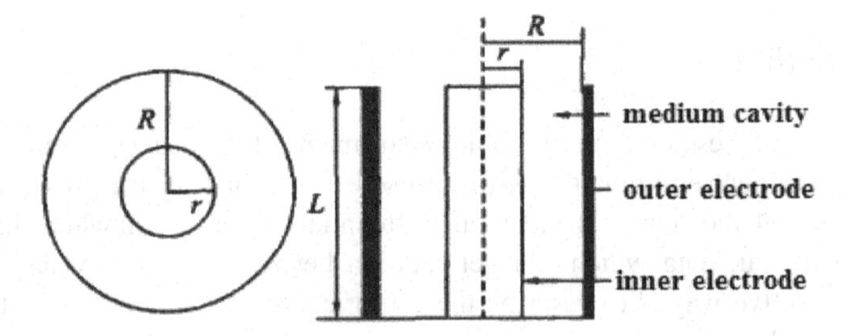

Fig. 1. Structure of capacitance sensor

Column capacitance is composed of two concentric metal cylinders as electrodes. The height of two electrodes is L; radius of inner electrode is r; radius of outer electrode is R. When L>>R-r, the edge effect of column can be ignored. Suppose electric charges carried in capacitor electrodes are +q and –q, which are distributed on inner and outer cylindrical surfaces. Absolute value of electric charges carried by every unit length of column is $\lambda(\lambda = \frac{q}{L})$. It can be drawn from calculation formula that:

$$C = \frac{\varepsilon_r}{1.8 \ln \frac{R}{r}} \times 10^{-10} \tag{1}$$

For grains with moisture content of M, their corresponding relative dielectric constant is ε_r. When moisture content of grain change is $M + \Delta M$, the relative dielectric constant change is $\varepsilon_r + \Delta\varepsilon_r$, thus, the capacitance change is:

$$C + \Delta C = \frac{2\pi L \varepsilon_r \varepsilon_0}{\ln \frac{R}{r}} (1 + \frac{\Delta\varepsilon_r}{\varepsilon_r}) \tag{2}$$

As a result of these, capacitance relative change $\frac{\Delta C}{C}$ will show linear relation with grain relative dielectric constant change $\frac{\Delta\varepsilon_r}{\varepsilon_r}$. The measurement of capacitance value change can be used to gain grain dielectric constant change. This paper utilizes oscillation circuit to solve capacitance sensor value. Putting capacitance sensor as part of RC oscillation circuit, capacitance value can be gained by measuring its output frequency.

2.2 Measurement of Environmental Parameter

Under constant environmental parameter, the capacitance change of the above-mentioned capacitance sensors will be related with loaded grain moisture content. That is linear relation. However, capacitance of sensor is also related with environmental parameter, such as environmental temperature, and compactness of loaded grain. Without the consideration of these, the measured moisture content is rough and inaccurate. Therefore, we need to measure the above parameters and find the relation of them with output. Normally, measurement of environmental temperature is very simple. By using thermistor to get voltage, temperature data can be gained through amplifying circuit and A/D conversion circuit.

The starting point of grain moisture experiment model is the design of portable moisture gauge. Structure of moisture data acquisition is shown in figure 2. In it, capacitance sensor test circuit is oscillation circuit. Front end data collection will use frequency measurement to gain capacitance change value. Temperature measurement circuit and pressure sensor circuit both transfer temperature and weight to voltage, and then, gain the number through A/D conversion circuit. Front end collection circuit will upload the three acquired variations to principle computer through RS 232 to save for the fusion research.

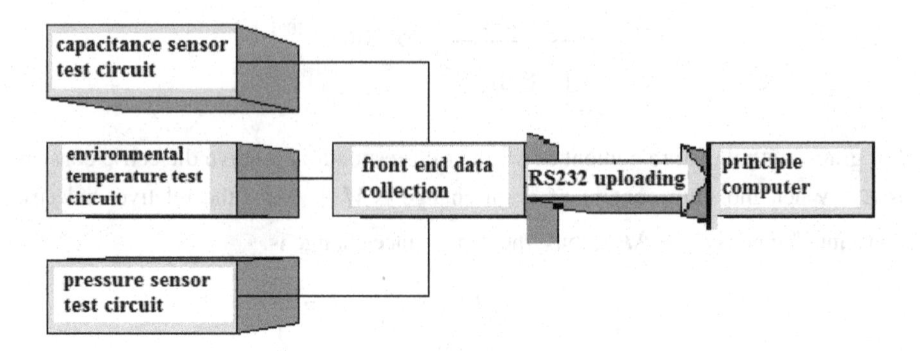

Fig. 2. Experiment model structure of grain moisture data acquisition

2.3 Collection of Sample Data

In order to know the input and output features of moisture sensor under various temperatures and compactness, calibrate them with following steps in lab.

(1) According to moisture measurement theory and operation process in GB /T3 543. 6-1995 *Crop Seeds Inspection* Norms, lab electric oven method is conducted. Then, use formula to calculate moisture percentage of wheat sample [5]:

$$\text{Moisture percentage} = \frac{M2 - M3}{M2 - M1} \times 100\%$$

In formula, $M1$ is weight of sample box and lid (unit: g); $M2$ is weight of sample box and lid and sample before drying (unit: g); $M3$ is weight of sample box and lid and sample after drying (unit: g).

(2) Take a certain amount of grains, and make up grain samples with different moisture content in line with grain moisture content formula: $M = \frac{W1 - W2}{W1} \times 100\%$.

(3) Seal the samples and put them into freezer. After that, set the temperature to -5℃.

(4) When moistures of all samples are basically absorbed, measure the moisture content of every sample with dry oven, and label them with vacuum package.

(5) Put samples with different moisture contents in moisture sensor. In order to have better data under various conditions, the experiment is done in incubators. Through the experiments, the relation between moisture content of different grains and measured capacitance value can be acquired. Besides, measured capacitance change caused by changed environmental temperature, grain weight, compactness and other factors will be gained. The large amount of data gained by experiments in designed circuit will be saved in Excel.

3 LIBSVM and Data Fusion

Support vector machine starts from solution of optimal hyperplane on linear division. Its principle is choose the normal direction that can make the interval largest. The corresponding two extreme lines are the optimal lines. Largest classification interval is actually the control on promotion ability. Meanwhile, get the minimum structure risk by adopting minimum training error and optimal generalization.

3.1 LIBSVM Algorithm [6]

Core algorithm of LINSVM is decomposition algorithm and sequence minimum optimization (SMO). The essence is to tackle a quadratic programming:

$$\left.\begin{array}{l} \min \frac{1}{2}\alpha^T Q\alpha - e^T\alpha \\ 0 \leq \alpha_i \leq C,\ i = 1,\ldots l \\ y^T\alpha = 0 \end{array}\right\} \quad (3)$$

In it, e is unit vector; C is upper bound of parameter; Q is positive definite or half positive definite matrix. Through decomposition, we solve this problem. The specifics are as follows:

1) Give number of elements in working set |B| = q ≤ 1 (q is even number) and accuracy requirement. Set the initial point 1 =, K=1.

2) If k is the optimal solution of issue, it will stop. Otherwise, re-search working set B {1,-l}, |B| = q. Define N = {1, -l}/B, and $_{B}^{k}, _{N}^{K}$ is sub-vector of vector k.

They are relatively B and N.

$$\left.\begin{array}{l} \min \frac{1}{2}\alpha_B^T Q_{BB}\alpha_B - (e^B - Q_{BN}\alpha_N^K)^T\alpha_B \\ 0 \leq (\alpha_B)_i \leq C, i = 1,\ldots q \\ y_B^T\alpha_B = -y_N^T\alpha_N^K \\ Q = \begin{bmatrix} Q_{BB} & Q_{BN} \\ Q_{NB} & Q_{NN} \end{bmatrix} \end{array}\right\} \quad (4)$$

3) Solve the quadratic form of B:

4) Set $_{B}^{K+1}$ as the optimal solution of formula (2) and set $_{N}^{K+1} = _{N}^{K}$.

Set K + 1 → K, and repeat 2). The above is decomposition, and sequence minimal optimization is similar to decomposition. Make B|=2, which means that choose two elements a time, to gain new 1,2. On the whole, LIBSVM adopts decomposition. After choosing working set B, use sequence optimal optimization to solve the quadratic form about B.

3.2 LIBSVM Parameter Choice

The most important in LIBSVM is kernel function and choice of its corresponding parameters. The common kernel functions are linear kernel, multinomial kernel and RBF kernel. Generally speaking, RBF kernel and radial basis function are chosen. They only have one undetermined parameter, and the larger the parameter, the faster the convergence rate is. Before the regression analysis by LIBSVM, firstly, parameter optimization has to be done on RBF function; then, interactive verification. In all parameters, the most important ones are c and g. c is the parameter (loss function) of C-SVC, e -SVR and v-SVR; g is the set of gamma function in kernel function. Three values bestmse, bestc, bestg given by LIBSVM respectively represent optimal mean square error, optimal c value and g value. The first optimization can gain the contour map of c and g value. Then, in line with contour map, ensure the range of c, g and by optimization, fine the suitable c and g value. The contour map of c, g fine optimization is shown in figure 3. From it, it can be drawn that convergence range of c is (-3, 3), and that of g is (-4, 4). By SVMcgForRegress function in LIBSVM, the suitable c, g values can be gained by c, g range.

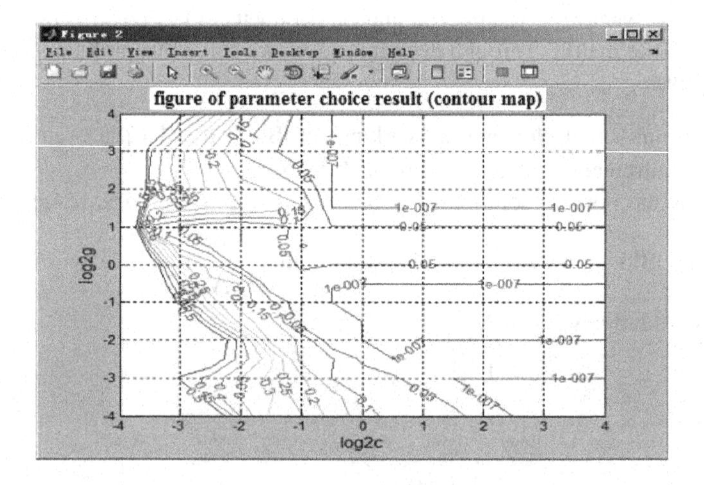

Fig. 3. c, g fine optimizing contour map

3.3 Normalization of Training Sample Data

Before fitting by SVR, generalization on extractive features is of great significance. The advantage of it is that in the establishment of classification hyperplane, avoiding the feature of large dynamic range overlapping small dynamic range. This provides them the same function. The other advantage is that feature of difficult large number calculation, which can cause computing overflow, can be avoided in inner product calculation of feature vector. Therefore, generalization on feature is necessary. After the process of generalization, the feature range is limited between [-1, 1]. This paper takes advantage of SVM classification to conduct generalization to produced data.

3.4 Regression Fitting of Experimental Data

This experiment adopts fine-quality RBF kernel and radial basis function to conduct parameter optimization in line with the parameter optimization principle. The rice fitting curve is shown in figure 4. In it, "+" is real value, and "o" is forecast value. The figure displays that function approximate error in the whole measurement process can be controlled in high level. The mean square error has reached 1e-0015. Therefore, LIBSVM has good function approximate capability.

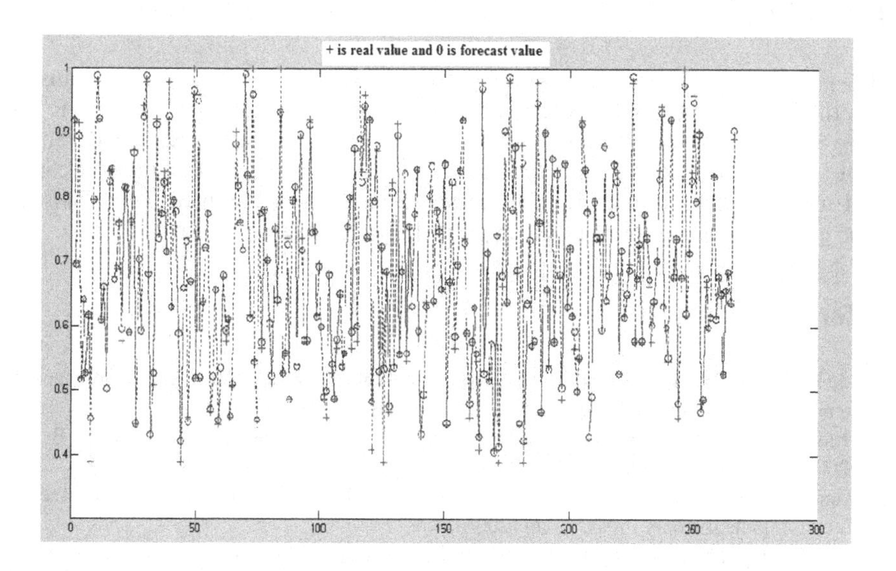

Fig. 4. LIBSVM regression fitting result

3.5 Error Detection

Table 1 is the correspondence relation between measured moisture content of rice by fitting curve and actual moisture content by standard calibration. Table 2 is correspondence relation between wheat measured moisture content and actual moisture content. The tables show that measurement errors of fitting curve are all less than 0.5%, 0.4% of maximum. So, it enjoys low error and accurate measurement.

Table 1. Correspondence table of rice measured moisture content and actual moisture content

Actual moisture content %	9.8	9.9	12.2	13.9	14.0	15.9	18.5	22.1	24.9	25.5	29.0
Measured moisture content%	9.7	10.0	12.2	13.8	14.2	15.5	18.6	21.9	24.8	25.3	28.6

Table 2. Correspondence table of wheat measured moisture content and actual moisture content

Actual moisture content%	7.8	8.7	9.6	12.9	14.5	15.7	16.5	17.1	18.9	20.5	21.0
Measured moisture content%	7.8	8.6	9.7	12.8	14.4	15.6	16.6	17.5	18.8	20.6	21.4

4 Conclusions

This paper analyzes the feasibility of utilizing LIBSVM to conduct grain moisture data fitting. Besides, it takes advantage of radial basis function to conduct SVR data curve regression fitting on grain moisture data. By these, the fitting curve under multi-input values is gained. The practice shows that curve fitted by this method can be used in practical measurement. The measurement error is less than 0.5%. This has some guiding meaning for research on portable grain moisture gauge.

References

1. Li, X., Wang, W., Lei, T.: Multi- sensor information fusion technology and its perspective in agricultural engineering. Agricultural Engineering (3), 73–76 (2003)
2. Huo, E.: Multi-sensor data fusion handbook, pp. 48–55. China Machine Press, Beijing (2008)
3. Kang, Y.: Data fusion theory and application, pp. 99–112. Xi' an Electronic Science and Technology University Press, Xi'an (2006)
4. Hou, G.: Test and sensor technology. Harbin Institute of Technology Press, Harbin (1998)
5. Wen, J., Zhu, C., Dezhi, L., et al.: Intelligence control system of circulated grain dryer. Agricultural Engineering 22(11), 89–92 (2006)
6. Vapnik, V., Golowich, S., Smola, A.: Support vector method forfunction approximation, regression estimation, and signal processing. In: 10th Conference on Neural Information Processing Systems, Denver, Colorado, pp. 281–287 (1996)

Optimized Location Planning of Logistics Park Based on Implicit Enumeration

Liu Xiaohui[1], Zhao Xueyu[2,*], Xue Yapei[1], and Zhu Cuijuan[1]

[1] College of Traffic and Building Engineering
Beihua University
Jilin, China
[2] College of Foreign Language
Beihua University, Jilin, China
liuxiaohui.1959@163.com, kobe_1978@163.com, aprilxue@qq.com

Abstract. AHP, Fuzzy Comprehensive Evaluation, Centre-of-gravity method, Cross Median Method, Kuehn—Hamburger Model、Baumol—Wolfe Model and Blson Model are the most used location methods on logistics parks. Those methods are obviously different from each other in many ways, such as what factors should be considered or ignored, the form and establishing method of mathematical model, the model's calculation methods、application and the advantages and disadvantages, and so on. On the basis of comparing and analyzing various location methods of logistics parks, this paper determines the most suitable scheme, establishes the mathematical model and solves it with Implicit Enumeration, and finally examines a site to determine its suitability for logistics parks.

Keywords: Logistics Parks, Mathematical models, Implicit Enumeration, Location schemes.

1 Introduction

Logistics Park is a site formed by many different types of logistics corporations with narrow distribution, which is specializing in logistics service and owning multiple logistics facilities [1]. Its infrastructure, complementary conditions and logistics service contain corresponding logistics service for industrial activity in cities or areas [2].

Research on the location of logistics parks is an important component of logistics system planning. It is also a complicated multi-objective decision making problem, and plays an important role in researching theories and methods of logistics parks location. Decision making of location has a great influence on the park's function and comprehensive benefits. A rational site selection for logistics parks favors a rapid and efficient product circulation, gives enterprises and consumers large vantage; It also favors logistics' socialization, meets the need of socialization production well, protects environment and reduce cities' traffic pressure. It can be favorable toward rationalizing the distribution of logistics resources, lowering logistics cost and investment

* Corresponding author.

H. Tan (Ed.): Knowledge Discovery and Data Mining, AISC 135, pp. 327–332.
springerlink.com

legitimately, lowering treasury and purchasing and selling cost, saving and accelerating the flow of funds, improving fund utilization rate, and realizing Zero inventory. And it will help new techniques' development and application, make operations more proficient. It can creates the overall techniques' development and application, make the operations more proficient. It can creates the overall effectiveness of society, promote the development of city economics, push forward Large numbers of associative businesses, advance industry project development and resources development ,thus expand the channels of employment and provide market demands for idle equipments and idle facilities.

Two main research technique have been followed, one being qualitative method, the other being fixed quantity method. In qualitative method, analytic hierarchy process (AHP) and Fuzzy Comprehensive Evaluation are combined to proceed index evaluation of programmes.

Fixed quantity method can be divided mainly into continuous model and discrete model. The continuous model believes that the site of Logistics Park can be any point in a plane. And two of representative methods are Centre-of-gravity method and Cross Median Method; While discrete model believes that the logistic park site is the optimum point among several limited feasible point, of which representative models are Kuehn—Hamburger Model, Baumol—Wolfe Model and Blson Model, etc.[3][4][5].

2 Steps of Locating Logistics Park

Collecting and Sorting Out the Materials

The locating method is usually obtained by calculation of net cost, that is, making model of transportation fee and logistics facilities cost .etc, establishing mathematical models according to constraint condition and objective function, thus seeking a minimum-cost programme.

1) Controlling the volume of business. It contains: Factory-to-logistics park transportation burden; logistics park-to-customers transportation burden; quantity kept in Logistics Park.

2) Controlling expenses. It contains: Factory-to-logistics park transportation fee; logistics park-to-customers transportation fee; fees related to logistics facilities and land as well as labor cost and operating outlay.

3) Others. Considering necessary vehicles, operating personnel, loading and unloading way, loading and unloading machine cost .etc, combined with cost analysis.

Constraint Conditions Analyzing of Location

While selecting location, there needs to make clear the necessity, purpose and meaning of establishing Logistics Park; regulatory systems: which areas are not permitted to build logistic park according to legal rules on destined land; requirements: containing the target clients——forecast on customers' present and future distribution, rate of growth in number of goods operations as well as transportation area coverage; land: available land or purchasing new land? If it is the latter, then how expensive it is? How is the land distribution at permit level of land cost? Transpiration service conditions:

reporting time of delivery, delivery cycles, and customers-to-logistics park distance calculated according to delivery time; Transportation conditions: near transportation nodes such as railway freight station, seaport and terminal bus station .etc, near transporters' office location; currency function conditions: is whether distribution function should be separated from logistics function? Does Logistics Park have the function of processing and circulation? If does, is it whether to limit locating area from a perspective of guaranteeing number of employees and convenience to pass through; others: different type of logistics has different special requirements.

Sieving Addresses

Preliminary locating area is determined, after fully classifying and analyzing materials achieved, considering into varies factors and forecasting the requirements, that is, inceptive appraisal locations are determined.

Quantitative Analysis

Select different model to calculate against different situation, and get a result.

B. *Result evaluating*

Combining market applicability, conditions of purchasing land and service quality .etc, evaluate the result, and judge if the result has the realistic meaning and feasibility.

Double-Checking

Analyze relative impact degree of other factors on the result, endow differ power, and double-check the result with weighted method. If our result follow the double-checking, then the result is final result; If not, then come back to third step to continue calculate until obtaining a final result.

Confirming Locating Result

After passing the double-checking by weighted method, then the calculating result can be regarded as final computed result and it is also optimum solution [6].

3 The Optimized Locating Scheme of Logistics Parks

Locating methods of Logistics Parks are obviously different from each other in many ways, such as what factors should be considered or ignored, the form and establishing method of mathematical model, the model's calculation methods, application and the advantages and disadvantages, and so on. Related data is obtained by investigating Logistics Park's present situation and understanding logistics demand, the beginning and the end sites, transportation and so on. On the basis of these data, we compare and analyze varies locating methods, determine the most suitable method, establish the mathematical model, and finally settle the locating problem of Logistics Park to fulfill the demand of social circumstance and geography traffic environment [7].

Establishing Mathematical Model for the Location of Logistics Park

There are eight alternative sites (points) Di (i=1, 2..., 8). We specify: in east area, select no more than one point between D_1 and D_2; in west area, select no more than one point between D_3 and D_4; in south area, select no more than one point between D_5 and D_6; in north area, select no more than one point between D_7 and D_8. The investment of D_1 site makes an estimate of b_1 one million Yuan RMB, and obtainable annual interest is estimated to be c_1 three million Yuan RMB. The investment of D_2 site makes an estimate of b_2 800,000 Yuan RMB, and obtainable annual interest is estimated to be c_2 2.4 million Yuan RMB. The investment of D_3 site makes an estimate of b_3 1.2 million Yuan RMB, and obtainable annual interest is estimated to be c_3 3.5 million Yuan RMB. The investment of D_4 site makes an estimate of b_4 1.1 million Yuan RMB, and obtainable annual interest is estimated to be c_4 3.2 million Yuan RMB. The investment of D_5 site makes an estimate of b_5 1.3 million Yuan RMB, and obtainable annual interest is estimated to be c_5 3.7 million Yuan RMB. The investment of D_6 site makes an estimate of b_6 900,000 Yuan RMB, and obtainable annual interest is estimated to be c_6 2.7 million Yuan RMB. The investment of D_7 site makes an estimate of b_7 1.4 million Yuan RMB, and obtainable annual interest is estimated to be c_7 4 million Yuan RMB. The investment of D_8 site makes an estimate of b_8 3 million Yuan RMB, and obtainable annual interest is estimated to be c_8 4.6 million Yuan RMB. The optimal location of Logistics Park is selected under the premise of maximum annual return.

Suppose variable $x_i = \begin{cases} 1, \text{When } D_i \text{ site selected} \\ 0, \text{When } D_i \text{ site unselected} \end{cases}$ (i=1,2,...,8), annual return z remarks the objective.

The mathematical model of Logistics Park is:

$$\max z = \sum_{i=1}^{8} c_i x_i$$

$$\text{s.t.} \begin{cases} \sum_{i=1}^{8} b_i x_i \le 460 \\ x_1 + x_2 \le 1 \\ x_3 + x_4 \le 1 \\ x_5 + x_6 \le 1 \\ x_7 + x_8 \le 1 \\ x_i = 1\text{或}0,\ i = 1,2,\cdots,8 \end{cases}$$

Solving Mathematical Models with Implicit Enumeration

$$\max z = 240x_2+270x_6+300x_1+300x_8+320x_4+350x_3+370x_5+400x_7$$

$$240x_2+270x_6+300x_1+300x_8+320x_4+350x_3+370x_5+400x_7 \ge 1130 \quad ◎$$

$$80x_2+90x_6+100x_1+100x_8+110x_4+120x_3+130x_5+140x_7 \le 460 \quad ①$$

$$x_1+x_2 \le 1 \quad ②$$

$$x_3+x_4 \le 1 \quad ③$$

$$x_5+x_6 \quad ④$$

$$x_7+x_8 \le 1 \quad ⑤$$

$$x_i = 1\text{或}0 \ (i=1,2...8) \quad ⑥$$

Alternative Site (x2,x6,x1,x8,x4,x3,x5,x7)	Constraint Condition						Objective Value
	◎	①	②	③	④	⑤	
(0, 0, 0, 0, 0, 0, 0, 0)	0						
(0, 0, 0, 0, 0, 0, 0, 1)	400						
(0, 0, 1, 1, 0, 1, 1, 0)	1320	450	1	1	1	1	1320
(0, 0, 1, 1, 1, 0, 1, 0)	1290	440	1	1	1	1	1290
(0, 1, 1, 0, 0, 1, 0, 1)	1320	450	1	1	1	1	1320
(0, 1, 1, 0, 1, 0, 0, 1)	1290	440	1	1	1	1	1290
(0, 1, 1, 1, 0, 1, 0, 0)	1220	410	1	1	1	1	1220
(0, 1, 1, 1, 1, 0, 0, 0)	1190	400	1	1	1	1	1190
(1, 0, 0, 0, 1, 0, 1, 1)	1330	460	1	1	1	1	1330
(1, 0, 0, 1, 0, 1, 1, 0)	1260	430	1	1	1	1	1260
(1, 0, 0, 1, 1, 0, 1, 0)	1230	420	1	1	1	1	1230
(1, 1, 0, 0, 0, 1, 0, 1)	1260	430	1	1	1	1	1260
(1, 1, 0, 0, 1, 0, 0, 1)	1230	420	1	1	1	1	1230
(1, 1, 0, 1, 0, 1, 0, 0)	1160	390	1	1	1	1	1160
(1, 1, 0, 1, 1, 0, 0, 0)	1130	380	1	1	1	1	1130

Optimum Solution $(x_1, x_2, x_3, x_4, x_5, x_6, x_7, x_8) = (0, 1, 0, 1, 1, 0, 1, 0)$, Optimum Value max z=13,300,000Yuan RMB [9][10][11].

The Optimized Locating Scheme of Logistics Parks

According to the above-analyzed-and-calculated result, 13.3 million Yuan RMB can be obtained by building logistics parks in east (D_2), west (D_4), south (D_5) and north (D_7) four sites. Thereby the optimal location is determined [12] [13].

4 Conclusion

The Implicit Enumeration is used to solve mathematical models of integer programming, and the rational and proposed locations of logistics parks are calculated. We hope that our research can provide some instructive ideas and methods [14] [15].

Acknowledgment. Jilin Province social science fund project (2008Bglx21)
Jilin city technological development fund project (201042307).

References

1. Lei, L.: Research on the Location Planning of Logistics Park on the basis of Rough Set and Delphi technique method. Logistics Technology (1), 37–40 (2008)
2. Chen, Y.: Synthetic Evaluate on the Location Planning of Logistics Park. Railway Purchase and Logistics (3), 16–18 (2006)
3. Ma, H.: Regional Study on Logistics Distribution Center Location, vol. (3), pp. 10–11. Southeast University (2005)
4. Fang, Z.: Overview on the Address-Selected Method of Logistics Center. Logistics Technology (9), 42–44 (2008)
5. Lei, Z.: Study on the Methods of Logistic Center Location in the West of China, vol. (9), p. 26. Shanghai Maritime University Press (2005)
6. Deng, A.: Logistics Engineering. China Machine Press, Beijing (2002)
7. Yu, C.: Jilin Province Logistics Requirement Analysis. Changchun University Journal (6), 56–57 (2008)
8. Qian, S.: Operational Research. Qinghua University Press, Beijing (2005)
9. He, F.: Research and Application of Logistics Park Locating Decision. Chungking University (2006)
10. Chen, T.: Research on Planning Schemes and Supporting Policies of Logistics Park. Jilin University (2008)
11. Guo, J.: Logistics Park Location and Layout. Master's thesis of Wuhan University of Technology, Master's thesis (2005)
12. Feng, Y.: Comprehensive Evaluation Methods of Logistics Park Location and Allied Studies. Central South University (2008)
13. Liu, D.: Research on the Location and Scale Development of Logistics Park in Wuhan City. Wuhan University (2005)
14. Li, H.: Application of Analytic Hierarchy Process on the Location of Logistics Park. Science & Technology Information (16), 23–25 (2007)
15. Wang, Z.: Modern Logistics Management. China Labors Press, Beijing (2001)

Preliminary Study on Remote Sensing Teaching in Environmental Engineering Specialty

Jinxiang Yang, Xiaolong Li, and Wenjuan Liu

Anhui University of Science and Technology Huainan Anhui, China
jxyang@aust.edu.cn

Abstract. With the intensify of global resource, energy and environmental issues, and the competition of digital earth strategy, the community's demand for professionals in remote sensing is growing, as well as students in relevant field need to master certain theoretical knowledge and basic skills of remote sensing. In this paper, combining the characteristics of environmental engineering specialty, teaching methods of environmental remote sensing are discussed from the three aspects which are the teaching content, teaching means and practice teaching.

Keywords: environmental engineering, remote sensing, teaching.

1 Introduction

Remote sensing (RS) is developed a new interdisciplinary since 60 years of the 20th century, which involves modern physics, space science, computer technology, mathematical methods and scientific theory and many other areas of the Earth, it is a sophisticated and practical detection technology.[1] As a modern spatial information science of the core technology, RS has been widely used in government departments and many industries in the economic field involved in land management, urban construction, forestry, environment, agriculture and other sectors. [2]

With the intensify of global resource, energy and environmental issues, and the competition of Digital Earth strategy, the community's demand for professionals in RS is growing, as well as students in relevant field need to master certain theoretical knowledge and basic skills of remote sensing.[3] In the Many courses of environmental engineering speciality, the status of RS is becoming more important, its advanced high-technology and a wide range of applicability make students increasingly be interested in this course. In this paper, combining the characteristics of environmental engineering speciality, teaching methods of the environmental remote sensing (ERS) are discussed.

2 The Choice of Teaching Content

As a monitoring technology and an application means, RS technology needs users to obtain information through RS data, and then the information obtained is used to solve the problems caused by economic activity. Therefore, the choice of teaching content is

H. Tan (Ed.): Knowledge Discovery and Data Mining, AISC 135, pp. 333–336.
springerlink.com

particularly important. ERS is the remote sensing technology, which purpose is to detect the phenomenon and the dynamic of the Earth's surface environment, and it detects and studies spatial distribution, time scale, nature, development trend, impact and damage degree of environmental pollution in order to take environmental protection measures or formulate remote sensing activities about eco-environmental planning.

For this reason, the teaching contents of ERS are mainly arranged for the following modules: Firstly, the principles of ERS; Secondly, image processing and interpretation of ERS; thirdly, principle and applications of water environment remote sensing; fourthly, principle and applications of atmospheric environment remote sensing; fifthly, principle and applications of eco-environment remote sensing; sixthly, the principles and applications of disaster environmental remote sensing. In each teaching module, the importance is that theoretical teaching is built on the basis of practice, and the specific cases are used to explain remote sensing applications in the environmental field; at the same time, keeping the latest developments in remote sensing technology to rich teaching content, to further the scientific achievements of teachers infiltrate into teaching, convert into teaching materials, and infuse into the classroom. For example, remote sensing images of eutrophication in the Gulf are introduced into the teaching of water environment remote sensing to classify and identify of the scope and level of red tide occurrence in the sea trough RS images fusion, which makes teaching content with cutting-edge and the research process transform into course content, as well as trains the students to analyze and solve problems.

3 The Reformation of Teaching Styles

3.1 The Reformation of Teaching Means

Teaching methods are mainly used multimedia. Multimedia can accommodate a large number of remote sensing satellite images and different remote sensing images of the environmental application field, etc. For example, a variety of crops and other contaminated plants growth are identified in the color infrared images, water pollution and thermal pollution of rivers, lakes, reservoirs and sea are identified in the scanned images.

These make course content more vivid, intuitive and easier to understand and accept by students. The capacity of multimedia teaching is large. Teaching content can be made a flash, which easily attracts the interest of students and seize the attention of students. Such as introducing the principle of the color addition and subtraction, it can be made a flash to vividly and clearly show the results of different colors adding. Accordingly, this method makes it easier for students to grasp and understand.

3.2 The Reformation of Teaching Methods

In a variety of teaching methods, how to enhance the interaction of teachers and students in teaching activities is very important. Teachers should establish a student-centered concept, guide the students to actively participate in teaching activities, encourage students to ask questions boldly and actively express their views [3]. Teachers may use timely problems heuristic method and discussion method to develop students' independent thinking and expression according the needs of teaching content. In the teaching process, teachers may use the methods such as clever design questions and group discussion to enhance the exchange and discussion among students, between teachers and students, improve student learning interest, and make the learning state from passive learning to active learning. Accordingly, it improves teaching effectiveness. For example, when teaching RS bands selection, pope curve of the same ground object in the different states are given, problem heuristic is used to make students discuss how to choose the bands to distinguish ground object under different conditions.

4 Enhancement of Practice Teaching

Environmental Remote Sensing course is a strong practical course. Practice teaching is the deepening and complement of theoretical teaching, and is the learning process apprehending the abstract theory knowledge through a combination of theory and practice for the students. Currently, training on the innovative spirit and practical ability of undergraduates has become the main content of the higher education reform in China. Therefore, practice teaching must be strengthened. Practical courses of RS is the key link of interactive learning between teachers and students, its importance is that plays the initiative of students. By the demonstration and explanation in projector, teacher may let students operate correction, mosaics, processing and classification of RS images with their own hands; accordingly it enhances the understanding of basic knowledge for students. While teachers should deepen the practice lesson content, increase appropriately the experimental class, reduce the validation experiments, increase the design and synthesis experiments, and introduce the scientific research achievements into the RS teaching to strengthen development of RS experiment skills. When completing the learning process, Students not only increase the skills of comprehensive analysis problems and problem-solving, but also inspire them to explore their own passion.

Concluding Remarks

The rapid development of RS technology and constant enlargement of application in environment field enable its course construction and teaching to show a certain dynamic. Therefore, the RS teaching process need to follow development frontier of the subject, and constantly adjust the teaching content and teaching methods with the times. In a word, combined with the school and professional characteristics, the teaching of ERS courses should carry out course construction and teaching reform from the demand for social application-oriented and innovative talents.

References

1. Dong, F., Wang, X.: An Approach to the System of Practical Teaching in Remote Sensing. Journal of Anhui Institute of Education 23(6), 47–51 (2005)
2. Qin, W.: Searching and Practice in the Teaching Reform of Remote Sensing Image Processing Course of Geography Major. Modern Surveying and Mapping 30(6), 45–48 (2007)

3. Zhao, Y.: Analysis of Teaching Reformation in the Course of Remote Sensing Principles and Applications. China Adult Education 10, 168–169 (2008)
4. Zheng, W.: Reformation of remote sensing practical course for geography major in normal universities. China Modern Educationa Equipment 105(17), 79–81 (2010)
5. Sha, J., Lin, Z., Chen, F.: Teaching Reformation in the Course of Remote Sensing. Fun Jian Geography 17(2), 27–28 (2002)
6. Chen, X., Tong, J., Liu, C.: Probe into the Teaching Reform of the Remote Sensing Principle in the Specialty of Surveying and Mapping. Minge Survying 6(2), 79–80 (2008)
7. Zhang, A., Yi, H., Cui, Q.: Searching and Practice in the Investigating Teaching of Remote Sensing Course. Bulletin of Survying and Mapping 12, 59–61 (2005)

Architectural Study on Undergraduate Teaching System of Hydrology and Water Resources Major

Xiaolong Li, Jinxiang Yang, and Xiaojun Zhang

School of Earth and Environment and Foreign Languages,
Anhui University of Science and Technology, Huainan 232001, China
xlli@aust.edu.cn

Abstract. Hydrology and water resource engineering major is a new-opened major in our university in 2008. So we need constantly study and research it on new major of basic education, practice education, modification and formulation of teaching plan, location of training objective and direction, and course offered.

Keywords: Hydrology and water resource engineering, undergraduate teaching system, architectural study.

1 Introduction

In 1998, in the list of degrees of the national ordinary institutions of higher learning issued by MOE, the former land and hydrology major of water conservancy is renamed Hydrology and water resource engineering major. It is not just a simple change of major name, it enlarges major connotation adjusting to era background, which makes the name of major more reasonable, scientific and standardized[1] [2] [3].

Besides China, the major universities which set up hydrology water resource department or hydrology water resource major in other countries are Russia meteorological and hydrological Leningrad University, Arizona University in America, Paris VI, Pierre et Marie Curie FR, University de Besancon. Besides there are also hydrology departments in some universities in Canada, Ireland, India, Czech. In order to reinforce water resource education around the world, The UNESCO sets up 32 graduate student courses of hydrology water resource in the world. The main nfluential ones are International hydrological graduate student class in National University of Ireland, The Dutch water environment engineering college graduate student class., Lausanne, Switzerland held federal institute of hydrological graduate student class.

Chinese famous professor, liu guangwen, set up in 1952 the first Hydrological professional in East China Institute of Water Conservancy (now Hohai University). It recruited postgraduates since 1955. It is one of the first courses, and got the gift rights of bachelor degree, master degree and doctor degree in 1981. There are 38 universities in china that set up Hydrology and water resources engineering undergraduate course, mainly are Hohai University, Wuhan University, CUG (China University of Geosciences), Chang'an University, JLU (Jilin University). Our university began to set up the major in 2007, and in September, 2008 began to recruit undergraduates of the major, 4-medical-year. The discipline nature is technology. Students are given bachelor's degree after graduation.

H. Tan (Ed.): Knowledge Discovery and Data Mining, AISC 135, pp. 337–341.
springerlink.com

2 Speciality Provision Background

Hydrology and water resources professional in our university was set up on the basis of Geological Engineering, environmental engineering, surveying and mapping engineering, Resources and Environment and Urban and Rural Planning , GIS(geographic information system). after 1958, our university set up several majors, such as radio geology, geophysical exploration , geophysical exploration , hydrogeololgy, and coal field geology. engineering takes the Inheritance and integration experience and history , Especially Mine hydrogeology direction in Geological engineering, it is on basic of Hydrogeological and engineering geological professional, accumulates abundant Running experience and Teacher Resource, and also delivered a great of excellent professionals engaged in Engineering and Technology.

In recent years, with the fast development of coal industry, our country attaches great importance to safe production of coal industry and prevention and cure of geological disaster. It extrudes day and day with the hydrogeololgy problem of coal mine safe production and Prevention and control of water disasters problem. It needs a great deal of professionals andtechnicals on hydrogeololgy and Prevention and control of water disasters. It can solve the problem of Professional talent shortage with the establishment of Hydrology and water resources professional. We can say that Hydrology and water resources professional has a good development. And our Hydrology and water resources engineering are set up under such a Favorable situation.

On the beginning of Setting up and launching our Hydrology and water resources engineering, it was a serious acid test in the following aspects, construction and development of professional teaching, Professional talent training mode, foster of professionals for the national economic construction. The newly-built Hydrology and water resources engineering faces a serious problem- late Started and high competition, though with a deep Teaching Basic and background. Since the publishing of the major in 1998, there is many this bachelor's degree programs in different universities in China. So it is of great competition. It is an Impact, a crossing and a test, and also a new challenge of Self-awareness and thinking.

Now, Higher education career in our country is on the turning course from exam-oriented education to quality-oriented education. So there are many Teaching research topics for us to constantly know ,think and solve, such as whether the new professional foundation education and professional practice education can reflect completely practical and innovative ability of the students, whether Software conditions and hardware facilities can satisfy the Teaching content requirements, and construction and practice in many aspects such as teaching plan establishing , training target and raise direction positioning and course offered. So, it is necessary for us to develop the study and research on Undergraduate teaching of hydrology and water resources engineering.

3 Professional Teaching System Setting Principles

Undergraduate teaching construction is a systematic engineering and can not be finished in a day. It needs a thorough teaching programme and matched course, but also emphasis. Build a good soft and hardware teaching system. Strengthen the foundation of professional teaching and experimental teaching construction and realize diversified teaching methods. Meanwhile, reasonably optimize and strengthen construction of teacher's troop. In contrast, there are many shortcomings in hydrology and water resources engineering of our university, so we need constantly perfect and embed our study and practice. Concrete Suggestions are as follows:

(1) Further teaching programme and curricula In-depth and optimization.

Undergraduate degree education emphasizes on "Broad caliber, thick foundation". This is the big direction and principle. It not only reflects vivid era characters, but also combines its own characters of our department. Take Geology and environment foundation course as the basic, highlights hydrogeololgy, Mine flood prevention and control, Water resources survey, hydrologic statistics, environmental engineering, exploration project. Cooperate with each other in a reasonable distribution teaching plan and teaching content.

(2) Further establish and fine stroke of the direction of professional development.

Nowadays, on the basic of Hydrology and water resources engineering training plan by our department, and with past Hydrological geology engineering geology professional foundation platform, combining with school characteristic of our university, we serve Mining enterprises and local enterprises and institutions. In subject setting, take deeply consideration market needs and combine with our Operating practices. In Specialized subject setting, we draw up two module directions: hydrological and engineering geology and water resources assessment. And its main solid bases are shown in table1.

Table 1. Hydrology and water resources engineering professional core courses

Course number	Course name	Credit	Total class hours	Inside class hours	Practice class hours	Semester Suggestions
0104001130	general geology	3	56	40	16	3
0104205130	General Hydrogeology	3	48	38	10	5
0104206130	hydrogeochemistry	2	36	36		5
0104207130	underground water dynamics	3	48	42	6	6
0104208130	Water resources assessment and management	3	48	42	6	6
0104209130	Soil water dynamics	2	32	32		5

These directions and courses can't satisfy market needs. It needs a deep study whether Professional and technical personnel can play a role in society. It must combine with our nation's policy, and get a deep understanding and constant adjustment with Graduates feedback.

(3) The renewal of the teaching means and teaching material system construction.

It is a Not -to -be -ignored subject. Along with the development of science and technology, the multimedia teaching has a large amount of information and flexible Form, therefore is a teaching method to be energetically advocated. For the newly-opened hydrology and Water resources engineering, it should take more Multimedia forms in the undergraduate teaching. Therefore, we need take a lot of large and detailed work, such as Gather information source, entry photo, and build model. It is also a serious problem to select teaching material for some specialized courses. And also, network, CAI courseware, construction and use of teaching model forms.

(4) Fully strengthen Practice education link

Teaching practice is an important link. We must completely strengthen Input and management. For the training scheme worked at present on Hydrology and water resources engineering, there are four important practice contents on Undergraduate course education stage: hydrological geological mapping practice, hydrologic exploration cognition practice, elementary surveying and the graduation stage practice and design. It is shown in table2.

It should have Clear purpose and mission for each residency, especially construction investment of practice base and the implementation of practice task. Practice base construction must reflect the characteristics of the times. The experimental teaching must also increase the design experiments and comprehensive experiment contents of proportion. Through the different contents of residency, raises student's beginning ability and innovative thinking.

(5) Continue to strengthen the construction of teachers' team, to optimize the structure of talent of teachers' team.

Age, degree and title structure of teachers' team play a key role to the quality of education and teaching. And at the same time, we should increase Professors lecture proportion, and make fine the job of "pass on, help and put in motion" of the young teachers. Now, there, in our department, are many problems of teachers' team to be solved in hydrology and water resources engineering. How to take measures to improve the situation is another top we will face.

Table 2. Professional practice curriculum

Course number	Course name	Credit	Practice class hours	Semester Suggestions
0104230140	elementary surveying	1	28.8	6
0104231140	Hydrological and geological cognition practice	1	28.8	3
0104232140	Hydrological geological mapping	4	115.2	7
0104233140	Water conservancy project cognition practice	1	28.8	4
0104234140	Graduation Practice	5	144	8
0104235140	Diploma Project and Thesis	13	370.5	8
0104236140	thesis defense			8
0104237140	Hydrology and water resources survey	2	57.6	6

4 Peroration

The development and study of undergraduate teaching is mainly to improve the quality of teaching of undergraduate education. For the study of undergraduate teaching construction of our Hydrology and water resources engineering, it should not just on the form, and take the model of other universities. It should give play to its own characters on the original advantages of hydrogeology and Prevention and control of mine water disasters. At the same, we should not to be too conservative. We should see clearly the situation, constantly broaden schooling method, and improve students' practical ability and innovation ability. We need to give birth boutiques and top students, and gradually establish our own brand. Also, through the identification of national engineering education for hydrology and water resources engineering professional, we can set the foundation for graduates for a global employment and supply large numbers of hydrology and Water resources engineering and technical personnel for our country and the society as a whole.

References

1. Li, D., Li, F.-B., Fang, X.-F.: Discussion on the Practice Teaching for the Specialty of Hdrology and Water Resources Engineering. Journal of Shijazhuang University of Economics 31, 127–129 (2008)
2. Zhou, X.: Some Ideas on the Modification of Undergraduate Programs of the Speciality of Hydrology and Water Resources Engineering. Chinese Geological Education (1), 56–581 (2005)
3. Wang, M., Gao, Z.-J.: Teaching and Reforming on Professional English for Hydrogeology Specialty. Chinese Geological Education (4), 162–164 (2009)
4. Dou, M., Zuo, Q.-T., Li, G.-Q.: The Course Construction and Teaching Study of Water Resources Planning and Management. Journal of Architectural Education in Institutions of Higher Learning (6), 91–93 (2008)

Appendix

Project Support: Young teacher fund of Anhui University of science and technology.

Exploration of Data Structure Experiment Teaching

WenJuan Liu, JinXiang Yang, XiaoLong Li, and DiLin Pan

Anhui University of Science and Technology Huainan Anhui, China
liuwj@aust.edu.cn, jxyang@aust.edu.cn

Abstract. In this paper, combining with experiment teaching experience of "Data structure", on the basis of the analysis of problems in the experiment teaching process and fully taking into account the needs of teaching and students' actual situation, the stratified experiment teaching model closely integrating with theory teaching is established in order to better play the role of "data structure" experimental course, which teaches students in accordance with their aptitude, centers on students, and pay attention to their interests and abilities.

Keywords: data structure, experiment teaching, exploration.

1 Introduction

Data structure course is an important professional core course of computer-related specialty, which has a linking role throughout the professional curriculum system[1,2]. But this course content is more abstract, students are often difficult to understand it, especially cannot master flexibly the application of the algorithm. Therefore, teaching of the curriculum should not only focus on students' mastery of theory, Should pay more attention to develop experimental ability of analysis problems and problem-solving. Experimental teaching is an important means which students verify, master and apply algorithms, and students can deepen understanding of the theory learning through experiments. Strengthening the experiment teaching and applying the knowledge taught in class to the experiment make students to deepen their theoretical knowledge when solving practical problems, these lay a solid foundation for professional learning.

2 Problems in the Experimental Teaching Process

The current experimental teaching effect is not satisfactory, and students often are lack of interest in the experiment or don't know how to start. I analyze the following reasons combining with teaching experience for many years.

(1) Foundation is not sturdy

High-level programming language is one of the leading courses for data structure; students' mastery degree of it is directly related to the teaching effect of data structures. Because the data structure is the prerequisite of core professional courses in computer-related specialty, most of the institutions set up the course after a high-level language programming. On the one hand, students are exposed less computer-related

H. Tan (Ed.): Knowledge Discovery and Data Mining, AISC 135, pp. 343–347.
springerlink.com

courses, and lack of deep understanding of the computer, thus these affect students' mastery of the data structure knowledge; the other hand, because setting of teaching hours is unreasonable, foundation of students' programming language is weak, and mastery is not sturdy.

(2) Curriculum itself difficulty is large

Content of Data Structures course is not only rich, but also its knowledge points are more, and the curriculum itself has strong logicality and high abstractness, there are certain requirements for prerequisites of high-level language programming and discrete mathematics. Therefore, when students learn abstract concepts and algorithms, they are often difficult to accept, and often behave that they can understand in class, but don't know how to do when designing algorithm. And students feel that they do not know what the use is and how to use after learning the course, they are quite at a loss and their enthusiasm is not high in the learning process.

(3) Arrangement of experiment teaching content is unreasonable

A variety of data structures in Data Structures course are self-contained and more independent, so knowledge points are more, rather complex, and more important, they all need to know and master. Therefore, in the experimental teaching, in order to use more data structure knowledge in the limited experiment link, there is polarization during subject designing. Some of the subjects emphasize practice of a single knowledge point; design of some subjects is too comprehensive, which make students do not know how to start and shrink back from difficulties.

(4) Lack of experiment teaching hours

Traditional teaching guiding ideology of "emphasis on theory, ignoring the experiment " is still very serious, most of institutions attach importance to theory teaching, it results that the experiment teaching hours are not enough to form a complete experiment teaching system[3]. For "data structure" of strong theory and high abstraction, students need to understand the using of different data structures in the experiment, shortening of experiment hours more limit students to master this course. A semester, students often feel that there is no gain in experimental course, and nor is there any practical ability to improve, the students have no confidence in programming. It easily creates the situation that students' experimental ability is low; they can test, but do not read the program and design the program, so it doesn't reach the training goal.

Taking into account many problems in experiment teaching of the course, it is necessary that the original experiment teaching is reformed and the stratified experiment teaching model closely integrating with theory teaching is established.

3 The Establishment of Experiment Teaching

3.1 Well-Designed Teaching Contents

Under the guidance of the course outline, according to knowledge point, a representative, typical algorithms of moderate difficulty is choose timely for students to program and work on computer debugging. A specific application circumstance is designed for each data structure to make the students know the application situation and the advantages and disadvantages of various data structures. At the same time, students are allowed to design their own data structures to meet the needs of the algorithm, so that they gain a deeper understanding of the advantages and application situation of the classic data structures.

(1) Stratified teaching, teaching students in accordance with their aptitude.

Because of uneven levels of students, experiment content is stratified in accordance with students' aptitude, mainly including three types: basic experiments, designing experiments and comprehensive experiments. Basic experiments mainly include representation of various types of data structures and implementation of basic operations. For these experiments, students should be to complete the corresponding content in the experimental classroom, the teacher need to give field guidance, and discover problems and solve them in time. Designing experiments aim at implementing a simple application mainly using a data structure, which require students master the logic representation, storage implementation and the corresponding algorithm of the data structure, and teachers can choose some practical issues to discuss and resolve according to the actual situation of the students. For example, linked list is used to implement one-variable polynomial four arithmetic operations. Comprehensive experiments aim at implementing a comprehensive application using one or several data structures, which require students gradually master ideas and methods of program design and development in the course of study. Students can be divided into groups and arranged a number of large-scale comprehensive subjects by teachers, and guided to solve practical problems by using integrated use learned knowledge. For example, two data structures (stack and queue) are used to achieve the parking problem.

(2) Enhancing the interest of subjects, improving motivation and self-confidence of students.

The premise strengthening the students' practical operation ability should be to improve their interest and confidence to dare to hands-on program code, which can improves their problem-solving abilities. When designing experiment subjects, strong interest subjects should be designed combined with practical; so that students understand that they apply what they have learnt in use, and it can enhance their learning fun and enthusiasm. For example, the application of stack structure can be designed to "maze path search"; the application of tree structure can be designed to "Huffman Tree and Huffman Coding" or "Tictactoe Game"; the application of graph can be designed to the problem of "Campus Tour Guide" or "Gobang" [4]. By solving a particular problem needed to design an algorithm, students are divided into groups to discuss, and students in the group are able to demonstrate their ability and inspired their team spirit, and also to stimulate their problem-solving confidence through different views of the debate.

3.2 Enhancing Interaction of Experiment Teaching and Learning

(1) Enhancing interaction of teachers and students, achieving phase progress of teaching and learning.

Experimental subject is students, so after teachers create a basic framework, thinking and practical space is given to students in order to achieve mission-driven teaching. Experimental subjects may be arranged for students every 1-2 weeks in advance, in order to make students have enough time to think and write basic code under the class, and go to class with questions.

(2) Enhancing communication between students, establishing a good learning atmosphere.

To make exchanges between students more active and effective, for the comprehensive subjects, the practice that all the students do the same experiment project can be break, and experiment grouping is taken. Students spontaneously formed study groups, each group selected topics according to their interests. The task is allocated within group; they help each other and co-operate together to accomplish the task.

3.3 Integration of Modern Resources and Materials

Network multimedia resources can provide rich three-dimensional teaching resources, during teaching teachers should take full advantage of the demonstration system, excellent course websites and other resources which can help students teaching and learning after class [5]. Each school can build its own the network teaching platform of data structure to provide multimedia courseware, source code of classical example, course video, online testing, and communication forums. Using multimedia technology to create a vivid, lively and interesting teaching situation, students can complete a series of learning process by self-study, self-test question and exchange. This can better assist the classroom teaching and develop students' independent learning and environment ability in a network environment; accordingly the efficiency and quality of teaching are effectively improved.

3.4 Enriching Examination Forms, Stimulating Learning Passion

Examination is one of the problems most concerned by students; the course experiment should hold focus on students' hands-on experimental and innovative ability. To this end, we should choose a variety of exam modes for students to test the level of experiment.

(1) Real-time evaluation. For simple basic experiments and designing experiments, the teacher may random check students' completion situation in the middle of a time in experiment course, and enable students to point out the idea and implications of the some source code achieved certain functions in classroom, which judge the extent of the students' truly mastering the knowledge.

(2) Reply-type report. For the comprehensive experiments, the form of packet completion is used. Within the specified time, each team submits a lab report, and replies the entire experiment (Subject analysis process, important data structure design ideas, operation methods and its optimization of algorithms). Trough three ways of answer questions by students, group comments and teachers' comments, teachers can check whether students achieve the intended learning objectives and the final learning results, so that students truly understand and master the knowledge points.

(3) Writing lab report. The report not only requires students to give experimental procedures and results, but also to write the knowledge preparation, design idea, the results of analysis. From the lab report, teacher can examine the logic organization capacity.

4 Concluding Remarks

Study of Any course, its ultimate goal is to apply. Therefore, reasonable arrangement of comprehensive and innovative experiment links is the best platform which ensures to transform theoretical knowledge into application ability. Experiment teaching is an important part of the teaching system in university, a reasonable and scientific experimental teaching is particularly important for consolidating, understanding the learned theory knowledge and improving their innovative ability. "Data structure" is a strong theoretical and experimental course; it needs the theory teaching results to guide the experimental teaching process, and more need the experimental teaching process to strengthen the theory teaching effectiveness. Therefore, the systemic and scientific arrangement of experimental teaching is the critical path which trains students' hands-on experiments to solve practical problems.

References

1. Han, L.: About Computer Specialized "Construction of data" High-quality Goods Curriculum Construction. Computer Knowledge and Technology (Academic Exchange) 11, 209–217 (2006)
2. Zhang, H., Zheng, H.: Discussion on Teaching of Data Structure. Modern Computer 15, 57–58 (2010)
3. Zhang, G., Li, H.: The Reformation and Exploration of Data Structure Experimental Teaching. Higher Education Forum 3, 103–104 (2008)
4. Chen, X.: Application of Case Teaching Method in Data Structure Course Teaching. Computer Era. 1, 50–51 (2011)
5. Chen, X., Liu, K.: Research and Practice of the Data Structures Excellent Course Construction. Journal of Nanyang Institute of Technology 1, 121–125 (2009)

Survey of Job Happiness of College Students in Enterprise

Xiujuan Yan, Shanshan Liu, and Jianfeng Hu

Job Consult Center of JiangXi BlueSky University
jxyxj1220@126.com,
lssandsunshine@163.com,
huguess@21cn.com

Abstract. The attention of College Students' employment is from difficulty to quality, job quality of college students is reflected from employment rate, employability, satisfaction to the positive experience of happiness. College students are no longer just to find a job, but get happiness experience from job. We explore factors which affect the job happiness of college students by questionnaire, and propose relevant recommendations, in order to better understand the employment situation of university students, enhancing the experience of job happiness, thus contributing to carry out career guidance for college students, solving the problem of employment of college students.

Keywords: College Students, Job Happiness.

Today, enterprises have invested a lot of money on recruiting each year, college students are the main target of enterprises' recruiting. One hand, many students find work after graduation, but many enterprises would not be able to complete recruitment, and students have been recruited is not stable, college students leave their job frequently. There are many reasons for college students to work unstably, which are physical and spiritual, internal and external, subjective and objective, etc. The main reason is that experience of job happiness which the students get from the work is not strong. Work is not only with survival, but also reflects a person's dignity, dignity is emotional need of people, it is an inner experience. The dignity can make people work with a positive emotional experience, thereby enhancing the quality of employment of college students.

Therefore, the students are not too much, not really useless, not without ability and not well adapted to the needs of enterprises, but lack of dignity and a sense of working people, working people do not have a positive emotional experience. How to get students to appreciate the joy from the work and well-being, positive experience is to improve the current employment situation of university students is an effective way. Well-being of students work to understand the situation, analyze the impact of employment of university graduates happiness factor is to help students adjust their concept of employment correct job attitude, expanding employment horizons, enhance professional competence and effective way.

H. Tan (Ed.): Knowledge Discovery and Data Mining, AISC 135, pp. 349–354.
springerlink.com

1 Job Happiness of College Students

Job happiness of college students decides to the quality of employment, the quality of employment affect job happiness of college students. College students' job quality depends largely on the student's subjective discretion, and Job Happiness Index is the core indicator of students' subjective quality of employment. Therefore the graduate students' job quality depends largely on the student's job happiness experience. And at the same time, the experience is influenced by the job quality. The students who have high-quality employment are more likely positive and those who have low-quality employment are more likely negative. Happiness is a kind of psychological experience. It is a fact judgment of objective conditions and state for life and also a value judgment of subjective meaning and satisfaction. It is shown as a positive psychological experience based on life satisfaction. Job Happiness Index is peoples' feeling for their own job status. Job happiness is an integration of satisfaction, happiness and sense of worth.

Students the quality of employment depends largely on the student's own level of subjective identity, employment, student well-being index reflects the core indicators of subjective quality of employment, so the quality of employment of college students' employment depends largely on the level of well-being experience. Meanwhile, the employment of university students the experience of happiness is also affected by the level of quality of employment, the employment of high quality students more likely to have a positive emotional experience, the work is easier to appreciate from the feeling of happiness; and low quality of employment is more likely to have negative feelings experience, easier to appreciate from work is not a happy feeling. Happiness is a psychological experience; it is both the objective conditions of life and determines which state a fact, but also for the subjective life satisfaction and a sense of value judgments. It showed on the basis of life satisfaction have a positive psychological experience. Employment, happiness can be understood as satisfaction, happiness and the sense of organic unity.

This happy state of employment of university graduates according to the findings from the perspective of well-being of college students to explore the work of well-being factors, and then analyze the factors affecting the stability of the work of students, we evaluated subjective and objective indicator. Subjective indicator includes job expectancy, interpersonal relation and career goal, and objective indicator includes salary, and Professional counterparts, work area and industry prospect.

2 Questionnaire

2.1 Production of the Questionnaire

The subject of questionnaire is divided into enterprises and employees, we find out working condition and job happiness of college students from the perspective of enterprises and employees. Enterprise questionnaire includes the contents of enterprises' listed situation, nature, industry classify, establishing age, scale, proportion of male and female staff, the main method of recruiting, the main job types of job and so on. We realize the basic situation of enterprises and find out the correlation between enterprises' characteristics and job happiness experience of college students. Employees questionnaire contains the contents of students' gender, only child or not, marital status, native place, age, education, work experience, post type, salary, professional counterparts, relationships, career goals and so on. We realize the employment situation and propose to build a evaluation system on the quality of graduate students' employment based on Job Happiness Index.

2.2 Distribution of Questionnaire

The research is proceeding in twenty enterprises in which locates in ChangShu of JiangSu Province, the amount of employees questionnaires we distribute is four hundred and fifty, and we returned four hundred twenty questionnaires, of which there are four hundred valid questionnaires, the efficiency of questionnaire is 95%.

2.3 Basic Situation of Questionnaire

The research refers to twenty enterprises, there are three listed companies, two foreign-invested enterprises, fourteen private enterprises, four Hong Kong, Macao, and Taiwan-invested enterprises; four hundreds staffs joined in the research, there are 172 male staffs and 228 female staffs.

3 Graduate Employment Happiness Factor

The research shows that students lack of employment happiness. Among the 400 in employees involved in research, about 1 / 4 of the employees score their sense of well-being 80 points of more. And nearly 1 / 6 of the employees score under 60 points. There are many influencing factors of graduate employment happiness such as individuals, enterprises, material and spiritual aspects.

3.1 Personal Traits

Personal information refers to basic situation and the feeling for their work, including gender, marital status, age, education, and if the work be in accord with the individual characters, interests, ideals ,career expectations, profession and the awareness of work, the sense of accomplishment. According to the research, the male employees have stronger feeling of well-being than the female employees. Among those who score their sense of well-being 80 points or more, 52 are male employees which takes 57.78% and 38 are females, takes 42.28%. However, the feelings of well-being between the only child and non-only child are almost the same, so as to the married employees and singles. Elders can feel happy better from than the younger.

The higher conformity of employment reality and professional expectations, the more counterpart of occupation and the profession, the more conformity of their work and their own career development goals, which make university employees have stronger sense of happiness.

This object of study is university students, who have independent values, emotional traits, and so on. They pay more attention to self-experience. In addition work solve the problem of existence, it also meet the spiritual needs. Employment of college students is increasingly concerned about the level of their inner experience and feelings. The gap between reality and expectations, the conformity of engaged in professional and self-development goals, job enrichment and autonomy, and from work in the sense of accomplishment, which are important factors of influencing employment of university students experience a sense of well-being.

In addition, since the University for a Later Expansion, the difficulty of college students' employment has become a reality. In the severe employment situation, many university students choose the employment guidelines of "first employment, after choosing". When a real role to work, university students find that work and their personality, interests and goals of their own career development are inconsistent. In addition, university students of accepted higher education are ambitious, and they want a career with a strong desire. With increasing number of graduates, the employers recruit standards are improving in due course, Especially the education standards. This has made many graduates to enter business later, and professional competence on the job can not fully play, and their own values can not achieve. This psychological gap is easy to make them a sense of loss and frustration, resulting in less work experience happiness.

3.2 Enterprise Factor

According to this survey, the future development of company, the culture and interpersonal atmosphere of company, the management systems and processes of company that are important factors of influencing employment of university students experience a sense of well-being. These factors are in second, third and fourth in a number of influencing factors. Enterprise is a carrier of university students' career; it will affect the happiness of their work experience. Organizational factors include the future development of company, the image size and level of company, the management systems and processes of company, and the culture and interpersonal atmosphere of company. The university students in the work want to achieve personal development and goals through enterprise organization. Therefore, enterprises have great significance to them. The future development of company, and size and efficiency of company influence their future development in the enterprise. The visibility of enterprise is their social status symbol, but also is a favorable basis for their re-career. The working conditions of enterprise directly influence their work feelings in the enterprise feelings.

3.3 Salaries

According to this survey, pay factor is a very important factor of influencing employment of university students experience a sense of well-being. 55% of the survey indicated that "the pay and benefits satisfaction" is the impact factor of their work happiness, and its' influence on many factors are the first. The pay of employees determines his economic and social status, and concerns his quality of life and space. For young people, pay and benefits concern whether they can maintain their good life. When they feel the remuneration paid by enterprises are not well on behalf of the value of their human resources, or can not correctly assess their contribution to the enterprise, they will have a negative emotional experience. They will lose confidence to work, and will feel lost and sad in the work process. From another perspective, the work can not make them feel happy.

3.4 Career Goals

According to this survey, we find that the employees of having clear objectives and occupation consistent with their career development goals have a strong sense of happiness, and the employees happiness of having clear objectives and occupation inconsistent with their career development goals is weak. Employees of participate in research have about 65% employees of clear objectives and occupation consistent with their career development goals, and their happiness level in more than 80 points are 70% of total of more than 80 points employees. Employees of participate in research have about 35% employees of clear objectives and occupation inconsistent with their career development goals, and their happiness level in more than 80 points are 10% of total of more than 80 points employees. For university students, they are in the process of exploring their own career development. In addition to the work of the corresponding material rewards, the more important is to find their career development opportunities to better achieve their career goals, which is a psychological reward. Therefore, the availability of career development opportunities and occupation consist with their career development goals, which are essential for them. If employees feel enterprises could not provide for their development of space and support and the occupations and their career goals more distant, it will reduce the work of passion and power, and it can not be better from the work to appreciate happiness. An important criterion of employment options in today's college students is to find professional development platform, to engage the occupations of better meet the development goal.

3.5 Interpersonal Factors

According to this survey, 10% of college employees are very satisfied with their current interpersonal relationship, and 42% of college employees are the employment happiness level in more than 80 points in this group. 71% of college employees are satisfied with their current interpersonal relationship, and 23% of college staff is the employment happiness level in more than 80 points in this group. Employees of the employment happiness level in more than 80 points in the two groups are 92% of total of more than 80 points employees. For employees, good interpersonal atmosphere is a psychological income. Compared with the campus relationships, interpersonal business will be relatively complex. Students' staff in enterprise need to obtain a certain material rewards, but also need others recognition and support. Harmonious interpersonal atmosphere is one of them. Only in a harmonious relationship, it makes college employees experience happiness at work. Therefore, college employees are not suited to the business environment, which will lead to the generation of college students leave.

Employment of college students experience a sense of well-being will be the important factor of influencing the employment stability of college students and the employment quality of college students. Through analyzing the influencing factors of employment of college students experience a sense of well-being, it can do for enterprises and institutions, social at the college students' employment for reference and borrowing. It assists enterprise to a better understanding of the college staff requirements, then enterprises can provide development platform. It helps colleges in the better carrying out the student employment guidance education, and enhances the professional competence of college students. It helps the community better relieve the employment pressure of college students, and solve the employment problem of college students.

Acknowledgments. This study was supported by 2010 Humanities and Social Science Foundation of Ministry of Education of China (10YJC880048) and Teaching Reform Project of JiangXi BlueSky University Educational Committee (JY1027). The authors are grateful to the anonymous reviewers who made constructive comments.

References

1. Lin, H.: On university Graduates' Employment Training System of Dalian, p. 11. Dalian University of Technology (2007)
2. Weng, J., Zhou, B., Han, Y.: Changes in Employment Stability of Developed Countries: Causes and problems. Journal of ZheJiang University of Technology (Social Science), 146–152 (June 2008)
3. Li, X.: The empirical Analysis and countermeasure research of the demission impacts of college students in enterprises. Journal of Guangdong College of finance and Economics, 41–46 (June 2008)
4. Lingwen, Fangli, Luofu: The study on the effecting factors and the accommodating factors of Chinese enterprise employees turnover intention. Journal of Xiangtan University (Philosophy and Social Sciences), 65–69 (July 2005)

Study on the Physical Education Reform of Colleges and Universities Based on the *Regulation on National Fitness*

Liying Ren, Xueming Yang, Xiaoping Xia, and Jinliang Shi

Sport Department of Southwest Petroleum University, Chengdu Sichuan 610500
shijinliang275@126.com

Abstract. College physical education plays an important role in promoting national fitness activities. The promulgation of *Regulation on National Fitness* has put forward new requirements for college sports culture and human resource development and management, is of vital significance for the Physical Education Reform of colleges and universities. The study suggested that college physical education must implement the spirit of *Regulation on National Fitness* and carry out reforms appropriately. Take college students' interests as the starting point, improving optional course system in physical education; take the students' participation as the core, optimize the evaluation system; take extra-curricular group activities as a useful complement to ensure smooth implementation of the *Regulation on National Fitness*.

Keywords: Regulation on National Fitness, colleges and universities, physical education, reform.

1 Introduction

School physical education is the foundation of community sports and family sports. Sports activities in colleges are the basis of national fitness activities. In college, Energetic young people are hungry for sports knowledge and they also have requirements for sports, so college physical education should be an educational, stage and long-term education for the purpose of promoting youth physical and mental development and increasing the body quality. Speeding up the construction of campus sports culture and conforming to the development of *Regulation on National Fitness* is the development direction of college sports in new area. How much the students master the sports knowledge, how much they love sports and the degree of their physical ability and etc will affect their performance in physical activity after work and this effect can not be ignored.

In the implementation of the *Regulation on National Fitness*, people should master the methods and means if they want to take exercise, while mastering the methods and means need education and time. College physical education enable college students receive good physical education. By planned training, foster students to engage in lifelong sports, so as to promote their love to sports and participate in sports independently and spontaneously. Therefore, not only they promote physical and mental development, but also enjoy the happiness of the sports, making sports culture

H. Tan (Ed.): Knowledge Discovery and Data Mining, AISC 135, pp. 355–360.
springerlink.com

becomes an indispensable part of life and work. And consciously fulfill obligations and protect the right of citizens to participate in national fitness activities.

2 New Requirements for College Physical Education According to *Regulation on National Fitness*

The promulgation and implementation of the *Regulation on National Fitness* make national fitness turn from "plan" to "regulations" which provides a more powerful political protection in promoting healthy and sustainable development of national fitness activities. The core of the *Regulation on National Fitness* is to carry out national fitness activities in terms of system and protect the legitimate rights and interests of citizens in national fitness activities. National fitness mainly focus on teenagers, improving their sports ability and fitness level as well as leading them to engage in life-long sports activities is new historic mission that the *Regulation on National Fitness* have entrusted to college sports work.college physical education should coordinate with the national economic and social development, accelerating its pace of reform, cultivating students' consciousness of physical activity, teaching them scientific training methods to improve sports ability and be physically active on a regular basis so as to lay the foundation for the integration of health education in school, society and family in the fervor of physical exercise. Consequently, all these will effectively promote the full implementation of *Regulation on National Fitness,* achieving the ultimate goal of national fitness and building a harmonious society.

2.1 Material Culture of College Sports Is the Material Basis to Implement the *Regulation on National Fitness*

Sports venues are important parts of campus sports culture and are Material guarantee to ensure the smooth implementation of campus sports cultural activities. It directly affects students' participation in sports and the development of the school sports work. In recent years, expanding college enrollment and dramatically increased students number make the existing sports facilities can not meet the sports needs of the students. Therefore, it is of vital importance to accelerate the construction of material culture of college sports and establish efficient and rational stadium. According to the *Regulation on National Fitness*, the construction of college sports venue must be combined with the reality of large-scale college students, reflecting the people's needs and to establish comfort, safety, modern and adequate sports venues and equipment that can meet a variety of requirements. Thus to stimulate students' desire to exercise and perform its social sports function in accordance with *Regulation on National Fitness.*

2.2 Human Resources of College Sports Is the Manpower Guarantee to Implement the *Regulation on National Fitness*

Human resource development is the key to the development of college sports culture, and we must pay attention to the educating and developing of the human resources of college sports at the core of the physical education teachers.Because teachers are the organizers of educational activities who play a leading role in students ' development. Actually, requirements of *Regulation on National Fitness* to college physical education are the requirements to sports organizers and PE teachers. Whether the renewal of teaching mind, or teaching content and teaching methodologies, depend on the quality of teachers. Therefore, it is of vital importance to increase the on the job training of PE teachers so as to make teaching ability and the concept of the PE teachers keeping pace with he times.

2.3 System Culture of College Sports Is the System Guarantee to Implement the *Regulation on National Fitness*

It is the basic requirement and purpose for the *Regulation on National Fitness* to let all college students participate in sports independently and spontaneously; it is also the ultimate goal of the construction of college sports culture. The survey found that enthusiasm and self-consciousness of college students participating in sports activities is still a serious problem. Therefore, it is imperative to establish campus sports and cultural organizations, to strengthen the organizational management of departments, to guarantee the school's sports and cultural activities going smoothly under the macro-control. Firstly, schools should set up a special working committee of sports, making overall decisions of school sports cultural development. Departments at all levels should also set up a special leading group of sports and the corresponding grass-roots organizations, and actively organize, conduct, manage the unit sports cultural activities, to ensure sports and cultural activities carrying out smoothly and effectively. Secondly, studying and formulation the related policies, measures and systems of sports cultural activities according to the spirit of *Regulation on National Fitness*, achieve the institutionalization and standardization of the sports cultural activities. Sports cadres in student union and sports activists are powerful forces that play an important role in cultivating the environment of campus sports culture. So, through various forms of sports grass-roots organizations, attract vast students involved in sports activities, cultivating their organizing and management capability, carrying out the student's self-organizing and self-management. In this way, not only can it arouse people's enthusiasm and initiative, but also it can promote democratic management of sports cultural activities, ensuring the thorough implementation of *Regulation on National Fitness* in colleges and universities.

3 Study on the Physical Education Reform of Colleges and Universities Based on the *Regulation on National Fitness*

In order to ensure the smooth implementation of *Regulation on National Fitness* in colleges and universities, making the fitness consciousness realized by all the population, promoting fitness activities in the whole society, the existing the existing PE curriculum and extra-curricular activities should reform appropriately, only in this way, can promote school sports being integrated into national fitness.

3.1 Take College Students' Interests as the Starting Point, Improving Optional Course System in Physical Education

Optional PE lesson, as a substantial sports reform, through 10 years of practice, featured by its wide range of contents, students with a certain right to choose, flexible teaching and organization more, have loved and accepted by students. However, due to some objective factors, the quality and effect of Optional PE lesson is not very satisfactory.

At present, the situation of teachers qualification, venues and facilities are the main factors that affecting the establishment of optional PE lessons, so students' hobbies are always conflicting with the volume of some optional lessons, therefore, only a small number of students really satisfied with their chosen sport.since the venues and facilities are important factors in ooptional PE lessons, it will not only affect the setting of optional PE lessons, limit the number of the elective lesson, but will also affect the student's elective motivation. If restricted by venues, students have to ignore their interest in choosing lessons and change another one. Badminton, tennis, table tennis, aerobics and other projects is more popular in students, however, since these items lack of qualified teachers and higher requirements in venues and facilities, they are not carry out extensively.

In order to fully reflecting the optional PE lessons to meet the requirements of students in choosing "class time, teachers and learning content on their own master", protecting the right of students to participate in national fitness activities proposed by *Regulation on National Fitness,* we must resolve the key problems, such as teachers qualifications, venues and equipment, the class setting should be centered on students.

Optional PE lesson satisfied students' autonomy, meanwhile, it also increases the difficulty of teaching unity. Since people have their own demands, so it will have a bad effect on the learning initiative of students if the outline and the content of courses were unified. Under the current optional course system, sstudents generally are divided into class based on the selected classes. Students come from different regions in a class, as the regional high school physical education are very different, so each class of students differ from each other in physical fitness and physical ability. The uneven acceptance caused various learning speed. In addition, their understanding of the physical object and purpose various a lot and there are considerable differences in students' learning requirements for , so a fixed course content is difficult to meet the different requirements of students. This will have a negative effect on their enthusiasm for learning, not conducive to the development of personality and ability of students, of course, the teaching result is not very optimistic. Therefore, setting courses in different levels in the same optional courses is an effective way to solve this problem.

3.2 Take the Students' Participation as the Core, Optimizing the Evaluation System

Currently, the sports score of the students are basically depend on rate of attendance, learning attitude, testing grade of National Student Physical Health Standard, special technology , theory test and other aspects. using uniform standards to measure students test scores is mainly focused on the horizontal comparison of their learning effect, ignoring the individual differences, played down the longitudinal evaluation of students' learning process .This competitive evaluation mechanism restricts the sound development of physical education, easy to dampen the enthusiasm of students, not conducive to the formation of lifelong physical consciousness, and are not consistent with *Regulation on National Fitness.* It is thus clear that guiding by the core of "participation" emphasized by *Regulation on National Fitness*; it is an important way to promote the sound development of physical education for colleges to develop an evaluation mechanism fit for university student's physical learning.

Under the guidance of the scientific concept of development, school sports should be centre around all students, adhering to the guiding ideology of "health first". The evaluation principles should reflect the student's interest in sport, emotional experience, personality differences, and fully mobilizing the enthusiasm of students, to enable students to develop good habit of exercising. The evaluation way should desalt the summative assessment, paying attention to the process of evaluation, student self-evaluation and mutual evaluation. In the formulation of evaluation standards, it needs to desalt the inherent genetic indicators, highlighting the acquired indicators, strengthening the function of feedback and incentives.

Regulation on National Fitness emphasizes the core of "participation", so in the process of establishing the evaluation methods, students' participation attitude should be considered as an important indicator in evaluating students' learning effect. The evaluation should also take full account of whether the students actively involved in the activities, whether they think proactively and practicing over and over again in order to achieve the goal as well as their collaboration with peers and self-perseverance to overcome difficulties in morning exercise, physical education, extra-curricular physical exercise, sports competitions and other activities. In this way, it can better reflect the student's attitude toward physical education, sports interests, sports habits, physical ability, and many other conditions.

3.3 Take Extra-Curricular Group Activities as a Useful Complement, Ensuring Smooth Implementation of the *Regulation on National Fitness*

According to the *Regulation on National Fitness,* the Schools shall ensure one hour of sports activities a day for students in school, so two hours per week of physical education can not meet the national fitness demands. In order to ensure the implementation of *Regulation on National Fitness,* colleges should carry out a series of extensive extra-curricular physical activity, so it is also particularly important to establish a sound supervision system of extra-curricular physical activity.

Extra-curricular physical activity should focus on guidance, developing the backbone of sports by promoting the training of student-athletes who will lead many students to participate in after-school exercise. In order to attract more students participating in activities, colleges should also carry out a variety of competitions, reducing the difficulty of the game with reality, expanding the number of participants, organizing large-scale events with high interest.

In addition, colleges should full play the role of the organization of various sports associations, to enable students to manage students, making the extra-curricular physical activity to be a long-term, spontaneous development.

4 Conclusion

Regulation on National Fitness is an important political step to improve the people's livelihood and it proposed a new culture demands to college sports works. The reform of college physical education should in accordance with the spirit of *Regulation on National Fitness,* building a harmonious, civilized and healthy campus sports culture, and so as to response to the *Regulation on National Fitness.*

Acknowledgements. Fund Project: University-level teaching reform project of Southwest Petroleum University "the study and practice of the implementation of *Regulation on National Fitness in* institutions of higher learning".

References

1. The State Council of the People's Republic of China. Regulation on National Fitness. China Legal System Publishing House, Beijing (2009)
2. Lian, H.: Analysis of Physical Education Reform of Colleges and Universities Should Be Accommodate With the Needs of the National Fitness Program. Acta of Hubei Radio and Television University (5) (2006)
3. Li, M.: The practice and Thinking of "Three-Independence" in Physical Education. Physical Journal (12) (2008)
4. Xia, X., Shi, J.: Study on the Physical Education Reform of Higher Vocational School under the Regulation on National Fitness. Career Forum 32 (2009)
5. Cai, Y., Zhang, Y., Li, W.: The Strategic Significance of the Promulgation of the Regulation on National Fitness. Acta of Beijing Sport University (9) (2009)
6. Zhai, H.: New Requirements to School Physical Education From the Regulation on National Fitness. Acta of Wuhan Sport University (1) (2011)

A Study on Problems and Countermeasures of China's Higher Vocational Education

Xiaozheng Liu

Journal Editorial Department, Henan College of Finance and Taxation, Zhengzhou, 451464
2580832012@qq.com

Abstract. As an important component of higher learning, higher vocational education in China has almost accounted for half of education industry: with the scale of operation rapidly expanded, managing skills constantly improved, schooling characteristics more and more specific, higher vocational education has become a big concern and primary focus of the whole society. At the same time, however, we must admit there still exists some long-standing mistaken ideas concerning about higher vocational education. In addition, the mechanism of some vocational colleges is rigid and tardy; the "dual-competency" teaching staff is deficient and the teachers' comprehensive quality remains to be improved; the personnel training lacks variety of characteristics; the graduates are comparatively not very competitive as expected in seeking employment, so on and so forth. All these problems will surely restrict the sustainable development of higher vocational education. Some measures must be taken in order to promote higher vocational education to develop faster and better. They include: abolish the prejudice against vocational education; strengthen its appeal; highlight its characteristics in personnel training; strengthen the construction of teaching staff and optimize their overall qualifications; strengthen the organic integration of production, education and research; and improve graduates' employment competitiveness, etc.

Keywords: higher education, vocational education, educational reform, personnel training.

1 Introduction

In recent years, with the application of opening and reform policy, China's higher education has been rapidly developing. As an important component of higher learning, higher vocational education also witnesses a rapid development: the scale of operation has been greatly expanded, managing skills constantly improved, schooling characteristics more and more specific. By 2009, institutions of higher education had 28.26 million enrolled students, including 12.80 million students from higher vocational colleges. After many years of development, higher vocational education has almost accounted for half of education industry and become a big concern and primary focus of the whole society. Some higher vocational colleges, with peculiar schooling characteristics of their own, have really established their self-identity in the society. Since China released National Plan for Medium and Long-term Educational

H. Tan (Ed.): Knowledge Discovery and Data Mining, AISC 135, pp. 361–368.
springerlink.com

Reform and Development (2010-2020) in 2010, vocational education has been much more highly evaluated and may even take wing in its accelerated development. But of course, there is also some hindrance to its further forward going.

2 Hindrance to Further Development of Higher Vocational Education

As an important component of higher learning, higher vocational education has experienced rapid development and made great achievements in recent years, but it is still not in compliance with the requirements of the times and society, as is shown in the following several aspects:

2.1 There Still Exists Some Long-Standing Mistaken Ideas Concerning about Higher Vocational Education

Nowadays, although higher vocational education has become an important component of China's higher education and reached an unprecedented developing level, however, due to traditional beliefs in education and talents training, people from all walks of life cannot have a clear and thorough understanding of higher vocational education, including talents cultivation mode and training characteristics. In their eyes, compared with regular higher education, higher vocational education is much inferior in many aspects, such as a lower level in teaching, a shortage of qualified teachers, a fickle atmosphere in study, a low quality of enrolled students, a lack of competitive advantage in seeking employment, etc. It seems that students who receive this low-level higher education are unable to be well-trained and become genuine talents, not to say a satisfactory job and a bright future waiting ahead upon their graduation. These mistaken ideas directly affect the further development of higher vocational education. Some higher vocational colleges are even faced with difficulties to fulfill their enrollment quota.

2.2 The Mechanism of Some Vocational Colleges Is Rigid and Tardy

Nowadays, China is in the period of readjusting and optimizing the industrial structure. Higher vocational education undertakes the arduous task of improving the quality of workers and training professionals, which concerns the future development of Chinese economy and society. For this reason, deepened reform of higher education is carried out, more higher vocational universities and colleges are founded. Some secondary vocational schools are incorporated and upgraded as higher vocational universities and colleges, so do some adult schools via transformation. Because of the habitual inertia effect, these newly emerged higher vocational colleges are likely to follow the old path in such respects as the type and level of its education, running characteristics, property, etc. Besides, since the internal body, the mechanism of management and teaching staff are generally not allocated as required by a standard higher vocational college, the introduction of talents, promotion of regular management and improvement of teaching level would be seriously affected.

2.3 The "Dual-Competency" Teaching Staff Is Deficient and the Teachers' Comprehensive Quality Remains to Be Improved

As we can see, in recent years, teaching resources in higher vocational colleges strength have been greatly enhanced, and their overall quality are also improved, but in a whole, there is still a certain disparity to the demand of higher vocational education development. Firstly, in present situation of enrollment expansion, teachers of higher vocational colleges are obviously insufficient in quantity. Secondly, teachers' structure is unreasonable, and many of them are inexperienced in social practice. Since China's higher vocational education starts relatively late, with some additional historical reasons, teachers of basic courses disproportionably outnumber those of specialized ones. Few versatile and so-called "dual-competency" teachers can be found, most teachers just don't know how to put their knowledge into practical use, which combine to make the further development of higher vocational education more difficult. Finally, teaching personnel in higher vocational colleges is unstable, there is not a smooth channel for teachers to be refreshed. Though governments at all levels attach great importance to higher vocational education, and provide big support to its development in terms of finances and policies, owing to various kinds of internal and external factors, to name a few, the non-standardized management in assessing teachers' professional titles and their employment as well as the low wages and poor welfare, result in teachers of higher vocational colleges to be unstable, some of them would go job-hopping. To make up for the loss of some experienced teachers, higher vocational colleges will have to employ graduates from regular institutions of higher learning. These newly recruited teachers, even if rich in theoretical knowledge, will need time to enrich their practical experience, so as to be more competent at their work. "dual-competency" teachers, who are qualified both in theory and practice, are really badly needed in higher vocational colleges. But since they are mostly from a trade or enterprise at the basic level, the currently implemented personnel system prevents them from obtaining some authorized positions in higher vocational colleges. In addition to idea differences and unreasonable specialities disposition, such soft underbellies in softwares are also increasingly highlighted as insufficient number of teachers, irrational team structure, impropriate teacher-student ratio, and low proportion of dual-qualified teachers to full-timer teachers.

2.4 The Personnel Training of Higher Vocational Education Lacks Variety of Characteristics

Quite different from regular higher education whose aim is to turn out academic and research oriented students, higher vocational education should be designed to train "skill and application type" talents according to the requirement of social economic development. But at present, many higher vocational colleges seldom carry on in-depth and prospective study on social demand for talents. Their training objective is obscure; their training scheme can not embody the characteristics of higher vocational education mode; In addition, curricula design is not focused on cultivating interdisciplinary talents with applicable abilities; curricula system which combines teaching of theory, practice and occupation capability has not yet been really set up; some courses are somewhat overlapping; Furthermore, cooperation of most higher vocational colleges with enterprise combining learning with research and production just rests on the superficial experiment and training level, to establish a more profound cooperative mechanism in which both vocational colleges and enterprise can coexist, get mutual reciprocity and make mutual benefits is still on the way.

2.5 Graduates from Vocational Colleges Are Comparatively Not Very Competitive as Expected in Seeking Employment

As higher vocational education is designed to train “skill and application type” talents, some higher vocational colleges attach too much importance to cultivating their students’ practical abilities. So teaching of specialized and practical training courses is overemphasized while many other basic courses take up only a small proportion. Overall development of vocational students is hindered, their knowledge horizon is generally partial and narrow. When our society calls for all-round developing and high-quality talents, compared with graduates from regular institutions of higher learning, graduates from higher vocational colleges are usually not competitive as expected, even if they can get a vacancy in the job market, they still have difficulty to have a wide space to develop themselves, and the rate is relatively slow too. With the development of times, we are now in an information era. “specialized personnel” is a little out-of-date and difficult to meet the need of the knowledge economy society. Therefore, higher vocational education should not only attach importance to the cultivation of students’ basic skills, but also give prominence to students’ overall development, so that we help them acquire abundant knowledge, be more competent and have higher quality, enhance their ability and broaden the dimension to make further progress.

3 Countermeasures against Problems of China’s Higher Vocational Education

As higher vocational education is an important component of higher education, we should observe its characteristics and spare no efforts to promote its further development revolving around the respects of running idea, running patterns, specialty set-up, curriculum system, teaching staff as well as cultivation objective, etc. Meanwhile, we must insist on scientific development concept, improve the quality of education and constantly strengthen the managerial benefit.

3.1 Abandon Social Prejudice against Higher Vocational Education and Try to Improve Its Appeal

It is suggested that higher vocational education should reform its teaching and enrollment mode. The related department should lay equal stress on academic and vocational qualification credentials so that students can get two certificates at the same time upon graduation. School authorities should also strive to push forward a bridging program between professional courses and vocational training in order to establish a perfect curriculum system for higher vocational education. Graduates from vocational schools are encouraged to take up in-service training courses so that a direct entrance system can be improved and channels for their continual study can be expanded. Meanwhile, the state should improve the social status and treatment of the craftsman, give more praises and awards to those skillful talents who have made outstanding contributions, create a good social atmosphere that every trade has its master, rectify people's false concept about higher vocational education, encourage more young people to study in higher vocational colleges, make good command of professional knowledge and skill, and develop into talented contributors of the country and society.

3.2 Highlight the Characteristics of Higher Vocational Education in Personnel Training

As its name implies, the aim of higher vocational and technical education is to cultivate interdisciplinary talents with applicable abilities. So, higher vocational education must stick to the following tenets: quality and service being highly evaluated; teaching being employment-oriented; promoting teaching reform and carrying on the talent cultivation mode based on combination of theory with practice, school-enterprise cooperation and post practice. In the teaching process, it is unnecessary to pursue the integrality and systematization of theoretical knowledge and topmost superiority of professional knowledge. The basic principle of instruction in higher vocational colleges is "fundamentality and adequacy". To construct students' rational knowledge, ability and quality structure is more indispensable. The ultimate aim of higher vocational education is to train high level technicians armed with both necessary theoretic knowledge and strong practical skills. We should have a reasonable collocation of different subjects so as to enable specialized courses, practical training and basic courses to be coordinated, interacted and supplemented each other very well and form an organic whole. In addition, we should try to help students to possess the pioneering consciousness, spirit and ability, foster their occupational morality, conscientiousness and cooperativeness, and promote their quality of ideology, ethnics, humanity and psychology.

3.3 Strengthen the Construction of Teaching Staff and Optimize Their Overall Qualifications

The level of teaching staff is an important demonstration of a school's development level. The construction of "dual-competency" teachers is of vital importance to the development of higher vocational education. Teachers in higher vocational colleges should not only have profound theoretical knowledge, but also should have strong practical ability to guide students in training practice and help them solve problems encountered in production practices. In the modern society when technology is constantly progressing and industry is frequently reorganized, teachers of higher vocational colleges should study constantly and optimize their knowledge structure. If necessary, they should go deep into the basic level to expand their vision and improve their practical ability. Higher vocational colleges should set down preferential policies and take effective measures to encourage teachers to receive an in-service training or pursue an advanced study and develop into "dual-competency" teachers. Teaching is learning, through closely cooperating with enterprises or companies or setting up training practice base, teachers can get access to the training base to improve their skills, so that they can "master many skills while specializing in one". Teachers are also encouraged to take up vocational skills qualification exams so that a big proportion of them can get corresponding credentials. We should bring up or bring in teachers with high-quality, high academic qualifications and rich experiences, and provide them with good working and living conditions. Some key engineers and technicians in enterprises and institutions can also be hired with high salaries as part-time teachers. If the related authorities can build and perfect a managing system to enable skillful experts in enterprises and factories to work part-time in vocational colleges, it might be better to share outstanding teachers, strengthen teachers' teamwork building and optimize the overall quality of teaching staff. In addition, we should also put emphasis to information technology since it has revolutionary influence on educational development. Higher vocational education should intensify the application of information technology so as to improve teachers' level in grasping information technology, upgrade their teaching conception, and finally improve the teaching method effect. Only by working along many lines, can teachers' academic credentials and teaching abilities be constantly improved; a team of "dual-competency" teachers who are good at both theoretical knowledge and practical ability can be formed.

3.4 Set Up an Organic Integration of Production, Education and Research and Improve Graduates' Employment Competitiveness

As an era full of opportunities and challenges in the 21st century, the whole world is becoming a global village. The flow of capital and labor service is accelerating. New technology, especially the information and communication technology, is developing vigorously. New technology and tertiary industry characteristic of intensified content of knowledge becomes a leading industry. As a matter of fact, the competition in era of knowledge economy is competition of knowledge and talents, and in the final analysis, the competition of education. At present, China's economic construction has run into a serious challenge. Being short of hi-tech talents, a good deal of advanced technology can't be easily turned into productivity; and the advanced technology introduced can't be well digested, absorbed and further developed. Under such background, higher vocational education, by setting up an organic integration of production, education, training and research, is likely to improve graduates' employment competitiveness, cope with changes of job market more flexibly, and turn out various kinds of hi-tech talents to meet the needs of China's economic construction. From successful experiences of higher vocational education in foreign countries, it is essential to reinforce practice teaching, because it directly ensures the teaching quality. As is pointed out in the document of Education Ministry entitled On Strengthening Talents Training Work in Higher Vocational and Academic Education: change the present situation of clinging excessively to theory teaching by exploring and setting up a relatively independent practice teaching system. Practice teaching should take greater proportions in teaching schedule, and absorb the latest social, scientific and technological development. Teaching content of experiment course should be reformed by reducing the demonstrating or validating phases, and giving prominence to polytechnic experiment as well as procedures of crafts and designing, so that a practice teaching system, with an organic integration of practical abilities and manipulative skills, technology application and professional skills, general practical abilities and comprehensive skills, can be gradually established. "The construction of laboratories and practice training base should be strengthened. In accordance with characteristics of advanced vocational and academic education, we should keep on renewing teaching instruments and facilities, improving their modern scientific and technological contents, and establishing more multi-functional laboratories in which teaching, scientific research and production are well integrated. Meanwhile, steady-going practicing or training bases should be built both in and outside higher vocational colleges". Thus, through school-enterprise cooperation, higher vocational students can not merely utilize resources and obtain proficient practical skills in enterprises, but also have a sharp sense of working requirements for their future jobs, so that they can make for their future careers good preparations in such aspects as basic knowledge, professional skills and mental quality, and meet what is required by the society and enterprises for talents of high quality and excellent skills. At the same time, this would be also very beneficial for higher vocational colleges to improve their graduates' employment rate and competitiveness in job-seeking, promote their quality of teaching to a much higher level, and gain better social recognition for themselves.

4 Conclusion

Higher vocational education composes an important part of our country's higher education, and takes a particularly important position in the strategy of making our country powerful and prosperous by the way of booming the science and education to cultivate more talents. It has broad development prospects, as is pointed out in National Plan for Medium and Long-term Educational Reform and Development (2010-2020), China will "spare no efforts to develop vocational education. Developing vocational education is an important way to promote development of economy and employment, improve people's livelihood and tackle the three farmer-related problems; It is also a key link of alleviating the contradiction of workforce's structure between demand-and-supply, so it must be put in a very prominent place. The main development goal of China's manpower resources is, by 2015, the number of people with schooling of higher education will reach 145 million, that is to say, the proportion of main workforce's population receiving higher education should be up to 15%, and proportion of the newly added laborers receiving senior-school education should be 87%; By 2020, the number of people with schooling of higher education will reach 195 million, with the proportion up to 20%, and proportion of the newly added laborers receiving senior-school education should be 90%. Such a prospect shows us a clear direction and clarifies the tasks for further promoting the development of higher vocational education.

Development of higher vocational education should be in accordance with the development of national economy and social development, and the state must give policy support and guarantee from macroscopic perspective. It is advisable to bring higher vocational education within socio-economic development and industrial development plan, establish and perfect an investing mechanism through many channels, increase investment in vocational education, impel the scale and specialty setup of higher vocational education to be in conformity with the demand of socio-economic development. At the same time, higher vocational colleges should adapt to educational reform and needs of the times, and strengthen connotation from such respects as reforming educational structure and teaching pattern, strengthening the quality of teaching staff, highlighting the typical features of this kind of education. Only in this way, can we set up in our country a new system of higher vocational education with an open and high-standard operational mechanism, performing together and communicating with formal regular higher education and linking up with the secondary vocational one; embodied with the characteristics of multiple educational bodies, different academic levels, complete educational functions, job-market oriented and fundamentally application-based.

References

1. National Plan for Medium and Long-term Educational Reform and Development (2010-2020). [DB/OL], `http://www.gov.cn`, `http://www.gov.cn/jrzg/2010-07/29/content_1667143.htm`
2. Yang, J., Cao, H., Zhang, Y.: Probe of Working and Learning Alternation Teaching Mode in Higher Vocational Colleges. Journal of Zhangjiakou Technological Institute (1), 17–19 (2011)
3. Lei, Z.: Developing Vocational Education: An Important Action of Implementing the Strategy of Revitalizing the Nation through Science and Education. Journal of Vocational and Technical Education (4), 5–7 (2007)
4. Bai, X., Chen, S.: Analysis of the Necessity and Means of University Students' Vocational Education. Study of Pursuing Further Education (5), 35–36 (2007)
5. Song, J.: Study of Institutional Construction Based on Connotation Development in Higher Vocational Colleges. Journal of Jiyuan Technological Institute (1), 1–3 (2011)

Research on Cost Accounting Informatization

Ping Cheng* and Xiongyu Zhou

School of Accounting, Chongqing University of Technology,
Chongqing, China
Chgpg2006@163.com

Abstract. Economic globalization makes cost accounting faced with challenge, but informatization gives cost accounting a chance to develop. Accounting informatization is a method which cost accounting use to answer the challenge. And cost accounting give accounting informatization a chance to play a role. This paper introduces the cost accounting and accounting informatization, and then analyses the relation between cost accounting and accounting informatization. At last it analyzes the influence which the accounting informatization gives to cost accounting.

Keywords: informatization, cost accounting, influence.

1 Introduction

There are two big trends which will not be blocked in the 21st century. They are informatization and economic globalization. The two trends will develop more and more quickly. Accounting informatization will be an inexorable trend which will not be blocked under the information. Accounting informatization has born great influence to accounting. It makes great difference in improving the enterprise management efficiency and management level. For one thing," Economic globalization swept across the world. Enterprise is facing the global market and the competition from many countries of similar enterprises. Market expands and opportunities increases. At the same time the market risk increase and the market competition becomes incentive. It brings serious challenges to Chinese enterprises' development and survival". The cost advantage for an enterprise survival is vital, and cost accounting is more and more paid attention to [1]. On the other hand, new economic environment has also brought a challenge for cost accounting. Under the informatization environment and economic globalization environment enterprise must learn to use accounting informatization technology means to deal with the challenge which the new economic environment brings to achieve cost advantage. Accounting informatization is important for cost accounting to deal with new challenges and achieve cost advantage. At first, accounting informatization and cost accounting are introduced in this article. Then it analyzes challenges which cost accounting in the new economic environment is faced with and the relationship between accounting informatization and cost accounting under the challenges. At last it analyses the influence which accounting informatization give to cost accounting.

* Corresponding author.

H. Tan (Ed.): Knowledge Discovery and Data Mining, AISC 135, pp. 369–374.
springerlink.com

2 Accounting Informatization

Account informatization is that enterprises take accounting information as manage information resources and Comprehensively use information technology of computer and network communication to get accounting information, process accounting information, transform accounting information and use accounting information. By this way, it provide enough, It provide enough, real time and comprehensive information for the enterprise management, control decision and economic operation. The development of accounting informatization is on condition of integration of economy and society, the digitization and network development and based on accounting computerization. It has the function of financial accounting and function of controlling and management. Meanwhile it has the characteristics of openness, diversity and Intelligence. Achieving accounting informatization let business operators and users of the information use enterprise accounting information for the future of the company financial situation to make reasonable forecast at any time and make right decision for management and development. It can directly read data from other systems. It also deals with, stores and transfers data. Meanwhile, with enterprise nets and external network realizing interrelation, the user of accounting information can get related accounting information at any time. With the comprehensive using of information technology, it greatly improved the timeliness of information and the predictive value information and feedback value of information. Then it will improve the level of management.

3 Cost Accounting

3.1 The Meaning of Cost Accounting

Cost accounting is the accounting that calculates the total and the unit cost of the product to account all production costs. The core content of the cost accounting is calculating the costs. Cost accounting is cost accountants assist enterprises to manage and control the management of the company, formulate long-term or strategic decisions, and formulate a favorable cost control method to reduce costs and improve quality. Cost is the product of the commodity economy. It is one of the commodity economy economic categories. And it is the main component of commodity value. Cost content often subject to administration's needs. In addition, due to the content of the economic activities are different, the meaning of the cost accounting is different. With the development of social economy and the improvement of requirements of management, cost concept and connotation are in constant development and change. We can feel the cost expand gradually. To sum up, the cost of economic content has two things which are common: one is the formation of the cost is at some goals for the object [2]. The other is cost is one kind of cost which is to achieve certain goal.

3.2 The Emergence and Development of Cost Accounting

Cost accounting experienced four stages: early cost accounting, modern cost accounting, contemporary cost accounting and strategic cost accounting. The way and theory system of cost accounting are different in different stages. In early cost accounting stage, Cost accounting began to produce. It mainly used process method or partial method. Cost accounting system was mainly about product cost calculation to determine cost inventory and cost of sales. In modern costing accounting stage, cost accounting entered a new stage of development. It mainly used the standard cost system and cost forecast to control the cost. In contemporary cost accounting stage, cost accounting was in a new stage. The focus of development of cost accounting became how to forecast the decision and plan cost instead of controlling, calculating and analyzing. In the strategic cost accounting stage, cost management concerned more about customer perceived value. Meanwhile, it concerns more about internal organization management. It is as far as possible to eliminate the internal cost of increasing customer value to achieve the advantage of the market [3].

3.3 The Function of Cost Accounting

The function of cost accounting is the function which the cost accounting has in management. As an important branch of accounting, the basic function of cost accounting is the same as accounting. It has the two basic function of reflect and oversight. And the reflection is the key function [4].

3.4 The Challenge of Cost Accounting

Under the situation of economic globalization being speeding up, new economic environment has appeared. And new economic environment bring big challenges to both enterprise and cost accounting." New economic environment bring the following challenges [5]:

(1) World competition become more and more fiercely.
(2) Industrial structure changes.
(3)The environment of production develops.
(4) Just-in-time production system appears.
(5)Total quality management has new requirements.
(6) Activity-based cost method is used.

4 The Relationship between Cost Accounting and Accounting Informatization

Accounting informatization is an effective method to resolve cost accounting's challenge. And cost accounting provides a platform for accounting informatization playing a role. Accounting informatization can help cost accounting to resolve challenges and develop. In general, new economic environment bring the following requirement: (1) It requires cost accounting to subdivide its function. (2) It requires cost accounting to be more accurate. (3) It requires cost accounting to improve its efficiency. (4) It requires cost accounting to change its old cost calculating method. At first, accounting informatization improves the efficiency of cost accounting. With this, enterprises will have more energy to find out the more accurate cost accounting methods and subdivides function based on requirement. With these problems being resolved, cost accounting method will be improved. Previously, accounting informatization was only involved with financial statements. The using in cost accounting will promote the using of accounting informatization in new field. And it will provide new space for the development of accounting informatization, rich connotation of accounting informatization and provide a platform for accounting informatization playing a role.

5 Accounting Informatization on the Impact of Cost Accounting

Accounting information on the impact of cost accounting mainly through the collation and processing of cost accounting information to help companies to reduce costs, according to the purchase, production to analysis of how accounting information to assist enterprises to improve management and efficiency to reduce costs: .

5.1 The Management of Procurement Cost

It is essential important to control the cost of purchasing for a company's achievement The reduction of procurement cost not only reflected on the reduce of business cash outflow, but also directly reflected on the decline in production costs, and increase profits, and enhancing the competitiveness of enterprises. the proportion of the costs of material in production costs are often more than 50% , therefore, control the cost of procurement and declining it is one of the most important means for a enterprise to reduce the cost and increase the profit . "There are two ways to reduce procurement costs: Firstly, to improve the efficiency of procurement staff to reduce labor costs; secondly, to research quality suppliers, and purchase excellent quality and reasonable price goods. Now, we will analysis how accounting information to help companies to reduce procurement costs:

(1) Accounting information to assist the establishment of vendor files.

The implementation of information technology is easily to organize the files of vendors , It can bring it up timely when need the information of vendors, , have improved the efficiency of procurement staff, in the mean times, comparing of suppliers to provide fast access to excellent quality and reasonable price products high-quality suppliers.

(2) Accounting information to help establish the price file.

The implementation of accounting information to facilitate business purchasing department to establish prices file for all material of purchase. In order to gain the material more quickly which has the superiority price, and reduce the purchasing cost?

(3) Accounting information affords ways to gain the information of manufacturer for enterprise, which help enterprises deal with manufacturers directly, and reducing the middle process and costs.

5.2 Manufacturing Cost Management

Accounting informatization reduce the enterprise's costs through two aspect on the produce:

Accounting informatization helps enterprises to establish a "JIT" and "TQC (total quality control)." Traditional cost reduction is achieved by cost savings, striving for not to waste resources in the job site and improving the working methods in order to save costs, which is a primary form to lower costs, but this cost reduction is a temporary solution, and real-time production systems through "zero inventory" form to avoid almost all of the inventory cost, totally control to “zero defects” form to avoid almost all of the maintenance costs and product failure caused by other costs. in-time production system require manufacturers to market requirements based on customer orders or the number, variety, quality standards and delivery times for production, procurement arrangements. the former production process must according to the latter production process’s required information about products, semi-finished products or components The number, size, quality and timing of demand for production [7] "," accounting informatization enhanced communication between the enterprise and customer, help enterprises to establish timely control system; total quality control is divided into four stages: plan , Do, Check and Action. Eight steps: find the problem, identify the factors, clearly the important factors, afford improvements, and implement measures to check the implementation of measures, standardize the good measures, and deal with legacy issues. Fourteen kinds of tools: the implementation of the plan and inspection phase, in order to analyze and solve problems, use of fourteen kinds of tools (methods): stratification, Pareto Law, cause and effect analysis, histogram method, control chart, correlation diagram method, check the graph, the diagram method, KJ method, system diagram method, matrix method, matrix data analysis, PDPC law and vector chart method. The first seven species in which traditional methods, the late seven species produced, called the new seven kinds of tools [8]. "Accounting informatization to help enterprises to establish a comprehensive quality control system in the following three areas: strengthening the communication of information through a comprehensive control system, four stages of contact, information gathering and processing through eight steps to improve the efficiency of a computer that tool for the comprehensive quality control systems provide an effective tool.

Accounting informatization is effective means of accurate product costing. Accurate accounting information in its accounting methods and the proper cost accounting methods to ensure the accuracy of corporate accounting, cost control on the right product is important to reduce costs. Because only to master the right product to calculate the cost of the profits of each product, so that enterprises can use limited resources to focus on manufacturing products with high level of profitability, which for the company's manufacturing strategy to provide direction.

6 Summary

Under the global economic environment, there is great challenge in cost accounting. But accounting informatization has provided effective means for the challenge. It have a great influence in cost accounting, which offers opportunity for the development of cost accounting, supplies a new stage for the further development of it as well. Accounting informatization must be made the best of it by enterprises to promote the development of cost accounting.

References

1. China Institute of International Economic Relations. The trend of globalization and China's Countermeasures, vol. 15. Shishi Press, Beijing (2007)
2. Luo, F.: Cost accounting, p. 25. Higher Education Press, Beijing (2009)
3. Xing, Q.: The new environment, the development of cost accounting analysis. Henan Business College 6, 15 (2005)
4. Yu, F., Li, L.: Cost accounting, p. 6. China Renmin University Press, Beijing (2009)
5. Tang, G.: The new economic environment, the development trend of China's cost-accounting major. Technology Consulting Herald 12, 14 (2007)
6. Li, M.: Control of the purchase cost. Management & Technology 9, 17 (2009)
7. Guo, X.: On-time production system, the impact of modern accounting. Finance and Accounting 3, 8 (2003)
8. Zhang, S.: On the importance of total quality control. Accounting Research 7, 15 (2007)
9. Wang, L.: Analysis of cost accounting. Guangdong Business College 4, 9 (2008)
10. Wang, L.: On our corporate cost accounting trends and countermeasures. Hunan Vocational College of Environmental Biotechnology 2, 39 (2008)

Analysis of Factors Affecting Audit Charge in China's Listed Companies

Juan Wang[1] and Jingya Zhao[2]

[1] Xi'an University of Architecture and Technology, School of Management, Xi'an
[2] Xi'an University of Architecture and Technology
juan_851030@163.com

Abstract. This paper uses univariaty frequency method to analyze the effecting factors to audit fees, according to the sample unit of SSE 380 Index Data. The results suggest that the audit fees are positively correlated with the corporate scale, the debt-to-asset ratio, and the scale of the accounting firm. Whereas negative correlation between the audit fees and current ratio can be concluded. Finally advices are presented according to the research results.

Keywords: Audit fees, effecting factors, SSE 380.

1 Introduction

As China's capital market develops, audit plays a pivotal role in ensuring its rational allocation of resources and maintaining the sound development of social economy. As an important part of audit procedures, audit charge indirectly affects the development of audit industry and improves the information transparency of listed companies. However, due to the imperfections of China's capital market and the fierce competition in audit market as well as other reasons, China's current audit charge still exists many unreasonable. Therefore, the analysis for factors of audit charge is extremely important in ruling the audit market.

2 Analysis of Factors Affecting Audit Fees

According to the degree of impact on audit charge, the affecting factors will be divided into the basic factors and other factors. Of which the basic factors affecting audit charge include: the audited unit's scale and complexity and audit risk.

2.1 The Audited Unit's Scale

The larger the unit being audited, the more the brokerage business and accounting issues, the CPA's work will be more difficult, the audit costs will increase, thus the audit charge will naturally increase.

H. Tan (Ed.): Knowledge Discovery and Data Mining, AISC 135, pp. 375–382.
springerlink.com

2.2 The Audited Unit's Complexity

When the audited unit is more complex, the possibility of its occurrence of affiliated transaction will increase. In previous studies, the accounts receivable and inventory to total assets ratio are often used as a measure indicator of the complexity of audited business, as the accounts receivable and inventory are the focus of audit, compared to other audit procedures, they needs to invest more human and material resources.

2.3 The Audit Risk

Audit risk refers to the possibility of CPA issues inappropriate audit opinion after the audit when material misstatement or omission exist in accounting statements. The current ratio is often used as an indicator to measure company's short-term liquidity, the higher the ratio, the stronger the liquidity of listed companies, the smaller the audit risk, the less the charge.

2.4 The Scale of Accounting Firm

The scale of accounting firm usually represents its audit quality and reputation, large scale accounting firms represent high quality audit services , their good reputation and independence can also ensure all company owners and potential investors obtain more useful accounting information, this will drives customers willing to pay higher audit charge.

In addition, the company's ownership structure, the impact of non-audit services, listed companies to change the scale of accounting firms and auditing term as well as geographical factors also play a underestimated role in affecting the pricing of audit.

3 Statistical Study of Factors Affecting Audit Charge in China

This study will use statistical methods to expound and prove the four basic factors affecting audit charge.

3.1 Sample Selection

The financial data and audit charge data of listed companies used in this paper are from CCER database. Select the sample stocks of SSE 380 Index as a research object, because it covers a wide range of industries and represents a strategic direction of national economic development and direction of economic restructuring. In view of this database has not yet disclosed the relevant data of listed companies in 2010, so the 2009 relevant financial data of SSE 380 Index sample stocks is selected as a research object. In addition, given the comparability of data and research needs, screen in accordance with the following four methods , 341 groups of samples are obtained : (1) Weed out the listed companies appeared as "-95" in the term of "audit charge";(2) Exclude the listed companies that have not yet disclosed audit charge; (3) For other indicators required in this paper, if not available, also need to be eliminated.

3.2 Assumptions and the Choice of Variables

This paper selected the total assets at the end of period to measure the scale of listed companies, and speculated that audit charge would increase with the expansion of the scale of audited units.

According to previous studies to select the accounts receivable and inventory to total assets ratio as indicator to measure the complexity of the business being audited, and assuming that with the increasing of accounts receivable and inventory proportion, the audit charge will increase.

Based on experience, this paper chose the two financial indicators of current ratio and asset-liability ratio to measure the audit risk. It can be inferred that the audit charge decreases with the increasing of current ratio and increases with the increasing of asset-liability ratio.

This paper considered the total annual income as a indicator to study the scale of accounting firm. According to Association of Certified Public Accountants released in 2009, China's "Top Ten" accounting firms are: PWC Zhongtian, Deloitte Touche Tohmatsu, KPMG, Ernst & Young, RSM, Ericsson, SHINEWING, Tianjian, Crowe Horwath, Daxin Certified Public Accountants. Assume that: the audit charge will increase with the expansion of the scale of accounting firms.

3.3 Statistical Analysis

The Descriptive Statistical Analysis of Audit Charge

In the 341 groups of samples, the maximum of audit charge is 10.726 million yuan, the minimum is 150,000 yuan, divide them into six groups (each group includes up limit but not low limit), obtained by the results in Table 1, Figure 1. It can be seen that China's listed company's audit charge basically presents as normal distribution in statistics, the whole sample's average value of audit charge is 732,300 yuan, and 77% of the listed company's audit charges are concentrated in the 200,000-800,000 million range, there is only a small number of extreme values.

Table 1. The group statistics table of audit charge

audit charge (ten thousand Yuan)	sample number	proportion(%)
0 - 20	3	0.88%
20 - 40	79	23.17%
40 - 60	114	33.43%
60 - 80	70	20.53%
80 - 100	34	9.97%
≥100	41	12.02%
total	341	100%

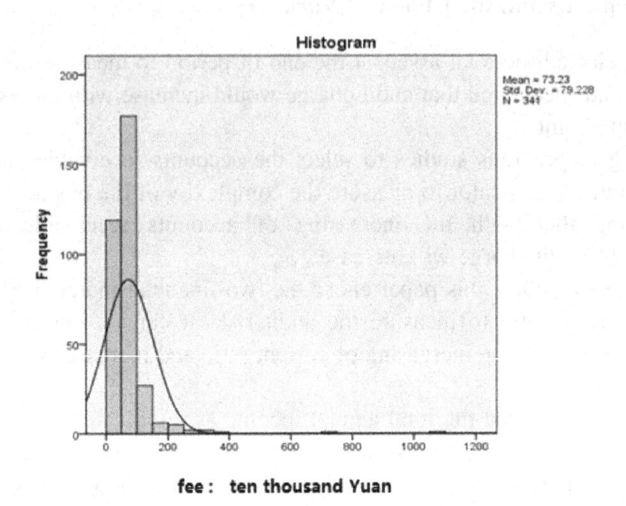

Fig. 1. The group statistics column of audit charge

We also screen the audit charge of 2005-2008 in accordance with the above method and then compare them with the audit charge of 2009, the descriptive statistical analysis table of 2005-2009 audit fee is as follows:

Table 2. The descriptive statistics of audit fees in 2005-2009 annual financial report

unit: ten thousand Yuan

year	sample number	minimum value	maximum value	mean value	median
2005	313	10	615	58.60	40
2006	286	10	1186	68.37	45
2007	286	15	1186	72.24	50
2008	306	15	693	67.63	55
2009	341	15	1073	73.23	58

It can be seen from Table 2 that the 2005-2007 average audit charge have been showing a gradual upward trend, affected by the 2008 global financial crisis, the payment of audit charge of corporate customers declined slightly, but in 2009 it began to improve.

2, The Descriptive Statistics Analysis of Factors affecting Audit Charge

(1) Company's Assets Scale and Audit Fees

According to the scale of year-end total assets of listed companies, the listed company is divided into five groups, the descriptive statistics for company's assets scale and audit charge is as follows:

Table 3. The descriptive statistics table for company's assets scale and audit fees

unit: ten thousand Yuan

assets scale	≤15	15~30	30 ~50	50 ~100	>100
sample number	63	120	68	52	38
average charge	57.94	53.79	77.02	80.56	164.14
median charge	51.00	53.00	65.00	75.00	99.00
minimum charge	20	15	20	30	32
maximum charge	90	178	300	350	1073

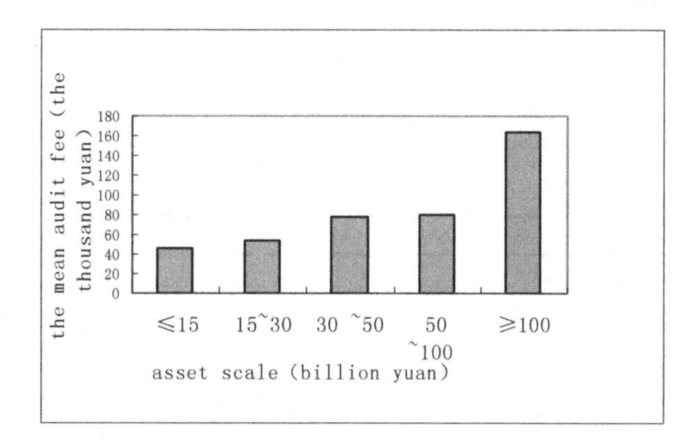

Fig. 2. The relation diagram for assets scale and audit charge of unit being audited

As is shown in Table 3 and Figure 2, the mean value of each group's audit charge gradually tends upward with the expansion of audited unit's assets scale and the trend is very clear, that is the audit charge increases with the expansion of audited unit's scale. The result is consist with those of China's accounting firms actually take the customer's asset as a basis, which is based on the audit charge standard drawn up by financial department in conjunction with price department.

(2) The Accounts Receivable to Total Assets Ratio and Audit Charge

According to the accounts receivable to total assets ratio, the sample is divided into three groups, the descriptive statistics of average charge and median charge in each group is shown in Table 4:

Table 4. The descriptive statistics for accounts receivable to total assets ratio and audit charge

accounts receivable to total assets ratio	≤0.05 below	0.05~0.1	>0.1
sample number	155	67	119
average charge	77.71	70.25	69.06
median charge	55.00	60.00	60.00
minimum charge	15	21	20
maximum charge	1073	350	300

It can be drawn from Table 3 that although the average audit charge gradually decreases with the increasing of total accounts receivable to total assets ratio, the trend is not obvious; the median charge presents a upward trend in the ratio of below 0.05 and 0.05 to 0.1, but the median charge in the ratio of 0.05 to 0.1 and above 0.1 are the same, which is not the same as assumed previously that the audit charge increases with the ratio. According to the results can be speculated that accounts receivable to total assets ratio is of no obvious relationship with audit charge. This showed that the CPA didn't consider the proportion of accounts receivable as the main factor in determining the audit charge or they may be affected by other factors and ignore the proportion of accounts receivable will impact the audit charge.

(3) The Inventory to Total Assets Ratio and Audit Charge

In accordance with the inventory to total assets ratio, the entire sample is divided into three groups, statistics for charge in each group is as shown below:

Table 5. The descriptive statistics table for inventory to total assets ratio in manufacturing industry and audit charge

inventory to total assets ratio	≤0.1	0.1~0.2	>0.2
sample number	119	110	112
average charge	79.29	78.1	61.99
median charge	60	58.5	55
minimum charge	20	15	20
maximum charge	1073	728	280

As can be seen from the above table, each group's average and median audit charge have no obvious relationship with the inventory to total assets ratio. But it may be due to larger differences of inventory ratio in various sectors, so we further select the manufacturing industry with larger inventory proportion to study the impact of inventory to total assets ratio on audit charge, but after analysis we still can't get the conclusion that the factor has a significant relationship with audit charge.

The reason for the conclusion may be due to accounting firms adopt with "supervision without counting" instead of supervision of counting in strict sense for listed companies when auditing ,and insufficient attention to the proportion of inventory in total assets when measuring the complexity of client's audit.

(4) The Current Ratio and Audit Charge

The sample's current rate is divided into three groups, the average and median charges of each group are as shown below:

Table 6. The descriptive statistics stable of current rate and audit charge

current ratio	≤1.0	1.0~2.0	>2.0
sample number	97	182	62
average charge	83.10	73.58	56.73
median charge	65.00	55.00	46.00
minimum charge	20	15	25
maximum charge	1073	728	200

From the data in the table can be seen as the current rate increases, the mean and median charges between each group all have emerged a trend of declining; Compare each group, obtained by the result that the median charge is lower than average charge. It can be concluded that the audit charge decreases with the increasing of current ratio, which is consistent with the expected hypothesis.

(5) Asset-Liability Ratio and Audit Charge

According to the asset-liability ratio, the sample is divided into three groups, after statistical analysis, the mean and median charges of each group are as shown in Table 7:

Table 7. The relationship between asset-liability ratio and the audit charge

asset-liability ratio	≤0.3	0.3~0.5	>0.5
sample number	52	112	179
average charge	60.41	66.00	81.47
median charge	50.50	53.00	60.00
minimum charge	21	15	20
maximum charge	200	728	1073

Found by observing Table 7, the mean and median audit charge of each group all present a tread of upward with the increasing of asset-liability ratio. Asset-liability ratio determines the business risk of company, the certified public accountants determine the audit risk of listed companies according to their business risk, thus to increase the audit charge for risk compensation. Therefore, the greater the asset-liability ratio, the higher the audit charge, which is consistent with the hypothesis in this paper.

(6) The Scale of Accounting Firms and Audit Charge

Divide the sample into "Top Ten" and "non-top ten" according to their own audit accounting firms. Choose ten listed companies with smaller total asset scale from the "Top Ten", and accordingly select ten companies with similar asset size from the "non-top ten", compare their audit charges and the results are as shown in Table 8:

Table 8. The relationship between the scale of accounting firms and audit charge

"non-top ten" accounting firms		"top ten" accounting firms	
total asset(billion yuan)	audit charge(ten thousand Yuan)	Total asset(billion yuan)	udit charge(ten thousand Yuan)
85.07	63	84.27	110
98.70	80	95.00	90
104.96	55	103.43	60
107.66	40	107.83	180
126.38	80	125.99	90
150.59	90	149.67	100
161.65	76	159.68	289
185.79	60	178.41	220
219.53	91.2	222.09	118
241.97	60	232.60	228

The above results fully prove that in the case of assets of similar size, since with a good reputation, technical level and higher risk control, if the scale of accounting firms is larger, they charge higher audit fees compared to other small and medium sized firms, this is also consistent with this hypothesis.

4 Research Conclusions and Recommendations

After the study, this paper concluded the following conclusions: the audit charge increases with the increasing of audited entity's size, asset-liability ratio and the scale of accounting firms; audit charge decreases with the increasing of inventory to total assets ratio and current ratio; and accounts receivable to total assets ratio has no significant relationship with audit charge.

In connection with the earlier study, we proposed the following countermeasures:

Although the China Securities Regulatory Commission issued the "Information Disclosure Q & A - No. 6," in late 2001, which calls for the listed companies will consider the compensation paid to accounting firms as an important matter to disclose in the annual report, in the process of removing the samples, we found the actual disclosure results are less than ideal. Therefore, improving the disclosure system of audit charge is imminent, we should further improve the existing regulations and strengthen the improvement, integrity, and mandatory of information disclosure, and make full use of a variety of information and communication carriers, combining with a variety of channels for information disclosure.

Accounting firms should combine with the audit risk of unit being audited and audit complexity to charge when auditing. The analysis of accounts receivable and inventory to total assets ratio showed that in the implementation of audit procedures for accounts receivable and inventory, the CPA audit should execute the "circularization" and “supervision of counting” procedures in traditional sense, and the two should be regarded as ignorable factors in audit charge.

When identifying fraud of listed companies, the relevant regulatory bodies, intermediaries can prove according to the relationship between company's annual audit charge and related factors.

References

1. Yang, H., Li, J.: The Empirical Research Review of China’s Factors Affecting Audit Charge. Accounting Communications (Academic Edition) 06, 65–67 (2008)
2. Li, Z.: Research on Factors Affecting Audit Charge - Based on China’s Experience Data in Securities A-Share Market [Master Thesis]. Fuzhou University, Fujian (2004)
3. Zhang J.: The Empirical Research on China’s Factors Affecting Audit Charge (Master Thesis). Southwest University, Chongqing (2009)
4. Titman, S., Trueman, B.: Information quality and the value of new issues. Journal of Accounting and Economics 8, 159–172 (1986)
5. Simunic, D.A.: The pricing of audit services: theory and evidence. Journal of Accounting Research 18, 161–190 (1980)
6. Qiu, C.: Research on Factors Affecting Audit Charge in China’s Listed Companies (Master Thesis). Central South University, Hunan (2008)
7. DeAngelo, L.: Auditor size and auditor quality. Journal of Accounting and Economics 12, 183–199 (1980)

The Research of Workflow Management System in Teaching Procedures

Qiongjie Kou and Quanyou Zhang

Xuchang University, Xuchang, Henan Province, China
kouqiongjie@163.com, sealaplace@yahoo.com.cn

Abstract. This paper analysis and design a management system for "poor understanding students" and analysis system module function based on workflows which fully consider education object diversity. It's easy to understand knowledge and technology for poor understanding students using that teaching systems.

Keywords: Workflow, Management System, Teaching.

1 Background

Informational teaching is becoming more and more common in recent years. In the practice of teaching we can know "poor understanding students" have a weak foundation, low level and slow of understanding etc. If I can design a system according to the characteristics, so that students will learn more knowledge and understand more clearly. But some companies at home and abroad have only developed the remote education systems, such as Dell. It had released the new Interconnected Class Solution on May 14, 2010. Join the interactive whiteboard and multi-purpose short focal projector together teachers can be in any place for interactive teaching which is supported by 3D projection. Another company Beijing Chengwei New Century Network Communication Co. his "Cenwave" Network Real-time Classroom and Counselling the Question's Answer System is a successful example. Through broadcasting the real-time video of teacher's sound, teacher's writing and teacher's lecture notes, the lifelike course is moved to the online. These systems in a certain extent remedy the traditional teaching defect. However it doesn't match every education object. We can't make those so-called "poor understanding students" whose understanding ability is poor improve their knowledge greatly in advanced teaching mode. In addition, some experts and scholars use workflows to design teaching management systems [1] and E-learning learning systems etc. These systems just abide workflows 3R rule called Router Rules and Roles [2]. It doesn't match any education object. Thus, we must be paid attention to "poor understanding students". The problem is how to make full use of information technology to make them quickly understand and nimbly used the knowledge in the same conditions. This paper studies workflows in the light of the characteristics of "poor understanding students". I design a management system in teaching procedures and analysis system module function. This system can help students to understand and expand student's knowledge in teaching procedures.

H. Tan (Ed.): Knowledge Discovery and Data Mining, AISC 135, pp. 383–386.
springerlink.com

2 Systems Function

There are several definitions of workflows at present such as the definition of "Workflow Management Coalition". Its definition is that business process is all or part of automation. Documents information or task is according to certain rules of procedure to circulate. Realize the coordination of organization between the members in order to achieve the integral goal of business [3]. This paper study is based on workflows principle. The systems of design is according to characteristics of "poor understanding students" having a weak foundation, low level and slow of understanding etc.

This system is mainly consisted of four parts: user interface, workflows engine, search engine and database. The user can browse different web by the application program whose activities is controlled by workflows engine. The content of data web is sourced from the question database. The key word or sentence is annotated in the application programe. put that annotation into the database of intelligent teacher which is provided data for search engines. The database of information analysis which data is sourced from the question database is used to analyses students learning factors and the knowledge of understanding.

The definition of process is deployed on the workflows database. The components of workflows engine according to the document of process definition drive the process flow. If found some activities need to invoke services, the system will send request information by the URL. Web service bus deal with affairs according to the command of manager which is the components of workflows, then, execute the next process or task. First user passes login authentications and chooses different learning styles such as homework, testing, online classroom, etc in the macro aspect. Second, students can accesses to different modules by his need. Every question and quiz appears on computer screen in workflows form. Each key words of question is treated as the search target and explained. Each key words of explanation is stored into the database of intelligent teacher. When meet new key words in exams, students can search information in database. The search results will be appeared on a separate web. According to actual condition, the content may be a speech, pictures, videos and text. If still do not understand the search result, students can continue to search new key words or sentences until understand key words or sentences, then return to the home web of question. By the workflows engine, control student can continue to do the next question, finally complete the study target and summarize the study. Besides can input text information, the key words met also select and mark key words which appear on computer screen in summary, which so that remind guided memory next time.

3 Systems Module

This system is divided into user login module database expansion module workflows module assignments module exams module and search engine module.

3.1 User Login Module

When users login into system, users will be identified to find the illegal users. There are many main technology of access control at present. Such as the dynamic access control (DAC), the mandatory access control (MAC) and the role based access control (RBAC) [4]. This system is based on Role-based access control (RBAC) by IP address. First identify users to filtrate illegal users. Then it verify role to implement access control.

3.2 Database Expansion Module

This system do a survey or questionnaire to record and store interactive data into the database of questions. This system transforms the data and store into the database of intelligent analysis to provide the data for intelligent analysis for the improvement of teaching.

3.3 Workflows Module

In the workflows module there are mainly two kinds: one kind is the special information for administrator, such as information of monitoring system operation state. Another kind is information of processing status. When receive files, we not only can see basic information and content of file, but also can see steps of before, current and next operation. Specific information will appear in next operation. If the file is filed, all circulation of document can be seen in the process tracking including the specific content of each step. If not complete documents in transfer process, the process tracking will display all step from start to terminate. The above information using is displayed by dynamic web technology.

3.4 Assignments and Exams Module

This module to workflows showcases assignments and exams, students user can only according to test process sequence completed. Students do exams and assignments which appear on computer screen on workflows form. If won't answer questions, Students can use search engine module to find information of key words until understand and answer questions. After submit paper, students make a personal summary. Summary web include all key words appeared for user to choose. Summary is convenient to be reviewed by day, week, month and year categorizing storage. Teacher can understand the students' learning progress and grasp the difficulty of exam, meanwhile summary gain and loss in the teaching process.

3.5 Search Engine Module

Base on the database of questions, it chooses various form note key words, such as text, speech, pictures and video. This information is inputted into the database of intelligent teachers, which physical storage structure is hash table. In the database of intelligent teachers, we build index and process for queries or application. In order to improve querying speed, we build relationships between the large amount of data table and little data table.

4 Summary

The paper studies workflows is according to the characteristics of students, based on workflows to design a search question information systems. We should analysis the system module function and the required information technology. This system can provide search function, help students to understand and expand student's knowledge in the process of answering questions. Otherwise the number of processes grows rapidly and complex data became more and more. How to make the searching efficient and quickly? How to make systems more perfectly? These will be required for further study.

References

1. Ni, C.: Design and Implementation of Web-Based Lab Teaching Information Management System. Research and Exploration in Laboratory (1) (2011)
2. Zhang, Z.: From Workflow to Learning Flow: A New Framework to Design and Develop E-learning System. Laboratory Science (1) (2011)
3. Cao, F.: Design and Implementation of Students Elective Courses System Based on the Workflow Technology. Modern Computer (10) (2010)
4. Deng, Q., Yao, J., Chen, C., He, W.: A Service Enterprise Automation Management System Based on Web and Workflow Technology. Experiment Science & Technology (5) (2010)
5. Liu, Y., Wu, J.: Integration of OA and ERP based on service-oriented architectures. Journal of Computer Applications (3) (2008)
6. Lin, F., Yuan, L.: Research and Development of a Teaching Management System Based on "workflow" mode. Journal of Wuhan University of Science and Engineering (1)

The Master of Engineering Program in Integrated Circuits and Systems: A Case Study in Tsinghua University

Hong Chen, Zhihua Wang, and Yangdong Deng

Institute of Microelectronics, Tsinghua University
Beijing, 100084, China
{hongchen,zhihua,dyd}@tsinghua.edu.cn

Abstract. With the quick development of Integrated Circuit (IC) industry in China, a big group of qualified engineers are needed urgently. In this paper the importance role of IC industry in China and the unbalance between the need and talent supply of the engineers is described. The Institute of Microelectronics of Tsinghua University (IMETU) has made significant efforts on education of master students majored in integrated circuits and systems to improve the quality and quantity of engineers to meet the requirement. Based on the analysis of the current system of the education of bachelor, master and doctor degrees in microelectronics engineering, the steps taken by IMETU are given. As a result, the master program in integrated circuits and systems of IMETU is attracting a smoothly increasing number of enrolled students. Meanwhile, a survey showed that the current employers gave a high rating on the quality of the graduates of this program.

Keywords: Master of engineering, Integrate circuits and systems, Institute of Microelectronics of Tsinghua University (IMETU), Student recruitment.

1 Introduction

With the quick development China has been obtaining and advancing technology in a manner and at a pace that has surprised most people. The high-tech semiconductor industry exemplifies this growth. In less than 10 years China went from being five generations behind in Integrated Circuit (IC) semiconductor processing, to being current with the state-of-the-art. With the help of Chinese engineers, Chinese companies will soon be producing the leading IC technologies for the world with the already existing infrastructure to incorporate semiconductors into advanced systems, and a large internal (and external) market [1].

The sale of IC in China increased gradually from 2002 to 2011 shown in Fig.1. According to the statistic data of the Department of Commerce of China, the total foreign trade value of the import of IC and microelectronics subassembly is about $127 billion dollars, which is 1.6 times of that of the crude oil in 2007. The IC field is playing a strategic and irreplaceable role in Chinese economy and politics. It's forecasted that the sale of IC industry in China will exceed $45 billions dollars and reach $54 billions dollars in 2012. China will be the most important base of the IC manufacturing in the world [2].

H. Tan (Ed.): Knowledge Discovery and Data Mining, AISC 135, pp. 387–396.
springerlink.com

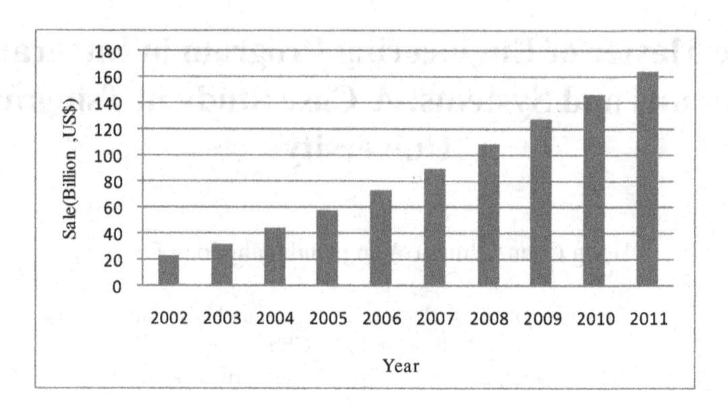

Fig. 1. Sale of IC industry from 2002 to 2011 in China (From CCID Consulting 2007 [2])

As a result, lots of qualified engineers are needed to meet the requirement of the rapid development of IC industry in China.

From the statistic data of custom, about 120,000 IC design engineers will be need in the IC field in 2011, and more than 200,000 IC engineers are lacking during the next ten years from 2010 to 2020. However, the number of the IC engineers in China now is far from meeting the requirement of the IC industry. So far only about fifteen universities (such as Tsinghua University, Peking University, Fudan University, and so on) in China have dedicated microelectronics programs, while the number of graduated students majoring in this program is only about 300 in each year [2].

Besides, IC engineers are expected to receive a higher level of training in analytical skills, practical ingenuity, creativity, communication and teamwork skills, business and management skills, high ethical standards, professionalism, leadership, including bridging public policy and technology, dynamism, agility, resilience, flexibility and should be lifelong learners [3].

It is thus challenging for universities in the world to bring up a big group of talented engineers in IC field. As the only institute entitled to award the master degree of engineering in integrated circuit and system in Tsinghua University, the Institute of Microelectronics of Tsinghua University (IMETU) has to address the challenge in a constructive manner. Accordingly, the following question has to be raised: What should engineering education be like so as to prepare the next generation of students for effective engagement in the engineering profession in 2020?

There has been a large body of work on identifying the problems of today's engineering education (e.g., [3]). Based on these results and our analysis on the undergraduate/graduate programs in IMETU, the major issues to be resolved are as follows.

1) When starting their career, graduated students do not have enough skills that should be taught in colleges. In fact, usually the employers have to pay the expense of training them.
2) The current education system is still very close what it was 50 years ago, although significant progresses have been made in our knowledge base.

3) Considering the complexity of current engineering practices, engineers are required to receive training in a wide range of areas. A master program seems to be necessary for students to deal with the heterogeneity of knowledge. In other words, the undergraduate and master programs should be considered as an integral process to provide the necessary skills and establish a well-round knowledge base. However, current undergraduate programs "still look more like 'preparation for a Ph.D. program' [3]."
4) Very often the faculty members do not have real industry experiences. As a result, they might not truly understand the fast-changing employer requirements.
5) Due to the rising operational expense of modern universities, it can be costly to offer engineering education programs.

Traditionally, the education system of eastern Asian countries put an emphasis on diligence and discipline, but lacks training in the ability of innovation. Many universities have realized this problem, but it is still a long way to go for cultivating such a culture.

2 System of the Education of BS, MS and Ph.D in Microelectronics Engineering

The structure of current BS/MS/Ph.D. degree programs is a legacy of the 19th century engineering education system. In other words, it is "vocationally oriented" and can be traced to the science components of liberal arts curricula. Obviously, such a structure is not fully consistent with today's requirements for a true "Professional Program" [3]. In this section, we use the microelectronics engineering programs of Tsinghua University as a typical example.

2.1 Bachelor Program

A four-year program is offered for undergraduates. The semester system consists of two regular 18-week semesters in the spring and fall seasons, as well as a 10-week summer semester.

The credit system requires a minimal 172 credits for an undergraduate to get Bachelor's degree, among which 140 credits are from spring and fall semesters with an average of 20 credit hours per week, 17 credits are from summer semesters, and 15 credits are from the diploma project/thesis in the last semester of the undergraduate program.

In the third year, the undergraduates will have a chance to select their majors as electronics and microelectronics. Every year around 60 students will choose the microelectronics major.

2.2 Master Programs

Masters in this major include Master of Science and Master of Engineering. The Master of Science program is tuition free, while the students of Master of Engineering program have to pay the tuition.

1) Master of Science (MS)

The MS students must complete basic required courses, major required courses, and other trainings. Graduate students have to finish 25 credits in total, including 21 credits from courses and 4 credits for required trainings. The detailed credit requirements are as follows:

-- Basic required courses (5 credits);

-- Required trainings (4 credits);

-- Major required courses, including mathematics and fundamental major courses;

--Elective courses recommended by advisor (at least 7 credits).

Candidates need to complete a research project for MS degree and submit a thesis with appropriate theoretical and experimental significance. Guided by their advisors, master candidates are required to complete a topic selection report before May 1st of the second semester. This report is then reviewed and should be approved by a panel of three to five faculty members or experts before execution.

After the report is approved, no less than one calendar year is required for the candidate to apply for oral defense.

At the end of the semester before oral defense, the candidate must pass the intermediate investigation which is organized by a reviewing panel.

MS candidates are required to have at least one peer-reviewed paper accepted by a core journal or conference.

2) Master of Engineering (ME)

Full-time ME candidates study courses for one year in the University following the credit system. Compared to MS, the training of ME is more oriented to practical application. Professional off-campus practice should be no less than six months or one year, depending on the working experience of the candidate. Duration for completing ME thesis should be no shorter than one year. The total length of study is two to three school years.

Full-time ME candidates need to acquire at least 30 credits in total. The detailed composition is

-- Courses (at least 27 credits):

-- Professional practice, 3 credits;

-- Master degree thesis.

2.3 Ph.D. Program

The Ph.D. program usually takes four to five years (for candidates with bachelor's degrees) or three to four years (for candidates with master degrees) to finish. Upon enrollment, Ph.D. candidates should schedule their programs including relevant courses, academic literature review, research topic selection, etc. The schedule must be approved by the academic degree committee of department.

Ph.D. candidates need to acquire a minimum of 18 credits for whom with master degrees, or 39 credits for whom with bachelor's degrees.

Oral qualifying examination is usually carried out two to three years after enrollment, together with a presentation of research proposal including innovative ideas and preliminary results. Qualification committee should consist of no less than ten members, including at least seven Ph.D. advisors. Qualified Ph.D. candidates need at least one year to do research and compose their dissertations. At least three months before the oral defense, a final academic report should be submitted to a review committee for approval. Candidates should send their dissertations to five to seven experts (at least one of them is anonymous) for review at least 45 days before oral defense. Oral defenses can be carried out only if it is approved by no less than five reviewers and all anonymous reviewers.

Ph.D. candidates are required to publish peer-reviewed papers satisfying either of the following requirements. a) Acceptance of at least four papers, including three journal papers with two indexed by SCI or EI. b) Acceptance of a full-length journal paper (excluding letters, abstracts and reports) with SCI impact factor higher than 3.0 (for Physical Electronics), 2.5 (for Information and Communication Engineering) or 1.0 (for Automation).

3 Efforts of IMETU to Improve the Microelectronics Engineering Education

3.1 Attracting Prospective Engineering Students

Due to fast increasing complexity of modern IC systems, a master program is necessary for many students to start their career in this area. However, many undergraduates would not pursue a master of engineering degree after receiving their bachelor degree. The major reasons are three: 1) Many students from other colleges, or even from non-electronic departments of Tsinghua University, do not know that they have the chance to be recruited by IMETU; 2) Many students cannot afford the tuition fee; 3) Students feel uncomfortable about the difference between the MS and ME programs, especially because the ME students have to pay the tuition.

IMETU has taken efforts to resolve the problems. Firstly, we designed documents and activities to attract the undergraduate students to further their study in IC field. Different outreach initiatives were implemented over the years 2008–2010. We have edited different materials in different format aimed at the universities. These include poster, multimedia presentations and books which described the degree offered by Tsinghua University and featuring profiles of working professionals. Presentations include pictures of the campus and facilities. The posters have been mailed to the top 20 colleges in engineering in China, aiming to let as more as possible undergraduates know our program of master of engineering in integrated circuits and systems.

Additionally, IMETU recently redesigned its web site, which allowed users to extract detailed information about the degrees offered, campus life, admissions, and faculties. Especially, the web tries to deliver the key information about what it means to be an engineer in IC field. Such information includes challenging and highly rewarding job opportunities, career paths, and the social, economic and environmental impact of their work. This effort increases the visibility of the engineers for prospective students and society as a whole [4].

Secondly, one of the main initiatives undertaken is the development of one-day workshops. We invite the renowned professors in the IC field to introduce the characteristics, the importance to the society and the bright future of the field to the undergraduates. The graduated students from the master program will also be invited to share their experiences in their studying and career in the field with the students, and answer questions to help the students to learn more about the field and the future working chances.

We set up various scholarships to help students with difficulty to pay their tuition. In addition, many of the students are be funded by their advisors.

Instead of being static strategies, these different outreach initiatives implemented over the years constantly adapt every year. For instance, recently we began to arrange the the outreach program by cooperating with the industry.

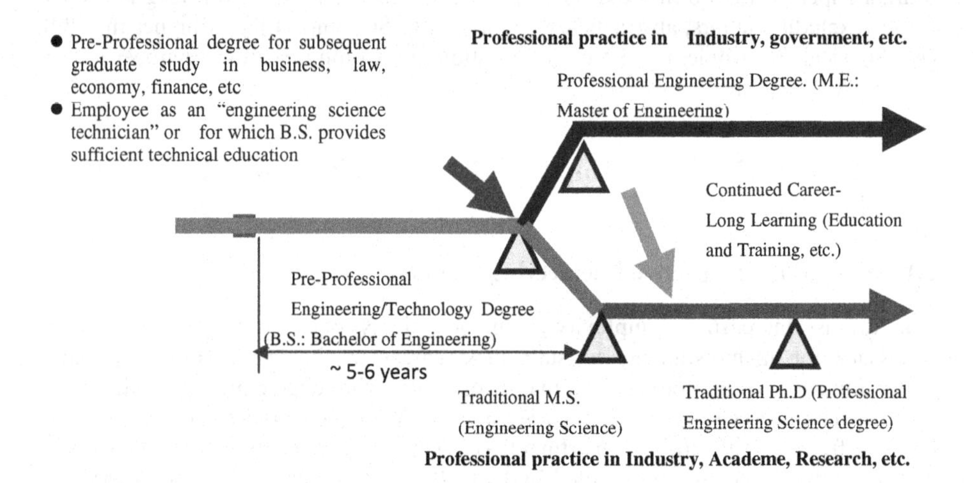

Fig. 2. A Possible Degree System of engineering (From [3])

3.2 Reformation of the Undergraduate and Graduate Program

We change the undergraduate program to for more chances of hands-on practice. Firstly, a new program, Students Research Training (SRT), was initialized in 2003 and soon became very popular among undergraduates who want to get exposure in real research projects. The undergraduate can apply the SRT program proposed by the faculties and get some funding from the university. The program covers almost all the research areas of IMETU including micro/nano electronics and micro system, system chip integration, micro/nano technology, micro/nano device and system CAD method. The program will last for one or two academic years for a project. So far about 30% to 50% undergraduate will join the SRT program in each semester, among which 10% students will select the SRT program offered outside their own departments. The SRT program is a good way to help the undergraduate to learn the cut-edging technologies in the research area and enhance their abilities in finding and solving problems in real projects.

Besides, we are trying to reform the undergraduate course as well as the graduate course of engineering. More experiments and projects will be added, and more breadth and depth courses will be given.

As shown in Fig. 2, the reformation will started from the bachelor course of IC engineering. At the undergraduate level, a modern "systems engineering" perspective is adopted so as to better determine what really needs to be presented as well as when and how to present it in our efforts to educate students to work in a modern engineering environment. Instead of creating courses to meet specific (and too often parochial) needs, we try to help students a basic understanding of the unity of the fundamental tools and concepts needed for engineering practice rather than providing them a vast bag of tricks for solving selected problems. As said in [3], any attempts to reform undergraduate programs could only be performed in a way that does not damage the quality of what we now have at the graduate level.

In educating future IC engineers, the quality is more important than the quantity even under the urgent need of the IC industry. In both undergraduate and graduate programs, "design-build-test" project experience has been increasingly reintroduced in many curricula as an effective means to bridge the gap between engineering theory and practice, and significantly enhance student learning, motivation and retention [3].

At the graduate level, many steps have been taken to improve the microelectronics engineering education.

First, we innovation the old courses by adding more practices and open new courses to widen the vision of the students including some culture, economics management , accounting and so on as well as some major courses. The main specialized courses are as follows: "Progress in Semiconductor Device Physics", "Digital VLSI Circuits", "Analog LSI Circuits", "High-level Synthesis of Digital VLSI Systems", "IC-CAD", "IC Manufacture Processing and Equipment", "Analysis and Design of CMOS RF Integrated Circuits", "IC Design Practice", "Cryptography and Network Security", "Digital Integrated System Design", "New Micro/nano Electronic Material and Devices", "PLL Design and Clock/Frequency Generation", "Semiconductor Memory Technologies", "Microprocessor Architecture and Design", "Introduction to VLSI Testing Methodology", "Microelectronic Packaging Technology", "Physics of Micro/Nano Fabrication", "MEMS (Micro electromechanical system)", "VLSI DSP (Digital Signal Processing)", "IC Manufacture and Processing Management", "Advanced Topics in Microelectronics", "Integrated Ferroelectrics", "Nano-electronic Devices" and "CMOS Integrated Circuit Manufacturing Experiment" and so on. These courses could be compulsory or optional depending on the students' research interest.

Second, more and more students without a microelectronics background get enrolled in our master of engineering program in recent years. Accordingly, we arrange special courses to help these students quickly pick up. The special courses are compulsory and cover the basic knowledge of multiple disciplines such as the microelectronics device and circuits, and signal and system. The new students should finish these courses in their first semester. Then students could smoothly catch up more advanced microelectronics courses.

Third, we provide practice chances to let the students to enhance their abilities in circuits design. For example, we have a course of "IC Design Practice" in the second academic year which provides full-customer IC design practice including a complete design procedure implemented from architecture consideration, floorplan, circuit design and skills, layout design and verification, whole chip analysis toward tapeout, all the way to packaged chip testing. The course teaches students how to realize a chip and requires each student to design a chip, tape it out and test the chips finally.

Moreover, the double master degree program between Tsinghua University and Katholieke Universiteit Leuven (KUL) is open to the master students of science as well as to the master students of engineering in integrated circuits and systems. In recent year more and more master students of engineering in integrated circuits and systems are interested in the program and have been selected to join the program. And delightfully, the students back from KUL are doing well and their evaluation is much higher.

Of course, a world-class teaching team is essential for our objective. We are now taking efforts to recruit high quality faculty members with diverse background, excellent practice-oriented teaching ability, and strong motivation to perform significant research. They will also serve as role models for their students as a group of engineers who pursue an academic career of life-long learning and team-based creative problem solving.

3.3 Reformation of the Laboratories of EDA and 1.0-Micron Fabrication Line

We have two main laboratories for students to do some research on devices or to design and fabrication chips. One is the EDA laboratory which possesses E5500 and E3000 servers, an IC test system, more than 60 workstations, approximately 200 PCs and the latest EDA software from Synopsys, Cadence, Mentor Graphics, etc. Another one is the clean rooms with an area of approximately 800 m2. There is a 1.0-micron fabrication line and a number of advanced micro- and nano-fabrication tools, such as double-sided alignment system (MA6), ICP, SPM, MOCVD, etc., for micro-/nano-devices and circuit chips. We also have an industrial VLSI process pilot line, the first and the only IC process line operated by a university in mainland China, and an important part of China's Northern Microelectronics R&D Base. The process line consists of a complete line of 5-inch sub-micron CMOS process facilities, with a throughput higher than 50K wafers per year.

The laboratories have served for the students and faculties in Tsinghua University for almost 30 years. It's worth mentioning that we established China's first Sub-micron VLSI Pilot Line in 1989; developed the first 1Mb Chinese ROM in 1993 and the first IC card with 1Kb EEPROM in China in 1995 with technologies successfully transferred to industry; developed a series of high power and low noise SiGe HBT devices and successfully delivered to production in 2002; jointly developed 0.18μm/4Mb embedded SONOS memory with integrated process and IP circuits in 2005. In 2004, IC design for the second generation residence ID card won the second prize of the Science and Technology Advancement Award of Beijing, and has been used widely in China.

To provide better service, we make lots efforts to reform the laboratories. Firstly, we enlarge the area of them to hold more people and decorate them for better condition. Secondly, we replace the old equipment with new ones. Moreover, new liberal policy to allow them could be open to students of researchers from other universities or research institutes not only to students and researchers from Tsinghua University. These facilities can meet the research and testing needs of nano devices, semi-conductors, photo-electronic devices and OEIC, and integrated circuit chips.

4 Outcome of the Cultivation of Master of Engineering

The effectiveness of these actions is objectively and quantitatively reflected in the official data of enrollment in the IMETU during the last few academic years. Figure 3 shows the number of engineering freshman enrolled in IMETU from 2004 majored in integrated circuits and systems. By 2010, there are 274 graduated master students of engineering in integrated circuits and systems from IMETU.

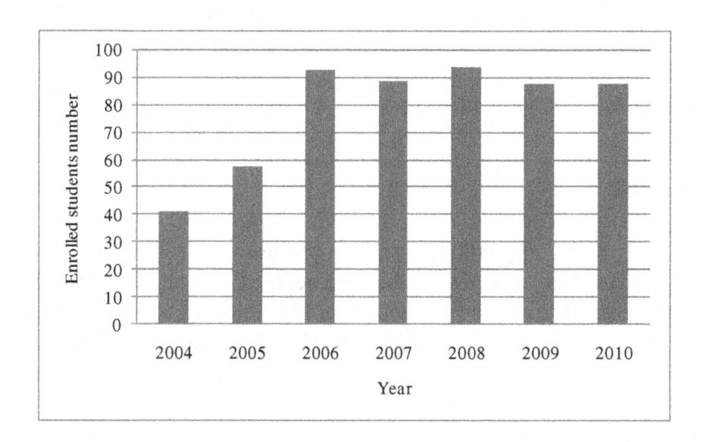

Fig. 3. Enrollment number of the master students in IC engineering

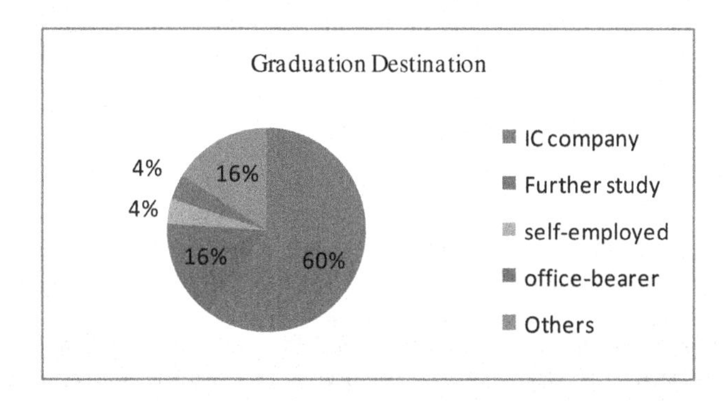

Fig. 4. Graduation destination of the master students in IC engineering

According to the recent Graduation Destination Survey (GDS), the results are shown in Fig.4, from which we can find about 80% graduated students with the master degree of engineering in integrated circuits and systems still are engaged in the same fields, in which 60% work for the famous IC companies (such as IBM, Motorola, Intel, Tongfang Microelectronics, Nvidia, Marvel Semiconductor, etc.) or institutes (for example Chinese Academy of Sciences) around the world, 16% further their study and 4% set up their own business in IC field (such as OmniVision, NetScreen). About 20% graduated students engage in different fields, in which about 4% work for government as office-bearer and 16% work in other field such as the finance, sale, software and so on.

The first group of graduated master students in integrated circuits and systems from IMETU has been working for 3 years in IC field from 2007. By 2010 the 4 groups of graduated students have earned a good reputation in the society from the feedback of the IC companies and institutes. In recent each year the employers will come to the IMETU to recruit new groups of the master students in IC engineering. Especially, they think there is little difference between the abilities of the master students of sciences and those of engineering. A positive feedback mechanism has been established between the IMETU and the society.

5 Conclusion

This paper surveys the essential role of IC industry in China and the unbalance between the demand and supply of the engineers. Considerable efforts have been made by IMETU to improve the master program majoring in integrated circuits and systems. We developed a series of outreach initiatives and workshops to effectively future students. Moreover, s systematic reformation is carried out on the bachelor program and master program to better prepare students for their career in IC industry. The results in recent years prove the effectiveness of the educational activities.

Acknowledgments. The authors would like to thank their colleagues: Xin Qian, Yuan Hu, Lei Guo for their hard working on the master engineering program in IMETU. This research was supported in part by the National Natural Science Foundation of China (No. 60906010).

References

1. Pecht, M.G.: How China is Closing the Semiconductor Technology Gap. IEEE Transactions on Components and Packaging Technologies 27(3), 616–619 (2004)
2. CCID Consulting, `http://www.ccidnet.com/`
3. McMasters, J.H.: Thoughts on the Engineer of 2020– An Early 21st Century [Aerospace] Industry Perspective. In: The Future of Engineering, Science and Mathematics: Who Will Lead? A Conference on the 50th Anniversary of Harvey Mudd College, Claremont, California, January 14 (2006)
4. Lopez-Martin, A.J.: Attracting Prospective Engineering Students in the Emerging European Space for Higher Education. IEEE Transactions on Education 53(1), 46–52 (2010)

Research on Risk Management of Project Schedule

Jinling Liu

Tianjin Beiyang Park Investment and Development Co. Ltd
Liujinlingtianjin@126.com

Abstract. This paper analyzed the risk of the project and its impact on the schedule of the project. It puts forward that the key to schedule control is duration control, and simulates the total period of project schedule. It sets up a risk analysis model of the project schedule according to the results of the simulation, and then it gets a schedule estimate of the engineering system and the risk distribution of the project total period. It provided an example to prove this model can be used for the analysis of schedule risks. At last it puts forward some measures to deal with the schedule risks.

Keywords: engineering project, schedule risk, schedule deviation.

1 The Analysis of Project Risk and Its Impact on the Schedule

1.1 The Concept of Project Risk

Risk management is a progress of ensuring a lower cost, minimizing the risk loss and getting high security. It uses scientific methods to identify, estimate, assess, reply and monitor the risks, then selects the best management measures to address risks [1].

The purpose of analysis of the project schedule risk is to identify and manage the weak link of the progress, to determine the statistical distribution in network planning project, to provide managers with the risk value of the progress to remind them to take measures and then to provide the necessary reviews and decision-making information.

1.2 Analysis of the Impact on Schedule Deviation

In the execution of the schedule, if the actual progress does not match the planned schedule, whether it is necessary to modify and adjust the original plan to adapt to the actual situation depends on the circumstances of the case deviation [2].

1) When the progress deviation is reflected in the actual work ahead of schedule, if we speed up the implementation of some work schedule, it can often lead to changes in using of resources.
2) When the progress deviation is reflected in the actual progress behind the planned schedule, if the work with progress deviation is the key work, due to the lag of work progress, it inevitably leads to the earliest start time of follow-up work delayed and

H. Tan (Ed.): Knowledge Discovery and Data Mining, AISC 135, pp. 397–404.
springerlink.com

the whole project duration extended, thus, we must take certain adjust measures to the original schedule.
3) If the work with progress deviation is not the key work, and the number of the day lags behind the schedule is beyond the total time difference, the delay of work progress can also lead to the earliest start time of follow-up work delayed and the whole project duration extended, thus, we must take certain adjust measures to the original schedule [3].
4) If the work with progress deviation is not the key work, and the number of the day lags behind the schedule is beyond its free float but not the total time difference, due to the work progress delay just causes the delay of the earliest start time of the follow-up work but has no effect on the whole project duration, at this time we take certain adjust measures to the original schedule just when the earliest start time of follow-up work should not be postponed [4].

2 The Risk Management of Project Schedule

Project risk management use scientific methodology to identify and assess the risk of project schedule and then use response and control measures to effectively control risk, at last achieve overall goal of the project reliably. The schedule control is a progress of supervision, inspection, guidance and corrective actions to the whole schedule in a limited work period according to a pre-decided, rational and economic project schedule plan.

The key to schedule control is duration control. Duration is the time required from the beginning to the completion of a series of construction activities. Its objectives include: the total duration objectives realized by the total schedule plan; the sub-schedule plans (procurement, design, construction, etc.) or the duration objectives realized by sub-item schedule; milestone objectives realized by each phase of the schedule. Schedule control is the process of repeated cycles. It reflects the dynamic processes of the schedule control system in controlling the construction progress [5]. In a range of limits (the minimum cost corresponding to the optimal duration), accelerating the construction schedule can achieve the purpose of cost reduction. But beyond this limits, accelerating the construction schedule will result in the increase of input costs. This paper will make simulations to project schedule plan, and put forward relevant measures to high-risk procedure in simulation results.

2.1 Time Parameters of Project Schedule

1) Duration of work is a time from work start to finish.
2) The duration refers to the time required to complete a task. It generally includes: calculating duration, required duration, planned project duration.
3) The earliest start time: under the restriction of predecessor activity, this may be the earliest start time.
4) The earliest completion time: under the restriction of predecessor activity, this may be the earliest completion time.
5) The latest finish time: without affecting the task completed on schedule, this must be the latest time to complete the task.

6) Latest start time: without affecting the task completed on schedule, this must be the latest time to begin the task.
7) Total difference: without affecting the duration, the mobile time can be use to a task.
8) The free float: without prejudice to tight work after the first, the mobile time can be use to a task.
9) Critical tasks and critical path: when the planned duration equal to the calculated period, the task is the key task if its total difference is zero; when the two are not the same, the task is the key task if its total difference is the smallest; when we use node calculation, if the planned duration equal to the calculated period, the work of key tasks are satisfied the following three conditions:

$$ET_i = LT_i \tag{1}$$

$$ET_j = LT_j \tag{2}$$

$$LT_j - ET_i - D_{i-j} = 0 \tag{3}$$

2.2 Simulation Analysis on the Total Project Schedule Duration

The total duration of the project schedule depends on the activity cycle on the critical path. Because the cycle of each progress is a random variable, so the total duration is composed of a series of random variables, which itself is a random variable. The most important is that the total duration is with statistical characteristics. We generally use triangular distribution to show the probability distribution of each activity cycle, which are respectively the most optimistic estimate value, the most likely estimate value and the most pessimistic estimate value. The mach easier way is to give estimates of the proportion of the three values: $K_1 : 1 : K_2$ K_1 = the most optimistic estimate value/ the most likely estimate value; K_2 = the most pessimistic estimate value/ the most likely estimate value.

The simulation runs to the step K, we get ach activity cycle $\{tik\}$ of step K plan through random sampling. Through the former methods, we calculate the time parameters of activities, find the critical path, calculate the total duration $\{TK\}$ and the sample mean and sample standard deviation of the total duration.

Sample mean: $\frac{1}{N}\sum_{K=1}^{N} T_K$

Sample standard deviation: $\sqrt{\frac{1}{N-1}\sum_{K=1}^{N}(T_K \frac{1}{N}\sum_{K=1}^{N} T_K)^2}$

$i = 1,2,..., N$ is the number of activities in the project schedule; $K = 1,2,..., N$ is the times of simulation. t_{ik} is the step K sampling value of the activity i; T_k is the total duration of the step K simulation. When the number of simulation reaches N, we make a statistical analysis for the simulation results of the total duration. We get the histogram of the total duration through grouping plan frequency, and regard it approximately as the density curve of the total duration, thus, we have a batter overall picture of the distribution regularity of the total duration. It can be used for decision-making.

$$T\min = \min_{k\in\{1,...,N\}} \{T_k\} \tag{4}$$

$$T\max = \max_{k\in\{1,...,N\}} \{T_k\} \tag{5}$$

$$T_g = \frac{T\max - T\min}{L} \tag{6}$$

So $T_1 = T\min + T_g; T_2 = T_1 + T_g; ...; T_m = T_{m-1} + T_g$

$T\min$ is the minimum total duration of N simulation; $T\max$ is the maximum total duration of N simulation; L is the number of packet groups; T_g is the distance between the groups.

The group results of the total duration T are $[T\min, T_1], [T_1, T_2], [T_2, T_3], ..., [T_{n-1}, T_n]$, we get the statistical frequency of the N times simulation results according to intervals, then we can get the approximate density function $f(T_j)$ of the total duration T. we can also get the approximate cumulative distribution $F(T_j)$ according to cumulative intervals of the total duration. At last, we will get the statistical distribution curve and the cumulative distribution curve of the total duration.

2.3 The Risk Estimates of the Total Duration

Schedule risk depends on technical risk, the reasonability of project scheduling, the adequacy of resources, personal experience and the management status of the enterprise, etc.

The risk distribution of the total duration can be estimated:

$$R(T_j) = 1 - F(T_j) \tag{7}$$

$R(T_j)$ is the risk value of construction work according to specified total duration T_j;
$F(T_j)$ is the approximate cumulative probability of the total duration T_j.

The approximate distribution density function of the total duration T is $f(T_j)$, we can use computer to process the approximate cumulative distribution $F(T_j)$ and the risk distribution $R(T_j)$ of the total duration into function curve and data tables. By consulting the distribution density function or data tables, we can observe and understand the distribution shape of the total duration T. Through the curve of cumulative function or the data tables, we can get the relationship between the total duration T and completion probability P and then can make schedule risk analysis. 1) if we input the specified total duration T* we can get corresponding designated completion probability P*. 2) if we input designated completion probability P* we can get corresponding specified total duration T*.

2.4 Examples

We use an example to explain the processing of project schedule simulation and total duration risk analysis. Node ① is connected to ② ③ ④, node ② is connected to ⑤, node ③ is connected to ⑤ ⑥, node ④ is connected to ⑥, nodes ⑤ ⑥ are connected to ⑦. Parameters table 1 is as follows:

Table 1. Project parameters

activity	Activity cycle	Cycle probability distribution
	——	——
① to ②	3	0.85:1:1.4
① to ③	2	0.7:1:1.2
① to ④	4.5	0.8:1:1.25
② to ⑤	5	0.85:1:1.3
③ to ⑤	0	0.75:1:1.25
③ to ⑥	7	0.65:1:1.2
④ to ⑥	8	0.73:1:1.35
⑤ to ⑦	8	0.73:1:1.35
⑥ to ⑦	6.5	0.75:1:1.3
ending node ⑦	——	——

Table 2. Statistical analysis of simulation results

activity (i,j)	Activity cycle		Earliest start time		Earliest completion time		Latest start time		Latest completion time	
	Mean	SD	Mean	SD	Mean	SD	Mean	SD	Mean	SD
beginning node ①	0	0	0	0	0	0	0	0	0	0
① to ②	3.021	0.247	0	0	3.29	0.342	2.745	1.836	6.039	1.809
① to ③	2.278	0.056	0	0	1.95	0.22	4.336	1.261	6.283	1.302
① to ④	4.291	0.389	0	0	4.57	0.372	0.079	0.337	4.651	0.47
② to ⑤	4.681	0.307	3.293	0.342	8.54	0.561	6.039	1.809	11.288	1.746
③ to ⑤	0	0	1.947	0.22	1.95	0.22	11.29	1.745	11.288	1.745
③ to ⑥	7.088	0.759	1.947	0.22	8.77	0.755	6.283	1.302	13.058	1.113
④ to ⑥	8.121	0.442	4.572	0.372	13	1.156	4.651	0.47	13.058	1.113
⑤ to ⑦	7.703	0.732	8.543	0.561	17	1.369	11.29	1.745	19.715	1.293
⑥ to ⑦	6.638	0.326	12.98	1.156	19.6	1.362	13.06	1.113	19.715	1.293
Ending node ⑦	0	0	19.715	1.293	19.7	1.293	19.72	1.293	19.715	1.293

Table 3. Statistical frequency distribution table of the total duration

group mean	frequency	probability	cumulative probability	risk value
16.9718	1	0.01	1%	99%
17.5357	5	0.05	6%	94%
18.0997	6	0.66	12%	88%
18.5637	12	0.12	24%	76%
19.2276	11	0.11	35%	65%
19.7916	17	0.17	52%	48%
20.3556	18	0.18	70%	30%
20.9195	10	0.1	80%	20%
21.4835	11	0.11	91%	9%
22.0475	3	0.06	94%	6%
22.6114	0	0.03	100%	0%

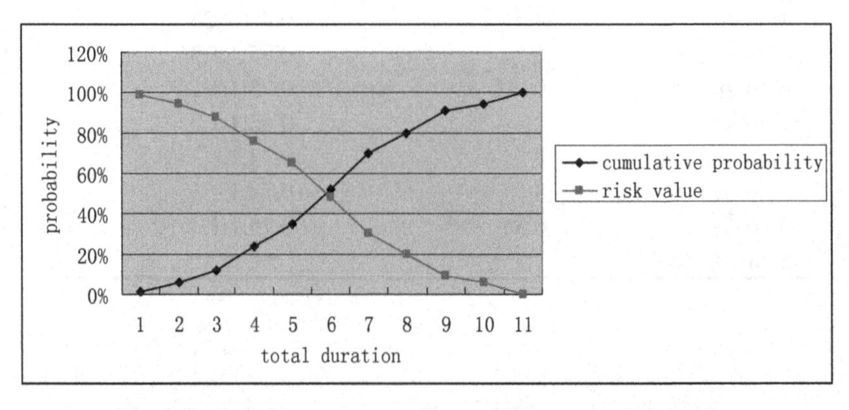

Fig. 1. Statistical frequency distribution figure of the total duration

The simulation times of the risk analysis processing of the total duration is 100; the mean of the total duration is 19.050; standard deviation is 0.273; minimum is 16.972,maximum is 22.611; average risk is 68.64%. It shows that the risk distribution of duration is more balanced, but the average risk 68.64%also shows that the duration is facing fairly big risks. We can determine which activity need to response to according to the risk values of each duration. That is the activity with the maximum value 22.611.

3 Response Measures to Project Schedule Risk

We get the high-risk steps in project schedule according to the simulation results and put forward certain appropriate measures for improvement.

3.1 Strengthen Management of Production Factors Allocation

Allocation of production factors including: labor, capital, materials, equipment, and the survey, summarizing, analysis, forecasting and control of its stock, flow and flow division. Rational allocation of production factors is an effect way to improve construction efficiency and increase management efficiency, and also the core and key point to dynamic control of project schedule nodes. In the dynamic control, we must attach great importance to the condition changes in the internal and external of the whole construction system, and track subjective and objective condition of site timely. We should stick to use a lot of time to know and study the people, materials, machinery and progress of the project. We also should do analysis and forecast to the resource requirements of each progress and the total resources, the actual progress conditions of the machine and project, the contradiction among the resource requirements, the total resources and the actual input resources, then make specifications of input direction and take adjustment measures to ensure realizing the schedule goals.

3.2 Strict Process Control

We should know well about the actual situation in the construction site. We record the start date, the work process and the end date of each progresses, its function is to provide the original information for the inspection, analysis, adjustment and summarization for the plan to implementation. Thus, the strict process control includes three basic requirements: first, we must track record; second, we must record accurately; third, we should record files through charts [6].

In addition, there are a few points need to be strengthened:

1) It should strengthen the organization and management. It requires various management departments and the managers that take part in the task do coordination work because the network plan is tight in time arranging. Thus, we should start from an overall situation to organize reasonably and arrange the labor, materials and equipments in a unified way, and in the organization, we should make the network plan into a technical documentation which every one must obey, then we can create conditions for the implementation of network plan.

2) We should make the control of hierarchical network planning more important to the duration to ensure the overall achievement of objectives. We should definite the responsibility strictly, make integrated control measures of the overall objective, phase objectives and node objectives according to management responsibilities. We also should study technology and organization, goals and resources, time and effect to find out the best combination points [7].

4 Conclusions

With increasing demanding to the schedule of project in nowadays, the management of project schedule risk becomes imperative. This paper analyzes the impact of schedule deviation firstly, then finds out the deviation and its reasons through the comparison between the objectives of planned schedule and actual implementation schedule, and takes corrective measures to adjust the deviation, thus achieves the control to the project schedule. It does simulations to the project schedule of the total duration, identifies and assesses the project schedule risks, and then on this basis it puts forward response and control measures in view of production factors, process, duration and economic responsibility system, thus we can control the risk effectively.

References

1. Antill, J.M., Woodhead, R.W.: Critical Path Methods in Construction Practice. A Wiley Inter-science Publication, New York (1982)
2. Davis, E.W.: Project Scheduling under Resource Constraints Historical Review and Categorization of Procedures. AIIE Trans. 5, 297–312 (1973)
3. Morse, L., Whitehouse, G.: A Study of Combining Heuristics for Scheduling Projects with Limited Multiple Resources. J. Comp. and Industrial Engineering 15, 153–161 (1988)
4. Wu, Z.: The Time Value and Risk in Network Planning. The Theory and Practice of System Engineering, 1–5 (1994)
5. Wang, C.: The Research on Scheduling Problem of Fuzzy Processing Time. System Engineering. 14, 238–242 (1999)
6. Chen, H.: A Fuzzy Model of Calculating Security Difference in Network Planning Model. System Engineering 16, 192–196 (2001)
7. Hu, J., Da, Q.: The Critical Path Analysis of Single-code Fuzzy Network. The Theory and Practice of System Engineering, 108–112 (1998)

Seizing the Big and Freeing the Small in Program Design Teaching

JunRui Liu and XueFeng Jiang

College of Computer Science, Northwestern Polytechnical University, Xi'an, China

Abstract. At present, the teaching of program design stays in the language level, and the student lacks the programming thinking and the ability of program designing. After the analysis of the core tasks of the developing course, the strategy of seizing the big and freeing the small is proposed. This strategy advocates restructuring major enterprises and relaxing control over small ones in the teaching of program design, its aim is to cultivate students with the ability of using language, the programming thinking and the idea of software development, to preserve the quality of the core teaching task, and to improve the students' comprehensive quality and the curriculum level. The results of my school practice show that the strategy of seizing the big and freeing the small can effectively improve the teaching effectiveness and enhance the students' skills of using language and program development.

Keywords: the teaching level, the core task of teaching, seizing the big and freeing the small, the ability of using language, the programming thinking.

1 Introduction

The programming courses is one basic computer course of China's institutions, There are a lot of people interested in the topic how to improve the teaching effectiveness and enhance the teaching level. The students only understand the lexical grammar of the language, without the ability of using language and the ability to solve problems. That is, the students who have learn the program design don't have the skills of program design. In this paper, the author combines with the practice of self and puts forward the teaching thought that the teachers should restructure major enterprises and relax control over small ones in the teaching of program design, its aim is to cultivate students with the ability of using language, the programming thinking and the idea of software development, to preserve the quality of the core teaching task, and to improve the students' comprehensive quality and the curriculum level.

2 Seizing the Big and Freeing the Small in the Stage of Making Teaching Target

In order to improve the teaching effectiveness and enhance the curriculum level, we should firstly check whether the curriculum objectives meet the requirements of the course trends and the training goals. With the popularization of computer applications,

H. Tan (Ed.): Knowledge Discovery and Data Mining, AISC 135, pp. 405–409.
springerlink.com

the use of computer science concepts to solve problems, design systems and the ability to understand human behavior is no longer just a requirement for the computer science students. The extensive application of computer in many areas requires more students with computational thinking. As a public computer based curriculum, the program design courses should change its training objectives, and can not stop programming language level, but to have students use computers to analyze and solve problems.

Therefore, we should consider the curriculum development when we formulate the course objectives, and understand that teaching the students the language is small while teaching the student to use the programming language as a tool for the computational thinking is the major training objectives. The curriculum goals should be emphasized that the ability to use language, the programming skill, the programming ideas, the ideas of software development and the skill of using the programming language solve practical problems.

While, in order to protect the new curriculum teaching objectives can be achieved, the other teaching stage should make the appropriate changes, such as the teaching content design, the theoretical teaching, the practice and the assessment areas. The change method of the teaching process is as following.

3 Seizing the Big and Freeing the Small in the Theoretical Teaching

The teaching material use the programming language as the background, has extensive knowledge of the language while ignores the using technology introduction of the language. In the teaching process, if the teacher has no the appropriate teaching methods, he will always be in high pressure and struggle to teach a number of knowledge points. So the course is a anticlimax. While, the students in the learning process has not got its priorities, are unclear about the course of study purposes, and thus gradually lost interest in learning the curriculum. Therefore, the adjustment of the theoretical teaching that clear teaching objectives and understand the instructional cues is necessary. The following introduce how to seize the big and free the small in the theoretical teaching.

(1) For the teaching content and the instructional cues, skill training is major while learning a language is small

In order to protect the program's new teaching objectives can be achieved, the curriculum content should be improved accordingly. The teaching content should be based on the language, and add the content which use to train the students' skills of using language, such as data structures, algorithm design and advanced programming, etc. The design of the teaching content should enable the students to master a language, to have the ability of complex data processing, to master the approach of comprehensive programming, and to learn much specialized application technology such as numerical calculation, interface programming and network programming.

Also, because the knowledge of the programming course is numerous, if the language is acted as the teaching cues, the teaching process will be messy and confuse. Then, the teachers and the students will pay much attention to the details, and thus diminish or ignore the use of language ability. Therefore, the programming courses should emphasize the skills development as the main line, and the goal is to solve the problem. The teaching process should not be too much emphasis on the language knowledge but to data representation and algorithms.

In the teaching process, the introduction of language knowledge is not based solely on the language level, but should be raised to the direction of skills development. Not only the functions and usage of the operators and expressions should be introduced, and the reason of learning them that the computer is essentially accomplished the work by calculating should be tell also. Such an approach allows students to know the use of each knowledge point of use, so the students take great interest in the curriculum.

(2) In the classroom teaching, the core nature of the knowledge points is major, and the branches and leaves is small

There is a lot of knowledge points in the program design course, all knowledge teach in a limited class is almost impossible, and this teaching mode will make the students confuse and with great pressure, then the learning effect will be weaken. Therefore, in the classroom teaching process, the teacher should seize the core content of each knowledge unit, explain this knowledge points succinctly through examples and metaphorical, so that students can fully understand and master the knowledge and have the heuristic thinking. The other secondary details are advocated to learn by self-learning. Only these will make the teaching process and the teaching clues clear, and improve the teaching effectiveness.

4 Seizing the Big and Freeing the Small in the Practice Teaching

In many domestic schools the curriculum's practical aspects is not very of attention, such as the teaching resources and the equipment are not adequate in practice time, the practice intensity and the practice quality can not be guaranteed. This is a great misunderstanding to the programming course. The survey result of U.S. National Training Laboratories show that the average retention rate of the learning content in classroom is 5%, while in the practice and teaching others is as high as 75% and 90%. From the survey results can be seen that practice is the best way to reinforce learning. Also, the program design is a strong practical course, so in curriculum design process practice should be pay more attention to improve the teaching effectiveness and enhance the teaching height.

(1) Programming in the actual programming environment is major, and the off-line training is small

In the past, because the teaching resources are limited, the off-line programming is very common. The students can not master the language accurately, and can not use the development environment masterly. When the teaching resource is abundant, the practices in the actual programming environment should be emphasized, and the off-line programming is no longer advocated. The students are asked to verify each program in the programming environments, master the use of development tools, learn the debugging skills, and have the ability to program design.

(2) The strength and quality of practice is major, and the form of practice is small

As the programming course is a practical course, and only a lot of programming can help the students master the use of the language, the strength and quality of practice must be ensured. In practice sessions, the students must be supervised to program fully, and their work should be tracked and guided. And because of the conflict between the teaching resources and the students, the training approach may no restrictions. The education software can provide students with the practice conditions anytime and anywhere, such as programming in the dormitories, training via mobile phone.

In addition, for the students in the initial learning, the quality of programming should not be emphasized, the students should be allowed to program with the help of the learning platform resources, interactive discussion among students, the Internet. That is conducive to the students learning confidence, the coexistence between students and the ability to cooperate.

Thus, in teaching practice, as long as the students complete a lot of programming training to acquire the language skills and programming methods, the teacher can ignore the specific forms of practice.

5 Seizing the Big and Freeing the Small in the Assessment

The traditional assessment of programming courses generally focus on knowledge of the language, which make the teachers and the students pay much attention to the language details in the teaching process, and neglect the ability of using the language. Thus, the assessment should make the appropriate changes. The assessment should be to examine students' language proficiency, the program design capacity, while the examination of the language knowledge should be relaxed.

Assessment areas should be in a real programming environment, and examine mainly the students the program design capabilities in the real programming environment. To enable the students to understand that they need not only to learn a programming language, but also master the use of skills, calculate their own thinking, the assessment form can only use the program design subject, or even replace the traditional assessment with the Curriculum design.

In addition, the opening of the learning resources in the examination process, such as the students can refer to the manual instructions for the use of library functions, could help the students to avoid spending too much energy in the memory of the language points, so that students can understand that the ability of using language for the solving of the practical problem is their major task, the language knowledge can be remembered through more practice rather than the focus and the burden of the learning process.

6 Conclusion

The strategy of seizing the big and freeing the small in this paper has been practiced in my school at least two years. The experimental results show that seizing the big and freeing the small can promote the curriculum development, ensure the teaching effectiveness and enhance the students' ability. Therefore, the authors believe that the strategy of seizing the big and freeing the small is worthy of popularization and application.

References

1. Liu, C.: How to motivate students to learn "C" interestingly in technical schools. Teaching and Research 9, 40 (2010)
2. Zou, G., et al.: Developing a varity of capacities of the students with the remote education resources. Computer Education 9, 21–23 (2010)
3. Guo, H.: The interesting teaching methods of the C language in the practice teaching. China Power Education 28, 148–149 (2010)
4. Liu, Z.: The research and the practice of the projects teaching pedagogy in the C language. Chinese Adult Education 4, 139–140 (2010)
5. Dai, L.: How to improve the teaching quality of the application-oriented university "VC ++ Programming" course. China Adult Education 21, 176–177 (2009)
6. Fu, Y.: The reform of the C ++ programming courses. Higher Education Research 6, 47–48 (2008)
7. Xia, M.: The teaching method of the Visual Foxpro programming. Higher Education Research 6, 40–41 (2007)

The Research of Teaching Mode in Basic Computer Courses

XueFeng Jiang and JunRui Liu

College of Computer Science, Northwestern Polytechnical University, Xi'an, China

Abstract. The current teaching methods of the programming course are advanced, but the teaching effect is poor. In view of this phenomenon, the author suggests the use of appropriate teaching modes to implement the teaching methods so that the progressiveness and the superiority of the teaching methods can be play. If the teaching mode is appropriate, the teaching effectiveness and the teaching height could be promoted. The results of the practice show that the appropriate teaching mode could promote the curriculum development, ensure the teaching effectiveness and enhance students' ability.

Keywords: teaching mode, precision teaching, vivid teaching, tracking and controlling, the reward and punishment system, discussing and exchanging.

1 Introduction

The programming course is a basic computer course in the higher education, many educators working on the teaching methods. Now, there are a lot of advanced teaching methods, such as the project pedagogy, the interests teaching, the happy teaching and the case teaching and so on. From the design strategy of these teaching methods, their implementation is useful for the improvement of the teaching effect and the promotion of the teaching height. But that is not the case, the teaching effect is still poor after many teachers used these teaching methods, and even some teachers in the teaching process can not successfully implement these advanced concepts of these teaching methods. The reason is the problem of inappropriate teaching methods. In this paper, in view of the characteristics of the course, the author puts forward a number of appropriate teaching modes.

2 Teaching Mode in Theory Teaching

There is a lot of knowledge points in the program design course, all knowledge teach in a limited class is almost impossible, and this teaching mode will make the students confuse and with great pressure, then the learning effect will be weaken and the students will lost the interesting in the course. Therefore, the use of appropriate teaching methods for ensuring the learning effect and the learning interesting of the students has a crucial role.

In view of the characteristics of the program design course, the author thinks there are two teaching modes can be used in the theory teaching.

H. Tan (Ed.): Knowledge Discovery and Data Mining, AISC 135, pp. 411–414.
springerlink.com

(1) Precision Teaching

There is a lot of knowledge points in the program design course, all knowledge teach in a limited class is almost impossible. Therefore, the teacher should analyze each teaching unit and pick out their core concepts and key points, then introduce them in the precision teaching mode.

The precision teaching mode requires the teachers to be familiar with the contents of each knowledge point and their position in the curriculum. If the teachers can not accurately pick out the core of the teaching knowledge point, the precision teaching mode not only fail to improve teaching effectiveness, but also misguided the study direction of this courses, then make the entire teaching objectives failed. Only the teachers pick out correctly the core of each teaching unit, and teach finely them by analogy and examples to enable the students to fully understand the knowledge point, know why them should be learned, and acquire the ability to apply the knowledge points. Only the students have a deep understanding of the knowledge points, the teacher could let students self-teach the other content of this knowledge unit. For example, the content of the pointer unit is quite rich, many teachers can not even be completed this teaching tasks within the allotted hours, the students are confusing with a wide variety of pointers applications, the teaching effect can not be guaranteed. However, if the teachers can extract the core knowledge of the teaching unit, and teach the concept of pointers succinctly through multi-faceted examples and multi-angle interpretation, so that the students can understand the nature of the pointer, then a variety of pointer applications can let students self-taught because the students can see the nature of the pointer from the phenomenon of the pointer application.

(2) Vivid Teaching

Computer programming language stands on the computer perspective and abstracts the problems, there is a great difference between computer programming language and natural language. The students often can not think on the computer thinking perspective, and in the learning process they has been unable to enter into the computer field, then they have a sense of failure and loss of interest in learning slowly. Therefore, in the classroom using the right, lively teaching modes to help the students to think the problem on the computer perspective is an important tool to ensure learning effect and enhance students' learning interesting.

The most effective way of vivid teaching is to change the teaching language, and replace the blunt professional language with the vivid language and the facile language, and the knowledge for students to accept difficultly can be given some life examples to help understand. Here are a few examples to illustrate that the proper expression language for ensuring teaching effectiveness is very important.

Example 1: For the teaching of the function definition and the return statement, the teachers can use the borrowing and returning relationship in the life to illustrate. The defining of the return value in the function header is borrowing while the return statement is returning. If you borrow someone something in life, you should have good habits to return, and the thing returned must be the thing borrowed. Similar to this, if the function has a definition of the return value, the return statement should be used to return the return value. Otherwise, the compiler will be angry.

Example 2: The student can not understand why 127 add 1 is -128 in the 8-bit computer, our colleagues used a very philosophical story to interpret of this phenomenon. If the number of the computer is compared to adult greed, everyone should have their own wealth, but if he I very greedy while he has wealth that he should own, for example at the status of 127, seizing anything that didn't belong to him will make him fall, that is at the status of -128. Through this interpretation, the students not only understand the computer number, but also learned the truth in life.

In addition, adding some body language in the teaching also makes the classroom to become interesting, the appropriate body language can help the students to understand and accept the knowledge, and recognize and grasp the classroom focus and so on.

In short, as the first of the teaching process, the appropriate teaching mode in the theory teaching can ensure the teaching effectiveness. Precision teaching and vivid teaching can also make the students to be interesting in the course and obtain a good learning effect.

3 Teaching Mode in Practice Teaching

In many domestic schools the curriculum's practical aspects is not very of attention, such as the teaching resources and the equipment are not adequate in practice time, the practice intensity and the practice quality can not be guaranteed. This is a great misunderstanding to the programming course. The survey result of U.S. National Training Laboratories show that the average retention rate of the learning content in classroom is 5%, while in the practice and teaching others is as high as 75% and 90%. From the survey results can be seen that practice is the best way to reinforce learning. Also, the program design is a strong practical course, so in curriculum design process practice should be pay more attention to improve the teaching effectiveness and enhance the teaching height.

The author thinks there are the following teaching modes can be used in the practice teaching.

(1) Tracking and controlling

In the practice session, the teacher is easy to take two extremes.

The first one is lack of management. The students in the experimental class can not be for guidance, and thus can not solve the problems encountered to give up. Both the teachers and the students ignore the importance of the practical aspects, and want to muddle through their work.

The second one is too spoiled to the students. The teacher guide the students everything, even help the students coding and debugging, and the teacher is a completely course nanny. The students lost the courage and the ability to solve problems completely.

In these two extreme modes of practice teaching, the teacher can not supervise the students to practice effectively, and can not develop the students ability to use language. Therefore, we should try to avoid the recurrence of these two extremes in the practice teaching. The teacher should guide the students in the way of inspire, master the practice of most students, and adjust the teaching activities according to the practice case. The teachers can use some educational software to do this work, such as the network teaching platform, the experimental systems. All students can be understand and controlled, and all students can be guided individually with the help of the online teaching platform.

(2) Rewards and penalties

Appropriate reward and penalties system is an effective means to urge the students to practice, stimulate their interest and enthusiasm in practice. And using the job management system to select the students which have complete the job excellently with the appropriate reward, and the students which have complete the job poorly with the appropriate reward, this can motivate the students to learn.

(3) Discussion and exchange

From the US National Training Laboratory's findings can be seen, discussion and teaching others is a good means to consolidate teaching effectiveness. Therefore, the teachers should urge the students to create a good discussing and exchange system, such as the course interest groups, the QQ discussion groups, and often design some projects which need the students to discuss and collaborate. In addition, the teacher can appoint the excellent students as the teaching assistants, and these teaching assistants can help other students to learn the program design. This is also a good means of communication.

These practical teaching modes are useful for strengthening the intensity and quality of the practice, and promoting the teaching effectiveness.

4 Conclusion

The teaching mode described in this paper has been practiced in my school. The experimental results show that the appropriate teaching mode can promote the curriculum development, ensure the teaching effectiveness and enhance the students' ability. Therefore, the authors believe that the appropriate teaching mode play a very important role in the teaching process, and these appropriate teaching mode are worthy of popularization and application.

References

1. Chen, G., et al.: The teaching mode and method of the C++ program design. Computer Education 11, 135–137 (2011)
2. Zhang, S., et al.: The teaching reform and practice of the program design course. The Adult Education in China 5, 137–138 (2010)
3. Zhong, D.: The reform and practice of the university public computer course teaching. Journal of Chongqing Arts and Sciences Academy (Natural Science Edition) 2, 70–72 (2007)
4. Li, J.: The reform of C language experimental teaching. The Civilian Battalion Science and Technology 9, 73 (2010)
5. Liu, Z.: The research and the practice of the projects teaching pedagogy in the C language. Chinese Adult Education 4, 139–140 (2010)
6. Ao, Z.: The research of the college computer course teaching reform. Journal of Hunan Science and Technology Institute 8, 78–80 (2010)
7. Guo, H.: The application of the interesting teaching method used in the C language practice teaching. China Power Education 28, 148–149 (2010)

Web Service Registry and Load Balance of Invocation Based on JXTA

Chenni Wu, Wei Zhang, and Shuangshuang Zhou

Software College of Northeastern University
110819 Shenyang, China
wcnbear@sina.com,
{weiz5535,shuangshuang.yaya}@gmail.com

Abstract. Web services are used to support large-scale distributed applications and interoperable technology. Two of the most important things about web service are its registry and invocation. Traditional registry strategy depends on UDDI protocol, which is an abbreviation of Universal Description, Discovery, and Integration. But UDDI has a typical shortcoming: it is based on central server and the server is easy to become the bottleneck. In centralized architecture there are lots of issues about reliability, performance, scalability and management scope. So in this article, we propose a registry mechanism based on JXTA where is a java'p2p technology; it can remove the central server through p2p technology. Then we propose a load balance of invocation method inside the p2p network to make our JXTA system more effective. We describe our task about this article to make it easy to understand. Finally, we narrate the shortage of this article and introduce future work.

Keywords: p2p, web service, JXTA, load balance, UDDI.

1 Introduction

A Web service is a method of communication between two electronic devices over a network. The W3C defines a "Web service" as "a software system designed to support interoperable machine-to-machine interaction over a network. Web service is widely used nowadays. It has become the one of the most popular technologies in recently years. Lots of related problem about web service appear. One of the most concerned issues is the web service registry. The traditional method to store and register web service is UDDI. Universal Description, Discovery and Integration (UDDI) is a platform-independent, Extensible Markup Language (XML)-based registry for businesses worldwide to list themselves on the Internet and a mechanism to register and locate web service applications [1]. But UDDI has a typical shortcoming; it is based on central server and the server is easy to become bottleneck. In centralized architecture there are lots of issues about reliability, performance, scalability and management scope. So first of all, we purpose a method to solve this registry problem.

We take use of P2P technology to implement web service registry, which can remove the central server. Peer-to-peer (P2P) computing or networking is a distributed

H. Tan (Ed.): Knowledge Discovery and Data Mining, AISC 135, pp. 415–422.
springerlink.com

application architecture that partitions tasks or workloads between peers. In Java area, JXTA is one of the P2P technology implementation [2]. It provides a mechanism named advertisement. It is a well-formed xml document, which store information owned by peer. So we thought we can put web service information into advertisement to construct "web service advertisement", and so that we can easily search web service in the P2P network through JXTA searching method.

Now we can easily search the web service advertisement and send invocation request to the peer node (which is also called edge node) which store the web service advertisement. But in most cases, a same web service can be stored in many edge nodes, and which to choose is difficult. So in this article we propose some load balance mechanism to choose a suitable edge node. Those mechanisms are named "RR"、"Random"、"Weighted Random". In the next part of this article, we will describe those mechanisms in detail.

2 Main Structure

The main idea of this paper is a simple but powerful mechanism to implement web service registry and load balance of service invocation. First of all, we will introduce the theory of P2P web service network, and how can we store web service information into JXTA advertisement. After the service information is stored, how can we find the web service we need. It depends on the mechanism of searching inside the JXTA technology. We will describe the searching process with flow chart in detail. After the storing and searching is done, then we can send service request to corresponding node that stored web service advertisement. The node will invoke the real web service with the parameters in the client request and return back invocation result to requester. The next problem we faced with is which node to choose to execute the web service, because we know that there are many edge nodes storing the same web service advertisement. We propose three load balance method, which is "RR"、"Random"、 "Weighted Random", to realize the load balance of web service invocation. Realizing load balance depends on the rendezvous peer which is the node storing advertisements. We describe the detail of rendezvous peer with structure chart, and then discuss the process of load balance with flow chart in detail.

The paper is structured as follows: Section 2 introduces the main structure of this article. Section 3 is devoted to the web service registry. Section 4 introduces the process of searching and invoking a web service. Section 5 shows how to realize the load balance mechanism. The final section closes the paper tracking also directions for future works.

3 Service Registry

In UDDI, the service description store in central server. But in JXTA, the P2P network, the service description can be stored in each peer node. Service will be inserted into JXTA advertisement. In JXTA, we can divide node into two kinds, one is edge peer and the other is rendezvous peer, they can be in the same peer group. In order to describe the JXTA network in detail, and we can look at the below chart firstly.

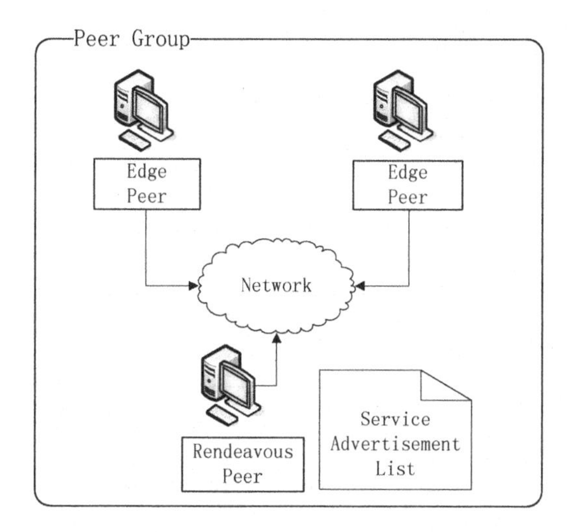

Fig. 1. JXTA Network Structure

We can find that in a same peer group, there is at least one rendezvous and one or more edge peers. All of the advertisements in the same peer group are registered into rendezvous peer. The advertisement list is called Service Advertisement List. Different peer groups connected through rendezvous peer. Just the same as router in IPV4 network, rendezvous peer is responsible to connect to each nodes within the network and the other rendezvous peers. So the whole network is just like IPV4's.

The structure of the network is known to us, and how the web service stored? In JXTA, the media between different nodes is advertisement which is a meaningful XML document. Divided by type, advertisement can be PeerAdvertisement, PeerGroupAdvertisement,PipeAdvertisement,PeerInfoAdvertisement,EndpointAdvertisement,RdvAdvertisement and so on [2]. They have special usage and such as PipeAdvertisement are used to identify the address of peer node so that the others can connect to it. PeerInfoAdvertisement is responsible to let others know what the node's basic information. Because of the XML's feature, we can easily to append our attachment to the advertisement, so we choose PeerInfoAdvertisement. Web service is described by WSDL document. WSDL is an XML format for describing network services as a set of endpoints operating on messages containing either document-oriented or procedure-oriented information [3]. In the WSDL document, we can find out the basic information about this service such as service name, service description, operation names, operation parameters, operation type and so on. So we can construct a XML attachment and then insert it into advertisement. The structure of the advertisement is just like below.

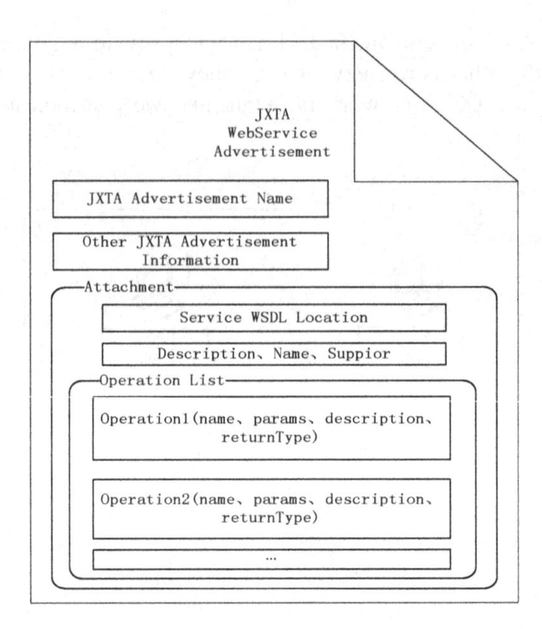

Fig. 2. Web Service Advertisement

We can find that the service information is attached with advertisement. It includes service WSDL location, service description (which is used by searching engine) and so on. So far, the service registry problem has been solved.

4 Searching and Invocation

The above part introduces how we can store and register web service in the JXTA network. In this part, the searching and invocation of web service will be solved. Searching and Invocation are the most important part of this architecture. The purpose of searching is to get the needed service from the vast P2P network. After we know the location of web service advertisement, users will need to invoke the specific operation of the web service. So user will send a request to the node that store web service advertisement, then the node will be responsible to invoke web service and turn the result back to requester.

First of all, we should solve the searching problem. Fortunately, JXTA provides perfect mechanism to search advertisement in P2P network. JXTA can search advertisement by name. The name is just the same as service name in WSDL, and the name can be just a regex, such as "hello*"、"hello?" So it is easy to search web service advertisement though JXTA searching mechanism. The searching structure of JXTA is just like below chart [4].

In the chart, there are some kinds of component. Different LAN (Local Area Network) in P2P communicates with the other LAN through rendezvous peer. The searching request is also dispatched by rendezvous peer.

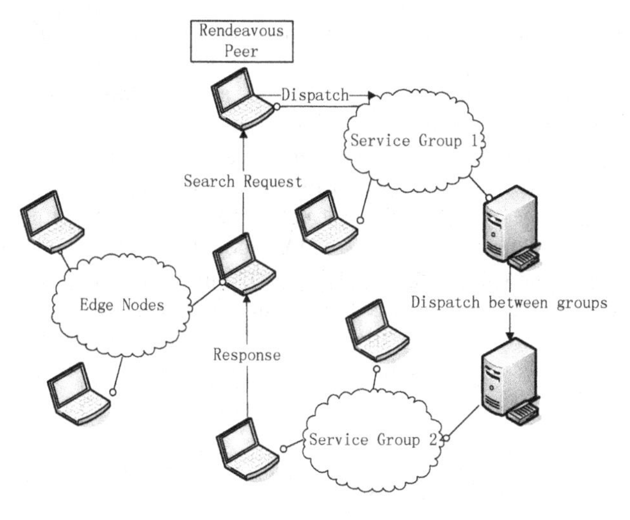

Fig. 3. The Mechanism of Searching in JXTA

Because of the searching process is not easy to explain. So we construct a flow chart. It stands just at the following.

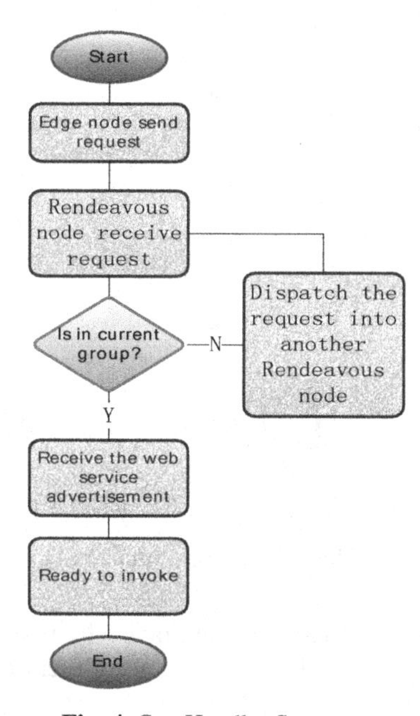

Fig. 4. Qos Handler Structure

In the searching process, edge node sends a searching request with service name pattern to the rendezvous node that is in the same LAN. The rendezvous node receive request and judge whether in the LAN has a service match the name pattern. If it hits, then rendezvous node receive the web service advertisement and return it back to the request node, and the node will be ready to invoke the service. If it does not hit, then the rendezvous node will dispatch the request to the other rendezvous nodes which are the neighbors, then the neighbor will execute the same process until finding the web service advertisement matching name pattern.

After the request node receives the web service advertisement, the request node can invoke the web service operation. According to the web service advertisement, the request module in request node will construct request document with the user input, and then send it to the target node. The SOAP module in target node will parse the receiving document and construct a SOAP message to the real web server. The result will be sent back to SOAP module. SOAP module will send it to the request module in request node, and after parsing the result SOAP, the user will receive the final result. The whole process is shown below.

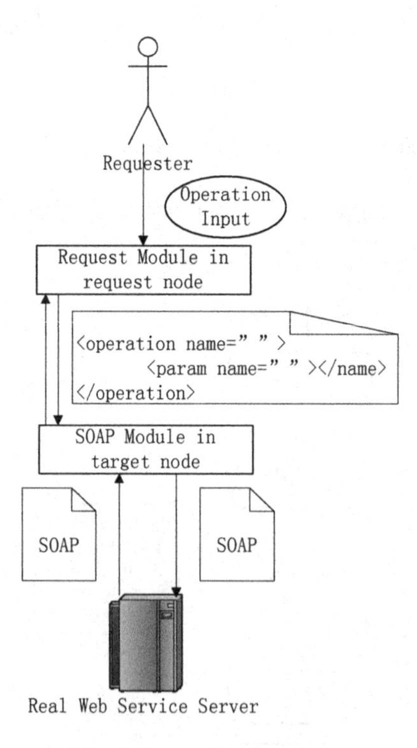

Fig. 5. Invocation Structure

After the above process, the searching and invocation is completed. The next part we will discuss how we can do the load balance of web service invocation.

5 Load Balance

In most cases, the invocations in the P2P network will be so much. Same web service advertisements can distributed across multiple edge nodes. If user invokes this kind of web service many times, we hope it can choose the most suitable edge node to execute the web service. Traditionally, there are three strategies to implement load balance.

First of all, it is called RR (round-robin), if A、B and C have the same web service advertisements. The selection of the node is by order that is one by one. Second, it is called Random, which is randomly choosing a node to execute web service. The last one is weighted random. Because we know different machine has different processing ability. Based on theory of able people should do more work, the higher capability of the machine is, the higher we choose it to execute web service probability is. So different edge node has a number of weights, which is the symbol of its processing ability [5].

In different situations, different strategy can be selected. Based on the selected strategy, we set up load balance structure, which is like below chart.

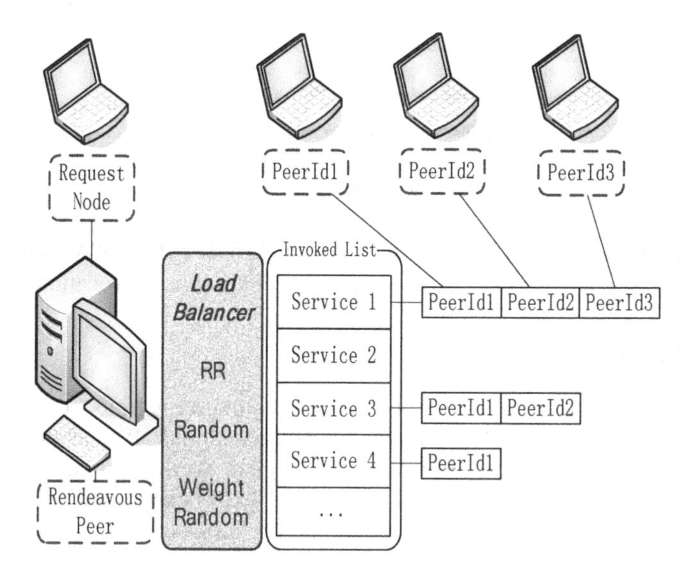

Fig. 6. Load Balance Structure

In rendezvous node, there is a module named load balancer. The balancer stores an invoked list, and each item in the list is a data structure which contains service name and a list of peer id. When a request is received by rendezvous peer, the load balancer will search from the invoked list. If found, we choose a peerId using the selected load balance strategy. A peer id is an UUID which can identify a node in the whole network. So we can directly request the target node with peer id. If there is not a service matching the name pattern, then rendezvous peer will boast the request, and if found, rendezvous peer will append it into invoked list. If the target node with target peer id has removed the web service advertisement, the load balancer will remove the node in the invoked list and then boast the request to find the other node which meets the requirement. The whole process can be described with a flow chart. It is shown below.

The load balance strategy has been tested from many times, and it is proved very effective. The load balance process is suitable for this environment.

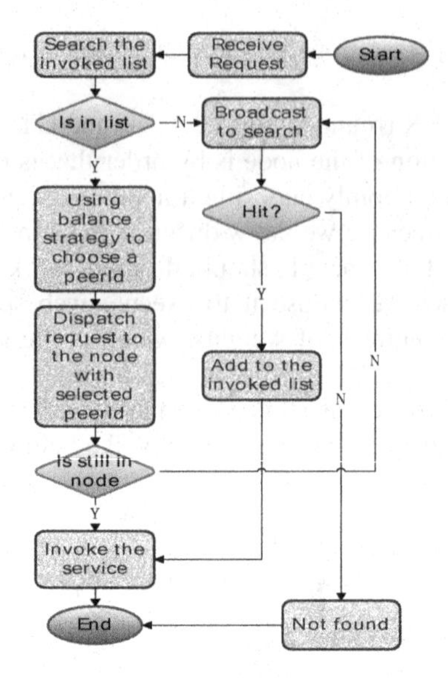

Fig. 7. Load Balance Process

After the above three parts, this article has finished. In this article we complete the load balance、service registry、service searching and service invocation.

6 Conclusions

In this article, we first analysis the shortage of the traditional model of web service registry and then we propose our solution method based on P2P technology. Next, we discuss the service searching and invocation method, so that we can really use the web service. Finally, we consider the problem of load balance and solve it through some load balance strategy. But those methods are still not perfect; we will do our best to make the method better and better.

Acknowledgments. This research work is supported by "Natural Science Foundation of Liaoning Province" (20092006).

References

1. UDDI, `http://en.wikipedia.org/wiki/Universal_Description_Discovery_and_Integration`
2. Oaks, S., Traversat, B., Gong, L.: JXTA Technology Manual, 2nd edn., pp. 64–69 (2004)
3. WSDL, `http://en.wikipedia.org/wiki/Web_Services_Description_Language`
4. Xing, C., Qi, Y., Zhang, Z.: JXTA network resources, research and implementation of the search model, pp. 2–3 (2007)
5. Peng, B.: JXTA-based P2P computing load balancing strategy, pp. 3–4 (2008)

Improved Dynamic Frame Slotted ALOHA Algorithm for Anti-collision in RFID Systems

Shian Liu and Xiaojuan Peng

Guang Zhou Maritime College, Guang Zhou, China

Abstract. The collision caused by tags is one of the most important problems which affect the efficiency of the RFID system. This paper presents the classical ALOHA algorithm in detail and estimates the real-time number of tags according to the collision so as to adjust the frame size of dynamically. Then a new algorithm based on estimating of the tag number is proposed. Analysis and simulation both show that the presented IDFSA algorithm is very effective compared to the conventional algorithms.

Keywords: RFID, ALOHA algorithm, anti-collision, dynamic slot.

1 Introduction

Recently RFID (Radio Frequency IDentification) attracts attention as an alternative to the bar code in the distribution industry, supply chain and banking sector. Nevertheless, RFID has disadvantages about the problem of identified data clearness, the slow progress of RFID standardization and so on. One of the largest disadvantages in RFID system is its low tag identification efficiency by tag collision [1]. Tag collision is the event that the reader cannot identify the data of tag when more than one tags occupying the same RF communication channel simultaneously. Therefore we must reduce tag collision for increasing tag identification efficiency. So far, several tag anti-collision algorithms have been proposed. Among them, the most widely used ones are framed slotted ALOHA algorithm and binary search algorithm. One of the popular anti-collision algorithms is ALOHA-type algorithms, which are simple and show good performance when the number of tags to read is small. However, they generally require exponentially increasing number of slots to identify the tags as the number of tag increases.

In the paper, we propose a new anti-collision algorithm called Improved Dynamic Framed Slotted ALOHA (IDFSA) which estimates the number of unread tags first and adjusts the number of responding tags or the frame size to give the optimal system efficiency. As a result, in the proposed method, the number of slots to read the tags increases linearly as the number of tags does. Simulation results show that the proposed algorithm improves the slot efficiency compared to the conventional algorithms.

H. Tan (Ed.): Knowledge Discovery and Data Mining, AISC 135, pp. 423–430.
springerlink.com

2 ALOHA Algorithm

ALOHA algorithm is a probability algorithm, which belongs to control algorithms of electronic label. When tag comes into reader range, it can send its serial number to reader automatically and communicate with reader subsequently. During one tag transmitting message to reader, the other one also transmits message, information collision will come into being [2].

2.1 Frame Slotted ALOHA Algorithm

Slotted ALOHA algorithm is the tag identification method that each tag transmits its serial number to the reader in the slot of a frame and the reader identifies the tag when it receives the serial number of the tag without collision [3]. A time slot is a time interval that tags transmit their serial number. The reader identifies a tag when a time slot is occupied by only one tag. The current RFID system uses a kind of slotted ALOHA known by framed slotted ALOHA algorithm. A frame is a time interval between requests of a reader and consists of a number of slots. A read cycle is tag identifying process that consists of a frame.

2.2 Basic Framed Slotted ALOHA (BFSA) Algorithm

BFSA algorithm uses a fixed frame size and does not change the size during the process of tag identification. In BFSA, the reader offers information to the tags about the frame size and the random number which is used to select a slot in the frame. Each tag selects a slot number for access using the random number and responds to the slot number in the frame [4].

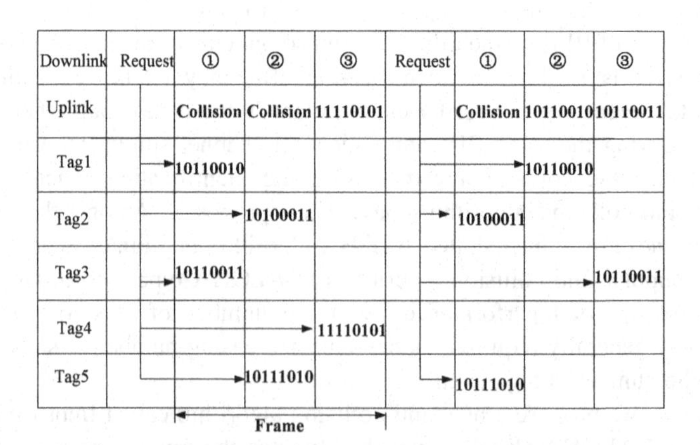

Downlink
Request
①
②
③
Request
①
②
③
Uplink
Collision
Collision
11110101
Collision
10110010
10110011
Tag1
10110010
10110010
Tag2
10100011
10100011
Tag3
10110011
10110011
Tag4
11110101
Tag5
10111010
10111010
Frame

Fig. 1. The process of BFSA algorithm

Fig.1 presents the process of BFSA algorithm. In the first read cycle, Tag 1 and Tag 3 simultaneously transmit their serial numbers in Slot 1. Tag 2 and Tag 5 transmit their serial numbers in Slot 2 respectively. As those are collided each other, i.e. tag collision, Tag 1, 2, 3 and 5 must respond next request of the reader. The reader can identify Tag 4 in the first reader cycle because there is only one tag response in the time Slot 3. In this example, the frame size is set to three slots.

Since the frame size of BFSA algorithm is fixed, its implementation is simple, however, it has a weak point that drops efficiency of tag identification.

2.3 Dynamic Framed Slotted ALOHA (DFSA) Algorithm

DFSA algorithm changes the frame size for efficient tag identification [5]. To determine the frame size, it uses the information such as the number of slots used to identify the tag and the number of the slots collided and so on. So DFSA algorithm can solve partially the problem of BFSA that is inefficient to identify the tag. DFSA algorithm has several versions depending on the methods changing the frame size. Among them, we will briefly explain the two popular methods appearing in [6].

DFSA algorithm can identify the tag efficiently because the reader regulates the frame size according to the number of tags. But, the frame size change alone can not reduce sufficiently the tag collision when there are a number of tags because it can not increase the frame size indefinitely.

3 Improved Dynamic Framed Slotted ALOHA Algorithm

3.1 Optimum Frame Size

The ratio of the number of identification slot to frame size is named as throughput of transmission channel. The larger the throughput of the system, the higher identification speed. To obtain the largest throughput, identification slot should be maximized. Assuming that frame size is N, the number of tag is n within reading and writing scope of readers, thus the probability which the one slot is identification slot can be written as

$$S = P_1 = \frac{C_n^1 (N-1)^{n-1}}{N^n} = (\frac{n}{N})(1-\frac{1}{N})^{n-1} \tag{1}$$

Where S is the throughput of transmission channel. N and n are variables in the equations, to obtain the most throughput, let's partially differentiate (1) in respect to N fixing n .The condition to maximize S is $dS/dN = 0$, it is therefore found that the expression $n = N$ is the maximum condition of S in respect to N.

$$S_{\max} = (1 - 1/N)^{N-1} \tag{2}$$

Resolving the equation (2) we have

$$\lim S_{\max} = (\frac{N-1}{N})^{N-1} = \frac{1}{e} \approx 0.368 \tag{3}$$

The above equality shows that RFID readers have the most throughput when frame size of request frame is equal to the number of tags, thus to obtain the number of tags can be the principal problem to be resolved.

3.2 Unread Tags Evaluating Algorithm

In RFID system, the information from tags to readers is uncontinuous, it is difficult to be defined as a random process. According to received information of frame size from reader request, tags can yield a random number within the scope of the frame size. The random number is evenly distributed within the scope of the frame size. According to it, we can estimate the existing number of tags from collision slot.

In the first place, it is assumed that, there are n tags which are located in efficient scope of readers. Thus the number distribution of tags selecting the ith slot can be expressed as

$$P_{k|i}(\text{x tags selecting the ith slotted}) = \frac{C_n^x (F-1)^{n-x}}{F^n} \tag{4}$$

Where F is frame size, if the collision occurs in the ith slot, it shows that the number of the tag in the ith Slot is two at least. Thus conditional probability distribution can be written as

$$P(\text{x tags selecting the ith slotted} \mid \text{collision}) = \begin{cases} 0 & k = 0 \text{ or } 1 \\ \dfrac{P_x}{1 - P_0 - P_1} & k \geq 2 \end{cases} \tag{5}$$

Conditional probability expectation of the number of tags during the ith slot can be expressed as

$$m = \sum_{x=2}^{n} x P_x^0 = \frac{\sum_{x=2}^{n} x \cdot C_n^x \cdot (F-1)^{(n-x)}}{F^n - (F-1)^{n-1}(F-1+n)} \tag{6}$$

According to derivation, if one slot is empty, the number of tags selecting the slot is zero; if message received accurately, the number of tags selecting the slot is one; if collision occurs, the number of tags selecting the slot is e; hence the number of tags which has not yet read correctly can be forecasted

$$N' = m \cdot N_c \tag{7}$$

Where N_c is the number of slot when collision occurring. By evaluating equation (7), the number of unread tags can be obtained, and adjusting the size of the next frame in term of the number of unread tags subsequently.

To calculate the number of unread tags, the initial number of unread tags should be known to us, during actual identification process, the number of tags which entered into the scope of readers may be unknown usually, program is also need to adjust the initial number of tags During implementing.

4 The Implementation of IDFSA

First, initializing the parameters: by letting $F = F'$, where F' is the initial values of system frame size in the light of actual conditions, Values is 16 generally. initializing the parameters during frame period: by letting $N_c = N_R = 0, S_c = F$, where N_C is the count of the collision slots, N_R is the count of the identification slots, S_C is the count of the readers frame slots.

(1) Send instructions with frame slots size, the RF tags entering into identification range of reader will make response to the event. Readers will detect all slots of one frame. It can be judged for three circumstances: ① data received correctly, by adding 1 to the count of the identification slots ,namely $N_R + 1$; ② Collision occurred, by adding 1 to the count of the collision slots, namely $N_c + 1$; ③ no acknowledge signal. in all three cases, by decreasing 1 to the count of the reader slots, namely $S_c - 1$.

(2) Judge the count of the reader slots whether it is 0, if it was equal to zero, and then judge the count of the collision slots whether it is 0, if it was also equal to zero, express that all RFID tags within identification range have been identified, then end this checking process. If it was not equal to zero, to call a subroutine which is used for adjusting frame size and reset the frame size of readers. If S_C was not equal to zero, the readers will send instructions unceasingly, detect the next slot and continue to step (1) and judge.

Flow of estimating the number of unread tag is shown in Fig.2.

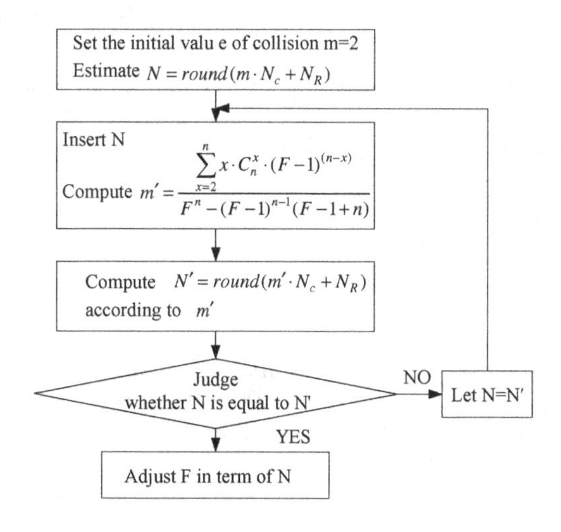

Fig. 2. Flow of estimating the number of unread tag

5 Performance Analysis and Simulation for IDFSA Algorithm

We can easily enumerate each combinations of N_c and N_R, where F is frame size and the number of tag is n.

If $N_c > 0$, count the next frame size using equation (7) and round a number to the nearest integer. Then we have

$$N(n,F) = F + \sum_{N_c=1}^{n/2} \sum_{N_R=0}^{n-2N_c} P(N_c, N_R | n, F) \cdot N(n - N_R, round(m_{(n,F)} \cdot N_c)) \tag{8}$$

For probability $P(N_c, N_R|n, F)$, recursive formula as below

$$\begin{aligned} P(N_c, N_R|n, F) = {} & P(N_c, N_R|n-1, F) \cdot \frac{N_c}{F} + P(N_c - 1, N_R + 1|n, F) \cdot \frac{N_R + 1}{F} \\ & + P(N_c, N_R - 1|n-1, F) \cdot \frac{F - N_c - N_R + 1}{F} \end{aligned} \tag{9}$$

Where

$N(n,F)$ is the number of slot which is in need of during every round (including the slot number of the current frame)

$P(N_c, N_R|n, F)$ is conditional probability, When N_C slots in a collision and N_R received correctly at the same time, where the current frame size is F and the number of tags is n.

1) according to equation (6), construct function m according to equation, input parameters is the tag number N and the current frame size F, output is collision coefficient m;

2) according to equation (9), construct probability matrix function P, input parameters is the tag number N and the current frame size F, the output is probability matrix Which is composed of N_C and N_R.

3) By using main function, construct the matrix solution of equation of first degree in term of equation (8), to obtain the values of collision coefficient and probability,we should call a subroutine m function and P function above, the variable of equation set is relative to demand.

4) Solve equations and save solution of equations which the number n of tag is from small to big, then we can obtain the solution of tag n+1 and Solution Matrix of tag n, subsequently extract the necessary data from Matrix and form figure.

Using MATLAB to simulate the process of reader identifying tag and analyze performance of the Improved DFSA algorithm. It is assumed that the largest number of entering into reading range is 70, the initial frame size F is 16, the frame size of FSA is 16, 32 respectively, system finish 200 cycles. The average of the total slot number was taken. Fig.3, Fig.4 plots the simulation curves respectively.

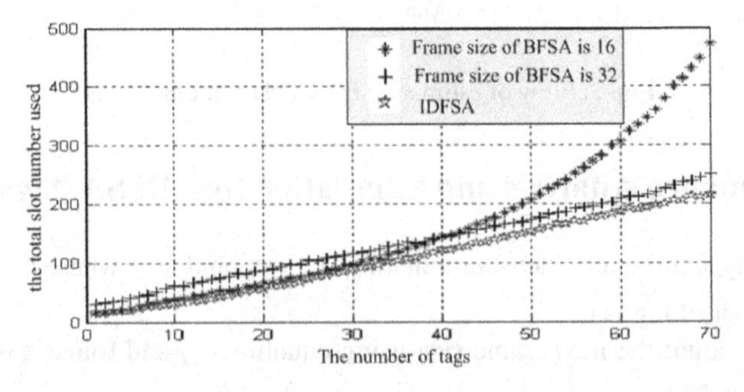

Fig. 3. Plot of slots needed to identify tag (FSA and Improved DFSA algorithm)

Fig.3 shows that when a certain number tag enter into RF FIELD, the number of slot is 16, the number of tags is greater than 45, the total slot number to identify all tags using FSA will rise quickly, and the number of slot is 32, the number of tags is smaller than 40, the total slot number to identify all tags using FSA will be more. If the number of tags is greater than 40, the total slot number to identify all tags using FSA will still rise quickly at speed above. The Improved DFSA will save slot in the whole range with respect to the two algorithms above.

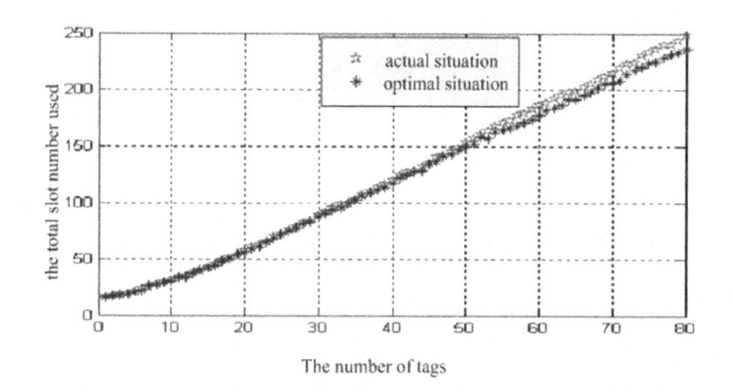

Fig. 4. Comparison between actual situation and optimal situation

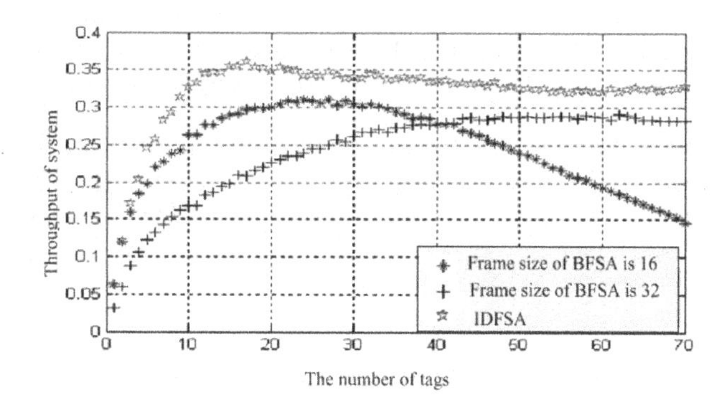

Fig. 5. Comparison of throughput between FSA and Improved DFSA algorithm

It is found that from Fig.4 the difference is not evident between changing frame size according to actual known number of tags and adjusting frame size by estimation of the number of tags. The Improved DFSA algorithm has good properties.

Now analyze performance of Improved DFSA algorithm in the aspect of throughput. It is assumed that the largest number of entering into reading range is 70, the initial frame size F is 16, the frame size of FSA is 16, 32 respectively, system finish 300 cycles. The average of the throughput was taken, as shown in Fig.5, which plots the simulation curve respectively.

As shown in Fig.5, Whether the number of tags is smaller than 40 or bigger than 40, the throughput of Improved DFSA algorithm is greater than the other two methods above, and in whole identification range the throughput of system is smooth highly.

6 Conclusions

In the paper, a new anti-collision algorithm called Improved Dynamic Framed Slotted ALOHA (IDFSA) has been proposed which estimates the number of unread tags first and adjusts the number of responding tags or the frame size to give the optimal system efficiency. As a result, in the proposed method, the number of slots to read the tags increases linearly as the number of tags does. Simulation results show that the proposed algorithm improves the slot efficiency compared to the conventional algorithms.

References

1. Finkenzeller, K.: RFID handbook, 2nd edn., pp. 195–219. John Wiley & Sons (2003)
2. Wieselthier, J.E., Ephremides, A., Michaels, L.A.: An Exact Analysis and Performance Evaluation of Framed ALOHA with Capture. IEEE Transactions on Communications 37(02), 125–137 (1989)
3. Roberts, L.G.: Extensions of Packet Communication Technology to a Hand Held Personal Terminal. In: Spring Joint Computer Conference, vol. 40, pp. 295–298 (1972)
4. Huang, X., Le, S.: Efficient Dynamic Framed Slotted ALOHA for RFID Passive Tags. In: The 9th International Conference on Advanced Communication Technology, vol. 1, pp. 94–97 (2007)
5. Lee, S.-R., Joo, S.-D., Lee, C.-W.: An enhanced dynamic framed slotted ALOHA algorithm for RFID tag identification. In: IEEE Proceedings of the Second Annual International Conference on Mobile and Ubiquitous Systems: Networking and Services, vol. 13, pp. 166–172 (2005)
6. Peng, Q., Zhang, M., Wu, W.: Variant Enhanced Dynamic Framed Slotted ALOHA Algorithm for Fast Object Identification in RFID System. In: 2007 IEEE International Workshop on Anti-counterfeiting, Security, Identification, pp. 88–91 (April 2007)

NC Skill Cultivation for Students with Specialty of Mechanical Design, Manufacturing and Automation

Xueguang Li and Shuren Zhang

College of Mechanical and Electrical Engineering,
Changchun University of Science and Technology , Changchun 130022
lixg_1979@163.com, srzhang@163.com

Abstract. The development of NC technology have a profound impact on manufacturing , the development of course of NC machine tool technology will help us understand and master the NC technology better . Along with adjustment to society's vocational needs, the construction of some courses has become more prominent. The necessity of the construction of the NC technology course is analyzed from the aspects of the actual needs of the community and students mastering the knowledge, and the course teaching method is discussed, and the corresponding data statistics for different teaching methods for NC technology course are done through a sampling survey of students.

Keywords: NC machine tools, NC technology, teaching methods.

1 Introduction

With the development of computer technology and electronic information technology, a revolutionary change has occurred in the manufacturing, NC machine tools as the representative manufacturing mode has become a symbol of the manufacturing industry. CAD/CAM/CAPP's widely use has further promoted the development of numerical control technology. At present, more than 90 percent of all industries use NC machine tools in mechanical manufacturing, involving aerospace, aviation, military, instrumentation, automobile, textile and other industries.

With the development of NC technology, the demand on the NC talent also increases, but most courses introduce CNC programming and CNC machine operation, the students don't know too much about CNC machine structure and characteristics, are not clear with possible problems in the use and corresponding measures, so the NC skill is difficult to obtain. Through the study of the course of NC technology , students can understand the internal structure of machine tools better, and then can make better use of NC machine tools. while the pressure of college graduate employment also increases gradually, higher education changes from the "elite education" to "public education" in the development process, through enlarging students' knowledge, also can further relieve employment pressure, make the students' employment range further expanded[1].

H. Tan (Ed.): Knowledge Discovery and Data Mining, AISC 135, pp. 431–436.
springerlink.com

2 Use the Process of Solving Fault to Reflect the Process of S Obtaining Skill

The course that cultivate NC skill involves many areas of knowledge , including mechanical, electronic control, computer, hydraulic, PLC and and many other branches[2] , and the course does not solves problems only a single aspect or areas, the likelihood is multidisciplinary questions, so that through the study of this course, students will learn multiple disciplines together, think about questions from the point of practical use , and consider maintenance from the angle of the design. Such as teaching the composition of NC machine tools, the students will not learn the contents profoundly when explaining the composition of the NC machine tools just from theory, but if explaining like pictures showed in Fig .1, combining with a certain fault to search according to the signal flow, will significantly enhance the understanding of the students and make them serve multiple purposes by analogy.

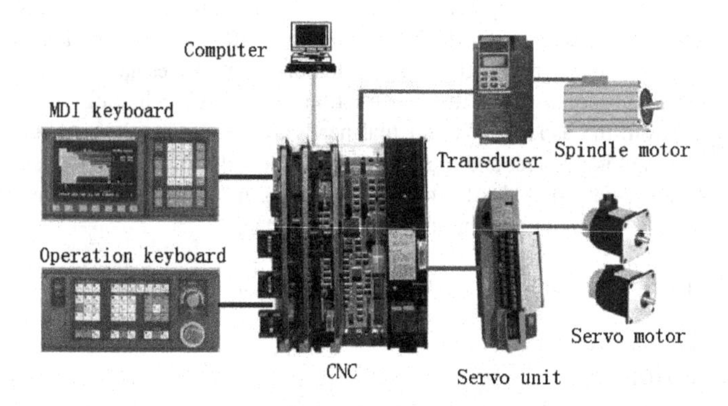

Fig. 1. Composition of NC machine tools

Now take a certain coordinate which is invalid to NC commands and do not move for example and do analysis ,possible fault positions contain axial plate of numerical control system, servo module, servo motor, electrical encoder, screw and nuts, etc. So during troubleshooting, we must firstly understand these parts very clear, understand the relationships between the parts and signal flow clear which involve mechanical and electrical together closely, at least understanding the structure of mechanical transmission parts involved in the movement of machine axes, such as the connection between screw nut and table, screw and the coupling, motor and coupling. In addition we have to understand the processes of control, such as the connection between axial plate and servo module, servo module and the servo motor, including control flow and wiring connecting methods of the enable signal and feedback signal, as shown in Fig .2. A invalid axes fault involves theory of mechanical and electrical control, interface, and other aspects, so when dealing with fault, analyze possible problems in machinery and control, we can eventually solve fault.

Through this case we can see that fault diagnosis of NC machine tool involves comprehensive use of various knowledge, including mastering some practical experience in the course of lesson, can make students comprehensive use professional knowledge which have learned[3] . So that both we can learn new knowledge, and knowledge can be applied to the concrete practice , apply the professional knowledge and the theory, and not just confined to theoretical level, because undeniably a large number of graduate students will put into some areas related to applied technology. So comprehensive use the knowledge theory of various aspects connected with NC machine to analyze and solve some actual problems during the concentration study in schools stage, will undoubtedly enhance the students' actual ability to solve problems.

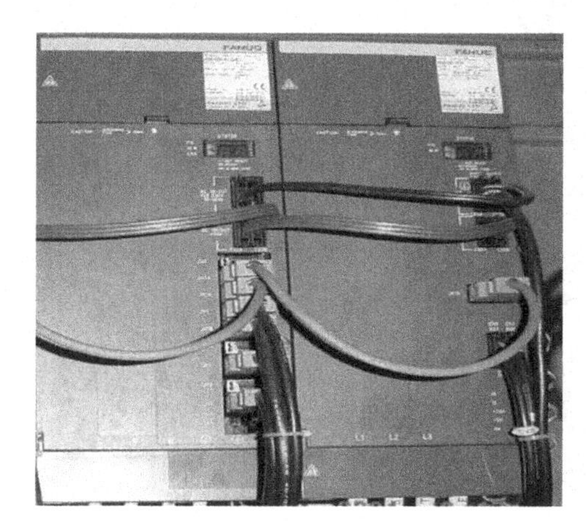

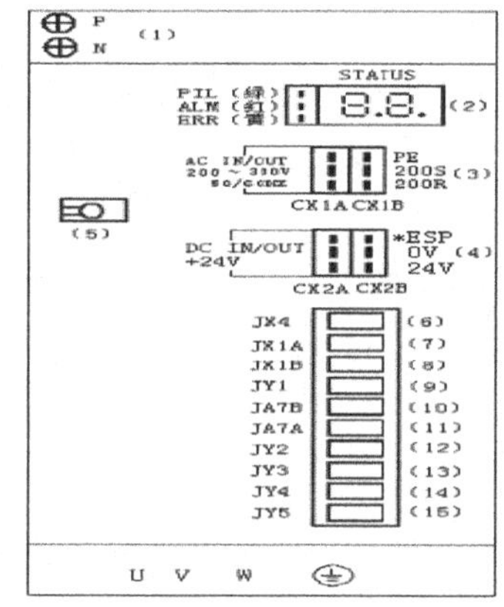

Fig. 2. Servo module and connection

3 Approach of Improving Student's NC Skill Level

NC skill, which involves both theories of fault diagnosis and maintenance of NC machine tool and the crossover of machine, electricity, liquid multidisciplinary and comprehensive utilization, so during teaching knowledge we should explain by combining theory with practice to improve the ability of solving practical problems mainly, while enhancing the comprehensive ability of students so that the students can validate and reappear the theoretical knowledge through actual case or intuitive media form.

So the course should be taught with more actual fault case and the actual image, instead of just the boring theoretical explanations. Because it is the fault diagnosis and maintenance of NC machine tool, so the structure of NC machine tools, function characteristics, the method of control must be very clearly understood. A single component or element may be known or studied, but an independent functional component assembled by several parts together are not very clear, for example we are familiar with machine table, motor, screw nut pair, coupling, bearing and bearing etc parts , but are not familiar with how to assemble a machine feeding transmission system, how to install the various components, and how to control the movement of worktable.

If providing a large number of actual pictures or multimedia material during the interpretation process, the students will have a deep understanding in sensory, and it is helpful to deepen understanding and enhance students' interest in learning,. The method of control and main transmission is shown in Fig.3,through combine with the actual picture, it will help the students enhance the understanding and mastering of the content.

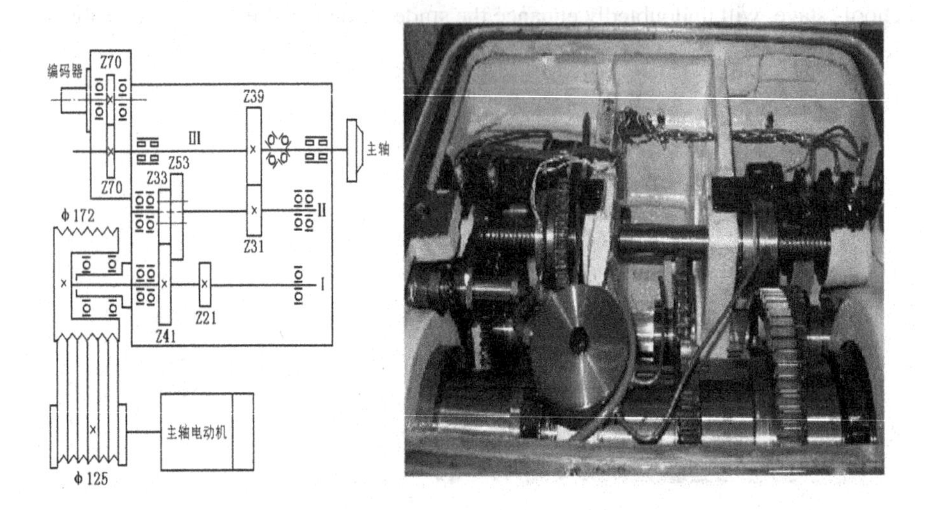

Fig. 3. Main transmission system

NC machine tool is an organic whole, its' fault usually involves several interrelated parts, so various disciplines cross and link together in the course, mechanical structure and control system structure must be manipulated in order to deal with the occurrence of fault capability. Because the time which the students contact machine is relatively small, and the students know little about the internal structure . So we should make full use of various multimedia form to aid teaching during the process of teaching, and present complex problems or structures to students in simple and intuitive way. It is better to let students see objects if having conditions , and explaining the theory after a direct impression of sense would be helpful for the students to understand and remember.

Secondly, consider solving problems from the angle of the design, if we are very clear of the structure or the control , if we can read the mind of the designer, then the same problems will have different solutions. The author took a random survey of students of several classes , the results is shown in Fig.4 ,some students were in favor of case analysis combined with theoretical teaching , some students were in favor of more scene teaching and the others are incline to case teaching. The survey data showed that students were more inclined to the learning way of combinating theory and experiment together.

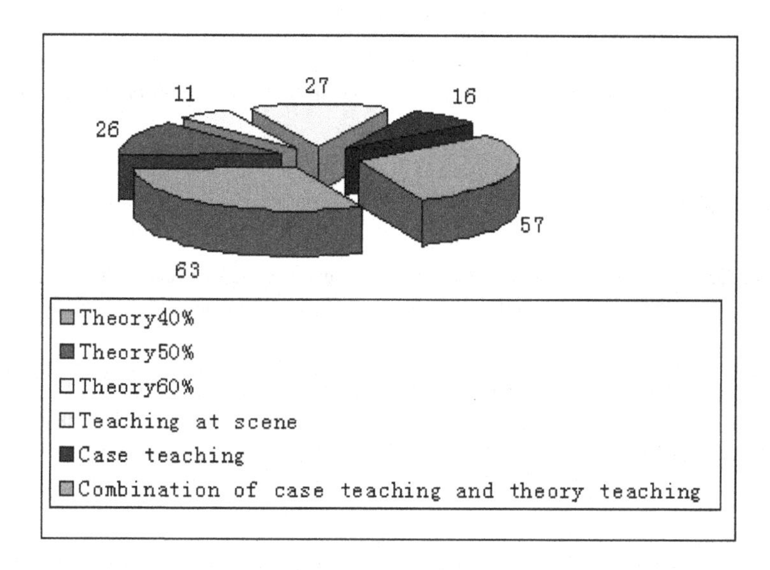

Fig. 4. Survey of teaching methods

4 Prospective Goal

NC technology is a strong comprehensive course, after the study of this course, the students can master the general methods of fault diagnosis of NC machine tool and comprehend the structure, control theory, connections between software and hardware of NC machine tool. Secondly, comprehensive use and further understanding of all kinds of mechanical, electrical components in an integrated platform will help students develop the ability of practical work. Fault of NC machine tool is not entirely the fault of machine in the traditional sense, the modern NC machine tool has applied the newest technology of power electronics, motor, testing and computer technology. Due to the rapid development of these technologies and their successful application in NC machine tools, new and completely different demands are put forward in the operation, maintenance and management of NC machine tools; whereas the students also can comprehensive use the knowledge of these disciplines through the study of the course of NC technology.

5 Conclusion

The development of the manufacturing industry puts forward new requirements of the development of NC machine tools and the NC technology, and the related knowledge structure create new changes with the development of NC machine tool technology, the building of the NC technology course is particularly important on the premise of the employment pressure and the increasing demand of NC technology talent. On the one hand ,through the study of this course the students can improve their comprehensive ability, enhance the understanding and application of the crossed multi-disciplinary knowledge, on the other hand, the students can further develop the range of employment, through the study and practice of this course the students with specialty of mechanical design, manufacturing and automation will be provided with a broader employment space.

References

1. Dai, Y.: Thoughts of offering undergraduate specialty of diagnosis and maintenance of NC machine tool. Journal of TIANJIN University of Technology and Education 15(3), 60–62 (2005)
2. Liu, J., Jiang, Q.: The partice explore in project about reform for course of diagnosis and maintenance of NC machine tool. Communication of Vocational Education 12, 34–35 (2005)
3. Zhang, T., Wang, Q.: The application of case teaching method in the course of diagnosis and maintenance of NC machine tool. Science & Technology Information 29, 612–615 (2008)

Research and Application of Heterogeneous Network Topology Discovery Algorithm Based on Multiple Spanning Tree Protocol

Dancheng Li, Chen Zheng, Chunyan Han, and Yixian Liu

Software College Northeastern University
Shenyang 110004, Liaoning, P.R. China
ldc@mail.neu.edu.cn, jackychen129@gmail.com

Abstract. Heterogeneous network topology discovery is significant for network management and network. According to the requirement of the real network environment ,in this paper we analyze the current mainstream data-link level topology discovery algorithms, including spanning tree topology discovery algorithms and forwarding-table topology discovery algorithm, discussing the advantages and disadvantages of each method and come up with a heterogeneous network topology discovery process based on multiple spanning Tree protocol including STP / RSTP, PVST +, MISTP / MSTP protocols, The algorithm has been implemented and tested in several enterprise-level networks, and the results demonstrate that it discovers the physical topology information, VLAN information , spanning tree logical topology information relatively accurate.

Keywords: Network Topology Discovery, Multiple Spanning Tree Protocol, SNMP, VLAN technology.

1 Introduction

With the network development, the enterprises` network become considerably large and complex, in addition, a various kinds of network equipments have been adapted in the network as long as the data-link level Spanning-Tree protocol is becoming complicated. Therefore, the algorithm suit for a heterogeneous network based on multiple Spanning-Tree protocols is accumulating important to the large enterprises.

Network topology discovery is quite a beneficial part to the whole network management software. It supervises the existing resources such as routers, switches, subnets, VLANs and Spanning-Tree Protocols. In today`s enterprises, a ascend number of different equipments and Spanning-Tree protocols have been set in order to fulfill the need of multi-VLAN network environment. For this reason, topology discovery algorithm has to be more suitable. In this paper, an efficient algorithm for heterogeneous network has been discussed and proposed.

H. Tan (Ed.): Knowledge Discovery and Data Mining, AISC 135, pp. 437–444.
springerlink.com

The remainder of this paper is organized as follows. Section 2 briefly presents the related works followed by section 3 which presents the system architecture. In Section 4, the proposed algorithm is described in detail. The implementation and testing are shown in section 5. Finally, section 6 gives the conclusions.

2 Related Works

Layer 3 topology discovery algorithm has been mature in network management algorithm, many approaches have been proposed such as in [1] and [2]. Layer 2 topology discovery algorithm mainly has two mainstream ways. In[3] is proposed to discovery the topology based on the tables of Spanning-Tree protocol(STP) information. Another important solution was came up by [4],it solely depends on the standard addresses forwarding table (AFT) information collected in SNMP MIB. Both algorithms are not perfect in some perspective, there exists a problem which can not be solved in theory. This article has given a considerably usable solution. Based on the solution, the article continues to study the Multiple Spanning Tree Protocol, including the [5] proposed the features of STP\RSTP, the [6] argued the characters of PVST+ and the relationship between MISTP and MST discussed in [7]. After have a deep understand of these protocols, it can safely draw a overall algorithm based on this results. The papers has no studied on the logical topology information. Considering the security of network, some enterprises' networks will block some ICMP protocol, such as ping and trace-route. Therefore the ICMP based algorithms are not always practicable for the enterprise-level network topology discovery. The approach [8] proposes a method based on the SNMP to build a packet in order to gain the topology.

Nowadays SNMP is supported in each router and managed switch and becomes a necessary tool for network management. Besides, multiple Spanning-tree protocol have been widely adopted by many companies. Considering the characteristics of enterprise-level networks, this paper propose a common-suitable network topology algorithm that can works well for network management. This algorithm is based on SNMP, and can discover the Layer 2 network topology accurately and the logical multiple Spanning-tree protocol topology information.

3 System Architecture

The architecture consists of three modules as illustrated in Figure 1, the Data Collecting Module, the Topology Analyzing Module and the Topology Visualizing Module.

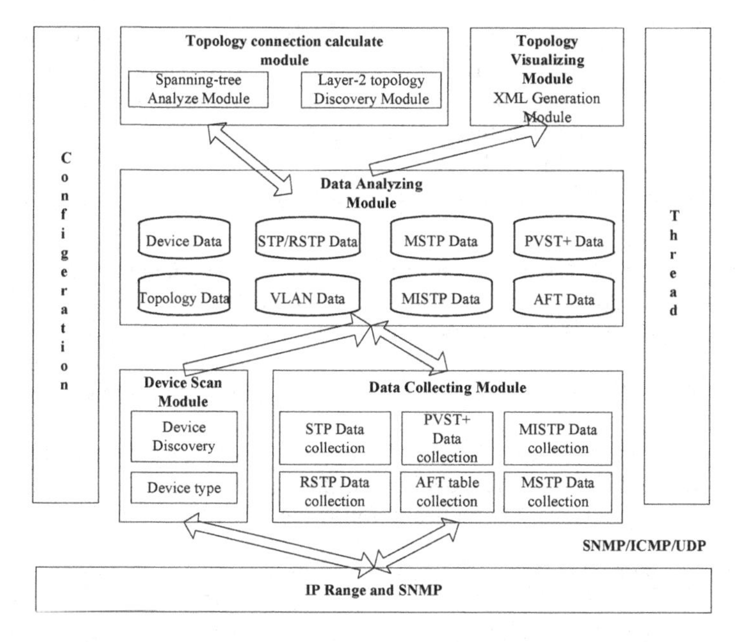

Fig. 1. System Architecture

Firstly, the Data Colleting Module discovers and detects types of the active devices among the IP addresses range of the enterprise network. Then, this module restores and selects the topology related data that are stored in MIBs of the routers and switches. The Topology Analyzing Module is responsible for analyzing the interconnections between the discovered devices and the logical connection information of multiple spanning tree through the collected data, and the result will be stored in the XML files. Finally, the Topology Visualizing Module visualizes the network topology through the data stored in the result files.

This paper is only focused on Data Collecting Module and Topology Analyzing Module. The remainder paper describes the two modules in detail.

4 Algorithms Used

4.1 Probing and Detecting Active Devices

In order to gain the Active devices in the management network, It should study the devices` ARP table to get the IP information, what`s more, the IP is not all active, It can use PING or other tools to finish the detection, such as FPING.

4.2 Collecting Topology Data from Devices

The algorithm mainly uses SNMP4J and SNMPWALK open-source tools to gain the information needed in devices. Because of VLAN information can not be gain by

SNMP4J using community@VlanID way, so the algorithm combines two tools to get information. The other problem is that the ARP table, MAC information is not static, it always changes, so, it must not only consider the gathering information but also use the incremental gathering method. The algorithm uses the timer to dynamically update the table to solve the problem.

The major basis tables used in this algorithm lie as following.

a. Device-MAC binging table

Table shows the device IP and all MAC information belonging to this device. The iftable in Interface group store the related information.

b. Device port-interface corresponding table

Table shows the virtual ports number related to real interface information. dot1dBasePortTable in dot1dBase group stores corresponding information.

c. ARP table

Table shows the MAC addresses related to the interface in device. The ipNetToMediaTable stores corresponding information.

The MIB table OID information lies as following.

TABLE	OID
dot1dBasePortTable	1.3.6.1.2.1.17.1.4
ipNetToMediaTable	1.3.6.1.2.1.4.22

4.3 Analyzing Layer 2 Topology

The Article adopts two algorithm to suit to the heterogeneous topology, algorithm based on down-port AFT uses to meet the multiple spanning-tree protocol, if the topology uses STP, it can only uses the algorithm based on STP. The specific algorithm shows below.

a. Algorithm based on down-port AFT

Firstly, it should collects the dot1dTpFdbTable to study the AFT information. Then, using the data to calculate the connection relationship, the procedure lies as follows.

Step1. According to the AFT information to decide which up-port and down-port is.

Step2. Decide the descendant ports.

Step3. Decide the parents nodes and parents ports.

Step4. Find the nodes who has zero descendant.

Step5. Find the parents ports.

Step6. Decide the connection between the nodes who has zero descendant and its father node.

Step7. Minus the parent nodes` descendants numbers

Step8. Repeat the Step5 and Step 6 until the nodes list is empty.

b. Algorithm based on STP(Spanning Tree Protocol)

Firstly, collecting the dot1dStpPortTable information, gaining ports, forwarding status, root switcher, designed switcher, designed port information.

Secondly, select ports whose forwarding status is 5, select the appropriate designed ports.

Thirdly, according to the dot1dbasebridgeaddresss to decide the connection relationship.

The MIB table OID information lies as following.

TABLE	OID
dot1dTpFdbTable	1.3.6.1.2.17.4.3
dot1dStpPortTable	1.3.6.1.2.1.17.2.15.1
dot1dbasebridgeaddresss	1.3.6.1.2.1.17.1.1

4.4 Analyzing VLAN Information

a. determine the equipment type

Based on the sysDescr in theMIB-II system group, it may include the enterprise information.

b. collect the switches` VLAN information

In H3C equipments, hwdot1qVlanMIBTable in the private MIB contains the VLAN list information. Specifically, it can gain the VLAN list by study the hwdot1qVlanIndex.

In Cisco equipments, the Layer-3 equipments should gain the VLAN list information through the cviRoutedVlanIfIndex in cviVlanInterfaceIndexTable. The Layer-2 equipments should gain the the VLAN list information through the ifDescr with the 'vlan' from the start.

a. calculate the VLAN ports

In H3C devices, check the hwdot1qVlanPorts, according to the principle that 8bits show that if the 8 to 1 and 16 to 9 ports belong to this VLAN, it can gain which ports belong to which VLAN. And the Trunk port is calculated by meeting all the ports bits.

In Cisco equipments, VLAN ports and Trunk ports information come from the vlanTrunkPortDynamicStatus and vmMembershipTable

b. VLAN information display

After gain the information of VLAN, based on the physical topology connection, it can display the topology information according to the VLAN information, a set of equipments belongs to the same VLAN can be shown in a unique topology.

The MIB table OID information lies as following.

Table	OID
sysDescr	1.3.6.1.2.1.1.1
hwdot1qVlanIndex	1.3.6.1.4.1.2011.2.23.1.2.1.1 1.1
cviRoutedVlanIfIndex	1.3.6.1.4.1.9.9.128.1.1.1.1.3
ifDescr	1.3.6.1.2.1.2.2.1.2
VlanPorts	1.3.6.1.4.1.2011.2.23.1.2.1.1 1.3
vlanTrunkPortDynamicStatus	1.3.6.1.4.1.9.9.46.1.6.1.1.14
vmMembershipTable	1.3.6.1.4.1.9.9.68.1.2.2

4.5 Analyzing Mutiple Spanning-Tree Topology

It needs to decide the multiple spaning-tree type, according to the type information to decide which table should be check and collect the information in order to gain the logical spanning tree connection information. Through the Dot1dStpVersion can gain the type information.

a. STP/RSTP

According to the Dot1dStpVersion, value equals 0 means STP, 2 means RSTP

When it is STP information has already gained when study the layer-2 physic information using algorithm based on STP.

When it is RSTP, it should check the stpxRSTPPortRoleValue in stpRSTPPortRoleTable to decide the alternate and backup ports information.

And then check the stpxRSTPPortRoleInstanceIndex value to decide the RSTP belongs to which Instance of PVST+.

At last, use stpxRSTPPortRoleValue to decide the true type of the ports.

b. PVST+/MISTP/MSTP

Firstly, check the stpxSpanningTreeType in stpxSpanningTreeObjects to confirm which type the spanning tree protocol is.

If it is 1or5, it means PVST+. It should check the stpxPVSTVlanEntry in stpxPVSTVlanTable, it can gain the VLAN Index and if the PVST+ is running on this VLAN. And then it has to check the cvbStpEntry in cvbStpTable to know which ports belong to which VLAN.

If it is 2, it means MISTP. It should firstly check the stpxMISTPInstanceNumber to know the instance number. Additionally, it can study the stpxMISTPInstanceEntry in stpxMISTPInstanceTable to make sure the relationship between VLAN and Instance in MISTP.

If it is 4, it means MSTP. Firstly, it should check the stpxMSTInstanceEntry in stpxMSTInstanceTable to gain the the relationship between VLAN and Instance in MSTP. In addition, it can obtain the ports working-status by studying the stpxMSTPortRoleEntry in stpxMSTPortRoleTable.

Till now, the mainstream multiple spanning-tree protocols information have been discussed above and the way to gain important logical topology information have been given above at the same time, which means the algorithm is temporarily over.

The MIB table OID information lies as following.

Table	OID
Dot1dStpVersion	1.3.6.1.2.1.17.2.16
stpxRSTPPortRoleValue	1.3.6.1.4.1.9.9.82.1.12.2.1
stpxSpanningTreeType	1.3.6.1.4.1.9.9.82.1.6.1
stpxSpanningTreeObjects	1.3.6.1.4.1.9.9.82.1.6
cvbStpEntry	1.3.6.1.4.1.9.9.56.1.1.1.1
stpxMISTPInstanceNumber	1.3.6.1.4.1.9.9.82.1.7.1
stpxMSTInstanceEntry	1.3.6.1.4.1.9.9.82.1.11.9.1
stpxMSTPortRoleEntry	1.3.6.1.4.1.9.9.82.1.11.12.1

5 Implementation

The topology discovery algorithm is implemented in Java and the Data Collecting Module is designed to work concurrently. The algorithm has been executed in several enterprise networks, such as networks in the banks and telecom companies. The experimental results validate the algorithm, demonstrating it can work efficiently and accurately. In an enterprise network which has 40 devices among the range with more than 100 IP addresses, the algorithm can complete discovering topology of the network within 4 minutes. Figure 2 shows the visualization of the topology including PC.

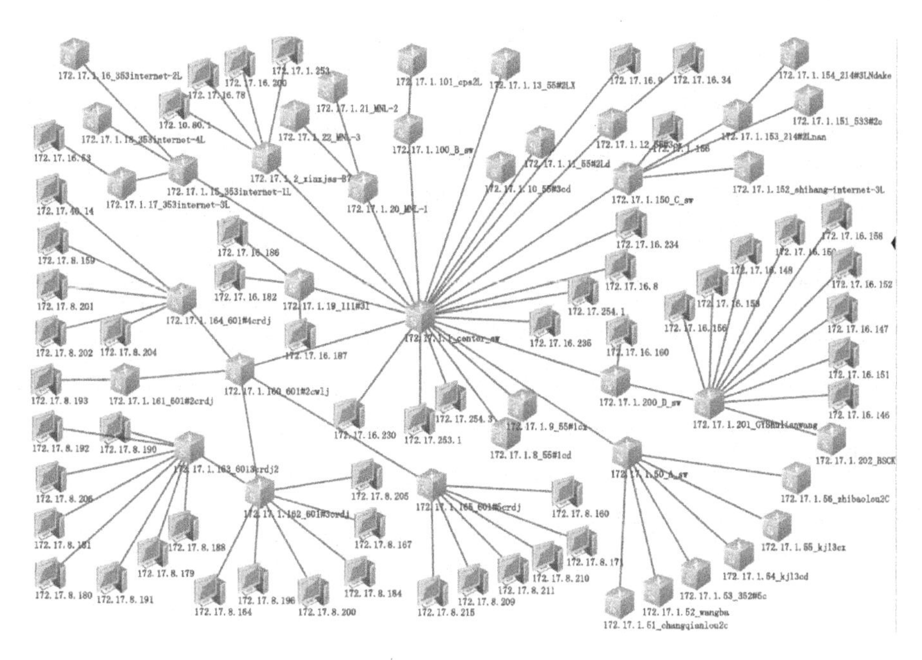

Fig. 2. Topology including PC

6 Conclusion

Heterogeneous Network Topology Discovery Algorithm Based on Multiple Spanning Tree Protocol plays an import role in real enterprises` network management. According to the requirement of many companies` requirements, this paper presents an efficient algorithm for heterogeneous Network Topology Discovery. The algorithm only need that the equipments in the network run SNMP. In the algorithm, For making sure that the AFT is complete, the article proposed a incremental discovery method, at the same time, the algorithm based on AFT and Spanning Tree both have been adopted to match different network environments. Based on the layer-2 discovery results, the logical connection information of heterogeneous multiple

Spanning-tree can be found at last. In multiple Spanning-tree procedure, VLAN information has been used necessarily, the algorithm has been implemented and tested in several enterprise-level networks, and the results demonstrate that it discovers the network topology accurately and efficiently.

Acknowledgement. This research work is supported by "Natural Science Foundation of Liaoning Province" (20092006).

References

1. Siamwalla, R., Sharma, R., Keshav, S.: Discovering Internet topology. Cornell Network Research Group Department of Computer Science Cornell University, Ithaca, NY 14853 (1999)
2. Stott, T.: Snmp-based layer-3 path discovery. Tech. Rep. ALR-2002-005, Avaya Labs Research, Avaya Inc., Basking Ridge, NJ (2002)
3. Bejerano, Y., Breitbart, Y., Garofalakis, M., Rastogi, R.: Physical Topology Discovery for Large Multi-Subnet Networks. Bell Labs, Lucent Technologies 600 Mountain Ave., Murray Hill, NJ 07974
4. Lowekamp, B., O'Hallaron, D.R., Gross, T.R.: Topology Discovery for large Ethenet Networks. In: Proceeding of ACM SIGCOMM, San Diego, California (August 2001)
5. SNMPbaidubaike (EB/OL) (March 2011) (in Chinese), `http://baike.baidu.com/view/2899.html`
6. PVST+ (EB/OL) (January 2011) (in Chinese), `http://yuxin.blog.51cto.com/869751/261249`
7. MSTP/MISTPbaiduwenku (in Chinese), `http://wenku.baidu.com/view/b55a997da26925c52cc5bfe3.html`
8. Lin, H.-C., Lai, S.-C., Chen, P.-W.: An Algorithm for Automatic Topology Discovery. National Tsing Hua University, HsinChu (December 1998)

Research of Photoelectric Position Acquisition System

Haima Yang[1,2], Siyong Fu[1], Yihua Hu[2], Jin Liu[1], and Jianyu Wang[2]

[1] School of Optical Electrical and Computer Engineering, University of Shanghai for Science and Technology, Shanghai 200093, China
[2] Faculty Office II, Shanghai Institute of Technical Physics of the Chinese Academy of Sciences, Shanghai, 200083, China
snowyhm@sina.com

Abstract. A high speed multi-channel photoelectric data acquisition system has been designed in this paper with FPGA as its master controller. This system is used in the application of PSD that measured the angel of space beacon light. In this system, it's capable of 4-route synchronized acquisition, 64 kHz max sampling, 12-bit real precision and 16-bit resolution. By the data analysis, the system's real resolution of ray point on the PSD is better than 10um, and the time response is less than 5ms.

Keywords: FPGA, Data Acquisition System, 2D-PSD, Space beacon light.

1 Introduction

It is well known that the data acquisition system is an important part in measurement-control system. The performance of DAQ will affect the whole system's accuracy and other parameters. In traditional DAQ system, MCU or DSP is used as the master controller to complete the task of data collection. But the MCU has low calculating speed and precision, it cannot meet the needing of high speed and high accuracy. Although the DSP's function is fast enough to satisfy the real-time field, it still cannot perform parallel operations or parallel processing [1]. Concerning to the advantage of FPGA: the good parallel calculation、 the high speed ability and the flexible processing, this paper focuses its point on FPGA chip as the master controller in the DAQ system.

2 System Architecture

This system consists of four modules: power supply, I / V conversion amplifier circuit, signal process circuit, the interface of FPGA and communication. The overall structure of the system is shown as in the figure 1. The PSD optical signal is amplified by I / V conversion circuit, then processed and filtered by the signal conditioning circuit, at last, changed to the differential signal that transmit to ADC. Taking into account the logic level of the chip ADC and the FPGA, it is need to increase the level conversion circuit to solve the problem of voltage matching.

H. Tan (Ed.): Knowledge Discovery and Data Mining, AISC 135, pp. 445–451.
springerlink.com

3 Measuring Principle

Position Sensitive Detectors (PSD) is a photoelectric device that is sensitive to the incident position of light. Its principle is based on transverse photoelectric effect. It is well known that sufficiently energetic radiation normally incident on a PN junction produces a photo-voltage across the junction. There is potential difference Φ in the direction of parallel PN junction at the condition of non-uniform irradiation [2], [3].

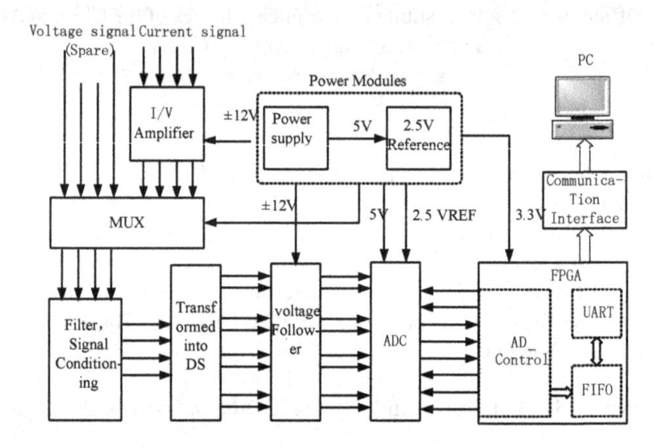

Fig. 1. System Architecture

$$\Delta^2\Phi(x,y,t) - \frac{J_s\rho_p}{W_p}[\exp(\frac{q\Phi(x,y,t)}{kT}) - 1] - \frac{\rho_p c}{W_p}\frac{\partial\Phi(x,y,t)}{\partial t} = -\frac{q\rho_p f(x,y,t)}{W_p} \quad (1)$$

This equation is known as Lucovsky equation. Where q is the magnitude of the electronic charge, k is Boltzmann's constant, T is the absolute temperature, J_s is the saturation current density of the junction, ρ_p is the resistivity of the p region, W_p the thickness of the p region, c is the barrier capacity, $f(x,y,t)$ is the hole electron pair separation function.

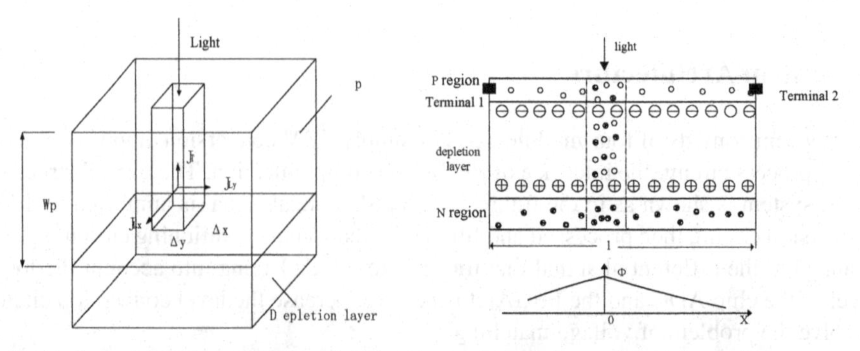

Fig. 2. Part of the three-dimensional PSD

Fig. 3. Cross-section of the PSD

If a one-dimensional PSD is reverse biased with zero load, then $\Phi << 0$, $\exp(\frac{q\Phi(x,t)}{kT}) \approx 0$, the equation (1) can be abbreviated

$$\frac{\partial\Phi(x,t)}{\partial t} - \frac{1}{rc}\frac{\partial^2\Phi(x,t)}{\partial x^2} = \frac{[qf(x,t) + J_s]}{c} \tag{2}$$

Where r is the resistance of P region, if $x = 0$ or $x = l$, then $\Phi = 0$, or if $t = 0$, then $\Phi = 0$, and $qf(x,t) >> J_s$, when this conditions are met, the formal solution of (2) is given by

$$\Phi(x,t) = \frac{2I_0}{cl}\sum_{n=1}^{\infty}\frac{rcl^2}{n^2\pi^2}\sin\frac{n\pi X}{l}\sin\frac{n\pi x}{l}\{1 - \exp(-n^2\pi^2 t / rcl^2)\} \tag{3}$$

Where I_0 is the density of photo-generated current, the instantaneous current output from the two electrodes of PSD is given by

$$i_1 = \frac{1}{r}\frac{\partial\Phi}{\partial x}\bigg|_{x=0} = I_0\frac{2}{\pi}\sum_{n=1}^{\infty}\frac{1}{n}\sin\frac{n\pi X}{l}\{1 - \exp(-n^2\pi^2 t / rcl^2)\} \tag{4}$$

$$i_2 = \frac{1}{r}\frac{\partial\Phi}{\partial x}\bigg|_{x=l} = -I_0\frac{2}{\pi}\sum_{n=1}^{\infty}\frac{1}{n}\cos(n\pi)\sin\frac{n\pi X}{l}\{1 - \exp(-n^2\pi^2 t / rcl^2)\} \tag{5}$$

By the t pass to the limit, it follows that

$$i_{1\infty} = \frac{1}{r}\frac{\partial\Phi}{\partial x}\bigg|_{x=0} = I_0\frac{2}{\pi}\sum_{n=1}^{\infty}\frac{1}{n}\sin\frac{n\pi X}{l} = I_0(1 - \frac{X}{l}) \tag{6}$$

$$i_{2\infty} = \frac{1}{r}\frac{\partial\Phi}{\partial x}\bigg|_{x=0} = -I_0\frac{2}{\pi}\sum_{n=1}^{\infty}\frac{1}{n}\cos\frac{n\pi X}{l} = I_0\frac{X}{l} \tag{7}$$

By combining (6) and (7)

$$1 - \frac{I_{1\infty} - I_{2\infty}}{I_{1\infty} + I_{2\infty}} = \frac{2X}{l} \tag{8}$$

If the origin of coordinates is set at the center of PSD, then the location of the incident light point is given by

$$x = \frac{l}{2}\frac{I_{2\infty} - I_{1\infty}}{I_{1\infty} + I_{2\infty}} \tag{9}$$

The equation indicates that the location of incident light point has a linear relationship with the current flow out the terminal.

4 The Logic Design of FPGA

This system uses EP2C8Q208C8 as the master controller chip. It has the following characteristics: High-density architecture with 8256 LE; embed 36 9X9-bit multipliers; support multi-Voltage I/O standard and the Nios II CPU, can meet most of the middle and small size system needs[4]. Analog-digital conversion controller is consisted of AD73360 which is produced by ADI Company. AD73360 is characterized by six-channel of A / D conversion function with 16-bit valid data, the master clock up to 16.384MHz, the rate of sample is up to 64KHz, and the gain amplifier of each channel can be set to 0dB to 38dB of eight levels, allowing a single power supply with 3V or 5V, the sampling frequency and the serial communication speed can be programmable to meet the requirements of synchronous acquisition system.

The logic design of FPGA is made up of three parts AD_control, FIFO and UART, the design of each module is shown as follows.

4.1 ADC's Interface Logic

The key design of system is the interface of ADC's logic, there is the problem of crossing clock domains and synchronization of asynchronous signals [5], [6], [7]. The chip AD73360 serial communication timing diagram is shown as Figure 4. the serial communication clock(SCLK) is divide from the master clock(MCLK), before entering a data byte, the serial input signal frame SDIFS need pulled a cycle, the valid data of SDI is locked at the falling edge of SCLK. While to transmit the data, the signal frame SDOFS is set with a cycle of SCLK, the output data is locked at the rising edge of SCLK. From the serial communication timing diagram, in order to build regular communication between FPGA and ADC, It have to deal with this problems of crossing clock domains and synchronization of asynchronous signals to ensure all the signal synchronized with the master clock of FPGA, otherwise there will be a problem of metastable caused by asynchronous signals. There is a method of 'Sample Synchronization Register Chain' to solve the problem shown in Figure 5, the signal of ADC pass through two D flip-flop to synchronize with the master clock of FPGA.

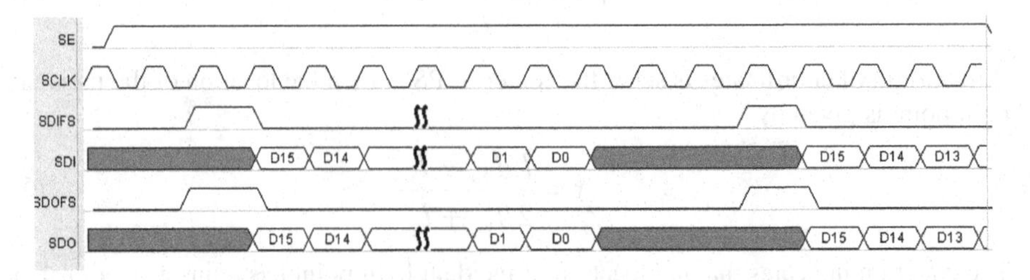

Fig. 4. ADC Serial Communication Timing Diagram

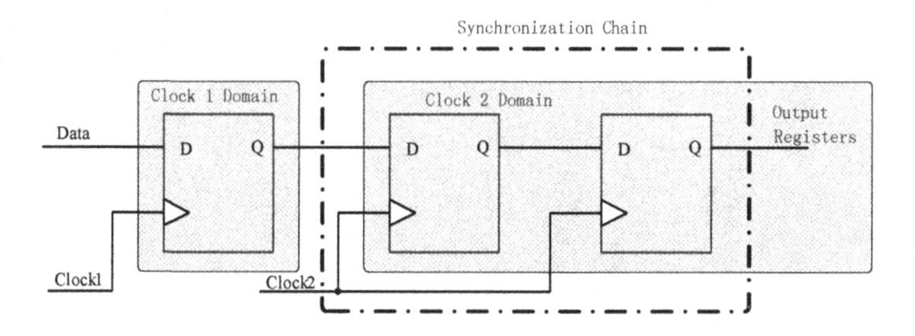

Fig. 5. Sample Synchronization Register Chain

The rising edge and the falling edge of SCLK and SDOFS are nabbed at the master clock 50MHz, meanwhile stored into the SDOFS_NEG, SDOFS_POS, SCLK_NEG, SCLK_POS registers. According the ADC serial communication timing diagram, Use Verilog HDL to program the interface of communication between FPGA and ADC, synthesized by QuartusII software, and downloaded into FPGA. The embedded logic analyzer (Signal Tap ii) is used to analyze the process of communication, from the Figure 6 which is nabbed by the Signal Tap ii, FPGA and the ADC has been successfully communicate.

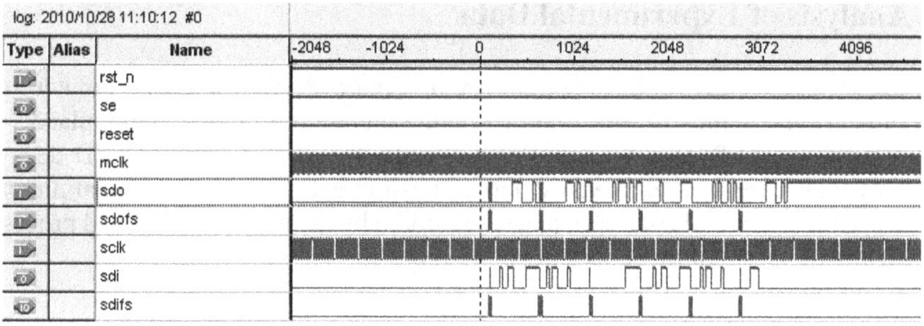

log: 2010/10/28 11:10:12 #0
Type Alias Name -2048 -1024 0 1024 2048 3072 4096
rst_n
se
reset
mclk
sdo
sdofs
sclk
sdi
sdifs

Fig. 6. The communication between FPGA and ADC

4.2 Design of FIFO and UART Transmit Module

Metastability and how to generate empty and full flag correctly is key in the design of asynchronous FIFO, FIFO is a memory of special functions, and it transmits data by the rule of First- in- First-out [8]. From Figure 7, the data read from FIFO read or write to FIFO is controlled only by the clock, does not need to read or write the address. According to the characteristics of FIFO timing diagram, the signal wrreq should coordinate with the data of AD73360 to ensure the communication is successfully.

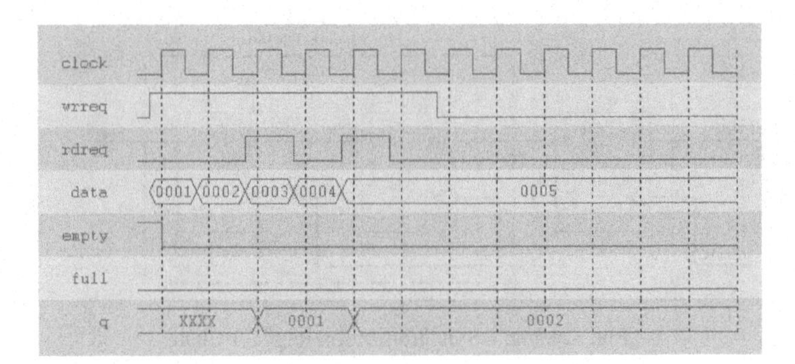

Fig. 7. FIFO timing diagram

UART interface protocol is composed of the start bit, data bits, parity bit and stop bits. The design of the UART module is divided into two parts, one is the baud rate setting and another is the data transmission. The nature of the baud rate setting module is a frequency divider. Taking into account external crystal of the FPGA is 50MHz. then just divide into 5027 to produce the baud rate 19200bps. In other words, when the count up to 2604, the send flag of the UART is set, then one bit of data can be transmitted.

5 Analysis of Experimental Data

The experimental environment as shown in Figure 8, fixe a Laser Pointer on a two-dimensional manually adjusting rack with regulation accuracy 10um, place some filters between 2D-PSD and Laser Pointer, make the photocurrent of PSD decay to 10uA by adjust the optical intensity, move 0.01mm every 30s in horizontal direction and keep still for 30s. The measurement data stored by the host computer and processed by software. One set of raw data shown in Figure 9.

Fig. 8. Experimental devices and measuring program

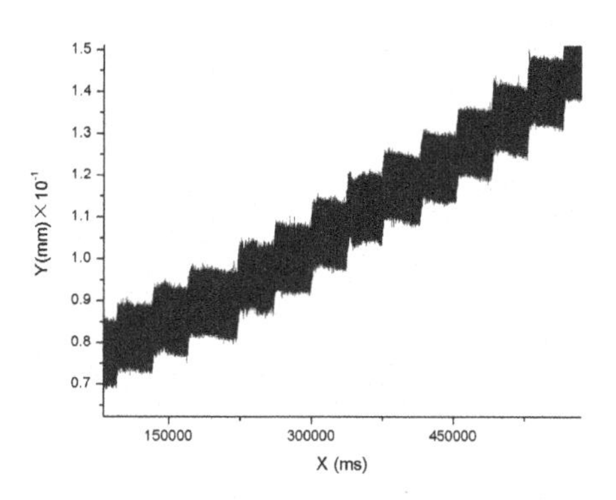

Fig. 9. Horizontal position measurements

Figure 9 shows that the real resolution of ray point on the PSD is better than 10um.

6 Conclusions

This paper describes a design of a data acquisition system for photocurrent of PSD, uses the chip AD73360 to achieve a design of A/D conversion with high accuracy, solves the problem how the single-ended level change to the differential level signal, and uses HDL language to drive the chip of ADC. The experiment data indicates that this system is effectively in the field of work, and can provide technical assistance for the further study on the application of 2D-PSD.

Acknowledgments. Foundation item: 1. Supported by Innovation Fund of SITP No. Q-ZY-14; 2. Supported by the Science and Technology Commission of Shanghai Municipality Project under Grant No. 10160501700. Corresponding author: Jianyu Wang, E-mail: jywang@mail.sitp.ac.cn, Tel: +086-021-25051000.

References

1. Guo, J.R., You, C., Zhou, K.: A 10 GHz 4:1 MUX and 1:4 DEMUX implemented by a Gigahertz SiGe FPGA for fast ADC. Integration the VLSI Journal 38, 525–540 (2005)
2. Lucovsky, G.: Photo-effects in Non-uniformly Irradiated P-N junctions. Appl. Phys. (1960)
3. Owen, R.B., Awcock, M.L.: One and two dimensional position sensing semiconductor detectors. IEEE Trans. Ucl. Sci., 290–303 (1968)
4. Altera Corporation: Cyclone FPGA family data sheet. Altera Corporation (2003)
5. Wang, X.Y., Lu, Y.H., Zhang, L.K.: Design and implementation of high-speed real-time data acquisition system based on FPGA. The Journal of China Universities of Posts and Telecommunications 13, 61–66 (2006)
6. Nurdan, K., Conka, N.T., Besch, H.J.: FPGA-based data acquisition system for a Compton camera. Nuclear Instruments and Methods in Physics Research Section 510, 122–125 (2003)
7. Joshua, M.J., Gerardo, O.V.: FPGA-based real-time remote monitoring system. Computers and Electronics in Agriculture 49, 272–285 (2005)
8. Andrade, R.P., Cumplido, R.: A versatile linear insertion sorter based on an FIFO scheme. Microelectronics Journal 40, 1705–1713 (2009)

An Improved Particle Swarm Algorithm Based on Cultural Algorithm for Constrained Optimization

Lina Wang, Cuiwen Cao*, Zhenhao Xu, and Xingsheng Gu

Key Laboratory of Advanced Control and Optimization for Chemical Processes
(East China University of Science and Technology),
Ministry of Education, 200237 Shanghai, China
caocuiwen@ecust.edu.cn

Abstract. This paper develops an improved particle swarm optimization algorithm based on cultural algorithm for constrained optimization problems. Firstly, chaos method is utilized in the initialization process of single swarm in population space to assure the searching breadth, and evolves with standard particle swarm optimization (PSO). Secondly, fixed proportion elites are selected from population space to construct the swarm of belief space through acceptance function. Then, the belief space updates its normative knowledge and situational knowledge according to the elite particles, and the elite-swarm in the belief space performs PSO operation according to the update knowledge and generates new particles. After that, the belief space renews the knowledge again, and passes down the new knowledge which has been updated twice to give better guidance to all the particles in the population space. The efficiency of the initialization strategy and the double evolving knowledge strategy are verified in six constrained optimization problems.

Keywords: particle swarm optimization, cultural algorithm, constrained optimization.

1 Introduction

In the last decade, evolutionary algorithm has gained more and more attention and been used in diverse application fields. In 1995, Kennedy and Eberhart [1] developed PSO which has many attractive characteristics, such as easy for understanding and implementation. However, a few deficiencies like trapping in local optimum and slow convergence which degrade the performance of PSO. In order to compensate these drawbacks, PSO is embedded into the framework of standard Cultural Algorithm (CA) which is proposed by Reynolds, R.G. [2]. Studies mainly focus on the improvements of parameters and the modifications on structure of population space [3][4]. In recent years, some researches begin to add some operations in belief space [5][6][7].

Inspired by their work, an improved particle swarm optimization based on cultural algorithm (IPSOCA) is proposed in this paper. Firstly, the chaos method is utilized in the initialization process to assure the searching breadth. Secondly, fixed proportion

* Corresponding author.

H. Tan (Ed.): Knowledge Discovery and Data Mining, AISC 135, pp. 453–460.
springerlink.com

elites are selected from population space to construct the swarm of belief space. Then, the belief space updates its normative knowledge and situational knowledge, and the elite-swarm performs PSO operation according to the update knowledge and generates new particles. After that, the belief space renews the knowledge again and passes down the knowledge which has been updated twice to give better guidance to all the particles in the population space. The initialization strategy and the double evolving knowledge strategy are verified in six constrained optimization problems.

The contents are organized as follows: Section 2 briefly introduces PSO and the PSO based on CA (PSOCA). Section 3 presents the methodology of IPSOCA we used. In section 4, experiment results and analysis on test functions are presented. The last section is the conclusions.

2 Overview of PSO and PSOCA

PSO[3][8] is a heuristic algorithm that characterizes constructive cooperation between particles, each particle represents a candidate solution of problem ($X_i = (x_i^1, x_i^2, \cdots x_i^D)$, i=1,2,...N), where N is the number of particles in population, D means the dimensions of each particle. Particles search global optimum through flight and update in position. The position of each particle is modified to better place by a new velocity which is based on its historical velocity, its personal best (*Pbest*) position and the global best position (*Gbest*). In the t th iteration, the i th particle update position and velocity in j th dimension following the bellow rule:

$$v_i^j(t+1) = wv_i^j(t) + c_1 r_1 (Pbest_i^j(t) - x_i^j(t)) + c_2 r_2 (Gbest_i^j(t) - x_i^j(t)) \tag{1}$$

$$x_i^j(t+1) = x_i^j(t) + v_i^j(t+1) \tag{2}$$

Where $x_i^j(t)$, $x_i^j(t+1)$ denote the current and updated position,. $v_i^j(t)$, $v_i^j(t+1)$, correspond to the velocity accordingly. c_1, c_2 are acceleration constants. r_1, r_2 are generated within the interval of [0, 1] randomly. w is the inertial weight.

Recently, a novel algorithm PSOCA composed of PSO and CA has been proposed, in which PSO is taken into population space and involves as standard PSO[4][8][9]. Ma, H.M. and Ye, C.M. [6] embedded standard PSO into the population space and the belief space simultaneously, the two spaces performed independent evolution in parallel and influenced with each other. They applied successfully this method to the knapsack problem. Wu, L.Y. et al.[7] adopted the framework of parallel evolution with one single swarm in the population space and the other in the belief space. The two swarms have the same particles in quantity, and evolved dependent and affected with each other. They showed good performance in unconstrained problems.

To assure the searching breadth, the chaos initialization method is as follows:

For the first iteration, initialize positions and velocities of all particles randomly based on chaotic sequence generated by logistic chaos mapping as rule(3)[10], which can improve the quality of initialized particles by means of regularity, ergodicity and intrinsic stochastic properties of chaotic motion.

$$\begin{cases} x_{n+1} = u \times x_n (1 - x_n) \\ x_0 = rand(), n = 0,1,2 \cdots n, u = 4 \end{cases} \tag{3}$$

Where x_{n+1} is the $n+1$th chaos variable mapped into variable region. *rand*() is random number generated within the interval of [0,1].

3 IPSOCA for Constrained Optimization

3.1 Description of IPSOCA

The brief framework of IPSOCA is depicted in Fig.1. Initially, particles are generated with the chaos initialization method of the PSOCA in Section 2, and evaluated in the population space and contributed fixed proportion elite particles as the initial particles of the belief space. The belief space updates normative knowledge and situational knowledge according to the elite particles, and then the elite-swarm in belief space performs PSO operation continually influenced by the knowledge updated and generates new particles. After that, the belief space renews the knowledge again and passes down the knowledge which has been updated twice to give better and further guidance to all the particles in the population space to find better solutions. On the one hand, the chaos initialization method assures the diversity of particles in the evolving process; on the other hand, the idea that the knowledge sources update twice per iteration accelerates the evolution process of knowledge.

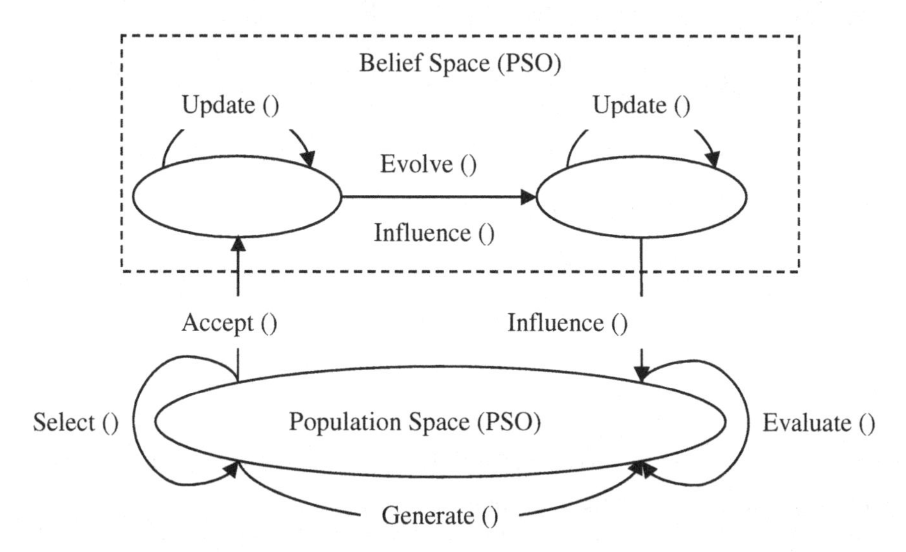

Fig. 1. Framework of IPSOCA

3.2 Design of IPSOCA

The design of IPSOCA covers three parts correspond to its framework: the population space design, the belief space design and the communication protocol design. The execution steps are introduced in detail as follows.

Ⅰ. Design of the population space:

For the first iteration, initialize positions and velocities of particles of the according to equation (3). For each of the following iterations, compare the fitness of each particle with the precious best particle, if current particle's fitness is better, it turns to be the global best particle. Update the velocity and position according to the equation (1) and (2) respectively. If some particles fly too rapidly or too slowly, it will result in locating outside the regions demarcated by the belief space. Therefore, we update their velocities and positions by rule (4) and (5) as follows:

$$if\,(v_i^j(t) > v_{\max}^j) \qquad v_i^j(t) = x_{\max}^j - x_{\min}^j \tag{4}$$

$$\begin{cases} if\,(x_i^j(t) > u^j(t)) & x_i^j(t) = x_i^j(t) - |\,rand() * (u^j(t) - l^j(t)\,| \\ if\,(x_i^j(t) < l^j(t)) & x_i^j(t) = x_i^j(t) + |\,rand() * (u^j(t) - l^j(t)\,| \end{cases} \tag{5}$$

Where $x_{\max}^j$ and $x_{\min}^j$ are the upper and lower boundaries of the j th dimension of particle, which are defined by problem, $u^j(t)$ and $l^j(t)$ denote the upper and lower limits respectively indicated by the belief space in t th iteration.

Ⅱ. Design of the belief space

In this paper, design of the belief space includes: update knowledge and elite particles evolve continuously on the basis of updating knowledge.

There are two kinds of knowledge sources adopted in the belief space, normative knowledge is updated through formula (6) ~ (9) as follow:

$$u^j(t+1) = \begin{cases} x_i^j(t), & if\ x_i^j(t) > u^j(t) \quad or\ fit(x_i(t)) > U^j(t) \\ u^j(t), & else \end{cases} \tag{6}$$

$$U^j(t+1) = \begin{cases} fit(x_i(t)), if\ x_i^j(t) > u^j(t) & or\ fit(x_i(t)) < U^j(t) \\ U^j(t), & else \end{cases} \tag{7}$$

$$l^j(t+1) = \begin{cases} x_i^j(t), & if\ x_i^j(t) < l^j(t) \quad or\ fit(x_i(t)) < L^j(t) \\ l^j(t), & else \end{cases} \tag{8}$$

$$L^j(t+1) = \begin{cases} fit(x_i(t)), if\ x_i^j(t) < l^j(t) & or\ fit(x_i(t)) < L^j(t) \\ L^j(t), & else \end{cases} \tag{9}$$

Where $U^j(t)$ and $L^j(t)$ are the maximum and minimum fitness, $fit(x_i(t))$ is the fitness of the i th particle, t is the current iteration.

Situational knowledge is composed of particles in the belief space, the size of which is 20% of the population space. The way to update situational knowledge includes two sections: (a) compare the top 20% particles of the population space with the particles reside in the belief space and select the elite particles; (b) reserve better elite particles after evolving them continually.

The way to update the velocities and positions is the same as that of the population space according to equation (1) and (2), while a different rule is adopted to influence particles that fly too rapidly or too slowly and the particles which is out of the regions demarcated by the updated knowledge, shown as bellow:

$$if\,(v_i^j(t) > v_{\max}^j) \qquad v_i^j(t) = v_{\max}^j / 2 \tag{10}$$

$$\begin{cases} if(x_i^j(t) > u^j(t)) & x_i^j(t) = x_{best}^j(t) - |rand() * (u^j(t) - l^j(t)| \\ if(x_i^j(t) < l^j(t)) & x_i^j(t) = x_{best}^j(t) + |rand() * (u^j(t) - l^j(t)| \end{cases} \tag{11}$$

Where $x_{best}^j(t)$ means the j th dimension of the best particle at t th iteration.

The velocity is designed to become slow so as to ensure the diversity in search space and avoid overflying. The particles out of the demarcated regions will be replaced by particles around the best one to add more effective information and speed up the evolution of knowledge, which can make the convergence process much shorter.

Ⅲ. Design of communication protocol

Communication protocol, a bridge for transferring information between the two spaces, plays an important role in the whole evolution process. One is accept() function, the other is influence() function. In this paper, the belief space accepts top 20% particles per generation because there is no complex updating mechanism. Consequently, it is not time consuming but updates knowledge better and faster. The population space perform evolution supported by the influence() function, in which the worst 20% particles are replaced with elites in the belief space to enhance the whole capability in searching best solution.

3.3 Constraints Handling

In our approach, penalty function is used to handle the constraints taking advantage of its simple principle and ease of implementation [11]. The violations of constraints are incorporated into fitness function so that the fitness is modified through the rule bellow to remain feasible solution and disregard unfeasible solution effectively [8].

$$fit(x) = \begin{cases} obj(x) & if\ g_i(x) \le 0 \\ obj(x) + q\sum_{i=1}^{n} g_i(x) & else \end{cases} \tag{12}$$

Where n is the number of constraints, q is punishing factor. In this paper, q is set to be 10000 for f1, f2 and 100000 for other test functions.

4 Experiment Results and Analysis

Aiming at evaluating the performance and effectiveness of IPSOCA for solving constrained optimization problems including maximization and minimization,, six benchmark test functions [4][12][13] are chosen to simulate and compare the accuracy and convergence speed with PSO and PSOCA.

For three algorithms, basic parameters is set to be same completely in order to enhance the comparability, the parameters related are set as follows: acceleration constants $c_1=c_2=2$; inertial weight w decreases from 0.8 to 0.2 linearly for f1, f6, and 1 to 0.4 linearly for other functions ; population size is 100; maximum number of iteration is 1000. Every test function using three algorithms is repeated for 200 times independently. The simulation results are summarized in Table 1 and shown in Fig. 2.

Table 1. Simulation results of PSO, PSOCA and IPSOCA

f(x)	algorithm	optimal	ε	best result	mean	std.dev	convergence ratio	Average convergence iteration
	PSO			-30662.74188474	-30623.45863495	3.151e+01	0%	1000
f1	PSOCA	-30665.539	1e−6	-30665.53867178	-30658.94398306	6.542e+01	71%	729.82
	IPSOCA			-30665.53867178	-30665.53867178	2.705e−10	100%	209.72
	PSO			-6834.47486964	-5356.69833320	9.000e+02	0%	1000
f2	PSOCA	-6961.81388	1e−6	-6961.81387558	-6949.21526413	3.104e+01	81.5%	505.92
	IPSOCA			-6961.81387558	-6961.81387558	2.264e−11	100%	306.69
	PSO			0.09582503	0.09582503	4.664e−17	100%	585.76
f3	PSOCA	0.095825	1e−6	0.09582503	0.09582503	2.927e−16	100%	41.62
	IPSOCA			0.09582503	0.09582503	2.646e−16	100%	28.06
	PSO			-9	-9	5.683e−11	100%	777.41
f4	PSOCA	-9	1e−6	-9	-9	0.000e+00	100%	201.52
	IPSOCA			-9	-9	0.000e+00	100%	129.63
	PSO			2.98863636	2.98863636	1.671e−08	100%	791.85
f5	PSOCA	2.9886364	1e−6	2.98863636	2.98863636	3.184e−11	100%	296.83
	IPSOCA			2.98863636	2.98863636	1.026e−14	100%	152.35
	PSO			-5.50801327	-5.50801327	4.452e−15	100%	80.45
f6	PSOCA	-5.5079	1e−6	-5.50801327	-5.50801327	5.268e−15	100%	36.65
	IPSOCA			-5.50801327	-5.50801327	7.643e−15	100%	22.10

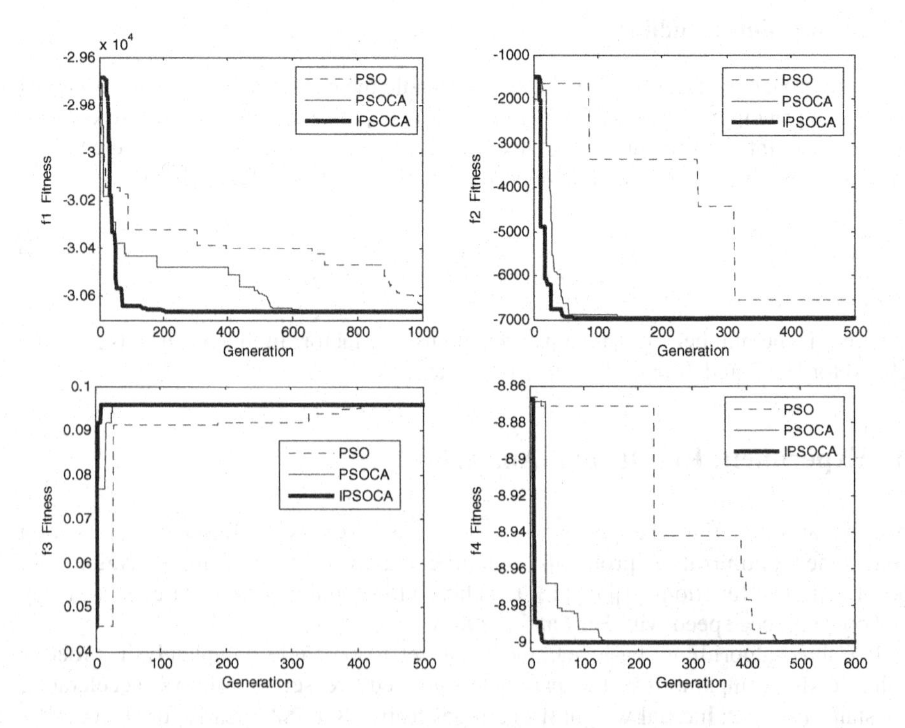

Fig. 2. Comparison of fitness curve for each test function

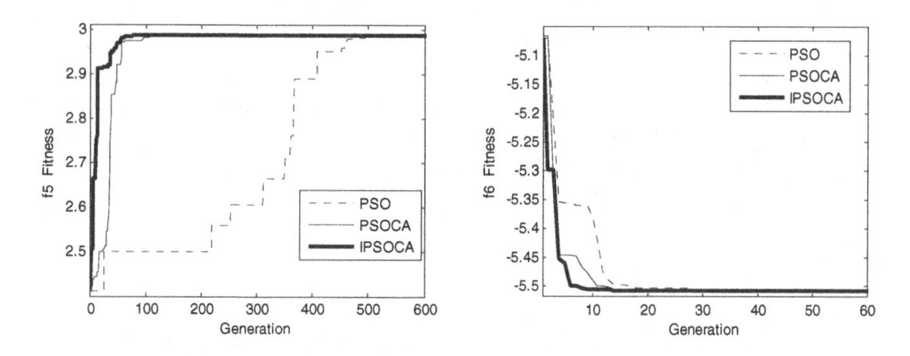

Fig. 2. *(continued)*

The performance of three algorithms is evaluated from five aspects. Here, ε is the convergence precision, the convergence ratio is defined as the percentage that the times converged to global best with the total runtimes (200), the average convergence iteration is the average iterations converged to best solution at the first time. As can be seen from Table 1 and Fig. 2, PSOCA obtains much better performance than basic PSO, which shows the outstanding advantages of CA model. Furthermore, the improved algorithm IPSOCA performs the better convergence speed and higher accuracy than PSOCA and PSO in all of the test functions. However, it's necessary to further analyzing the standard deviation in function f3 and f6.

5 Conclusion

In this paper, we present IPSOCA for constrained optimization problem, which integrates PSO into the framework of CA both in population space and belief space. Firstly, the chaos method is utilized in the initialization process of the single swarm in population space to assure the searching breadth, and evolves with standard PSO. Secondly, fixed proportion elites are selected from population space to construct the swarm of belief space through acceptance function. Then, the belief space updates its normative knowledge and situational knowledge according to the elite particles, and the elite-swarm in the belief space performs PSO operation according to the update knowledge and generates new particles. After that, the belief space renews the knowledge again and passes down the knowledge which has been updated twice to give better and further guidance to all the particles in the population space. The improved mechanism makes the IPSOCA more effectively, not only increases the diversity of PSO to achieve the global best solution, but also accelerates the convergence velocity. The simulation results of the six constrained functions verify the validity of the algorithm.

Acknowledgment. Financial support from the National High Technology Research and Development Program of China (863 Program) (No. 2009AA04Z141), Fundamental Research Funds for the Central Universities, Doctor Foundation of Ministry of Education of China (Grant No. 200802510010), Shanghai commission of Nature Science (No. 10ZR1408300) and Shanghai Leading Academic Discipline Project (Grant No. B504) is grateful appreciated.

References

1. Kennedy, J., Eberhart, R.C.: Particle swarm optimization. In: Proceedings of the IEEE International Conference on Neural Networks, Piscataway NJ, pp. 1942–1948 (1995)
2. Reynolds, R.G.: An Introduction to Cultural Algorithm. In: Proceedings of the 3rd Annual Conference on Evolutionary Programming, pp. 131–139. World Scientific, Singapore (1994)
3. Daneshyari, M., Yen, G.G.: Cultural-Based Multiobjective Particle Swarm Optimization. Systems, Man, and Cybernetics, Part B: Cybernetics 41, 553–567 (2011)
4. Gao, L.L., Liu, H., Li, T.X.: Particle Swarm Based on Cultural Algorithm for Solving Constrained Optimization Problems. Computer Engineering 34, 179–181 (2008) (in Chinese)
5. Huanng, H.Y., Gu, X.S.: Cultural Binary Particle Swarm Optimization Algorithm and Its Application in Fault Diagnosis. Journal of Donghua University 5, 474–481 (2009)
6. Ma, H.M., Ye, C.M.: Parallel Particle Swarm Optimization Algorithm Based on Cultural Evolution. Computer Engineering 34, 193–195 (2008) (in Chinese)
7. Wu, L.Y., Sun, H., Bai, M.M., Li, M.: New parallel Particle Swarm Optimization based on cultural algorithm. Computer Engineering and Applications 45, 44–46 (2009) (in Chinese)
8. Dos Santos Coelho, L., Mariani, V.C.: An Efficient Particle Swarm Optimization Approach Based on Cultural Algorithm Applied to Mechanical Design. Evolutionary Computation, 1099–1104 (2006)
9. Lin, C.J., Chen, C.H., Lin, C.T.: A Hybrid of Cooperative Particle Swarm Optimization and Cultural Algorithm for Neural Fuzzy Networks and Its Prediction Applications. Systems, Man, and Cybernetics.Part C: Applications and Reviews 39, 55–68 (2009)
10. Li, B., Jiang, W.S.: Chaos Optimization Method and Its Application. Control Theory and Application 14, 613–615 (1997) (in Chinese)
11. He, Q., Wang, L.: A Hybrid Particle Swarm Optimization with a Feasibility-based Rule for Constrained Optimization. Applied Mathematics and Computation 186, 1407–1422 (2007)
12. Thomas, P.R., Yao, X.: Stochastic Ranking for Constrained Evolutionary Optimization. Evolutionary Computation 4, 284–294 (2000)
13. Song, S.B., Cai, H.J., Kang, Y.: Genetic Algorithm Solution for Constrained Optimization. Journal of Northwest A&F University (Natural Science Edition) 33, 150–154 (2005) (in Chinese)

The Comparison of Program Implementations of the Acquisition of Vibration Data Based on LabVIEW

Wen Li[1], Changjing Wu[2], and Huimin Zhao[1]

[1] Software Technology Institute
[2] Electric and Information Institute
Dalian Jiaotong University, Dalian, Liaoning Province, P.R. China
lw6017@vip.sina.com, wuchangjing_2008@163.com, hmzhao@126.com

Abstract. A system of motor control and acquisition of vibration data built based on LabVIEW and NI data acquisition module is introduced. The motor control problem caused by collecting and saving data and the optimization of LabVIEW program are discussed. The influence of three programs on control performance are tested and results show that better control performance can be obtained when TDMS is used to realize data acquisition and storage.

Keywords: optimization of LabVIEW program, data acquisition and storage, motor control, sampling frequency.

1 Introduction

LabVIEW is powerful graphical programming software of National Instruments Corporation (NI) and is widely applied in various fields such as measurement and control, simulation, data analysis and so on. Its friendly program interface and intuitive program block diagram bring great convenience for users and programmer [1-2]. Various tasks of control and test can be implemented by NI software and hardware conveniently. It is important to design LabVIEW program under the same condition of system hardware. Different programming methods of control and data acquisition used will affect control performance and data acquisition efficiency.

What is discussed in this paper rises from the procedure of the actual motor speed control system and the acquisition and storage of vibration data. The vibration data of different positions of motor-generator is gathered using acceleration data acquisition card of NI USB-9234 of which sampling frequency is set at 25.6kHz when the motor is controlled. The speed of motor becomes unstable when the vibration data acquisition was added in control process. The computer configuration to operate this system is that CPU is Intel Pentium processor 1.86GHz and memory is 798MHz 2GB. When the computer is replaced by a higher performance one of which configuration is that CPU is Intel Core2 Duo T6570 2.1GHz and memory is 1.19GHz 2.0GB, the influence on motor resulted from the operation of data acquisition program is more or less decreased. In order to ensure that data acquisition program can not influence on speed control in terms of no computer's performance improvement, the measure to improve data acquisition program is taken. In this paper the execution efficiency of

H. Tan (Ed.): Knowledge Discovery and Data Mining, AISC 135, pp. 461–466.
springerlink.com

data acquisition and storage from different kinds of program implementation is studied, through comparison of their influence on control process.

2 The Parameter Selection of Vibration Data Acquisition

The motor of this experimental system is controlled by a converter. When converter is running some harmonics must be generated. Harmonics can lead to high frequency vibration of motor. From the vibration spectrum of the motor speed control system, there are strong vibration signals near of 800Hz and 4kHz. To capture and restore these vibration signals, the sampling frequency should be set at no less than 8kHz in the light of Shannon sampling theorem. To restore vibration signal perfectly, the practical sampling frequency is usually 3-5 times higher than the frequency of sampled signals. The following equation [5] provides the available sampling frequency of NI 9234:

$$f_s = \frac{f_M \div 256}{n} \tag{1}$$

where n is any integer from 1 to 31 and the internal master timebase f_M is 13.1072MHz. According to this equation the sampling frequency is set at 25.6kHz in this study.

To verify the efficiency of different programs, running time of different programs would be recorded which have the same sampling frequency and the same sampling number.

3 The Acquisition and Storage of Vibration Data

3.1 The Scquisition and Storage of Vibration Data Using Express VI

Express VI is more powerful than other VI, because of its dialog box of parameter configuration, in which some data can be set rather than in program block diagram. Then the program block diagram is simplified. Tasks of data acquisition, display and storage can be completed by Express VI which includes "DAQ Assist", "Waveform Graph" and "Write LabVIEW Measurement File". In "DAQ Assist" dialog box of parameter configuration the acceleration option of analog input can be selected. And physical channel, sampling frequency, conversion and so on also can be selected. According to creating "Write LabVIEW Measurement File", file name, file format and other parameters can be set in it automately.

Because it is simple for programing to use Express VI, the beginners can design some simple programs and some programs of lower demanding running speed. However, there are a lot of applications in which some functions can't be used in our program. These functions not only occupy memory space but also affect the running speed of program. It is the reason that the data acquisition and storage program influence on the motor control program.

In order to test the low operation efficiency of Express VI, a program (Fig.1) is designed. In this program, data acquisition and storage of 4 channels will be completed. The program running period can be test by timing the actual time needed of data acquisition and storage.

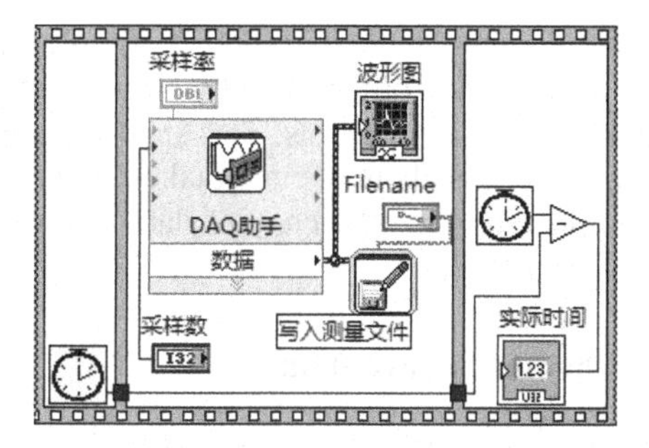

Fig. 1. Timing acquisition program of Express VI

It is found that although the program actual running time is 1793± 30ms (because the data of several operations is different, and its value changes in certain range.) If "Write LabVIEW Measurement File" is deleted and only the data acquisition and display program is operated, its actual running time is 1150± 10ms. This shows that the consuming time of finishing the task of data storage is about 640ms.

3.2 Creating DAQmx Channel to Acquire Vibration Data

In order to realize efficient system control, the time of data acquisition and save should be short to the greatest extent. So the DAQmx VI is used to complete data acquisition. The program operation efficiency is observed under different methods of vibration data acquisition. The channel of data vibration acquisition is created by DAQmx VI. Firstly, create 4 channels of "DAQmx Create Virtual Channel" and set up related parameters. Secondly, create "DAQmx Timing" to set up sampling mode and frequency. Then, create "DAQmx Start Task" and "DAQmx Read" to read and save data. Finally, create "DAQmx Clear Task" after out of cycle. Only one "DAQmx Create Virtual Channel" is showed in Fig.2. Comparing with "DAQ Assist" which integrates various functions and operates multiple task channels each program running, the efficiency of data acquisition using DAQmx is apparently higher than "DAQ Assist".

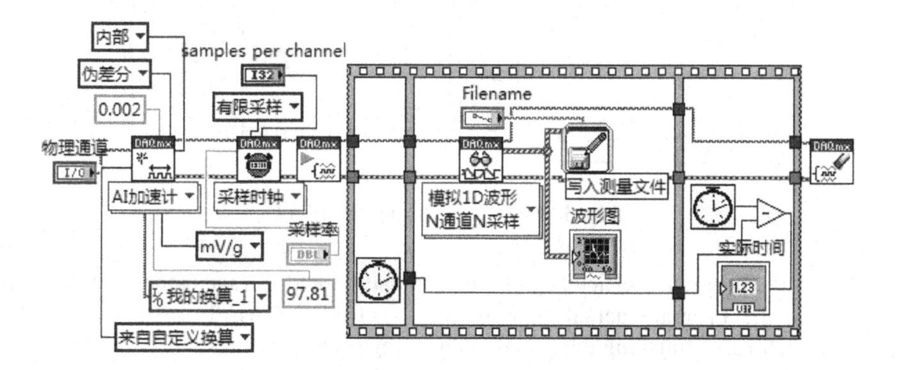

Fig. 2. Collect data using DAQmx's VI

The program actual running time is 1660 ± 50ms. If "Write LabVIEW Measurement File" is deleted, the program only carries out data acquisition and display, its actual running time is 1004 ± 1ms. This shows that the consuming time of finishing the task of data storage is about 656ms(it takes about 30 ms that to finish the task of data storage). Apparently the efficiency of data acquisition using DAQmx is much higher than "DAQ Assist".

3.3 Creating TDMS Channel to Save Data

"Write LabVIEW Measurement File" is an Express VI as the same as "DAQ assist". There is problem of low efficiency in program running process. The data is saved in the file format of ".lvm" by "Write LabVIEW Measurement File". This file content is decimal data which can be opened in Excel. Inside the computer the data should be processed in binary form. Because transforming the binary data into the decimal data will occupy the CPU resource, the file with suffix ".lvm" will need more time for data storage and make efficiency of data storage lower. However "TDMS Write" saves data in binary format and without the transformation, the data storage time must be shorter than the former. So TDMS's VI is adopted to complete the task of data storage. There are three steps for the programming. Firstly, create "TDMS Start Task" to set up the path of data storage, file format and so on. Secondly, create "TDMS Write" to save data in program cycle. And finally, create "TDMS Close" after out of cycle. The program block diagram of the data acquisition and storage using TDMS'VI is showed in Fig.3. It improves the process of data acquisition and storage.

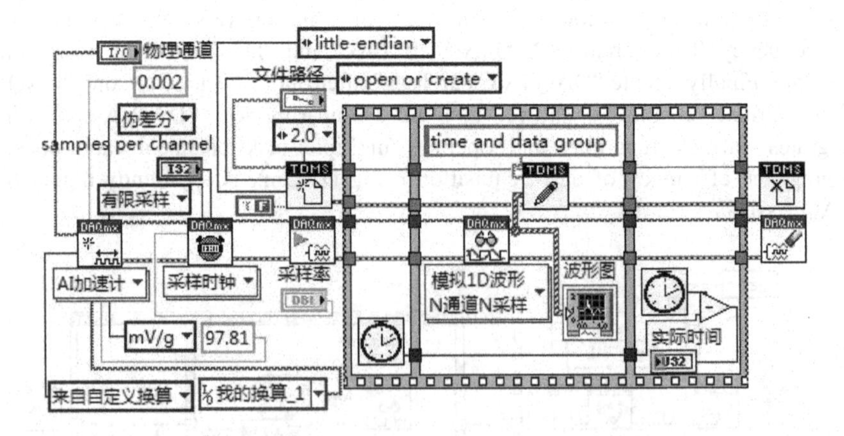

Fig. 3. Data storage using TDMS'VI

To test the program's efficiency of data acquisition and storage using TDMS'VI to save data, the program actual running time is 1040 ± 2ms. This shows that the consuming time of finishing the task of data storage is about 36ms.

3.4 The Comparison of Data Acquisition and Storage with the Two Method

In order to facilitate the comparison, some results of data acquisition and storage obtained with the two method are listed in Table 1.

The difference of program running efficiency between whether or not using Express VI is showed in Table 1. The actual running time of three programs decreases successively. The actual running time of program showed in Fig.1 is the longest both data acquisition and data storage. So its efficiency is the lowest on comparing with other two programs. The data acquisition program is replaced by DAQmx' VI in the program showed in Fig.2. Its actual running time is almost the same as set running time. After deleting program of data storage, its actual running time is entirely the same as setting running time. Since the program of data storage isn't improved, the efficiency of program showed in Fig.2 is lower than Fig.3's. So actual running time of program showed in Fig.3 is the shortest and its efficiency is the highest one.

Table 1. The comparison of data

Program \ Data		Sampling frequency	Actual running time	Time of data storage
Program of Fig.1	Program of Fig.1	25.6kHz	1793±30ms	640ms
	Delete data storage		1150±10ms	—
Program of Fig.2	Program of Fig.2		1660±50ms	656ms
	Delete data storage		1004±1ms	—
Program of Fig.3			1040±2ms	36ms

4 Conclusion

From above comparison of different methods of data acquisition and storage program, a method of minimal influence on control program is found. By operating motor control program and data acquisition and storage program in actual motor control system, the influence on motor speed control showed in Fig.1-Fig.3 is analyzed and discussed. The results of experiment show that when the program showed in Fig.3 is operated with motor control program in original computer the phenomenon that the data acquisition and storage program influence on the motor control program disappears. It can eventually come true that Motor control program and data acquisition program can be simultaneously run in one PC perfectly without improving computer's performance.

Acknowledgement. This paper is supported by National Natural Science Foundation of P. R. China (60870009).

References

1. Ruan, Q.: LabVIEW and I. Beijing University of Aeronautics and Astronautics Press, Beijing (2009)
2. Wang, L., Tao, M.: Proficient in LabVIEW8.0. Electronic Industry Press, Beijing (2007)
3. Long, H., Gu, Y.: LabVIEW8.2.1 and DAQ Data Acquisition. Tsinghua University Press, Beijing (2008)
4. Zhu, M., Li, Y., Bu, X.: Processing and Analysis of Test Signal. Beijing University of Aeronautics and Astronautics Press, Beijing (2006)
5. Operation Instructions and Specifications of NI 9234, National Instruments Corp., Hungary (2008)

Study the Social Adaptability of University Students and the Relationship with Mental Health and Personality

Bingwei Wang, Qiuping Guo, Qiuyun Liu, Jia Gao, and Shuangfeng Wei

Henan Institute of Science and Technology, Xinxiang, Henan, China
wangbingwei@yahoo.cn, guoqp@163.com, lqy@hist.edu.cn,
gaojia5834@163.com, 459447397@qq.com

Abstract. Objective: To study the social adaptability of university students and the relationship with mental health and personality and carry out pertinence management. Methods: The conditions of the social adaptability, psychological health and character of 441 persons in grade 2010 were investigated by using Social Adaptability Self-report Questionnaire (SASQ), Symptom Checklist 90(SCL-90) and Eysenck Personality Questionnaire (EPQ). Results: Those students have significant different social adaptability in gender (P=0.002), different educational background (P=0.000) and different majors (P=0.015). The students' social adaptability of art discipline is lower than the other majors' students statistically (P<0.05). The total score and ten factors'score of SCL-90 correlated with social adaptability negatively(r=-0.217～-0.337). P N items correlated with social adaptability negatively(r=-0.176, -0.401). E L items correlated with social adaptability positively(r=0.370, 0.21). Regression analysis shows, PHOB (β=-0.186) , PAR (β=0.100) , E (β=0.222) , N (β=-0.237) , L (β=0.095). So the results are useful to prognosticate the social adaptability. Conclusions: The university students' social adaptability is in the middle level. We should consider the adaptability difference in the different groups, and take more consideration to the female students, junior college students and student's major in art. We need continuously optimized the personality character as a basic training.

Keywords: university students, social adaptability, mental health, personality, relationship.

1 Introduction

It is considered that social adaptability(S-A) is the one that individuals have to learn, choose and avoid in their life, study and communication in order to adapt to the outer environment, culture and to meet the demands of their own physical and mental development. Furthermore, it also reflects the individual's ability to manage daily life and take the social responsibility independently, and whether one has attained the ability in accordance with his age and the expectation of social culture. The essence of social adaptability is the representation of social intelligence.

Mental health refers to a kind of good state based on the normal development of intelligence, when internal psychological coordination complies with external individual behavior. In psychology, the personality usually refers to the unique

H. Tan (Ed.): Knowledge Discovery and Data Mining, AISC 135, pp. 467–472.
springerlink.com

psychological reflection that is the total of the different trending psychological characteristic.

At present, it has been generally accepted that some college graduates couldn't better adapt themselves to society, conform to the demands of times, meet all kinds of social needs and keep Pace with the development of society harmoniously. The causes are related to graduates themselves, society, and talent training modes of colleges and universities and so on.

To explore their social adaptability in university students, and the relationship with mental health and personality, to provide the basis for student management, to strengthen students management and make it effective, we have had investigated in 441 freshmen of 2010 grade in Henan Institute of Science and Technology, studying their social adaptability, the relationship with mental health and personality. The reports are as follows.

2 Objects and Methods

2.1 Objects

Here are 441 freshmen aged 17-22 years old of 2010 grade. There are165 male students and 276 female students. Among of them, 51, 75, 72, 72, 142, and 29 students are majoring in agricultural disciplines, science, arts, economics management, art discipline, and sports discipline respectively, 372 of them are undergraduates, 69 of them are junior college students.

2.2 Tools and Conducting Methods

2.2.1 Social Adaptability Self-report Questionnaire

"Social Adaptability Self-report Questionnaire" is applied to investigate the social adaptability of adolescent students which from 14 to 28 years old, it includes 15 questions, there are five to seven alternative answers in questionnaire for each questions. This questionnaire is helpful for the tester to learn their psychological maturity level, that is to say, they can learn how is their social adaptability developing level, to help them to make themselves perfect quickly and adapt the social life well. The higher score means the more high psychological maturity level.

2.2.2 Symptom Checklist 90 (SCL-90)

The Symptom Checklist-90 (SCL-90) is a relatively brief self-report psychometric instrument (questionnaire) published by Pearson Assessment. It is designed to evaluate a broad range of psychological problems and symptoms of psychopathology. Responses were made on scales from 1 (never/not at all) to 5 (always/very much), which involve ten Symptom Checklists such as, somatization (SOM), obsessive-compulsive(O-C), interpersonal-sensitivity(I-S), depression(DEP), anxiety(ANX), hostility(HOS), phobic anxiety(PHOB), paranoid ideation(PAR), psychoticism (PSY), additional items(ADD). The higher the score is, the more serious psychological problem is.

2.2.3 Eysenck Personal Questionnaire(EPQ)

Eysenck Personality Questionnaire (EPQ) is a questionnaire to assess the personality traits of a person. EPQ includes Extroversion/Introversion (E), Neuroticism/Stability (N), Psychoticism/Socialisation (P), Lie (L)4.

2.3 Data Analyses

Statistical analyses were performed using programs available in the Statistical Package for Social Sciences (SPSS for Windows release 11.5). The significance of observed associations and/or differences between variables was tested using the Student's t test, analysis of variance (ANOVA, F test), Pearson correlation coefficient, and Regression. A difference was considered to be statistically significant if P<0.05.

3 Results

3.1 The Different Social Adaptability in Gender, Different Educational Background and Different Majors

Those students have significant different social adaptability in gender (*P*=0.002), different educational background (*P*=0.000) and different majors (*P*=0.015). The students' social adaptability of art discipline are lower than the other majors' students statistically (*P*<0.05).

Table 1. The different social adaptability in gender, different educational background and different majors ($\bar{x}$ ±s)

variable	basic document	social adaptability	*t/F*	*P*
Gender	Male n=276	96.61±10.80	3.043	0.002
	Female n=125	93.22±11.58		
Educational background	Undergraduate n=372	95.32±11.2	3.6111	0.000
	junior college student n=69	90.00±11.49		
Major	Agricultural n=51	93.49±11.90	2.87	0.015
	Science n=75	96.53±12.12		
	Arts n=72	95.74±9.53		
	Economic management discipline n=72	96.18±11.31		
	Art discipline n=142	91.83±11.43		
	Sports discipline n=29	96.69±11.09		
Total	441	94.49±11.40		

3.2 The Relationship between Social Adaptability and Mental Health

The total score and ten factors' score correlated with social adaptability negatively(r=-0.217~-0.337).

Table 2. The correlation coefficient of SCL-90 and social adaptability total score (r)

factor	TOTAL	SOM	O-C	I-S	DEP	ANX	HOS	PHOB	PAR	PSY	ADD
S-A	-0.310	-0.217	-0.252	-0.313	-0.297	-0.262	-0.227	-0.337	-0.236	-0.274	-0.238

Note: In this table, all indexes have been inspected by statistic, P < 0.001.

3.3 The Relationship between Social Adaptability and Personality

P N items correlated with social adaptability negatively(r=-0.176, -0.401). E L items correlated with social adaptability positively(r=0.370, 0.21).

Table 3. The correlation coefficient of EPQ Factors and social adaptability total score (r)

factor	P	E	N	L
S-A	-0.176	0.370	-0.401	0.210

Note: In this table, all indexes have been inspected by statistic, P < 0.001.

3.4 Regression Analysis Affecting Mental Health and Personality Factors

Take college students' social adaptability as the dependent variable, ten factors of mental health and four factors of personality characteristics as the prediction variables, we have forecasting analysis by using Backward Multiple Regression. The result is shown in table 6.

Table 4. Coefficients and values of the Regression of SCL-90 and EPQ on social adaptability total score (a)

Independent variables	Unstandardized Coefficients		Standardized Coefficients	*t*	*P*
	B	Std. Error	Beta(β)		
Constant	89.354	3.993		22.377	0.000
PHOB	-5.007	1.543	-0.186	-3.246	0.001
PAR	2.402	1.414	0.100	1.698	0.090
E	0.612	0.129	0.222	4.731	0.000
N	-0.514	0.120	-0.237	-4.268	0.000
L	0.317	0.159	0.095	1.988	0.047

a Dependent Variable: social adaptability total score.

Model Summary: R=0.49, R^2=0.24, ΔR^2=0.23, *F*=28.65, *P*=0.000.

4 Discussion

The university students' Social Adaptability is in the middle level, it needed to be improved in the future. Most of the testers can handle everything skillful and easily and cope everything flexibly either in the daily life or in the study and interpersonal communication. This ability is benefit for their developing in the future. But some of students showed the poor social adaptability, it played a bad role in their following college life.

The social adaptability of the female students was worse than that of male students; this point was also confirmed by the results of Nie Yangang (Adolescents' Social Adaptive Behaviors and Their Influencing Factors, 2005). The social adaptability of the junior college students was worse than that of undergraduates. Because the junior college students' educational background was poor, they have worse achievements and a comparatively speaking lower self-confidence; all these factors affected their social adaptability. The social adaptability of students in different majors are different, especially the students major in art showed a worse social adaptability than those of students in other majors, this point maybe the students major in art pursued the personality too much, they would not to change themselves to adapt the social environment actively. In a word, the gender, educational background and major will play a very important role in the student's social adaptability.

The social adaptability is positive correlated to the mental health level; it means that someone with the better social adaptability, his/her mental healthy level is higher. The social adaptability has a highest correlation to the interpersonal sensitivity and paranoid ideation specially. Those, very sensitive to interpersonal relationship and very paranoid ideation, are difficult adapt the social. The personality study model think that the social adaptability behavior are close related to the character formation and manifestations, the stable adaptability behavior is the personality character. All these data showed that the social adaptability behavior is close correlated to the personality character, among them, it showed the highest correlation to the Extroversion/Introversion (E), Neuroticism/Stability (N), those with a stable mood and outward are easier to adapt the social.

The regression model also showed that phobic anxiety, paranoid ideation, extroversion/introversion, neuroticism/stability, and lie forecast the university student s' social adaptability together.

5 Conclusions

When we try to improve the social adaptability of university students, we should consider the adaptability difference in the different groups, and take more consideration to the female students, colleague students and student's major in art, guide them to adapt the colleague life as soon as possible. Since the close relationship between the social adaptability formation and the personality and mental health level, when we improve the university student's social adaptability, we need continuously optimized the personality character as a basic training, and train them getting along with others in a trust, respective, friendly and sympathy manner, not in a suspicion, jealousy, fear and hostility manner and other negative attitude, therefore they can built a good interpersonal relationship. We also need guide them to accept their advantage and disadvantage, and don't strict self too much, and don't pay more attention to other's attitude to them, to have self-confidence and self-esteem, so they can appreciate and understand others openly and good at cooperate with others. They also need to be train how to deal with and cope with the pressure and frustration from the family, university and social life, how to good at use the resource from the inner and outer environments. We suggest the colleague students take part in all the social life, observe and think over everything carefully, and accumulate the life experience; therefore they can control themselves and improve the ability to handle some problems. The students also need to develop the ability how to control and catharsis the emotion. The students should grasp lots of methods to adapt the emotion, when they face the pressure and frustration, they can't control by the bad mood, keep the reason and control the adverse emotion accumulation at the same time, they can catharsis the adverse mood by talking out, sports and sublimation and keep vigorously fighting will. They should chasten the firmly, self-control, bravely and decisively will quality. The good will is close relative to the goal and ideal of life. Suggesting the university students set up the goal of life early; chasten the will in the life struggle process because those with the strong will remain the invincible position in any abominable environments.

Note: This is parts of research results of the project "Study of the colleague student's social adaptation under perspective of alleviating the employment pressure" (SKL-2011-2373) from Federation of Henan Social Science. 2011.

References

1. Wang, X., Wang, X., Ma, H., et al.: Rating scales for mental health. Beijing: Chinese Mental Health Journal, 33–37 (1999)
2. Gong, Y.: Eysenck Personal Questionnaire Handbook, vol. (2), p. 31. Hunan Medical College Publishing Company, Changsha (1993)
3. Austin, F.J.: Relationships beteen ability and personality: Does intelligence contribute positively to personal and social adaptability? Personality and Individual Differences (32), 1391–1411 (2002)
4. Searle, W., Ward, C.: The prediction of psychological and social cultural adaptability during cross-cultural transitions. Intercultural Journal of Intercultural Relations 4, 449–464 (1990)
5. Lee, R.M., Keough, K.A., Sexton, J.D.: Social connectedness, social appraisal, and perceived stress in college women and men. Journal of Counseling and Development: JCD 80, 355–361 (2002)
6. Williams, K.L., Galliher, R.V.: Predicting depression and Self-esteem from Social connectedness, support, and competence. Journal of Social and Clinical Psychology 25, 855–874 (2006)

The Study and Analysis on Psychological Commissioners' Mental Health and Personality

Bingwei Wang, Li Guo, Qiuyun Liu, Yongduo Wang, and Shuangfeng Wei

Henan Institute of Science and Technology, Xinxiang, Henan, China
wangbingwei@yahoo.cn, gl@hist.edu.cn, lqy@hist.edu.cn,
269188468@qq.com, 459447397@qq.com

Abstract. Objective: To study the mental health and personality of psychological commissioners and provide the timely help and pertinence training the healthy psychology and good personality in order to make them competent for the job and develop it smoothly in the future. Methods: 4531 persons which are included 337 psychological commissioners in grade 2010 freshmen were investigated by using Symptom Checklist 90 (SCL-90). 3540 freshmen which are included 240 psychological commissioners were investigated by using Eysenck Personality Questionnaire (EPQ). Results: In SCL-90, the total score and different factors values of those psychological commissioners were markedly lower than the other freshmen. In addition to somatization, hostility, the contrasts of the other scores are statistical ($P<0.05$). In EPQ, In E items, the score the psychological commissioners got is higher than the other freshmen. While in L item, the score is lower ($P<0.01$). As those psychological commissioners' mental health, only the contrast of phobic anxiety is statistical between male and female ($P<0.01$). To the students of different majors, those scores have no obvious differences ($P>0.05$). In EPQ , As those psychological commissioners' personality, the score in P item that the female got is lower than that the male got, while in E N L items, the scores are higher than those the male got. Score in L is statistical ($P<0.01$). Those scores in P E N L have no obvious differences in different majors ($P>0.05$). Conclusions: The psychological commissioners' mental healthy level is obviously better than other students. The personality characteristics of the psychological commissioners are suitable for the mental health education. There is no big difference of mental health and personality characteristics of psychological commissioners among different group members. All these results remind us that the choosing standard is very important when we choose the psychological commissioners. Furthermore, we should strengthen the psychological commissioners training.

Keywords: psychological commissioner, mental health, personality.

1 Introduction

Psychological commissioners systems are a new thing of the mental health education in China University. Since 2004, Tianjin University introduced and established a crisis intervention and quick response system based on the psychological commissioners, it

H. Tan (Ed.): Knowledge Discovery and Data Mining, AISC 135, pp. 473–479.
springerlink.com

has been popularized and Implemented in .Universities rapidly. The recently survey showed that the 61.8% universities had setup the class psychological commissioners. One male and female class psychological commissioners were setup in one class respectively in our university. Psychological commissioners systems exhibited the really benefits of the mental health education in China university. For example, it made up for the deficiency of psychological counseling professionals, enhanced the mental health education team strength, accelerated the information collection, increased the student's warning and prevention ability against the psychological crisis events, made the psychological crisis intervention network more widely and made the propaganda of the mental health knowledge more popularized.

The psychological commissioners as the go-between for the students, it built a good bridge for the students that need help and the mental health prevention system. The special role of the psychological commissioners required that they should own the basically quality is the better psychological quality. Therefore, in this investigation, we compared the mental health condition and personality characteristics of the psychological commissioners and normal students in order to better understand it. Find the psychological problems and deficiency of the personality of the psychological commissioners, and provide the timely help and pertinence training the healthy psychology and good personality in order to make them competent for the job and develop it smoothly in the future. The reports are as follows.

2 Objects and Methods

2.1 Objects

Here are freshmen aged 17-22 years old of 2010 grade in Henan Institute of Science and Technology. 4531 freshmen which are included 337 psychological commissioners were investigated by using Symptom Checklist 90. 3540 freshmen which are included 240 psychological commissioners were investigated by using Eysenck Personality Questionnaire. They are majoring in agricultural, science, arts, economic management discipline, art discipline, sports discipline, and so on.

2.2 Tools and Conducting Methods

2.2.1 Symptom Checklist 90 (SCL-90)

The Symptom Checklist-90 (SCL-90) is a relatively brief self-report psychometric instrument (questionnaire) published by Pearson Assessment. It is designed to evaluate a broad range of psychological problems and symptoms of psychopathology. Responses were made on scales from 1 (never/not at all) to 5 (always/very much), which involve ten Symptom Checklists such as, somatization (SOM), obsessive-compulsive(O-C), interpersonal-sensitivity(I-S), depression(DEP), anxiety(ANX), hostility(HOS), phobic anxiety(PHOB), paranoid ideation(PAR), psychoticism (PSY), additional items(ADD). The higher the score is, the more serious psychological problem is.

2.2.2 Eysenck Personal Questionnaire(EPQ)

Eysenck Personality Questionnaire (EPQ) is a questionnaire to assess the personality traits of a person. EPQ includes Extroversion/Introversion (E), Neuroticism/Stability (N), Psychoticism/Socialisation (P), Lie (L)4.

2.3 Data Analyses

Statistical analyses were performed using programs available in the Statistical Package for Social Sciences (SPSS for Windows release 11.5). The significance of observed associations and/or differences between variables was tested using the Student's t test, analysis of variance (ANOVA, F test). A difference was considered to be statistically significant if $P<0.05$.

3 Results

3.1 Comparison of the SCL-90 Factor Score between Psychological Commissioners and the Other Freshmen

The total score and different factors values of those psychological commissioners were markedly lower than the other freshmen. In addition to somatization, hostility, the contrasts of the other scores are statistical ($P< 0.05$). Please refer to Tab.1.

Table 1. Comparison of the SCL-90 factor score between psychological commissioners and the other freshmen ($\bar{x}$ ±s)

factor	psychological commissioners (n=337)	the other freshmen (n=4 192)	*t*	Sig.
SOM	1.21±0.28	1.24±0.35	-1.776	0.076
O-C	1.70±0.50	1.79±0.58	-2.760	0.006
I-S	1.51±0.48	1.64±0.58	-3.754	0.000
DEP	1.34±0.38	1.44±0.50	-3.584	0.000
ANX	1.33±0.44	1.37±0.45	-3.445	0.001
HOS	1.23±0.50	1.37±0.45	-1.732	0.083
PHOB	1.33±0.40	1.37±0.45	-4.379	0.000
PAR	1.36±0.31	1.41±0.46	-2.129	0.033
PSY	1.34±0.34	1.40±0.42	-2.599	0.009
ADD	1.30±0.37	1.38±0.43	-3.231	0.001
TOTAL	123.37±29.37	130.49±36.53	-3.489	0.000

3.2 Comparison of the EPQ Factor Score between Psychological Commissioners and the Other Freshmen

The results showed that the original scores in E L items, which are got by investigated psychological commissioners, are very different from the other freshmen ($P<0.01$). In E items, the score the psychological commissioners got is higher than the other freshmen. While in L items, the score are lower. Please refer to Tab.2.

Table 2. Comparison of the EPQ factor score between psychological commissioners and the other freshmen ($\bar{x}$ ±s)

factor	psychological commissioners (n=240)	the other freshmen (n=3 300)	t	Sig.
P	3.41±2.21	3.37±2.24	0.256	0.798
E	15.53±3.40	14.21±4.12	4.840	0.000
N	8.14±5.15	8.64±5.29	-1.415	0.157
L	12.92±3.66	13.60±3.48	-2.922	0.003

3.3 Comparison of the SCL-90 Factor Score between Male and Female of Psychological Commissioners

The score the female got in somatization, obsessive-compulsive, interpersonal sensitivity, depression, anxiety, phobic anxiety, psychoticism, additional items, the total score is higher than that the male got. But only the contrast of phobic anxiety is statistical (P<0.01). While the score the male got in hostility, paranoid ideation is higher than the girls got, it has no obvious difference (P>0.05). Please refer to Tab.3.

Table 3. Comparison of the SCL-90 factor score between male and female ($\bar{x}$ ±s)

factor	male n=159	female n=178	t	Sig.
SOM	1.21±0.29	1.21±0.28	-0.102	0.508
O-C	1.68±0.51	1.73±0.50	-0.891	0.824
I-S	1.51±0.48	1.52±0.48	-0.299	0.854
DEP	1.31±0.37	1.37±0.39	-1.602	0.068
ANX	1.33±0.34	1.37±0.39	-1.056	0.195
HOS	1.33±0.43	1.32±0.36	0.157	0.203
PHOB	1.19±0.29	1.27±0.37	-1.974	0.006
PAR	1.37±0.44	1.34±0.39	0.678	0.403
PSY	1.33±0.34	1.35±0.34	-0.418	0.683
ADD	1.28±0.37	1.32±0.37	-0.901	0.894
TOTAL	121.97±29.56	124.62±29.23	-0.828	0.582

3.4 Comparison of the SCL-90 Factor Score among Psychological Commissioners of Different Majors

The scores those psychological commissioners in different majors got have no obvious differences (P>0.05). Please refer to Tab.4.

Table 4. Comparison of the SCL-90 factor score among psychological commissioners of different majors

factor	Sum of Squares	Mean Square	F	Sig.
SOM	0.54	0.108	1.359	0.240
O-C	1.83	0.366	1.451	0.205
I-S	0.97	0.195	0.849	0.516
DEP	0.97	0.194	1.335	0.249
ANX	0.65	0.129	0.944	0.452
HOS	0.90	0.180	1.138	0.340
PHOB	0.58	0.116	1.028	0.401
PAR	0.60	0.121	0.697	0.626
PSY	0.16	0.033	0.281	0.923
ADD	0.62	0.123	0.891	0.487
TOTAL	4398.16	879.632	1.020	0.406

3.5 Comparison of the EPQ Factor Score between Male and Female of Psychological Commissioners

The result shows that the score in P item that the female got is lower than that the male got, while in E N L items, the scores are higher than those the male got. Score in L is statistical ($P<0.01$). Score in P E N item have no difference ($P>0.05$). Please refer to Tab 5.

Table 5. Comparison of the EPQ factor score between male and female ($\bar{x}$ ±s)

factor	male(n=110)	female(n=130)	t	Sig.
P	3.71±2.13	3.16±2.26	1.919	0.056
E	15.39±3.38	15.65±3.43	-0.595	0.552
N	8.10±5.04	8.17±5.25	-0.104	0.918
L	11.87±3.58	13.80±3.50	-4.205	0.000

3.6 Comparison of the EPQ Factor Score among Psychological Commissioners of Different Majors

The result shows that the scores in E N items have no differences ($P>0.05$), the scores in P L have differences ($P<0.01$). Please refer to Tab 6.

Table 6. Comparison of the EPQ factor score among psychological commissioners of different majors

factor	Sum of Squares	Mean Square	F	Sig.
P	77.63	19.41	4.167	0.003
E	48.08	12.02	1.038	0.388
N	356.51	89.13	3.506	0.008
L	95.97	23.99	1.816	0.126

4 Discussion

The psychological commissioners' mental healthy level is obviously better than other students. The working nature of the psychological commissioners asked them have the more healthy psychological quality, the results showed that their psychological quality meet this work. On the one hand, because they are outstanding as class cadres and have more chance to training the psychological quality, on the other hand, it also showed choosing the psychological commissioners is more scientific, the other reasoning is the mental healthy training was helpful to improve their psychological quality.

The personality characteristics of the psychological commissioners are suitable for the mental health education. In comparison, the psychological commissioners are outward and optimistic, more stable mood and more brave face to true self. All these qualities make them easily to communication, help and affect the classmates around them.

The difference of mental health and personality characteristics of psychological commissioners among different group members is not clear. In this study on the all the student's mental health and personality characteristics, we found that there are obvious difference in the different gender, resource and majors, but there is no difference among the psychological commissioners. The reasons may be that they are the outstanding group, they had breakthrough the limitation on the gender, resource and majors, and overcome the character shortcoming , and can untangle the stress and adjust the physical and mental well, so different psychological commissioners groups' mental health level and personality characteristics are similar.

5 Conclusions

All these results remind us that when we choose the psychological commissioners, the choosing standard is very important, we should select which are optimistic, outgoing, outward and good mental health condition, which be good at communication and have some good language expression ability, which are interesting in psychology and love the psychological education, which have the mass basis among the students and love the class working, and have the service consciousness.

Furthermore, we should strengthen the psychological commissioners training and improve their psychological health level, perfect the personality characteristics and clear the working responsibility, learn the content of psychological health education, experience the psychological health group activity's charm. At the same time, we should guide the psychological commissioners to study the psychological health, crisis and prevention methods, and helpful to find the classmate's psychological problem and learn how to deal with some normal psychological crisis events.

References

1. Du, F.: The Survey on the Status of the Psychological Commissioners in University. Education and Vocation (4), 36–38 (2009)
2. Huang, Q.: The Research and Analysis on the Status of the Psychological Commissioners in University. China Journal of Health Psychology, 1149–1152
3. Kochanska, G., Katherin, D., Goldman, M., et al.: Maternal reports of conscience development and temperament in young children. Child Development 65(3), 852–868 (1994)
4. Wang, X., Wang, X., Ma, H., et al: Rating scales for mental health. Beijing: Chinese Mental Health Journal, 33–37 (1999)
5. Gong, Y.: Eysenck Personal Questionnaire Handbook, vol. (2), p. 31. Hunan Medical College publishing company, Changsha (1993)

Queueing Analysis of the Decoding Process for Intra-session Network Coding with Random Linear Codes

Yuan Yuan, Zhen Huang, Shengyun Liu, and Yuxing Peng

National Laboratory for Parallel and Distributed Processing,
National University of Defense Technology, Changsha, 410073, China
{yuanyuan,liushengyun,huangzhen,pengyuxing}@nudt.edu.cn

Abstract. Efficient designs for intra-session network coding based practical applications largely rely on a better understanding on its queueing behaviors. However, few work devote on this topics. In this paper, we build a multi-channel batch service queueing system ($M^N/D^m/1$) with control feedbacks to describe the decoding process of intra-session network coding with random linear codes and try to answer several fundamental questions, including for example, how to analyze braking redundancy? Under what condition is the system stable? How's quantitative relationship between the inter-decoding delay and the generation granularity?

Keywords: Intra-session network coding, Segment granularity, Control feedbacks, Queueing analysis.

1 Introduction

Network coding has enjoyed much popularity within the research community since it was first introduced in its seminal paper by Ahlswede et al. [1]. With network coding, intermediate nodes encode multiple data packets algebraically together by using, for example, bitwise XOR or linear combination in Galois field, while destination nodes perform decoding operations to recover the original data. The benefits of network coding include improving network throughput, reducing energy consumption, enhancing network reliability, and much more.

According to the operations on sessions, network coding can be classified into inter-session and intra-session [2]. Intra-session network coding only allows intermediate nodes to encode the packets within the same session, which has found its success in peer-to-peer systems for content dissemination [3, 4]. Since with random linear codes, the coding coefficients can be chosen randomly and independently, and can be implemented in a fully distributed fashion, it is often utilized as the practical form for intra-session network coding [5].

In the practical applications, to maintain a tractable computation overhead, each session is split into multiple generations with equal number of packets. The number of packets in each generation, as the intra-session coding range, is called *generation granularity*. During the data dissemination from multiple upstream nodes, each destination has to hold its decoding operation until the number of coded packets

H. Tan (Ed.): Knowledge Discovery and Data Mining, AISC 135, pp. 481–488.
springerlink.com

buffered in its queue achieves the generation granularity. Once the decoding operation can be activated, destinations send braking messages to their upstream nodes in order to switch the data transmission to the next generation. This kind of dissemination protocol leads to highly resilient data transmissions in a multiple-sources fashion without the needs of explicit cooperation.

Most related work on generation based intra-session network coding focus on protocol design and system implementation [3, 4]. An efficient design of these practical applications largely relies on a better understanding on the stochastic behavior of the decoding process for intra-session network coding, which can be summarized as many unsolved fundamental questions, including for examples, How to analyze braking redundancy? Under what condition is the system stable? How's quantitative relationship between the inter-decoding delay and the generation granularity? Nevertheless, few of existing researches try to explore these problems.

To fill this vacancy, in this paper, we build a multi-channel batch service queueing system ($M^N/D^m/1$) with control feedbacks to describe the decoding process of intra-session network coding with random linear codes. Other than traditional batch queues [6], the control feedbacks complicate our analysis. To obtain explicit solutions, we first conduct rigorous discussion on the distribution of the braking redundancy and its expectation. Under the practical settings, we then utilize an approximation technique to compute the steady state probabilities of the queueing system. Finally, a full characterization for the departure process is provided.

2 Preliminary

In this section, we review the principle of practical intra-session network coding with random linear codes. Suppose that the generation granularity is denoted by m, and the original packets in generation i are denoted by vector $\mathbf{O}_i = \{O^i_1, O^i_2, \ldots, O^i_m\}$. The random linear codes are assumed to be defined in Galois field $\mathbf{GF}(2^q)$, where q is assumed to be large enough and $q/8 = \alpha(\alpha \in \mathbb{Z}^+)$. $\beta = \lceil s_o / \alpha \rceil$.

For a source, before sending out a packet, it first randomly and independently selects m coefficients from a Galois field $\mathbf{GF}(2^q)$, which are denoted by $\mathbf{G} = \{g_1, g_2, \ldots, g_m\}$. $\mathbf{G}$ is called the global encoding vector (GEV), since it acts on the original packets. Then, it encodes the original packets with the coefficients by using linear combination to derive a coded packet of generation i, say C_i, which can be expressed as: $C^i = \mathbf{G} \cdot (O^i)^T = \sum_{1 \le k \le m} g_k O^i_k$. $\mathbf{G}$ will be piggybacked in the header of the coded packet C_i and sent to the downstream nodes.

For an intermediate node, if the piggybacked GVE of a newly received packet is linear dependent with those of coded packets already buffered, which is no use for decoding, and will be discarded immediately. We call these discarded coded packets as *linear dependent redundancy*. Otherwise, are called *innovative* packets. Suppose that there are h independent coded packets of generation i have already been buffered, which are denoted by vector $\mathbf{C}^i = \{C^i_1, C^i_2, \ldots, C^i_h\}$, and suppose that their piggybacked GEVs are $\{\mathbf{G}_1, \mathbf{G}_2, \ldots, \mathbf{G}_h\}$, respectively. The relay first randomly and independently selects h ($1 \le h \le m$) coefficients from Galois field $\mathbf{GF}(2^q)$, denoted by vector $\mathbf{L} = \{e_1, e_2, \ldots, e_h\}$. $\mathbf{L}$ is called the local encoding vector (LEV), since it only acts on the coded packets. Then, the relay linearly combines the h coded packets with coefficient vector $\mathbf{L}$ to obtain a new coded block $\widetilde{C^i}$, which is:

$$\widetilde{C^i} = \mathbf{L}\cdot(\mathbf{C}^i)^T = \sum_{1\le j\le h} l_j \cdot \sum_{1\le k\le m} g_{j,k} O_k^i = \sum_{1\le k\le m} (l_1 g_{1,k} + l_2 g_{2,k} + \cdots + l_h g_{h,k}) \cdot O_k^i \tag{1}$$

Suppose $\widetilde{g_k} = e_1 g_{1,k} + e_2 g_{2,k} + ... + e_h g_{h,k}$, such that $\widetilde{C^i} = \sum_{1\le k\le m} \widetilde{g_k} O_k^i = \widetilde{\mathbf{G}} \cdot (\mathbf{O}^i)^T$, where $\widetilde{\mathbf{G}}$ denotes the GEV of $\widetilde{C^i}$ and will be piggybacked in the transmission.

For a destination, once an innovative packet of a certain generation is received, it uses *Gauss-Jordan elimination* for progress decoding. When the number of innovative packets received exceeds the generation granularity m, it sends braking message to its upstream nodes to start the data transmission of the next generation, and then can obtain a generation of the original data simultaneously by decoding.

3 Queueing Model of the Decoding Process

As shown in Fig. 1, the decoding process at a destination can be modeled as a multi-channel batch service queueing system ($M^N/D^m/1$) with control feedbacks. The specific configurations are listed as follows:

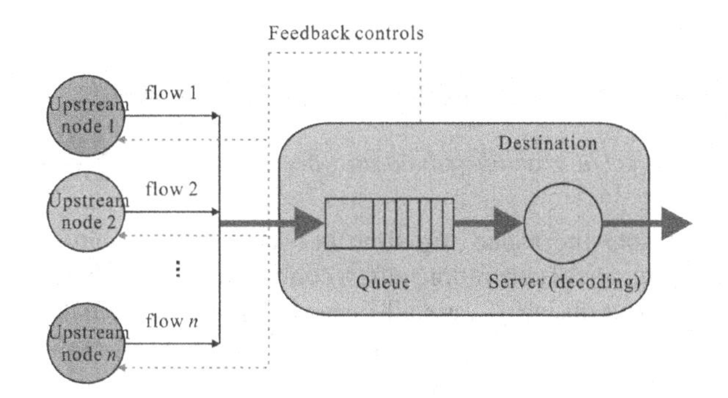

Fig. 1. Queueing model for decoding process

- **Traffic arrivals:** N Input flows from N upstream nodes are assumed to be independent, and can be classified into K groups according to their different upload capacities. The inter-arrival times of a group i flow a_i are independent identically distributed (i.i.d) random variables which follows an exponential distribution with mean arrival rate λ_i. Suppose N_i denotes the number of flows in group i, such that $\sum_{i=1}^{K} N_i = N$ and $\sum_{i=1}^{K} N_i\lambda_i = \lambda$. Based on the superposition property of Poisson arrivals, the N input Poisson flows aggregating together also form a Poisson flow with the aggregating intensity of λ.
- **Service discipline:** Server keeps idle until more than m coded packets are buffered. Each service once removes m coded packets from the queue.

- **Service time:** By using Gauss-Jordan elimination, the service time can be approximated as the processing time of the incoming m-th innovative packet with other m-1 code packets already processed. Since m is fixed, the service time, denoted by T_s, can be regarded as a constant. We will discuss the explicit expression of T_s later in the analysis.
- **Queueing discipline:** The destination maintains a single FIFO queue with infinite queue size.
- **Control feedbacks:** All the channels are assumed to be duplex and the transmissions of control feedbacks are independent with the data transmissions. The inter-arrival times of the feedbacks b_i are i.i.d. random variables which also follows an exponential distribution with mean arrival rate δ_i. At the instant that the m-th coded packet is received, the server sends out N feedbacks to its upstream nodes.

4 Queueing Analysis

Since parts of the coded packets will be discarded during the transmissions of the braking messages, to analyze this queueing system, an important breakthrough is to derive the expected braking redundancy.

4.1 Braking Redundancy

Base on the above configurations, we have the following theorem:

Theorem 1. *The expected braking redundancy per generation is* $\sum_{i=1}^{K} \frac{N_i \lambda_i}{\delta_i} + N$.

Proof. Let T^i_{rtt} denote the round trip time between the destination to its group i upstream nodes. For each generation, after control feedbacks being send out, the coded packets received during T^i_{rtt} are all linear independent redundancies. It is clear that the value of T^i_{rtt} equals to the sum of the two independent random variables, a_i and b_i, thus the CDF of T^i_{rtt} can be obtained as follows:

$$\begin{aligned} F_i(t) &= \mathbf{Pr}\{a_i + b_i \le t\} \\ &= \iint_{x+y\le t} \lambda_i e^{-\lambda_i x} \cdot \delta_i e^{-\delta_i y} dxdy \\ &= 1 - e^{-\delta_i t} - \frac{\delta_i}{(\lambda_i - \delta_i)} (e^{-\delta_i t} - e^{-\lambda_i t}). \end{aligned} \tag{2}$$

Thus, the PDF of T^i_{rtt} is

$$f_i(t) = \frac{\lambda_i \delta_i}{(\lambda_i - \delta_i)} \cdot (e^{-\delta_i t} - e^{-\lambda_i t}) \qquad (t \ge 0) \tag{3}$$

Let r_i denote the number of coded packets sent by the group i flow for one generation during a T^i_{rtt}. Suppose that χ_i = $\mathbf{Pr}\{r_i{=}k\}$, by using the law of total probability, we have:

$$\chi_i = \int_0^\infty \mathbf{Pr}\{r_i = k, t < T_{rtt}^i < t + dt\}dt = \int_0^\infty \mathbf{Pr}\{r_i = k \mid t_i = t\} f_i(t)dt$$
$$= \int_0^\infty \frac{(\lambda_i t)^k e^{-\lambda_i t}}{k!} f_i(t)dt \tag{4}$$

To find the explicit solution of the distribution and the first moment of r_i, we then define the z-transform for r_i as

$$\begin{aligned} R_i(z) &\triangleq \mathbf{E}[z^{r_i}] \triangleq \sum_{k=0}^{\infty} \mathbf{Pr}\{r_i = k\} z^k \\ &= \sum_{k=0}^{\infty} \int_0^\infty \frac{(\lambda_i t)^k e^{-\lambda_i t}}{k!} f_i(t)dtz^k \\ &= \int_0^\infty e^{-\lambda_i t} \left(\sum_{k=0}^{\infty} \frac{(\lambda_i tz)^k}{k!} \right) f_i(t)dt \\ &= \int_0^\infty e^{-(\lambda_i - \lambda_i z)t} f_i(t)dt \end{aligned} \tag{5}$$

Since the property that the first order derivative of z-transforms evaluated for $z = 1$ gives the first moment of the random variable r_i, we further have:

$$\mathbf{E}[r_i] = \left.\frac{dR_i(z)}{dz}\right|_{z=1} = \lambda_i \left(\frac{\lambda_i}{(\lambda_i - \delta_i)\delta_i} - \frac{\delta_i}{(\lambda_i - \delta_i)\lambda_i} \right) = \frac{\lambda_i}{\delta_i} + 1 \tag{6}$$

Since the N input flows are independent, we could derive the expected braking redundancy per generation $\bar{r}_{gen}$ as follows:

$$\bar{r}_{gen} = \mathbf{E}[\sum_{i=1}^{K} N_i r_i] = \sum_{i=1}^{K} N_i \mathbf{E}[r_i] = \sum_{i=1}^{K} \frac{N_i \lambda_i}{\delta_i} + \sum_{i=1}^{K} N_i = \sum_{i=1}^{K} \frac{N_i \lambda_i}{\delta_i} + N \tag{7}$$

□

4.2 Service Time

Since Gauss-Jordan elimination is a variant of Gaussian elimination, such that the computation time of the formal one can be approximated by the later one. The whole decoding process of Gaussian elimination can be divided into two phrases: computing the inverse of a $m \times m$ coefficient matrix, and multiplying the inverse matrix with m coded blocks. Let Δ_p denotes the processing time for a unit multiplication operation or a unit addition in $\mathbf{GF}(2^q)$, thus the total decoding time T_{dec} can be expressed as:

$$\begin{aligned} T_{dec} &= ((m+1)m(m-1) + m)\Delta_p + (m+1)m(m-1)\Delta_p + (m\Delta_p + (m-1)\Delta_p)m\beta \\ &\approx 2\Delta_p m^2 (m + \beta). \end{aligned} \tag{8}$$

Since the weight of the processing time for m-th innovative packet w_s can be expressed as $w_s = (m-1) \Big/ \frac{(m-1+1)(m-1)}{2} = 2/m$. Thus, we have

$$T_s = T_{dec} \cdot w_s \approx 4\Delta_p m(m + \beta). \tag{9}$$

4.3 Stability

From the above analysis on braking redundancy, the proportion of the coded packets which are allowed into the queue per generation can be calculated as $m/(m+\bar{r}_{gen})$. Based on stability theorem of queueing system, the following theorem about the utilization factor can be obtained.

Theorem 2. *The queueing system has steady state probability, if generation granularity m satisfies the following condition:*

$$0 < m < \left\lfloor \frac{(1-\varphi\beta)+\sqrt{(1-\varphi\beta)^2+4\beta\bar{r}_{gen}}}{2\varphi} \right\rfloor \tag{10}$$

Proof. Let ρ be the utilization factor of this queueing system. Based on the definition of utilization factor, we have:

$$\rho = \lambda\bar{x} = \left(\lambda\frac{m}{m+\bar{r}_{gen}}\right)\frac{T_s}{m} = \frac{4\lambda\Delta_p m(m+\beta)}{m+\bar{r}_{gen}} \tag{11}$$

Since whether the steady state probabilities exist is decided by the stability of the queueing system, which requires $\rho < 1$. Let $\varphi = 4\lambda\Delta_p$, thus, we obtain a quadratic inequality as: $\varphi m^2 + (\varphi\beta - 1)m - \bar{r}_{gen} < 0$.

It is easy to obtain two roots from the above inequality, those are

$$x_{1,2} = \frac{(1-\varphi\beta)\pm\sqrt{(1-\varphi\beta)^2+4\beta\bar{r}_{gen}}}{2\varphi} \tag{12}$$

Since $m \in \mathbb{Z}^+$, to maintain a stable queueing system, the generation granularity should satisfy the following conditions: $0 < m < \left\lfloor \frac{(1-\varphi\beta)+\sqrt{(1-\varphi\beta)^2+4\beta\bar{r}_{gen}}}{2\varphi} \right\rfloor$ □

4.4 Steady State Probability

For non-Markov service time queueing systems with Poisson arrivals, most of the traditional approaches use embedded Markov chain technique with z-transforms [7] to obtain steady state probabilities. However, through these techniques, one needs first to compute the roots of an m-degree equation, and then solves an m-dimensional linear equation set. Since in real-world applications, m is often on the order of hundreds, the computation overhead becomes undesirable.

Fortunately, from the practical configurations of intra-session network coding based systems, Gauss-Jordan elimination significantly reduces the service time, which makes the utilization factor of our queueing system sufficiently low. This observation helps us find an alternative approach. According to the literature [9], in a batch service queueing system, if the batch size is high ($m \geq 60$) and utilization factor is small ($\rho < 0.5$), it is reasonable to assume that the post-departure queue length probability $\mathbf{Pr}\{\tilde{q} = k\}$ equals to the probability that the number of innovative packet arriving in the queue during a service period $\mathbf{Pr}\{\tilde{v}=k\}$, since there will approximately be no queueing. Let $\tilde{a}$ denote the number of coded packets sent out from the N upstream nodes, which has the mean arrival rate of λ. We thus need to find out the relationship between $\tilde{a}$ and $\tilde{v}$. Using the expected braking redundancy per generation, we could derive the following approximate formulas:

$$\mathbf{Pr}\{\tilde{q}=k\}=\mathbf{Pr}\{\tilde{v}=k\}\approx\begin{cases}\mathbf{Pr}\{\tilde{a}\le\overline{r}_{gen}\} & (k=0)\\ \mathbf{Pr}\{\tilde{a}=k+\overline{r}_{gen}\} & (0<k<m)\end{cases} \tag{13}$$

We ignore the case that $k\ge m$, since the probability is very small.

4.5 Departure Process

To characterize the departure process of our queueing system, we need to analyze the inter-departure time.

Let d be the random variable denoting the inter-departure time of the original packets decoded. Let $\tilde{r}$ denote the number of discarded packets during a service period. We conduct our analysis at a batch departure instant, which involves two cases:

- $\tilde{q}=0$: Upon a batch departure instant, if the queue is empty, suppose that r_s redundant packets have already been discarded, the time till the next batch departure is the sum of $m+(\overline{r}_{gen}-\tilde{r})$ inter-arrival times and a service time.
- $0<\tilde{q}<m$: Upon a batch departure instant, if the number of coded packets buffered is less than m, then the time till the next batch departure is the sum of $m-\tilde{q}$ inter-arrival times to form a batch and a service time.

Let $d(t)$ and $D(t)$ denote the PDF and the CDF of the inter-departure time at this case, respectively. According to the above analysis, we thus have:

$$\begin{aligned}D(t)&=\mathbf{Pr}\{d\le t\}\\
&=\mathbf{Pr}\{\sum_{i=1}^{m+\overline{r}_{gen}-\tilde{r}}t_i+T_s\le t\,|\,1\le\tilde{r}\le\overline{r}_{gen}\}+\mathbf{Pr}\{\sum_{i=1}^{m-\tilde{q}}t_i+T_s\le t\,|\,1\le\tilde{q}\le m-1\}\\
&=\sum_{l=1}^{\overline{r}_{gen}}\mathbf{Pr}\{\tilde{r}=l\}\cdot\mathbf{Pr}\{\sum_{i=1}^{m+\overline{r}_{gen}-\tilde{r}}t_i+T_s\le t\}+\sum_{k=1}^{m-1}\mathbf{Pr}\{\tilde{q}=k\}\cdot\mathbf{Pr}\{\sum_{i=1}^{m-k}t_i+T_s\le t\}\\
&=\sum_{l=1}^{\overline{r}_{gen}}\mathbf{Pr}\{\tilde{a}=l\}\cdot\mathbf{Pr}\{\sum_{i=1}^{m+\overline{r}_{gen}-l}t_i+T_s\le t\}+\sum_{k=1+\overline{r}_{gen}}^{m+\overline{r}_{gen}-1}\mathbf{Pr}\{\tilde{a}=k\}\cdot\mathbf{Pr}\{\sum_{i=1}^{m+\overline{r}_{gen}-k}t_i+T_s\le t\}\\
&=\mathbf{Pr}\{\sum_{i=1}^{m+\overline{r}_{gen}}t_i\le t\}\qquad (t>T_s)\end{aligned} \tag{14}$$

It is well known that if the input flow is Poisson process, the sum of $m+\overline{r}_{gen}$ inter-arrival times follows Gamma distribution, therefore, $d(t)$ can be expressed as:

$$d(t)=\lambda e^{-\lambda t}\frac{(\lambda t)^{m+\overline{r}_{gen}-1}}{(m+\overline{r}_{gen}-1)!}\qquad (t>T_s) \tag{15}$$

Based on the above equations, we could compute the expected inter-departure time as follows:

$$\begin{aligned}\mathbf{E}[d] &= \int_{T_s}^{+\infty} t\cdot\lambda e^{-\lambda t}\frac{(\lambda t)^{m+\overline{r}_{gen}-1}}{(m+\overline{r}_{gen}-1)!}dt \\ &= \frac{1}{\lambda(m+\overline{r}_{gen}-1)!}\cdot\int_{\lambda T_s}^{+\infty} x^{m+\overline{r}_{gen}}d(-e^{-x}) \\ &= \frac{(m+\overline{r}_{gen})e^{-\lambda T_s}}{\lambda}\sum_{i=0}^{m+\overline{r}_{gen}}\frac{(\lambda T_s)^i}{i!} \\ &= \frac{(m+\overline{r}_{gen})}{\lambda}(1-\varepsilon),\end{aligned} \tag{16}$$

where $\varepsilon = \sum_{i=m+\overline{r}_{gen}}^{+\infty}\frac{(\lambda T_s)^i}{i!}\Big/\sum_{i=0}^{+\infty}\frac{(\lambda T_s)^i}{i!}$ is small value, and decreases as m increases.

5 Conclusion

In this paper, we conduct queueing analysis for the decoding process of intra-session network coding, which can be build as a multi-channel batch service queueing system ($M^N/D^m/1$) with control feedbacks. Our analysis can help better understand the critical parameter *generation granularity* in intra-session network coding based systems and is useful for the future system design and implementation.

References

1. Ahlswede, R., Cai, N., Li, S., Yeung, R.: Network information flow. IEEE Trans. Information Theory 46(4), 1204–1216 (2000)
2. Ho, T., Lun, D.: Network Coding: An Introduction. Cambridge University Press (2008)
3. Gkantsidis, C., Rodriguez, P.: Network coding for large scale content distribution. In: Proceedings of IEEE INFOCOM (2005)
4. Wang, M., Li, B.: Lava: A reality check of network coding in peer-to-peer live streaming. In: Proceedings of IEEE INFOCOM (2007)
5. Chou, P.A., Wu, Y., Jain, K.: Practical network coding. In: Proceedings of the Annual Allerton Conference on Communication Control and Computing (2003)
6. Chaudhry, M.L., Templeton, J.G.C.: A First Course in Bulk Queues. John Wiley & Sons
7. Kleinrock, L.: Queueing Systems: Theory, vol. I. John Wiley & Sons
8. Briere, G., Chaudhry, M.L.: Computational Analysis of Single-Server Bulk-Service Queues, M/GY/1. Advances in Applied Probability 21(1), 207–225 (1989)

The Teaching Reform and Practice of the Improvement of the Application Ability of Mechanical and Electrical Transmission Control

Yunlong Yuan, Weimin Zhao, and Jie Hu

College of Mechanical Engineering, Ningbo University of Technology, Ningbo, 315016

Abstract. The existing problems of the traditional teaching of "Mechanical and Electrical Transmission Control" course have been analyzed in this paper. From the perspective of improving the application ability of students' engineering practice, some corresponding teaching reform measures have been proposed in this paper, which have been also applied in the teaching practice. According to the results of teaching practice, the teaching reform has promoted the improvement of this curriculum and enhanced the interest and the motivation of students for learning this course. At the same time, the improvement of the application ability of engineering practice and the innovation ability of students has also achieved good results.

Keywords: mechanical and electrical transmission control, application ability, teaching reform.

1 Introduction

With the advance and the development of technology, the modern enterprise is becoming more and more automated. Therefore, it has put forward higher requirements for the electro-mechanical engineering and technical personnel. As a graduate in the department of mechanical manufacturing and automation, he should not only possess the solid and strong mechanical knowledge and disciplines as well as the related theoretical knowledge of electrical control automation, but also have the strong application ability of engineering practice. Only in this way can they adapt to the requirements of modern enterprises for the engineering and technical personnel and be qualified for the design, the development and the maintenance of modern automation system equipment.

The "Mechanical and Electrical Transmission Control" is a professional course in the department of mechanical manufacturing and automation, which is also the basic course of the related automatic control theory. Based on some pre-courses about the electricity, it has taught the traditional relay-contactor control technology, the motor control technology, the programmable controller and some other techniques according to the application status of automatic control technology in modern enterprises. It is an important course which has a close integration of theory and practical application and that is also closely related to the automatic control in modern enterprises.

H. Tan (Ed.): Knowledge Discovery and Data Mining, AISC 135, pp. 489–496.
springerlink.com

2 The Shortcomings of Traditional Teaching

However, in the long-term teaching practice, we can find that the traditional lecture-based teaching method has the following drawbacks, which has been not suited to the highly practical teaching requirements. So it is required to reform this traditional teaching method so as to adapt to the development of modern control technology and the demands of curriculum reform of "Mechanical and Electrical Transmission Control".

2.1 The Teaching Content Is Redundant and Obsolete

The "Mechanical and Electrical Transmission Control" is a practical and comprehensive course which should be learnt after completing some pre-courses about the electrical engineering and other circuit knowledge. However, with the development, the adjustment and the reform of some related course content, on the one hand, a part of content in original syllabus is overlapped with other courses. For example, the content about the velocity modulation of DC and AC motor has overlapped with the "electric drive" or the "electrical engineering" courses, which has caused the repeat explanation in teaching process. On the other hand, with the development of modern control technology, the new control method and the control devices have been constantly introduced, such as the programmable controller technology and the application. The explanation of this new knowledge is absent.

2.2 The Method of Theoretical Teaching Is Monotonous and the Content Is Boring

The theoretical teaching is generally performed in classroom, which always employs the verbal forms and whose form is single. Meanwhile, the partial content of this course is closely related to the mathematics, the physics, the electricity and other courses. The theory is complicated and the concept is abstract. The students are difficult to understand them except rote. Then it is difficult to stimulate the learning interest of students. For example, when the teachers explain the working principle of the electrical control circuit, the schematic diagrams of various electrical equipments are numerous and the drawings are similar. The students can't contact with the entity and don't have the perceptual knowledge, so the students will feel boring and have difficulty in understanding them in the learning process. There also exist the same problems in explaining the instructions of programmable controller and the programming practice.

2.3 The Theoretical Teaching Is Disjointed with the Practice Teaching

In the teaching process, the traditional teaching method refers that the theoretical teaching comes first and the practice teaching follows subsequently. However, there are different teachers for the practice teaching and the theoretical teaching. And after a certain time, sometimes the teaching schedule will be not synchronized and the practice teaching will lag behind, which causes the mismatch between theory and practice. Meanwhile, as the theoretical study is not solid, the students will not effectively use the theoretical knowledge when training, let alone employ the theory to guide practice. Therefore, the mismatch between the theoretical teaching and the practice teaching has increased the difficulty of theoretical teaching. And the practice teaching without the guidance of theory may become the simple repetition, which is monotonous and boring so that the students can't effectively grasp the practical skills. Then the solid and effective practical teaching effectiveness will not be achieved.

2.4 The Assessment Method Is Single and Rough

The traditional assessment method always employs the summative evaluation system in which "a course is taught in the end and a score will set life and death". The form is single so that the assault phenomenon before exams is serious and the students may also prone to the idea of cheating. Meanwhile, the main content of examination is usually the exam of knowledge points, which can't fully reflect the comprehensive ability and the self-learning ability of students. Especially the experimental evaluation, the result is just determined by an experimental report. This assessment form is difficult to mobilize the learning enthusiasm of students.

3 It Is Required to Reform the Teaching Content and Method as Well as Improve the Teaching Effectiveness of Courses and the Engineering Application Ability of Students

In order to improve the teaching effectiveness and enhance the engineering practical application ability of students, it is necessary to keep abreast of the times to reform the teaching and abandon the irrational teaching programs. The "Mechanical and Electrical Transmission Control" is a highly practical course. It is required to reform it from the aspects of optimizing the teaching content, updating the methods and means of theoretical teaching and practice teaching, reforming the assessment methods and some others. Then the good teaching effectiveness will be obtained.

3.1 It Is Required to Explicit the Practical Application Requirements and Optimize the Course Content

The teaching is serving for the application. The school should actively strengthen the links with enterprises, organize the teachers to carry out the in-depth research in enterprises and understand the knowledge competency requirements of modern enterprises for the engineering and technical personnel. According to the situation that the content of "Mechanical and Electrical Transmission Control" course is more widely and it has overlapped or duplicated with the related course content, the teaching and research group has closely contacted with the actual production, highlighted the engineering application and carefully studied the course content on the basis of the personnel training requirements. At the same time, they have also modified the original syllabus and optimized the teaching content according to the principle that "the theory is sufficient and the practice is strengthened". In the new syllabus, the content of the frequently-used electrical components has been introduced. The original requirements for the basic aspects of circuit and the requirements for the design of the traditional relay-contactor control system have been maintained. Meanwhile, the explanation of the control diagram of the typical machine electrical control system has been also eliminated. Finally the lectures about the principle of programmable controller as well as the design and the development process of the control system have been significantly increased, because the programmable controller system has become a major force of the modern automated control system, which has been in line with the requirements of modern enterprises for the control engineering and technical personnel. After the modification, the teaching content has not only maintained the understanding and the mastery of basic content as well as avoided the overlap with the related courses, but also expanded the understanding and the application of the new knowledge.

3.2 It Is Necessary to Improve the Teaching Methods and Means and Promote the Students to Understand and Master the Knowledge

Having the good draft recognition ability of electrical schematic diagram and the analytical design capability is the basic quality for a mechanism automated engineering and technical personnel. However, the analysis process of electrical schematic diagram is a very abstract process. The circuit diagram is difficult for the students to understand in the teaching process. The mastery of the design and the analysis of electrical schematic diagram is the basis to improve the electrical automated analysis and design ability as well as the engineering practice ability of students. Therefore, the teachers are required to take various measures to improve the teaching methods and means so that the teaching process can be simple but profound as well as very interesting. Then the students will actively participate in the teaching process and the teaching effectiveness will be improved.

1) It is necessary to use the modern education methods and give full play to the advantages of multimedia courseware. There are a lot of electric schematic diagrams in the "Mechanical and Electrical Transmission Control" course which need to be explained and analyzed. And the working process of some circuit diagrams is very complex and "dead", so it is difficult for the students to understand them. The teaching and research group should use the multimedia technology to make the current "flow" in circuit diagrams. Then the working process of circuit diagram can be intuitively displayed so as to help the students analyze and understand the working principle of circuits.

2) It is required to combine with the psychological characteristics of students and flexibly use the various teaching methods. In the classroom, the teachers should combine with the psychological features of students and the appropriate knowledge to change the teaching methods and means according to the characteristics of various elements in curriculum. The students should be taught in accordance of their aptitude, in accordance of questions and in accordance of individuals. For example, when the author designs the specific project instances, he will always employ the "situational" approach. The students will be considered as the technical personnel and the teachers will be regarded as the customer or the chief engineers. They should guide the students to analyze, discuss and communicate with others as well as "force" the students into the role. In the specific problem situations, the students are guided to initiatively think and try the best to solve the "problems". This teaching method can

greatly stimulate the "business sense" and the learning interest of students, which can promote the improvement of self-learning ability and the innovation ability of students.

The different teaching methods can be employed according to the different content, such as the heuristic teaching method or the interactive teaching method with discussion sessions. When the method of discussion session is employed, the topic can be designated by teachers. They should try to select the course content which the students are more concerned with and are more interested in. Generally speaking, there is no standard answer, which can fully stimulate the imagination of students. As the subject of discussion, the students can be divided into groups or participate personally. The discussion topic will be announced ahead of time so that the students can have the opportunity to search the materials and can be fully prepared. Finally they will be reviewed by teachers.

3.3 It Is Required to Reform the Teaching Content and the Teaching Methods as Well as Enhance the Practical Application Ability of Students

The experimental teaching is an important part of the course teaching, which has played an important role in consolidating the obtained knowledge of students, understanding the actual environment of engineering application and improving the application ability as well as the innovative ability of practice. In order to improve the effectiveness of practice teaching, the teaching and research group has broken the traditional experimental teaching method and carried out a series of experimental teaching reforms based on the individual, the application-oriented and the creative talent-training objectives.

1) It is required to optimize the selected experimental projects and improve the practical operation ability of students. In order to improve the practical ability, the practical operation ability and the innovative ability of students, the demonstrative and validation experimental projects have been eliminated in the practice teaching. Some innovative and designable experimental projects have been increased. At the same time, the practice opportunities including letting the students design the circuit themselves, the assembly, the wiring, the commissioning and some others have been also increased.

2) It is necessary to use the open experimental teaching method. It is required to employ the open experimental teaching method which has combined with the curricular and the extracurricular as well as integrated the time before class, in class and after class so as to break the restrains of traditional experimental teaching in the aspects of time, space and the amount of knowledge information. At the same time, the range of experimental teaching in time and space will be also expanded and the flexibility of experimental teaching will be also enhanced. What's more, the initiative and the autonomy of students in the experimental process will be also given full play so as to make them receive the comprehensive training of knowledge and ability. During the experiment, the students are the body of experiment teaching and the teachers will guide, inspire and evaluate them throughout the process.

3) It is required to employ the multi-objective training teaching method which has combined with the global situation and the individuation. As for the basic experiments and most of the comprehensive experiments, it is necessary to employ the whole staff

teaching and require each student to participate. As for the design of the comprehensive experiment and the study of the innovative experiments, the students can independently take the elective courses according to their interests and abilities. It is necessary to make the students fully discuss their own design ideas, the programs and the experimental results as well as improve the learning autonomy and initiative of students to a large extent. However, as for some innovative and extra-curricular technology innovation projects with larger difficulty, it is required to select some excellent students to participate in it through the optimal way and assign some teachers to guide them. Through the diverse teaching method, it not only ensures the students to master the necessary knowledge and skills, but also fully exploit the potential of students.

4) It is necessary to set up the large-scale design comprehensive and research-based training programs. The teaching and research group can combine with years of practical teaching experience to independently design the large-scale hydraulic comprehensive training platform according to the requirements of training the application-oriented talents. We can carry out various control systematical design and development trainings such as the relay-contact control, the microcomputer control, the PLC control and the IPC control on it. The controlled objects may be the modular machine tools, the excavators, the injection molding machines, the robots, the various three-phase stepper motors and the DC motors, etc. The students can independently select the programs and design the whole system according to the requirements. Then the assembly, the wiring, the programming and the debugging will be followed subsequently. After that, the whole training process will be completed. This platform has provided a good practice environment for the students to carry out the development and the training of Mechanical and electrical transmission control system, which has also greatly enhanced the innovative design awareness and the practical application ability of students.

5) It is required to actively explore the teaching approach with a combination of “production, learning and research”. The school should cultivate the high-quality engineering applied-oriented talents and try to provide the factory training conditions, the training jobs and the sufficient practice time for the students. The unilateral power of the school is difficult to achieve it. And the “school-enterprise integration” is a good cooperation mode. Through this mode, the talent-training program can be developed by school and enterprises. The enterprises can provide the internship positions for the students and the school can further improve the quality of teaching as well as cultivate some high-quality personnel to meet the requirements of modern enterprises.

3.4 It Is Required to Reform the Assessment Methods and Evaluate the Whole Process of Learning

The assessment is the important method to test the teaching reform and the teaching effectiveness, which is also the important feedback link in the teaching quality control system. The traditional single assessment method must be reformed. So it is necessary to use the various, flexible and scientific assessment tools and methods. As for the assessment of the “Mechanical and Electrical Transmission Control” course, the teaching and research group has employed a new examination method named “an open book". The students are only allowed to bring a sheep of A4 paper into the examination room. The content which the students are relatively unfamiliar can be recorded in this paper for inspection. This method has broken through the original simple closed book mode, which has promoted the students to take the initiative to review as well as find and record their weaknesses in their knowledge from the textbook. Therefore, it focuses on assessing the practical analysis application ability of students for the knowledge. As for the assessment of practical aspect, it is not allowed to only test an experimental report of students. It is required to focus on the assessment of the participation in the experiment and the comprehensive experimental ability of students, including whether the experimental program is reasonable, whether they make good use of the instruments, whether the experimental data is correctly handled and whether the experimental process has been given a correct evaluation, etc. It is necessary to employ the mode with a combination of process assessment and objective assessment and an integration of qualitative assessment and quantitative assessment. The "test" should be throughout the whole learning process. As for the final results of students, it should be given according to the proportion of the normal operating results, the scores of final examination, the laboratory results and other aspects.

As for the assessment of large-scale comprehensive trainings such as the course design, firstly, the training topics should come from the engineering practice. Secondly, the assessment should be set up the phase examination nodes and each node has a certain percentage of scores. As the answers of the engineering practice questions may be different, so it will be OK if the design program is reasonable and the performance can meet the requirements, which helps the students to give full play to their imagination and expand the thinking. At the same time, the plagiarism as well as the phenomenon that the operations are strikingly similar has been also avoided among the students to some extent.

The practice has shown that the reform of assessment methods has guided the changes of learning ways for students. Meanwhile, it has also mobilized the learning enthusiasm and initiative of students and made the combination of theory and practice become more consciously, which has greatly improved the teaching effectiveness and promoted the development of teaching reform.

4 Conclusion

In short, there are some methods of teaching but the methods are not fixed. Through the active exploration in the teaching of "Mechanical and Electrical Transmission Control" course, the teaching and research group has constantly reformed and summarized. Finally they have got the conclusion that "the course content is the guarantee to develop and improve the application ability teaching and the development of various practical training is the measure to improve the application ability teaching". Through the teaching reform, the improvement of course content has been promoted, the learning interests and the initiative of students have been enhanced, the teaching quality has been improved, the construction and the development of curriculum has been promoted and the optimal effectiveness of teaching and learning has been achieved. At the same time, the engineering practical application ability and the innovative ability of students have been also greatly enhanced and it is helpful to cultivate some high-quality application-oriented engineering and technical personnel for the society.

References

1. Lei, Q., Zhao, M.: Analysis on the Training Objective of Higher Engineering Education. Studies in Higher Education (11) (2007)
2. Zhong, X., Qu, Z.: The Exploration and Practice of the Training Program of Building the Modern Mechanical Engineering Application-oriented undergraduates. China University Teaching(11) (2005)
3. Shi, M.: Trend of Reformation and Development of the Engineering Education in USA. Researches in Higher Education of Engineering (5) (2002)
4. Hu, Z., He, H., Hao, H.: Exploration of Experiment Teaching Reform of Electro-Mechanical Drive and Control. Journal of Guangdong University of Technology (Social Science Edition) (z1), 153–154 (2006)
5. Li, H.: Study on the Formation and the Development of Quality-oriented Education and Examination-oriented Education. Social and Psychological Sciences (4), 126–127 (2009)
6. Zhang, H.: The Teaching Reform and the Practice of "Modern Electrical Control Technology". Beijing Education 12, 44–45 (2008)
7. Liu, G., Liu, G., et al.: The Exploration and Practice of the Training of Application-oriented Talents, vol. (9). National Defense University Press, Changsha (2007)
8. An, Q.: The Teaching Model that the System Develops the Innovative Ability. Researches in Higher Education of Engineering (1), 77–79 (2004)

An Analysis of the Application of Granular Computing in Private Data Protection

Zhuqing Yang

Jiangsu College of Information Technology, Wuxi, Jiangsu, China
zhuqinyang2011@126.com

Abstract. With the development of computer technology and the Internet in recent years, privacy issues of user private data receive much attention. Faced with complex and vast database and new database systems that are constantly emerging, any single tool, module or method would be unable to process data information faster and more effectively, not to mention the protection of private data information. Based on the further development and application of granular computing in recent years, the author focuses on the application of granular computing in data processing and gives an analysis on functions of data mining modules with the integration of granular computing, providing certain reference for establishing new data mining model and protecting private data.

Keywords: granular computing, protection of private data, data mining.

1 Introduction

With the further development and application of computer technology and Internet technology, network information sharing enhances the speed of globalization several times. However, with the development of network information technology, problems of personal private data leakage draw more and more attentions. As for data analyst, personal information is extremely important. The pursuit of a balance between publishing privacy and keeping crucial content secret has become the focus of many scientists and analysts. In recent years, new concepts and computing model based on granular computing have been attached importance to in terms of privacy protection and publishing. The application of granular computing in private data will be briefly discussed in this paper.

2 An Overview of Granular Computing

2.1 Origin and Developing Situation of Granular Computing

The concept of granular computing was first proposed in 1997, being considered as a brand new field involving multiple disciplines. With the deepening of granular computing study, research groups specialized in granular computing have been developed internationally in recent years. A number of scholars carry out researches on disciplinary application of granular computing from a variety of theoretical perspectives, involving linguistics, clinical medicine, mathematics, etc.

H. Tan (Ed.): Knowledge Discovery and Data Mining, AISC 135, pp. 497–502.
springerlink.com

L.A.Zedeh proposed the general framework of granular computing, which means constructing and defining granular by using generalized constraint. This method is called granular computing; Y.Y.Yao and others applied granular computing to data mining, knowledge discovery, machine learning, etc. Supported by acquired data of network information, the research mainly has the following characteristics: 1. Describe aggregate granules with the language of logical decision; 2. Synthesize neighborhood system, interval analysis, Rough Set theory and granular computing to carry out in-depth research on granular information computing; 3. Solve the problem of consistent classification with grids structured by all divisions. T.Y.Lin conducted intensive researchs on structure, expression and application of granular computing, and studied granular model of binary relation. In addition, such aspects as the granular quotient space theory and extension set were also discussed deeply.

2.2 The Concept of Granular Computing

Granular computing is a new concept, as well as a new model, of information processing, which solves problem on different granular layer. Granular computing theory incorporates a variety of computer theories like rough set, fuzzy set, artificial intelligence, which can be regarded as a new breakthrough in the area of private data protection. L.A.Zedeh pointed out that there are three basic concepts of human cognition, namely granulation, composition and causation. Granulation refers to decomposition of a whole into parts. Composition means merging into a whole from parts. Causation refers to relation between cause and effect. He further put forward granular computing based on this theory, indicating that granular computing is the granulation of fuzzy information, the superset of rough set theory and interval computing, as well as the subset of granular mathematics. Granulation, measuring problems in detail at different levels, is the process of splitting effectively the existing whole large object into small granular set.

2.3 The Application of Granular Computing

(1) Data security

Using granular computing method to conduct researches on conflict analysis of security policy and gaming model was primarily based on building up impenetrable firewall of data set between competing companies. The basic idea is that, in defence process, data without conflict should be kept in one side of the "wall" and conflict relation should be established, being recorded as CIR. However, conflict relation is neither reflexive nor transitive, so it is impossible to acquire separated CIR class when all data are classified with CIR. As a result, this approach cannot be successfully applied. Lin Zaoyang revised it on the basis of granular computing theory, which is security protection wall of China Great Wall. Overseas researches on data security are relatively early, but theories of rough set granular computing are rarely seen.

(2) Graphic and image processing
This theoretical framework was initially put forward by Zedeh, which is appropriately reducing precision processing and solving image analysis system by taking advantage of granular computing based on rough set theory. Afterwards, lots of scholars also joined in and discussed the problem of how to granulate internal information of image, as well as image compression method based on fuzzy relation. On the basis of these researches, image segmentation based on the granular synthesis principle and specific interpolation method based on fuzzy granular structure of image were successfully designed.
(3) Data mining
It means the process of discovering useful, novel and available data information from a large number of incomplete data stored in database or storage, which will then be converted into knowledge. Because of large amount of data and vast network distribution, traditional data mining approach fails to meet current needs. From the perspective of data mining, association rule is one of the important types. Combined with digital mapping technology, Lin proposes using bitmap representation of information granule to conduct association rule mining; Wang Guoyin and others suggest a method by combining the rough set theory that process information with granular computing to obtain approximate information; Yao and others carry out in-depth research about the application of granular computing in data mining field.
(4) Complicated problems
Mainly the analysis approach from rough granule to fine granule, avoiding complex computational process and acquiring a better solution.

In short, granular computing theory has been widely used in various areas and will continue to be discussed intensively.

3 Current Situation of Privacy Protection Research

With the in-depth development of computer network technology, such multiple needs as data sharing, privacy protection, private data maintenance and others received much attention from universities and scientific research institutions at home and abroad, and became a research hot spot in the area of data discovery and information security. There are mainly two aspects of private data protection and data mining: Firstly, processing and protection of initial data like name, ID number, age, gender, etc., which avoid leakage of personal private data. Secondly, how to obtain required data through data mining without malicious mining of private data. Based on the two points, private data protection of data mining at present mainly includes the privacy protection of centralized data mining and the privacy protection of distributed data mining.

Privacy protection technology of centralized data mining contains such approaches as data conversion, data blocking, data perturbation, K-anonymity, data set structuring through reflection, etc. According to data fragmentation, privacy protection technology of distributed data mining contains the privacy protection technology of vertical data fragmentation, which can be also divided into the privacy protection technology based on two parties of security and the one based on multiple parties of security.

4 Applications of Data Mining and Private Data Protection Based on Granular Computing

There is a huge amount of data information in database or data warehouse. It always wastes time and energy to discover useful information knowledge. Certain achievements of data mining technology have now been made, yet faced with complex and vast database and new database systems that are constantly emerging, any single tool, module or method would be unable to process data information faster and more effectively, not to mention the protection of private data information.

In information processing, user query relevant information in database with SQL of inquiry system and get the incomplete system So; then process the incomplete system So through information anonymization based on granular computing; and finally display the incomplete system So (without decision attribute), which is obtained through information anonymization, in the form of query results. The process is shown in the following diagram.

Compared with traditional data mining, this method has the following advantages:

First, when data inclusion is incomplete, uncertain or fuzzy, approximate solution can be acquired by granular competing; Second, data mining of granular computing can get approximate knowledge and relatively accurate information on lower cost; Third, it is able to abstract and simplify problem so as to improve the efficiency of data mining; Fourth, high dimensional data can be converted into polynomial time solution; Last, it helps to make reasonable decision.

4.1 Data Preprocessing Module

Data preprocessing refers to the organization of existing business data by making use of new module, which is mainly embodied in the following three aspects:

(1) Certain requirements for data set that initial data cannot be used for direct mining.
(2) The workload of processing is larger than that of sheer mining.
(3) Great influence on mining algorithm, which might cause serious interference or misguided results.

Data preprocessing mainly includes data cleansing, integration, abstract conversion, reduction, etc. This is a complex project, which needs to convert data into a form that is suitable for mining. As for mining algorithms like neural network, the accuracy of mining can be only achieved through data normalization. Commonly used methods of normalization include maximum and minimum normalization, zero-mean normalization, ten bases conversion normalization, etc. Besides, switching, rotation and projection in data warehouse can be also used for converting data and generating knowledge base. Uncertain attribute value only needs to be assigned with a discrete value and finally granulated.

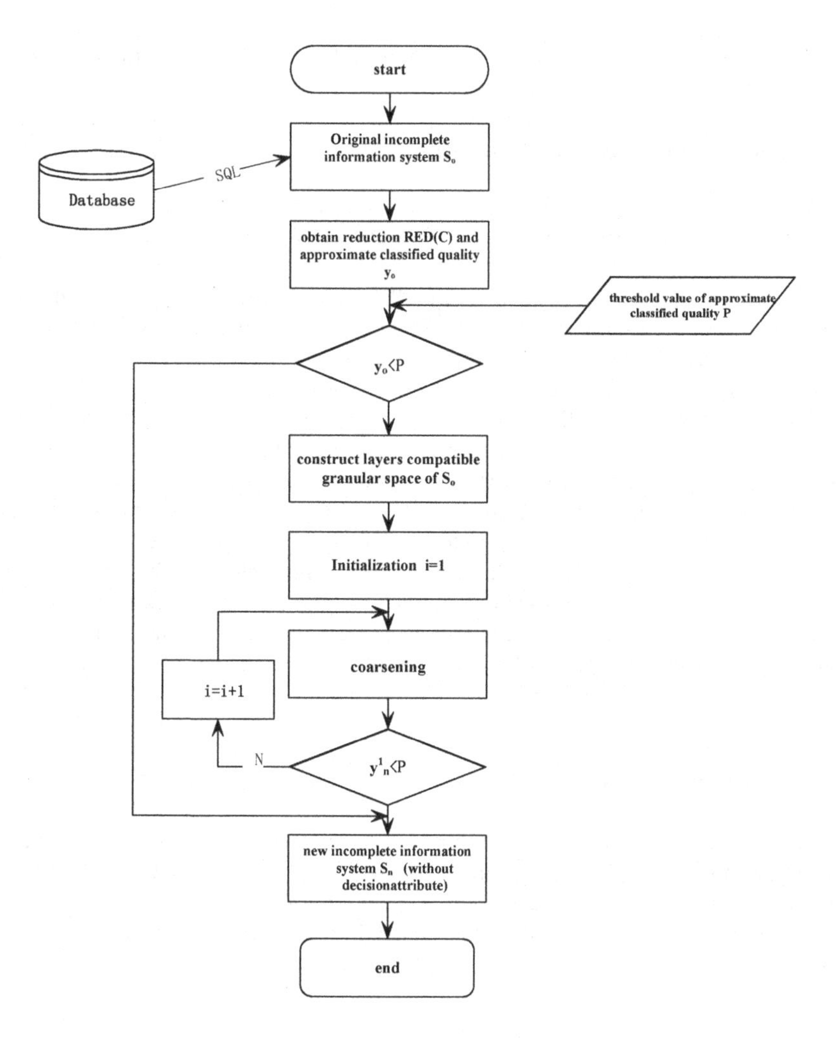

Fig. 1. The process of granular computing

4.2 Reduction Module

Eliminate redundant data through clustering or deleting redundancy features, so as to acquire simplified data subset. Main approaches are: 1. Attribute reduction. Reduce data scale through eliminating attributes that are irrelevant to mining task or redundant, so that the probability distribution of new data subset can be approximate to the probability distribution of original data as much as possible. 2. Data block reduction. Replace original data with smaller data representation. There are mainly parameterized method and parameterless method. Parameterized method means evaluate data with a model instead of storing actual data. 3. Discretization and concept hierarchy. Use concepts of higher layer to substitute for initial data or concepts of lower layer. Divide attribute threshold value into several intervals to achieve continuous attribute discretization.

4.3 Mining Module Based on Association Rules

Set mining rules as association rules, so that there is a relationship between descriptive data items. General objects are transaction databases. Association rules refers to identifying relevance between transactions, which can be generally divided into two subproblems, namely discovering all frequent itemsets, of which support is greater than or equal to minimum support threshold, and generating association rules with frequent pattern which satisfy credibility threshold.

Association rules require support and certainty factor of rules from user so as to find out rules that satisfy all requirements of support and certainty factor, namely automatic generation of rules. Decision-making requirements or conditions can be also defined in accordance with user needs so as to find out the minimal rules that satisfy requirements, namely custom decision mining and custom condition mining. Finally interpret rules with the original meaning of attribute. The classic mining algorithm of association rules is the Apriori algorithm, namely obtaining frequent itemsets. Generate smaller frequent itemsets with the method of iteration so as to produce larger frequent itemsets. Determine repeatedly which itemsets are frequent through querying transaction database until candidate sets cannot be produced, which means data are completely mined.

5 Conclusion

In a word, establish new data mining module through analysis on function implementation of each data mining module and the combination of granular computing, and conduct calculation on association rules of data mining, so as to provide optimal information for data analysis and provide data support for control and decision-making.

References

1. Zhi, H., Lin, Y.: Data Warehouse. China Machine Press, Beijing (2003)
2. Xu, F., Zhang, L.: An Analysis of Uneven Granules Clustering Based on Quotient Space. Computer Engineering (3) (2005)
3. Zhang, C., Li, C., Zhangling: Realizing the High-precision Fuzzy Control Based on the Theory of Quotient Space(Methods of Granular Computing). Computer Engineering and Applications (30) (2009)

P Systems with Proteins Working in the Minimally Parallel Mode

Chun Lu[1,2], Xiao-jie Chen[3], and Xiao-long Shi[1,2,⋆]

[1] Key Laboratory of Image Processing and Intelligent Control
[2] Department of Control Science and Engineering
[3] School of Mechanical Science and Engineering
Huazhong University of Science and Technology
Wuhan 430074, China
luchun.hust@gmails.com, chenxiaojie75@163.com,
shixiaolong@mail.hust.edu.cn

Abstract. Membrane systems are distributed and parallel computing devices inspired from the structure and the functioning of living cells, called P systems. Most variant of P systems have been proved to be universal in the model of maximally parallel mode. But this hypothesis does not have biologically realistic support. In order to construct a more "realistic" system, we introduce the minimal parallelism into the protein P systems. Minimal parallelism is a way the rules of a P system used: from each set of applicable rules associated to the same membrane, at least one must be applied. We investigate the computing power of P systems with proteins on membranes working in the minimal parallelism mode. Such systems are shown to be computationally complete even with only three membranes and one protein on each membrane.

Keywords: membrane computing, protein membrane, minimal parallelism, universality.

1 Introduction

Membrane systems (P systems) were introduced by Gh. Păun as distributed parallel computing devices, based on inspiration from biochemistry, especially with respect to the structure and the functioning of a living cell [1]. The cell is considered as a set of compartments enclosed by membranes which are nested one in another. The basic model consists of a hierarchical structure composed by several membranes, embedded into a main membrane called the *skin*. Membranes divide the Euclidean space into *regions*, that contain multisets of *objects* (represented by symbols of an alphabet) and *evolution rules*. Using these rules, the objects may evolve and/or move from a region to a neighboring one. Most variants of P systems have been proved to be Turing equivalent [2] and computationally efficient [3].

P systems with proteins on membranes (MP systems) [4] is a variant of P systems bridging the gaps between membrane computing [5] and brane calculi

⋆ Corresponding author.

H. Tan (Ed.): Knowledge Discovery and Data Mining, AISC 135, pp. 503–510.
springerlink.com

[6]. Both of them start from the same reality as the living cell, but they develop in different directions and have different objectives. Membrane computing tries to abstract the computing power of biologically inspired models in the Turing sense, whereas brane calculi works only with objects – called proteins – placed on membranes, while the evolution is based on membrane handling operations, such as exocytosis, phagocytosis, etc., in the framework of process algebra.

Various operations on membranes appear in both areas. A few related attempts in this direction include [7]. The power of some of the operations individually ($3ffp$) and in combination ($2res$ and $4cpp$, $2res$ and $1cpp$, $1res$ and $2ffp$) has been looked at in [4] and universality results obtained with arbitrary number of proteins (in case of $3ffp$, $1res$ and $2ffp$), and with 2 proteins for the other two combinations. Using the trade-off between space and time by the means of biologically inspired operation as the membrane division, it is able to solve NP-complete problems [8] and PSPACE problems [9].

All papers mentioned above apply the rules in the maximally parallel mode: in each computation step, the chosen multiset of rules cannot be extended anymore, i.e., no further rule could be added to this chosen multiset of rules in such a way that the resulting extended multiset still could be applied. Obviously, it is not "realistic" from a biological point of view. Actually, the number of rules used in a chemical reaction is not equal the number of rules applicable, but usual less than.

As membrane systems are inspired by living cell behavior, a continuous research topic in this area was looking for as bio-realistic computing models as possible. In this direction, another strategy of applying rules in parallel is introduced recently, the so-called *minimal parallelism* [10]: in each computation step, the chosen multiset of rules to be applied in parallel cannot be extended by any rule out of a set of rules from which no rule has been chosen so far for this multiset of rules in such a way that the resulting extended multiset still could be applied (this is not the only way to interpret the idea of minimal parallelism, e.g., see [11] for other possibilities). Minimal parallelism relaxes the condition of using the rules in a maximally parallel way.

In this work, *non-determinism minimal parallelism* is introduced into the P systems with proteins on membranes and the computational completeness of the systems in this mode is also considered by simulating register machines.

The paper is organized as follows. In section 2, we introduce the minimal parallelism for MP systems, the recognizer MP system, the register machines and some necessary preliminaries. In section 3, it is provided their universality by simulating a register machine with only one protein on each membrane. Conclusions and some final remarks are given in section 4.

2 MP Systems in the Minimally Parallel Mode

The reader is assumed to be familiar with basic elements of membrane computing and computability, from [12]. So only a few notions and notations are mentioned, which are used through the paper.

As usual, for a given alphabet V, V^* denotes the set of all finite strings over V, with the empty string denoted by λ, and the membrane structures are expressed by correctly matching labeled parentheses. The family of recursively enumerable sets of natural numbers is denoted by NRE.

In the P systems this paper focuses on, we use two types of objects, *proteins* and usual *objects*; the former are placed **on** the membranes, the latter are placed **in** the regions delimited by membranes. A protein p on a membrane (with label) i is written in the form $[_i p\ |]_i$. Both the regions of a membrane structure and the membranes can contain multisets of objects and proteins, respectively.

A P system with proteins on membranes of the initial degree $m \geq 1$ is a device of the form

$$\Pi = (O, P, \Sigma, \mu, \omega_1/z_1, \ldots, \omega_m/z_m, E, R_1, \ldots, R_m, i_o)$$

- m is the degree of the system (the number of membranes);
- O is the set of objects;
- P is the set of proteins (with $P \cap O = \emptyset$);
- μ is the membrane structure consisting of m membranes enumerated by $1, \ldots, m$;
- $\omega_1, \ldots, \omega_m$ are the (strings representing the) multisets of objects present in the m regions of the membrane structure μ;
- $z_1, \ldots, z_m$ are the multisets of proteins present on the m membranes of μ;
- $E \subseteq O$ is the set of objects present in the environment (in an arbitrarily large number of copies each);
- $R_1, \ldots, R_m$ are finite sets of rules associated with the m membranes of μ;
- i_o is the output membrane, one membrane from μ.

$a[_i p\ \|]_i \to b[_i p'\ \|]_i$, $a[_i p\ \|]_i \to b[_i p'\ \|]_i$	$1cp$	$1res$
$a[_i p\ \|]_i \to [_i p \mid a]_i$, $a[_i p\ \|]_i \to [_i p' \mid a]_i$	$2cp$	$2res$
$[_i p \mid a]_i \to b[_i p'\ \|]_i$, $a[_i p\ \|]_i \to [_i p' \mid b]_i$	$3cp$	$3res$
$a[_i p \mid b]_i \to b[_i p' \mid a]_i$	$4cp$	$4res$
$a[_i p \mid b]_i \to c[_i p' \mid d]_i$	$5cp$	$5res$

The actions and the notations of rules are considered for handling the objects and the proteins; in all of them, a, b, c, d are objects, p and p' are proteins, and i is a label (*cp* means *change protein*, if $p = p'$, then the rules of type *cp* become rules of type res which means that protein is *restricted* during the evolution):

$1cp$: object evolution rules, associated with membranes and depending on the label and the protein on the membrane, a possible object in the membrane or the upper membrane is involved; $2cp$: movement rules, a possibly object is introduced in or sent out of a membrane without change by the protein p on the membrane i; $3cp$: modified movement rules, a possibly modified object is introduced in or sent out of the membrane by the protein p on the source or target membrane; $4cp$: interchanging rules, depending on the label of and the protein on the membrane, a possibly object is exchanged with one object outside; $5cp$: modified-interchanging rules, one possibly modified object is sent out of the membrane whereas another possibly modified object outside is introduced in the membrane; in all cases $1res - 5res$, the protein is not evolved, it plays the role of a catalyst, just assisting the evolution or the movement of objects.

In what follows, the rules are applied in a *non-deterministic minimally parallel* manner: in each step, from each set R_i ($1 \leq i \leq n$) we use at least one rule (without specifying how many) provided that this is possible for a chosen selection of rules. The rules to be used, as well as the objects and proteins to which they are applied, are non-deterministically chosen. More specifically, we assign non-deterministically objects to rules, starting with one rule from each set R_i. If we cannot enable any rule for a certain set R_i, then this set remains idle. After having one rule enabled in each set R_i for which this is possible, if we still have objects and proteins which can evolve, then we evolve them by any number of rules (possibly none).

We emphasize an important aspect: the competition for objects. It is possible that rules from two sets R_i, R_j associated with adjacent membranes i, j (that is, membranes which have access to a common region, either horizontally or vertically in the membrane structure) use the same objects, so that only rules from one of these sets can be used. Such conflicts are resolved in the following way: if possible, objects are assigned non-deterministically to a rule from one set, say R_i; then, if possible, other objects are assigned non-deterministically to a rule from R_j, and thus fulfilling the condition of minimal parallelism. After that, further rules from R_i or R_j can be used for the remaining objects, non-deterministically assigning objects to rules. If no rule from the other set (R_j, respectively R_i) can be used after assigning objects to a rule from R_i or R_j, then this is a correct choice of rules. It depends on the first assignment of objects to rules in order to make applicable rules from each set.

If, at one step, two or more rules from the same set can be applied to the same objects and proteins, then only one rule will be non-deterministically chosen. Moreover, type involving a membrane h form a set R_h. Moreover, if a membrane h appears several times in a given configuration of the system, then for each occurrence i of the membrane we consider a different set R_{h_i}. Then, in each step, from each set R_{h_i}, $h \in H$, from which at least a rule can be used, at least one rule must be used.

At each step, a P system is characterized by a *configuration* consisting of all multisets of objects and proteins present on the corresponding membranes. For example, $\mathcal{C} = \omega_1/z_1, \ldots, \omega_m/z_m$ is the initial configuration, given by the definition of the P system. By applying the rules in a non-deterministic maximally parallel manner, we obtain transitions between the configurations of the system. A finite sequence of configurations is called *computation.* A computation *halts* if it reaches a configuration where no rule can be applied to the existing objects and proteins.

Only halting computations are considered successful, thus a non-halting computation will yield no result. With a halting computation, we associate a result in the form of the multiplicity of objects in region i_o in the halting configuration. The set of numbers generated in this way by a system Π is denoted by $N(\Pi)$ and the family of such sets, generated by systems having initially at most n_1 membranes, is denoted by $N_{mp}OP_{n_1}(types - of - rules)$, with the subscript "*mp*" indicating the "minimal parallelism" used in computations, and types-of-rules indicating the allowed types of rules.

In what follows we give several universality result for MP systems working in the minimally parallel mode.

3 Computational Completeness

In this section, we prove the universality of MP systems working in the minimally parallel mode; as it is the case also for the maximally parallel mode, only the rule of type $2cpp$ is used – as in [4].

We start now to investigate classes of MP systems with rules as above which are computationally complete, able to characterize *NRE*, and begin by considering systems in which only one type of rules is used. Without the ability of creating objects in the system, we need a supply of objects with the skin membrane containing object a_r for $1 \leq r \leq m$, in arbitrarily many copies.

Register machines is an important tool for characterizing *NRE*, hence the Turing computability. Let us recall their definitions from [13].

An n-register machine is a construct $M = (n, l_0, l_h, I)$ where:

- n is the number of registers;
- I is a set of labeled instructions of the form $(l : op(i), l', l'')$; $op(i)$ is an operation on register i of M; symbols l, l', l'' belong to the set of labels associated in a one-to-one manner with instructions of I;
- l_0 is the initial label;
- l_h is the final label.

The instructions allowed by an n-register machine with an output tape are:

- l : $(\mathrm{ADD}(i), l', l'')$ – add one to the contents of register i and proceed to instruction l' or to instruction l'';
- l : $(\mathrm{SUB}(i), l', l'')$ – jump to instruction l'' if register i is empty; otherwise subtract one from register i and jump to the instruction labeled by l' (these two cases are often called *zero-test* and *decrement*);
- l_h: HALT– finish the computation. This is the unique instruction with label h.

If a register machine $M = (n, l_0, l_h, I)$, starting from the instruction labeled by l_0 with all registers being empty, halts with values n_j in register j, $1 \leq j \leq m$, and the empty contents of registers $m+1, \ldots, n$, then it generates the vector $(n_1, \ldots, n_m) \in \mathbb{N}^m$. The result of a halting computation of a register machine with an output tape is a sequence of symbols written on that tape.

It is known that register machines with $m+2$ registers can generate all recursively enumerable sets of m-dimensional vectors (we can also require that the only instructions associated to the output registers are increment instructions). Moreover, register machines with 2 registers and an output tape can generate all recursively enumerable languages.

Theorem 1. $N_{mp}OP_3(pro_1; 2cpp) = NRE$.

Proof. We consider a register machine $M = (3, l_0, l_h, R)$ without direct loops in the ADD instructions and construct the system with degree 3.

$$\Pi = (O, P, [_0[_1]_1[_2]_2]_0, \lambda/l_0 p, E, R_1, R_2, R_3, 1)$$

with the following components :
$O = \{a_r \mid 1 \leq r \leq 3\} \cup \{c_l \mid l \in R\} \cup \{c, d\}$,
$P = \{l, l', l'' \mid l \in R\} \cup \{p, p', p''\} \cup \{p_l \mid l \in R\}$,
$E = \{a_r \mid 1 \leq r \leq 3\}$,
and the following rules, where protein l_i corresponds to register machine instruction label l_i and the number of object a_r corresponds to the contains of register machine r.

1. For an ADD instruction l_i : (ADD(r), l_j, $l_k \in R$), we consider the rules
   $c_l[_1 l_i | \rightarrow [_1 l'_i | c_l$,
   $a_r[_1 l'_i | \rightarrow [_1 l''_i | a_r$,
   $[_1 l''_i | c_l \rightarrow c_l[_1 l_j |$,
   $[_1 l''_i | c_l \rightarrow c_l[_1 l_k |$.
   With the presence of label-protein l_i on membrane 1, object c_l centers into the membrane 1 and attracts the simulation of l_i. In the next step, the primed protein l'_i introduces the object a with the correct subscript r into the membrane 1, and gets one more prime. With double primes, label-protein l_i sends out the object c_l and changes into l_j or l_k non-deterministically and lose the primes. In ADD instruction, there is no rule can be used for membrane 2, and in each step, there is only one rule can be used for membrane 1.
2. For a SUB instruction l_i : (SUB(r), l_j, $l_k \in R$), we consider details of decrement/zero test instructions in the following two cases. In each configuration, only the membrane, the proteins and the objects relevant for the simulation are specified.

step	R_1(with a_r)	R_2
1	$c_l[_1 l_i \vert\rightarrow [_1 l'_i \vert c_l$	–
2	$[_1 l'_i \vert c \rightarrow c[_1 l''_i \vert$	–
3	$e[_1 l''_i \vert\rightarrow [_1 l'''_i \vert e$	$c[_2 P \vert\rightarrow [_2 P' \vert c$
4	$[_1 l'''_i \vert a \rightarrow a[_1 l'''_j \vert$	$d[_2 P' \vert\rightarrow [_2 P'' \vert d$
5	–	$[_2 P'' \vert c \rightarrow c[_2 P' \vert$
6	$c[_1 l'''_j \vert\rightarrow [_1 l''_j \vert c$	$[_2 P' \vert d \rightarrow d[_2 P \vert$
7	$[_1 l''_j \vert e \rightarrow e[_1 l'_j \vert$	–
8	$[_1 l'_j \vert c_l \rightarrow c_l[_1 l_j \vert$	–

Suppose that there are at least one a_r in membrane 1. With protein l_i on membrane 1, we apply the rule $c_l[_1 l_i | \rightarrow [_1 l'_i | c_l$ which changes the protein l_i into l'_i and moves promotor object c_l inside. This means the beginning of the simulation of instruction. Then, protein l'_i sends out the object c which will return in step 6 and gets one more prime. In the third step, object e

enters membrane 1 with protein l_i with double primes, while object c from membrane 1 is introduced into membrane 2. In membrane 1, these happen in both cases irrespective whether object a_r exists or not in it. In the forth step, label-protein l_i with three primes sends out object a with correct subscript r and changes to l_j with the same number prime. At the same time, object d in *skin* membrane enters membrane 2 with protein P' on membrane 2 and adds one more prime to the protein P'. In the fifth step, there is no rule can be used in R_1, and the protein P'' sends out object c and loses one prime. In the sixth step, the object c from membrane 2 goes back membrane 1 and removes one prime of protein l_j. In the following steps, protein l_j sends out object e and c_l, which came from the skin membrane, orderly and loses one prime every time.

step	R_1(without a_r)	R_2
1	$c_l[_1 l_i \mid\rightarrow [_1 l'_i \mid c_l$	–
2	$[_1 l'_i \mid c \rightarrow c[_1 l''_i \mid$	–
3	$e[_1 l''_i \mid\rightarrow [_1 l'''_i \mid e$	$c[_2 P \mid\rightarrow [_2 P' \mid c$
4	–	$d[_2 P' \mid\rightarrow [_2 P'' \mid d$
5	–	$[_2 P'' \mid c \rightarrow c[_2 P' \mid$
6	$c[_1 l'''_i \mid\rightarrow [_1 l''_k \mid c$	$[_2 P' \mid d \rightarrow d[_2 P \mid$
7	$[_1 l''_k \mid e \rightarrow e[_1 l'_k \mid$	–
8	$[_1 l'_k \mid c_l \rightarrow c_l[_1 l_k \mid$	–

If there is no a_r in membrane 1, no rule can be used from step 4-5. Membrane 1 will stay idle with protein l'''_i until object c is sent into skin membrane. In step 6, object c enters membrane 1 with protein l'''_i and changes it into l''_k. In the following two steps, protein l''_k also have to send out the object a and c_l to introduce the corresponding label-protein l_k.

3. When the halt label l_h is present on the membrane, c_l is introduced in it and never sent out. Because, no further simulation can start without c_l in skin membrane, the number of copies of a_r in membrane 1 is equal to the value of register r of M.
   $c_l[_1 l_h \mid\rightarrow [_1 l'_h \mid c_l$.

Therefore the register machine is correctly simulated independently of the time of execution of the rules involved in the simulation.

4 Conclusion

The main contribution of this paper is the use of the minimal parallelism in the framework of P systems with proteins on membranes. The universality result was proved in this framework, for one type of rule.

As a further study of the computing power of MP systems in minimally parallel mode, it is necessary to obtain fast solutions to computationally hard problems in this bounding way of using rules. In this field, cell division and cell separation are used widely in P systems to solve the NP-complete problems [14] even the QSAT problems [15]. It is a opening problem to design a MP system with cell division or cell separation working in the mode of minimal parallelism, which solves the NP-complete problems, even the PSPACE problems.

Acknowledgments. The corresponding author of this paper is Xiao-long Shi. This work was supported by National Natural Science Foundation of China (61033003 and 60971085), the allocated section of the Basic Fund for the Scientific Research and Operation of Central Universities (2011TS006), the opening Fund of Key Laboratory of Image Processing and Intelligent Control, Ministry of Education (200905).

References

1. Păun, G.: Computing with membranes. Journal of Computer and System Sciences 61, 108–143 (2000)
2. Pan, L., Zeng, X., Zhang, X.: Time-free spiking neural P systems. Neural Computation 23, 1320–1342 (2011)
3. Zhang, X., Wang, S., Niu, Y., Pan, L.: Tissue P systems with cell separation: attacking the partition problem. Science China Information Sciences 54, 293–304 (2011)
4. Păun, A., Popa, B.: P systems with proteins on membranes. Fundamenta Informaticae 72, 467–483 (2006)
5. Dassow, J., Păun, G.: On the power of membrane computing. Journal of Universal Computer Science 5, 33–49 (1999)
6. Cardelli, L.: Brane Calculi. In: Danos, V., Schachter, V. (eds.) CMSB 2004. LNCS (LNBI), vol. 3082, pp. 257–278. Springer, Heidelberg (2005)
7. Krishna, S.N.: Universality Results for P Systems based on Brane Calculi Operations. Theoretical Computer Science 371, 88–105 (2007)
8. Păun, A., Popa, B.: P Systems with Proteins on Membranes and Membrane Division. In: Ibarra, O.H., Dang, Z. (eds.) DLT 2006. LNCS, vol. 4036, pp. 292–303. Springer, Heidelberg (2006)
9. Sosík, P., Păun, A., Rodríguez-Patón, A., Pérez, D.: On the Power of Computing with Proteins on Membranes. In: Păun, G., Pérez-Jiménez, M.J., Riscos-Núñez, A., Rozenberg, G., Salomaa, A. (eds.) WMC 2009. LNCS, vol. 5957, pp. 448–460. Springer, Heidelberg (2010)
10. Ciobanu, G., Pan, L., Păun, G., et al.: P Systems with Minimal Parallelism. Theoretical Computer Science 4618, 62–76 (2007)
11. Freund, R., Verlan, S.: A Formal Framework for Static (Tissue) P Systems. In: Eleftherakis, G., Kefalas, P., Păun, G., Rozenberg, G., Salomaa, A. (eds.) WMC 2007. LNCS, vol. 4860, pp. 271–284. Springer, Heidelberg (2007)
12. Păun, G. (ed.): Membrane Computing-An Introduction. Springer, Berlin (2002)
13. Minsky, M.: Computation–Finite and Infinite Machines. Prentice-Hall, Englewood Cliffs (1967)
14. Niu, Y., Pan, L., Pérez-Jiménez, M.J., Font, M.R.: A tissue P systems based uniform solution to tripartite matching problem. Fundamenta Informaticae 109, 179–188 (2011)
15. Ishdorj, T.-O., Leporati, A., Pan, L., Zeng, X., Zhang, X.: Deterministic solutions to QSAT and Q3SAT by spiking neural P systems with pre-computed resources. Theoretical Computer Science 411, 2345–2358 (2010)

Interactive Research of Traditional Aesthetics and Teaching of Art and Design

Jinxiang Ma[1], Xuguang Yang[1], and Mingyu Gao[2]

[1] Environmental Management College of China, Qinhuangdao, 066004, P.R. China
[2] Northeast Agricultural University, Haerbin, 150030, P.R. China

Abstract. Traditional Chinese painting aesthetics and traditional Chinese culture are closely linked; with the accumulation of national culture, the aesthetics has strong spiritual elements of national culture and cultural concepts. Today, with the development of technology and society, art and design promote the development of China's designing industry, and during the college art teaching, art and design take the heritage the concept of traditional aesthetics; at the same time, esthetic thoughts are excavated and re-interpretated and applied to modern art designing, so that art and design open up a new realm in the field of art designing and teaching, promote national cultural heritage and development, and display a design style with Chinese unique characters The essence of traditional aesthetic elements is to learnand experience in continuous. It requires college students to integrate aesthetic ideas into the design of the modern product, which is an integration of spiritual civilization and cultural sentiments, to convey the cultural connotation that the design works carries.

Keywords: traditional esthetics, art and design, interavtive, national culture, education.

1 Art and Design Education and Teaching

In China's higher education, western art educating theories have exerted great impacts on Chinese design [1] and art teaching; the constant introduction of a variety of Western art education ideas, theories, and teaching experience has greatly promoted the progress and development of the theory and practice of Chinese art education. However, with the introduction and absorption, we have to pay full attention to and strengthen traditional Chinese culture education of the students, and combine with our own esthetics culture and characteristics of Chinese ethnic philosophy in the design, so that Chinese art education can truly display its unique artistic connotation and we can fundamentally improve the humanity, artistic and cultural accomplishment, and aesthetic sensibility of students majored in design.

Traditional Chinese esthetics thoughts are a great component of traditional Chinese culture, which could date back to the concepts of "day" and "place" in the farming community. For example, the early geometric patterns on painted pottery, figures, birds, fish, frogs and other totem decorations, they all had a profound impact on national culture. From sculptures and portraits of the Han Dynasty, paintings of Tang Dynasty and Song dynasty, to porcelains of Ming Dynasty and Qing Dynasty, they all

H. Tan (Ed.): Knowledge Discovery and Data Mining, AISC 135, pp. 511–515.
springerlink.com

contained deep philosophy and imagery. China's dragon and phoenix totem which existed very early in the folklore, is the most auspicious and the most sacred decorating pattern, and consist of strange and mysterious animal images imaginarily created by the ancient people. Through creative practice of generations of artists and craftsmen, they enriched the artistic expression and manifestation of this decorating pattern. This aesthetic concept "image" has melted into the art of painting, has developed a unique painting system with an abstract, symbolic, distorted manner and appearance. It expresses the natural images with the distinctive ways of thinking and integrates of Confucianism, Taoism and Thoughts of Chan, as well as fully demonstrates their understandings of beauty. This unique sense of aesthetics shows our Chinese unique artistic style. When the aesthetic ideas come from one's heart, aesthetic experience of the design work is often in a particular environment, instantly touches people's soul.

2008 Beijing Olympic medals are designed to organize these elements with ethnic characteristics to modern designing. The front of the Medal is the unified patterns, in accordance to the provisions of the International Olympic Committee --Victoire de Samothrace and Panathinaiko, while the back is inlaid with nephrite disk shaped from the ancient Chinese dragon decorative pattern; on the meatal center on its back is engraved Beijing Olympic Games emblem. The hook of the Medal is evolved by traditional Chinese jade twin-dragons pattern. The medal as a whole is elegant and rich of national characteristics, and the box of medals, ribbons released with the medal, also possess the aesthetic charms of traditional Chinese culture, highlighting the distinct Chinese characteristics and styles. Beijing Olympics' medal box is wooden lacquer box made by traditional Chinese craft, square shaped. The edges of the cover are slightly curved, implied with the meaning of harmony of the whole universe. The ribbon is woven in red with cloud- pattern, with blessing of happy and fortune. This design is a perfect expression of the unique interpretation of the Olympic spirit by the wisdom of the Chinese people.

Therefore, in the teaching of design, it is of importance to fully mobilize the students' thinking pattern to intrgrate traditional technique of symbol expression into modern art design, with the combination of traditional aesthetics and modern aesthetics, so that they can penetrate with each other, to display to people a wide range of designing styles, and this state of art and design can attract people with its endless charm.

2 Chinese Painting under the Cultural Heritage of Traditional Chinese Aesthetics

Chinese paintings are required to be vivid, neat, natural in the creation, and to be lively and charming in the expression. They pay great attention to the expression of the spirits of the images and subjective feelings. Their artistic quality is to describe natural images through a distinctive way of which is thinking images thinking and images creating. Same with Chinese paintings, the Chinese calligraphy also focuses on charming, and through its presentation skills, conveys its unique "character" "mood" and "temperament". However the art of calligraphy is more "impressionistic". Its Aesthetics creation does not only display the "image", but also has to express the realm of the calligraphy [2].

The art of calligraphy and art of painting tend to permeate, influent and blend with each other, which is an important aesthetic characteristic in the form of literati paintings. It is a way of expressing philosophy world view and aesthetic sensitivity of a particular stage of social development. The decorating, graphy and aesthetics it presents all indicate the image creation that is characterized in oriental painting aesthetics. What's more, Confucianism, Taoism and Emptiness ideology of Chan have also become the soul of art of Chinese painting and calligraphy creation, the creative process of which is dominated by emotions and Chinese spirits, and to describe natural images through images thinking and creating. This way and manifestation of painting and writing forms Chinese painting system, the deep philosophical meaning it contains shows a clear spiritual image. We apply these aesthetic ideas of calligraphy and painting to the modern art design, the representative designs are: well-carved wooden craft, beautiful ceramics, murals, furniture imitating the styles of Ming Dynasty and Qing Dynasty, classical and simple purple-sand tea set, and so on. They are adhering to the aesthetic srandard of "natural", and combine with colorful Chinese painting and fascinating calligraphy, so that the charm and artistic image of the product can be more expressive, as well as demonstrating to the world the broad spiritual culture of Chinese ancient civilization.

3 Packaging Design of Modern Products and Its Elements

To take images, calligraphy, colors as visual elements to decorate products, and to neatly apply those elements in designing, they can make people better appreciate the aesthetic meaning it contains. "DaoXiangCun", a famous food brand in China, it has a traditional food culture. The design of its mooncake packaging is very fascinating. Clusters of peony with ethnic-style and a full moon are set against the scarlet red background, give people a strong feeling of warm and luck with the images of blooming flowers and the full moon. "Three characters of "Daoxiangcun" which are written with running style, penetrate the essence of the Chinese art of calligraphy with so much elegance. Packaging is designed mainly in red, in match with yellow, gold, silver, black and white, giving people a sense of antique, as well as endowing the product with a traditional charm. Another example is "XingHuaCun" which represent the culture of "Fenjiu", a famous drink in China. On its package, it quot a poem called "Qing ming", whose writer is a well-known poet in Tang Dynasty named Du Mu --- "Where can a winehouse be found to drown his sadness? A cowherd points to Almond Flower (Xing Hua) Village in the distance" as its text decoration, echoing with the picture of the shepherd who blowing reed pipe by sitting on a cow's back. This design fascinates people greatly. The subtle integration of these traditional patterns and calligraphy enriches Chinese designing art, giving the packaging a strong ethnic scent [3], spreading Chinese aesthetics, as well as expressing the national sentiment.

Traditional culture of China is the inexhaustible treasure of modern art and design. "China" is originated from china, the Porcelain, which indicates that Chinese ceramic culture can best represent our Chinese culture. Blue and white porcelain of China is the one with the most national characteristics, which began in Tang and Song dynasties, the technology had been mature in Jingdezhen in Yuan Dynasty. "Blue and white porcelain Olympic gold and silver medals"in 2008 presented blue and white porcelain, the most typical symbolof Chinese culture in front of people. Its lush green and blue pattern, clean and clear, bright and elegant, combined with creative paper cutting and the Olympic spirit as a whole, is one of the products with the most Chinese characteristics. The traditional blue and white pattern has also been emerging in the fashion show, it sents off a special kind of beauty in cooperation with the modern fashion designs. Those features of blue and white porcelain also can be applied into industrial designing. With more and more traditional Chinese pattern blended the into modern aesthetic elements, people could sketched out classic green cloud patterns of blue and white porcelain on the surface of electrical products as smooth as porcelain, blue and white well matched, simple and elegant vividly displayed by the product. This simple and soft feeling contains more and more traditional Chinese elements [4].

However, pottery is known for its simple and elegant style, the prominent features of which are displayed throuth ceramic art. There are a variety of potteries, some of the best are Paleolithic gray pottery, terracotta from magnetic mountain culture, painted pottery from Yangshao culture, Dawenkou eggshell pottery, white pottery of the Shang Dynasty, hard pottery of the Western Zhou Dynasty, as well as the The Terracotta Army of Qin Dynasty, glazed pottery of the Han Dynasty, Tang-dynasty trio-colored glazed pottery, trio-colored pottery of Song and Liao Dynasties, purple clay tea pot of Ming and Qing Dynasties, coloured glaze, etc. They all have delicate patterns, solemn and exquisite, shining with a unique charm. In design teaching, we must first comprehend and master the characteristics of ceramic art and aesthetic features, then, apply the deep connotation of Chinese traditional culture to the creation and decoration of the product. The use of ancient Chinese charaters, ancient pottery, folk culture, religional culture, calligraphy, seal cutting, poetry and other literary elements of the traditional cultural [5] themes endows the pottery with inner vitality, reflecting the charm of Chinese aesthetic culture--- “vivid and lively” , “Nature and Humanity”.

4 Chinese Traditional Designing Patterns, Cultural Elements

Firstly, music, chess, poetry and painting, literature, drama, food, clothing, architecture, traditional medicine, religion, philosophy, arts and crafts, flora and fauna, legends, myths, agricultural culture, so on so forth, all of which require college students, in the process of learning how to design, to understand the humanity environment in various regions, living habits and religious beliefs, clarify the characteristics, types and uses of the design product. Second, with the continuous development of the times, people's aesthetic habits will also be changing, so it is necessary to investigate the aesthetic habits and the acceptance ability of the consumers in different regions, which requires the designer to keep up with the aesthetic pursuit of the times.

Therefore, Professors need to stimulate students to think actively, to inherit the essence of traditional culture, and to learn and experience in life, so that they can make traditional aesthetic culture melt into modern art design [6]. Only when traditional aesthetic culture integrates with modern art design, can the designer's thoughts, artistic accomplishments and concepts be reflected. Only in this way, can people design a piece of work that forms the mainstream with both characteristics traditional cultural and modern times.

References

1. Lang, Y.: Outline of the History of Chinese Aesthetics. Shanghai People's Publishing House, Shanghai (1999)
2. Yi, P.: Chinese Aesthetic Cultural History. Shandong Pictorial Press (2000)
3. Zhang, W.: Exploration of Esthetics of Confucianism, Taoism and Buddhism. Chinese Social Science Press (1991)
4. Yang, Y.: Chinese Culture Implied in Olympic Medals – Chinese Stamp and nephrite disk. Zhejiang University Press (2008)
5. Xu, W., Xu, L.: Chinese Traditional Auspicious Patterns. Tianjin Yangliuqing art community Press (1994)
6. Zheng, J.: Art of Chinese Auspicious Patterns Design. People's Fine Arts Publishing House (2010)

The Analysis of "BRAND" Image and Performance Tendencies in Space Design

Jinxiang Ma[1], Jun Liu[1], and Mingyu Gao[2]

[1] Environmental Management College of China, Qinhuangdao, 066004, P.R. China
[2] Northeast Agricultural University, Haerbin, 150030, P.R. China

Abstract. In the 21st century, people attach great importance to experience, image and emotion. The design of modern commercial space and industrial development leads to the quality of products and service that meet users' needs. It is reflected in the design of personalization, and a variety of meaningful visual orientation. As for space, it embodies in the complementary relationship. And with such features as the central we studied the spatial structure in brand store, to find out that the brand has its own popular features, and interior design tend to appear in brand image, retail and store.

Keywords: brand image, Users, Consumption and way of life, Performance tendencies.

1 Brand Image and New Consumption Tendency

1.1 Brand Image Analysis

Modern society is an age to create and show personality. Therefore, enterprises promote their brand to the world through popularizing its brand personality, and then, they produce and sell its products as well as its service to create a natural brand. Brand image [1] is a compound concept that can make us gain a series of organizational intuition, and form loyalty and brand in the mind of consumers, including emotions, attitudes, imaginations, etc. The brand image that based on consumer trends and product is not simply to convey the visual impression of the item only, but requires corporate planners to control the brand's reputation, its value is to make others feel the presence of the brand. Therefore, brand image reflects the enterprise value.

1.2 The Brand Character Analysis

Brand feature which is regarded as the enterprise logo in commercial space and the association of ideas that reflects on the desperations of the consumers [2], including the core features of all elements, refer to the overall impression of the users on the brand.

2 The Tendencies of Marketplace and PRADA Brand Features

2.1 The Features of Store Design

From a negative point of view, it refers to the user as the object, including the induced formation of trafficking. Therefore, the store design is related to the interior design,

H. Tan (Ed.): Knowledge Discovery and Data Mining, AISC 135, pp. 517–523.
springerlink.com

visual products, marketing activities, including lighting, color and other diverse areas. Users will choose their own user model with the best image and their favorite stores which is determined by intuition. Therefore, the enterprises through the store, product personalization and customer management approach of the new form of trafficking, in which penetrates a designer's creative space, showing the main features of retail space design.

2.2 Store Design Performance Tendencies

Through the recent retail popular way and the architectural decoration design, brand attracts users and provides new experience. PRADA clothing brand take this design as the tendency, reflecting the phenomenon of this age and cultural characteristics, and form interactions base on the various shape. Therefore, the significance of performance tendencies is constantly changing and diversifying, such as expressing empty space, flat surface and open space, the pursuit of the comparison and purity of space, the use of transparent material, translucent material, to show off twists, enter, reflection effects, etc.

2.3 PRADA Brand Features

Prada is designed for the ordinary women, and with an elegant brand image, it pursue an extraordinary nature in an popular way, in order to induce female clients to participant in, and to provide marketing strategies which is emotional and cognitive experience [3], also, to strengthen the management of brand image, to improve sales rate, reflecting on valuing the brand as the company's core strategic.

3 Analysis of the Brand Image and the Tendency in PRADA Retail Stores

3.1 Case Investigation and Analysis

Analysis base on the brand image and PRADA retail performance tendency, and investigate the scope of cases in the United States, Japan, PRADA stores.

3.2 PRADA Store Brand Image and Performance Tendency Analysis

(1)PRADA sales store brand image
PRADA modern store design is beyond the practical purpose of goods on display and on sale, It strengthen the brand image and establish a user-based brand strategy, customers can choose a brand that can meet their real needs as well as to show their identity, brand image gives certain social characteristics.

(2)PRADA stores appear performance tendencies
Store's performance also reflects the tendency of the phenomenon of the times, the decorative effect and space structure of inner and outer space of PRADA stores in New York's are beyond the limits of time and space to convey the PRADA brand value.

Building internal and external refraction of light forms a pitch. Through transparent and opaque material to show the effects of twists and turns, entering into, firing .Give a sense of vision open by links of the expansion of time and view.

Stores in L.A perform light transmission and non-material through the visual projection of space, also imply that stores express opaque materials through the shape of space. From the space material form and image, eliminate the sense of material. So this performance orientation, together with the information of digital environment, media, and the ubiquitous environment, show the image of light.

(3) the brand examples of PRADA stores
The United States (New York 2001) Ram Koolhaas OMA

First, strengthen the brand personality, cutting-edge technology [4], in order to demonstrate the understanding of ubiquitous displays and a varieties of screen toward new cultural forms, and to reflect the brand space and user interaction. These are given in Table 1.

Table 1. The comparison of PRADA stores tendencies (New York 2001)

location	United States (New York), 2001, Koolhaas design	
keyword	Diversity Spatial structure of the future direction	
Brand image	Continuous variable cultural and exhibition space, for a project to expand the diversity of the brand personality tendency to provide a unique exhibition space.	Screen media, cutting-edge IT technology application, build a customer point of interest and future of space management system. Experience, fashion brand
Keywords	Fuzzy without boundaries Elimination of internal and external space	
Performance orientation	Athletics track constitutes a novel display method and display presentation, the dual use of space technology, vertical interior space connected with the flow structure, the results of the twists and turns and space, the building of the variant type field to complete the spatial difference	Space of uncertainty, the ceiling of the substance inside the tendency to eliminate irregular inside and outside the boundaries of space to build a ubiquitous interface environment

Japan 2003, Herzog de Meuron (fashionable, symbolic, and build common environment - connectivity between the urban environment, structure, spatial integration.).

The decorative form borrowed concave, convex space, "look, can look, have seen, show ", to convey the information of "image, sound, light", and the diversity of the space will be integrated into the design of commercial public space. These are given in Table 2. United States (L.A 2004), Ram Koolhaas OMA (to eliminate the boundaries of inside and outside space and to achieve the interaction of diversity space) Variable space design compose a whole, including the various features of commercial space designs. These are given in Table 3.

(4) PRADA store case analysis

Expand the details of the characteristics of the investigation and analysis of PRADA brand image, brand personality, developmental and variability which is based on localization and environmental factors, The works of Herzog & de Meuron are the examples of Semantic construction, symbols become the rhythm of events, is precisely to establish communication between the environment: the rhythm of social structure and material structure linked together [5]. These are given in Table 4.

Table 2. The comparison of PRADA stores tendencies (Japan 2003)

location	(2), Japan Aoyama, 2003 Herzog & de Meuron design	
keyword	Building its own deep-floating, fashion, symbolic Common environmental structure	
Brand image	Fashionable, floating of deep, symbolic buildings, provide an internal space laboratory project, using culture marketing to keep intimacy with customer	Importing cutting-edge IT technology, the pursuit of continuously varying levels of Horizontal pipe and vent pipe constitute a variety of information, ideas of, and build common environment, and strengthen the future direction and visual diversity
Keywords	the urban environment connectivity uniforms in structure, space and outfits	
Performance orientation	The structure, space, positive performance of the integration of the tendency to have the dynamic properties of glass, giving the visual openness and continuity of its landmark building the image generated	The formation of a structure of space, the structure become the loop space grades, using internal and external spaces into each other, in order to produce maximization expansion of space the concept of Scenery type square.

Table 3. The comparison of PRADA stores tendencies (L.A 2004)

location	United States (L. A), 2004 Koolhaas design	
keyword	Interior and exterior boundaries to eliminate (point of "folding")	
Brand image	Interior and exterior boundaries to eliminate (point of "folding")	A positive step together with the products displayed at the bottom, provides a role of bridge to accommodate the topography of spatial continuity of the project fold expansion of the interaction between customers
Keywords	Vagueness and Diversity Project Maximizing the mutual	
Performance orientation	Try to form a conception of space with the concept of slope deformation on the ground., suggesting that poured into the space	Cognitive space connected to the internal flow structure, a continuous variable-type space to the center display showing the connection project, together with the visual openness, inducing symmetrical connectivity.

Table 4. Proposals to PRADA brand image and performance orientation after research

Case	PRADA brand image and performance orientation	
2001 New York (Koolhaas) PRADA stores 2001	Brand	Diversity Spatial structure of the future direction
	Performance orientation	Fuzzy without boundaries Elimination of internal and external space
Herzog & de Meuron 2003	Brand	Building its own deep-floating, fashion, symbolic Common environmental structure
	Performance orientation	the urban environment connectivity uniforms in structure, space and outfits
L. A 2004	Brand	Interior and exterior boundaries to eliminate
	Performance orientation	Vagueness and Diversity Project Maximizing the mutual penetration of space

Various types of culture influence consumption patterns of consumer; meanwhile, it plays a variation on the clothing culture. The era in which the culture was guided by the goods has passed away. Stores develop through Mutual integration [6] of architecture and clothing, while integrating the world of designer works into space design, to form a metaphor for the design of visual brand image. After 2000, the brand relies on stores, strengthening the corporate brand image and personality, reflecting brand integrity. Store brand promise and explains the requirements of brand strategy, store is beyond the practical purposes such as commodity display and sale, increasing the brand image and awareness of culture ,shaping the brand personality and emphasizing the store a variety of selling strategic, structuring the cognitive of the value of the brand image.

4 The Follow-Up Study of PRADA Brand Image and Performance Tendencies

The interior design, tend to occur in PRADA brand stores, is out of the basic design, and analyze the characteristics of stores, clothing flagship stores, performance orientation, brand image, and personality, to explore the strategy and direction that can beat its rivals [7].

The analysis is based on user needs and the development and creativity of future PRADA store design, maintain the shared cultural marketing, interior design, cutting-edge IT and other basic concepts, explain the reason of difference from design, to develop the new selling strategic systems of clothing and space, using brand image that can be perceived in interior design, and bring about the rise in apparel brand image finally.

Although the performance of interior design of stores tend to be different from each other, it is better to pursue the constantly updated image than to be content with the basic brand image. At the same time, provide and experience a new concept which is different from traditional marketing, but it also implies that the development of brand image and personality tend towards the performance of detailed, differentiated, future development, ranging from visual design to create a new brand image.

That is to say, the combination of sensibility and culture demonstrate the functional, symbolic of the performance trend. Also in art and commercial building design, consider the fuzzy point of brand image and the performance trend in internal and external space. And put forward the space differentiation. Therefore, the design trend and brand image not only create clothing trade, but also explain the difference between external manifestations of culture and the different inner meaning.

This is the effective props to communicate for commercial space; it also can be the development of interdependent and complementary relationship. Later, I think of doing research on the symbol of the cognitive and the future brand value.

References

1. Li, M., Jin, S.: Visual, garment terminal sells display planning. China textile press, Shanghai (2007)
2. Su, J.: Research on the tendency of the interior design of stores. Korean Construction Association Proceedings 209, 147–148 (2006)

3. Jin, Z.: Research on the communication space design of the brand in which consumer interested. Hongik University Dissertation 109, 36–38 (2005)
4. Piao, A.: The research on space experience appeared in the store. Konkuk University Dissertation 9, 26–27 (2003)
5. Pohlman, A., Mudd, S.: Image as a function of group and product type. Applied Psychokgy 26, 36–38 (1973)
6. Bingham, N.: The New Boutique Fashion and Design. Merrell Publishers Limited, San Francisco (2005)
7. http://baike.baidu.com/view/5049126.html?fromTaglist#sub5054547

Application of Regression Analysis in Data Processing of Physical Experiment of College

Jianxin Peng and Yigang Lu

Department of Physics, School of Science, South China University of Technology, Guangzhou, Guangdong, China
phjxpeng@163.com

Abstract. Using regression analysis to process college physical experimental data may reduce tedious and complex manual calculations by data processing software and improve the accuracy of the experimental data processing results. The application of linear and nonlinear regression analysis, especially piecewise regression analysis was discussed in college physical experiment of data processing using Excel and Origin software through some examples in this paper. The detail procedure for college physical experiment of data processing using the regression analysis was given and the processing results were demonstrated by figures with a more direct and visual way. The advantages and disadvantages of regression analysis using different software for physical experimental data processing were also discussed. Applying regression analysis to process college physical experimental data through using computer application software can deepen the student to master physical experiment of college, boost the teaching quality for physical experiment of college and enhance the efficiency of student in learning.

Keywords: data processing, regression analysis, physical experiment of college.

1 Introduction

College physical experimental data processing is an important part in college physical experimental teaching. The experimental data processing method from most college physical experimental textbooks usually includes the following: list method, mapping method, graphical method and least squares method [1]. However, if the experimental data is analyzed by hand or calculator in college physical experiment course, it is very slow, low accuracy and quite boring. Moreover, this kind of method is very serious obstacle to students' enthusiasm, initiative, creativity and scientific thinking formation. It could improve the students' ability of solving practical problems if we can combine data processing with software applications with help of computer. Excel and Origin are commonly data processing software which can be used for data statistic analysis, processing and data mapping. Using these software for data processing can avoid the complexity of programming as C, C++ and Matlab language, and has the features of perceptual intuition, easy learning and using. It can obtain experimental results and laws quickly when Excel and Origin are used. Using these softwares for data processing can also overcomes the defect of time-consuming and easy making error by manual calculation [3].

H. Tan (Ed.): Knowledge Discovery and Data Mining, AISC 135, pp. 525–531.
springerlink.com

It is often to analyze the relationship between two physical quantities which have a function relation and then determine the relationship according to many groups of observed data of them. The method that using mathematical analysis finds a best fit curve from a group experimental data is known as the regression equation [4]. Regression analysis mainly analyze a specific cause and effect relationship between two variables, use mathematical models to represent their specific relationship and further get the relationships between variables in quantities (i.e. the regression equation) [5]. Regression analysis is a frequently-used method in data processing. In this paper, the method of linear regression and nonlinear regression analysis, especially in piecewise regression analysis in college physical experimental data processing are discussed with some examples using Excel and Origin software.

2 Regression Analysis Using Excel

The initial data from experiment measurement should be input into Excel and necessary calculation is conducted with Excel when regression analyzing is applied. Table 1 shows the relationship vibration frequency v and tension T from string vibration experiment. It requires students to find the relation between v and T with lgv as horizontal axis and lgT as vertical axis [6]. In order to get the relationship between v and T, firstly, we draw the scatter-plot according lgv and lgT, the result is shown in figure 1(a) which can be obviously seen that it is linear relationship. Right click any data point with mouse and click "Add the trend line" in the popup menu, then choose "linearity" in the "type" box and click "option" to choose "Display formula" and "display R^2", at last click "yes", a regression line would be obtained as shown in figure 1(a). The regression equation and related coefficient R^2 are provided. We can obtain the relationship between v and T from regression equation shown in figure 1(a).

$$\nu = 12.717 \lg T^{0.4776} \tag{1}$$

Table 1. The vibration frequency v and tension T of string

T	v	lgT	lgv
9.8	38.2	0.99	1.58
19.6	52.0	1.29	1.72
29.4	63.2	1.47	1.80
39.2	73.8	1.59	1.87
49.0	82.2	1.69	1.91

We could also draw scatter-plot directly instead of the conversion of v and T, the steps are same to figure 1(a) mentioned above except for choosing "exponentiation" in the "type" option. It can be found that the relationship between v and T is consistent in both methods.

According to the relationship $\nu = (n/2L)\sqrt{T/\rho}$, the relation curve between v^2 and T can be obtained the method mentioned above and shows in figure 1(c). The string destiny ρ can be obtained according to regression equation if wave number n and string length L are known.

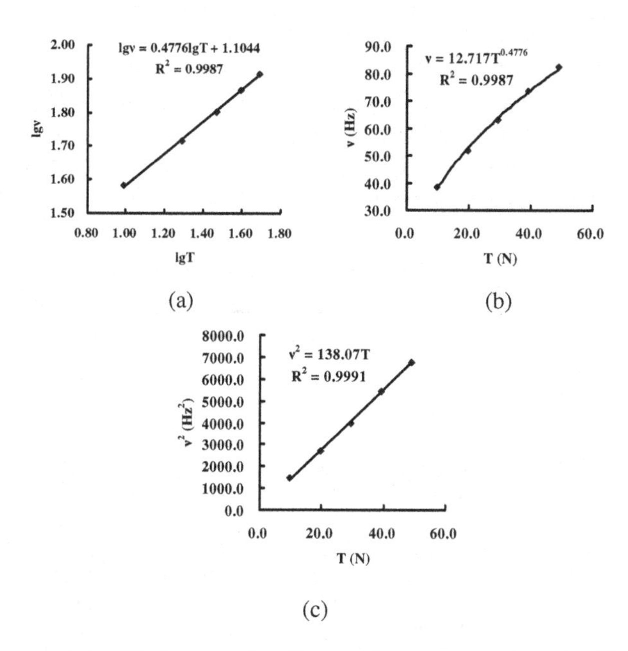

Fig. 1. The relationship between the vibration frequency v and tension T of string

Now we use piecewise regression analysis to process experimental data in Excel. Table 2 shows the measured data of magneto-resistance effect. The magnetic induction intensity B, magnetic resistance R and the relative rate of change Δ R/R(0) are calculated in Excel, R(0) is magnetic resistance when B equals zero. It requires students to plot B versus Δ R/R(0) using B as horizontal axis and Δ R/R(0) as vertical axis. When magnetic field is weaker, Δ R/R(0) is proportional to the B2 and it is proportional to the B in high magnetic fields, so a piecewise regression fitting is required for data processing.

The detail methods of piecewise regression using Excel software are demonstrated below. Firstly, we draw the scatter-plot with the horizontal axis B and vertical axis Δ R/R(0); then find demarcation point of piecewise regression fitting. B is proportional to the Δ R/R(0) when B>0.12T in this example, so linear regression can be conducted firstly. One click the demarcation point and read the data, then find the position in Excel and right click, choose "source data" and change X and Y to be linear data in the popup window, choose any data with right click, click "Add the trend line" in the popup menu, choose "linearity" in the "type" box and then click "option", choose "Display formula" and "Display R^2", at last choose "yes", Regression analysis of linear parts are completed. When the magnetic field is weaker, Δ R/R(0) is proportional to B^2, but there is no square trend line available in Excel, so we can calculate B^2 value when B is smaller than 0.12T, click "tools", "statistic analysis" "regression", then input B^2 and Δ R/R(0) in the popup window "X value input area" and "Y value input area", set up the relevant option, the related-coefficient between B^2 and Δ R/R(0), the coefficient in regression equation and the predicted values of Δ R/R(0) are calculated. Finally, we click right in the drawing area, then choose “initial data”, “series”, “add”, add two series of data and plot figure. One plot shows calculated B versus ΔR/R(0), another is B versus ΔR/R(0) obtained from linear regression analysis. Figure 2 shows the relationship between B and ΔR/R(0) after piecewise regression. The nonlinear and linear equation are ΔR/R(0)=59.768B2 and ΔR/R(0)=2.923B+0.377, respectively and related-coefficient are 0.9898 and 0.9994.

Table 2. The measured data of magneto-resistance effect

U_1 (mv)	I_1 (mA)	U_2 (mv)	I_2 (mA)	B(T)	R(Ω)	ΔR/R(0)
0	2.22	799.2	2.22	0	360.0	0
3.1	2.21	800.5	2.20	0.0082	363.9	0.0107
6.6	2.14	800.1	2.14	0.0179	373.9	0.0386
10.0	2.09	800.5	2.08	0.0278	384.9	0.0690
12.8	2.01	800.3	2.00	0.0370	400.2	0.1115
15.4	1.94	799.3	1.94	0.0462	412.0	0.1445
18.9	1.72	799.6	1.72	0.0639	464.9	0.2913
22.2	1.49	800.2	1.49	0.0866	537.0	0.4918
25.2	1.31	800.8	1.31	0.1118	611.3	0.6980
32.8	1.20	799.5	1.20	0.1589	666.3	0.8507
39.6	1.12	800.0	1.12	0.2056	714.3	0.9841
45.8	1.05	799.9	1.05	0.2536	761.8	1.1161
51.3	0.99	799.3	0.99	0.3013	807.4	1.2427
55.8	0.94	800.6	0.93	0.3451	860.9	1.3913
59.8	0.89	799.5	0.88	0.3906	908.5	1.5237
63.5	0.85	800.2	0.84	0.4343	952.6	1.6462

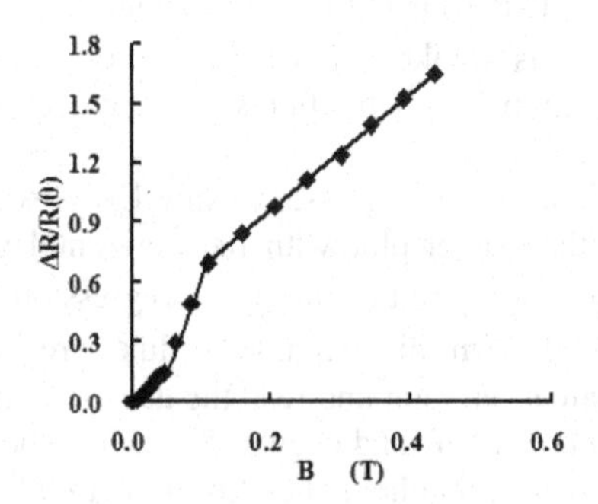

Fig. 2. The relationship between ΔR/R(0) and B

3 Regression Analysis Using Origin

Starting Origin software, a data worksheet window is displayed and the default is two columns, which are A(X) and B(Y). We click the first column header and choose “properties” in the pull-down menu after right clicking, the data display format, column name and columns marks, etc. can be set. The data waited for processing should import or duplicate from excel to data workbook. For example, we can copy lgT, lgv in Table 1 to A (X), B (Y) column and choose the two column data with mouse, then click "tools" in Main Interface. There is a popup window when choosing "linear fit" in pull-down menu after setting relevant options. Then we click "fit" to obtain the scatter diagram of initial data, linear regression line, the correlation coefficient and the coefficient of regression equation. We can get same regression result like figure 1(a) when modifying the horizontal and vertical coordinates of property, title and other relevant settings.

Regarding the nonlinear regression of experimental data, the data firstly should be inputted or duplicated to worksheet, such as duplicating T and v in table 1 to A(X), B(Y) column, then choose these two column data, click "analysis" in main interface, and choose the "Non-Linear Curve Fit", "Advance Fit Tool..." in the pull-down menu, and then choose the regression class and function in popup window. For this example, we choose "power" the class and "Allometric1"(which equation is $y=Ax^b$) the function, then click "fit", "100 Iter", "Done" successively. The regression curve, equation, coefficient and related coefficient R would be obtained. These results are same to those in figure 1(b).

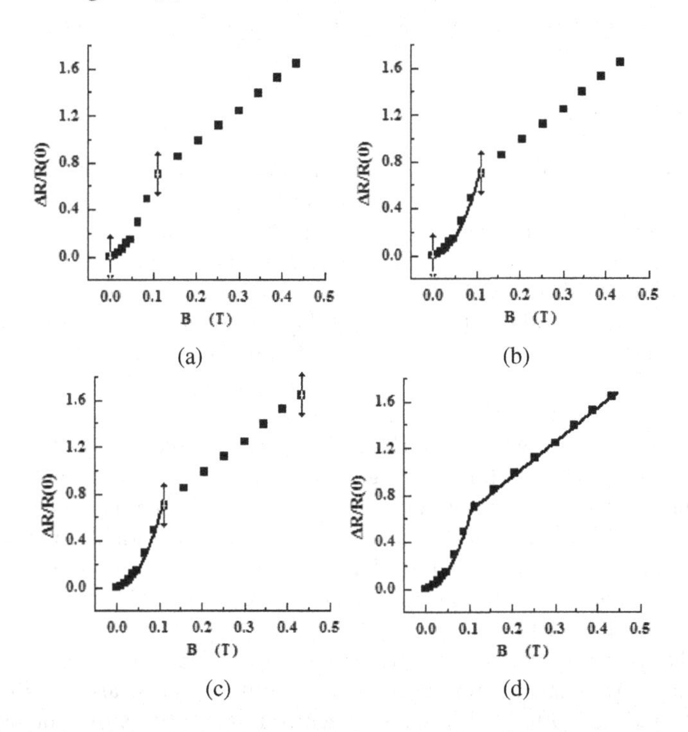

Fig. 3. The relationship between Δ R/R(0) and B using piecewise regression fit

It is easier to use Origin to carry out piecewise regression analysis than Excel in processing college physical experimental data. Duplicating or inputting the data of B and Δ R/R(0) in table 2 to worksheet of Origin and choose the two column data, then click the "plot" menu in main interface which shows the order of many graphs in origin such as straight line graph, scatterplot, histogram, pie chart, vectogram in the pull-down menu. A drawing window and shown the graph of the relationship between B and Δ R/R(0) as shown in figure 3(a) after choosing "scatter". The software would automatically produce data identifier at the both ends of scatterplot track when we click "data selector" in the toolbar of drawing window. We can move the end identifier to the boundary of piecewise fitting data with the press-button "Data selector" as shown in figure 3(a) and carry out the fitting of the first part. Firstly, we click "Analysis" in main interface, choose "Non-Linear Curve Fit", "Advance Fit Tool..." in the pull-down menu, then choose "Polynomial", "Parabola", and then click "fit" and set A=B=0 together with removing the "√" in "vary" option. Finally, when we can click "100 Iter..." and "Done", the corresponding nonlinear regression curve would be obtained. When we carry out the fitting of the next part, we firstly click the "Data selector" button similarly, the data identifier in head of scatterplot track should be moved to the second inflection point as shown in figure 3(c), the linear regression fitting method is same as the regression fitting of lgT and lgv mentioned above using origin. The result is shown in figure 3(d). The nonlinear and linear equation are $\Delta R/R(0)=59.800B^2$ and $\Delta R/R(0)=2.923B+0.377$, and the square of the correlation coefficient values are 0.9816 and 0.9994, respectively.

4 Discussion

Excel has the powerful function of data calculation, analysis and statistics. It is convenient to operate and easy to master even though one who does not have computer professional knowledge. The complicated mathematical analysis and statistics could be carried out in college physical experimental data processing with the function provided by Excel software. The relationship and the change tendency among data in the worksheet can be display by figure. The parameters of the regression equation, standard deviation and fitting line can be obtained together through linear regression analysis of experimental data with linear characteristics. These actions transform complex data analysis into simple one [7]. However, there are certain restrictions when use Excel for curve regression. Only exponential, logarithmic, power and multinomial regression analysis of experimental data can be carried out. When other curve regression analysis of data is applied, we must transformed nonlinear data into linear one first through the variable substitution, and then carry out the regression analysis for data after transformation. The function of graphics processing and analysis for excel is inferior to that for origin which is simple, quick and powerful.

Origin has powerful function of data analysis (including calculations, statistics and curve fitting) and graph plotting and has the merits of quick and flexible for data processing and easy to use [1]. It can be seen from the process of data analysis that origin is beneficial to carry out the regression fitting with complex physical experimental data and to verify complex physical phenomenon and law by experiment. Using origin to plot fitting curve can avoid the experimental error brought by hand plotting in coordinate paper traditionally. Using origin, we can perform both linear regression and nonlinear regression analysis. While, using Excel, nonlinear regression analysis need be transformed into linear through the variable substitution. If the experimental data need further calculation, such as B and Δ R/R(0) in table 2, we can calculate them in origin directly or import them to Origin from Excel, but the calculation of data in excel is easier.

It needs point out that Matlab provide powerful numerical calculation function. Matlab can handle data quickly and effectively and can demonstrate results with corresponding graphics. It is intuitive, convenient and vivid to illustrate physical phenomenon and law [3,7]. But there is no Matlab course in junior college and the use of Matlab is restricted.

From mentioned above about the process of regression analysis, it is known that Excel and Origin could carry out effectively regression analysis for college physical experimental data. They are intuitive, easy operating and can avoid the complex and boring manual calculation and improve the accuracy of experimental data processing results. It can deepen the students' mastering of the university physical experiment, consolidate the students' understanding college physicals knowledge and improves students' learning efficiency. It not only promotes university physical experimental teaching quality, but also improves students' hands-on ability and comprehensive quality.

References

1. Liu, F., Wang, A.F., Sun, D.P., Yang, T.L.: Application of Origin Software in College Physics Experiment Data Processing. Experiment Science & Technology 8, 19–21 (2010)
2. Chen, Y.L., Ding, L.G., Zhang, L.: Application of Excel to Data Processing of Physics Experiment. Research and Exploration in Laboratory 26, 63–65 (2007)
3. Wan, H.J., Luo, X.B., Yang, J.P.: On the Enhancement of MATLAB-based Data Processing Capacity of College Physics Experiment. Journal of Changchun Normal University (Natural Science) 29, 59–61 (2010)
4. Li, X.Y.: Physical Experiment of College, pp. 18–32. China Higher Education Press, Beijing (2004)
5. Zeng, W.Y.: Statistics, pp. 210–248. Beijing University Press, Beijing (2006)
6. Ni, X.L.: Physical Experiment of College, pp. 139–143. South China University of Technology Press, Guangzhou (2005)
7. Yuan, A.J., Wang, J.Y.: The Compared Software For The Data-Process of Physics Experiment. Physical Experiment of College 20, 82–85 (2007)

Importance and Challenges of Digital Construction for Academic Library

Mingfeng Xu

Library of Weifang University, Weifang, Shandong, China
wdhy1998@163.com

Abstract. Digital construction brings great impact and influence to library and it has great importance of strengthening digital construction of the library, the challenge that its face determines the survival and development of library. The article states the importance and features of digital library, analyses its the challenge and inevitability for development, and puts forward the strategy to these challenges.

Keywords: academic library, digitalization construction, challenge.

1 Introduction

After 1990's, with the rapid development of computer technology, especially the universal network technology, digital storage and transmission technology, it makes people have the new requirement on the document information about processing, storage, inquires and use, etc. Therefore, the Digital Library (Digital Library, referred to as "DL") comes into being. DL is the complete knowledge positioning system, but also for the future development of the Internet information management. It can be widely used in social culture, life-long education, mass media, business consulting, e-government, and all other social organizations of the public information transmission.

2 The Significance of the Establishment of the Digital Libraries

The development of the modern information technology and the construction of the digital library, which makes the content and form of library have larger changes than traditional library. Comparing with the traditional library, DL is virtually based on the network environment and sharing of the extensible knowledge network system. It also is a large scale, distributed, easy to use, no time limit as well, can be achieved across the library seamless and intelligent search across the knowledge center. Digital library is established with the following important significance.

2.1 Expanding the Range of Collection

The networked libraries can be connected to a variety of commercial electronic document supply centers, on-line retrieval centers, electronic magazines centers and all other the Internet network. These external information resources, although not part

H. Tan (Ed.): Knowledge Discovery and Data Mining, AISC 135, pp. 533–537.
springerlink.com

of the museum its own resources, but the network can connect or retrieve them and be available to users. This resource virtually becomes a part of the library, which is the "virtual collection". At the same time, the digital libraries take up the relatively small amount of physical space, due to the large amount of digital information stored in the countless in disk storage.

2.2 Achieving Large-Scale Resource Sharing

The digital library is the foundation of the national and global library network. In the digital library under the premise of achieving network, it can truly achieve a wide range of data resources sharing, the higher the level of the network, the greater the range of resource sharing. The network digital library of information resources sharing will become a reality, once the national and even global integrated. It will radically change the traditional mode of operation, such as no paper, all-weather and all-round coverage for the network to provide the reader fast and efficient quality service anywhere. The reader will use a terminal never leave home through all over the world book, and known world affairs.

2.3 Meeting the Reader Much More Needs

The digital library collections digital form of information also includes other digital information, in addition to the material via printing on books, Such as video, audio data, computer programs, etc, to meet the diverse needs of readers.

2.4 The Function of High-Speed Information Retrieval and Comprehensive Information

Compared with the traditional library, the digital library is a revolutionary change, which is high speed, accurate information retrieval and comprehensive information. As long as there is a title or key words, through the digital library of powerful search engine, the reader can accurately access to the required books in a very short period of time, at the same time, it can ensure that the readers never leave home, according to their own needs to obtain the required information.

2.5 To Implement Low Cost of Operation

The digital library realizes information sharing, which means that each library collection into the circulation of library books for all networking, and to ensure that every book can be successfully "shelves. "It is impossible for the traditional libraries; it can be achieved only after the networking digital libraries.

2.6 Effective Communication and Protection of the Chinese Culture

The construction of digital library will quickly reverse the lack of information about the status of Chinese on the Internet and form the Chinese culture in the overall advantages of the Internet. The digital Library was the best means that preserved and continued the development of national documentary heritage; all the precious material is processed by the digital, the original will be stored in a more appropriate environment.

3 The Characteristics of the Digital Library

The digital library has the following features:

Ⅰ.Opening: the digital library based on the network information technology is an open information platform and has a wide range of readers.

Ⅱ. Integration: to make a variety of information services and multimedia applications of digital libraries highly integrated which can provide single point transmission, and can also provide more transmission, achieve more computer information transmission between as well; And because of its integration can also be used to transfer control information to achieve a variety of remote control.

Ⅲ. High efficiency: the application of network information technology improves the efficiency of information processing. The digital library will connect many computers together via the internet to continue to access and transmit information, which makes the literature information resource efficiency greatly improved.

Ⅳ. Real-time: the digital library makes us understand the resources of information promptly; reflect the principle of quick, public and fairness of information.

It is because of the distinctive characteristics of the digital library that makes the library's digitalization construction become the important competition weights in the industry in future.

4 Facing Challenges of the Digital Construction

The essence of library competition is the dispute of information collection, processing, the transmission, services and use, is the dispute of competing for users, show the characteristic, compare benefits, measured the value. In order to have a place in the future competition must do the following:

4.1 The Construction of Hardware Conditions

In the modern information society, it is difficult to imagine a library has backward technology and equipment, the library can provide satisfactory service for the reader, similarly, the backward of technology and equipment also can make the library lose the core competitive power. Therefore, college library should increase investment of strength in the information processing equipment, keep up with the development and make the universities library information resources satisfy the need of reader in time.

The improvement of hardware conditions should be highlighted create the open information environment; the key point is to seize the following two aspects: ① setting up a library network and service system with high performance and availability. For now, the campus local area network is been building in the colleges and universities around the country, suggest building the backbone network by using Gigabit Ethernet, also fit out switch and switch integrated device with the high-speed Ethernet module.② Setting up Multimedia Reading Room. Multimedia reading room building should be based the high starting point, with good conditional can be equipped with interactive multimedia network system, at the same time, readers can scan, copy and print other series service facilities, also help teachers implement network-teaching.

4.2 The Construction of Software Conditions

Under the network environment, who had the resources advantage, and form the irreplaceable advantage in a certain range, realized I have not, I have the more, I best, who is to obtain advantages and pole position in competition. To play a good foundation of resources under the environment of network, we should not only pay attention to the professional and characteristic of matter, also pay attention to the variety of carrier form; not only pay attention to the construction of literature resources, the more attention to the disposition and combination of digital information resources; not only to explore the sources and content of information, the more to improve the quality of information, development with the proliferation of new products; not only to speed up the speed of information transmission, simplify the transfer process, increase the order way, and improve the ability of information obtain, analyze, access, so as to establish effective information resources advantage.

4.3 The Introduction and Cultivation of Excellent Talents

An excellent talent includes the professionals and technical with high responsibility and the administrative staff with a strong affinity. Librarian's quality decides the development level of library. At present, there are three broad the relation of librarian's quality and library's development: One is advanced type. That is the librarian's quality higher than the need of library practical work; make librarianship has the more broad space of the library's development. The second is adaptive. That is the personnel quality and the library's development to suit basically, both balanced development. The third is the lag librarians. Librarian quality behind the need of library's development hindered the process of library's development. University library need advanced type and adaptive librarians, phased out lag type librarians. As society progresses and new technologies have emerged, university libraries are need of professional technology personnel in urgent, adapt to the new situation, and improve the service level.

4.4 The Competition between the Libraries

The competition between the libraries is in fact that is to fight for readers and information markets competition in the future. While the environment of the university library's digital network setting up, especially not in a "wall" barriers and constraints of time, the libraries face the fierce competition. It is Who has more books, thick foundation resources, speedy transmission, more ways to retrieve, high quality service, good social efficiency, more advanced means that has a competitive advantage. Therefore, we must have the resources and strengths over competitors in part or the whole of the value chain activities. The most important thing is to build the brand and highlight features.

4.5 The Issues of Digital Copyright

The legal issues of digital libraries are actually the copyright and intellectual property issues in the use of the Internet age, which has become more prominent in the construction of digital Libraries. How to both protect the author's intellectual property rights, while allowing all kinds of cultural, scientific and technological achievements of civilization into the digital library services for more people to create greater value, this is a serious problem. At present the international community has no single solution, national laws vary. In addition, our people's awareness of copyright is still relatively weak, they can not be better to avoid copyright issues during the construction of the digital library and use of information resources resulting in violations have occurred.

4.6 Improve the Performance of Macro-management

According to a variety of factors that affect the digital libraries to improve the effectiveness of macro-management can formulate many corresponding strategy and measures to improve the effectiveness of macro-management, for example: to deepen the traditional library system, innovate the macro-management system and management mechanism of the digital library, reduce the cost of macro-management of digital libraries, improve the quality of macro-management of digital libraries, improve the efficiency of macro-management of digital libraries improve the effectiveness of macro-management of digital libraries, etc.

5 Summary

The important practical significance to construct the digital libraries and the characteristics of the digital libraries have been discussed. From six aspects, including the construction of hardware and software conditions, the introduction and training of high-quality personnel, competition between the library, the issues of digital copyright and the improvement of the effectiveness of macro-management, we investigated the challenges and measures that the future digital library be faced with, which has some guiding significance for the construction and development of the future digital library.

References

1. Tang, X.M.: Discussion of core competitiveness of University Library. Library Theory and Practice (3) (2008)
2. Xie, C.Z.: The marketing strategy and fashion of digital library. New Technology of Library and Information Service (1), 7–9 (2003)
3. Wu, Y.F., Hu, Y.L.: Competition and development of library under the network environment. Library Theory and Practice (2) (2008)
4. Liu, D.Z., Cao, X.Q.: On construction and development of digitized libraries in college. Library Theory and Practice (7) (2009)

The Research on Physical Education Teaching in Sino-foreign Cooperation in Running Schools

Baoling Ma[1], Xiangdong Wang[2], and Qiang Li[3]

[1] E&A College of Hebei Normal University of Science Technology, Hebei, Qinhuangdao, China
[2] The College of Physical Education, Hebei Normal University of Science Technology, Hebei, Qinhuangdao, China
[3] Hebei Vocational College of Foreign Languages, Heibei, Qinhuangdao, China
mabaoling528@163.com

Abstract. College Physical Education is an important part of the whole higher education. In order to optimize Sino-foreign Cooperation in Running Schools PE curriculum setting and Reform College PE teaching in Sino-foreign Cooperation in Running Schools.The paper studies PE documentation, teaching practice, teaching subjects, teaching-led teaching environment of Sino-foreign cooperative education institutions in china by literature review, questionnaire and interview. The result shows that Teaching the basic integrity of the document, standard, but does not reflect the characteristics of Sino-foreign cooperation in running schools sports;Education was the main motivation for diversification trend sports, sports a soft spot for foreign;Teaching-led structure has its own characteristics, but the school is not conducive to sustainable development;Barely meet the physical education field devices to work with students on the sports needs were quite different.And analysis of this study, presented by "international" training as the goal of foreign cooperation in running schools sports development ideas, to Chinese and foreign cooperation in running schools provide physical education to carry out the work of reference.

Keywords: Sino-foreign co-operational, Physical Education teaching, Future development.

1 Introduction

Sino-foreign cooperation in running schools is an important part of higher education. From primary exploration and formation to rapid development, the position of Sino-foreign cooperation in running schools is finally established in China's higher education. Sino-foreign cooperation in running schools refers to the activity of educational institution with Chinese citizens as the main enrollment target, which is co-sponsored legally by Chinese and foreign educational institutions in China. With the goal of "introducing international courses, selecting international faculties, establishing international schools and cultivating international talents", the dominant idea of not going abroad and the means of bilingual teaching, it combines excellent

H. Tan (Ed.): Knowledge Discovery and Data Mining, AISC 135, pp. 539–546.
springerlink.com

resources, ideas and outcomes of foreign education with concrete practice of China, improving the school-running quality and talent-cultivating level of China's higher education. There are differences between colleges and universities of Sino-foreign cooperation and regular colleges and universities in terms of educatee, education property, cultivating objective and education concept, which will inevitably lead to differences of education content and teaching management methods. However, physical education is an important part of higher education that cannot be ignored. For this reason, the study is aimed at carring out an investigation on physical education system of Sino-foreign cooperative educational institutions currently existed in China. The research findings not only help to improve the integral level of physical education of Sino-foreign cooperative educational institutions, but also provide a useful theoretical support for the overall development of Sino-foreign cooperative educational institutions as well as a scientific decision-making basis for the future development of physical education of Sino-foreign cooperative educational institutions.

2 Research Objects and Methods

2.1 Research Objects

Physical education in secondary college of Sino-foreign cooperation is the research object. Relevant leaders, principals of PE, full-time PE teachers and college students from 18 secondary colleges of Sino-foreign cooperation are respondents. Interview leaders and principals, and carry out questionnaire survey on 90 full-time PE teachers and 600 college students of subordinate colleges.

2.2 Research Methods

With regard to content involved in this topic, search by Journal Net and WanFang Database, consult and collect research literatures at home and abroad related to this paper. Design questionnaire according to research purpose, conduct questionnaire survey and interview on respondents, and make mathematical statistical analysis on obtained data through Microsoft EXCEL2003 and the statistical software SPSS13.0.

3 Research Findings

3.1 Physical Education Documents of Sino-foreign Cooperative Education

Sino-foreign cooperative educational institutions attach much importance to creating physical education documents. More than 80% colleges have teaching syllabus and materials, and more than 90% teachers have teaching plans and programs. In this study, foreign language condition is investigated as a feature of physical education documents of Sino-foreign cooperative educational institutions. The results show that only two schools (11.11%) compiled syllabus bilingually. The interview record suggests the reason why schools and teachers do not use bilingual teaching documents: most PE teachers of Sino-foreign cooperative educational institutions graduated from domestic higher PE colleges, who are not familiar with purposes, requirements and objectives of Sino-foreign cooperation in running schools. They think that it is only a way of domestic higher education and rarely study the difference between the two. Thus, some colleges simply borrow physical education documents of regular colleges. There is no uniform requirement of the nation on physical education of Sino-foreign cooperative educational institutions, and the compilation of syllabus is not put on agenda. So far, there is still no special material for physical education of Sino-foreign cooperative colleges.

3.2 Implementation of Physical Education of Sino-foreign Cooperative Education

Teaching contents of Sino-foreign cooperative colleges involves mainly competitive sports while less traditional sports (44.44%) and recreational fitness sports (27.78%), as well as much less foreign sports (16.67%). With respect to curriculum model, 83.33% of the institutions have the main model of combining general courses with special elective courses and 27.78% of the institutions choose the teaching model of sports club. Physical education evaluation of Sino-foreign cooperative colleges places emphasis on the evaluation of students while ignores the evaluation of teachers. The evaluation of students in terms of "learning" mainly focuses on results evaluation, which accounts for 72.22%. Colleges that attach importance to process evaluation take the smallest proportion, which is only 33.33%. Colleges that combine results evaluation and process evaluation account for 22.22%. As for the evaluation of teachers in terms of "teaching", most colleges adopt only students' evaluation of teaching. Evaluations from experts and peers are largely formalism. Individual schools evaluate those with no teaching accident and being praised for teaching as excellent, those with teaching accident as unqualified, and the rest as ordinary. Based on the above investigation results, physical education of Sino-foreign cooperative educational institutions and that of regular colleges are basically the same with each other in terms of teaching content, curriculum model and evaluation system. There is no project with foreign features introduced to physical education content nor curriculum model. And evaluation system is completely the domestic evaluation system without any school-running characteristics of Sino-foreign cooperation.

3.3 Physical Education Subjects of Sino-foreign Cooperative Education

Motivations for participating in sports activities of students of Sino-foreign cooperative colleges show a diversified trend. Students of Sino-foreign cooperative colleges, who choose improving physical fitness, learning knowledge skills and mastering body-building methods as their motivation for participating in physical exercises, take a higher percentage. However, the investigation of 2007 shows that the number of students who choose improving physical fitness, recreation and interest as their motivation for taking part in extracurricular exercises is only 34.9%, which is significantly lower than the result of this survey. The percentage of students in Sino-foreign cooperative colleges who choose passing exams of physical education as the motivation for PE exercises is lower than 40%. Yet the percentage of college students in Hubei Province who choose this motivation is 61.1% in 2007, which is remarkably higher than the result of this survey. From the nature of motivation, students have a higher level of recognition in terms of common functions of PE, as well as the recognition of special value orientation of PE. Girls have a higher level of requirements than boys, which are more stronger. In spite of better situation in terms of getting credit, they still show a passive attitude towards learning.

Sports interest. From categories of sports selection, students of Sino-foreign cooperative educational institutions have a wide range of sports interest. Besides, there is such a large difference between boys and girls that they have different interest in 60% of the sports activities. From characteristics of sports, boys prefer ball games with strong competitiveness and antagonism, while girls prefer sports that are entertaining, less antagonistic and non-contact with body directly. Meanwhile, it can be seen that boys and girls have a common characteristic that they prefer foreign sports taekwondo and yoga. It can be informed from student interviews that they have great curiosity about foreign sports which are not quite popular in China, such as baseball, rugby football, racing car, curling, etc. Results of the survey suggest that the physical education content of Sino-foreign cooperative education should choose some foreign sports properly so as to show characteristics of the physical education content of Sino-foreign cooperative education and meet the needs of students in terms of sports.

3.4 Physical Education Leader of Sino-foreign Cooperative Education

The age structure of PE teachers in Sino-foreign cooperative educational institutes shows the trend of getting younger. Education background is mainly undergraduate and most of them have the education background of undergraduate, as well as the professional title of intermediate professional title. The results of this survey are similar to higher independent college in terms of education background and professional title structure of PE teachers. From the analysis of PE teacher source, full-time PE teachers in Sino-foreign cooperative educational institutes are quite few, and most part-time teachers are depending on their parent schools. External teachers from other colleges are mostly course teachers, retired teachers, graduate students of physical education colleges or professional athletes. The English level of PE teachers are relatively low, most of whom are below the college English level. The research content of Sino-foreign cooperative college teachers is mainly based on domestic sports research while hardly involves foreign physical education. There is a decreasing trend in the liquidity of full-time PE teachers in Sino-foreign cooperative colleges year by year, but it is higher than that of parent schools.

3.5 Sports Filed and Equipment of Sino-foreign Cooperative Colleges

Evaluate sports field and equipment of Sino-foreign cooperative colleges in accordance with the standard in undergraduate teaching evaluation of Ministry of Education that the C-level indicator of sports field and equipment is that the area per student is larger than or equal to $3m^2$. The evaluation result shows that sports field and equipment in Sino-foreign cooperative colleges are insufficient. Therefore, in questionnaires and interviews of teachers and students, once there are more than 80% of the teachers and students who think sports field and equipment are sufficient, it can be concluded that sports filed and equipment in the school are able to meet the requirements of teaching and extracurricular sports activities. The investigation indicates that there are very few Sino-foreign cooperative colleges can meet the needs of teaching and extracurricular physical exercise. Sports field and equipment of most colleges can only satisfy the requirements of teaching, and even a small number of colleges fail to meet the requirements of teaching. There are might be such reasons: insufficient educational investment; shortage of teacher resources; PE lessons are arranged in extracurricular activities time conflicting with both of the time and space of extracurricular physical exercise. Thus the limited filed and equipment becomes increasingly inadequate.

4 Analysis and Discussion

4.1 The Impact of Decision-Making Level on Physical Education of Sino-Foreign Cooperative Colleges

From such aspects as the use of teaching documents, provision of PE teachers and investment of teaching situations, it can be seen that leaders of Sino-foreign cooperative colleges pay no attention to whether bilingual teaching is used in physical education. They attach too much importance to whether professional knowledge skills can be geared to international standards while ignored the main part of higher education——whether physical education can be geared to international standards. Nowadays, there are different degrees of difference between China and foreign countries in terms of both physical education and leisure sports, which are still not valued by leaders and will finally, lead to the fact that teaching documents, teacher training and construction of teaching situation in Sino-foreign cooperative colleges fail to get rapid international development. Thus, it is really difficult to achieve the target of not studying abroad.

4.2 The Effect of PE Teachers on Sino-foreign Cooperative Colleges

PE teacher is one of the main parts that are engaged in physical education activities, so that structure of teaching staff is directly related to the implementation of teaching task and the overall quality of teachers has a direct impact on teaching quality. Studies show that PE teachers training of present physical institutes and normal colleges is basically disconnected with the current new curriculum standard of middle and primary schools and even universities. However, this study result indicates that most PE teachers in Sino-foreign cooperative colleges are young teachers. They are active in thinking, outgoing and close to students, but lack teaching experience. Loose class management and poor classroom organizing capacity will inevitably lead to the fact during teaching process that the "seriousness" of PE lesson is ignored. Low scientific research level and few studies on foreign physical education will hinder the sustainable development of physical education of Sino-foreign cooperative colleges towards scientific, rational and international direction. It is not merely a problem of Sino-foreign cooperative colleges that the English level of PE teacher is relatively low. However, the requirements of teacher's English level in Sino-foreign cooperative colleges should be higher than that of regular colleges and universities. Otherwise, the normal implementation of bilingual teaching would be seriously affected. Sino-foreign cooperative colleges have such defects as short running time, fast development, hurried employment, rough selection and unsound property of school-running and management system, leaving troubles for teachers with high education and high professional title who worry about instable foundation of personnel relationship. This will finally lead to the high liquidity of teachers, instability of teaching team and disadvantage of teaching management, as well as hesitation and delay in terms of PE teachers training of Sino-foreign cooperative colleges.

4.3 The Effect of Students on Physical Education in Sino-foreign Cooperative Colleges

Most students of Sino-foreign cooperative colleges have more superior family conditions yet poor scores than regular college students. Besides, it can be known from teachers' interview that students of Sino-foreign cooperative colleges have active thoughts, unique concepts, vigorous personality and colorful extracurricular life. Compared to culture courses, this is seemingly drawback yet has a good simulative effect on physical education. The motivation for physical exercises in Sino-foreign cooperative colleges has been greatly improved. Phenomena of passive exercises for getting credits are becoming less. Acquiring knowledge skills and improving physical and mental health become the main motivations of students for physical exercise, coming before all the other motivations. It has something to do with good sports lifestyle of students in Sino-foreign cooperative colleges. Most PE teachers indicate that students of Sino-foreign cooperative colleges have a higher sports interest level as well as higher enthusiasm for taking part in extracurricular physical exercises. Sports exercises are tend to be international, modernized and fashion-oriented. This is a favorable phenomenon for physical education. However, the result of psychological investigation shows that students of Sino-foreign cooperative colleges contain more factors of stubbornness and hostility than regular college students. The result indicates that PE teachers in Sino-foreign cooperative colleges should be fully familiar with psychological differences between students of Sino-foreign cooperative colleges and regular college students. Instructional education concerning psychological aspects like stubbornness and hostility should be strengthened during teaching process, so as to fully show the physical exercise function of improving mental health level. The choice and organization of physical education content should be in line with the characteristics of physical and mental development of students.

4.4 The Effect of Teaching Situation on Physical Education in Sino-foreign Cooperative Colleges

Sports field and equipment are necessary material guarantees for physical education, extracurricular sports activities, extracurricular sports training and normal competition, which demonstrate the strength of running school, and are also important factors and conditions of attracting students. Some Sino-foreign cooperative colleges do not have the 400m standard track and field playground, not to mention sports stadium. Most students think that they are paying high tuition fees and deserve teaching equipment. High tuition fees should be used for Sino-foreign cooperative colleges to introduce foreign curricular system, as well as improve teaching environment and teaching equipment. Yet these are not reflected in aspects like sports field facility and sports equipment, which are used for keeping up appearance of colleges. Obviously, Sino-foreign cooperative education is extremely stingy about investment in physical education. Therefore, sports facilities in Sino-foreign cooperative colleges present a seriously highlighted problem of physical education, which will finally have impact on the internationalization of physical education in Sino-foreign cooperative colleges.

5 Conclusion and Suggestion

5.1 Conclusion

(1) Teaching documents are basically complete and standard, but bilingual teaching is not taken seriously. Innovation strength of physical education is not strong enough, which mainly borrows the physical education pattern of regular colleges and fails to show the physical education characteristics of Sino-foreign cooperative education. (2) Motivations for sports of students in Sino-foreign cooperative colleges show a diversified trend. Motivations of taking exercises only for getting PE credits have improved markedly. The great majority of students show special preference to foreign sports and have great curiosity about foreign sports which are not quite popular in China. They are quite enthusiastic about physical exercises. (3) It is not conducive to the sustainable development of colleges that the age structure of teachers is showing the trend of getting younger, professional titles and education degrees are low, teaching experience is insufficient, the proportion of full-time teacher is relatively low, teaching team is instable with high liquidity and teachers training is hesitating and lagging. (4) Inadequate sports field facilities. It can be seen from teaching documents, teachers training and sports field facilities that physical education is ignored in Sino-foreign cooperative schools.

5.2 Suggestion

Improve the system and standardize management, so as to reinforce the macro-management of physical education in Sino-foreign cooperative colleges. Construct a PE teacher's team with high comprehensive quality by improving preferential policies and introducing professionals with high titles and education degrees from home and abroad. Encourage and support PE teachers to participate actively in sports science research and foreign language learning, cultivating a group of teachers with superior teaching ability, high level of English and intense atmosphere of scientific research. Create a good environment for physical education by making rational planning and increasing investment. Implement the policy for education in China and introduce advanced educational concepts from foreign countries, so as to the characteristic system of physical education of Sino-foreign cooperation in running schools. Based on our own country as well as foreign countries, offer more rich and international curriculum systems that are entertaining, practical, bodybuilding and self-determined, achieving the higher education goal of international cultivation.

Acknowledgments. Foundation project: research subject of social science development of Hebei Province in 2010 "The Research on Integral Optimization of Physical Education of Sino-foreign Cooperative Colleges" (Project number: 201003119).

References

1. Zhang, X.: On Innovation of Physical Education Teaching Content from the Evloution of Physical Education Teaching Content Abroad. Journal of Sports and Science 23(2), 10–13 (2002)
2. Qin, B.: Comparative Research on Sense of Physical Education between Foreign Countries and China. Journal of Sports and Science 28(6), 97–101 (2007)
3. Shi, Y., Lei, A.-L.: Comparative Study on Chinese and Foreign Teaching Models of P.E. and the Imperfect Construction of Chinese Ones. Journal of Shenyang Physical Education Institute 24(5), 89–93 (2005)
4. Gong, X., Yin, J.: A Comparative Study of Leisure Sport Cultures between China and Western Countries. Journal of Capital Institute of Physical Education 21(5), 547–550 (2009)
5. Zhou, H.: University College of Physical Education Curriculum Reform of the Status quo and Development Trend of research. Wuhan Institute of Physical Education, Master's Degree Thesis (2007)
6. Zhao, P.: A Research on the Influences of Differential Treatment on Improving College Students'Fitness——A case study of the students in the Taiyuan University of Technology. Journal of Beijing Sport University 33(6), 101–104 (2010)
7. Zhang, Y., Liu, Q., Liu, R.: A Comparative Study and Analysis of Contents and Programs of Chinese and Foreign P.E. classes. Journal of Beijing Sport University 25(3), 385–388 (2002)

Using EDA Technology to Design Comprehensive Experiments of Digital Circuit Course

Shimin Du, Xiangsheng Yang, and Runping Yang

College of Science & Technology, Ningbo University,
315211, Ningbo Zhejiang, China
{dushimin,yangxiangsheng,yangrunping}@nbu.edu.cn

Abstract. The Electronic Design Automation (EDA) technology represents the development direction of modern electronic design technology. This paper analyzes the problems of digital circuit experiment in our university, and proposes a new way which employs EDA Technology to design digital circuit comprehensive experiments to help students improve their comprehensive application abilities of the knowledge of digital circuit. Then, a comprehensive experiment named digital frequency meter is designed by combining the integrated applications of the common MSI such as Multiplexer, decoder, counter and register, which involves a number of knowledge points of digital circuit. The detailed module partition scheme and the implementation circuit of main modules are also given. The actual experiment results indicate that designing digital circuit comprehensive experiments by the use of EDA technology is helpful to improve students' interest in this course and improve their comprehensive application ability of digital circuit knowledge.

Keywords: EDA, MAX+PlusII, digital circuit, comprehensive experiment.

1 Introduction

Digital circuit is an important fundamental course of many electrical information majors such as communications engineering, electronics information science and technology. Digital circuit experiment plays a vital role in helping students to understand the knowledge of digital circuit and grasp its theory and method. However, the digital circuit experiments in most colleges consist mainly of some simple experiments such as voting machine, counters and others mainly constructed with the 74-series components. This approach can help students more easily understand and grasp the knowledge of various circuit units, but it is difficult for students to establish links between them and integrate flexibly them in various practical applications[1][2]. Secondly, because of the restrictions in course hours and chip resources available, the current digital circuit experiments are generally dominated by replication experiments, and the comprehensive experiments are still lacking [3][4][5].

In view of the above problems in digital circuit experiment teaching, we incorporate the modern EDA technology in the experimental teaching process to fully make use of existing experimental devices and EDA software. Several comprehensive digital circuit experiments have been developed based on the MAX + Plus II software

H. Tan (Ed.): Knowledge Discovery and Data Mining, AISC 135, pp. 547–554.
springerlink.com

which is widely used in colleges because of its simplicity and powerful features. Here we take the traditional comprehensive experiment "digital frequency meter" as example to illustrate our exploration in integrating the modern EDA technology into digital circuit's experiment.

2 MAX+Plus II

MAX+Plus II development tool, the 3rd generation of PLD development system of Altera Corporation in the United States is mainly for the design of new devices and large-scale CPLD/FPGA. Designers can adopt their familiar design methods (such as schematic or hardware description language or waveform inputs, etc.) to establish the design, and MAX+Plus II will automatically convert these designs into the final desired formats. Users can easily complete a variety of digital circuits or systems by taking use of the powerful features provided by the MAX+Plus II software in editing, compilation, simulation, synthesis and chip programming, and the design speed is very fast. MAX+Plus II also provides a rich components and model library containing all the 74-series components, which facilitates students to learn the circuit schematic design. In addition, because of MAX+Plus II's powerful digital circuit simulation and timing analysis capabilities, it is also very convenient for designers to perform functional simulation and timing analysis on the developing digital circuits or systems. Finally, with the help of download cable and EDA experimental development system, we can perform hardware testing or verification on the developing digital circuits or systems.

3 Circuit Design of Digital Frequency Meter

As its name suggests, a digital frequency meter is used to measure the frequency of an input signal. In this experiment, the input signal to be measured is generated from a laboratory signal generator. Its amplitude varies between ±5V, and its waveform can be sine, triangular or rectangular. The frequency range of input signal is 0.1 kHz~10MHz. We need to design a circuit to achieve the measurement of input signal' frequency and display the measured results on the digital tubes in a dynamic scanning manner. The frequency measurement method adopted is counting [8], that is, the frequency of input signal is obtained by calculating the number of pulses in one second. According to design requirements, we divide the entire system into 6 modules, and they are pulse generation module, control module, signal shaping module, pulse counting module, storage module, and dynamic display module. Fig. 1 gives the block diagram of entire system and relationships between these modules.

In Fig. 1, the signal shaping module is used to convert various input signal waveforms (such as triangle wave, sine wave, and so on) into the rectangular wave with the same frequency. The pulse generation module produces different frequency clock signals required for other modules. The control module is used to generate the various control signals required for subsequent modules, such as the enable and clear signals for pulse counting module, and the clock signal for storage module. The clock signal of pulse counting module comes from the output of signal shaping module.

The counting result will be delivered to the storage module after the counting is finished. The dynamic display module send the counting result to digital tubes in a dynamic scanning way.

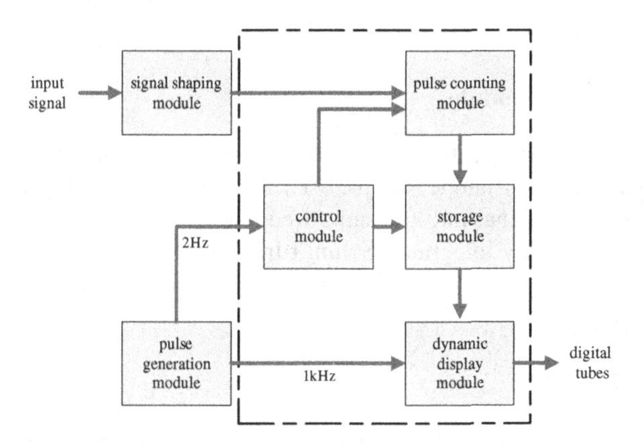

Fig. 1. Block diagram of digital frequency meter

In order to make full use of existing EDA experimental conditions and to save chip resources and design time, the modules in dash-dotted frame will be implemented on EDA experiment development system while other modules (including the signal shaping and the pulse generation modules) will be built by students themselves through choose appropriate chips. In this way, the experiment contents will be more comprehensive, and more helpful to train students' integrated application capabilities and practical operation capabilities.

3.1 Pulse Generation Module

As mentioned above, this module is used to generate various frequency clock signals required for other modules, including the 2Hz time base for control module and 1 kHz scanning signal for dynamic display module. Its circuit diagram consists of a 14-stage ripple carry binary counters CD4060, a crystal oscillator (the crystal frequency is 32768Hz) and several RC components, as shown in Fig. 2.

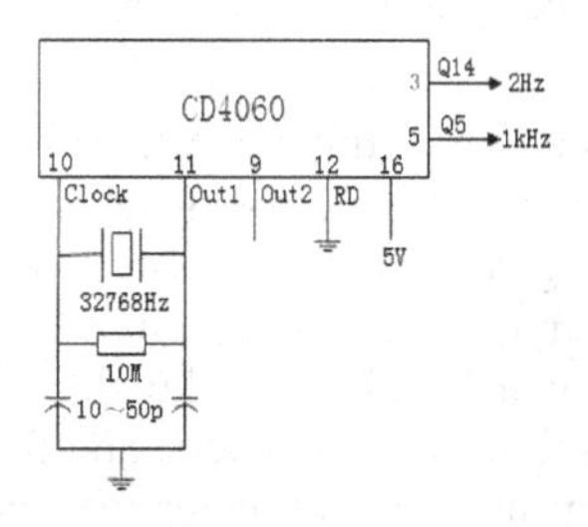

Fig. 2. Circuit diagram of pulse generation module

As the 2Hz time base, the clock signal of control module, has a high demand for frequency accuracy, the oscillation signal is generated by a crystal oscillator because crystal oscillator has very high frequency stability. The oscillation signal is divided multi-stage by CD4060, and the required 2Hz and 1 kHz clock signals can be obtained at pin 3 and pin 5 of CD4060 respectively.

3.2 Signal Shaping Module

According to design specification, the waveform of input signal be measured can be non-rectangular (such as triangle, or sine). Therefore, it is necessary to transform this non-rectangular-wave signal into the same frequency rectangular wave signal. In this paper, this is achieved by integrated Schmitt trigger 7414 chips. The output of 7414 will be an ideal rectangular wave and it can be delivered to pulse counting module as input clock signal.

3.3 Control Module

The method of frequency measurement adopted in this paper is counting. The frequency of input signal is obtained by calculating the number of input signal pulse per second. Therefore, some control signals need to be generated in this module. Firstly, we needs to generate a 1 sec pulse width signal, and take it as count-enable input signal of pulse counting module. Secondly, after the counting process is finished, a latch clock signal needs to generate to store the counting result into the registers of storage module to keep the stability of display data. In addition, after storing the counting results, a clear signal is required to reset all counters of pulse counting module to prepare for the next counting. Accordingly, we can get this module's working timing sequence, as shown in Fig. 3.

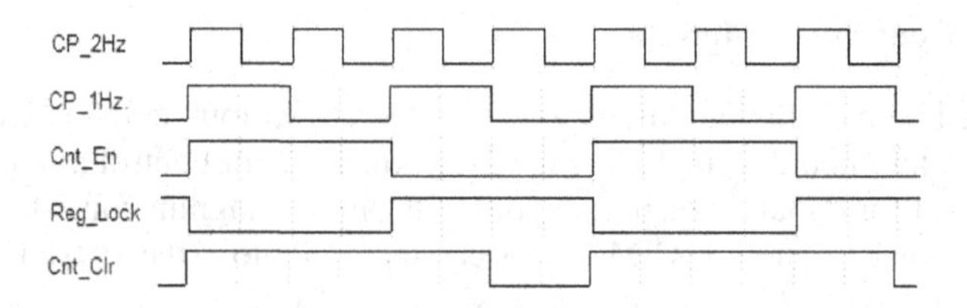

Fig. 3. Working timing sequence of control module

Fig. 4 shows the circuit diagram of this module's. Where the input signal *CP_2Hz* is the input clock signal generated by pulse generation module, the output signal *Cnt_En* is the count enable signal with 1sec pulse width and is obtained through dividing the *CP_2Hz* twice by frequency-halving circuit which consists of a D flip-flop and an inverter, the output signal *Reg_Lock* is the clock signal of storage module and used to store the counting result into the registers, and the output signal *Cnt_Clr* is the reset signal of pulse counting module. The latter two output signals can be obtained by simple combinational logic operations with *Cnt_En* and *CP_1Hz*.

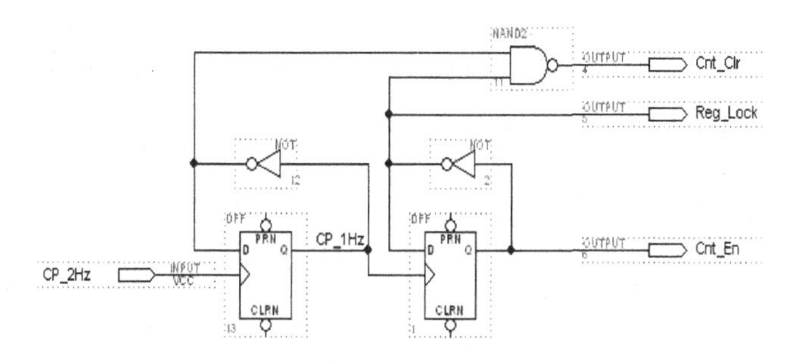

Fig. 4. Circuit diagram of pulse generation module

3.4 Pulse Counting Module

This module is used to count the input signal pulses after shaping. It consists of eight decade counters 74160 in synchronous cascade way. The start/stop signals of these counters are controlled by the control modules output *Cnt_En*. When the *Cnt_En* signal is high, these counters begin to count, and vice versa. When the counting process is finished and the count results are stored, the signal *Cnt_Clr* from control module will clear all counters to prepare for the next counting.

3.5 Storage Module

This module is used to store the counting results of pulse counting module, and consists of four 8D registers 74273 with an asynchronous clear signal. These registers are clocked simultaneously and droved by the *Reg_Lock* from the control module. When the counting process is finished, the control module will drive the output *Reg_Lock* a low-to-high transition and the counting results will be locked into 32-bit registers. As pulse counting module and storage module circuit diagrams are relatively simple, they are not given in detail due to limited space.

3.6 Dynamic Display Module

The circuit diagram of this module is shown in Fig. 5. Where input *Scan_Clk* is input clock signal, input *D[31..0]* is the counting result from the storage module, output signals *BitSel[0..7]* and *Seg[0..6]* are 8 bit selection signal and the 7-segment display code respectively. It can be seen from the figure that the entire circuit consists of an octal counter which is made up of an NAND gate and a 74160 counter, four 8-line to 1-line multiplexers, a 3-8 decoder and a 7-segment display decoder 7448. The octal counter used to generate the address signal for the 3-8 decoder and four multiplexers. Four multiplexers is used to select 4-bit data from the 32-bit input *D[31..0]* and output them to the 7-segment display decoder 7448. Decoder 7448 converts the input 4-bit data into corresponding 7-segment display codes. When the octal counter counts periodically from the "000"-"111", *BitSel [0..7]* which drives the bit select lines of digital tubes will output its each bit to low level in turn, and *Seg [0..6]* outputs the corresponding 7-segment display codes to digital tubes at the same time. Thus, the frequency measurement results can be dynamically displayed on the digital tubes.

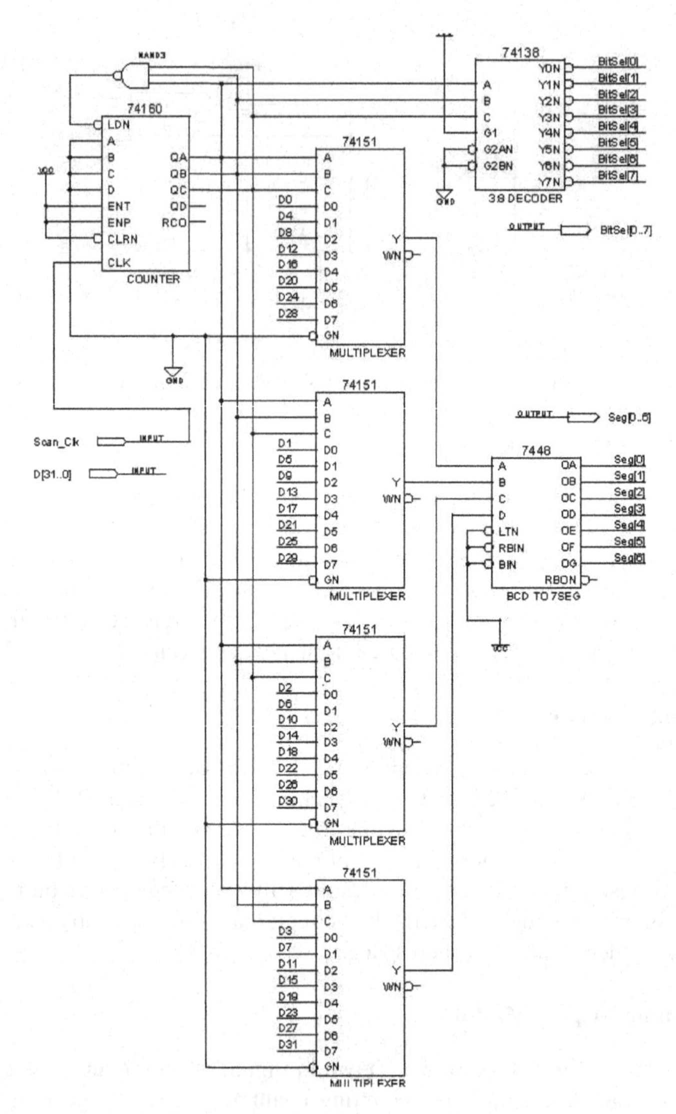

Fig. 5. Circuit diagram of dynamic display module

3.7 Whole Circuit of Digital Frequency Meter

After finishing the design of these modules mentioned above in MAX+Plus II, we can assembly them into a whole system according the block diagram shown in Fig. 1. The whole circuit of digital frequency meter is shown in Fig. 6. Where, *Fn* is the input signal to be measured after shaping by signal shaping module , *CP_2Hz* and *CP_1kHz* are two input clock signals generated by pulse generation module, and ctl, *counter, reg32b and scan_disp* represent control module, pulse counting module, storage module and dynamic display module respectively.

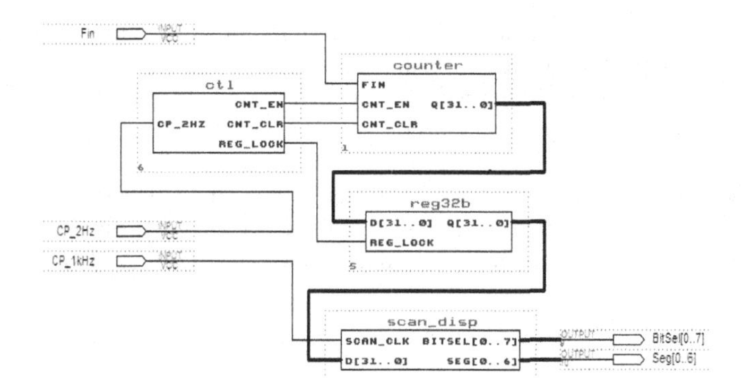

Fig. 6. Whole circuit of digital frequency meter

4 Software Simulation and Hardware Verification

MAX+Plus II features powerful simulation and analysis capabilities. Therefore, during each phase of the design process, we can exploit MAX+plusII simulation tools *Simulator* to carry out functional or timing simulations on our designing circuits to ensure logical correctness. This is one of the significant advantages of EDA technology. Fig. 7 shows the simulation waveform of dynamic display module. It can be seen from figure that the designed circuit can correctly implement the desired logic function. As other module's simulation are similar, and no elaboration is given here.

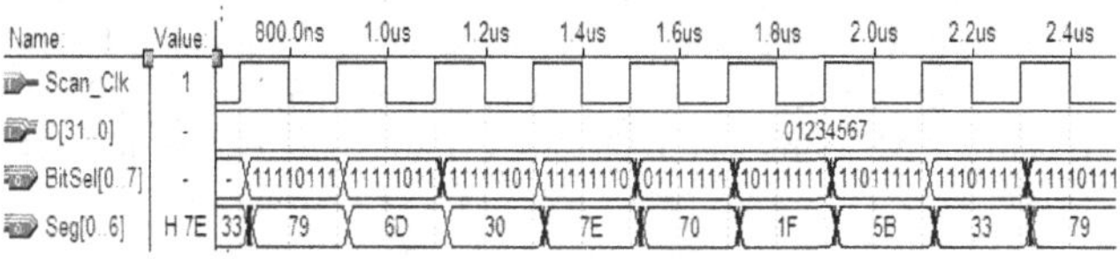

Fig. 7 Simulation waveforms of dynamic display module

After the software simulations of all the modules in MAX+Plus II have been finished, we can perform hardware verification on the designed circuit. The first step of hardware verification is to assignment a pin for each I/O port of the top-level circuit shown in Fig. 6. Then, we need to download our circuit into the EDA experimental development system through the download cable. After that, the signal to be measured is introduced to our designed circuit through port *Fn*. The actual circuit verification shows our designed digital frequency meter runs normally and achieves the desired requirements.

5 Conclusion

Exploiting EDA technology to design comprehensive experiments of digital circuit is a new experimental teaching method of modern digital circuits. We have gradually established this new experimental teaching pattern by integrating the EDA technology into the teaching practice in recent years. This pattern can effectively improve students' abilities in exploiting comprehensively their knowledge to solve practical problems, because many commonly-used components and design methods will be used during the design process. It also can help students become familiar with and grasp the most advanced circuit design methods and skills. The actual student experiment results also indicated that the comprehensive experiment based on EDA technology not only deepen their understanding of the course contents, but also promote their interest in this course.

References

[1] Tian, J.-Y., Xia, L.-Y.: Teaching Practice of EDA-Based Electronic Technology. Theory and Practice of Education 25(6), 54–55 (2005)

[2] Jiao, J.-S.: Applications of EDA Technology in the Education of Digital Electronic Course. Modern Electronic Technology 30(10), 168–169, 173 (2007)

[3] Xu, X.-M., Guo, Z.-L., Xing, X.-F., et al.: Reform of Digital System Course Design Based on EDA. Experimental Technology and Management 24(3), 25–27 (2007)

[4] Huang, Q.-Y.: Design Type of Experiment Study on Digital Circuits by Using EDA Technology. Semiconductor Technology 31(1), 19–22 (2006)

[5] Cui, G.-W., Li, W.-T.: Exploration and Practice of a New Digital Electric Technology Course Design Mode Based on EDA Technology. Experimental Technology and Management 25(1), 123–125 (2008)

[6] Zhang, Z.-L., Cai, Y.-Q., Wang, Y.: Design and Realization of Digital Frequency Meter Based on FPGA. Process Automation Instrumentation 27(11), 10–13, 17 (2006)

[7] Tang, X.: Practice of Open Comprehensive Experimental Teaching Using Modern Electronic Technology. Experimental Technology and Management 24(9), 35–39 (2007)

[8] Lin, Z.-J.: Electronic Measurement Technology. Publishing House of Electronics Industry (2007)

[9] Yarbrough, J.M.: Digital Logic Applications and Design. China Machine Press (2002)

Improve Practice Engineering Teaching Ability of Young Teachers by School-Enterprise Cooperation

Dong Guo, Lei Mao, Huilan Sun, Bo Wang, Xuezheng Zhang, and Yunfeng Yao

College of Material Science and Engineering,
Hebei University of Science and Technology, Shijiazhuang, 050018, China
guodongbill@yahoo.com.cn

Abstract. Lack of practice engineering teaching ability of young teachers is quite common in many colleges. It is useful to select school-enterprise cooperation improve the ability by school-enterprise cooperation, and this mode has a great effect on individual and colleges. Opportunities should be created by colleges to enable young teachers achieve the training. On the other hand, young teachers should locate themselves appropriately and understand their responsibility sufficiently.

Keywords: school-enterprise cooperation, practice engineering teaching ability, young teachers.

1 Introduction

In 1987, national education committee clearly proposes that practical teaching should be enhanced, and promotes comprehensive reform of teaching content and teaching methods. But today there is still a long way to go. Now students training objectives in many colleges were defined as "applied technical personnel". It requires a solid theoretical knowledge as base of practical ability. It is particularly important for engineering specialties that has higher professional and practical requirements is for. In recent years as the sustained and rapid development of economy, the demand for high quality engineering and technical personnel continued to increase. It is pointed out in "The outline of reform and development on China's education" that it is relied on education that we develop our nation, and it is relied on teachers that we develop our education. Teacher is educational root; the level of quality of teachers comes to a decision the teaching mass of the school and the development mass of the student. Colleges and universities in China developed rapidly in recent years. Scale of school-running expanded rapidly. A large number of young teachers entered teachers' troops in the high school, and become main force gradually. As the major force in teaching, the level of quality of young teachers relates to the training quality of talents in colleges and universities and even the university's development and future. It has great significance to enhance the building young teacher team. School-enterprise cooperation could adapt the development needs of economic and social to higher education, it is effective

H. Tan (Ed.): Knowledge Discovery and Data Mining, AISC 135, pp. 555–560.
springerlink.com

teaching methods to cultivate high quality talents with innovation spirit and practical ability, it is the only way to serve society and promote social development for higher education, and it is the only way to improve their teaching of scientific research for young teachers as soon as possible.[1] For education ability of young teachers in colleges and universities, this article started with the improvement of young teachers education capacity from the education of in colleges and universities, offered to combine industry, learning and research to promote development practice engineering ability of young teachers.

2 What Is the Practice Engineering Ability?

The definition of practicing ability is very broad; in brief it could be understood as the ability to practice. In particular, refers to the ability of subjective purpose, consciously transform object. Subject is a person with initiative; object is subject to awareness or modification of an object. Practice is formed and reproduced in the practice. That means only through practical activities could teacher's practical ability show.

3 Current Situation of Practice Ability of Young Teachers in Colleges and Universities

At present young teachers in colleges and universities mainly born in the 1970 later and the 1980s. They grew during China's reform and opening up. They are excellent part of young people. They are inclined to pragmatism, ideologically independent activity. Orientation of their values on life is diversified. Their lives are full of personality, and their demands are varied. Its main features: first of all, they were in their prime and full of vigor. They occupied an absolute time and space advantages and were in a period of memory, understanding and strength of the best in life. Secondly, their psychologies are tending to maturing, that senses of occupational and professional attitudes are maturing. Thirdly, they have pioneering and innovative intellectual characteristics. [2] Relative to the middle-aged teachers, most of young teacher have a master and doctor degrees background, solid expertise and reasonable knowledge structure. They have active mind, strong learning ability and accept new knowledge fast. They have a strong sense of identity to students groups. They are easy to communicate with students. They could master and use modern information technology expertly. But on the other hand, there are clear disadvantages for young teachers. Most of them are only a "home-school-school", namely, the "hss" teacher. Although most of them have higher degree, seldom of them has thorough experience of normal education. They are lack of experience to industrial enterprises to participate in engineering design, technological transformation, accept re-education. They are not deep enough understanding on strong practical courses. They are difficult to manage the classroom atmosphere, professionalism needs to be improved.

4 School-Enterprise Cooperation Is Effective Way to Improve Practical Teaching for Young Teachers and Scientific Research Level

School-enterprise cooperation refers to professional teachers to enter the joint venture, participating in the production, research, product promotion, and service all aspects of engineering practice. This is profitable whether for enterprise or university, both from a process or the result is a win-win. Teachers provide technical support for enterprise, improving the competitiveness of products. Teachers could overcome weaknesses of "separation of theory and practice" through participation in practice and scientific activities at different levels. They could adjust teaching content timely, improving the quality of its academic quality and teaching. School-enterprise cooperation plays an important role in improvement the teaching ability of young teachers. [3]

4.1 Improve the Ability of Practice Teaching of Young Teachers

Young teachers should familiar with the practice process of work-integrated. It could guarantee the quality of practical teaching. So it could be accomplished that teaching tasks of practice and training targets of practical technical talents. Young teachers carry out technical research and technical service and other forms of research projects with enterprise. It could make them always in the forefront of industry. They would bridge the gap itself in terms of theory with practice, raising the level of technology and scientific research. They participate in the actual business of product development, design and other activities, improving their level of education and teaching theory and practice, so as to achieve the purpose of improving teaching ability. [4]

4.2 Cultivating Excellent Talents with Creative Ability

Mode of training young teachers is the premise and foundation of student training model. Young teachers cultivated by school-enterprise cooperation must have strong research and practice of teaching ability. They could be competent for complete practical training of technical personnel well. During young teachers and enterprise contacts, the links of their personal as well as schools and enterprises could be enhanced. It could lay a good foundation for students outside of practice teaching, and create good places for students' practical skills and more practice opportunities. It is certainly helpful to alleviate contradictions of the disjointedness of theory and its practice in the undergraduate teaching process currently widespread. It is very beneficial for the cultivation of creative talents.

Teachers can access to different levels of scientific research via school-enterprise cooperation. And it would enable students to understand frontier knowledge and developments via teaching process.

4.3 Broaden Channels for Student Employment

Young teachers could understand dynamic development of industry demand and demand for personnel via the process of participation practice exercises. They could adjust direction and focus of training students purposefully and make the profession more features, increase the students' employment chips. At the same time, teachers can introduce the best graduates to the enterprise, to broaden channels of employment for students.

5 Colleges Actively Create Conditions for Growth of Young Teachers

Young teacher the most lacking is experience, and experience must be accumulated through practice exercises bit by bit. Therefore, in the process of training young teachers should take "wide cultivation and fine selection" ideas. On this issue colleges should be willing to long-term investment, not a quick success, for growth of young teachers to create a relaxed environment.

Firstly, colleges should cooperate with related enterprises according to their professional and characteristics and help young teachers into the enterprise to work. At the same time, colleges should alleviate burden of job evaluation of young teachers, encouraging them to get out of the school and approach enterprise and production site, maintaining regular communication and contact, establishing ties and communication channels for school-enterprise cooperation. Young teachers carried out social investigation of industry to understand their current status and trends in technical, technology and equipment, so that supplement new technology and new processes of production site. [5]

Secondly, colleges should give teachers more freedom on relevant curriculum development. For example, encourage enterprise experts to involve in syllabus modification and preparation of teaching materials, so it could be ensured that practicability, progress and integrity of course content. We take competency standards in different stages as teaching goals and adjust timely teaching order according to actual situation. So we could organize production practice pertinently.

Thirdly, colleges should take active measures to encourage young teachers out of school, for enterprise and actual production situation. They carry out initiative in technology services, take on enterprise research projects. So theoretical knowledge should be turned into practical productive forces. Some young teachers could accumulate more professional solid theoretical foundation, provided with a certain research and development capabilities.

6 Young Teachers Should Correctly Deal with School-Enterprise Cooperation

Colleges contact enterprises to create conditions for cultivation of young teachers. But whether it could be long-term sustainability depends primarily on young teacher's own conditions and level of effort. This training model has higher requirements for teachers, young teachers should be strict with themselves, to correctly understand and treat school-enterprise cooperation.

Firstly, young teachers who are in business training process must not arrogant, they must actively devote themselves to thinking, down to earth in enterprise learning, promote self-training.

Secondly, young teachers should start from enterprise, find research projects from market and put their focus on research applications. Young teachers consult engineers and technicians for advice with some subjects in teaching. So they could improve their abilities of popularization and application of new technologies, research and development capabilities.

Thirdly, young teachers should establish extensive contacts with enterprises, keeping information contact with enterprises at any time and accessing business information timely. Therefore, students could learn the latest knowledge and information in classroom.

Fourth, young teachers should make every effort to make enterprise more closely cooperation with colleges. Young teachers should understand their aims not only raising their personal quality during school-enterprise cooperation, but the link between colleges and enterprises. They should always keep in mind what they are the image of college representatives. It requires young teachers solve real problems for the enterprise under support of college and obtain enterprise's trust. Gradually enterprise's dependence on intellectual and human resources of college is increased. Ultimately truly achieve mutual integration of school-enterprise has been built. This is also the key that school-enterprise cooperation could be sustained.

Enterprise is both a place for using and fostering talents. School-enterprise cooperation is an effective way to train for young teachers on duty. It is significant to colleges and enterprises to build a long-term platform for cooperation. It should promote the formation of long-term working mechanism and technology innovation. It has also important significance to cultivate young teachers and students in innovative areas. In order to strengthen job training and research work of young teachers, teachers training program on duty should be planned actively. Stable research base and platform for school-enterprise cooperation should be created. It should be established that a sound operational mechanism, evaluation mechanism, protection mechanisms and incentives mechanisms. Policy levers should be carried out to encourage teachers to tempering in enterprises, research institutes, government departments and other part-time practical training, this experience can be gradually considered as important basis of a teacher for job appointment, evaluation.

7 Conclusions

Under the background of increasing demand for high quality engineering and technical personnel in current community, development orientation of college students has been "application-oriented technical personnel". This requires colleges to change the current practice of teaching ability of young teachers are generally weak in reality, and school-enterprise cooperation is an effective way to change this situation. Colleges should create conditions in the process of training young teachers and establish communication channels actively. Young teachers should connect with their own characteristics, improve their practical teaching ability. In this process, young teachers should clear their positions and responsibilities. Through this effective way young teacher could achieve their aim to enhancing practical teaching ability and improving research capacity as well as its position in colleges.

References

1. Yang, S.Z., Zhang, F.R.: Foundation of innovation lies in practice. Researches in Higher Education of Engineering 2, 9–12 (2001)
2. Zou, M.: Study on comparison between middle-aged and young-aged teacher in college. Journal of Leshan Teachers College 11, 30–33 (2008)
3. Zhang, H., Ren, C.Y., Zhang, B.: Young teachers to enhance practical ability to explore. Journal of Social Science of Jiamusi University 27(4), 88–89 (2009)
4. Jiang, Q., Zhu, J.M.: On the construction of young college teacher's education capacity. Journal of Hefei University of Technology (Social Sciences) 23(6), 28–31 (2009)
5. Zhang, X.P., Yu, Y.C.: An integration of research and industry into instruction to promote young teachers at-work training. Journal of Ningbo University (Educational Science Edition) 32(4), 100–103 (2010)

On Countermeasures and Dilemma for Higher Engineering Education Infiltrating Humane Education in China

Xiaoping Li[1], Yan Gao[1,*], Huilin Chen[2], and Lifang Cheng[3]

[1] Training Department, Air Force Radar Academy, Wuhan, China
[2] Department of Politics, Air Force Radar Academy, Wuhan, China
[3] Basic Courses Department, Air Force Radar Academy, Wuhan, China
ga0yan@163.com

Abstract. Aiming at the internal requirements of humane education in engineering education, this paper reveals the problems and causes in the fusion process of current college engineering education and humane education. Besides, some new thoughts and approaches are also presented on how to deepen the practical hot issue about the fusion process of current college engineering education and humane education from the following aspects: transformation of viewpoints on teaching and knowledge, innovation of administrative system in college teaching, construction of curriculum system with a cross of engineering and humanity, enhancement of humanity makings of teachers in engineering courses, and establishment of campus culture with abundant inside information on humanity.

Keywords: Engineering education, Humane education, Fusion.

1 Introduction

In recent years some large-scale engineering construction is not harmonious with the environment, and even destroys the environment engineering construction, causing shortage of resources, resulting in cultural recession, fully exposed engineering and technical personnel in scientific practice, pursuing benefit and being eager to success. And they pay little attention to engineering ethics and sustainable development. The awareness is weak in the humanities and social sense of responsibility. Therefore, science and engineering university must attach great importance to the humanities education, natural science, engineering technology with humanities and social science, professional education in fusion And at the same time the training of various development of humanities and social for science students is needed, so that the university can train qualified engineers who adapt to the development of contemporary engineering in the future.

* Corresponding author.

H. Tan (Ed.): Knowledge Discovery and Data Mining, AISC 135, pp. 561–566.
springerlink.com

2 The Characteristics of the Humanities Education in Project University Education

Shuzi Yang points out, "Without scientific, humanism will be incomplete; also without humanities, science foundation and Jane; also, no raw human science, is also incomplete science, science is not perfect. Science must be attached to humanistic spirits and it seeks truth, certain... must have the sense of responsibility. The humanities; for good, there must be a good sense of responsibility for both which will undoubtedly form the sense of responsibility promoting social progress "[1]. Higher engineering education of humane education has both the shape the spiritual world, puring people's feelings and providing people the value realization methods of the humanities education with the generality and its characteristics.

First, it is the humanistic education of higher education. It has the resources of the need to develop potential, and transformation of higher education, which can use the particularity and the advantages of the humanities education, such as through setting text, science, engineering, other disciplines mutual infiltration and interdisciplinary curriculum to use the comprehensive interdisciplinary research activities, to diversify campus culture and cultivate students' humanistic qualities.

Secondly, it is the project education of humanistic education; it is the way of permeability, which should be different from other types of higher education of the humanities education. Engineering education of humanistic education, should be focused on engineering problems, and engineering talent on the quality of the training altogether.

Third, its effects of the humanities education is gradual, is mainly focus on the problem of how to do things, which have a vital role in comprehending the development of the man, but because this effect in colleges are not obvious, to a degree it will be blurred by the achievements in study, so that the real effects can not be easily showed out which makes teachers and students' humanities social sciences of knowledge and learning passion become decreased. Colleges and universities must overcome blundering thoughts, resolutely abandon that only for immediate interests regardless of the short-term and long-term needs of humanity, narrowing views. They are obliged to attach importance to humanistic education unremitting.

3 The Problems and Reasons of Present Project College Education in Humanities Education

At present, project education in humanities education concept basicly establishes in the university. Though engineering colleges integrate the science and technology education with the humanities education in teaching practice, but the main form is to open some special lecture and elective courses (such as science), or combined with a scientific theory, the short time to explain human to help students to understand the cultural background knowledge of science and engineering college students, and so on, contains still exist in the humanities deficiency of the outstanding problems. Refering to the surveys in Chinese IT industry, manufacturing, construction, transportation, post and telecommunications, education research and many industries, there are 44.0% of employers consider they hired engineering graduate are lack of scientific attitude and spirit (such as rigorous, careful attitude, dare to doubt, criticism and innovative spirit, etc), 40.2% think that its engineering consciousness (such as economic management consciousness, ethical consciousness, moral consciousness, etc) and innovation consciousness, 38.0% think the graduate are lack of recognition, formation and they can not solve engineering problems. In addition, they think engineering graduate can't use engineering practice and the necessary technology and modern tools effectively, they don't have proficient operation experiment and analysis of the data and explanation.Their independent working ability is not strong, language and written communication ability is weak, lifelong learning literacy is not beautiful, active learning enthusiasm can not last for long .The percentage of choosing and employing persons also accounted for a significant portion [2]. Visibility, deepening humanistic education, improving the overall humanistic quality comprehensively is Chinese important mission in engineering education currently.

Currently, humanities education in engineering education of university do not have substantial effects for many reasons, the main reason is:

First, there is the "demand" but not "security". In 1998, the ministry of education issued: the higher school should strengthen cultural quality of education, and universities have introduced corresponding measures of enhancing humanistic education. However, in the teaching courses whether educating humanities depends on teachers' personal interest. And most universities don’t have systems and clear requirement, the ability level of humanistic education, and haven’t included humanistic education in the engineering course syllabus, also not in the teacher evaluation system. Moreover, incentive mechanism is not perfect, restricting the teachers for reforms, restricting enthusiasm and creativity, the determination to explore the humanities education practice for teachers.

Then there's "enter" but not "integration". Humanistic education now has entered "into" engineering education, in the idea, the requirements, lesson plans, interpretation, but embodied, this reflects is only as "adornment ", and the course content and teaching process relatively separated, which is not really fused into teaching, the humanities education mostly stay in the level of the teachers, and their teaching management knowledge and the level of idea, without reaching the depth of "practice". Some engineering college students are influenced by the thought of social utility and by the pressure of employment situation, the study presents the obvious "possession of sexual orientation", paying attention to the function of the value, neglecting the connotation of knowledge, even the humanity of knowledge is ignored.

Third, there is only the "pursuit" and not "method". Because the individuality of teachers is very strong, they have their own autonomy in teaching design and teaching implementation. As how to strengthen the education of humanistic education project, the key depends on the penetration of the pursuit, interest and ability of teachers. Many teachers are pursuing for the real engineering education embodying humanity education, however, because of the theoretical guidance is weak, especially the method is not enough, many teachers are difficult to combine concepts into practice, they are difficult to embody humanity education in the engineering education [3].

4 Deepen the Fusion of University Education and the Humanities Education Strategy

To truly realize the fusion of project education and the humanitic education, strengthening the education and culture in teaching, universities have to completely change the condition of university that teaching activity of technology education and the humanitic education are separated, constructing a holistic, organic teaching activities to develop the teaching level system, methods and strategies of research.

4.1 Change the View of Teaching and Knowledge

In order to provide overall human nature "nutrition" with the scientific knowledge engineering education for university students, we must think from the view of teaching understanding, view of scientific knowledge and college course and so on to overcome the tendency of ideology and one-sided science, in order to establish a complete harmonious environment of education.

First of all, teaching understanding view should transform from the "scientific epistemology" to "life understanding". The orientation of "scientific epistemology" teaching method will make humanities generated resources lose, evaporate and make the composition course come into a single objective logic system. Therefore, universities must change their narrow scientific understanding in teaching method, establishing a comprehensive view of "life understanding ".

Secondly, the scientific knowledge view must refer to both "the dominant knowledge" and "the tacit knowledge". Knowledge or knowledge composition have "dominant" and "hidden" points, this "hidden" knowledge is proposed by the famous British scientist and philosophers Polanyi, he proposed "tacit knowledge" or "personal knowledge", "general people always think conversation knowledge is all human knowledge, but in fact it is just a huge iceberg out of the water, and the little spire tacit knowledge is hidden in the grand part." The dominant components in scientific knowledge, apparently, is not directly related to the humanities composition, its humanistic ingredients often belong to "sense" and "hidden" knowledge category.

4.2 The Reform of Teaching Management System

The improvement of college teaching management system, making it is beneficial for the development of humanitic education in engineering education. Firstly, we should perfect credit system, in order to make credit as the learning tie, promoting and strengthening the fusion of scientific education and humanistic education. Credit is the calculating unit in curriculum teaching. This is a kind of suitable educational system for the advanced education in teaching management system, credit system teaching plan is the embodiment of the fusion of professional education and humanities education. Second, implement and gradually improve the Lord minor system and the double major system. Such as MIT encourages conditional students to study engineering and humanism as across content and minor study. Third, implement the second bachelor's degree, engineering students study across the engineering and humanism,so that they can get a second bachelor's degree or double a bachelor's degree. Fourth, in the management for students, universities should break the traditional management and conduct of activities, such as MIT arranges accommodation across different subjects, organizes activities for students. It create necessary conditions for the interact communication for all students.

4.3 Build the Cross System of Engineering and Humanistic Curriculum

There exists two drawbacks in course structure system for a long time: a unilateral emphasis on training specialist which "is for training professionals," they set the professional class, making the smaller proportion of humanitic educational courses. The second aspect is that the course system structure is in a "closed the body", which not only separate professional education and the humanities education, but also close up the professional course structure system [5].

Colleges and universities must actively promote the penetration of arts and projects, supporting the development of the fusion of promising humanism and nature of science in the edge subject and emerging profession. In the settiing of curriculum, they should put the humanistic education and engineering education in the equal palce, they must reflect the common characters of university education but also reflect the professional characteristics in curriculum system. Through the blend of various disciplines, restructuring and adjustment of subject, the integration of professional talent, rationally distributing the lessons, setting a reasonable proportion of professional curriculum and humanistic education course, guiding students choose suitable humanistic courses according to their own professional content.

4.4 Improve the Humanistic Quality of Engineering Teachers

Only engineering teachers possess a deep knowledge of humanism, professors infiltrate humanistic knowledge to the student, and spread good moral quality, noble personality charm and perfect personality strength exerts a subtle influence on students. Therefore, the school must attach great importance to the humane quality of engineering science teacher cultivation and training work, and must actively explore the way and mechanism to improve the quality of university teachers. In addition, the teacher must explore the humanistic connotation in professional courses and the teachers should penetrate the spirit of humanism. Such as the professional teachers in building subject, must not only teach students the construction technology, but make them understand the requirements of people, understand the relationship between the building and the surrounding environment design. They need to know build a warm building full of humanism can not leave the humanistic quality.

References

1. Yang, S.: Walking out of Half-man Age—Discussion on General Education and Cultural Quality Education by Scholars from Mainland Taiwan and Hongkong. Higher Education Publishing Company, Beijing (2002)
2. Li, Q., Zhao, Y.: An Investigation on Practical Abilities of College Students: Visual Angle of Employers and Graduates. Development and Evaluation of College Education (1), 106–115 (2008)

3. Li, X.: A Report on Study of Fusion of Science and Technology Education and Humane Education in Science Teaching (2008)
4. Huang, R.: Scientific and Humane Approaches of Polanyi. Natural Dialectics Communication (2), 30–37 (2000)
5. Liu, Y.: Investigation on Implement of Humane Education in Higher Engineering Education—A Comparative Study of MIT and HUST. Dissertations of Master's Degree in Huazhong University of Science and Technology (2007)
6. Zhang, M., Chen, F.: Develop STS Education to Promote Fusion of Liberal Arts and Science in Education within Universities of Science and Engineering. Higher Education in China (19), 19–20 (2001)

Discussion on Training of Engineering Students' Ability of Professional English Application

Yufeng Duan[1] and Zhaoxia Fu[2]

[1]School of Material Science and Technology, Hebei University of Science and Technology, Shijiazhuang, Hebei, China
[2] Office of Science and Technology, Hebei University of Science and Technology, Shijiazhuang, Hebei ,China
duanyufengz@yahoo.com.cn, fzx68@hebust.edu.cn

Abstract. This paper puts forward a reform model for engineering college English teaching considering that present college English teaching is weak in training of engineering students' ability of professional English application in China. Present college English course repeats high school English course a lot in teaching content, teaching goal and teaching requirements. The most important evaluating indicator for college English teaching and learning is the passing rate of CET (College English Test) -4 and CET-6. But that indicator is far from evaluating the ability of professional English. The results of questionnaire indicated that there is a great gap between the learned skill in college English and the requirements of practical work. It is suggested that college English teaching for engineering students be shifted to training students' ability of professional English systematically in whole four years college days from mainly focusing on basic English in first and second years of college.

Keywords: Engineering students, ability of professional English application, college English teaching, reform model.

1 Introduction

"Which course took the most time and energy during your university time?" If you ask this question to a person who graduated from a engineering university in mainland China, we are sure that the answer is English course. But if you ask them if their English skill comes in handy in the career, we are afraid that most of them will shake their heads. Present college English courses take a lot of teaching and learning resources, but the training for English application ability of the students, especially the ability of professional English application, is ralatively weak. It seems that the only goal of university English teaching and learning is to pass CET-4 or CET-6. Such a situation has existed for many years and needs to be reformed urgently.

As the global economic integrating more and more tightly, the English for specific purpose (ESP) education attracts increasing attention[1][2] . Some Chinese famous scholars also called for reform in college English teaching [3-6]. But the reforms still are ineffective by now, especially to engineering college English teaching.

H. Tan (Ed.): Knowledge Discovery and Data Mining, AISC 135, pp. 567–572.
springerlink.com

2 Present Situation of Engineering College English Teaching

At present, the English courses that engineering students take in university period consist of three parts: basic English, specialized professional English and specialized courses by bilingual teaching. The schedule of present engineering college English curriculum in 4 years is shown in Table1. It can be seen that the class hours of university English teaching mainly focus on basic English course. At this stage, the most important index of measuring teachers' teaching level and students' learning degree is passing rate of CET-4 or CET-6. Excepting college English courses finished in in low grades, there is only a very few English courses during the university time. The training aiming at the application abilities of professional English is very weak. Class hours of specialized professional English are very few. What is more, there is nearly no reformations of teaching content and teaching methods in specialized English course in the past decades. By now, the main task of specialized English course seems to be trainning students' ability of English-Chinese translation concerning professional information.

Table 1. Schedule of present engineering college English curriculum

Courses	Class hours	Opened time
Basic college English	240	The first and second year
Specialized professional English	30	The third year
Specialized course by bilingual teaching	30-50	The third year

In 2001, China's Ministry of Education promulgated a policy that 'Suggestions on strengthening teaching work of undergraduate courses and improving teaching quality'. They suggested to introduce original edition of foreign teaching materials and to use foreign languages in common courses and professional courses. Now, however, bilingual teaching is still in limited scale although ten years past. Generally, one major only has one bilingual teaching subject at present university and the class hours also are few. So there are a lot problems existing in the current college English teaching on how to train students' practical ability of professional English.

3 Existing Problems in Engineering College English Teaching

3.1 A Lot of Repeats Comparing with High School English

The present university basic English course repeats high school English course a lot in teaching content, teaching goal, and teaching requirements[7]. For example, the goals of college English teaching and high school English teaching both are to cultivate students comprehensive ability using forein language. The English vocabulary requirement for high school graduates is 3300-4500 and for undergraduate is only 4500 [7]. As the high school graduates' English ability improves year by year, this problem will become increasingly serious.

Since several years ago, some experts have called for college English content transferring from English for General Purposes (EGP) to English for Specific Purposes (ESP) [3][4][6] , but the situation has no obvious difference by now.

3.2 There Is a Great Gap between College English Teaching and Practical Work Requirements

On jobs, the occasions of engineering college graduates use English mainly include (1)reading technology materials, (2) translating professional materials from English to Chinese or from Chinese to English, (3)writing English technology reports, (4) written or oral communication about professional information, and (5) daily life communication in English. For an engineering and technical personnel, the English should be a convenient tool helping them to solve problems in the professional work. Obviously, in present basic college English course, the training of professional English ability to students is significantly deficient although a lot of teaching resources are costed. For example, the vocabulary of basic college English is very different from that a professional technical personnel should learned. Table 2 and Table 3 are results of a questionnaire from 250 third-year students who took bilingual teaching specialized courses, *Science of polymer interface*, in Hebei university of science and technology from 2006 to 2010. The students are asked to single out 1 choice from 4 options in the questionnaire.

Table 2. The results of questionnaire for "What is the most difficult thing you have encounted in the specialized bilingual course?"

Difficulty	Respondents	
	Numbers	%
Understanding of pecialized knowledge	53	21.2
Specialized English new words	179	71.6
Reading of specialized English materials	10	4.0
Others	8	3.2

It is clear that there is a great gap between the leraned skill in college English and the requirements of practical work. Hebei university of science and technology is a local key university and that survey results perhaps are a valuable reference for the colleges with similar teaching conditions.

Table 3. The results of questionnaire for ''Which kind of new words you have encounted most frequently in the specialized bilingual course?"

Difficulty	Respondents	
	Numbers	%
A. Specialized English new words	132	52.8
B. General English new words	46	18.4
C. Both A and B	61	24.4
D. Others	11	4.4

3.3 Present English Proficiency Test and Evaluation System Restrict the Reform of College English Teaching and Learning

Indeed, college English test(CET) band 4 and band 6 once played a big role in the process of promoting college students' English level. The scale and influence of CET-4 and CET-6 have expanded a lot since the test originated in 1980 s. Now the scores of CET-4 and CET-6 have become an important index of assessing teachers' teaching level and students' English ability. And even when enterprises recruit staff, score of CET-4 or CET-6 is a necessary information of candidates.

But, the negetive effects of CET-4 and CET-6 have not been paid enough attention. The arising problems also have not been solved effectively.The CET-4 and CET-6 only test common English ability of studens. So the present college English teaching need not link up with professional application. As a result, the students after graduation are not competent in jobs concerning specialized professional English [6]. Being faced with that reality, the authors hold that the college English teaching must be reformed. It is necessary to change the teaching goal, curriculum plan, and teaching content to meet the requirements of practical work that the students will take after their graduation.

4 Design about the Reform of Engineering College English

For the situation of engineering college English teaching in China, experiences of professional English teaching in other countries may be useful references [7-9]. The authors proposed a model for innovation of engineering college English teaching (Table 4). It is suggested that engingeering college English teaching be shifted to training students in professional English systematically in whole four years college days from mainly focusing on basic English in first and second years of college. In present teaching mode, most English teaching resources are used to complete basic English learning in the low grades but the class hours of subject English or specialized bilingual courses are relatively few, although these two courses are useful to train students practical ability of application.

Thus the authors suggested that more English teaching hours be arranged for training students' professional English ability. From the results of the survey (Table 2 and 3), it is the most difficult thing for student to understand the specialized words related subject. The authors think that if the great deal of specialized English vocabulary is merged with the basic English reading course, the training for studens' English reading aility and training for professional English ability maybe both achieved.

Table 4. Schedule of a reform model for engineering college English curriculum developed by the authors

Courses	Class hours	Opened time	Innovation
Basic college English	180	The first and second year	Add reading professional materials
Specialized courses by bilingual teaching	90	The third year	Add training of oral and written English
Specialized courses by full English teaching	30-50	The fourth year	Improvement of ability

In the new model, the teachers should select some reading materials easy to be underestood in subject knowledge, considering the background knowledge of the low grade students is weak. At the same time, the training of English writing ability in science and technology should be strengthened. The authors also suggest that every major open more than three bilingual teaching courses, in order to let the students face true original professional materials as possible.

In a present basic English class, there may be hundreds of students in many universities, so there is rarely oral English training in the class. Therefor it is a precious opportunity to the students to be trained in oral English in the bilingual teaching courses. After bilingual study is finished, at least one full English specialized class should be carried out to further improve studens' professional practical ability. The authors belive that students' professional English application ability will be significantly improved by taking the new reform model, while the class hours of English teaching nearly no change.

For implement of this new model, the most difficult thing to most university is the lack of qualified teachers at present. Usually, at present, most professional English teachers don't understand the students' professional knowledge while most specialized professional teachers' English background is limited. The authors suggest to change this difficult situation by training professional English teachers in some specialized courses or/and by training bilingual teachers and professional teachers in voice training and oral English training. On the other hand, English teachers and professional teachers may cooperate to finish the teaching work. For example, part of basic English reading can be taught by professional teachers to introduce some specialized content, while the bilingual course and specialized course can be taught by professional teachers in English. As the English level of new students and young teachers gets more and more high, the authors believe that the suggestd teaching model is feasible in the next few years.

International learning experience is imporant to the teachers who teach bilingual courses or specialized English courses. Colleen Willard-Holt[10] studied the influence of short international experience on bilingual teacheres. The research data indicated that international experience has positive and lasting effects on buildings of teachers' global consciousness, self-confidence and positive attitude towards life. That experience also is very helpful for the teachers to understand foreign culture. The postive effects will be transfered to students year after year by the teachers. With the rapid development of Chinese economy, most colleges have had the material conditions of arranging teachers to study abroad.

5 Conclusions

Although present college English teaching costs a lot of teaching resources, the training of students' professional English ability is very deficient. The goal of present college English teaching seems to be passing CET-4 or CET-6. The present college English course repeats high-school English course a lot in content, curriculum, vocabulary and etc. However, there is a great gap between what the student learned in college and what the graduates have to solve in work. Obviously, present college English teaching need to be reformed to meet the requirement of society. It is suggested that college

English teaching for engingeering speciality be shifted to training students in professional English systematically in whole four years college days from mainly focusing on basic English learning in first and second year of college. By reducing basic content from college English teaching, increasing class hours of bilingual teaching and English teaching of specialized courses, and combining part of professional reading and technology writing into basic English teaching, it is expected that the specialized English abilities of engineering students be significantly improved.

Acknowledgement. This research is supported by Education Science Foundation of Department of Education of Hebei Province (No. JYGH 2011020) and Social Science Foundation of Hebei Province (No. HB11GL009).

References

1. Julia, H., Ute Smit, S., Barbara, M.L.: ESP teacher education at the interface of theory and practice: Introducing a model of mediated corpus-based genre analysis. System 37, 99–109 (2009)
2. John, H., Maria, A.A.C.: Sustainability and local knowledge: The case of the Brazilian ESP Project 1980–2005. English for Specific Purposes 25, 109–122 (2006)
3. Cai, J.: Characteristics and solutions of college English teaching in transition. Foreign Language Teaching and Research 39(1), 30–31 (2007)
4. Cheng, Y.: Reform of English teaching against a background of China's entry WTO. Journal of Foreign Languages 42(6), 10–12 (2002)
5. Cai, J.: Factors affecting the shift of the focus of college English teaching in China. Foreign Languages Research 120(2), 40–45 (2010)
6. Zhang, Z.: Talking about reform in English teaching in China. Journal of Foreign Languages 146(4), 1–6 (2003)
7. Cheng, A.: Understanding learners and learning in ESP genre-based writing instruction. English for Specific Purposes 25, 76–89 (2006)
8. Chia, H., Ruth, J., Chia, H., Olive, F.: English for College Students in Taiwan: A Study of Perceptions of English Needs in a Medical Context. English for Specific Purposes 18(2), 107–119 (1999)
9. Sanner, S., Wilson, A.: The experiences of students with English as a second language in a baccalaureate nursing Program. Nurse Education Today 28, 807–813 (2008)
10. Colleen, W.H.: The impact of a short-term international experience for preservice teachers. Teaching and Teacher Education 17, 505–517 (2001)

The Construction of the Secondary Vocational Teachers Training System Based on 'Cooperation between School and Enterprise'

Chen Lidong, Ma Shuying, Shi Lei, Li Guofang, Zhang Liang, and Zheng Lixin

College of Mechanical and Electronic Engineering,
Hebei Normal University of Science and Technology, Qinhuangdao, China
chentian-940308@163.com, ma_shuying@126.com

Abstract. Vocational education teachers' training is important for vocational education development. Although China' secondary vocational education teachers' training system has been set up and made considerable progress, there are some problems that can not be neglected during the teachers training. In this paper, the existing problems of the secondary vocational teachers' training are introduced, and a new training system for the secondary vocational teachers based on 'Cooperation between School and Enterprise' is put forward. The system consists of three training levels of 'teacher's professional qualifications training', 'teacher improving training' and 'the backbone teacher's training'. The training system emphasizes the training methods of the three trainings. The system accords with the current vocational education development situation, and it is helpful for sending the teachers' comprehensive ability.

Keywords: secondary vocational education, teacher training, training system, cooperation between school and enterprise.

1 Introduction

With the development of economic construction in China, she needs a large number of high quality skilled personnel, whose training relies on the development of vocational and technical education, especially secondary vocational and technical education. In the recent 20 years, China's secondary vocational education has obtained a conspicuous success for transporting a large number of professional technical talents. However, the secondary vocational teachers come from different sources, some of them are university or college graduates; some of them are transferred from basic education schools, and also some come from the enterprises. So their basic knowledge of culture is weak or they are lack of knowledge of education theories or practical ability. Therefore, we must improve their professional quality to meet the need of secondary vocational school education, and the establishment and improvement of our secondary teacher training and vocational training system is very urgent theoretical and practical issue [1].

H. Tan (Ed.): Knowledge Discovery and Data Mining, AISC 135, pp. 573–578.
springerlink.com

Therefore, a training system of secondary vocational teachers based on "cooperation between school and enterprise" is introduced in this paper, which is not only useful for strengthening the construction of "dual-qualification" teachers, also helpful for improving the education quality of secondary vocational education.

2 Problems of the Teacher Training for the Secondary Vocational Education

To a certain extent, our current teacher training system in secondary vocational education has met the needs of secondary vocational teacher training at the present stage, and it has made its contribution to promote secondary vocational education in China. However, some problems of the system can not be ignored. Firstly, most young teachers do not have the chance to training in secondary vocational school, but the training demand is much larger than the current limited training capacity; secondly, the human and financial resources of the affiliation who has training teachers can not meet needs; thirdly, education authorities have make a greater effort, but education incentives and secure financing terms for the teachers' training can not be provided by the government; fourthly, training provided by the training base can not satisfy the teachers because of their different professional level; in addition, most training base mainly focus on theory teaching, lack of practical skill training, which goes against the healthy development of secondary vocational education [2]. Therefore, in order to enhance the teacher's quality, we should establish a suitable training system to strengthen secondary vocational education.

3 Construction Principles of Training System

3.1 'Application' as the Guiding Ideology, Emphasis on the Improvement of New Technologies and New Theories

The professional teachers of vocational school mostly come from colleges or universities, and they are lack of appropriate practical skills, ability of theory with practice and teaching skills. Thus we should pay attention to the teacher's actual demands, and offer some practice opportunities, at the same time, when organizing training we should deal well with all aspects of the theory teaching, practice operations, inspections, investigations, thesis writing and other training activities. With the faster update knowledge in modern society and training materials will inevitably lag behind the development of the situation, so the training must be forward-looking to adapt to teaching demands.

3.2 Inspiring Learning Enthusiasm of Trained Teachers and Improving the Quality of Training

Professional technique is the teaching basis for vocational teachers, and the improvement of the professional skills will promote the trained teachers to learn independently. Whether the choice of teaching content or the updating of teaching methods, it should take stimulating the teachers' learning enthusiasm as the reference object. Only in this way, it can stimulate and maintain the professional interests of trained teachers to active learning. In addition, it is the effective guarantee to achieve the training objectives of the training programs.

3.3 Centered by the Trained Teachers, Emphasizing the Dominant Position of Them

From course design to evaluation aspects, we should always place the comprehensive development of the trained teachers on the center. During teaching activities the trained teachers are training subject, and they should be guided to change the role--from the "appointing my training" into "I want to train." [3] Training departments should put forward specific requirements, and guide the trainee to adjust their psychological consciously, conversion task, change the role and focus into learning [4].

3.4 Concerned about Individual Differences of Trained Teachers

Because of the trained teachers' individual difference, such as knowledge reserve, interests and hobbies, practical skill and so on, training department should make study objects and evaluation methods according to it. In addition, they should propose the proper teaching suggestion which ensure that almost all trainees can complete their training objects. It will bring them a sense of fulfillment. Therefore, we need to combine teaching, advice, guidance together to meet the teachers' various requirements. For instance, we can adopt multi-level training method for different level teachers. Training methods can use the type of combination of mass teaching and individual guidance, self combined with face to face.

4 Framework and Content of Training Program

4.1 Training Program

Summarizing the successful experience of training base construction and teacher training, referring to advanced education concept, in this paper the training program based on the improvement of teachers' comprehensive vocational ability is designed, which take "double qualification teachers" for the training goal, and it adopts modular teaching and the way of "corporation between school and enterprise". The training system is shown as Fig. 1.

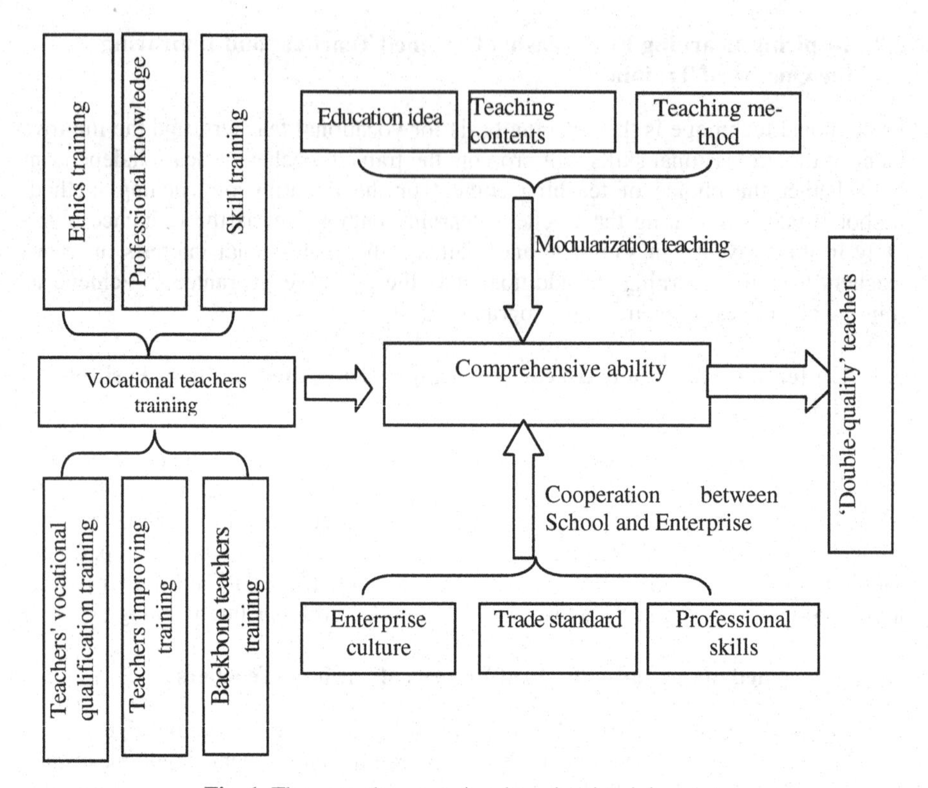

Fig. 1. The secondary vocational teachers' training system

4.2 Training Content

Teachers training content is the guarantee for the quality of the secondary vocational education, it affects the teachers' participation enthusiasm in the training. In the training system, the content is divided to teachers' vocational qualification training, teachers improving training and the backbone teacher training.

Teachers' Vocational Qualification Training. Teachers' vocational qualification training is mainly to train the specialty teachers for secondary vocational schools, and make them possess qualifications (including undertaking practical teaching individually, achieving professional qualification certificate, and so on). The training emphasizes basic teaching theory and the professional difficulties, and the key of the training is practical teaching based on the "corporation between school and enterprise". In the teachers' training, it should pay attention to the teaching ability and professional ability.

Teachers Improving Training. We should abide by the integration of knowledge, ability and quality when establishing training target of the secondary vocational. Moreover, we need to pay more attention to the teachers' urgent knowledge and ability to enhance, but also the comprehensive quality of the teachers should to be enhanced [5]. Therefore, we should design the personalized course systems based on requirement of vocational education. The mode that training organization and trainees develop the training courses together should be established. We can put up a training course supermarket to meet with trainees' individual demands, in which the prescriptive training contents can combined with the trainees' optional training contents. By the training, the trainees can obtain the new knowledge, new technology, new methods, new process and core skills, and their practical experience can be further enriched. In one hand, vocational teacher training base develops some training schemes by means of school-enterprise joint school-running to enhance the teachers' practical ability; in the other hand, vocational school should take full advantage of enterprise practical system, and make teachers training implement project to strengthen the training of school teachers practical ability. This training can make the trainees familiar with the modern production conditions, technical level, talent demand information, production process and the operational norms, which can further enhance the teachers' understanding of the importance of the school teaching and enterprise practice. The training will enhance the professional skills and operational capacity, and make the trainee be core of teachers who have mastered professional theory and the practical ability.

Backbone Teachers' Training. Backbone teacher training is to train the enhanced teachers to cultivate them to be a accomplishments who posses rich teaching research experience and teaching practical ability. They will undertake more than three specialty courses and the practical teaching, and they can guide students the comprehensive professional practice training and graduation project. They have the expertise of new technologies, new theories of knowledge, understanding of the professional development trends, technician or senior technician with the level of practical skill level of technician.

5 Implementation of the Training System

From July 2008 to August 2009, the research group selected three pilot sites-Jiangsu Teachers University of Technology, Qian'an Vocational Education Center and Hebei Normal University of Science and Technology to experiment the training system on the basis of the different levels of training programs, according to the relevant professional level competency standards to implement by the four modules as professional class, technology class, educational and training projects, in which experts seminars and exchange visits were peppered. The training is major on practical teaching, and theory seminars are supplement. The training can establish the concept of advanced vocational education and master the advanced teaching methods and modern educational technology and means to improve vocational education.

Through training, on the one hand, it brings participants' further strengthen teaching ability and makes them understand their curriculums and teaching reform; at the same time, their professional standards are further improved. In addition, by this means, the trainees can grasp more comprehensive professional technology in the field of new knowledge, new technology, new methods new technology and the core skills, and be familiar with the production process and the post operational norms.

On the other hand, the trainees have mastered the professional teaching methods, technological means of modern education and further enriched practical experience.

5 Results

This paper proposes a secondary vocational teacher training system and its principle based on ‘cooperation between school and enterprise’ to adapt to "double-quality" teacher training, starting from the present status of secondary vocational teacher training and the urgent need for ‘double-quality’ teachers. The system includes ‘teachers' vocational qualification training’, ‘teachers improving training’ and ‘backbone teachers training’, which accords with the development of vocational teacher training situation.

References

1. Yan, Y., Zhang, Y.: Problems and strategies in training vocational guidance teachers in secondary vocational schools. Journal of Tianjin University of Technology and Education 20(1), 56–58 (2010) (in Chinese)
2. Niu, H.L.: A study on teacher education and training for secondary vocational and technical education in China, in Master Paper. South West University, Chongqing (in Chinese)
3. Zhao, B.Z., Chen, R.H., Zhao, X.L.: Thoughts on the training of secondary vocational and technical education teachers. Journal of Hebei Normal University of Science and Technology 4(1), 44–46 (2005) (in Chinese)
4. Feng, Z.Y.: Research on faculty training system of China’s vocational education, in Master Paper. Hebei Normal University, Shijiazhuang (in Chinese)
5. Yang, X.X., Li, N.: Innovative teacher training mode to promote teachers’ professional development. Journal of Changchun Education Institute 21(1), 5–7 (2005) (in Chinese)

Research on High-Precision Time Synchronization Algorithm for Wireless Sensor Network

Chunxiang Zheng [1] and Jiadong Dong [2]

[1] School of Computer and Information, Anqing Teachers College, 246011, Anqing, China
[2] School of Physics&Electrical Engineering, Anqing Teachers College, 246011, Anqing, China
dongjiadong@sina.com

Abstract. Time synchronization is of great importance in local area networks for many applications. A wireless sensor networks (WSNs) is typically made up of numerous sensors which have different clock accuracies. Time synchronization is crucial to maintain data consistency, coordination, and perform other fundamental operations. In this paper, based on wireless sensor network a high-precision time synchronization algorithm are studied and the error analyses are given. In the whole network, we discuss the average synchronization error of all the nodes through simulation.

Keywords: time synchronization algorithm, wireless sensor network, distribution.

1 Introduction

In recent years, the wireless sensor network, has received much attention of researchers. Wireless sensor networks consist of a large number of tiny low-power devices capable of performing sensing and communication tasks. Time synchronization is a critical part of infrastructure for distributed system. Time unification is the most essential requirement. The reasons are as follows: the coordination and collaboration of sensor nodes need a common timescale; the nodes must sleep and wake up at the same time, so they need synchronization. In order to realize time unification, the time synchronization is the most important method.

Time important point in wireless sensor network is energy efficiency, which is possible with algorithms and techniques, not physical technique. That is time synchronization. Time synchronization algorithms started in distributed system. In general, time synchronization problem appears in many kinds of networks, for example, Internet and LANs. And, those traditional synchronization technologies used in them such as NTP, GPS, etc. In the case of NTP and GPS method, when the timestamp in messages is broadcasted to exploit as a global time, the sending time and the access time of messages incurs time delay problem for time synchronization. In the traditional time synchronization strategy, the time synchronization is considered, and the cost of time synchronization drops to smallest. However in actual distributed system, not only has the different time respectively in each node, but also their time carries on by different speed. Moreover, there are time network transmission delay and wrong node reference time.[1]

H. Tan (Ed.): Knowledge Discovery and Data Mining, AISC 135, pp. 579–583.
springerlink.com

2 Time Synchronization Model

There are some factors that affect the accuracy and precision achievable by existing time synchronization algorithms. These include the instability of the clock frequency and variable delays of message exchange. [2]

2.1 Instability of Clock Frequency

The clocks on different sensor nodes do not run at exactly the same rate. The phase difference between the clocks on two nodes will change over time due to frequency differences in the oscillators. The frequency instability introduces errors into the time synchronization algorithms. These errors are shown as points off the fitting line.

2.2 Variable Delays of Message Exchange

Existing time synchronization algorithms typically work by exchanging messages. The source of message latency has four distinct components.

Send Time. The time spent at the sender to construct the message. The send time also accounts for the time required to transfer the message from the host to its interface.

Access Time. The delay incurred waiting for access to the transmit channel. This is specific to the MAC protocol in use.

Propagation Time. The time needed for the message to transit from sender to receivers once it has left the sender. When the sender and receiver share access to the same physical media, this time is very small as it is simply the physical propagation time of the message through the medium.

Receive Time. The processing required for the receiver's network interface to receive the message from the channel and notify the host of its arrival.

With the development of computer network technology and master-slave system, compared with the centralized processing, distributed processing is widely used in many fields. Because the time synchronization is all foundation for application in distributed system, it can be used for wide range, such as Data Acquisition, Monitoring, Performance Analysis, Operating Optimization, and Fault Diagnosis for Power Plant, Electronic Commerce as well as Internet or the Intranet Network Time Service and so on. Two kinds of main technology of time synchronization in distributed system are the direct-connect time synchronization and based on NTP network time synchronization. [3]

Regarding high precision time synchronization system, direct-connect time transmission technology can be used. In order to receive and maintain time information, this method needs to install bus technique timing module, then the computer may directly connect and synchronization to the standard time source (GPS or central timing system). These modules may maintain the time independently without influence of master node, and crystal oscillator precision in these modules may match the time source. This synchronous mode can provide a high synchronous precision for single machine in the distributional system, but it cannot make full use of the network resources, and it has high cost input.

3 Time Synchronization Strategy

3.1 Distributed System Time Synchronization Fault-Tolerate Strategy

The distributed system time synchronization fault-tolerant strategy may be divided into three types: The master and servant type time synchronization strategy of time server, the Byzantium protocol type time synchronization strategy, the convergent function type time synchronization strategy [4].

The time synchronization strategy of the master and servant type, it is direct, simple widely used in the distributional system. Its strategy is to establish time synchronization to have the high confidence level and the availability time server. The other nodes penetrate the correspondence network and the server to pick up directly the correct time, and to revise respective time, then to achieve time synchronization in various nodes of the overall system. The merit of the synchronized strategy is the principle and the equipment very are all simple, but the shortcoming is that the fault-tolerant ability is low, and easily lead to a bottleneck effect of centralized management.

The Byzantium agreement type time synchronizes the strategy, it is another basic, also important fault-tolerant technology in the distributional system fault-tolerant strategy , its algorithm is that node acts as the transmitter, it transmits a specific information to all points, after as soon as the non-wrong node receives this information to pass through processing to be able correctly to broadcast again it for other surplus nodes, each node penetrates the exchange information by each other, it may know which nodes have the mistake by the majority decision way. Afterwards various nodes transmit the exchange information in turn, until all error nodes were found. The synchronize strategy has fault-tolerant ability, but it requires a lot to exchange information, and it aggravate the network load. It applies to a specific exchange of information. The need for time synchronization in the exchange of non-specific reference time information, there are application constraints.

The convergence function type time synchronizes strategy, the aim is to search every node reference time in the system, through an effective convergence function, from in the reference time which collects and decides the synchronizing time. Based on the results, every node adjusts its time. It basic step gathers as follows:

- Collection reference time value from other node.
- Estimate these time estimates, consideration influence factor, and obtain a correct time value.
- Calculus these time estimate using a convergence functional, and obtains a correct time value.
- Revises respective time according to the correct time value.

3.2 Time Synchronization Algorithm

The sliding window calculating method (Sliding Window Algorithm, was called SWA) is published by two scholars (Manfred J.Pfluegl and Douglas the M.Blough) in May, 1995, in IEEE Journal of Parallel and Distributed the Computing periodical. It belongs to a convergence functional time synchronization strategy. Comparing with other similar time synchronization algorithms, both in accuracy and fault tolerance, it has a better performance. After comparison, this paper will use the idea of the algorithms and make amendments, and time synchronization in wireless sensor networks will be applied.

The sliding window calculating method and the general convergence function type time synchronize strategy is similar. It uses the mechanism of the periodic time synchronizes. Suppose R on behalf of time synchronization cycle, The next time synchronization of cycles before the arrival, namely iR-p (i representatives of i time synchronization, p as the reference time information can indeed be served on time), namely first obtain the reference time value t_m, and then estimate the time estimated value which the cycle arrives $\tilde{X}_J = t_m + t_s - t_r + e_{mean}$, J representatives from node j to obtain the reference time value, t_s expresses the next time synchronizing time (i.e. ts=iR), t_r indicates the time which receives reference time value , e_{mean} expresses average error . The diagram of propagation time is shown in Figure 1.[5]

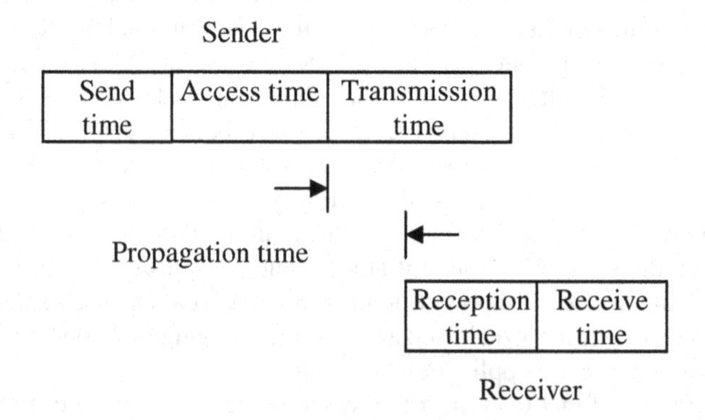

Fig. 1. Components of packet delay

4 Conclusion

In this paper, We proposed a new time synchronization method. This effect resulted in the decrease in the amount of power consumption so that the lifetime of sensor nodes could be prolonged. Each sensor node has synchronized its local time by interpolating with the time difference achieved from two messages.

The primary cause of influence system precision has two main reasons, one is the time information maximum delay value Δ_{max}, and when the overall system starts or the new client side joins. It is possible that it may influence precision because of tremendous Δ_{max} value. However, after several time synchronization cycles, it should be able to converge to a certain value, therefore the Δ_{max} value impact would be stabilized, It is little variability. Another time synchronizes the precision influence factor is network transmission delay ε_{max}, It is produced by main reasons: The TCP/IP network communication protocol influence network transmission delay; Network load and the machine processing of the busy state of the network transmission delay also influence network transmission delay and so on. Accuracy of the above-mentioned factors can be measured by the time delay of communication protocols. For other information and communication protocols delay, it can be used to estimate the minimum reference value, as close as possible at the time delay in communication protocols.

Acknowledgment. This work is partially supported by Natural Science Foundation of Education Department of Anhui Province (ID: KJ2011A199).

References

1. Romer, K., Blum, P., Meier, L.: Time Synchronization and Calibration in Wireless Sensor Network. In: Handbook of Sensor Networks: Algorithms and Architectures, pp. 199–237. Wiley and Sons (October 2005)
2. Gao, Q., Xu, B.: Time Synchronization Improvement for Wireless Sensor Networks. In: 1st International Symposium on Pervasive Computing and Application, pp. 805–810 (2006)
3. Ren, F.Y., Dong, S.Y.: A time synchronization mechanism and algorithm based on phase lock loop. Journal of Software 18(2), 372–380 (2007)
4. Akyildiz, I.F., Su, W., Sankarasubramaniam, Y., Cayirci, E.: Wireless sensor networks: a survey. Computer Networks 38(4), 393–422 (2002)
5. Elson, J., Romer, K.: Wireless sensor networks: A new regime for time synchronization. ACM SIGCOMM Computer Communication Review 33(1), 149–154 (2003)

able to converge to a certain value, therefore the Δ [illegible] value [illegible] could be stable [illegible] the precision [illegible] network transmission delay [illegible] The PCPAT [illegible] communication process [illegible] network transmission delay. Network load and [illegible] of the network transmission delay also influence [illegible] and so on [illegible] accuracy of the above mentioned [illegible] protocol. For other information and communication [illegible] delay [illegible] used to estimate the [illegible] the [illegible] communication [illegible].

[illegible]

[illegible]

1. [illegible] Wiley [illegible]
2. [illegible] Time Synchronization [illegible] Networks [illegible]
3. [illegible]
4. [illegible]

The Social Value and Development Strategies of Modern Distance Education Based on Humanistic Utility

Shurong Yan and Min Gao

School of Economics and Management, Chang'an University, Xi'an,710064, P.R. China
376052129@qq.com

Abstract. Recently the modern distance learning expands rapidly in our country, and it has made a huge progress. It's most remarkable social value is providing the material base for the establishment a studied society. It also can obviously improve the survival condition of the minority groups and promote the social justice, and train the new talented person for the society at the same time. But for the new educational reform,we must determine the developmental strategy based on a new point. The objective research is also a newer task to educate the current worker for a healthy and sustained development of the distance learning.

Keywords: distance learning, social value, developmental strategy.

1 The Concept of Distance Education

Distance education is a new teaching form.Students ,teachers, and educational organizations will be adopted for the systematic teaching multimedia and communication links. Modern distance education is based on computer networks,satellite communications and telecommunications networks. The word "modern" here mainly refers to the modernization educational thought, content , teaching methods and technology.

2 The Development of Distance Education in China

Its occurrence and development is always closely linked to the information technology, and education technology. Its development in China mainly has three stages.

2.1 Correspondence Education

Correspondence education originated in the United Kingdom University campaign in the 19th century 1860s, all capitalism countries began set correspondence school in 1980s. In China, early correspondence education generated with the modern Chinese printing industry and the development of the postal industry. In 1953, RenMin University of China and Northeast Normal University began correspondence, then gradually expanding to science, engineering, agriculture, medicine..., all types of colleges and universities. The main representatives are independent sets of correspondence schools and traditional universities in distance education.

H. Tan (Ed.): Knowledge Discovery and Data Mining, AISC 135, pp. 585–592.
springerlink.com

2.2 Tele-education

In 1980s its second generation rose in China. It worked by radio and television, recording, video, telephone, telex and other mass media , the main form of tele-education was radio and television. In China, in 1977, comrade Deng Xiaoping agreed Ministry of Education, the Central Radio and Television Bureau jointly organization to the Central Radio and TV University, then. China had the largest Radio and TV University in the world.

Its first and second teaching method mainly made use of the "one-to-many" and "one-way" communication, whether correspondence by printed materials or radio and TV recordings. Therefore, its first and second generation did not work well for the two-way communication between teachers and students or students, which could communicate primarily by postal, transport and the service system of correspondence instruction. It had a longer operation period, and less efficiency especially to correct and assess on student.While the phone was a very effective two-way communication tool, but it was not popular, and its communication costs was relatively higher in a long period, and its application was not sufficient. Thus, the fact that teachers and students could not communicate effectively was the first main problem in distance education.

2.3 Modern Distance Education

Distance education is a new educational form which using the network, multimedia, and modern information technology. It is the network which based on modern electronic information communication. Students, teachers, and educational institutions , all of them mainly used a variety of media and interactive tools for teaching and communication.

Relative to the other education, with the third generation technology, a teacher can transfer a large number of more complex information to students by emails ,chatting rooms and bulletin board. Using that,computer-assisted teaching, computer simulation, disk, CD-ROM and the internet and other channels of electronic resources to further demonstrate is the characteristic of distance education.

3 Humanity Utility Theory and Its Enlightenment

There are two major characteristics in utility theory of western traditional economics: the first is subjective utility, the second is individual utility.

Utility theory of western traditional economics emphasizes individual subjective feelings. It was established on the basis of individual utility, trying to judge and change the high or low social welfare by the utility of personal preference. Because of the variety of individual preferences, it is impossible to find which one could meet every individual preference, then the effectiveness of individual subjective research also leads to agnosticism. For a time the research of welfare economics turned into a dilemma.

Humanism emphasize that the improvement living quality of human is the most fundamental requirement in social development.

In the theory of modern utility, humanity utility, that is to say social group could promote the social development. There are fundamental differences between the traditional utility of economics and humanity utility. Humanity utility analyze different effect of social progress which from different social groups, and thinks that different community groups have different social progress to contribution of society. It is exploring that how to configure resource to ensure that the resources'supply and training for social groups, and thus it is more conducive to the promotion of social progress and civilizations in a long term. The "utility" here is referring to groupment and objectivity, so it is the social economics of developing, and it is breakthrough and innovation to the traditional utility theory.

According to wealth creation, there are three groups in social groups, which are forming group, created group and old-aged group. In the natural growth process, the consumption of resources of the three groups is different. The forming group is purely processes of resource consumption; creating group is processes of consumer resources and wealth creation. The creation is greater than consumption, and the extra wealth is used for raising forming groups and old-age groups and reproduction; the old-age group is also a resource consuming group, not wealth creating group.

Resources are including natural and social resources. Natural resources are including land, mineral resources and other original objective production. Social resources are including human, technical, information and other subjective producing capacity, so both of them combine as a source of wealth, which could create social wealth. Wealth is able to bring the material and spiritual things to enjoy, so either resources or wealth has the same meaning; because owning resources means for owning the wealth. According to the definition of theory of humanistic utility, wealth plays a facilitating role in human civilization process, so the effect of wealth or resources is called humanity utility in this paper.

The survival of the fittest is the law of evolution of society, the inevitable demand of social civilization and progress of mankind. However, if the competition starting point is unequal in the natural resource, such as education and production material from the outset, it is difficult to ensure the final distribution of income fair; since the low-income people is short of equal competition conditions from the outset, which cause the final allocation results will be less. That's "initial state" unfair which Rogers stressed in the Theory of Justice, and this led to increasing social gap between the rich and the poor, growing social discontent of the crux of the problem resulted from the system of distribution according to one's performance based market allocation mechanism.

In a country mainly based on market economy, the fundamental way to solve the problem of income distribution for government is to ensure the fairness to create a starting point for everybody. That is to say government should guarantee that each person accepted equal education in forming process, and ensure that every person own equal production resources in creating process under the macroeconomic regulation. The improvement of education on forming group can be achieved primarily by inclination configuration of national distance education resources, and the rural condition of education for creating group is primarily improved by of the public resources.

Distance education resources are an important part of public resources, and it plays an active role in meeting the equality of resources and solving the income gap.

4 The Social Value of Modern Distance Education

The most significant social value of modern distance education is to promote educational equity, and social equity after educational equity.

4.1 Distance Education Provides Material Basis for Building a Learning Society

As we known, nowadays society is a union of knowledge unprecedented prosperity, and information increase with an explosive form. A famous research report was noted that mankind had entered into a lifelong learning society. For everyone, including the students in university, still does not meet the real-life work, and needs to continue learning. The human has come to a life-long learning society. Thus, lifelong learning is more and more acceptable. The contradiction between the rapid increasingly learning needs and scarce educational resources become sharp, so only traditional educational institutions is not likely to meet. This requires that the family, society, education sector, as well as non-education sector must be able to provide more learning opportunities, and take participate in educational activities. That is to say, we should build a learning society, so that people can be learning freely at any moment. However, the traditional education is affected by time, space, and manpower, material resources, and it is difficult to meet the requirements of the learning society.

Distance education is not affected by time, space and geographic limitations, so it can bring the school by the network in office, home or even in every corner of the world.Then we can join work and learning together. Through the network, everyone can choose when and where to study. So it can help to whom one already has job and lost conditions or opportunities of all levels of education. We can say that internet distance education provides practical material basis for establishing a learning society.

4.2 Modern Distance Education Can Significantly Improve the Living Conditions of Vulnerable Groups and Promote Social Equity, and It Can Be Regarded as a Important Tool for Social Equality

As in occupies social resources and accessing to opportunities and interests often have many differences by different layers of population in the social structure, but education may be an important reason. For a long time, accessing to education or higher education is almost the only channel to change personal social identity. In this condition, education demand would be very strong. The main driver that the world held distance education is to provide a platform of accessing to education and success to the society members who have disadvantaged social and economic identity. Therefore, the social function and value of education are improving social status ultimately after the educated education.

Modern distance education has a broad and far-reaching influence for most remote learners. Such learning is a process of that learners carry out self-study, take participate in the collaboration necessary to achieve the knowledge, improve skills and enhancing the value of labor by the supporting and assistance of distance education institutions. Distance education is a multi-level, popular education, including education and non-academic continuing education, etc. It can provide more community members with professional and technical education, which breaks occupational, class, etc. Even celebrates staff, laid-off workers through training, to some extent, it can also change its disadvantaged status of the rights expression and the protection.

4.3 Modern Distance Education Develop New Social Talents

In the future, the society is full of information and it has the new requirements for students. The employer requires students not only to master professional knowledge, but also to master information technology and have good innovative ability. However, the traditional model cannot nicely solve these problems. Students are faced with numerous competition and pressure to their careers, so they require to be able to learn a skill on campus in order to develop their own various potential.

The network itself is the cornerstone of that the information age can produce and develop, and the students trained by distance education should be consistent with the requirements of the information society. Professor Xie Weihe in Beijing Normal University believed that the talented person trained by the network distance learning should havethree strong points at least:

First of all, they should have strong capabilities to obtain and select information, which are exactly essential knowledge and skills that modern society and information society need. The characteristic of the network distance learning is that it provides the students with this kind of ability.

Secondly, they should have a strong network capacity. In the book "The study –which wealth hidden among it", it shows that survival learning to survive, study, work, cooperate and so on in the network space. Therefore, a person who has strong networking capability also means that he has the quite strong ability in real life.

Finally, they should have a strong ability of self-construction, which includes the construction of new ideas and knowledge and so on. Because different knowledge and information centralized from each aspect unifies in together in the high-frequency. The students need to conform these knowledge and information, then discover and obtain the value and the significance, and establish the new significance.

Therefore, the raising of the new talented person is incorrect if only depending on the traditional education. So we must unify the network with distance learning at least, then train the talented person that the information society needs.

5 The Development Strategy of Distance Education

5.1 Exploring and Summing Up Seriously the Regulation Standards and Norms Which Both Suits the National Condition and the Market Rule

After three years of exploration, our modern distance education has presented the following characteristics and features: first of all, the market of China's education shows strong market demand and a huge market potential power, but there are still very far distance from 30% of the popular destination accepting higher education. ; second, with the aging of related technologies such as multimedia, database, data transmission and Internet application development technologies, and there are wider coverage and more convenient environment and conditions to achieve a high level; the third point , with the increase of the intelligent community and the decrease of the network tariff, Internet user groups from the Center City to all submissions in remote mountainous areas are growing rapidly. According to the projection of concerned experts, the average rate of increment of domestic modern distance learning will achieve 6.4% in near three year yearly.With the significant progress being made in our modern distance education development, there are indeed some undesirable problems, such as the management authority between Administrative department of education and the host institution are unclear, and the difficulty of a number of districts and colleges or filing, duplication and waste of teaching resources and so on. Times calling out as soon as the standards and norms of the law which have suitable conditions and in line with market to ensure healthy development of modern distance education.

1) Exploring the dynamic transmission technology standards based on computer networks, satellite television networks, telecommunications networks, broadband multimedia communication education in the public media, such as the Internet, satellite, cable TV, video conference system, wireless broadcast method of composite application and its corresponding specification system, etc..
2) Exploring the standards of tele-education resource description, pack, interactive interface between learning content (i.e. the software) and learning management system, descriptions for learners and character, and other aspects.
3) Exploring the effective and cooperative mode of learning management systems and content providers, educational services, the late running behavior supervision, education degree management between host institutions and contractors, educational administrative departments.
4) Exploring the models of online education resources cooperation, development, and delivery, including courseware, virtual simulation of a giant, experimental practice Library, also including in which provide, who will provide and how we provide, how many, etc.
5) Exploring the system of quantitative indicators and responsibility for the education site contractor qualification, such as the nature, conditions, qualifications, responsibilities, obligations, penalties, and other violations.
6) Exploring the shared principles and practices of national universal teaching resources, which could be a breakthrough in property rights relationship. Such as use, loan, rent range and so on.
7) Exploring the management system of tracking, monitoring and evaluation for teaching quality, such as who will monitor and how to monitor, control, and so on.

The starting point and end result of these exploration cognition and summary of laws and rules are breaking the barrier, putting the relation, optimization, complementing each other. Through the construction rules, it can form an effective macro-control mechanism and market mechanisms, and construct new patterns of an open modern distance education with Chinese characteristics. is the basic guiding principle needed to establish that we should adhere to the specification for the education practice, and promote innovation and improve the specification in the continuous development. So that it is conducive to the health of the distance education itself, but also conducive to full liberalization in the pilot of the development of distance education in the whole Chinese education market.

5.2 Develop and Construct Media Resources of Remote Education Positively

To develop the high-quality and large-scale modern remote education, network building is a premise. Speeding up steps and intensify the degree to develop and construct rich multimedia teaching resources is strategic focus to make sure that remote education is dynamically, and persistent and have long-term development advantages. At present, the rapid development of modern remote education and the pinch of teaching resources have become the main contradictions of restriting their own development. Modern remote education system is an perfect combination of three parts, hardware support, software support and teaching resources, the realize of the educational function of which shows the organic construction and combination of three levels, information-teaching ,procedure-teaching and culture education. Among them,information-teaching relies primarily on material and advanced courseware, made technology, procedure-teaching interactive teaching, and culture education virtual campus technology. And it also needs to fully exploit the characteristics of digital ,hypermedia ,no distance and economical , construct highly intelligent network courses with pictures and teletexts , network test bank, network experiment station, network library ,network base for answers ,discussions ,exchanges and practices. First, overall planning, put grasp and release together, focus on keys, reduce repetitive work. That is, in some relatively stable base disciplines which are generally demanded by community , led by the Ministry of education , we can organize expert groups to make a number of high-quality, universal boutique courseware's, and free of charge to society for universities sharing; At the same time , the courseware resources the existing 45 hosting universities develop should be on a regular basis to assess, accept and contest, the winner of which could be awarded as the standard curriculum resources and is required courseware and script tutorial among similar courses. For the professional and dynamical courses we can play the advantages of each university to make its own construction. Second, we can introduce a quantity of mature even entire English environment courseware resources from foreign universities, using them for network education, full-time education, adult education and even non-academic training and education in the common universities. The third is to take example by the joint experience of the RenMin University of China and the Central University, promoting co-operation between universities and establishments, collaboration, introducing market operation mechanism with commercialization for industrialization. Striving after three to five years of efforts, built to overwrite an existing higher education disciplines and courses of 60-70% of network multimedia courseware, support and services to increasingly large remote education network system, with continued development and expansion of vitality.

5.3 Walks the Path of International Cooperation Development

China has joined the WTO, and taking effect with the relevant provisions, foreign universities and educational institutions entering Chinese market has become the trend of the times. They could have walked the path of modern distance education, or independent office, or the Liaison Office, or use chain of all schools inside and outside, credit recognized, the connecting way of school record trail. In the new situation of cooperation at the same time with competition, we must be awarded and prepared.

In the face with the growing open education in China, we should encourage universities actively to seek multiple forms of cooperative education with foreign universities and educational institutions, and introduce, absorb, digest, transform and utilize outstanding educational resources overseas including curriculum, faculty, academic degree, particularly curriculum resources of economy, finance, trade, financial accounting, foreign language, law, IT, arts to meet the national multi-level needs. According to the existing practice in Tsinghua University, RenMin, Hunan University and other universities, we can predicted that it will have a broad prospect to run international professional college by strong cooperation. We should gear the internationally vocational training certificates, foreign academic and non-academic education, and continuing education to the international conventions in aid of advanced tools of modern distance education, and fully revitalize the education market, finally serving for the personnel training and modernization of our education.

References

1. Gao, Y.: Development and insufficient of distance learning of China. Test Weekly 50(11), 202 (2010)
2. Cai, H.: Teaching reform for improving the quality of distance education. Guangdong Radio and TV University Journal 40(6), 2–4 (2008)
3. Ying, Y.: The teaching innovation of adult education in the context of network. China Higher Education 30(24), 49–50 (2008)
4. Liu, M.: The construction of modern distance education management system based on quality management. The China Educational Technology 23(5), 48–50 (2005)
5. Ji, W., Luo, L.: Model and development strategy of modern distance education. The Education of Tsinghua University 45(2), 96–101 (2002)
6. Zhang, Y.: Social value and trend of modern distance education. The Reform and Development 35(2), 156–157 (2009)
7. Zhao, X., Cui, X.: The concept, characteristics and trends of modern distance education. Economic Market 30(4), 124–125 (2010)

Research on Cultivation Mechanism of Enterprise Ability for Undergraduates on the View of Employment

Shurong Yan

School of Economics and Management, Chang'an University, Xi'an, 710064, P.R. China
376052129@qq.com

Abstract. It is needed that to cultivate undergraduates to have a innovation spirit, pioneering consciousness and engineering ability in addition to the professional knowledge and skills. So the "four spiral" model is to set up to cultivate students. The "four spiral" relationship is formed by four kinds of force including government, enterprises, universities, enterprise students, which showing cross nonlinear, promoting the students' ability. To solve the present problem, this paper proposed that we must construct the operational mechanism and curriculum system of innovating the initiative capacity for these undergraduates.

Keywords: four spiral, entrepreneurship, undergraduates, innovation, mechanism.

1 Introduction

The development strategies' implementation to improve and expand enterprise employment is the purpose of "the national long-term education reform and development plan outline"(2010-2020), and it is also the higher requirement which nation and community put forward for the higher education[1], which means improving students' creativity and practical ability to adapt to complex , ever-changing environment initiative and respond to the future challenge of the enterprise education, it is a new mission in new era for universities. To meet this challenge, a series of effective research should be carried out on enterprise education.

1.1 The Concept of Enterprise Education

At the end of the 1980 s, the United Nations Educational Scientific and Cultural Organization(UNESCO) put forward the new concept "enterprise education" at the conference of international education development trend of 21 st century. UNESCO pointed out: generally speaking, the aim of enterprise education was to cultivate pioneering personal. It required the high school to take the enterprise skills and enterprise education as the basic goal, and made enterprise have a equal position as professional education.

Enterprise education is refer to the education activities which could train the career consciousness, quality and entrepreneurial skills for students. That is to say it train students how to adapt to the society, improve survival ability, as well as ways and means to create enterprise. Therefore, entrepreneurship education is a kind of

H. Tan (Ed.): Knowledge Discovery and Data Mining, AISC 135, pp. 593–599.
springerlink.com

subjectivity education and a higher quality education, and it is even a education of beyond. Its basic connotation is to develop and improve the students' quality, and train students' initiative and the spirit of innovation, and the ability engaged in a career. It means to develop the comprehensive ability which includes the spirit of creation , adventure , the skills of technologic or social, and the management.

1.1.1 Enterprise Education Is the Unity Tool to Cultivate Psychological Quality and Practical Ability

Enterprise education as a part of ability cultivating system, is subject to cultivate the personality psychology quality. According to the system theory, it will inevitably cultivate students' spirit and personality. As the shape system of individual mental character tendentiousness (ideals, beliefs, outlook, values, interest, etc), enterprise education would play a role of motivation and guiding.

1.1.2 Enterprise Education Is a Core System to Cultivate Students' Higher Comprehensive and Intelligence Ability

The ability ,which cultivated by the enterprise education, involves the various factors of rational knowledge inside and outside of oneself , this ability involved sensitivity, attention, memory, imagination, thought, etc, belongs to intelligence, constitutes the core ability of the enterprise education. The enterprise education divides into three levels, namely the professional ability, management and operation ability and general ability, in such a large ability system, the intelligence run through them all.

1.1.3 Enterprise Education Is a System to Educate Students' an Exploration Ability, Which Thinking and Doing Some Things Creatively and Changeable

One of characters of the enterprise education is to educate students' creative thinking as its' fundamental rule. Enterprise education requires individuals constantly developing their potential, exceeding their minds, keeping pace with the times, constantly transcend the reality and themselves in the social practice.

1.1.4 Enterprise Education Is the System Which Made Students Doing Very Strong Social Practice

Entrepreneurship practice activity is the soil, a stage and a appraisal standard to develop students' career ability.

1.2 The Type of Enterprise Education

Entrepreneurship education developed in China for more than ten years. Since 1997, some colleges and universities began to exploring it. For example, Tsinghua University carried out some education discussion practice, which attracted the attention of the Ministry of Education after exploring in some colleges. In March 2002,the conference was held in Beijing University of Aeronautics and Astronautics, and it confirmed nine schools including Tsinghua University, Beijing University of Aeronautics and Astronautics and so on as practicing colleges of enterprise education. In October 2003 and May 2004, Ministry of Education entrusted the training college of enterprise of Beijing University of Aeronautics and Astronautics to hold twice training course about the enterprise education. Entrepreneurship education was going into a period of diversified development under guide of government. The practicing institutes began to explore enterprise education; it has five types in the following.

1.2.1 The Entrepreneurship Education Model Education Courses-oriented

Renmin University of China offered the "entrepreneur spirit "," venture capital "and other courses for students, which combined the first class and second class focusing on training entrepreneurial awareness of the students for building entrepreneurial knowledge structure in entrepreneurial education.

1.2.2 The Entrepreneurship Education Entities Model Organization as a Carrier

Such as College of Aeronautics and Astronautics of Beijing, it established business park which emphasized on improving entrepreneurial knowledge and skills of the students by the method of entities functioning, teaching students how to entrepreneurship, and funding for student venture financing and advisory services.

1.2.3 Integrated Entrepreneurship Education Model Take Innovation as the Core

Tsinghua University, on the one hand, take the innovative education as the foundation of entrepreneurship education and payed attention to the basic quality of students in the course of professional knowledge. In addition, Tsinghua University established the basic framework and fundamental content of training system of innovation focusing on the overall quality' improving of students. In the whole school, they opened many education courses in the innovation and entrepreneurship. On the other hand, Tsinghua Univerisity holded the social activities in order to promote entrepreneurship education. For example, the implementation of the "business plan competition", setting up a "Science and technology innovation fund", funds for scientific and technological innovation activities for students. School had also established a special science and technology innovation centre to offer students the entrepreneurship and innovation activity instruction, consultation and evaluation.

1.2.4 Entrepreneurship Education Model Which Takes Students as Center

That is represented by Fudan University. This model take entrepreneurship, practice skills, team spirit training, entrepreneurial guidance, business team supporting for undergraduates.

1.2.5 Entrepreneurship Education Model Based on the Practice Activity

As Heilongjiang University, its character of innovation and entrepreneurship education is that, on the one hand, it take full advantage of its resources to build entrepreneurship practice foundation such as the use of research laboratories at university, it opened to students ,so that students can understand the disciplines of cutting-edge, high-tech equipment, display presentation of the latest achievements in scientific research. While at the same time they accepted the part of the outstanding students to participate in research projects, school used internal resources to organize students' venture experience, by operating the services companies, by establishing campus supermarket, bookstalls and tutors center, which could also management independent. Students could experience the whole process and thereby develop their entrepreneurial spirit guided by their teachers. On the other hand, campus entrepreneurial practice base could also be built. For example, it could establish the education model to integrate industry, academia, research, and the enterprise collaboration together, which could enable students to the scene, in-depth post and the atmosphere of entrepreneurship, growth their venture skills. The universities also could organize students' in-depth factory, rural areas, offer technology assistance, legal assistance, special survey, making students receive practice knowledge and competence in summer social practice activities. The universities can encourage students to practice on post taking advantage of their professional knowledge, especially selecting the excellent PhD, Masters to post in the social and business, thus it not only wide their training ways, but improve their overall entrepreneurship.

These five forms of entrepreneurship education are based in science education which fosters entrepreneurship and entrepreneurial capacity of students inevitably. Students are in the position of subordination, being educated in this system, which lagged to new training mechanism of entrepreneurship. The newer one which lets students' initiative or give the first place to students. So this article proposes and discusses construction the "four spiral" model to raising the employment rate of students '.

2 Design of the Model

Based on the purpose to promote the employment of college students, regarding college students as equally important behavior agents, the students constitute a part of "four spiral" model associated with the other three behavior agents.

2.1 The Basic Model

In this paper, the "four spiral "model is from the three spiral models of innovation which originated in biology. Richard Lewontin, a American geneticist, Harvard' professor considered that organisms could not only adapt to the environment, but also select, create and change living environment. The relationship of genes, the organization and the environment is a dialectical relationship which likes three spiral windings, constituting a causal relationship between each other. Professor Henry Essen (Etzkowitz) [2] from the sociology department of American New York State University and professor Roy, Lauder Randolph (Leydesdorff) [3]) from technology development Institute of Amsterdam in the Netherlands, introduced the model into economics, which was used to explain interdependent interactive relationship between the university, government and the enterprises with the development of knowledge economy. The university, government and enterprise' knowledge sharing is the core of the innovation system. The unit of triangular cooperation is the important factor to promote knowledge sharing and innovation. In the process of changing knowledge into productivity, each participant jointly promotes the rise of innovation spiral, realizing goals of the knowledge innovation.

The four spiral model is cultivating the ability of university students in order to improve employment, fully expressing the four powers including the enterprise, government ,university students and entrepreneurial. And it is based on market mechanism of cultivating the ability of university student, showing the cross nonlinear relationship of a "four spiral".

In the model above, the four spiral objects promote the interaction between knowledge information sharing and resource integration by the interaction of these four helicoids. The college students are the main actors, the implementers, organizers and leaders in the whole entrepreneurial process. The university is the culturist to develop the ability of students, and its main government function is providing policy and innovation environment for students and the enterprise is the guiders and imitator for them.

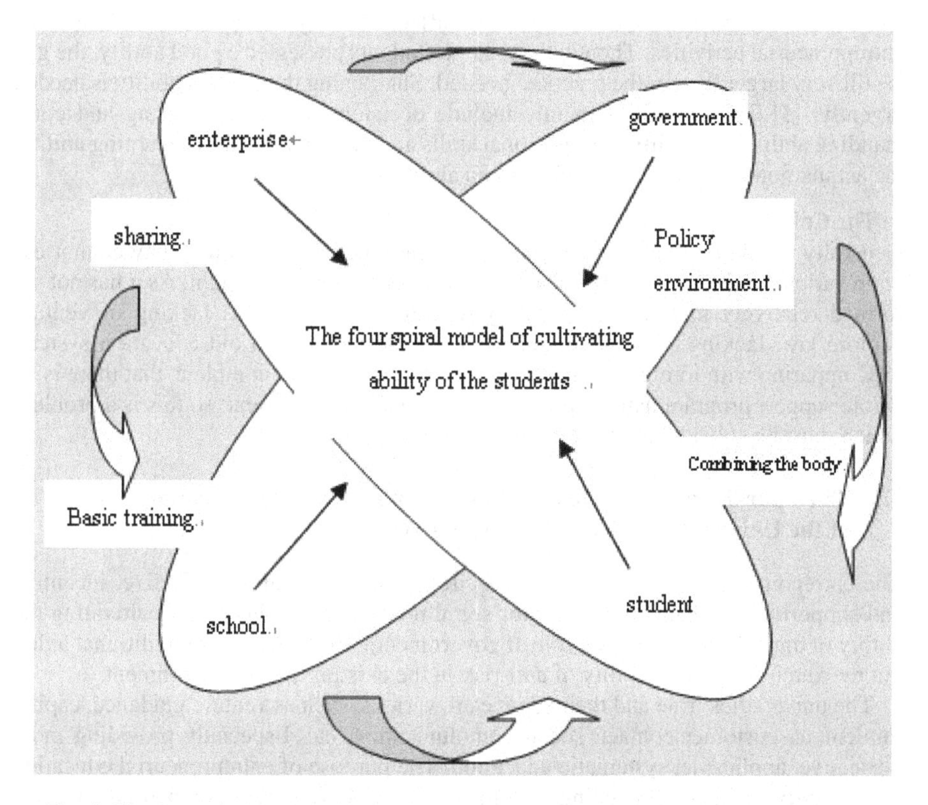

Fig. 1. The four spiral model of cultivating ability of the students on view of employment

1) The Enterprise

The main subject of market is enterprise which deploys technology resources and promotes the knowledge sharing by using the strength of market which is just a driving to help students in business ability training as the main body and the link to the enterprises when they realize the optimized allocation and effective development of human and technology capital for university students. By strengthening and pioneering college students, government, university of knowledge information sharing and cooperation, the various aspects of resources together, with the orientation of market provides the students' business ability training and provides certain experiment field.

2) The Government.
Fulfilling its public service function of government includes the provision of public goods and service (such as infrastructure, etc.). Government departments should provide focused service to the university students by the integration of resources and perfect the policy when students facing with difficulty in financing, expensive site, skills and service problems. They should set up more "green channel" and departments of industry and commerce, taxation, personnel and so on for college students as fast as they can.

3) The College Students Who Want Start an Undertaking.
Ability and self-cultivation of entrepreneurs affects directly the success or failure in the entrepreneurial activities. Though the college students possessed a good ability, the gap is still very larger than in the practice' needed. Sharpening themselves ability is needed urgently. [5] These abilities mainly include decision-making, organizing and commanding ability, flexibility, interpersonal skills and presentation skills, learning ability, the acquisition and processing information ability.

4) The College.
Generally speaking, the entrepreneurship education of Chinese university cannot explain various problems of entrepreneurial activity in theoretical height. As it has not yet formed relatively mature theoretical system and framework, as lacking of venture explore law, lacking of practice and having a shortcoming in old education system. [6]Comparing with foreign business environment mature and complete, that there is no whole support program nationwide on venture in China at present, so this is a problem which must be solved that faced by the colleges.

2.2 The Four Spiral Model for the Purpose of Cultivating the Ability of the University Student's Pioneering Work

The entrepreneurship of the students requires society to establish effective, incentive and supporting mechanism. In the four spiral model for the purpose of cultivating the ability of university student, the role of government is no longer direct traditional aider, but to exercise students' ability of anti-risk in the existing social environment.

The universities' role and their service are various, such as venture guidance, capital implement, customer contact, social communication, etc. Especially providing more normative, applicable, systematic education. The purpose of entrepreneurial education in university is various, including thinking consciousness and entrepreneurial concepts for college students. On the other hand, it can cultivate the various entrepreneurial comprehensive skills, and eventually students will have certain undertaking abilities.

In fact, the enterprise can provide college students with vast experimental site, and the latest technical information from the four interactive organizations, dealing with the market risk and changing technology rapidly at the same time.

3 Main Countermeasures

In order to solve the problem, the main matter is to construct the operation mechanism of the innovation ability cultivation for college students, and training curriculum system to improve students' innovation ability.

First, in order to construct the operation mechanism of innovative ability, we should abandon the single system that universities play a key role, and conduct the all-round system of supporting by the three institutions, improving the system function, fulfilling the role of system of innovative and entrepreneurship education. Then we can construct a good operation mechanism analyzing with the economics system and ensure the long-term execution of students' cultivation in innovative and entrepreneurship ability.

Secondly, for the perfection of cultivation course system of students majored in financial and economics, we should establish a complete cultivation course system for their innovative and entrepreneurship ability. In the process, we need to apply the concept of systems engineering in innovative and entrepreneurship education, linking the entrepreneurial theory and the subject courses. So, it can develop fully the education of innovative and entrepreneurial, and improve competition of college graduates in employment.

References

1. Shi, L.: Discusses ion of raising innovative ability of university students. Management and Science and Technology of Small and Medium-sized Enterprise 24(6), 222 (2010)
2. Etzkowitz, H., Chunyan, Z.: Regional innovation initiator: The entrepreneurial university in various triple helix models, National University of Singapore. In: Triple Helix Conference Theme Paper, vol. 23(6), p. 24 (2007)
3. Leydesdoff, L., Etzkowitz, H.: The dynamics of innovation: from national systems and "Mode 2" to a triple helix of university industry government relation. Research Policy 23(29), 111–112 (2000)
4. Ma, Y., Wang, S.Y.: The triple helix model of Universities, Government and Enterprise. Higher Engineering Education and Research 45(5), 29 (2008)
5. Zhu, X., Qian, C.: College entrepreneurship culture of necessity and the means. China Metallurgy Education 34(50), 31–33 (2004)
6. Cai, W., Wang, X., Huang, X.: Research of entrepreneurial ability system of students based on financial crisis. Fujian Forum (Social Science Education Edition) 56(10), 121–122 (2009)

Teaching Reform Exploration of Automobile Examination and Diagnosis Based on Ability Training

Liang Zhang, Shuzhen Li, Lei Shi, Guofang Li, Lidong Chen,
Shuying Ma, and Lixin Zheng

School of Mechanical and Electrical Engineering,
Hebei Normal University of Science & Technology, Qinhuangdao, Hebei, China
lzhang7608@163.com

Abstract. Aiming at the problem in the course teaching of Automobile Examination and Diagnosis, in order to adapt teaching request under the new situation, and guarantee the quality of teaching, this article carries on the reform exploration from the course content, the teaching method, and the inspection way and so on three aspects. In teaching, the theoretical teaching and the practical teaching are combined organically, the goal is to develop the independent and innovative spirit of students, and raise student's project practical ability, and improve the comprehensive abilities of handling practical problems.

Keywords: automobile examination and diagnosis, teaching reform, teaching method.

1 Introduction

Automobile Examination and Diagnosis is a professional course which the students of agricultural mechanization and its automation specialty are required in our college. Through the teaching of automobile engine and chassis of the detection and diagnosis, the detection and diagnosis of bodywork and accessories, automotive instrumentation and lighting system, the detection of automobile emission and noise detecting content of teaching, students can master the basic knowledge of the automobile detection and diagnosis technology, the correct use of automobile testing instruments and equipment, and master the inspection and diagnosis of automobile method. The curriculum content and automotive technology service occupation post demand are very close, the course teaching not only emphasizes theories, more emphasizes the practice teaching, and cultivates students' ability of analyzing and solving practical problems. In order to improve the quality of teaching, based on a summary of teaching practice of this course, this article carries on the reform exploration from the course content, the teaching method, the experiment teaching and the inspection way and other aspects.

2 Present Situation Analysis of the Course

The current course system of teaching content and teaching method has not adapt to the modern education's development needs, the reformation has to be made. The problems in course teaching are as follows:

H. Tan (Ed.): Knowledge Discovery and Data Mining, AISC 135, pp. 601–605.
springerlink.com

(1) Rapid development of automotive technology has led the automobile check-out facility's renewal and services technical the progress, therefore the instrumentation, the equipment, the examination method, and the examination method has undergone tremendous changes, but education on this change has the hysteresis quality, the curriculum content renews slowly, and falls behind in the new technological development step.

(2) The teaching mode takes the teachers as the center; the force-feeding teaching method is mostly adopted to give a lesson. In the teaching process, the knowledge is paid special attention to, the students' participation and between teachers and students, the interaction between students are of very few, and the learning activity of the students is neglected. It is not good for the enhancement of the students' initiative and enthusiasm, and is not good for the training of student's project practical ability.

(3) The experiment teaching still takes the teachers as the center, students are subsidiary role. There are too constraints for students in the experiment teaching. Before the experiment, students are requested to read the extremely detailed experimental instruction and carried on the experiment strictly in accordance with the experiment procedure and details of the experiment instruction. Students are completely passive participation, so their independent thinking and initiative are bound. Finally, they know its result but not know its reason.

(4) In teaching, it overemphasizes the student's theory result and testing scores, but neglects the student practical ability inspection.

3 Exploration of Teaching Reform

Aiming at the problem in the course teaching of Automobile Examination and Diagnosis, in order to adapt teaching request under the new situation, and guarantee the quality of teaching, and arouse student's study enthusiasm fully, and raise student's project practical ability, this article has carried on the reform exploration from the course content, the teaching method, the experiment teaching and the inspection way and so on, and has made the good teaching progress.

3.1 Teaching Content Reform

In course content selection, it highlights the emphasis, and makes the teaching content and the professional characteristics combined, and emphasizes learning ideas and knowledge utilization. It broadens students' academic horizon and stimulates students' interest in learning with the project background case. With increasing the new technology and the new knowledge promptly, the teaching content is enriched. The course content which reflect the latest scientific developments of the new testing method and analysis means consciously is set, so that students are always able to grasp the latest developments and the development trend in the automotive testing technology in the learning process, like this, it ensures the knowledge of the forward-looking and practical, also helps to stimulate the students' learning enthusiasm.

3.2 Teaching Method Reform

In the teaching, the teaching and learning enthusiasm is made full use of, "the teacher as the center" is transformed for "taking the teacher as the leadership, the student as the main body". The modern teaching method of the students' active participation and creation takes place of the traditional one in which the teacher and textbook are centered with one-way instruction and lecturing. The various teaching methods are adopted to give full play to students' initiative, arouse students interesting for study.

(1) Heuristic interaction teaching method
While teaching, teachers start the classroom instruction by certain questions, guide the student to follow the question to ponder, and make the students' thinking active. It actives classroom atmosphere, promotes students to ponder independently, ask questions and answer questions, stimulates the student potential, and transforms the student passive learning into the active learning by enhancing teacher-student interaction. For example, in the teaching of cylinder sealing detection and diagnosis analysis, which components does guarantee air cylinder's leak-proof quality is put forward firstly, then teacher organizes students to review related knowledge in automobile structure; secondly, which does the air cylinder leak-proof quality's evaluating indicator have is put forward, then the teacher and students review the related knowledge in automobile theory together, after the student have the right key, it can continue to in-depth analysis. Which problems do the building block have, the air cylinder leak-proof quality can worse is given continuously; finally, students are guided to identify all the influence factors of cylinder seal variation. Through the layers of the elicitation, it can help students to link automobile structure, automobile testing and fault diagnosis knowledge, cultivate the students' knowledge system, and train the students' ability to analyze and solve problems[1].

(2) Case analysis teaching method
The teaching goal of Auto detecting and fault diagnosis technology is to make students master the basic knowledge of the automobile detection and diagnosis technology, the correct use of automobile testing instruments and equipment, and master the inspection and diagnosis of automobile method. When teaching, teachers give the typical cases to introduce lectures. It can stimulate student's academic motivation and the intellectual curiosity, encourage the activity and initiative of the students to study, enhance the teachers and students interaction in the classroom, cultivate the awareness of the students to apply and create in study, and develop the ability of the students to analyze and solve problems[2]. Through this process, students not only learn troubleshooting skills, but also master the fault causes and relevant theoretical knowledge.

(3) Scene teaching method
The course teaching not only emphasizes theories, but also emphasizes practice teaching, and cultivates students' ability of analyzing and solving practical problems. Relying on the training room, the separation teaching mode of theory teaching and experiment teaching is changed, on-site teaching new pattern is carried out[3]. In the training room, through combining teaching form with teaching, learning, doing, examination and evaluation, it make the teachers teach, students learning operation,

practice and training, evaluation throughout the whole teaching process, learning in practice, and doing in study, it achieves the integration of teaching and learning. So it cultivates students' occupation accomplishment, improves the students' abilities, the teaching mode that the original classroom theory and professional classroom training of science are respectively carried out is changed, it realizes the teaching process of integration of theory and practice, and improve the teaching effect.

3.3 Teaching Means Reform

(1) Using the multimedia technology
With the rapid development of modern educational technology, multimedia technology and network technology are widely used in the field of education, teaching methods are diverse, the introduction of multimedia teaching technology in teaching can change the traditional teaching mode by simple blackboard, chalk type classroom teaching which lacks image and the vivid characteristics. It makes figure, text, image, audio and other information combined together, the class information and interesting are also increased greatly, and the shortage of traditional education is made up. The use of multimedia teaching means that introduce various detection equipments and knowledge to students makes the contents of course easy, simple and convenient for students to master knowledge and memory.

(2) Using the real teaching means
Using the material object to teach, the teaching intuitive can be strengthened. For example, in the teaching of electronic engine control system testing and diagnosis, it enhances students' perceptual knowledge that teachers adapt physical teaching means, and use multimeters and other testing instruments to test auto parts, such as sensor, fuel injector, electric steam pump etc, at the same time, they give presentation. Thus the abstract theory and practice are organically integrated, so that students recognize and understand from the three-dimensional fully.

3.4 Inspection Way Reform

Exam is used to judge students' learning effects and means of teaching quality. The practicality of this curriculum characteristic is very strong, so the assessment methods should be the combination of process evaluation and final evaluation, knowledge and skill evaluation, emphasize the importance of outstanding skills and process assessment. The sole inspection way of the curriculum is changed, and the inspection way that the written, peacetime result and experiment examination are combined is adopted to comprehensively evaluate the degree of the course content that students master.

4 Conclusion

Practice proves that it may greatly mobilize the learning initiative of students and improve the teaching effect with the reform of teaching method, the experiment teaching and the inspection way, teachers from the past teaching dominance into classroom teaching organizer, director, while students from the original in the class into the classroom active exploration.

References

1. Sang, N.: Teaching Exploration about Automobile Examination and Fault Diagnosis Course. Journal of Jiangsu Teachers University of Technology (Natural Science Edition) 14(3), 36–39 (2008)
2. Liu, G.B.: Teaching Exploration about Automobile Examination and Fault Diagnosis Technology Course in Application Colleges and Universities. The Science Education Article Collects 4, 127–128 (2008)
3. Automobile comprehensive fault diagnosis of Xingtai vocational and technical college, http://www1.xpc.edu.cn/dept/yzk/zdjc/index.html

Research on Teaching Reform of Mechanism and Machine Theory

Xiulong Chen*, Xiuli Qi, and Suyu Wang

Shandong University of Science and Technology,
Department of Mechanical and Electronic Engineering, Qingdao, 266510, Shandong, China
cxldy99@163.com

Abstract. In order to improve students' engineering qualities, engineering practice ability, innovative ideology and innovative ability, enhance students' learning interest and learning efficiency, establish a teaching program for engineering technology students, a series of reforming practice, which includes course architecture, teaching content, teaching method and means, course design and bilingual teaching, has been conducted. The teaching practice of mechanism and machine theory shows that the significant achievements has been made by the teaching reform. This research can provide the important practice of teaching reform for different engineering courses.

Keywords: mechanism and machine theory, teaching reform, practice, innovative ability.

1 Introduction

As a mechanical backbone course of the professional technical basis, mechanism and machine theory studies the common problems of the mechanical principle, kinematics and mechanical properties and design methods, and has strong engineering nature, practicality, applicability. Through the study of mechanism and machine theory, we can not only lay a solid theoretical foundation to learn further course of mechanical engineering professional, but also strengthen the adaptability of the student in the mechanical technology work, cultivate the students' mechanical design ability, innovation ability and practice ability which is power source for technology reform and theory innovation of students. On one hand, the basic requirements of mechanism and machine theory is to enable students to master the basic theory, basic knowledge and basic skills of modern mechanical engineering work. on the other hand, it's to expand the students' knowledge, acquire new knowledge, and track the development of the discipline, inspire students' ideas, and broaden students' horizons. In recent years, with the rapid development of modern science and technology, the concept of machinery has been developed, and mechanical engineering discipline has appeared a big change. With teaching reforming and deepening of mechanism and machine theory, the curriculum system of mechanism and machine theory is being continuously optimized, and the teaching content is being updated and increased (mainly displays in

* Corresponding author.

H. Tan (Ed.): Knowledge Discovery and Data Mining, AISC 135, pp. 607–613.
springerlink.com

the students' ability of innovation and engineering practice ability). In this new opening situation, the demand for students' English level is also being improved with the increasing frequency of international academic exchange. Therefore, in order to adapt to new teaching requirements, the reform of mechanism and machine theory, which includes centering on the course system, teaching contents, teaching method, teaching means, course design, bilingual teaching and so on, is imperative.

2 Existing Problems in Teaching Reform of Mechanism and Machine Theory

Some developed countries, such as the USA, Germany, Japan, attach great importance to the teaching reform of mechanism and machine theory course, they take the training courses as one of the most important course to develop the mechanical engineering students' innovation ability and practical ability. In order to realize the goal put forward in 1990s that higher engineering education should return to the project, many foreign universities and colleges carry out the teaching reform and practice research of mechanism and machine theory in light of their own actual strength, and make some theoretical and practical results[1] .

At present, the domestic universities also have done lots of beneficial research and practice around construction and teaching reform of mechanism and machine theory. Now many researchers present that the traditional teaching system of mechanism and machine theory ,which can not adapt to the requirement of talent cultivation in twenty-first Century, is mainly to guide mechanical analysis. Xiaohong Yu from Beijing University of Science and Technology has studied from three aspects such as optimizing the curriculum, integrating course content and improving teaching conditions, improving teaching methods and attaching importance to the construction of teaching faculty, and also gave some good suggestions. In view of the traditional mechanical experiment problems, Ping Mou from University of Electronic Science and Technology of China, has excavated innovation points from the existing experimental device, conducted mechanical principle experiment reform, passed on the improvement of the experimental device, developed new experiments for the purpose in designing, aroused students' experimental enthusiasm, and improved the quality of experimental class in view of the traditional mechanical experiment problems. Mu Tarif Ahmed from Xinjiang University has integrated with the course characteristics and teaching requirements of mechanism and machine theory, discussed and summarized from curriculum designing, the optimization of curriculum system, teaching and learning as well as the optimization of course examination, homework and score, and at last put forward a new knowledge and experience of the teaching reform of mechanism and machine theory. Lianqing Ji from Zhengzhou University of light industry has combined with mechanism and machine theory course design characteristics, reformed the content of the curriculum design and system for cultivating students' innovation consciousness in the curriculum design, and put forward some preliminary ideas and practice. Chunhua Fei from Ningbo University, Hairong Fang from Northern Jiaotong University, weidong Guo from Beihang University, xinme Yuan from Yangtze University and Dayi Yang [2-5] from Changchun Institute of Technology, have researched on the teaching reform of mechanism and machine theory, put forward reasonable suggestions, and accumulated valuable experience which play an important role in mechanical professional talent training quality especially the students' relevant ability.

In summary, the research of this field is still existing the following problems: ① The research surrounding mechanism and machine theory course construction and reform are still preliminary, it mainly reflects in the curriculum teaching that the relatively backward content, weak practicality, relatively single teaching method, low the teaching efficiency, the old content of curriculum design, weak professional characteristics, but there is urgent need for more intensive study; ②Teaching practice is few, and most of teaching reform results have not been completely applied to teaching practice; ③The research of this field only reforms certain aspect of the course, not totally reform engineering curriculum accord with the systematic principle.

According to the position that the mechanical engineering major in our college cultivating senior engineering and technical personnel, we establish the aim of mechanism and machine theory is to cultivate students' engineering quality, enhance students' innovation ability, improve the ability to analyze and solve problems. At present, the difference of our students' overall quality is bigger, and new contents of mechanism and machine theory are constantly enriched, teaching requirements are continuously improved, but the teaching hours of mechanism and machine theory are continually reduced. Therefore, the contradictions of teaching hours and teaching contents are outstanding.

In view of these problems and the actual situation of our school, we must carry out the research on teaching reform of mechanism and machine theory, reform the teaching content and teaching methods, use modern teaching means, improve the curriculum design and practical teaching conditions, perfect curriculum teaching system, improve the teaching effect, cultivate students' ability of innovation and practice, strengthen the curriculum construction, and form curriculum implementation scheme suiting to cultivate senior engineering undergraduate personnel.

3 The Goals of Teaching Reform of Mechanism and Machine Theory

Mechanism and machine theory, which is the key of mechanical innovation design and a most effectively course to cultivate the students' innovation ability and engineering quality in the mechanical courses, takes the machine and mechanism as the research object, mainly researches the basic principle and kinematic scheme design of mechanical product. Through the study of teaching reform of mechanism and machine theory, we stimulate the student to study the mechanism curriculum the enthusiasm, cultivate students' ability of mechanical kinematic scheme design and improve students' project quality, the practical ability, the innovative ideology and innovation ability. The goals of teaching reform of mechanism and machine theory are illuminated as follows.

(1) To build a new curriculum system " the main line is designing and the base is analysis and the end-result is the design of mechanical system plan", on the basis of the new curriculum system, we complete the optimization of teaching content,

compose a Chinese electronic teaching note of the mechanical principle and complete a set of Chinese multimedia course ware.

(2) We establish a multimedia material library of mechanism and machine theory, perfect theory of multimedia teaching software and enhance students' learning interest and teaching efficiency.

(3)We establish a curriculum design library of mechanism and machine theory, bring the computer technology into the curriculum design and train the students' ability to proceed with mechanism design and analysis by modern means.

(4)We compose a english electronic teaching note of the mechanical principle and complete a set of english multimedia courseware and english three-dimensional teaching material.

4 Main Content of Teaching Reform of Mechanism and Machine Theory

We should closely combine course characteristics with teaching requirements of mechanism and machine theory and study the teaching reform of mechanism and machine theory from the teaching content and course system, teaching method and means, the curriculum design and curriculum construction of bilingual teaching.

4.1 The Optimization of Curriculum System and the Renewal of Teaching Content

According to the guiding ideology that the foundation is the analysis and the core is the design and the computer-aided design and analysis should be strengthened, we explore on the motion design of mechanism system and dynamic design of mechanical system to organize the thought of course sections, integrate the courses and optimize curriculum system of mechanism and machine theory with by design, re-organize teaching content and reinforce engineering-thinking and mechanical specialty.

When we organize the curriculum system, we discuss curriculum chapters in line with the design and strengthen mechanism design elements. According to the motion design and dynamic design of mechanism system, mechanism and machine theory system is divided into two parts: ①The part of kinematic design of mechanism system, that mainly includes mechanism structure, the velocity instantaneous center method, the velocity and acceleration analysis and calculation, planar linkages and design, cam mechanism and design, gear mechanism and design, common institutions and combinatory mechanism and design, mechanical system design. ②The part of dynamic design of mechanical system, that mainly includes force analysis of mechanism, mechanical efficiency, balance of machinery, mechanical system dynamics. And the research on kinematics of mechanism system by analysis and statics study is increased, and a number of easily understood content is arranged, and that the mechanical friction, mechanical efficiency and mechanical system dynamics teaching are enhanced.

In order to enable students to know developmental direction and frontier field of mechanism and machine theory, we add our scientific research items are closely related to the contents of this course in recent years. Some researches are introduced in the undergraduate teaching classroom, such as the China postdoctoral science foundation project "the basic theory research of parallel coordinate measuring machine ", the science and technology project in Shandong colleges and universities " nonlinear elastic dynamics modeling and dynamical behavior analysis of high-speed spatial parallel machine" and enterprise project "the development of servo mechanical press", these scientific research items broaden the students' horizons, enrich the students' knowledge and obtain good teaching effect.

4.2 The Reform of Teaching Methods and Means

On the basis of select and supplement to the teaching content, we can apply modern teaching technology, such as virtual reality, the network teaching, multimedia, VCD video clips and on-line question-answering, the integration of varying teaching methods, and pursue the teaching method reform to improve the students' learning interest and initiative and the teaching efficiency.

We can increase the number of experiments, which includes the visit of mechanical observing laboratory and innovation laboratory, to improve students' perceptions of typical machines and develop students' interest in learning mechanism and machine theory. According to the physical models (such as rod mechanism, cam mechanism and gear mechanism), we can explain the importance of theoretical knowledge in mechanism and machine theory, and encourage students to learn the content deeply. Through the physical model, we can increase the visual perceptions of students, save their time and improve their learning efficiency. we also can introduce some of the machine, such as machine tools, presses, cars, sewing machines, to stimulate students' thirst for knowledge.

4.3 The Educational Reform of Curriculum Design

In order to cultivate of the students' creativity, practical ability and the tight control of the design process, we must carry out the educational reform of curriculum design. The reform of curriculum design includes setting up a number of the topics of the course design that close contact actual enterprise and have no great difficulty and reflect the characteristics of mining machinery and stimulate students' interest in learning, exploring a new model of curriculum design which combine with students' technological innovation activities and various types of college students' innovative product design competition, cultivating students' basic ability of engineering design by applying computer technology(the softwares such as UG, PRO/E, ADAMS are introduced into the curriculum design of mechanism and machine theory).

On the basis of understanding and coping the experience of teaching reform which are gained in curriculum design of mechanism and machine theory, a number of the topics of the curriculum design, which have the enterprise practical application background and mining machinery speciality, can be developed and accumulated. Then the library of the topics of the curriculum design that can avoid the routine design content is established. We can also introduce into a variety of teaching methods to improve the students' ability to solve the practical engineering problems(For example, we tried to introduce comprehensive experiment and innovation experiment in curriculum design, and conduct students to certify the validity and rationality of mechanism design by using the experiment.). We combine with curriculum design and students' technological innovation activities, and pay attention to stimulate students' creative potential, encourage and guide the students to actively participate in the innovative design competition, such as the competition of challenge cup of science and technology. We can also promote the application of computer technology in the curriculum design, and introduce the technology of virtual prototype into the curriculum design.

For example, combining with the production practice of my school students in forging machinery plant, the practical topic proposed is to design a multi-linkage pressure machine. By the process of mechanisms synthesis and selection, kinematic scales design and performance analyses, the practical topic encourage students to know a mechanical system as a whole, consolidate the theoretical knowledge learned, and improve srudents' quality and innovation. Besides, after the mechanism design of multi-linkage pressure machine is completed, students must use ADAMS software to verify the feasibility and correctness of analysis results above. This course can stimulate students' interest in learning and improve students' ability of computer application.

4.4 The Construction of Bilingual Teaching of Mechanism and Machine Theory

In order to narrow its gap with advanced world standards, realize the oversea latest research, provide students with a entirely new knowledge platform and culture students' level of professional foreign language, the construction of bilingual teaching of mechanism and machine theory, which includes writing English electronic teaching materials, establishing english multimedia teaching courseware, constructding stereoscopic teaching materials as well as improving methods and means of teaching english must be enforced. We maintain a small group setting (each class has no more than 30 students) in bilingual class' teaching. The students choosed to take part in bilingual class must have relatively high-levels of english, strong ability in reading english, some good learning ability and creative potential. Combined with the basic requirement of courses and the practical experience of our college, the teaching materials are recompiled on the basis of introducing teaching materials and multimedia courseware of mechanism and machine theory from the key colleges in USA.

5 Conclusion

According to the characteristics and existing problems of mechanism and machine theory course construction, we undertake a series of reform on the curriculum system and teaching content, teaching methods and means, and curriculum design and curriculum construction of bilingual teaching. The apparent effect have been made by the reform. The reform can meet the needs of course constructionour of machine theory of our school, cultivate and improve students' engineering quality and innovative ability. And the research has a great significance in the teaching reform of engineering specialty course in universities.

Acknowledgments. The work is supported by the project of starts of "SDUST", the teaching research foundation of Shandong university [Grant No. 2009223], and the project of specialty professional in Shandong province.

References

1. Guo, W.D., Liu, R., Li, J.T.: Reform and Practice of the Series and Content of the Course of Theory of Machines and Mechanisms. Journal of Taiyuan University of Technology (Social Sciences Edition) 26, 7–10 (2008)
2. Fang, H.R., Fang, Y.F.: On the Effect of Reform in Practice Education Programs in the Course of "Principle of Machinery" on Cultivating Students Creativity and Overall Quality. Journal of Northern Jiaotong University (Social Sciences Edition) 2, 63–66 (2003)
3. Guo, W.D., Liu, R., Li, J.T.: Reform and practice of the teaching method and means of theory of machines and mechanisms course. Journal of Taiyuan University of Technology (Social Sciences Edition) 26, 50–60 (2008)
4. Yuan, X.M., Hua, J., Zhou, C.X.: Teaching reform exploration on course exercise in the theory of machines. Journal of Yangtze University (Natural Science Edition) 5, 345–346 (2008)
5. Yang, D.Y., Chen, L.M.: Research and practice on the course system reform in mechanical theory. Journal of Changchun Institute of Technology (Social Sciences Edition) 9, 86–88 (2008)

Acknowledgments. [illegible] work is supported by the project of [illegible] of SDUST [illegible] teaching research foundation of Shandong University [illegible] (Grant No. [illegible]), and the project of Specialty [illegible] in Shandong province.

References

1. [illegible] Y. [illegible]: [illegible] the Curriculum [illegible] of the [illegible] Machines and Mechanisms [illegible] of Taiyuan University of Technology [illegible] (2006)
2. [illegible] Y. [illegible] on Teaching [illegible]
[illegible]

Research on Cultivation Model to Full-Time Postgraduate of Engineering Major

Xiulong Chen*, Yu Deng, Xiuli Qi, and Suyu Wang

Shandong University of Science and Technology,
Department of Mechanical and Electronic Engineering, Qingdao, 266510, Shandong, China
cxldy99@163.com

Abstract. In order to enhance cultivation quality and cultivation efficiency of full-time postgraduates of engineering major, a series of reforming practice, such as redesigning of teaching goals, making individualized training plan, implementing the postgraduate tutors system, developing international academic exchanges and lecture of professional knowledge, constructing postgraduate culture base and establishing evaluation quality system of postgraduate education, has been conducted. The significant achievements has been made by the reform. This research can promote the development and improvement of cultivation model to full-time postgraduates of engineering major.

Keywords: full-time postgraduates of engineering major, cultivation model, reform.

1 Introduction

As the basis of cultivating scientific and technological innovation talents, full-time postgraduate education of engineering major is the key to promote one nation's comprehensive strength. For a long time, our country's full-time postgraduate of engineering major have been cultivated in accordance with the "academic" type talents [1], but at present, the request of domestic employment market to full-time postgraduate of engineering major is the mixture of academic talents and applied talents instead of original pure academic talents. In recent years, college graduates of full-time postgraduate of engineering major have a portion to choose to work directly after graduation, some postgraduate choose further study for doctor's degree. For those who choose to work some go to the enterprise, some work in the scientific research institutes, colleges and universities. Therefore, according to the request of society to full-time postgraduate of engineering major, we must change cultivation system of the traditional academic postgraduate, re-explore the cultivation model for engineering postgraduates to really realize the request of society to them from the cultivation point of view to further improve the training quality of them.

The paper regards full-time postgraduate of engineering major as the study object, which intends to systematically study a series of reforming practice, such as redesigning the teaching goals, making individualized training plan, implementing the system

* Corresponding author.

H. Tan (Ed.): Knowledge Discovery and Data Mining, AISC 135, pp. 615–620.
springerlink.com

of supervisors guiding the graduate, developing international academic exchanges and lectures of professional knowledge, constructing multi-postgraduate culture base and establishing quality evaluation system of postgraduate education, and finally achieve the purpose of the classification of cultivation to postgraduates according to different cultivation goal and model, so as to enhance cultivation quality and cultivation efficiency of full-time postgraduates of engineering major in our school.

2 The Existing Problems in the Cultivation of Postgraduate of Engineering Major

At present, the european countries simultaneously cultivate postgraduate by academic type who are search-based talents and occupation type who are applied talents, especially pay more attention to occupation postgraduate [2]. The United States not only divides the postgraduate education into four types, including affiliated postgraduate education, occupation development postgraduate education, apprenticeship postgraduate education and community-centered postgraduate education, but also uses different cultivation model according to the different types of postgraduate education.

So far many colleges have begun to study postgraduate cultivation model and a lot of achievements have been obtained. The academician Hao Jiming pointed out that the postgraduate cultivation goals should be changed with the social demand, and we should enhance professional practical ability and occupation ability, along with the postgraduate enrollment proportion enlargement. Zheng Xiaoqi from Beihang University think that postgraduate cultivation should set out from our country's social economic development needs, selectively develop academic type, application type, professional talents. In the postgraduate training orientation and postgraduate educational system seminar held in Harbin Institute of Technology, the delegates agreed that we should establish a pluralism of postgraduate cultivation model, execute classification to education, separate education-oriented research and academic postgraduate from employment-oriented curriculum and applied postgraduate.

In conclusion, foreign countries have been relatively perfect studied engineering postgraduate cultivation model in the theory and practice research, and made great progress recently, while China has just started the relevant research. Though all the experts of colleges and departments of education have been fully aware of the necessity and urgency, but not study the relevant content in depth and formulate the relevant scheme. At present the main existing problems are pointed out as follows:

(1) The relevant theoretical research about cultivating model reform of postgraduate of engineering major is still lack, just stay on the preliminary understanding of basis; and the research content, which includs redesigning of teaching goals and studying problems of cultivation model in the different teaching goals, has rarely carried out.
(2) We are not sufficiently knowing foreign postgraduate educational system and cultivation model. and we regard less of fully use the relevant experience of other countries for reference.
(3) We study on more theory of cultivation model, but rarely research on implementation scheme of postgraduate cultivation model.

Therefore, it is necessary for us to rely on the relevant research results, intensively conduct research on problems of full-time postgraduate of engineering major cultivating model reform in theory and practice. This research can promote the development and improvement of cultivation model to full-time postgraduates of engineering major.

3 Reform the Target of Engineering Postgraduate Cultivation Model

The paper encloses three aspects to study the reform of full-time postgraduate of engineering major cultivation model such as designing of teaching goals, establishing the different training objectives of cultivation model and cultivation model research and practice. The specific goals are explained as follow: (1)We should combine with foreign well-known university full-time postgraduate of engineering major training experience to find out the deficiencies and problems of our country to be urgently solved, put forward to fit our country social development of engineering postgraduate cultivation target, formulate postgraduate of engineering major cultivation model according to different teaching goals. (2)We should take the mechanical and electrical engineering in our school as example to carry out practice research on reforming cultivation model.

4 Reform of Postgraduate of Engineering Major Cultivation Model

4.1 Redesigning of the Teaching Goals

The full-time postgraduate of engineering major are divided into academic degree and the application of professional master's degree, and they are respectively cultivated according to academic cultivation model and practical cultivation model. For the academic postgraduate, we should stress to develop their academic ability, improve their theoretical innovation level. So we can choose a number of academic postgraduates who have a broad theoretical basis and strong research strength to adopt a "2+3" scheme, which is that the postgraduate should finish all master's degree courses within 2 years, and then start the doctorate in the following 3 years. The mutual integration of studying in master stage and doctoral stage enable students to have a more focused time engaged in scientific research. For the applied postgraduate, we should focus on cultivating students' practical ability, research ability and the ability to solve practical problems, endow them with applicability, wide range of knowledge, high quality, strong applied talents and enhance their competitiveness in employment and work. Postgraduate may be required to take 1 year to finish their courses, take the remaining 1 year to 2 years to finish degree papers with better appliance in the projects.

4.2 Making Individualized Training Plan

Combining with the master's professional orientation, employment preference and hobbies, we make individualized training plan according to the different training objectives. We strive to build a variety of different combinations of training plan, such as academic postgraduate and applied postgraduate using different training plans, the postgraduate guided by studying the theory or oriented by solving the practical problem adopting different training plans, the postgraduate given priority to solve practical problems in business or publish academic papers using different training plans and so on. Cultivation of full-time postgraduate of engineering major should pay more attention to closely integrate selection in the study life and planning in the employment, postgraduate courses offered and students' academic tendency and basic skills. The courses offered should reflect the theory and application. We should increase the proportion of elective courses in the curriculum and encourage postgraduate and teachers in other colleges to study and research on interdisciplinary courses.

For instance, the training plan of full-time postgraduate of engineering major is divided into academic major and application major in mechanical engineering discipline of our university. In the academic training plan, basic theoretical courses include matrix theory, numerical analysis and elasto-plastic mechanics. Professional courses include modern design theory and method, computer control and interface technology and mechanical and electrical system analysis and design. The professional elective courses, which include CNC processing theory and programming, differential geometry, plastic forming theory, computer integrated manufacturing, finite element method and so on, are largely offered (account for about 80% of the courses). While in the application training plan, the basic theoretical courses include engineering mathematics and modern design theory and method. Professional courses include mechanical electronics engineering, hydraulic control, numerical control technology and CAD/CAM. The professional elective courses, which include PLC control technology, the principle and application of DSP, finite element and its application, digital signal processing, the design of single-chip microcomputer application system and practice, automobile design, mechanical reliability design and so on, are largely offered (account for about 85% of the courses), and are closely linked with the practical engineering.

4.3 Implementing the Postgraduate Tutor Group System

We should strictly implement the selection of tutors for full-time postgraduate of engineering major, simultaneously stress tutors' educational background, academic level and academic training, carry out tutorial annual examination and appointment system and stoutly cancel the unqualified tutors. We should change single tutorial system (one tutor to one student and one tutor to many student) to tutor group system (many tutor to one student and many tutor to many student). The teachers of tutor groups have different academic background, knowledge structure and different research directions. Therefore, adopting tutor group system to guide postgraduate can increase students' knowledge, broaden their specialty areas, enable students to master a wide professional knowledge, facilitate interdisciplinary cross and enhance cultivation quality of postgraduate. For the applied postgraduate, we should employ one or several tutors in order to enhance their practical operated ability and increase their competitiveness.

4.4 Developing International Academic Exchanges and Lectures of Professional Knowledge

In order to let the postgraduate know the latest research results of their subject at home and abroad and widen their knowledge, the construction of international academic exchanges system has been strengthened. We positively organize, guide and support postgraduate to walk out of school, participate academic exchange activities at home and abroad to build a multiple level, channel, form, perspective of high level of academic exchange platform, establish postgraduate academic exchanges system and finalize the funds of postgraduate academic exchange activities. We should invite domestic and overseas scholars and academic master to open professional advanced interdisciplinary academic lectures, lead postgraduates to communicate with distinguished scholar academic masters face to face and create a free and equal, rich and healthy academic atmosphere to arouse their intense learning desire and creative mind and create desire, open postgraduate academic vision, enhance the research and cultivation quality of postgraduate.

4.5 Constructing the Multi-cultivate Base for the Postgraduate

In order to cultivate engineering postgraduate practice ability and innovation ability, the multi-cultivate base for the postgraduate has been constructed. On the basis of establishing laboratory and practice base on campus, the culture base for postgraduate are vigorously expanded out of the school. We encourage postgraduate to walk into enterprises in the stage of graduation thesis, participate the development period of the new technology and new product, mainly study enterprise's practical problems in their graduation theses, and make full use of the good software and hardware resource to solve the enterprise's technical problems. It not only cultivate postgraduates' engineering and scientific research quality, but also broaden their employment channel. For example, our school has successively established training bases for postgraduate in Haier Group, Aucma Group etc. We regular organize postgraduate to those enterprises to study, carry out technical training and participate the development period of actual projects, so that the majority of postgraduate have very good exercise in practice.

4.6 The Establishing Quality Evaluation System of Postgraduate Education

The establishing quality evaluation system of postgraduate education,which is the reliable safeguard to enhance cultivation quality of full-time postgraduates of engineering major, can mobilize activity of postgraduate to study. In the quality evaluation system of postgraduate education, we should not only pay attention to the research ability of students, but also pay close attention to their academic morality and willpower. We conduct a comprehensive inspection of postgraduate study and research by a number of indexes, such as theoretical courses, participating in research projects' level and contribution, academic papers, the level of foreign language, computer level, the national patent, academic conferences and social practice. To applied postgraduates, we do not advocate the master of publish research papers, mainly evaluate them by the importance, difficulty and contribution of participating subjects of enterprises or research units, encourage them to participate practical subjects to study for the purpose of application and cultivate their ability of engineering practice.

5 Conclusion

According to existing problems researched on cultivation model to full-time postgraduates of engineering major, a series of reforming practice, such as redesigning of teaching goals, making individualized training plan, implementing the postgraduate tutor group system, developing international academic exchanges and lectures of professional knowledge, constructing multi-postgraduate culture base and establishing evaluation quality system of postgraduate education, have been conducted. The significant achievements have been made by the reform. The reform has the vital significance to cultivate engineering postgraduate innovative ability and creative ability, and enhance the cultivation quality and cultivation efficiency for postgraduate of engineering major.

Acknowledgments. The work is supported by creative education plan for Shandong university of science and technology postgraduate, the teaching research foundation of Shandong university [Grant No. 2009223], and the project of specialty professional in Shandong province.

References

1. Liu, G.: Study on Cultivation Model of Practice-oriented Talents in Local Universities. China Higher Education Research (5), 7–9 (2006)
2. Wu, Y., Ding, X.M.: Comparative study on the management system of postgraduate education in America, Britain, Russia and Japan. China Higher Education Research (2), 50–53 (2006)
3. Min, L.: Research on degree types and academic systems of postgraduate in Europe and its significance to China. Academic Degrees & Graduate Education (12), 38–41 (2003)
4. Zhang, Y.N.: Research of master graduate cultivation mode reformation and practice for engineering course in china. Dissertation for the Master Degree in Management, Harbin Institute of Technology (2007)
5. Zheng, X.Q., Xie, L.X., Xie, S.L.: The orientation and development of postgraduate education in research-oriented universities. Academic Degrees & Graduate Education (2), 10–14 (2006)

Discussion on Measures of Training Innovative Talents in Science and Technology in Regional Colleges

Zhaoxia Fu[1], Yufeng Duan[2], and Chen Wang[1]

[1] Department of Science and Research, Hebei University of Science and Technology, Shijiazhuang, Hebei, China
[2] School of Material Science and Technology, Hebei University of Science and Technology, Shijiazhuang, Hebei, China
fzx68@hebust.edu.cn, duanyufengz@yahoo.com.cn, kychu2008@163.com

Abstract. Based on analysis of status quo of training innovative talents in science and technology in regional colleges, this paper puts forward several measures to solve existing problems. Regional colleges should devote much more attention to the cultivating of innovative talents in science and technology. Some effective measures should be taken to help outstanding young teachers with potential talent in research become outstanding innovative talents. Several measures were discussed in this paper, such as financially supporting outstanding young teachers for research project, building a platform of research for young teachers with the aid of construction of key subjects, and offering opportunities of international training and exchange to outstanding young teachers. Reasonable incentives mechanism for innovative talents should be established including using the incentive resource effectively.

Keywords: innovative talents in science and technology, regional colleges, training system, incentive pattern.

1 Introduction

It forms a huge demand for talents in science and technology because the science and technology of rapid development has a huge boos. Science and technology innovation is the source of economic growth and determines the strength of national competitiveness. China has been the largest country in scientific and technological human resources in the world, but far from a powerful country in human resources. The overall innovation capability of talents in science and technology determines the scientific development level of the country. So cultivating and training innovative talents in science and technology is the key to implement talents strategy for powerful nation. In academicians meeting of the Chinese academy of sciences and Chinese academy of engineering in 2006 president HU Jintao stressed that:" The talents are the key to build an innovative country, especially innovative talents in science and technology. Without a large team of innovative talents in science and technology as support it is impossible to achieve the goal of building an innovative country. The competition in overall national strength all over the world is the competition of talents, especially innovative talents in the final analysis." Recognizing the importance of innovative talents in

H. Tan (Ed.): Knowledge Discovery and Data Mining, AISC 135, pp. 621–626.
springerlink.com

science and technology, the whole country is implementing talent strategy. Regional governments of many developed cities successively unveil a lot of efforts to introduce high-level talents and many famous schools have provided substantial financial funding for introducing talents. In the fighting of the introduction of talents in science and technology regional colleges are at a disadvantage due to the relatively weak in academic foundation and material conditions. On the one hand, colleges carry the teaching task to cultivate students to be innovative talents, on the other hand, also pay attention to the training of innovative talents in science and technology in the teachers. So for colleges it has great significance to self train innovative talents in science and technology.

2 The Current Problems in Cultivating and Training Innovative Talents in Science and Technology for Regional Colleges

With the further implementation of talents strategy for powerful nation the introduction and cultivation of innovative talents in science and technology has became a top priority. The quantity and quality of innovative talents in science and technology determines the competition of colleges. Now colleges have unveiled munificent policy to introduce talents in science and technology, but there are many questions such as stressing introduction, lighting management and neglecting culture and so on. It lacks of strict supervision system how to further develop under the new environment early for these talents in science and technology who have been introduced and lacks effective policies how to make them maintain further creative energy. Usually we very positively introduce talents and take base engineering and digital engineering of introducing talents as political achievement. But it lacks of scientific, long-term supervision, incentives, management strategies how to utilize innovative talents introduced in science and technology and train them to further enhance the level, realizehigher value, and do more innovative results. On the other hand, it is easy to neglect the discovering and developing of existing talents within colleges. In fact, discovering and developing existing talents within colleges and taking them early growth innovative talents in science and technology have more important strategic significance for regional colleges that have been at a comparative disadvantage in the competition of introducing talents.

Human studies show that innovative talents in science and technology have a certain time scope, namely history dynamic. In other words, the majority of them only maintain academic leading or technology leading in a certain period. So on the one hand, we must take appropriate strategies to supervise and support existing innovative talents in science and technology to make them maintain longer leading level. On the other hand, we must actively discover, continually support, systematically cultivate innovative talents in science and technology of different ages in the existing young teachers. Continuously posses innovative talents in science and technology is the basic condition to maintain the academic and research vitality of colleges and is the necessary guarantee of sustainable development of colleges.

3 Measures of Cultivating Innovative Talents in Science and Technology for Regional Colleges

3.1 Stressing Autonomous Culture of Young Teachers and Promoting Hidden Innovative Talents to Grow Dominant Innovative Talents

Innovative talents in science and technology are these talents who have innovative consciousness, innovative spirit, innovative thought, innovative ability and can obtain innovative achievements in the field of scientific research. Human studies show that talents can be divided into several steps. First, prospective talents, who have the essential condition of quality demand of talents. Second, latent talents, namely, high-level talents who have obtained leading innovative achievements, but not be recognized by social or peer now. Third, dominant talents, whose innovative achievements have been recognized by peer in a certain range, and at a leading in the field and are playing the lead role. At present the whole country is competing to introduce innovative talents in science and technology. In the fierce competition of introducing dominant talents regional colleges whose academic base and economic foundation are relatively weak are clearly at a disadvantage. So, it is particularly important to discover the backbones and support and cultivate them in existing young teachers and researchers. Most innovative talents in science and technology once had been unknown and gradually achieved remarkable innovative achievements after hard work. So college should seize the golden age of talents growth and take sustained, practical, effective measures to assist and promote prospective talents with potential as quickly as possible to grow latent talents or dominant talents with remarkable innovative achievements.

3.2 Cultivating Innovative Talents in Science and Technology on a Platform of Key Discipline Construction

The cultivation of innovative talents in science and technology needs to continuous innovation practice. Generally key disciplines of regional colleges have a good academic base, academic teams and hardware facilities and academic groups bear many great subjects, but also adequate scientific research funds. These hardware and software provides a good platform for cultivating innovative talents in science and technology. Although young teachers who graduated from school recently, including young teachers of doctoral students graduation, may have the potential to become innovative talents in science and technology, there are many questions such as how to adapt to the environment as soon as possible and look for opportunities to take orders from a good student into a outstanding scientific worker with undertaking positive role, working independently and team cooperative spirit. Facing the experimental conditions of relatively weak in regional colleges, some doctors have not good ideas to work because research environment can not achieve the desired. So the management departments of science and technology of colleges should play a very important linking role for novice young teachers to lead them to join the appropriate academic team and combine with the scientific research process of academic team and the cultivation of innovative talent teams in science and technology so as to lead them to take responsibility in key positions and create the opportunities of practice for them. Professional development is an important way to continuously improve innovation ability and achieve innovation value. Joining the practice of relatively wealth research projects will help them to further enhance the level and become dominant talents as earlier as possible.

3.3 Providing Financial Aid for Outstanding Young Teachers to Preside Research Projects to Improve the Innovative Ability in the Process of Research and Practice

Innovative talents in science and technology, especially innovative leading talents in science and technology, don't be determined by administrative assignment, but form by a lot of research and practical in long time. However, for young teachers of regional colleges it is very difficult to apply to chair the project due to qualifications, job title, and academic standards and so on. As the leader of project he needs to complete a series of research practices from application project, project establishment, implementation to completion project. The process can improve the ability to have academic frontier problems, develop academic ideas, chasten tackling volition and coordinate all works. So it is the necessary process to become innovative talents in science and technology. It is also an effective measure to cultivate innovative leading talents in science and technology that regional colleges provide financial aid for outstanding young teachers to preside research projects to improve the innovation ability in the process of research and practice.

The growth of innovative talents in science and technology has its own special laws. Their growth process and becoming talent have stage. Supported in the best age and putting up the basis platform in international research frontiers for them they can stand out and make outstanding performance as early as possible. In the process of selecting funded young teachers we must pay attention to justice and science and not only proof the frontier and innovation, but also consider the quality of scientific research and cultivation future of the people who received the financial aid. In the implementation process of the project we must closely track, take effective supervision measurement and actively guide.

3.4 Providing International Exchange Opportunities for Talents in Science and Technology to Expand Innovation Ability

Innovative talents in science and technology must understand the latest frontier developments and have an international perspective. Due to rapid knowledge renewal in information era some ability that talents in science and technology have had may also be degraded with the change of time or conditions. So the ability delay restricts talent development. To go abroad and directly communicate with international counterparts have played an irreplaceable role on expanding the academic horizons of talent, understanding advanced research ideas, learning advanced research methods. Learning and communication in the world can expand the academic view of talents in science and technology and their research ideas and thinking mode. After returning to China they are active in the first line of teaching and research. At the same time their words and actions also affect teachers and students around. In addition, for knowledge renewal and the maintenance of academic vigor of talents it is essential measures to strengthen academic communication with home and abroad, regularly invite domestic and foreign experts to hold an academic report and create a strong academic atmosphere.

3.5 Establishing an Efficient and Reasonable Incentives Mechanism of Innovative Talents in Science and Technology

Establishing the incentives mechanism of innovative talents in science and technology should be based on value principle, market principles, justice principles, system principles and sustainable development principles. The incentives mechanism includes both material and spiritual dimensions. At present, there are many tangible incentives ways, such as the greater degree on the allocation of housing issues, allocation of research funding, arrangement of spouses' employment and their children studying at school and so on. But it lacks of efficient measures for how to reuse talent and promote them to make innovative contribution as quickly as possible. In order to reflect the recognition for talents in science and technology arranging important administrative positions have become one of common modes. Affected by official standard thought some talents in science and technology are happy or ask to take important administrative duties. However, this often brings some negative results. On the one hand, busy administrative works spread their energy and time, dissipate their outstanding genius of scientific and research, delay, or even, end their innovative achievements in the process, which means reduce, or even, kill valuable scientific life of these talents. On the other hand, some talents may have extraordinary innovative ability in the field of scientific research, but lack administrative experience and skills; the result is to affect both scientific research and management. One might be a master in his own special field, and generalists is few after all. After years, seeing exceptional innovative talents to become mediocre administration officials, almost no contribution, whose sin is it? From this, it needs to a logical and efficient mechanism what strategy we take can reflect a high degree of reuse for talents in science and technology and inspire their sense of mission, responsibility and urgency from the spirit, as early as possible to make a greater innovative contribution. Management College by experts and Management College by academic can't be reduced to take expert with high academic level directly as certain level administrative duties and can't take administrative level as an incentive for expert. For talents we should make full use of their advantages, such as supporting the talents to construct academic organizations, giving full human rights and property rights, enjoying a relatively high political honor and material benefits, possessing full suggestion right and veto power for the development and operation of colleges. By above measures we can inspire these talents to work hard so as to complete the innovative performance. This will also help colleges create a free, relaxed academic atmosphere, and the works of colleges can be supervised by experts in time.

Technology innovation is a complex work which combines mental work and physical labor. So the evaluation and motivation behavior should follow the study law of innovation achievements and the growth law of talents for innovative talents in science and technology. Modern brain research has shown that Innovation is often driven from intuitive thinking drove by passion. It is the basic law of innovative thinking from image thinking which takes conjecture as starting point to associative thinking of a large span to logical thinking which takes demonstration as end point. Therefore, the incentive mechanism of management should avoid quick benefits and create a relatively loose and free regulatory environment so that talents in science and technology create significant innovation achievements. Currently, for the reward mechanism of talents there are still the questions of lacking of coordination between government departments and colleges so that the incentive resources can't be efficiently used. The multiple awards for certain performance or individual mean to lack a due incentive for other performance or individual. So rewards are inequitable. For the incentive of innovative talents in science and technology we should also fully consider the efficiency and benefits of resource using, and strive to maximize the effectiveness of incentive resources to promote more people to make innovative performance. An instance is very inspiring for how to efficiently use incentive resources. There is a special permanent parking space in the front of laboratory building of chemistry department of University of Southern California in the United States. The guideboard in the front parking spaces shows that it is a reward for a professor who obtained Nobel Prize-winning. What a clever idea! Colleges cost very little, but make the award-winning professor enjoy the honor and benefits, and students learned a lot when passing. For the management departments of regional colleges it is very important to expand horizons and ideas so as to make limited resources create the greatest benefits.

4 Conclusion

With the economic globalization and rapid development of science and technology the development of China and the world has entered a new stage. At this stage innovation ability becomes the core competence, which highlights the importance of innovative talents in science and technology. Regional colleges are in a weak position in the competition of talents introduction because relatively weak discipline base and funds. So self-training innovative talents in science and technology has important strategic significance for them. We can effectively help young talents in science and technology create achievement as early as possible so as to successfully complete the transition from latent talents to dominant talents by the measures, such as funding outstanding young teachers chair research project, putting up the research platform with the key disciplines, providing opportunities for international communication and so on. Colleges should establish an efficient and reasonable incentive mechanism for innovative talents in science and technology and improve the employment model and orderly use incentive resources.

Acknowledgement. The research is supported by Social Science Foundation of Hebei Province (No. HB11GL009).

References

1. Zhao, Y., Ren, M.: An Analysis on Leading Talents of Scientific & Technological Innovation. Journal of Zhejiang Gongcheng University 103(4), 70–74 (2010)
2. Liu, B.: The Idea of an Innovative talent: A Comparative Perspective. Comparative Education Review 156(45), 7–11 (2003)
3. Rao, S., Han, J., Liu, J.: The Concepts and Practice on S&T Innovation System Building of East China Normal University. R&D Management 20(2), 114–117 (2008)
4. Wei, B.: Discipline Construction and Talent Production of a Research University. Journal of Higher Education Management 4(4), 7–9 (2010)
5. Xu, J.: Study on establishment of motivating pattern for innovation talent. Modern Business Trade Industry 15, 39–41 (2010)

Construction of the Training System in Ascendant Helix for the Computer Application Courses in Higher Vocational Colleges

Qiming Tian

Dept. of Computer, Wenzhou Vocational & Technical College,
Wenzhou 325035, China
tqm78@126.com

Abstract. The irrational Practical teaching arrangements are common in the computer application courses of the higher vocational colleges. Aiming at this situation, this paper firstly presents that the computer application courses in the higher vocational colleges should have the regional trait, the practice trait and the trait of the higher education. Secondly, this paper proposes a novel Practical teaching system in ascendant helix. Thirdly, this paper further expounds how to organize the steps of practical teaching and how to test the practice skills. Practice has showed that the Practical teaching system in ascendant helix could help students to develop the practice skills and professional qualities required by the enterprises.

Keywords: higher vocational education, computer application, the curriculum reform, the practical teaching system.

1 Introduction

From the Eleventh Five-Year Plan, the Ministry of Education has specially released the document of <Several Opinions on Comprehensively Promoting the Quality of Teaching in Higher Vocational Education from the Ministry of Education>, and put forwards the urgency and importance of promoting the teaching quality of higher vocational education. Meanwhile, the Ministry of Education has proposed specific requirements on various aspects including teaching system, teachers, teaching condition, teaching reform, teaching appraisal and infrastructure, and made clear that it has been an important channel to promote teaching quality by improving condition of practice test and strengthening the construction of Practical teaching base and reform of teaching mode [1]. For the computer courses in higher vocational education with guidance of application, the effect of Practical teaching has directly influenced whether the students can really understand and master relevant computer courses, which is directly related to the practical ability and professional skill of the students, and related to the employment rate of the majors.

Currently, most courses in computer majors of higher vocational education are application courses including software design, animation design and production, plane image processing, and webpage design and production. These difficulties shall be

H. Tan (Ed.): Knowledge Discovery and Data Mining, AISC 135, pp. 627–633.
springerlink.com

solved in the computer application courses in vocational education including how to construct the Practical teaching system, how to arrange each Practical teaching link reasonably, and how to perform feasible appraisal on Practical teaching skill of students.

2 Current Situation of Practical Teaching System for Computer Application Courses of Higher Vocational Education

According to the survey, the irrational Practical teaching arrangements are common in the computer application courses of the higher vocational colleges, with mainly the following problems: (1) The emphasis on Practical teaching link is far lower than the emphasis on theoretical teaching link, and the class hours of practice are far more lower than the class hours of theoretical teaching, which goes against the development of practical skills. (2) The theoretical teaching link is out of line with the Practical teaching link. The practice course after the theoretical course is only used to verify the theories studied in previous stage, so the teaching mode integrated with theory and practice is not realized. (3) The practice course has only rested on the practical teaching level of single skill, and lacks of comprehensive, which can not develop students with systematic and comprehensive practical skills. (4) The Practical teaching link has only rested on class level, and the content of Practical teaching is limited to small case designed by teachers. The design and development of Practical teaching content and organizational link is not cooperated with the companies, which is not good to develop students with real work skills. (5) The faculty advisers of Practical teaching link are front-line teachers of schools who graduated from university or above without company experience, so the effect of Practical teaching will be influenced in certain degree.

According to the above phenomenon, the author has suggested that the computer application courses in higher vocational education shall meet the requirements of the three traits to really develop students with practical skill of market demand.

3 Requirements of Three Traits in Computer Application Course of Higher Vocational Education

The higher vocational education shall strengthen the regional trait, practical trait and trait of higher education [2]. Therefore, the computer application courses shall reflect the three traits.

3.1 Strengthen the Regional Trait

With orientation of application, the computer application courses in higher vocational colleges shall develop students with practical skills in real work by considering market application demand of their location in the selection of teaching content, jointing with work process of companies in the teaching arrangement, constructing simulated or real work environment with companies during the construction of practical teaching system, and constructing a teachers' team having double teaching qualities with company.

3.2 Extrude the Practical Trait

Compared with other majors and other courses, the computer application courses in higher vocational colleges shall further extrude the practical trait of the courses, and develop students with stronger practical capacity. The development of this practical capacity can be completed through three channels: (1) Training of basic skills: Realizing the integration of Practical teaching room with classroom, the teachers can give lessons through demonstration, and students can learn from the practice. (2) Training of comprehensive skill: Realizing the integration of Practical teaching and production, the students can attend the production and practical teaching in the training bases both inside and outside the school to develop their position capacity. (3) Training of creative conscious: Realizing the integration of graduation project and development service, the students can develop the application capacity to serve society during the comprehensive graduation practice by selecting real projects from companies as the title for the comprehensive practice of graduation project as the title for the comprehensive practice of graduation project.

3.3 Reflect the Trait of Higher Education

Different from the computer application courses in secondary vocational schools, the computer application courses of higher education shall develop students' capacity to perform scientific development and serve society. Therefore, the computer majors in higher vocational colleges shall set up students' studio or development research institution to undertake technical service orders from companies or institutions in this region and complete the regional service.

4 Construct the Training System in Ascendant Helix

All classes in computer application courses of higher vocational colleges are suggested to take in the Practical teaching room. With case orientation or project orientation, the knowledge and skill is integrated in every lively and direct case and project for teaching, resulting in the combination of teaching, learning, and practicing, and the integration of theory and practice.

4.1 Design of Training System in Ascendant Helix

The theory of Piaget's constructivism has suggested that the knowledge is neither the objective matter (empiricism) nor the subjective matter (Vitalism). Knowledge is a result constructed during the process of interaction between individuals and environment. Therefore, in the practical teaching of computer application courses of higher vocational colleges, the training system in ascendant helix can be designed as figure 1. The professional capacity and quality development has been the first place. With the interaction between work environment and practices including core skill practice, simulated project practice, order project practice and position practice, students can not only set up confidence in the easy-learning environment, but also promote their practical skills gradually in the happy-learning environment.

The training system in ascendant helix is consisted of four stages of circulations, as shown in figure 1. The skills of students increase along with the stages. The four stages are: (1) Core skill practice: it is arranged after the common case study or chapter teaching to develop the core single skill of this courses. (2) Simulated project practice: The simulated project practice will be introduced when single core skill of students has achieved certain level. As a result, students can apply the skill learned from previous stage comprehensively, and perform practice of single core skill for the next stage. (3) Order project practice: The students will enter the students' studio when they have certain comprehensive skills. The students' studio can undertake certain order projects and organize students to perform the order project practice. The core skill practice, simulated project practice and order project practice is performed in cycle. (4) Position practice: With certain accumulation, students can perform position practice in the practice bases both inside and outside the school to experience real work environment and work task.

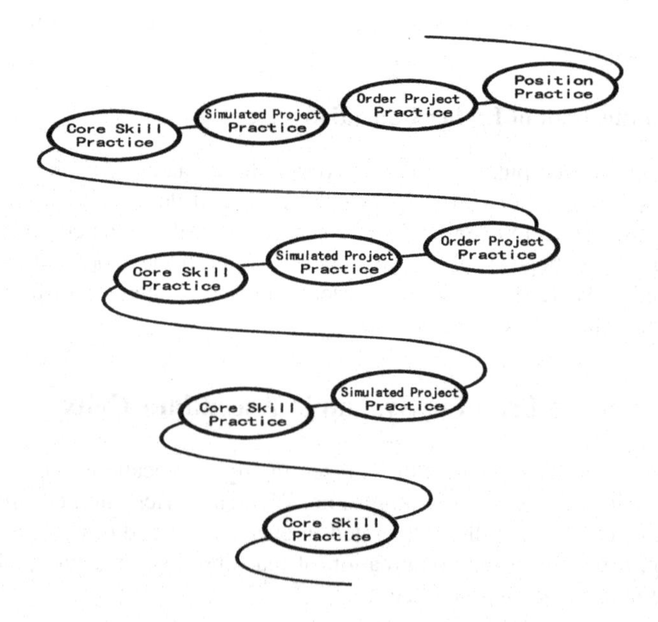

Fig. 1. The training system in ascendant helix

In the first state of training system in ascendant helix, students will attend the core skill practice when they finish a case study (or a chapter). The students at that time will complete training fake questions (not real work tasks, fake questions) covering core skills designed elaborately by teachers in class of the school (not real work environment, fake practice). When students master certain core skills, they begin to perform simulated project practice (complete classic work task in school class), which is fake practice for real questions. When students have mastered certain skills, based on the students' studio, the order practice is performed in the situation education by organizing students and introducing the order tasks from institutions and companies in the market gradually according to actual company operating system. As a result, students can contact real production task. Finally, based on the practice bases both inside and outside the school, students are organized to perform position practice to experience real development work in companies and enterprises. Both in the order project practice and the position practice, students can complete real tasks in real work environment, as shown in figure 2.

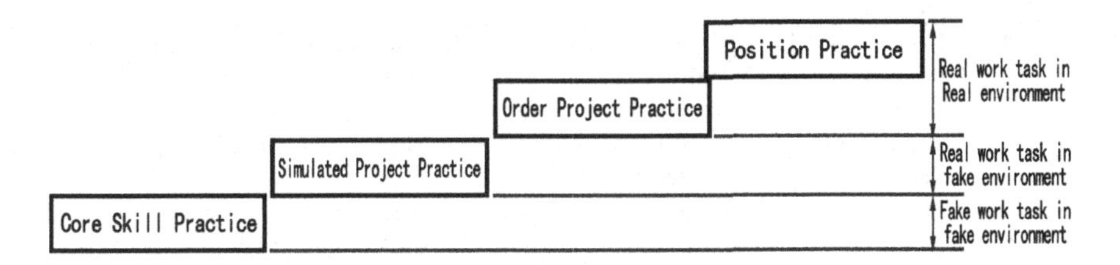

Fig. 2. Design of the steps of practical teaching

Computer application courses in higher vocational colleges are suggested to apply training system in ascendant helix to realize the alternate teaching mode integrating with work and learning (shown in figure 3): learning-work (simulated project practice)-learning-work (order project practice)-learning-work (position practice). As a result, students can learn from the practice and perform practice from learning. The professional work capacity and professional comprehensive quality can be developed. During the order project practice and position practice, based on the students' pioneering studio set up by students and practice bases both inside and outside the school, the real work environment can be set up. With teaching mode integrating with production and study, students can exercise and promote their capacity to analyze problem and solve problem through the technical service when they undertake project social technical service. In addition, the course goal of serving region can be fulfilled.

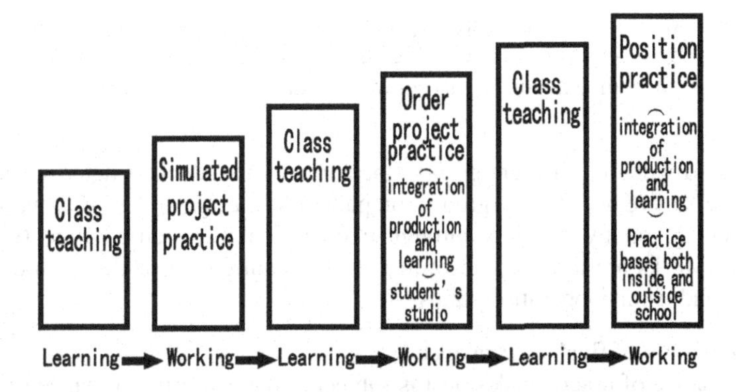

Fig. 3. The teaching mode of working and learning alternation

4.2 Implementation of Practical Teaching Process

The specific arrangement of the steps of practical teaching in training system in ascendant helix is shown in table 1.

Table 1. Arrangement of the steps of practical teaching

No.	Practical teaching link	Implementation of courses	Guidance method of teachers
1	Core skill practice	A person is a group Case drive	Concentrated demonstration Personal coaching
2	Simulated project practice	Team cooperation (3 to 4 persons a group) Simulated project drive	Project guidance Monitoring of development process
3	Order project practice	Work team (3 to 5 persons a group) Cooperative design order	Virtual company system Project guidance Monitoring of development process
4	Position practice	The practice company organize the work team and engineering process according to the actual work demand	Company project system Apprentice guidance Integration of monitoring and guidance

4.3 Appraisal of Practical Skills

The appraisal of practical skills has not only tested the professional comprehensive skills and knowledge of the students, but also tested whether the students have the professional capacity and quality required by companies including teamwork skill, creative skill, and word expression skill. The plan for the appraisal of practical skills is as follows.

(1) Personal test
Each student develops a simulation project with certain comprehensive difficulty interdentally according to project effect given by teachers. Assessment has been given to personal performance according to certain standard for evaluation.

(2) Team test
Students are divided to different groups (the amount of students is determined according to scale of orders). Every group completes the design order of comprehensive development given by teachers through cooperation in specified time (such as 1 week). The works are assessed according to the requirement of design order, works originality and team cooperation status.

(3) Determination of final performance
The performance of practical skill test is subject to the lower performance of personal performance and team performance, which has reflected that personal individual work capacity and teamwork capacity is a whole.

5 Conclusion

According to the current shortages in Practical teaching link of computer application courses in higher vocational education, the paper has not only suggested that the computer application courses in higher vocational colleges meet the requirements of three traits, but also designed a set of training system in ascendant helix. Furthermore, the paper has also put forwards suggestions on implementation of Practical teaching and appraisal plans of practical skills. The training system in ascendant helix has been implemented in some main courses of computer majors in Wenzhou Vocational and Technical College. According to the implementation result, it is much easier to develop students with required practical skills and professional quality in this training system.

Acknowledgment. This work was supported by 2007 project subsidy of Zhejiang Education and Science Planning (SC152).

References

1. Dong, L., Wang, H.-J., Sun, B.: Research and Exploration on Teaching of Computer Physical Training in Advanced Occupational Physical Education. Modern Computer, 81–82, 91 (2009)
2. Ding, J.: Research and Practice of Sustainable Development of Higher Vocational Education based on Three Traits. Journal of Higher Education, 72–77 (2010)

On Sino-foreign Investment College PE Curriculum Setting from the Perspective of International View

Shujuan Yuan[1], Zhanguo Fu[2], Hongwu Yang[3], and Yu Xi[3]

[1] E&A College of Hebei Normal University of Science &Technology, Qinhuangdao, Hebei, China
[2] Hebei Vocational & Technical College of Buiiding Materials, Qinhuangdao Heibei, China
[3] Qinhuangdao Institute of Technology, Qinhuangdao, Heibei, China
yuanshujuan2004@163.com

Abstract. By the means of literature datum, questionnaires and interviews with experts, we analyze and study the rationality, actual effect and scientific result of the PE curriculum and teaching content arrangement in Sino-foreign cooperation running college. Then proposed the "based on their own, facing the international, people-oriented "features of physical education curriculum model for Sino-foreign cooperation running college curriculum arrange and teaching content.

Keywords: Sino-foreign cooperation running schools, PE status quo, courses setting, teaching content.

1 Introduction

Sino-foreign cooperation running schools is a product of the internationalization of education, which is characterized by the introduction of foreign high-quality educational resources and advanced educational concepts, training high-quality compound talents adapting to economic globalization, [1]. After many years of working experience and nearly one-year research on the status of physical education of some of China's foreign cooperative education institutions, the authors find the introduction of foreign high-quality educational resources (curriculum, teaching content) doesn't performe obviously, the effect is not ideal that the characteristics of Chinese-foreign cooperation running is prominent. PE curriculum in Sino-foreign cooperative education institutions has not an exact conclusion, this paper proposed new curriculum model with the characteristics of Sino- foreign cooperation running schools on the basis of full investigation, according to actual physical education in Sino- foreign cooperation running schools, and the project content to meet students sports needs, which provide a reference for the reform of physical education in Sino-foreign cooperation running schools.

H. Tan (Ed.): Knowledge Discovery and Data Mining, AISC 135, pp. 635–642.
springerlink.com

2 Research Objects and Methods

2.1 Research Objects

18Sino-foreign cooperation running schools institutions; 90 physical education teachers; 600 freshmen and sophormores in schools.

2.2 Research Methods

(1) Literature datum According to the content of the topic, retrieve, collect and access to international and domestic research literature related to this article by Mirror Journal Net, Wanfang database, including Sino-foreign cooperation running schools, both domestic and foreign school physical education. Based on the education-related laws, regulations on Sino-foreign cooperation running schools education released by the Ministry of Education, the author analyzed and summarized the data.

(2) Questionnaires Three questionnaires are designed according to the contents of the study "Survey on Sino-foreign cooperative education institutions of physical education - expert questionnaire, teacher questionnaire and student questionnaire. Test the validity of three questionaires by Delphi method, the indicators identified by experts found rates of 89% or more; in order to test reliability of the questionnaire, 10 subjects are taken for two measurements, the time is interval week, after calculating the correlation coefficient between two measurements, "expert questionnaire" R = 0.88, P0.01, "Teacher Questionnaire" R = 0.85, P0.01, "student questionnaire" R = 0.83, P0.01, that the index system of the questionnaires has a higher reliability.18 expert questionnaires were issued, 18 were recovered, the recovery rate is 100%; 18 valid questionnaires, 100% efficiency. 90 teacher questionnaires were issued, 88 were recovered, the recovery rate is 97.78%; 85 valid questionnaires, 94.44%efficiency. 600 student questionnaires were issued, 584 were recovered, the recovery rate is 97.33%; 571 valid questionnaires, 93.17% efficiency.

(3) Interviews interviewing three Sino-foreign cooperation running education experts, seven experts and sports teachers of physical education by telephone on the status and development of physical education in Sino-foreign cooperative education.

(4)Mathematical statistics Using Microsoft Excel statistical software to statistically analyze survey data.

3 Results and Analysis

3.1 The Status of Physical Education in Sino-foreign Cooperative Educational Institutions

(1) PE class arrangement status in Sino-foreign cooperative colleges

Table 1. PE class arrangement section and the curriculum survey tables in Sino-foreign cooperative college n = 18

	Commercement section				
	Grade one	Grade two	Grade three	Grade four	
Number	18	18	2	--------	
%	100	100	11.11	--------	
	Course arrangement				
	Basic course	Common course	Profession course	Heallth course	Elective course
Number	2	11	18	5	2
%	11.11	61.11	100	27.78	11.11

Table 1 shows, from the commencement time, the existing Sino-foreign cooperation running schools sports commencement rate is 100%, mainly concentrated in the first and second grade, the third grade is less, only 11.11%, the fourth grade doesn't ibasically open. Analyzing from type of the commencement, PE general elective courses are mainly taught in sino-foreign cooperation running schools, 100% and 61.11%, schools offer basic courses, health courses and elective courses are less than 30%. Learning through the investigation, the basic course primarily is arranged in the first semester, common course is usually arranged in semester 1, 2, special classes are mainly in 3, 4, but some schools also commence special classes in the first and second semester. Health lesson for some of the sick, disabled, injured, frail students is set up and arrange in semester 1 to 4. Elective course set up for students above third grade. The results suggest that the rate is high that the first and second year commenced physical education in sino-foreign cooperative education institutions, physical education isn't set up basically in third and fourth grade, which does not meet the requirements of "National College Physical Education Curriculum Guidelines" that the studentsin the third grade and above set up physical education [2]. In Sino-foreign cooperation running schools, sports curriculum types are less, only the special and general courses are recognized. This curriculum model does not reflect the characteristics of Sino-foreign cooperation running schools, which are basically the same as college sports programs form [3], there was no signs that foreign physical education curriculum is introduced. Analyzing the reasons, the author found that in Sino-foreign cooperative education institution, full-time teachers are less, part-time teachers are mostly dependent on their own schools and local headquarters of PE teachers, retired teachers, their research is based on domestic sports, the concept of education about Sino-foreign cooperation running schools is vague , sports education and research in Sino-foreign cooperation running schools is minimal, they think physical education in Sino-foreign cooperation running schools is no qualitative difference from college physical education.

(2) The status of physical education content in Sino-foreign cooperative education institution.

Learning through the investigation, in Sino-foreign cooperative educational institutions, physical education contents mainly include project-based sports, traditional sports (44.44%), fitness and entertainment sports (27.78%) are less. Foreign fashion sports are (16.67%) less. The findings suggest that physical education contents in Sino-foreign cooperation running schools are no qualitative difference from that of college physical education.

Which has much difference from that foreign universities promote the physical education content connected with life, diversity, flexibility and native [4]. The reason is that, space, equipment is the main reason for restricting physical education; followed by physical education teachers are relatively scarce and lack of innovation in teaching practice on the research on teaching of foreign school sports, leading to lack of physical education programs, the content is not sufficient.

3.2 Research on Reasonable Arrangement of Physical Curriculum Model in Sino-foreign Cooperative Education Institutions

(1) Investigation on reasonable arrangement of physical curriculum model in Sino-foreign cooperative education institutions

Table 2. PE course survey in Sino-foreign cooperative college

	Common course		Profession course		Health course	
Term	Expert	Teacher	Expert	Teacher	Expert	Teacher
1	77.78	62	22.22	23		
2	55.56	36	44.44	49		
3			66.67	66	16.67	19
4			44.44	34	33.33	51
5\6					38.89	29
7\8					16.67	11
	Fitness course		Elective course		Not commercement	
Term	Expert	Teacher	Expert	Teacher	Expert	Teacher
1	18	18				
2	18	18				
3	18	18				
4	18	18	22.22			
5\6	18	14	55.56	35	5.56	21
7\8	18	3	72.22	27	11.11	47

Table 2 data shows that, most experts believe that in PE curriculum arrangement of Sino-foreign cooperative education institutions, common course:1,2 semester, special coursse 3, 4 semester, elective course: 5-8 semester; most teachers think that a reasonable PE curriculum: common course is commerced in ,the first semester,compulsory specifical elective course is taught in 2, 3 semester, health classes are taught in 4 semester, sports electives are in 5, 6 semester, 7-8 semester does not offer physical education. From the research, experts and teachers believe that general tutorials are arranged for a semester to help students adapt to college sports as early as possible ,to restore physical function and quality because of the entrance, develop interest in sports, in order to lay the foundation for the future sports learning; In the 2,3 semester special classes are taught to achieve the requirement in "National Universities Physical Education Teaching Guidelines" that each student grasp 1-2 motor skills more systematicly; physical education and health classes in the fourth semester are regarded as a physical feature of the design in Sino-foreign cooperation running schools, which is different from the existing health courses for the sick, disabled, injured, frail in colleges and universities offered by the students, but a required course open to all students, design for health status of current students physically and mentally [5], desire to physical health care and life-long sports knowledge, skills, and lack, as well as foreign sports and health needs [6]. It is no doubt to open the elective sports classes are taught to the students inGrade 3 and above, also in line with the regulations in 'University and College Physical Education Curriculum Guidelines "; physical education teachers do not agree to open physical education students in the fourth grade, primarily consider students' other courses study and prospective employment needs.

(2) Investigation of teaching content in Sino-foreign cooperative education institutions

① Reasonable choice survey of teaching content.

Table 3. Physical Education Content reasonable choice survey tables in Sino-foreign cooperative college

Order	1	2	3	4	5
T	Basketball	Football	Aerobics	Table tennis	Wushu
S(m)	Basketball	Table tennis	Football	Wushu	Boxing
S(f)	Aerobics	Shuttlecock	Body	Yoga	Table tennis
Order	6	7	8	9	10
T	PE games	Voleyball	Hike	Shuttlecock	Badminton
S(m)	Taekwondo	Badminton	Climbing	Skating	Billiards
S(f)	Hike	Wushu	Taekwondo	Badminton	Hip-hop

Table 3 data show that teachers and male and female students have different opinions on physical education content building. Teachers mainly consider the school's venues and equipment, teachers and students interests in the elective content, college sports are basically regular projects and outdoor projects Lord, even the indoor projects are those equipment space requirements of the project is not too high ,such as aerobics, table tennis and badminton. The majority of students like the indoor items, boys 6, girls 7, some projects open more difficult in many Sino-foreign cooperative education institutions, mainly not have the venue and facilities. Promption: the schools increase investment to facilities and equipment to meet students' interest, improve school physical education environment, and respond to it, teachers reform and reasonably lay out the existing space and equipment, increase utilization to meet the needs of teaching [7].

② foreign fashion sports included in the teaching content of the questionnaire,

Table 4. Physical Education Content Survey implanted foreign sports vision statistics

	Much hope	Hope	Option	Not hope
Expert	2	8	6	2
N=18	11.11	44.44	33.33	11.11
Teacher	14	27	30	11
N=85	16.47	31.76	35.29	12.94
Student	223	181	96	71
N=571	39.05	31.71	16.81	12.43

In the content of physical education of Sino-foreign cooperative education institutions, the introduction of foreign fashion sports is sponsored on the basis of the study of Sino-foreign cooperative education theory and practice by research group. Our proposed foreign fashion sports are more traditional and popular sports for foreign partners, such as: South Korea's taekwondo, wrestling in Japan, the United Kingdom rugby, squash, hockey in Canada, tennis and billiards in France, skating in Netherlands, Indian yoga, the American hip-hop, pilates and rock climbing, etc., the introduction of these elements are designed to enrich the content of physical education, train students interests , understand foreign sports, lay out the foundation for study abroad and international exchanges. Table 4 shows that most experts, teachers and Students agree, and hope to grasp or understand some foreign fashion sports at the university level, laid the groundwork to physical fitness after the students go abroad and enjoy sports ,and provide a platform to understand the foreign sports for those who do not go abroad.

3.3 Optimizement and Construction of the Physical Education Model in Sino-foreign Cooperative Education Institutions

The twin goals of physical education curriculum in Sino-foreign cooperative implementation, common goals and special goals. Common goal is to implement the spirit of "National Universities Physical Education Teaching Guidelines", special goal is formulated according to the provisions "Sino-foreign Cooperation Running Schools". To this end, Sino-foreign cooperation running schools physical education should be based on the domestic,open to the international, people-oriented, to reflect the characteristics of introduction of foreign high-quality educational resources and advanced educational concepts, training global high-quality compound talents (see Table 5).

Table 5. The reasonable setting of Physical Education in Sino-foreign cooperative college

Term	Course type	Teaching content	Teaching objectives
1	Compulsory Common Course	the main content: martial arts, the mass aerobics, the assistant content sports games and athletics	Develop students' basic skills, enhance physical fitness of students, develop interest in sports, lay the foundation for future learning
2-3	Compulsory professional elective course	S(M):Balls,Wushu,Taekwondo,Skating S(F):Ball, Wushu, Aerobics, PE dance, Yoga, Hip-hop	Master the basic knowledge, technology, skills of one or two sports, , meet the sports needs of students, lay the foundation for lifelong physical
4	Compulsory health course	Sports, Leisure sports, Fitness sports,etc PE.games	Improve body shape, develop physical function, relieve psychological stress, adjust attitude, improve their physical and mental health;
1-6	Fitness elective course	Health education, PE health knowledge and technology, Wushu,Taichi, Qigong, Yoga, Balls, Pegames, Hike	master the knowledge, skills and exercise prescription of physical health care, learn medical supervision and rehabilitation fitness evaluation, physical illnesses, enhance physical fitness;
5-6	Sports elective course	Wushu, Balls, Yoga, Body, Common asrobics, Pilates, Mountaineering, Hike, Pegames.	strengthen our inheritance and development of traditional sports culture, sports fashion enhance students to learn and master foreign fashion sports, meet students' demand for foreign fashion sports

4 Conclusions and Recommendations

4.1 Conclusions

(1) In Sino-foreign cooperative education institutions,high rate of commencement for freshmen and sophomores on physical education, low rate of commencement for junior ,zero rate of commencement for senior. Course type and content is basically identical with those in our colleges and universities,whether high-quality educational resources and the introduction of foreign advanced educational philosophy of curriculum weren't found.

(2) Sino-foreign cooperative education institutions insist on commencing general physical education curriculum and special elective courses, strengthening health physical education and elective courses, increasing the health class.

(3) Increasing traditional sports in the content of physical education teaching is the need to inherit our traditional sports; increasing the entertainment and fitness program is the needs of "health first", "lifetime sports" and students; the introduction of foreign sports fashion items is the needs of Sino-foreign cooperation running school education and students.

(4) Not offer physical education for students in fourth grade of the foreign cooperative education institutions was more supported by the surveyed experts, teachers and students, more conformed to the overall college teaching arrangement, which is reasonable.

(5) Carefully study and summary of teaching experience based on the survey results, we optimize and set out special physical education curriculum model in the Sino-foreign cooperative education institutions.

4.2 Recommendation

(1) Sino-foreign cooperation running schools physical education should adhere to the guiding ideology: "student-centered," "health first", based on the domestic, open to the international, multi-commence international curriculum included entertainment, practice, fitness, self and other rich contents, achieve the training objectives of "internationalization" of higher education.

(2) Sino-foreign cooperation running schools sports curriculum should start from our reality introduce foreign high-quality educational resources and advanced educational concepts, selectively introduce foreign successful teaching experience, absorb the sport for China, rich China's content of physical education, lay the foundation for students to study and physical exercise abroad.

(3) The purpose of Sino-foreign cooperation running schools is the introduction of foreign advanced teaching resources and educational philosophy, so that students in the country can accept foreign university education. Sino-foreign cooperative education institutions not only build the necessary sports facilities, purchase adequate teaching aids, but also learn some foreign experience to develop and construct multi-functional stadiums.

Acknowledgments. Foundation project: research subject of social science development of Hebei Province in 2010 "The Research on Integral Optimization of Physical Education of Sino-foreign Cooperative Colleges" (Project number: 201003119).

References

1. Yang, H., Luxi, H., Feng, G.: Sino-foreign cooperation in running schools and Countermeasures. Hebei Normal University of Science and Technology (4), 60–63 (2008)
2. Ministry of Education. Of Physical Education Curriculum Guidelines. Ministry of Education, Office files, August 12 (2002)
3. Sui, X.: Compare of the PE curriculum in Sino-USA University. Physical Education (3), 61–65 (2008)
4. Song, J.: The Enlightenment of Foreign Physical Education Physical Education Curriculum Theory to the development of Domestic PE Curriculum. Heilongjiang Higher Education Research (7), 150–152 (2008)
5. Ding, X.: Undergraduates Mental Health and Thinking in New Undergraduate College. Journal of Preventive Medicine (4), 703–705 (2010)
6. Yuan, L.: An analysis of Examination Results and University Students Health Education. Chinese People' s Health (5), 321–324 (2006)

Automatic Summarization of Web Page Based on Statistics and Structure

Shuangyi Zheng[1] and Junyang Yu[2]

[1] Engineering College of Management, University of Huazhong University of Science & Technology Wuhan, Hubei 430074, China
[2] Net Center of Henan University, Kaifeng 475001, China
Zhengsy@mail.scuec.edu.cn

Abstract. This paper discusses the automatic text information extraction, and presents the automatic summarization based on analysis of HTML tags and statistics. This method combines the summary extraction and Web structure. We get structure levels of the document using HTML tags and calculate the weight of sentences using text structure and the statistics of word frequency, in order to extract summary. Our experimental results indicate that this method accurately and completely specifies the main content of the document, having high accurate rate and recall rate.

Keywords: Automatic summarization, Web page summary, Web page structure, Statistics.

1 Introduction

Summarization is to provide the broad outline of information without comments and additional explanatory, and it states the important content concisely and accurately. Automatic summary is the process of automatically compiling and generating the summary using the computer. So far, the existed automatic text summarization system can be divided into two categories: statistic-based text summarization and knowledge-based text summarization[1]. The statistic-based text summarization has simple method and realizes easily, but it generates unsatisfactory result[2]. The knowledge-based text summarization is based on understanding the text information, which gets better result, but it is too difficult. For the Web page, Web document summarization should be used for all kinds of users not limited by fields. It should have certain content coverage which states the outline of Web page accurately and completely, and at the same time the generated summary must reach certain speed, in order to meet the requirement of processing a large number of Web documents.

HTML tags divide the web page into different structures, which represent different degrees of importance. For example, <title> is the headline of the Web page; <body> is the main content; <p> is the paragraph tag; <meta> contains keywords or the abstract. For the words in the Web page, if they are composed by normal types, they are not as important as the words which are composed by special types. We depend on the analysis of HTML tags in the Web page, combine with calculating the weights

H. Tan (Ed.): Knowledge Discovery and Data Mining, AISC 135, pp. 643–649.
springerlink.com

of keywords, and design a method to extract the summary of Web page. This paper automatically summarizes the web document, with a combination of HTML tags analysis and statistic. First, we analysis HTML tags of the document and get the paragraph information and all levels of subhead information; then we extract the document's keywords and key sentences using the statistic method and heuristic rules; at last, we generate the summary of the document after eliminating redundancy among key sentences with concept distance.

2 The Analysis of Web Page Structure

1) Web page cleaning.
Compared with ordinary text documents, the Web document contain much other information except the text, which includes Script, ad links, navigation links, and copyright information and so on. The automatic summarization of Web document needs the text message of the document, while other information such as hyperlinks, copyright information, date and time tagging is useless. We call such information as "noise", and its existence will affect the speed and quality of the summary. We should filter it before summarization, in order to get clean and orderly text information [3]. For getting text information quickly, our method is: first establish a list of expression chunks, and extract the least text chunk through matching expressions; then reject the "noise" in the text chunk and divide the text into paragraphs, combining with models match and heuristic rules; finally get the text information composed with paragraphs.

2) Extracting the Title.
The division of the article hierarchical structure is very important in the research of automatic summarization of Chinese Web page. The identification of titles is beneficial for leaning the overall framework of an article, while the title itself is the highly abstract. So the correct identification of titles can improve the quality of the abstract to a certain degree. Considering that there are subtitles with signs and subtitles without signs, we extract subtitles with the method of combining models match and heuristic rules. In order to extract various ranks of subtitles, we use the following method: to begin with, we extract the first subtitle, and judge the ranks of titles according to the specialty and consistency of fonts in HTML. Generally speaking, words with the same fonts belong to the same rank, and the larger font, the higher rank.

Then according to the subtitles, we divide the text into some text chunks, and regard every text chunk as a document. Repeat the process of extracting subtitles and dividing into text chunks, until the text chunk can't be divided or no subtitles. So the first title of text chunks in different levels can compose various subtitles of the document.

3) Extracting the Title.
HTML tags divide the whole document into several sections. We identify the title and subtitles; count the number of paragraphs, the number of sentences and positions of every sentence; get the information of HTML tags as showed in Fig. 1. So the Web page structure is mainly divided by HTML tags.

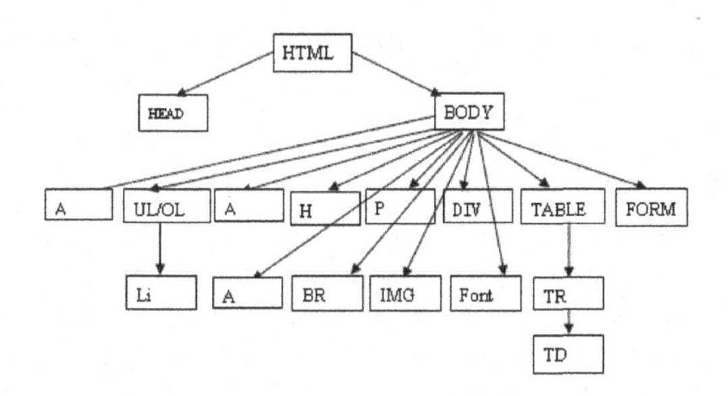

Fig. 1. Web Page structure Figure

3 Calculate the Words Weight

The collection of keywords usually can reflect the main contents presented in the article. Different from common word weight calculation formula and the classical TF-IDF, the weight value not only consider word frequency, but also word distribution. The word which occurs more often and distribute equally in every paragraph should have greater weight, so we introduce the Eq. 1.

$$W_{tk} = \lambda \times \beta \times \frac{tf_k \times \log_2\left(\frac{N}{n_K} + 0.2\right)}{\sqrt{\sum_{k=1}^{n} (tf_k)^2 \times \left[\log_2\left(\frac{N}{n_k} + 0.2\right)\right]^2}} \tag{1}$$

Where tfk is the frequency of the word tk in the document D; N is the number of paragraphs in D; nk is the number of paragraphs which contain tk; n is the number of words in the sentence,; Wtk is the weight of tk in the document D; nk/N can be explained as the paragraph density of word tk, but this is the reciprocal of the paragraph density; β is the adjustment coefficient of word frequency, and it is usually 1.

λ is the position scale factor, because words appeared in some special positions have special importance. This paper reweights words appeared in titles and subtitles, multiplying the scale factor in calculating Wtk. If words appeared many times, the maximum scale factor can be only considered. After repeated experiments and comparisons, this paper set the weight scale factor of titles as 2, and the weight proportion of subtitles declines as follows:

$$\lambda = 1 + \frac{0.4 * 1}{n + 1} \tag{2}$$

Where n is the hierarchy of subtitles. This paper records the weight of all words to calculate the weight of sentence.

4 Calculate the Weight of Sentences and Generate Summary

The weight of every sentence in the Web page should consider the weight of words and the weight of sentence structure.

1) Calculate the weight of sentence.
The automatic summarization based on statistics and text structure generally select a certain number of most representative sentences to compose the abstract[7]. In order to quantitatively measure the importance of sentences, every sentence (s_k) in the document should be weighted as $w(s_k)$. The primary weight is decided by the statistic result of word frequency in the sentence. The following factors are considered when confirm $w(s_k)$:

The importance of words contained in the sentence. The sum of weights of words is larger, and the importance of the sentence may be more.

The position of the sentence in the article. For example, the sentence in the first paragraph, the last paragraph, the beginning of the paragraph, the end the paragraph, title and subtitle usually summarize the main content of the article. The sentence in those positions should be weighted more.

The sentence contains indicative phrases. If the sentence contains general words such as "this paper states", "this paper presents", "this paper discusses", "in a word", "to sum up" and so on, it can summarize this paper and it should be weighted more. To synthesize the above factors, this paper use the following formula to calculate weights:

$$w(s_k) = \frac{LC(s_k) * CC(s_k) * EC(s_k) * \sum t_{kj}}{\|s_j\|_T} + \frac{p_i^2}{q_i} \quad (3)$$

Where $\sum t_{kj}$ is the sum of weights of key words contained in sk; $\|sk\|T$ is the length of sentence sk; LC(sk) is the position scale factor. It is specified as: weight sentences of the first paragraph and the beginning of paragraphs separately, the weight value of the first paragraph is 1.5, the first sentence of every paragraph is 1.3; CC(sk) is the scale factor of indicating words weight in sentences, and it is 1.5; EC(sk) is the scale factor of detail words weight in sentences, and it is 0.5; others are 1. pi is the number of important words in the sentence ; qi is the total number of words in the sentence; p_i^2/q_i is the measurement value of the sentence. The sentence containing various words generally has better measurement value, so adding the k makes the sentence has higher weight.

2) Select the key sentence.
The size of summary can be stipulated by the proportion or specific figures. Considering the importance of the first section (fc) and the last section (lc) , the ratio of exacting the summary from the two sections is higher than the rest of the article. The number of sentences as summary taking from the two sections is sat as $\|Abst(c)\|_s$:

$$\|Abst(c)\|_s = \left[1.5 * R * \|c\|_s\right] + 1 \quad (4)$$

or

$$\|Abst(c)\|_s = \frac{1.5 * SN * \|c\|_s}{\|d_i\|_s} + 1 \tag{5}$$

Where R is the specified ratio of summary; SN is the specified number of sentences in the summary; $\|c\|_s$ is the total number of sentences contained in the first section; $\|d\|_i$ is the total number of sentences contained in document d_i; “[]” is the bracket function, the left summary will be assigned to other sections. The number of sentences taken from other section c except section fc or lc is $\|Abst(c)\|_s$.

$$\|Abst(c)\|_s = \frac{\left(R * \|d_i\|_s - \|Abst(fc+lc)\|_s\right) * \|c\|_s}{\|d_i - fc - lc\|_s} + 1 \tag{6}$$

or

$$\|Abst(c)\|_s = \frac{\left(SN - \|Abst(fc+lc)\|_s\right) * \|c\|_s}{\|d_i - lc - fc\|_s} + 1 \tag{7}$$

Where $\|Abst(fc+lc)\|_s$ is the number of key sentences contained in the summary of the first section and last section. $\|di\text{-}fc\text{-}lc\|s$ is the number of sentences contained in document di except the first section and last section.

In order to keep the original logic, after selecting candidate sentences of the summary, the summary is formed by the output of these sentences.

5 Experimental Analysis

The main index evaluating the summary is the accurate rate and the recall rate[10]. The sentences summarized by many people are acted as a set. The evaluation criterion adopts the accurate rate and the recall rate of information retrieval, and the higher, the better. Supposed that the set of automatic summarized sentences is x and the set of artificial summarized sentences is y, the accurate rate and the recall rate will be calculated with the following method:

(1) The accurate rate(P): $P = (x \cap y)/x$. It is the ratio of the number of sentences in the automatic summarization which are supposed to be extracted to the number of all automatic summarized sentences.

(2) The recall rate(R): $R = (x \cap y)/y$. It is the ratio of the number of sentences in the automatic summarization which ought to be extracted to artificial summarized sentences.

(3) The compression rate (C): C=the length of summary / the length of article. It is the rate of the summary to the whole article.

At present, the acquired experimental samples are mainly from Web pages from BAIDU’s search system. The 5000 Web pages are from comment, news and scientific papers.

This paper mainly tests the integrity and conciseness from the coverage and the accurate rate. The summary result is quantitative analyzed by the accurate rate (P) and the recall rate (R). Experiments use the method sated in this paper to summarize automatically.

The hardware environment of experiments is: Pentium IV 2.4GHz CPU, 512M RAM computer, and the software environment is: Visual Studio 2005, SQL SEVER 2005 Enterprise Edition.

To fully test the effect of the summary, we download all kinds of 300 documents related news, science and so on. Then we exact the summary respectively at the compression rate of 10%, 15%, 20% and 25%. and exact sentences at the corresponding rate as "the ideal summary". We compare the traditional mechanical summarization based on word frequency statistics with the summarization based on topic segmentation and dynamic adjustment of the sentence weight stated in the paper.

Table 1. Resulting data of experiment

Subject Matter	Summarization System	Compression Rate	Accurate Rate	Recall Rate
Comment	In This Paper	10%	0.6428	0.5082
		15%	0.6819	0.5415
		20%	0.7126	0.6139
	Mechanical Summarization	10%	0.6009	0.4973
News	In This Paper	10%	0.6510	0.5269
		15%	0.6983	0.5615
		20%	0.7190	0.6221
	Mechanical Summarization	10%	0.6123	0.5112
Scientific papers	In This Paper	10%	0.6630	0.5435
		15%	0.7120	0.5693
		20%	0.7221	0.6312
	Mechanical Summarization	10%	0.5961	0.5125

As can be seen from Table 1, for various articles, especially comment documents, the accurate rate and the recall rate of the summary is higher than that of traditional summarization. It states that the method of topic segmentation and dynamic adjustment of the sentence weight gives great consideration to both the coverage rate and the accurate rate. The advantage embodies more obvious when the compression rate of the article is moderate.

6 Conclusion and Prospect

This paper presents the method of extracting the summary based on the text structure, and tests this summarization method to indicate the outline of contents accurately and comprehensively through experiments. Meanwhile, it has high accurate rate and recall rate, helping users to evaluate the value of Web page quickly and effectively. The summarization system based on this method can deal with the favored summary, and the accurate rate of this system is higher than that of other summarization systems.

But for search engines, because the number of internet Web pages is huge, the efficiency of exacting summaries is also the important factor. The method stated in this paper need be improved in this aspect.

References

1. Wang, Z., Wang, Y., Liu, C.: On Automatic Summarization and Its Classification. Journal of the China Society For Scientific and Technical Information 24(2), 214–221 (2005)
2. Morris, A.H., Kaspcr, G.M., Adams, D.A.: The Effects and Limitations of Automated Text Condensing on Reading Comprehension Performance. Information Systems Research 3(1) (1992)
3. Li, L., Zhong, Y., Guo, X.: The Application of Comprehensive Information Theory in Automatic Abstract System. Computer Engineering and Applications 36(1), 4–7 (2000)
4. Zhou, Y., Wang, J., Zheng, G.: Research and Implementation of Web Page Cleaning. Computer Engineering 28(9), 48–50 (2002)
5. Xu, X.: Recognize Subtitle from Chinese Web pages. Computer Science 31(22) (2004)
6. Larroca, N.J., Freitas, A.A., Kaestner, C.A.A.: Automatic Text Summarization Using a Machine Learning Approach. In: Bittencourt, G., Ramalho, G.L. (eds.) SBIA 2002. LNCS (LNAI), vol. 2507, pp. 205–215. Springer, Heidelberg (2002)
7. Minel, J-.L., Nugier, S., Piat, G.: How to Appreciate the Quality of Automatic Text Summarization. In: Proc. of the ACL/EACL 1997, pp. 25–30 (1997)
8. Herlocker, J., Konstan, J., Borchers, A., et al.: An algorithmic framework for performing collaborative filtering. In: Proceedings of the 1999 Conference on Research and Development in Information Retrieval, pp. 230–237. ACM Press, New York (1999)
9. Saggion, H., Lapalme, G.: Concept Identification and Presentation in the Context of Technical Text Summarization. In: Proc. of the Workshop on Automatic Summarization, pp. 1–10. Association for Computation Linguistics, New Brunswick (2000)
10. Fu, J., Chen, Q.: A New Evaluation Method for Automatic Text Summarization. Computer Engineering and Applications 18, 176–177 (2006)
11. Jin, X., Yang, B.-R., Jian, Z.-G.: Analysis of Automatic Abstracting Methods. Application Research of Computers 21(9), 5–6 (2004)

Design and Application of New Structure of Prefabricating Multi-Column Wall

Yu Rui and Cheng Ming

Building Department of Civil Engineering College, Tongji University, Shanghai 200092, China
badaqiao@163.com

Abstract. New structure of prefabricating multi-column wall (PMCW), the vertical members such as shear wall apply prefabricated type, and horizontal members such as floor and balcony apply congruent type, as well as that the grouting anchorage connection is adopted as vertical member joints, and horizontal and vertical components nodes are connected by concealed beam to form an overall structural system. Taking a high-rise residential for example, the basic component elements and construction process are introduced. Furthermore, corresponding conceptual connection measures are proposed. Results indicate that: the structure not only has the advantage of quick construction and energy conservation, but also can be connected in a whole as cast structure, so it is worth to promote and research.

Keywords: residential industrialization, prefabrication, assembly, component connection.

1 Introduction

PMCW is a new industrialization residential structure which is put forward by Prof YING Huiqing, and it mainly consists of precast concrete hollow shell, cast-in-situ column, precast composite floor and prefabricated stair. PMCW is an improvement to China's residential industrialization: compared with traditional concrete structure, it is more efficient in energy saving and environmental protection; compared with the existing fabricated structure, it is more security in seismic performance as joint connection is improved; as to economy, it integrates some kinds of construction, saves a large number of templates and speeds up construction progress. So, PMCW is a structure worth to research and apply.

Vertical ultimate capacity test and horizontal quasi-static test have been applied in early research, which proved that PMCW had good seismic behavior. Based on the research results, this article describes the technical characteristics of building approach and construction process.

2 Industrialized Production

2.1 Prefabricated Wall

Prefabricated wall is composed of precast concrete hollow shell, side columns, middle columns and concealed beam, as shown in figure1. Precast concrete hollow shell is the

H. Tan (Ed.): Knowledge Discovery and Data Mining, AISC 135, pp. 651–658.
springerlink.com

most important component, it is first manufactured at the factory, second transported to fabricating yard until strength meets design requirements, third lifted to construction floor and held in place, fourth reinforcement cage is placed in hollow, upper and lower reinforcement are connected in reserved binding hole, last concrete is poured.

Precast concrete hollow shell shows a certain strength and stiffness, which has capability to bear the shake and pressure as lifting and pouring. Meanwhile, it can improve the construction speed and reduce template installation, scaffolding erection and wet operation. Moreover, the shell is set as a permanent part of the wall; some sub-projects (such as heat insulating work, doors and windows work, facade decoration work, etc.) can be integrated in the shell processing, which could not only shorten the building period, but also ensure the quality of sub-projects. In addition, all of the structure nodes and main bearing skeleton are connected by cast-in-situ concrete; these settings can ensure a good overall performance and solve the faults of weak connection and seismic behavior in prefabricated structure.

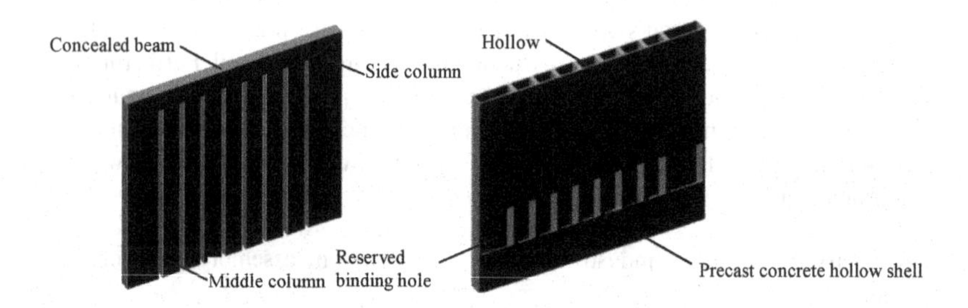

Fig. 1. Schematic diagram of fabricated wall

2.2 Precast Composite Floor and Balcony

The reinforcement of precast composite floor and balcony adopts truss type, which could ensure the exact location of steel bar in cast-in-situ concrete, and improve the strength and stiffness of joint surface. Total thickness of precast composite floor is designed to 180mm, and the prefabricated part is 80mm while cast-in-situ part is 100mm. Thickness of prefabricated part is determined by load-carrying capacity and hoisting stiffness. Meanwhile, thickness of cast-in-situ part should not only consider the steel bar arrangement and equipment line, but also meet the minimum thickness requirement of prefabricated structure in Chinese Code [1]. Design mentality of precast balcony is the same as precast composite floor, which could ensure the effective height of upper steel bar and improve the building quality.

2.3 Prefabricated Stair

Prefabricated stair is divided into two components, and each of them is composed of half platform board, half floor panel and a layer of step plate. Components are sustained in frame beam of layer and stair beam of rest platform. Gaps between two components are filled with cement mortar.

3 Construction Process

3.1 Manufacture

Precast concrete hollow shell is produced by special steel mold, and poured concrete in the way of loosing core. After that, external insulation, painting and window embedding are preceded gradually in the external wall, the production technology can reference the precast concrete composite wall[2]. As to the precast composite floor, precast balcony and stair, they can be manufactured according to conventional practice.

3.2 Transportation

Precast concrete hollow shell is transported by vertical fixing equipment, and trucks are equipped with special plane and line system. Meanwhile, trucks should be slow started and kept uniform; in order to avoid slant, speed also should be slow down when turning [3]. Precast composite floor and prefabricated stair could be transported by conventional way, but we must pay attention to avoid collision in the process of transportation.

3.3 Lifting

The lifting of prefabricated components needs a special hoisting tool, which is named "horizontal beam of adjustable lifting point" [4], as shown in figure2. The tool can adjust hoisting position according to different components, and ensure the vertical constant between lifting hook and component gravity. Not only that, the tool can efficiently prevent component slant in the lifting process, and conveniently fix position. As to the choice of tower crane, the lifting torque takes the first place, component weight and lifting distance should also be involved.

3.4 Steel Binding and Anchoring

After the completion of building foundation or lower layer, corresponding position should be set aside consistent reinforcement, which must meet the lap length. Upper reinforcement can be prefabricated first, and then lifted in hollow, reserved binding hole should be taken advantage of connecting upper and reserved lower steel; next, steel of concealed beam, composite floor, prefabricated stair and balcony can be bound and anchored; finally close template. Figure 3 displays the lifting and connection of prefabricated concrete hollow shell.

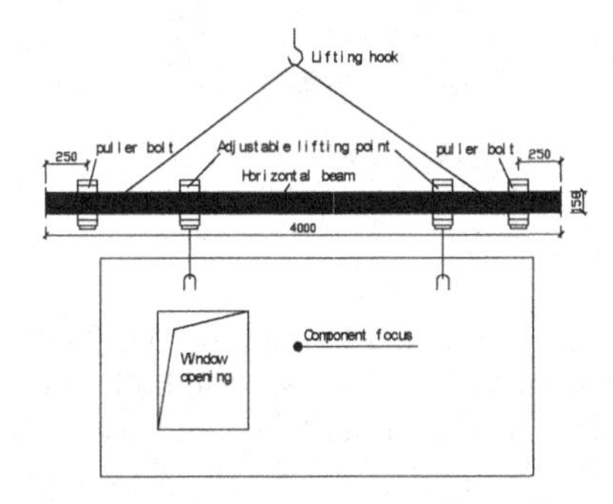

Fig. 2. Horizontal beam of adjustable lifting point

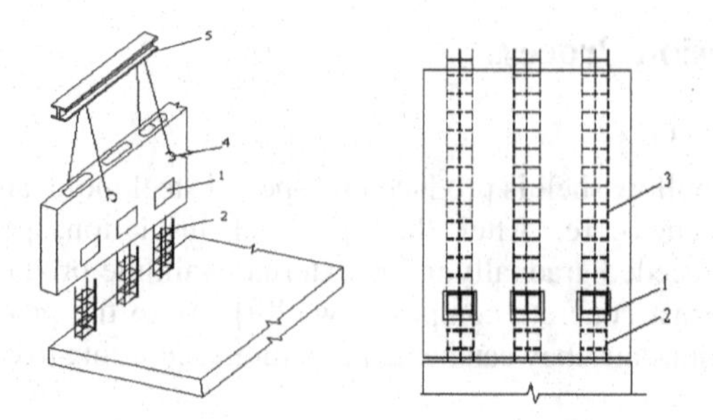

Fig. 3. Lifting and connection of prefabricated hollow shell
1-Reserved binding hole ; 2-Lower steel ; 3-Upper steel ;
4-Lifting hook ; 5- Horizontal beam of adjustable lifting point

3.5 Fixing Position

Prefabricated concrete hollow shell should be positioned by high-precision fixture after lifting. The high-precision fixture is composed of adjustable device and a series of embedded parts, which can achieve the regulation of three directions. The adjustable device includes elevation bolt, level and slant adjustment screw, as shown in figure 4.

One elevation bolt and level adjustment screw are set on both sides of the component bottom. Adjusting screw height can control the elevation of precast components; rotating level draw bars can push or pull the component bottom, which takes the role of level regulation. Meanwhile, one slant adjustment screw is also set on the both sides of component centre, and all the screw ends are fixed in structure floor. Rotating slant draw bars can push or pull the component top, which plays the role of verticality regulation [5]. As to the fixing of composite floor, prefabricated stair and balcony, there are many mature technologies in practice, so they are not described again.

Fig. 4. High-precision fixture

3.6 Pouring Concrete

Pouring concrete can be carried out after fixing position. The process is shown in figure 5, the middle and side columns of precast concrete hollow shell must be poured first, while concealed beam and composite floor be poured last.

There is something must be paid attention in the pouring process:

① Due to the smaller size of columns, pouring hose of diameter 200mm or below should be used. Internal vibrator must be pulled into the bottom of columns by the way of "quick plugging and slow pulling", and also tried to avoid colliding with concrete shell, which maybe lead to component migration. At the same time, in order to ensure the safety during construction, border protection must be focused too.

② To reduce the adverse influence of vibration and lateral pressure, and avoid component displacement or deformation, pouring speed should be slowed down to ensure construction quality.

③ If auto pump is adopted, concrete placing booms should be controlled carefully to avoid colliding with fixed structures. If fixed pump, high-precision fixture must be stagger with pump support system, and it should be special monitored during pouring process.

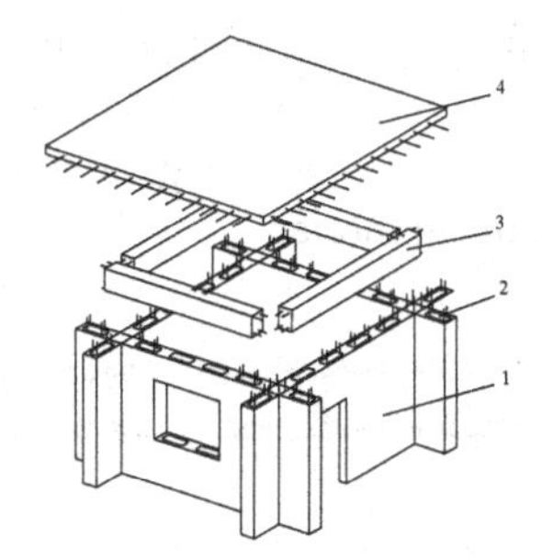

Fig. 5. Process of concrete pouring

1-Prefabricated wall ; 2-middle column ;
3-concealed beam ; 4-composite floor

4 Node Design of Prefabricated Wall

Node connection performance has a marked effect on the safety and durability of structure. According to the concrete connection pattern, earthquake resistant code and related fabricated structure design code, this article puts forward the connection measures of PMCW [6][7].

4.1 Vertical Connection

As to the vertical connection of precast hollow shell, lower wall is 500mm higher than floor; connection between upper and lower walls adopts tongue-and-groove join, whose inside and outside sewing are sealed by elastic glue, as shown in figure 5. Reinforcement is embedded in lower wall, and the length is 500mm, overlap join between upper and lower layer is completed in the reserved binding hole. This vertical connection can form rigid nodes.

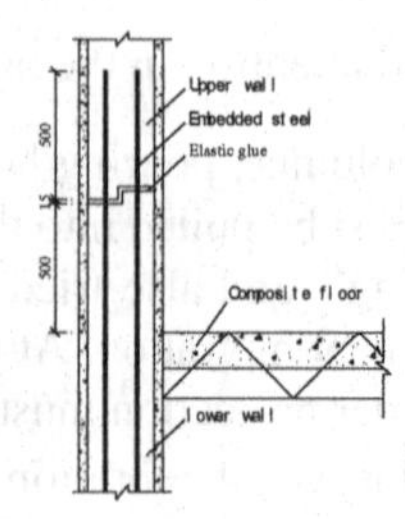

Fig. 6. Vertical connection of hollow shell

4.2 Horizontal Connection

The horizontal connection can be divided into three forms: L type, I type and T type, as shown in figure 7. As to L and T type, reinforcement is reserved in joint between lengthways and transverse wall and mutual anchored, the joint and column are poured simultaneously to form rigid connection. As to I type, when applied in external wall, vertical juncture should be sealed by elastic glue; when applied in interior wall, juncture should be sealed by cement mortar.

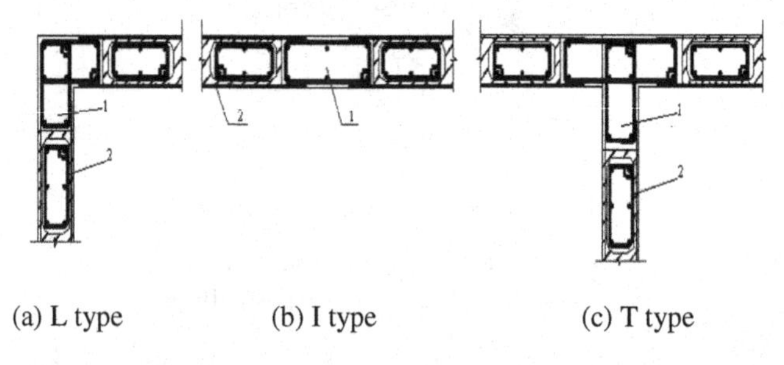

(a) L type (b) I type (c) T type

Fig. 7. Horizontal connection of hollow shell

1-On-site concrete ; 2-Precast concrete hollow shell

4.3 Connection with Foundation and Floor

In figure8, when connected with foundation, reinforcement is embedded in foundation top, and the length is 500mm; one steel waterstop is reserved in precast concrete hollow shell, which is more than 300mm form the foundation top. When connected with composite floor, upper reinforcement of floor should be bound in hollow shell, and the anchor length is more than $1.2l_{ae}$.

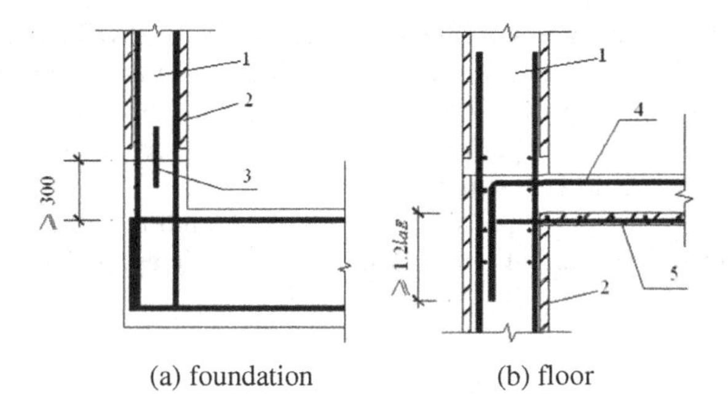

(a) foundation (b) floor

Fig. 8. Connection of hollow shell with foundation and floor

1-On-site concrete ; 2-Precast concrete hollow shell ;
3-Waterstops ; 4-Upper reinforcement ; 5-composite floor

4.4 Connection with Beam

The connection can be divided into three forms: side wall of middle storey, side wall of top storey and middle wall of middle storey, as shown in figure 9. Reinforcement between hollow and beam are mutual anchored, which must meet the limit of anchoring length.

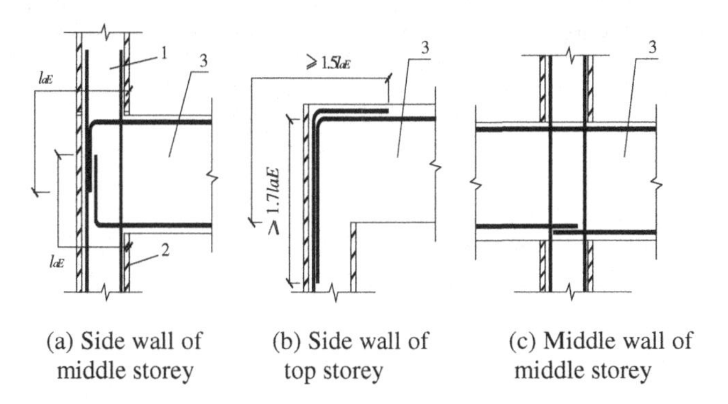

(a) Side wall of middle storey (b) Side wall of top storey (c) Middle wall of middle storey

Fig. 9. Connective joints of hollow shell with beam

1-On-site concrete ; 2-Precast concrete hollow shell ; 3-beam

5 Conclusion and Suggestion

The application shows that: first, the PMCW has high degree of industrialization, and construction progress can be sped up in a short term; second, crossover project is convenient and orderly, each process can be checked accuracy as equipment installation, so the quality can be guaranteed; third, construction takes up less field and site materials, moreover, the wet operation and noise of construction machinery are also declined; fourth, construction site is civilized and made less interference to around residents, which is good for environment protection; last, it saves lots of consumption of water and electricity, and achieves the goal of energy saving and emission reduction.

Various connections of PMCW are poured with on-site concrete, so that the carrying capacity, ductility and seismic performance can be guaranteed. Hence, the PMCW can be connected in a whole as cast structure.

Some structure practice, such as sealing of reserved binding hole and water repellent treatment, etc, must have further detailed research to guide application practice.

References

1. Cao, Z.: General theory of real estate economics, p. 321. Beijing University Press, Beijing (2004)
2. Ying, H., Zeng, J., Tan, Z., Wei, H.: Construction. Tongji University Press, Shanghai (2009)
3. Gong, H.: Construction Technology of Prefabricated house. Building Science (22), 81–82 (2010)
4. Gu, Z.: Accuracy control of prefabricated structure. Building Construction (10), 655–656 (2008)
5. Liu, B., Zhang, Y.: Seismic behavior of pretressed concrete frame. Building Structure (April 2005)
6. Liu, Z.: The theory and experiment research on Multi-Columns Wall with prefabrication construction method. PhD thesis of Tongji University (September 2010)
7. Jiang, L., Guo, Y.: Behavior of a new type of Self-locked joint of Concrete-filled steel tubular column and steel beam. Construction Technology (March 2009)

Discussion on Six-Step Approach of Programming Courses in Vocational Colleges*

Shaojie Du[1,2] and Xiaoju Zhang[1]

[1] Information & Engineering College, Binzhou Polytechnic, Binzhou, Shandong, China
[2] Institute of Vocational & Adult Education, East China Normal University, Shanghai, China
ducare@126.com

Abstract. Because of the abstract of programming courses, the weak learning ability of vocational college students and the shortage of learning resources, it is difficult to purely apply action-oriented teaching in programming courses nowadays. This paper provides an improved action-oriented teaching based on the characteristics of the course and the student and the existed learning resources. The new method is named IDCEAG, which follows the six steps of "importing a case→designing its algorithm→coding its function→explaining its knowledge→applying by self→guiding its summary". By practice, IDCEAG can achieve the teaching aims on training programming capability. The method is also suitable for the other integrated courses of theory and practice included in the computer related major.

Keywords: action-oriented teaching, programming, six-step approach, IDCEAG.

1 Introduction

Action-oriented teaching has been advocated in vocational education circles to train the professional ability recently. During studying, it requires not only with brain, but also with heart and hand. Action-oriented teaching makes the student more interested and innovative in learning, and then trains the professional ability in the activities [1]. Under the guidance of action-oriented teaching, vocational colleges carried out a bold reform. With the practical experience in curriculum reform of computer software major in Binzhou vocational college, the article provides an improved action-oriented teaching approach named IDCEAG, which fits for programming courses and vocational student.

2 Course Analysis

As an important curriculum in major program of computer software, it is the main aim of programming course to train the students' programming skills. The skills include

* The research is supported by department of education, Shandong Province Grant #J08WG60 to X. J. Zhang.

H. Tan (Ed.): Knowledge Discovery and Data Mining, AISC 135, pp. 659–665.
springerlink.com

the ability to use IDE, to code, to debug and to solve practical problems by programming with advanced language [2]. It requires to design an algorithm, and to build a mathematic model through observing, analyzing, inducing and reasoning in the process of solving practical problems. The process is creative exploration and constantly thinking beyond oneself [3].The thinking way is so abstract and different from usual problem-solving that the students are rusty and hard to accept. So to train program-thinking is the first difficulty of the course [3]. At the same time, grammar rules of programming language are vast, complex and closely connected with each other. It's not enough to study by reading and reciting. There is no vividly way for the students to know how the program runs. These determine that grasping the rules is another difficulty for programming teaching [4].

3 Teaching Objects Analysis

Some researchers carried out a survey about the learning psychology of the students in vocational college to improve teaching effect, whose result showed that the students were poor in learning, self-discipline, independent thinking, expressing, document writing and summarizing [5]. Nowadays most of the students want to learn under employment pressure. They like operation and practice, but tend to give up when meeting difficulties during studying [6].

There are many vocational students come from technical secondary school in the related major, basic theoretical knowledge of them is very poor. Because some local authorities lead a tendency to expand senior secondary school education, a lot of graduates of junior middle school entered technical secondary school when they couldn't go into senior school [7]. And at this stage, it stresses on grasping professional theoretical knowledge and training technical skills. Those make the students weak in basic theoretical knowledge. The other side, many vocational students come from technical secondary school have grasped some technical skills in his major, even have got the certificates of technical qualification [8].

4 Problems in Teaching

There is two parts in the teaching of Programming currently, first is explaining knowledge and the second is practicing. When explaining knowledge in multimedia classroom, shows knowledge and then gives an example related it. Teaching process includes explaining knowledge by PPT, showing an example and analysing the code [9]. The example is interested and practical to motivate the students. But to the students in the vocational college, the time to learn knowledge is too long to pay attention to it during the whole process. In addition, the distance between teacher and student will come into being in the process of learning from projection; the students can't grasp the emphasis and can't keep up studying if the speed of PPT playing is a little fast. Finally, the students can't practice in the multimedia classroom, when they practice in the computer room, it is possible that some knowledge learned is forgotten and the teaching effect discounted [10].

When practicing, the students program and code the task given by the teacher independently, and the teacher will solve and show the generality errors [11]. There are some problems also.

First, because the knowledge has been taught previously, it will not be taught again except emphases and usually it is forgotten. Without the knowledge and the example, it is difficult for the students to program and code by themselves; they will be dazed and don't know how to do. The generality errors will be solved when all of the students has finished, it is too long from the errors came, so the effect will not be good.

5 Idceag Method

Following the process of "getting information→planning→making a decision→carrying out→checking→evaluating", action-oriented teaching has been advocated in vocational education recently [12]. It emphasizes the independence of the students on getting information, planning, carrying out the plan and evaluating with the guidance of the teacher.

During practicing, the students will master vocational skills and professional knowledge, further construct their own experience and knowledge system [13].So, action-oriented teaching is a method of self-studying. However, the ability of self-studying is lack for vocational students. To use this method in vocational colleges, it needs enough teaching recourses and related technical materials, and leading-paper should be provided also. With the help of teaching resources and technical materials, the students can conceive their work results, and with the help of learning-paper, they can study in self-control [14]. However, the teaching resources and leading-paper mentioned above are in developing [15].Pure action-oriented teaching should be improved. In this paper, IDCEAG is provided based on the existed resources and the characteristics of programming courses. It follows six steps.

Step one is importing a case. It means to introduce an actual problem which needs to be solved by programming. The case must be interesting, low starting point and practical. "Interesting" means that the problems should be interesting and familiar with students, such as small games, password testing, etc. "Low starting point" means that the cases should not be too big or too difficult. "Practical" means that the function of the case should be applied frequently in real software development, such as data query. Such case will make the students interested in studying and contact studying with their future jobs.

Designing its algorithm is second. It follows the three steps below. Firstly, analyzes the interface and the data should be inputted. Secondly, analyzes the algorithm, such as how to handle the data, what kind of program structure to be used, how many variables demanded and their functions. Thirdly, analyzes what information should be feedback to customers and how to feedback, and then draws the flowing chart according to the analysis results. The second step is a key to transform actual problems to program-thinking, and it is an important way to train the ability to solve actual problems through programming [16]. Unknown problems solved with reference of the known, it meets Situated Cognition not only, but also meets From Quantitative Change to Qualitative Change.

The third step is coding its function. Complete code will be given to the students in this step, so they can implement by themselves. It is a key step to train practical ability. By coding, the ability to use IDE, to debug and to find the bug in codes will be developed.

The fourth is explaining its knowledge. The students could have experienced the process of the function as achieved and been interested in it. It's time to tell the students why and how it achieved, especially the new knowledge related. By explaining, it can make the students "know that something is so and why is so", at the same time it makes theory and phenomenon merged each other. New knowledge should be taught gradually, let easy one come first and not be limited to the knowledge in this case. The knowledge related and often used in programming should be introduced also. During explaining, errors can be set deliberately, so that students can deepen their understanding through debugging and revising them.

The fifth step is applying by self. Programming with advanced language is just like writing in English, which should not be only limited in reading the codes. It is more important to build the cognitive structure by reading, and then to code by themselves [17]. So, in this step, a related case with condition changed or function improved will be implemented by the students independently. They can discuss, use help documents and search on net. As analyzed above, the students in vocational college are poor in exploring and innovating, whether or not they can code the new function of the case is the key to go on studying.

There is a theory named cognitive migration in the field of cognitive psychology. Migration means the effect of a study on the other [18]. Thorndike, a famous psychologist and educator, believes the reason of migration among studying is that there are common elements in them, what are co-stimulation and reflect linking. In another word, the former study includes the same elements of the latter [18]. This theory supports the possibility to code independently. Broadly speaking, cognitive construction of writing is very similar with that of reading in content and organization. The two constructions are very similar but not same and closely link with each other, and fulfill migration conditions. The third and fourth step has made up a clear cognitive construction about the codes. The students have understood how the function is implemented, and how the statement and syntax worked. Further more, it is possible for the students to program by themselves, because new functions to be implemented are similar with or related to the one learned and practiced previously.

For example, in the course of 《Visual Basic.NET programming》, to learn "branch structure", the case "to judge if pass a test or not" is be imported. The knowledge about "IF statement" is included in the case. When explaining, the knowledge about "nested IF statement" and "Select Case statement" will be involved with expanded case "rating the scores", and new functions bellows are demanded to implement by the students independently:

① to judge the score inputted is between 1-100 or not, and give the logical expression;

② the program above ran by buttons. To run the program by pressing "Enter" after the scores inputted;

③ to judge the name inputted is "zhangsan" or not, the password is "123456" or not, if yes, "welcome" will be shown; If not, "invalid user" shown;

④ to input two values and operator, and then show the result, using "nested IF statement";

⑤ to compile the program, using "Select statement".

Having a foundation on "branch structure", the students are demanded to complete other functions using "IF statement", or to change the form of the program, so as to construct a real acknowledge structure.

Guiding its summary is the last step. The students should be asked to sum up their studying in order to consolidate the result. Because of their poor ability in writing, it is hard for them to write practice summary reports independently, or the quality of the summary is hard to guarantee [19]. So the effect of revision can not be achieved. Guiding-paper is provided to students, they can fill contents in the blanks designed in advance. Contents of the leading-paper include the four parts.

Part one is answering questions. According to the important knowledge in the cases, some questions are designed for students to answer, such as the effects of one property, the function of one statement, the premise of using an active controllers etc.

Part two is collecting faults. Bugs encountered in the program debugging and the solving methods should be written down, so as to avoid the same errors in programming in the future, and improve the ability in programming and debugging.

Function storing is the third part. Students are guided to change codes to functions consciously, thus work experience are accumulated with studying. The stored functions will become a resource after the course, which will lay a foundation for programming later after graduation. The functions, parameters and return values of the functions should be remarked, so as to use the functions correctly in future, and the functions are given more universally as much as possible [20].

The last part is self-reflection. Students should write down their feelings and experiences, and think about the inspiration. According to the branch structure mentioned above, the contents of the guidebook including:

① why should translate the contents to type of value in the processing of judging the score legal or not? Otherwise in judging usernames and passwords?

② the difference of expressing conditions between Select Case and IF statement;

③ Logic symbol OR was used above, and are there some ones, and the meaning of its respectively;

④ what debug errors encountered and how to solve in the processing of implementing the programs;

⑤ write down the harvest of the experiment simply.

6 Conclusions

Without enough learning resources, it is hard to carry out action-oriented teaching purely at present. It was improved to IDCEAG method according to the existed learning resources. Being guided strategically, the students explore initiatively, and build the knowledge structure gradually in the learning process. Then trained the ability of programming and accumulate programming experience. IDCEAG has been used in the past two grades, its effectiveness were very good. IDCEAG is not only suitable for programming courses, but also for other courses integrated theory and practice in computer software major, such as web design, dynamic website develop, web programming, etc.

References

1. Yi, Y., Wu, L.G.: On Programmers' Cultivating Objectives and Ways of Computer Software Technology. Vocational Education & Economic Research (1), 13–16 (January 2007)
2. Deng, J.H.: The Psychology of Coding and Programming, pp. 230–257. Tsinghua University Press, Beijing (2003)
3. Wu, D.Q.: A Research of Programming Learning Mind and Teaching Policy for College Students, pp. 15–21. Yangzhou University, Yangzhou (2006)
4. Jiang, Y., Sun, Y.J., Li, Z.W.: Talking about the Influence of Psychological State upon Teaching Effect in Vocational College. Liaoning Higher Vocational Technical Institute Journal 7(2), 140–141 (2005)
5. Xu, H.: Characteristics of Field-based Learning Curriculum in German Vocational Education. Research in Educational Development (1), 69–71 (January 2008)
6. Li, X.P.: Discussion on the Developing Model of the Cooperation between School and Enterprise in Chinese Vocational Education. Journal of Jiangsu Teachers University of Technology (3), 6–10 (March 2008)
7. Wu, Q.L.: Cognitive Psychology of Instruction, pp. 387–401. Shanghai Education Publishing House, Shanghai (2000)
8. Shen, J.Y.: A Reforming Experimental Teaching in Programming Design. Journal of Beijing Institute of Planning Labor Administration 11(1), 57–58 (2003)
9. Xiao, Y.: Discussion on Teaching Method of Java Programming in Vocational College. Modern Enterprise Education (8), 37–38 (August 2007)
10. Ai, C.C.: Thinking about Classroom Teaching of Programming in Vocational College. Gansu Science and Technology 24(7), 175–177 (2008)
11. Qiu, H.Z.: Analysis on learning Psychological and ability of vocational students. Journal of Jiangxi Finance College 19(12), 147–148 (2009)
12. Chen, T., Liu, H.: A Research on Self-learning Ability of Vocational Students on net. Education and Career (20), 155–156 (July 2007)
13. Zheng, Z.J.: An Analytical Study of the Teaching Practice in the Introductory Remark Approach. Zhuhai City Polytechnic College 13(1), 39–41 (2007)
14. Luo, Z.L.: Talking about the Reform of Programming Teaching in Vocational College. Journal of Hezhou University 23(3), 119–121 (2007)
15. Kong, C.L., Tian, G.: Based on Algorithm Design C Programming Language Instruction Method. Journal of Jilin University (Information Science Edition) 23(8), 5–7 (2008)
16. Xia, Y.: A Teaching Method in Program Design Based on Double-Center Mode. Journal of Sichuan College of Education 25(11), 109–111 (2009)

17. Lian, Z.Z., Wang, H.Z., Li, J.Y.: Research on the Teaching Model of Curriculum Design of Program Design. Journal of Science of Teachers' College and University 29(6), 95–97 (2009)
18. Tong, Q.: Constructivism Based Education Mode in Programming language Teaching Practice. Journal of Hunan University of Technology (5), 74–76 (May 2006)
19. Lu, M., Zou, Q.M.: Teaching Research of Programming Fundamentals Course Based on Capability Cultivation. Computer Era (8), 64–68 (August 2009)
20. Peng, Z., Li, D.Y., Geng, Z.L.: Research on the Algorithm-centered Teaching model of Programming Courses. Journal of North China Institute of Science and Technology 6(2), 90–92 (2009)

Inspiration of Cooperation between School and Enterprise in Germany and USA to China

Shaojie Du[1,2] and Shan Su[1]

[1] Information & Engineering College, Binzhou Polytechnic, Binzhou, Shandong, China
[2] Institute of Vocational & Adult Education, East China Normal University, Shanghai, China
ducare@126.com

Abstract. Because of the lack of corresponding policies and regulations and the short consideration to his employees in enterprise, cooperation between school and enterprise is skin deep in China. Though the cooperation modes are different in Germany and USA, both of them have corresponding policies and regulations, and special organizations. Enterprises in them, who take part in vocational education, will get allowance, enjoy tax relief and rise their famous degree. Taking the foreign experience as reference, the government should specify the responsibilities of enterprises for participating in vocational education in law, and motivate them by way of tax relief and publicizing the cooperation widely at the same time. In order to affect production order as less as possible, the vocational college should strengthen the management to the students when they practicing in enterprise.

Keywords: cooperation between school and enterprise, the role of government, the motivation of enterprise.

1 Introduction

Cooperation between school and enterprise, a training mode of skilled talents, is widely recognized and accepted in vocational education circle in China, and practiced energetically [1]. According to information, the first twenty-eight national demonstrative schools have cooperated with more than 5000 enterprises since 2006, which shows the rapid development of cooperation [2]. For late starting, there are no mature mode and corresponding policies and regulations about how to manage cooperation. The enthusiasm of enterprises to participation in vocational education is very low in fact. Generally, oral cooperation or surface cooperation with no stable relation is more than 75%. The breadth, depth and actual effect of cooperation were all far below the expectation [3].

The mode of cooperation between school and enterprise are different in Germany and USA, but both have achieved good results, and promoted economic development efficiently [4]. It mountain of stone, could besiege jade. Vocational education developed vigorously, we should study and reference carefully on the successful experience in developed countries, which will make the development of vocational education continuous and healthy in our country [5]. The paper analyzes the mode of

H. Tan (Ed.): Knowledge Discovery and Data Mining, AISC 135, pp. 667–672.
springerlink.com

cooperation, the role of government, the motivation of enterprise in Germany and USA, and the bottleneck in China. And then provides some measures to deepen cooperation according to our own conditions.

2 Dual-script System in Germany

2.1 Dual-script System

One is vocational school and the other is enterprise in dual-script system. The main task of school is to offer theory basis and training the practical skills is that of enterprise [6]. Vocational colleges don't enroll new students from community, and not contact with the enterprises directly. First, students select an enterprise and contract with it, and then deployed to a vocational college by State Ministry of Culture and Education [7]. Training funds are bared by enterprise entirely. Enterprise provides training venues, facilities and training instructors along to the training contracts for students, and pays them training allowance.

2.2 The Role of Government

The behaviors of enterprise and school are mostly constrained by legislation and coordinated in dual-script system. On legislation, there is a perfect legal system of vocational education in Germany [8]. Enterprise behavior in training is mainly constrained by federal laws, such as 《Laws of Vocational Education》, 《Protection Laws of Youth Labor》, 《Laws of Labor Promotion》 and 《Regulations of Handicraft Industry》 etc. Obligations of enterprises, qualifications of trainers and the procedure of training are prescribed in 《Laws of Vocational Education》 specifically, and school behavior are constrained by 《Laws of Schools》 issued by states [9].

Government is a bridge between enterprise and school besides legislation in Germany. State Ministry of Culture and Education deploys vocational college for the students enrolled in enterprise, and State Ministry Culture devises teaching plan of major according to the demand of enterprise, which implemented by the vocational college, so as to guarantee the training target of the two consistent [10].

2.3 The Motivation of Enterprise

Dual-script system is a mode that is mainly carried out by enterprise from the analysis above. Besides the external factors, such as power legal protection, social environment which advocating technology etc, there are still some internal factors cannot be ignored when talking about the reason that enterprise takes part in vocational education in Germany.

Firstly, enterprise is profitable. The expenses for the training are be borne by enterprise, but 50%-80% is supported by government. If the career trained is in line with the development trend, the supported can reach to 100% [11]. In addition, enterprise will enjoy tax relief. So it is worthwhile for enterprise.

Secondly, it can rise famous degree of enterprise. In Germany, enterprise which meets the requirements can enrol trainee, but not all. Training students shows the level of the enterprise, it is not only a kind of honour but also is a kind of publicity for enterprise, and increases its popular renown in society.

3 Cooperating Educational Mode in USA

3.1 Cooperating Educational Mode

Combined teaching with practice or alternated study with work, Cooperation education in USA is a mode of school-based [12]. Teachers seek for appropriate enterprise according to the major and interesting of the students, and make cooperation educational plan. Enterprise offers appropriate place and condition to practice base on its need and ability. Then school contracts with enterprise to determine the task, responsibilities, work time and remuneration of the students [13].

3.2 The Role of Government

In the respect of legislation, 《Laws of Higher Education in 1965》 passed by U.S. Congress. It brought out some regulations such as financial allocation, public support and setting up cooperation educational fund etc. Those strengthened the support of government and social sectors to cooperation education [14]. By the 1990s, America issued variety of encourage policies and measures, such as setting up special fund, providing low-interest loans, reducing or exempting some taxes and founding Science and Technology Park Zones etc [15].

In the respect of organization, national specialized agencies were established, whose task was to strengthen the management of cooperation education, to publish and coordinate the development of cooperation education. The school that carried out cooperation education had special department and staff to coordinate cooperation between school and enterprise [16].

3.3 The Motivation of Enterprises

Different from the legislative constraints in Germany, many enterprises participate in the vocational education due to the purpose of social welfare, and half past of them mainly due to charity. Research showed that there are some other references, such as enhancing its reputation, reducing labour cost and recruiting employee etc [17].

4 The Bottleneck for Cooperation in China

4.1 Specify the Responsibilities and Obligations of Enterprise Legally

An important reason for enterprise participating in the vocational education is that the government issues a perfect and operable law system in Germany, which makes cooperation being well a foot. It needs a law system to guarantee enterprise participating in vocational education in China. 《Trade and enterprise participate in vocational education executive way》 is suggested. It will state the status, function, task of the trade and enterprise in vocational education, and the organization, condition, responsibility, method, teacher and trainer, especially raising and managing fund. Guaranteed in law, it can make cooperation in real earnest.

4.2 Motivate the Enterprise

Government should offer corporate support policies to encourage enterprise to participate in the vocational education. Such as incentive plan, means that the government should apply tax relief or prior credit to those enterprises who take part in vocational education actively and effectively. Government should fund cooperation, award honour and bonus to effective enterprises by evaluating. And also, it should vigorously publicize the active enterprises to make them popular with people through news media.

4.3 School Cooperate Actively to Guarantee the Profit of the Enterprises

Considering the profit of the enterprises, vocational college should actively cooperate with enterprise and benefit them. First, vocational college should revise major and curriculum program, and make the students achieve the demand of the enterprise as much as possible, thus enterprise can save money on training new workers. Secondly, vocational college would help enterprise to train his employee in business, and thus induce the training cost. Before the students practise or train in enterprise, the school should lay out a detail plan for them together with enterprise. With the plan, the effect to production order will be as less as possible in time and place. At the same time, the school should guide the students to create profits for enterprise with strong professional responsibility during practice.

4.4 Publicize and Public the Social Responsibility of Enterprise for Participation in Vocational Education to the Whole Society

Vocational education is unique. It is successful or not is associated with multiple factors and mutual penetration, so it requires integration of interdisciplinary thinking, and also the support of the whole society, especially participation and support from enterprise. Corporate' social responsibility is manifold: in the creation of profit, be accountable to shareholder interests, meantime we must also take the responsibility on our employees, customers, partners, community, and industry and environmental. As a responsible business enterprise, including the skilled and highly skilled personnel, we should not only to pursue economic efficiency, but also focus on the technological advances and social long-term development, undertake the responsibility together with schools to cultivate highly skilled personnel and high quality workers. Government should make decisions based on the national macroeconomic policy, starting from the regional development perspective to introduce specific policy guidance, support, regulation and protection of enterprises to participate in vocational education, establish and coordinate a smooth channel for enterprises and schools to train skilled personnel, so to ensure that project of work and study completely implementation, then to form a government-led, community supervision, industries and enterprises to work closely with the school that is a new pattern of vocational education.

Vocational College faculty should update their views to have a sense of ownership, initiatively to seek opportunities of enterprise and school cooperation, actively participate in the technological transformation of enterprises and project development, and actively help companies train their employees to solve the urgent need to address the technical, production and management problems, so that enterprises could feel the mutual benefit of school-enterprise cooperation.

5 Conclusions

Cooperation between school and enterprise is a new demand to the development of school and enterprise. Taking the foreign experience as reference, the vocational college should make cooperation enlarge its scale, extend its depth and rich its content, and share profits with enterprise. So promote cooperation between school and enterprise in China to go ahead.

References

1. Peng, W., Song, X.J.: Try to discuss Industry Support of Education Product in Higher College. Journal of Shanxi Radio & TV University 10(5), 33–34 (2005)
2. Ma, Q.F.: China's Education Research & Review, pp. 486–523. East China Normal University Press (2010)
3. Hu, C.R.: A Study on the School-Enterprise Cooperation & Order Type Training in the Vocational Colleges. Journal of Chongqing Industry & Trade Polytechnic (1), 23–25 (2009)
4. Cai, Z.: Talking about Dual System Vocational Education. Hunan Nationality Technical Institute Journal (6), 90–93 (2008)
5. Liu, H.D.: Cooperation in America and Its Inspirations for China. Researches on Higher Education (10), 91–92 (2002)
6. Wang, Z.Q., Liu, X.H., Zhao, M.W.: Exploration on Dual-System Vocational Education of Germany. Journal of Shanxi Polytechnic Institute 5(2), 75–78 (2010)
7. Li, J.: Drawing on the Experience of Dual System Model for the Reference to Chinese Vocational Education. Journal of Xian Vocational Technical College 3(2), 55–57 (2010)
8. Hu, G.: On System Factor of Germany Success in Vocational Education. Journal of Nantong Textile Vocational Technology College (1), 84–86 (2010)
9. Lin, G.H., Kan, L.: Exploring the Road of Chinese Higher Vocational Development from the German Dual-system Vocational Education. Liaoning Higher Vocational Technical Institute Journal 11(2), 10–13 (2009)
10. Yi, Z.Y.: Successful Elements of Cooperation between School and Enterprise of Dual-system in Germany and Its Revelations to China. Vocational and Technical Education (17), 98–100 (2006)
11. Feng, X.F., Li, H.Z.: The Reason of Enterprises Participating Vocational Education in Germany and Its Inspirations for China. Education Exploration (1), 60–63 (2009)
12. Xu, P., Xu, J.Z.: An Analysis on the Mechanics of Cooperation in USA and its Significance. Heilongjiang Researches on Higher Education 1, 35–36 (2007)
13. Li, X.X.: Simply Talking about the America Community Colleges and Its Reference for Higher Vocational Education in China. Journal of Liaoning Agricultural Vocation-Technical College 12(1), 41–43 (2010)

14. Gao, H.: American Vocational Education Legislation and Its Revelation to China. Journal of Hunan University of Technology (2), 134–136 (2010)
15. Miao, Y.Y.: Development and Enlightenment: Vocational Education of US Community Colleges. Journal of Yangtze University: Social Sciences 32(2), 114–116 (2009)
16. Zhou, W.J., Wu, X.Y.: The Reform of Vocational Education Management System in the Transitional Era of the USA, vol. 24(9), pp. 30–35 (September 2009)
17. Zhang, F.J., Chen, L.G., Luo, Y.B.: Analysis of Incentive and Impediment of Employers' Involvement in Vocational Education in America. Comparative Education Review (5), 86–90 (2008)
18. Hu, Y.X., Cao, L.S., Liu, Y.H.: Bottleneck and Solutions about Cooperation between School and Enterprise in China. Higher Education Exploration (1), 103–107 (2009)
19. Gan, Y.L., Yang, R.: The Analysis on the Bottlenecks Encountered in Vocational and Technical Education School-enterprise Cooperation. Technological Development of Enterprise 29(3), 176–178 (2010)
20. Song, H.X.: Higher Vocational Education Development Bottlenecks Countermeasures under Domain of Educational Elements. Journal of Anyang Teachers College (1), 116–120 (2009)

Building Computer Graphics into an Independent Sub-discipline

Wei Haitao, Lu Hanrong, Xiong Jiajun, Gao Yan, Hu Jinsheng, and Wu Caihua

Air Force Radar Academe Fourthly Department, Wuhan, Hubei, China
weihaitaogood@126.com

Abstract. Conventionally the international education circle for computer graphics education insists on that computer graphics is based on graphical standards for displaying graphics. This mainstream point of view entails quite a lot of difficulties in computer graphics education. To resolve these problems, this paper proposes a new method of teaching that the fundamental methodology for realizing 3D graphical displays is done by a combination of computer programming and computer simulation. The discipline connotation of computer graphics is distilled in three aspects of research area, methodology and theory architecture. The position and role of computer graphics is re-established in computer science. We expect the new paradigm could solve the problem in computer graphics education that puzzled the international computer graphics education world for so many years. At the same time, the connotation of computer education and the subject framework of computer science are distilled effectively.

Keywords: Computer Graphics, Discipline Properties, Teaching Model, Programming, Computer Simulating.

1 Introduction

As a fundamental application course of computer science, the teaching of computer graphics has become a pressing education issue on a global scale. In particular, ACM SIGGRAPH Education Committee and Eurographics have shown profound concern about the problem in computer graphics education for 12 years [1-11], and expect the publication of an instructive approach that can solve the construction and teaching of this course[2, 11].

The reason for the confusion surrounding the teaching of computer graphics is that the content of traditional computer graphics courses is based around the construction and use of graphics standards[3,12~14]. However, graphics standards have no concept of geometric modeling (instead, geometric modeling is taught in traditional CAD courses), and no concept of software systems. Furthermore, they cannot of themselves automatically generate 3D graphics (because graphics standards are the driver of display card and the collection of subroutines), and their teaching has led to the discipline properties blurring of computer graphics. It is important that research results and development laws of computer graphics cannot be used to summarize in

H. Tan (Ed.): Knowledge Discovery and Data Mining, AISC 135, pp. 673–680.
springerlink.com

graphics standards. Merely studying graphics standards, beginners are unsure of the basic programming approaches to contemporary computer graphics application software like CAD and digital design, computer animation, 3D games, and virtual reality; this is the critical issue that needs to be addressed in reforming computer graphics courses. Since computer graphics is simply confined to graphics standards, there exist some deficits, which cannot be settled theoretically.

2 Reforming the Teaching Models of Traditional Computer Graphics Courses [15]

2.1 Adjusting the Content of Computer Graphics Courses

2.1.1 Utilizing 2D Graphics to Conceptualize Software Systems

All applications of computer graphics are result of programming and using the programming tools.

Using the simplicity of 2D graphical models for point, line, and surface to explain the implementation of software systems can allow students to acquire a complete picture of the use of computers to automatically generate graphics.

Software systems refer to large scale programs with complete and dynamic structures that follow system flow requirements in solving problems, coding the implementation of multiple models that describe the four processes of data and command input, storage management, operation processing and output display, in order to reach the desired objectives of auto-running software.

2.1.2 The Use of 3D Graphics: Constructing Physical Models for Objects, Lighting and Cameras, and Simulating Light Transmission to Generate 3D Graphics

Actual photographs involve light transmitted in the natural world and within camera models, and images are then generated and displayed in the camera lens. Computer-generated 3D graphics is a simulation of the physical process of light transmission and the generation and display of images. For example, ray tracing algorithms, radiosity algorithms and projections of 3D geometric models act in this way, and so the texture mapping process can be interpreted as "mapping photographs to the polygons in the screen, to simulate the reflective display effect generated from the point-by-point illumination of light on object surfaces." Based on this we can use a new idea of the simulation of light transmission, that gives us a unified interpretation of the generation of 3D graphics. The teaching content of 3D graphics can then be rearranged and ordered according to the theoretical framework of computer simulation: "systems, modeling, simulation algorithms, and evaluation".

The generation of 2D lines can be explained as a depiction and simulation of the point-by-point projection (simulation of light transmission) of the position (non geometric shape) of lines in space onto a display screen. This is a mathematical concept and proposition; it is not a process of actual physical change, so that realistic 3D graphics do not include this type of graphic display; but straight lines, curves and other graphics have important applications in mathematical modeling of 3D graphics.

3D graphics describes physical models of objects, lights and cameras through the construction of multiple mathematical models as the principal line. For example, geometric models that describe an object's physical shape (for example, the winged-edge data structure of a plane object, general cutting operation and Boolean operation of plane objects.), material models that determine the ability of an object surface to reflect and transmit light, and texture models that display details of the object surface; ray geometric models, color models, illumination models, radiosity algorithms and ray tracing algorithms etc that describe the physical characteristics of light; camera models displaying all kinds of graphical effects, e.g. the geometric transformation, clipping, projection, polygon filling (involving polygon's flat rendering, Gouraud rendering, Phong rendering, bump rendering and shadow rendering etc.) and texture mapping ,and image fusion algorithms of geometric models. From this we get multiple objects constructing and simulating the visible physical environment of the natural world, as well as simulating the transmission of light through this virtual visible environment and in camera models, and the concept of dynamically generating 3D images in the camera lens, as in 3ds max software using ray tracing algorithms for the process of rendering animated image files.

3D graphics software systems are built through actual programming practices under the guidance of concepts of software system, modeling and simulation. By inputting descriptive data of graphical models and commands into this operating software, realistic 3D graphics just like photographs can be automatically generated by computer.

Movements or deformations of objects, lighting changes, and camera movements can all form computer animation. Utilizing computer animation and games (operation commands + real-time animation + sound), we can carry out applied research on visual simulation, or the simulation of human behavior and evaluating capacity for action. Furthermore, numerical control machine tools are composed of multiple parts, hydraulic and electrical equipment; if these machines are endowed with the capacity for processing other mechanical parts, the relevant parts of the machine then also need to be capable of the necessary mechanical movement and interaction. This type of mechanical motion and simulation in computers can also constitute a type of computer animation with a specific effect; and the inherent geometric model of the mechanical motion and simulation of these machine tool parts has even higher precision requirements, because it has to control the geometric model precision of the processed parts, which is more complex than graphical display.

Simulation systems in computer graphics refer to that all visible physical phenomena involved in light transmission are the objects of study in computer graphics, thus determining the scope of research in this discipline. By evaluation, we mean that by comparing the differences between 3D computer-generated graphics and actual photographs, new mathematical models or simulation algorithms can be constantly proposed and used, and incremental improvements or updates applied to existing computational models, thus enabling the mathematical calculation process of computer graphics to move ever closer towards lighting , camera and the processes of real life physical changes for the construction of real object models (including rigid bodies, soft bodies, objects with multi-joint movement, fluids, gases, and fields), their movement and evolution, collision detection and deformation, and the display of object reflective effects.

2.2 Constructing a New Teaching Model for Computer Graphics

The above theoretical education and programming training for constructing 3D graphics software systems is consistent with the "implementation prerequisite of computability" requirement of computing theory for program design: "problems to be solved are described in the form of systems and models; this description can be converted into algorithms; complexity of these algorithms should be reasonable." This leads instruction models of teaching computer programming towards scientifically oriented teaching models; the case of 3D animation systems in computer graphics can be used to guide students in mastering the working methods of scientific research: "① accepting tasks and asking questions, or freely choosing topics: exploring and discovering new objects of study from the natural sciences, that are not part of the research scope of computer graphics, along with asking questions are prerequisites of thinking deeply about tasks. Freely selecting topics requires establishing the creativity, objectivity, targetability, feasibility, and scientific basis of research, and sorting out avenues for submission; for example the problem of exactly what kind of discipline is computer graphics is a dilemma that has occupied the attention of the international computer graphics education sector for many years[1~11]; ② analyzing problems: real photographs are defined by the three key factors of object, lighting and camera; and from this refining the key teaching points required for describing these physical models; ③ proposing solutions to problems: firstly using 2D graphics to establish the concept of software systems, and then establishing the concept of generating 3D images by simulating light transmission, based on multiple mathematical models required for describing the physical models of objects, lightings and cameras; ④ doing experiments to solve the problems: selecting the data structure, designing algorithms, writing source code and debugging the program in order to build 3D graphics software systems; ⑤ promoting the results and reaping the rewards: carrying out incremental improvements or updates to existing scientific theories and methods, publishing academic papers to demonstrate and report new developments, applying for new project grants and promoting the steady forward growth of the discipline. For example, when we completed the first demonstration that computer graphics can be classified as a branch of computer simulation studies, it laid the foundation for the writing of this paper." This is the basis of a complete education in programming. At the same time the various teaching methods of this course were formed, including instructive, problem exploration, training, and practical techniques.

2.3 Constructing a New System of Computer Graphics Courses: Towards Mature Teaching

After improving and rationalizing the theory of computer graphics systems, the principles and use of OpenGL graphics standard drawing tools can be explained. OpenGL involves concepts about physical models of cameras and lighting, texture models and material models etc, but has no concept of ray tracing algorithms, radiosity algorithms, geometric modeling and software systems. By transferring geometric parameters to OpenGL, students can easily utilize OpenGL programming to implement the display of 3D graphics. When combined with practical computer programming content, this results in the formation of a new type of computer graphics course with three subject forms: "theory (software systems, modeling, simulation and programming), tools (OpenGL) and applications (various applications for displaying graphics and their operation)". This is consistent with the development model of human society i.e. "explore the unknown world → form theoretical systems such as independent disciplines → make tools with the help of theoretical guidance → use the tools to remold the world", while the teaching theory system of computer graphics in China is the first of its kind in the world to achieve maturity.

3 The Discipline Connotation of Computer Graphics (It as a Teaching Course and Knowledge Classification, Original Innovation)

Through curriculum reform, the discipline connotation of computer graphics [6, 7] is distilled in three aspects of research area, methodology and theory architecture.

Independent research area: research area in computer graphs is independent; its fundamental question is "how to use computers to automatically generate images similar to those acquired by human observation of the world." The physical method of obtaining image is to take photos with camera or use artificial painting. So the way of generating this image by computer program automatically is only to simulate the transmission of light through the natural world and in camera models, and generating 3D images in the camera lens; Or simulating the process of brush drawing. This then determines that the scientific properties of computer graphics are steeped in computer simulation studies.

Methodology: since computer graphics belongs to the realm of computer application software, it follows that the basic approach established by computer application development i.e. "physical models (chemical models, biological models, artificial intelligence models, social models) → mathematical models (description of natural language and function that can be implemented through programming) → program structure(function structure of input, storage management, operation processing and output display) → program implementation", the working methods of scientific research, the method of abstraction (for instance, distilling the discipline connotation of computer graphics and the concept of software system), constructive mathematic methods represented in the form of recursive, inductive and iterative techniques, along with scientific paradigms[16] (used to define what should be studied, what questions should be raised, how to raise doubts about a problem and what kind of rules to follow in interpreting the answers we obtain, i.e. paradigms express the inherent laws and structure relations of scientific development), are all important parts of the methodology of computer graphics, and are fundamental ways of generating knowledge in the discipline of computer graphics.

Theoretical architecture: the theoretical architecture of computer graphics is described by the implementing prerequisite of computability (problems to be solved are described in the form of systems and models; this description can be converted into algorithms; complexity of these algorithms should be reasonable) and the theoretical framework of computer simulation (systems, modeling, simulation algorithms, and evaluation). The subject framework of computer graphics is expressed by the three subject forms of "theory (software systems, modeling, simulation and programming), tools (OpenGL, Direct3D, Java3D, ACIS, VRML, CUDA, OpenCL, OptiX, 3D games engine software package, PhysX engine etc.) and applications (various applications for displaying graphics and their operation, e.g. contests in computer animation, games and CAD; applications of computer 3D animation software that can produce a various of artistic pictures and effects of art.)". Along with the "theory, tools and applications" of computer science, this neatly shows the application of recursive techniques in computer science.

4 Re-establishing the Position and Role of Computer Graphics in Computer Science, as Well as the Connotation of Computer Education

Computer graphics is an important basic course for reforming and popularizing the education of computer programming, because it can highly distill multiple applications of mathematical methods in various fields due to the fact that linear and quadratic lines are the simplest math models. Furthermore, computer graphics has built bridge for intercommunion of the information world and the physical world through math models and visualization. And enabling computer education comes back to development orbit of science research, because computer simulation is a basic method of science research.

Data computation and simulation, data storage and retrieval, and data network communications have become the three characteristic components of 3D online games and other modern computer applications, and the corresponding typical courses are computer graphics, database and network communication. The teaching of computer graphics can seamlessly link up with the teaching of control software, because its approach to problem solving is ideologically interlinked with the approach of automatic control disciplines (i.e. first model, and then implement). As a category of visual and physical simulation application software, computer graphics fills a gap in the core curriculum of traditional Chinese computer courses and Computing Curricula 2005(CC2005) [17] in regard to the teaching of computer simulation principles.

A comprehensive education encompassing mature computer graphics and traditional computer courses allows the teaching of the three subject forms (i.e. the connotation of computer education and the subject framework of computer science) "theory (computer expertise and system theory; theory needs to answer: What can be effectively calculated automatically? What method should be used to study the proposition and what complete theoretical systems formed? What are the prerequisite conditions for implementing computability? How can this kind of automatic computation be implemented, how can the security and stability of computing devices and the accuracy of computing results be ensured?), tools (computer systems, operating systems, compiling systems) and applications (data calculation, reasoning and simulation; data storage and retrieval; data network communications; automatic data acquisition, and data output expression and control such as multimedia techniques)" to be fully showcased, enabling balanced development in computer education. They recur again to form the basis of multiple growth directions for computing majors, both specific (such as internet fire protection wall and antivirus—enhancing the security and stability of computer's operating systems; network computing and cloud computing; network data storage and retrieval for example Google; website building and web design and development, web browsers; HLA and instant communications; streaming media technology and media players; The Internet of things - it acts more like an expanded open computer network control system, regulating the automated input and output of data describing numerous controlled object models by means of sensors and controllers; office software, image processing, geographic information systems etc.) and synthesis (such as 3D online games); or in researching and developing computer hardware. The importance lies in that, through the teaching of computer graphics, we are able to elaborate the fundamental approaches to developing application programs base on the parading: physical models (chemical models, biological models, artificial intelligence models, social models) → mathematical models (description of natural language and function that can be implemented through programming) → program structure (function structure of input, storage management, operation processing and output display) → program implementation. This has transformed the thinking and outlook behind the teaching of computer graphics courses.

5 Conclusion

The teaching methods of traditional computer graphics (graphics standards) can be interpreted in the same way as the idea of simulating light transmission to generate 3D images. But this makes it difficult to conduct the teaching of computer graphics. The new teaching methods of computer graphics can be interpreted in the same way as the idea of simulating the changes in physical properties of objects, lighting and cameras, and simulating light transmission to generate 3D animation images. We expect the new teaching methods could solve the problem in computer graphics education that puzzled the international computer graphics education world for so many years. The education of computer graphics can be divided into two major branches. One is computer animation or CAD, the other is graphics standards or computer game. Both of them have a background for supporting industry. The principle of computer graphics education would involve various academic branches, as well as the geographic information system, which can better facilitate people's acceptance of it.

Thanks professor Marc J. Barr for his help and guidance when preparing this paper.

References

1. Cunningham, S.: Report of the 1999 Eurographics/SIGGRAPH Workshop on Graphics and Visualization Education, Coimbra, Portugal, July 3-5 (1999)
2. Cunningham, S.: Computer Graphics Education, July 6-7 (2002)
3. Cunningham, S.: Computer Graphics: Programming. Problem Solving, and Visual Communication (2003)

4. Cunningham, S., Hansmann, W., Laxer, C., Shi, J.: A report from the working group on computer graphics in computer science. In: 2004 SIGGRAPH/Eurographics Computer Graphics Education Workshop, June 2-6. Zhejiang University, Hangzhou (2004)
5. Bourdin, J.-J., Cunningham, S., Fairén, M., Hansmann, W.: Report of the CGE 2006 Computer Graphics Education Workshop, Vienna, Austria, September 9 (2006)
6. Alley, T.: Computer Graphics Knowledge Base Report, September 19 (2006)
7. Orr, G., Alley, T., Laxer, C., Geigel, J., Gold, S.: A Knowledge Base for the Emerging Discipline of Computer Graphics, July 9 (2007)
8. Barry, R.: SIGGRAPH Education Committee 2007 Annual Report (2007)
9. Scott Owen, G.: ACM SIGGRAPH Annual Report for FY (2008)
10. Cunningham, S., Cunningham, B.: Teaching Computer Graphics in Context Computer Graphics Education 2009 Workshop, Munich, Germany, March 31-April 1 (2009)
11. Barr, M.J.: SIGGRAPH Education Committee 2010 Annual Report, Los Angeles, July 25-29 (2010)
12. Cunningham, S.: Computer Graphics: Programming in OpenGL for Visual Communication. Prentice Hall, Upper Saddle River (2006)
13. Hill Jr., F.S., Kelley, S.M.: Computer Graphics Using OpenGL, 3rd edn. Prentice-Hall, Inc., Upper Saddle River (2006)
14. Angel, E.: Interactive Computer Graphics: A Top-Down Approach Using OpenGL, 5th edn. Addison Wesley, Boston (2008)
15. Wei, H.: Computer Graphics, 2nd edn. Publishing House of Electronics Industry, Beijing (2007)
16. http://en.wikipedia.org/wiki/Paradigm
17. Shackelford, R., Cross II, J.H., Davies, G., et al.: Computing Curricula, September 30 (2005)

New Design for Water Quality Early Warning System

Tian Jing, Zheng Shuyin, Zhang Guangxin*, Hou Dibo,
Huang Pingjie, and Zhang Jian

Department of Control Science & Engineering, State Key Laboratory
of Industrial Control Technology, Zhejiang University,
Hangzhou 310027, China
{TJ080052,autorabbit,gxzhang,houdb,
huangpingjie,askya}@zju.edu.cn

Abstract. Traditional water quality early warning systems are often static and the algorithms built in are difficult to modify while the system is in operation. So a changing environment has brought great challenges to the systems. To overcome the shortages, a new water quality early warning system is designed; it is based on the idea that the algorithms codes can be compiled dynamically and executed automatically. The algorithms codes can be updated in real time and executed as jobs. And thread-pool is used to execute the jobs in parallel and the system is designed to manage jobs for fault-tolerant. Finally, the realization of the framework is given and some rules to make the system stable and high fault tolerant are discussed.

Keywords: water quality, early warning, compiled dynamically, thread-pool.

1 Introduction

Recently, there are more and more fatal water pollution accidents [1] with rapid development of national economy and increasingly expanding urbanization in China. The accidents have heightened awareness concerning potential hazardous threats to Chinese today's water quality early-warning systems and emergency management system [2]. Lots of types of hazardous substances such as arsenic, fluoride, nitrate and organic chemicals have made the quality of many drinking water sources significantly degraded [3].

Nowadays, some traditional water quality early warning systems in China are static and they always work according to the early-warning algorithms already integrated in the systems. The algorithms are difficult to modify while the system is in operation. But water environment is complex and changing and it takes great challenges to the traditional early warning systems. Water quality early warning systems should reliably detect sudden contamination of the water supply and adapt to the changing water environment. So flexibility, stability and reliability are the important characteristics for water quality early warning systems.

* Corresponding author.

H. Tan (Ed.): Knowledge Discovery and Data Mining, AISC 135, pp. 681–685.
springerlink.com

In this paper, aiming at the shortages of the traditional water quality early warning systems, a new water quality early warning system is designed. The system can compile the algorithms codes dynamically by GNU Compiler Collection (GCC) of Linux operating system while the early warning system is in operation. And while the executing time and period is assigned to the algorithms, a job is generated and will be executed automatically as executing time and period. Thread-pool is used in the system to execute jobs in parallel and some rules is designed to manage the jobs for fault-tolerant to make the system stability.

The remainder of this paper is organized as follows. Section 2, is the framework of water quality early warning system designed in this paper and its mechanism. Section 3, is the realization of the framework and jobs' management is the key content in this part. In the final section conclusions are drawn and future work is discussed.

2 The Framework of New Water Quality Early Warning System

The water quality early warning system designed in this paper has two main functions: compile early warning algorithms codes and manage the jobs' execution processes. Early warning system should reliably detect sudden contamination of the water supply and also relay timely information on unusual threats to water systems [4]. While for an emergency management system, it should take treatment to response to pollution accidents as soon as possible.

2.1 Modules of the Framework

The framework is proposed in this paper, as shown in Figure 1. It is comprised of six modules including Job Manager, Job Info Database, GCC, Thread Pool and Executable File Library. It also defines the interface between users and the system.

Job Manager is the controller of the system. It receives the users' requests, accesses jobs' information to Job information database and generates a new job or updates the old one. Then it calls GCC to compile algorithms codes and updates the executable file library. The manager will check jobs' status in real time and control their operation state using thread pool.

Job Info Database is designed to record jobs' information, including job name, job execution time, job execution period, job status, early warning algorithm codes, executable file name, and etc. Job manager uses job execution time and period to determine when the job should execute. Early warning algorithm codes are used to be compiled by GCC and generate executable file. Executable file name is used to find the executable file and execute the job by thread pool. Job status is used to check the job status and control the job's execution.

GCC (GNU Compiler Collection) is a compiler system produced by the GNU Project supporting various programming languages [5]. It has been adopted as the standard compiler by most modern operating systems such as Linux. And it was extended to compile C program, C++ program, JAVA program and so on.

Executable file library includes the .out files which are generated by compiling algorithms codes. While job manager checks that a job's execution time is coming, the job manager will use thread pool [6] to generate a new thread to execute the file of the early warning algorithm.

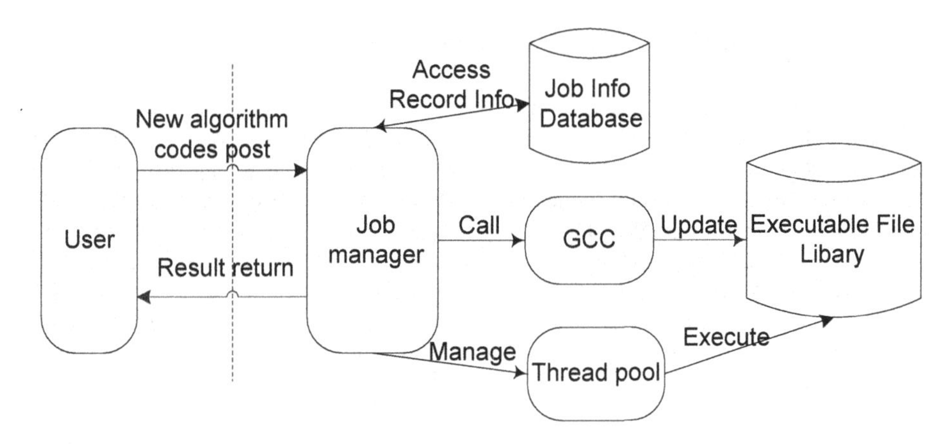

Fig. 1. The Framework of new water quality early warning system

2.2 The System's Mechanism

Based on GCC and thread pool, the mechanism of the system has two main sub processes: algorithm codes compilation sub process and job's auto execution sub process. The detailed mechanism of the system is as follows:

While the user posts codes of a new algorithm or updates codes of an old algorithm to the system and related information, job manager receives the request and accesses the request information to the job information database [7]. And job manager calls GCC to compile the codes and generate executable file in executable file library. If the process of compilation is completed successfully, a job is generated and the success information will return to the user. But if not, job will not be generated and the compilation error information will return to the user to try it again.

When a job is generated, it is in sleep status and does not run until its execution time is coming. Job manager checks jobs' status and calls thread pool to generate a new thread to start the job whose execution time is coming while the computer resources such as memory and CPU utilization are enough. Then the job is in run status.

The job executes the early warning algorithm and accesses the water quality data from the database. Then it analyzes the data according to the algorithm and determines if water quality is normal. While detecting pollution or threat pollution that maybe happen, it will access the analysis results into database and inform related staff to confirm the pollution situation. Once the pollution is confirmed, emergency response process will start and the related specialists will compose the emergency response plan according to national environmental emergency plan, local environmental situation and the local government's environmental emergency plan dynamically [8]. And the system designates emergency response tasks to the related environmental protection departments to execute.

3 The Realization of Jobs' Management

There are lots of jobs in the system. They run automatically while started by job manager. Because of dynamic compilation or the other reasons, some jobs may run into loop status and can not stop and some jobs may generate wrong results. So job management is very important to the system. So job manager is designed for this.

Job manager is designed to have a thread pool pointer to manage a thread pool. And job thread is a class which extends from a thread generated by thread pool. And it realizes the interface method and the entire jobs' algorithms should realize the method so that job threads can execute jobs. A job has five statuses: ready, suspend, sleep, run and abnormal. The relationship of these statuses is proposed in this paper, as shown in Figure 2.

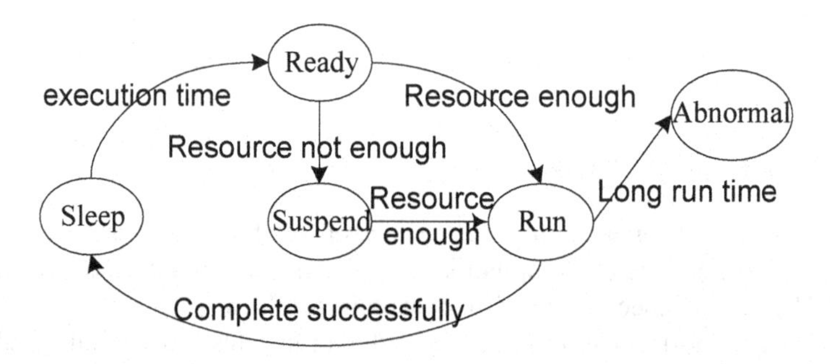

Fig. 2. Status Transition of a Job

While a job's execution time does not come, it is in sleep status. And the job does not do anything. Once its execution time is coming, the job's status will become ready. But the job does not do anything at all because maybe the number of thread pool's threads in "busy" status reaches maximum limit and computer memory is full. The job manager monitors the resources of thread pool and if there are enough resources, the status of the job in ready status will become run.

A job which is in run status is monitored by job manager. Job manager checks the status, the start time and the maximum execution time of a job by accessing database in real time. The problem job is in run status for very long time and its execution time exceeds the maximum execution time. Then job manager will report the state of jobs to users and apply for checking the source codes of the algorithm of the job. The status of the job will become abnormal and it will not be executed until user modifies the source codes of the algorithm and updates the job.

If there are not enough resources, the status of a job in read status will become suspend. The job in suspend status does not do anything at all. But the job has the permission to have a thread if the computer has enough resources. So the job is waiting for the other threads to complete their jobs and release their resources. If there are enough resources, the job's status will become run. And the job's status will become sleep while it is completed successfully.

Besides, in order to excluding wrong results cased by program logic which are difficult to find, the results will return to users in the first time when the job is executed and does not run until users make sure the results are correct.

4 Conclusions

Nowadays, more and more pollution accidents have brought great challenges to China's traditional water quality early warning systems. To deal with complex and changefully water environment pollution accidents, the stability and flexibility of water quality early warning system is very important.

The framework of the new water quality early warning system is designed to compile algorithm codes dynamically and execute the algorithm automatically. So in this paper the flexibility of the system has been improved much and jobs' management improves the stability too. However, dynamic compilation brings lots of conveniences to the system such as program debugging. Problem jobs are only killed and wait for users to modify algorithm codes. So it may bring lots of hidden danger to the system. The research in the future should be done to solve these problems.

References

1. Qu, J., Fan, M.: The Current State of Water Quality and Technology Development for Water Pollution Control in China. Environ. Sci. Technol. 40, 519–560 (2010)
2. Li, J., Lv, Y.-L., He, J.-Z., et al.: Spatial and Temporal Changes of Emerging Environmental Pollution Accidents and Impact Factors in China. Environ. Sci. 29, 47–51 (2008)
3. Zeng, W., Chai, Y., Wei, J.: Construction of emergency management system for fatal environmental pollution accidents in China. Asia-Pac. J. Chem. Eng. 4, 837–842 (2009)
4. Chapman, H., Owusu Yaw, A.: Rapid, State-of-the-Art Techniques for the Detection of Toxic Chemical Adulterants in Water Systems. IEEE Sens. J. 8, 203–209 (2008)
5. Gough, B.J.: An Introduction to GCC. Network Theory Ltd. (2004)
6. Xiao, H., Han, W., Jia, Y., et al.: Research and realization of efficient thread pool model in StarCCM2.0. C. E. 31, 67–69 (2005)
7. Moad, M., Orgun Mehmet, A.: Efficiently querying dynamic XML documents stored in relational database systems. International Journal of Intel. 5, 389–408 (2011)
8. Zhang, X., Chen, C., Lin, P., et al.: Emergency Drinking Water Treatment during Source Water Pollution Accidents in China: Origin analysis, Framework and Technologies. Environ. Sci. Technol. 45, 161–167 (2011)

Sliding Window Calculating Method of Time Synchronization Based on Information Fusion

Chunxiang Zheng[1] and Jiadong Dong[2]

[1] School of Computer and Information, Anqing Teachers College, 246011, China
[2] School of Physics & Electrical Engineering, Anqing Teachers College, 246011, China
dongjiadong@sina.com

Abstract. Time synchronization is very critical requirement at all four layers of proposed wireless sensor network architecture namely, Application layer, Network layer, Node layer, Sensors layer. At application Level, time synchronization is required for consistent distributed sensing and control. In order to realize united time, the united time service system or the time server must be established. This paper proposes Sliding Window Calculating Method for time synchronization based on information fusion. Fault-tolerance strategy of time synchronization is studied and the error analyses are given. It is necessary to add the functions of network clock synchronization to the information fusion to ensure accurate, efficient, and automatic operation of information fusion system.

Keywords: time synchronization, Sliding Window Calculating Method, information fusion.

1 Introduction

As a distributed system, time synchronization is an important issue for wireless sensor networks in information fusion. In general, time synchronization problem appears in many kinds of networks, for example, Internet and LANs. And, those traditional synchronization technologies used in them such as NTP, GPS, radio ranging, etc. can't be used directly in wireless sensor networks because of limited resources of energy, high density of nodes, etc. So, more research works have been done to design synchronization algorithms specifically for wireless sensor networks.

Based on recently the amazing growth of subminiature, low-cost and low-powered hardware, the wireless sensor network is strongly researched for the various application areas.

In this paper, we propose Sliding Window Calculating Method of Time Synchronization Based on Information Fusion. First, a spanning tree of all the nodes in the network is created. Then, it is divided into multiple sub trees. In each sub tree, we perform sub tree synchronization through two algorithms. After all the sub trees are synchronized are synchronized, time synchronization of the whole network is achieved. [1]

H. Tan (Ed.): Knowledge Discovery and Data Mining, AISC 135, pp. 687–691.
springerlink.com

The organization of the paper is as follows: In Section II , we review related work. In Section III, we propose Sliding Window Calculating Method of time synchronization for information. Finally, we present our conclusions in section IV.

2 Related Work

2.1 Network Time Protocol

Network time protocol is a time synchronization protocol which is widely used for computer networks such as the Internet. Computers exchange timestamps with a time server for a few times to average out the time drift. Then time is adjusted and synchronization is obtained. In the process, Manzullo's algorithm is adopted to find the time drift in a noisy and unstable communication. The precision is about one millisecond. Since power efficiency is not the major concern, NTP is not suitable for wireless sensor networks. [2]

2. 2 Analysis of Time Delay for Synchronization

The biggest enemy of precise time synchronization of sensor network is non-determinism. The synchronization time can be decomposed the packet delay into the following 4 components.

Send Time. The delay is spent in assembling a packet and delivering the packet to MAC layer in sender. It depends on the system call overhead of the operation system and the load of processor.

Access Time. The delay incurred waiting for access to the transmit channel. It is the least deterministic part of packet delay in the precondition that the state of wireless receiver chip on in each node is certain.

Propagation Time. The time needed for the message to transit from sender to receivers once it has left the sender. When the sender and receiver share access to the same physical media, this time is very small as it is simply the physical propagation time of the message through the medium.

Receive Time. The processing required for the receiver's network interface to receive the message from the channel and notify the host of its arrival.

With the development of computer network technology and master-slave system, compared with the centralized processing, distributed processing is widely used in many fields. Because the time synchronization is all foundation for application in distributed system, it can be used for wide range, such as Data Acquisition, Monitoring, Performance Analysis, Operating Optimization, and Fault Diagnosis for Power Plant, Electronic Commerce as well as Internet or the Intranet Network Time Service and so on. Two kinds of main technology of time synchronization in distributed system are the direct-connect time synchronization and based on NTP network time synchronization. [3]

Regarding high precision time synchronization system, direct-connect time transmission technology can be used. In order to receive and maintain time information, this method needs to install bus technique timing module, then the computer may directly connect and synchronization to the standard time source (GPS or central timing system). These modules may maintain the time independently without influence of master node, and crystal oscillator precision in these modules may match the time source. This synchronous mode can provide a high synchronous precision for single machine in the distributional system, but it cannot make full use of the network resources, and it has high cost input.

3 Sliding Window Calculating Method

3.1 Time Synchronization Fault-Tolerate Strategy

The distributed system time synchronization fault-tolerant strategy may be divided into three types: The master and servant type time synchronization strategy of time server, the Byzantium protocol type time synchronization strategy, the convergent function type time synchronization strategy [4].

The time synchronization strategy of the master and servant type, it is direct, simple widely used in the distributional system. Its strategy is to establish time synchronization to have the high confidence level and the availability time server. The other nodes penetrate the correspondence network and the server to pick up directly the correct time, and to revise respective time, then to achieve time synchronization in various nodes of the overall system. The merit of the synchronized strategy is the principle and the equipment very are all simple, but the shortcoming is that the fault-tolerant ability is low, and easily lead to a bottleneck effect of centralized management.

The Byzantium agreement type time synchronizes the strategy, it is another basic, also important fault-tolerant technology in the distributional system fault-tolerant strategy , its algorithm is that node acts as the transmitter, it transmits a specific information to all points, after as soon as the non-wrong node receives this information to pass through processing to be able correctly to broadcast again it for other surplus nodes, each node penetrates the exchange information by each other, it may know which nodes have the mistake by the majority decision way. Afterwards various nodes transmit the exchange information in turn, until all error nodes were found. The synchronize strategy has fault-tolerant ability, but it requires a lot to exchange information, and it aggravate the network load. It applies to a specific exchange of information. The need for time synchronization in the exchange of non-specific reference time information, there are application constraints.

The convergence function type time synchronizes strategy, the aim is to search every node reference time in the system, through an effective convergence function, from in the reference time which collects and decides the synchronizing time. Based on the results, every node adjusts its time. It basic step gathers as follows:

- Collection reference time value from other node.
- Estimate these time estimates, consideration influence factor, and obtain a correct time value.

- Calculus these time estimate using a convergence functional, and obtains a correct time value.
- Revises respective time according to the correct time value.

3.2 Sliding Window Calculating Method

The sliding window calculating method (Sliding Window Algorithm, was called SWA) is published by two scholars (Manfred J.Pfluegl and Douglas the M.Blough) in May, 1995, in IEEE Journal of Parallel and Distributed the Computing periodical. It belongs to a convergence functional time synchronization strategy. Comparing with other similar time synchronization algorithms, both in accuracy and fault tolerance, it has a better performance. After comparison, this paper will use the idea of the algorithms and make amendments, and time synchronization in wireless sensor networks will be applied.

The sliding window calculating method and the general convergence function type time synchronize strategy is similar. It uses the mechanism of the periodic time synchronizes. Suppose R on behalf of time synchronization cycle, The next time synchronization of cycles before the arrival, namely iR-p (i representatives of i time synchronization, p as the reference time information can indeed be served on time), namely first obtain the reference time value t_m, and then estimate the time estimated value which the cycle arrives $\tilde{X}_J = t_m + t_s - t_r + e_{mean}$, J representatives from node j to obtain the reference time value, t_s expresses the next time synchronizing time (i.e. ts=iR), t_r indicates the time which receives reference time value , e_{mean} expresses average error.

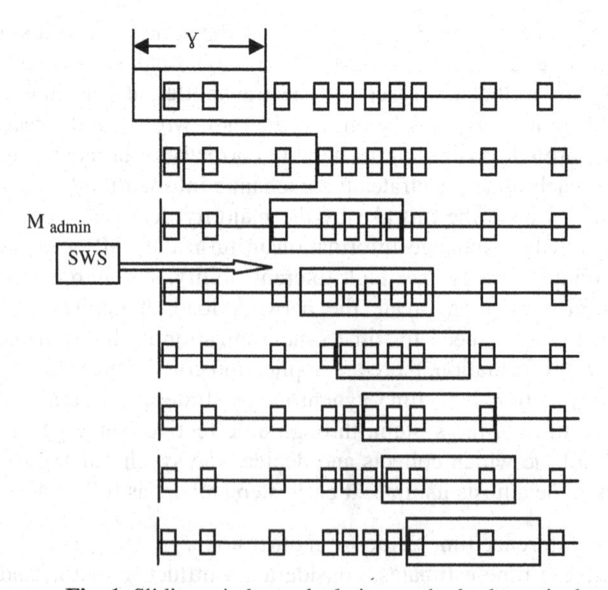

Fig. 1. Sliding window calculating method schematic drawing

After calculating each reference time estimated value, and then convergence function is adopted to calculate the synchronization time. Using the sliding window algorithm, synchronize time is obtained. The calculating method step is as follows:

First, sliding window program to be executed, inputs n and γ. (n for the reference time estimate of the number, γ for the window length)

Second, standard window Sample is calculated. (Containing the largest number of estimated value of the time window for the sample is as a standard windows sample. M_i).

Third, calculate M_{min}.(If the estimated value that contains most of the time window for the number of samples are non-unique, then select the smallest one among the group of the standard deviation of the sample as a standard window, the window sample M_{min} .)

Finally, find M_{admin}.(If the standard deviation of the smallest window of samples are non-unique, then select the first window sample for standard window sample M_{admin}.)

Sliding window algorithm is shown in Figure 1, the horizontal axis represents the time estimate timeline.

4 Conclusion

In this paper, we explored a new time synchronization improvement approach that can be applied to information fusion system, in which sensors monitor each other's behavior and achieve accurate detection of attacks in a distributed approach. Our work utilizes Bayes theorem to reduce the uncertainty of message exchange delay to improve time synchronization accuracy.

Accuracy of the above-mentioned factors can be measured by the time delay of communication protocols. We tested the algorithm only for one level of hierarchy (one Cluster head and two client nodes). At one level the global time variation is coming in microseconds. Further analysis of algorithm for higher level of hierarchy is required.

Acknowledgment. This work is partially supported by Natural Science Foundation of Education Department of Anhui Province (ID: KJ2011A199).

References

1. Claasen, T.: An Industry Perspective on Current and Future State of the Art in System-on-chip (soC) Technology. Proceedings of the IEEE 94, 1120–1137 (2006)
2. Gao, Q., Blow, K.J., Holding, D.J., Marshall, I.W.: Analysis of energy conservation in sensor networks. Wireless Networks 11(6), 787–794 (2005)
3. Ren, F.Y., Dong, S.Y.: A time synchronization mechanism and algorithm based on phase lock loop. Journal of Software 18(2), 372–380 (2007)
4. Zheng, J., Liu, W.: Time synchronization strategies in real time distributing system. Journal of Liaoning Technical University 23(1), 92–93 (2004)

Study on Evaluation Method of University Library Website

Jian-shan Gao[1] and Yan-ling Zheng[2]

[1] Library, Hengshui University, Hengshui, 053000, P.R. China
[2] Scientific Research Department, Hengshui University, Hengshui, 053000, P.R. China
gjszyl@163.com, zylgjs@126.com

Abstract. In order to scientifically, objectively and reasonably evaluate university library website, this paper proposes a method based on Interval Analysis Hierarchy Process (IAHP), Linear Programming (LP) and Gray Clustering. Firstly, we calculate the interval-weight of evaluation index system by IAHP. Secondly, we construct linear programming model and work out the optimal weight of the assessed object by LP. At the same time, we apply the grey clustering method to evaluate and analyze the assessed object. Finally, we made an empirical analysis on ten university library websites, which verified the effectiveness of this method.

Keywords: Grey Clustering, Interval Analysis Hierarchy Process, Linear Programming, University Library Website, Evaluation.

1 Introduction

University Library website has become the important platform that library users utilize library information resources. Enhancing the effectiveness and efficiency of library website, improving service quality and user satisfaction are primary challenges. Scientific, objective and reasonable evaluation to library website is an important means of library building an important way to achieve library reform and development [1]. Based on the importance of university libraries and the urgency of scientific evaluation, this study prepare to start from meet the needs of users, by empirical analysis on ten university library websites of Hebei province, proposed a grey clustering website evaluation method based on IAHP and LP. Library website construction quality is affected by many factors, well-considered all the factors, access to all the information is not realistic, meanwhile, website evaluation is obtained from user, it is also inevitably affected by personal knowledge, experience, many other known and unknown or uncertain factors. Therefore, website quality has the typical grey system characteristics, grey clustering evaluation method has strong applicability to evaluate library website.

2 The Research Method: Grey Clustering (GC)

GC belongs to Grey System Theory (GST). GST was pioneered by Deng Julong in 1982, which is suitable to deal with uncertainty question in less data little sample,

H. Tan (Ed.): Knowledge Discovery and Data Mining, AISC 135, pp. 693–700.
springerlink.com

incomplete information and devoid of experience. GC can divide into Grey Relational Clustering and Whitening Function on GC, its theoretical basis is Grey League Whitening function, according to the description of grey league set, compile and summary the specific survey or statistical data, in order to enhance understanding of things [2].

Generally Whitening weight function is divided into three basic types, the mathematical model is as following:

Hi stand for the Minimum value of high league;

Zi stand for the Median value of medium league;

Li stand for the Maximum value of low league;

Let d_{ij} be sample from i-th object for j-th indicator. The algorithm on Grey Clustering Whitening weight function is as follows:

$$\text{H,}\quad f_1=\begin{cases}1, & d_{ij}\ge H\\ \dfrac{d_{ij}-Z}{H-Z}, & Z\prec d_{ij}\prec H\\ 0, & d_{ij}\le Z\end{cases} \tag{1}$$

$$\text{Z,}\quad f_2=\begin{cases}0, & d_{ij}\ge H\\ \dfrac{H-d_{ij}}{H-Z}, & Z\prec d_{ij}\prec H\\ 1, & d_{ij}=Z\\ \dfrac{d_{ij}-L}{Z-L}, & L\prec d_{ij}\prec Z\\ 0, & d_{ij}\le L\end{cases} \tag{2}$$

$$\text{L,}\quad f_3=\begin{cases}1, & d_{ij}\le L\\ \dfrac{Z-d_{ij}}{Z-L}, & L\prec d_{ij}\prec Z\\ 0, & d_{ij}\ge Z\end{cases} \tag{3}$$

The procedure of GC evaluation as the following :

(1) Construct sample matrix
(2) Define polarity of each index
(3) Define Category boundaries of each index
(4) Construct Whitening function of each index
(5) Calculate grey clustering weight
(6) Work out comprehensive weight coefficient matrix
(7) Judge grey clustering level of each sample
(8) According to the comprehensive weight vector score, sort samples
(9) Result analysis.

3 Grey Clustering Evaluation of University Library Website

3.1 Construct the Evaluation Index System of Website

Refer to the constructed evaluation index system of the literature [3], from website features, web content, and organization to analyze the construction level of library website quality. Specific indicators as following: website features: Search functions C_1, Navigation C_2, Interactivity C_3, Response capacity C_4, Download speed C_5; Web content: Richness C_6, Characteristics C_7, Authority C_8, Timeliness C_9, Personalized C_{10}; Organization: Design style C_{11}, Directory and link structure C_{12}, Operability C_{13}(see table 1 shows).

Table 1. The evaluation system of university library website

Target layer	rule layer	Index layer	Weight interval
The evaluation of university library website (A)	Website features (B_1)	Search functions (C_1)	[0.138, 0.189]
		Navigation (C_2)	[0.235, 0.315]
		Interactivity (C_3)	[0.040, 0.054]
		Response capacity (C_4)	[0.027, 0.035]
		Download speed (C_5)	[0.057, 0.085]
	Web content (B_2)	Richness (C_6)	[0.126, 0.173]
		Characteristics (C_7)	[0.033, 0.053]
		Authority (C_8)	[0.033, 0.047]
		Timeliness (C_9)	[0.018, 0.026]
		Personalized (C_{10})	[0.017, 0.027]
	Organization (B_3)	Design style (C_{11})	[0.012, 0.016]
		Directory and link structure (C_{12})	[0.041, 0.056]
		Operability (C_{13})	[0.052, 0.075]

3.2 User Satisfaction Survey of Website Construction Quality

Select ten university library websites in Hebei Province as assessment object; investigate among students and teachers by the way of questionnaire. The processed survey data as follow table 2.

T_i stand for the assessed site, i=1,2,…10.

Table 2. The survey data of the assessment websites

	C_1	C_2	C_3	C_4	C_5	C_6	C_7	C_8	C_9	C_{10}	C_{11}	C_{12}	C_{13}
T_1	7.9	6.9	6.8	6.6	6.4	5.7	6.4	6.2	6.9	6.7	7.2	7.0	7.8
T_2	6.7	8.7	8.4	8.1	8.3	8.7	8.4	7.8	8.4	7.3	8.6	8.1	8.0
T_3	5.7	7.4	6.1	5.6	6.4	6.5	6.4	5.5	5.9	6.7	6.1	6.3	7.4
T_4	5.1	7.3	4.9	4.4	6.4	5.7	6.4	4.6	5.4	4.8	5.6	5.3	7.5
T_5	7.5	8.5	8.7	8.8	7.9	8.3	8.6	8.1	8.4	8.2	8.8	8.7	8.2
T_6	6.7	7.0	6.1	5.7	6.4	6.5	6.4	6.3	7.0	7.1	6.6	6.2	7.8
T_7	6.7	6.0	6.8	6.6	4.7	6.5	5.6	6.2	7.0	7.6	7.2	7.1	7.9
T_8	8.3	8.6	8.4	7.8	7.8	8.3	7.9	7.8	8.6	6.8	8.2	8.3	8.0
T_9	5.4	5.3	6.8	6.6	4.5	4.2	4.4	5.8	6.5	6.6	6.7	6.9	7.9
T_{10}	6.8	6.0	7.3	7.2	5.7	6.5	5.7	6.3	7.1	5.8	7.3	7.1	7.8

3.3 Calculate the Weight of the Evaluation Index System

There are a variety of methods when calculating the weight in a multi-index evaluation system. However, these methods can only get a fixed set of weights. IAHP uses Interval numbers instead of point value when calculating the weight; finally get a weight range in which each weight is reasonable. Usually because the advantage of each assessed object is different, its weight is different. There is always a weight to reflect the most advantages in the weight range of each index, which is the most suitable, dynamic weight for assessed object [4].

This paper works out the interval-weight of each index by using IAHP. See Table 1 shows.

Construct linear programming model and calculate the optimal weight of the assessed websites [5,6].

Let x_j stand for the optimal weight of the assessed websites.

Let a_{ij} stand for the value of each index which is the survey data of assessed website (table 2).

The linear programming model is : $\max Z = \sum_{j=1}^{n} a_{ij} x_j$

$$
s.t.\begin{cases}
0.14 \le x_1 \le 0.19, 0.24 \le x_2 \le 0.32 \\
0.04 \le x_3 \le 0.05, 0.0265 \le x_4 \le 0.03 \\
0.06 \le x_5 \le 0.09, 0.13 \le x_6 \le 0.17 \\
0.03 \le x_7 \le 0.05, 0.03 \le x_8 \le 0.05 \\
0.02 \le x_9 \le 0.03, 0.02 \le x_{10} \le 0.03 \\
0.01 \le x_{11} \le 0.02, 0.04 \le x_{12} \le 0.06 \\
0.0515 \le x_{13} \le 0.0747 \\
\sum_{j=1}^{n} x_j = 1, j = 1,2,3 \ldots\ldots n; i = w_1, w_2, \ldots w_{10}; n = 13
\end{cases}
$$

After calculating we get the optimal weights. SeeTable 3 shows.

Table 3. The optimal weight of each assessed site

	T_1	T_2	T_3	T_4	T_5	T_6	T_7	T_8	T_9	T_{10}
C_1	0.19	0.14	0.14	0.14	0.14	0.19	0.19	0.15	0.19	0.19
C_2	0.13	0.17	0.17	0.15	0.15	0.13	0.17	0.17	0.13	0.17
C_3	0.31	0.32	0.32	0.32	0.32	0.32	0.24	0.32	0.26	0.24
C_4	0.04	0.05	0.04	0.04	0.05	0.04	0.05	0.05	0.05	0.05
C_5	0.03	0.03	0.03	0.03	0.03	0.03	0.03	0.03	0.03	0.03
C_6	0.06	0.06	0.06	0.09	0.06	0.06	0.06	0.06	0.06	0.06
C_7	0.03	0.05	0.04	0.05	0.05	0.03	0.03	0.03	0.03	0.03
C_8	0.02	0.02	0.03	0.02	0.02	0.03	0.03	0.02	0.03	0.02
C_9	0.06	0.04	0.04	0.04	0.06	0.04	0.06	0.06	0.06	0.06
C_{10}	0.02	0.03	0.02	0.02	0.03	0.03	0.03	0.03	0.03	0.03
C_{11}	0.02	0.02	0.01	0.01	0.02	0.01	0.02	0.01	0.02	0.02
C_{12}	0.03	0.03	0.03	0.03	0.03	0.03	0.03	0.03	0.05	0.04
C_{13}	0.07	0.05	0.07	0.07	0.05	0.07	0.07	0.05	0.07	0.07

4 Grey Clustering Evaluation of University Library Website

4.1 Construct Sample Matrix

Construct sample matrix D according to the website survey data (Table 2).

$$D=\begin{bmatrix} 7..9 & 5.7 & 6.9 & 6.8 & 6.6 & 6.4 & 6.4 & 6.7 & 7.0 & 6.9 & 7.2 & 6.2 & 7.8 \\ 6.7 & 8.7 & 8.7 & 8.4 & 8.1 & 8.3 & 8.4 & 7.3 & 8.1 & 8.4 & 8.6 & 7.8 & 8.0 \\ 5.7 & 6.5 & 7.4 & 6.1 & 5.6 & 6.4 & 6.4 & 6.7 & 6.3 & 5.9 & 6.1 & 5.5 & 7.4 \\ 5.1 & 5.8 & 7.3 & 4.9 & 4.4 & 6.4 & 6.4 & 4.5 & 5.3 & 5.4 & 5.6 & 4.6 & 7.5 \\ 7.5 & 8.3 & 8.5 & 8.7 & 8.8 & 7.9 & 8.6 & 8.2 & 8.7 & 8.4 & 8.8 & 8.1 & 8.2 \\ 6.7 & 6.5 & 7.0 & 6.1 & 5.7 & 6.4 & 6.4 & 7.1 & 6.2 & 7.0 & 6.6 & 6.3 & 7.8 \\ 6.7 & 6.5 & 6.0 & 6.8 & 6.6 & 4.7 & 5.6 & 7.6 & 7.1 & 7.0 & 7.2 & 6.2 & 7.9 \\ 8.3 & 8.3 & 8.6 & 8.4 & 7.8 & 7.8 & 7.9 & 6.8 & 8.3 & 8.6 & 8.2 & 7.8 & 8.0 \\ 5.4 & 4.2 & 5.3 & 6.8 & 6.6 & 4.5 & 4.4 & 6.6 & 6.9 & 6.5 & 6.7 & 5.8 & 7.9 \\ 6.8 & 6.5 & 6.0 & 7.3 & 7.2 & 5.7 & 5.7 & 5.8 & 7.1 & 7.1 & 7.3 & 6.3 & 7.8 \end{bmatrix}$$

4.2 Change the Index Polarity

We can see all the indexes are positive polarity, without the need to change polarity.

4.3 Define Category Boundaries of Each Index

Set the evaluation league of university library Web site is "excellent", "good" and "poor". According to the distribution of the data Define the Limit values of each index for each gray league. SeeTable 4 shows.

H_i stand for the Minimum value of high league
Z_i stand for the Median value of medium league
L_i stand for the Maximum value of low league (i=1, 2,…,13)

Table 4. The Limit value of each gray league

	C_1	C_2	C_3	C_4	C_5	C_6	C_7	C_8	C_9	C_{10}	C_{11}	C_{12}	C_{13}
H_i	7.4	7.2	7.6	7.4	7.3	7.0	7.2	7.4	7.6	7.5	7.8	6.9	8.0
Z_i	6.0	5.7	6.4	6.2	5.9	5.8	5.8	6.6	6.5	6.5	6.7	5.7	7.7
L_i	6.7	6.5	7.0	6.8	6.6	6.4	6.5	7.0	7.0	7.0	7.2	6.3	7.8

4.4 Construct Whitening Function of Each Index

Index C_1" Search functions", the Whitening weight function is:

$$f_1=\begin{cases}1, & d_{ij}\geq 7.4\\ \dfrac{d_{ij}-Z}{H-Z}, & 6.7\prec d_{ij}\prec 7.4\\ 0, & d_{ij}\leq 6.7\end{cases}$$

$$f_2=\begin{cases}0, & d_{ij}\geq 7.4\\ \dfrac{H-d_{ij}}{H-Z}, & 6.7\prec d_{ij}\prec 7.4\\ 1, & d_{ij}=6.70\\ \dfrac{d_{ij}-L}{Z-L}, & 6.0\prec d_{ij}\prec 6.7\\ 0, & d_{ij}\leq 6.0\end{cases}$$

$$f_3=\begin{cases}1, & d_{ij}\leq 6.0\\ \dfrac{Z-d_{ij}}{Z-L}, & 6.0\prec d_{ij}\prec 6.7\\ 0, & d_{ij}\geq Z\end{cases}$$

In the same way, we can construct Whitening weight function of other indexes.

4.5 Calculate Weight Coefficient of Each Index

Get the survey data on C_1 into the Corresponding Whitening weight function; we can get the weight coefficient matrix R_1. In the same way, we can construct weight coefficient matrix of other indexes.

$$R_1 = \begin{bmatrix} excellent & good & poor \\ 1.00 & 0.00 & 0.00 \\ 0.03 & 0.97 & 0.00 \\ 0.00 & 0.00 & 1.00 \\ 0.00 & 0.00 & 1.00 \\ 1.00 & 0.00 & 0.00 \\ 0.03 & 0.97 & 0.00 \\ 0.03 & 0.97 & 0.00 \\ 1.00 & 0.00 & 0.00 \\ 0.00 & 0.00 & 1.00 \\ 0.12 & 0.88 & 0.00 \end{bmatrix}$$

4.6 Calculating Comprehensive Weight Coefficient Matrix

Set the Corresponding values of weight coefficient matrixes obtained weighted summary according to the optimal weight that table showed. Finally we get the comprehensive weight coefficient matrix M.

Calculate comprehensive weight coefficient:

$\eta_K(j) = \sum w(ij) \bullet f_K(ij)$,
k="excellent", "good", "poor"
i=1, 2, 3, …, 13
j=1, 2, 3, …, 10
$\eta_K(j)$ Means the grey clustering weight for *k*-th in *j*-th assessed website
$f_K(ij)$ Means the grey league weight coefficient for *k*-th in i index of *j*-th assessed website
$w(ij)$ Means the optimal weight for *i* in *j*-th assessed website

Finally, we get the comprehensive weight coefficient matrix M.

4.7 Evaluation Results

Can be seen from the obtained comprehensive weight coefficient matrix M the different value in each line correspond to "excellent", "good", "poor" grey category descript the size of strength that each website belongs to each grey category. The final evaluation results for each site were taken in the corresponding row vector corresponding to the maximum grey category.

As a result, the final evaluation results of the ten websites is T_2,T_5 and T_8 is excellent; T_1,T_6 ,T_7 and T_{10} is good; T_3,T_4 and T_9 is poor.

The integrated order is $T_5 > T_8 > T_2 > T_6 > T_1 > T_7 > T_{10} > T_3 > T_4 > T_9$.

$$M = \begin{bmatrix} excellent & good & poor \\ 0.19 & 0.62 & 0.19 \\ 0.86 & 0.14 & 0.00 \\ 0.25 & 0.34 & 0.41 \\ 0.19 & 0.25 & 0.55 \\ 1.00 & 0.00 & 0.00 \\ 0.02 & 0.82 & 0.16 \\ 0.09 & 0.57 & 0.33 \\ 0.98 & 0.01 & 0.01 \\ 0.03 & 0.19 & 0.78 \\ 0.17 & 0.49 & 0.34 \end{bmatrix} \begin{matrix} website & evaluation \\ T_1 & good \\ T_2 & excellent \\ T_3 & poor \\ T_4 & poor \\ T_5 & excellent \\ T_6 & good \\ T_7 & good \\ T_8 & excellent \\ T_9 & poor \\ T_{10} & good \end{matrix}$$

5 Concluding Remarks

This paper used the grey clustering method based on IAHP and LP to empirically analyze and evaluate ten university library websites, the research results show this approach is an effective and feasible method for website quality assessment. The comprehensive weight coefficient matrix M provides a wealth of information to comprehensively understand the assessed object. It provides the theoretical and practical basis guarantee for our country in the comparative study of the university library website and other relevant evaluation research.

References

1. Nankai University Research Group. Evaluation of Websites of Provincial Libraries in China. Journal of The National Library of China (3), 37–44 (July 2009)
2. Deng, J.: The Basic Theory of Grey System. Huazhong University of Science and Technology Press, Wuhan (2002)
3. Zheng, Y., Shi, B., Gao, J.: A Study on the Evaluation Method of University Library Web Sits Based on the IAHP. Journal of The China Society for Scientific and Technical Information 27(1), 127–132 (2008)
4. Wu, Y., et al.: Interval Approach to Analysis of Hierarchy Process. Journal of Tianjin University (5), 700–705 (September 1995)
5. Hu, Q., Wei, Y.: Linear Programming and Its Application, pp. 20–26. Science Press, Beijing (2004)
6. Zhang, G.: Linear Programming, p. 2. Wuhan University Press, Wuhan (2008)

The Construction of Online Video Resource Library with Streaming Media

Tongming Wang

Center of Modern Educational Technology, Hengshui University, Hengshui, 053000, China
wtm.2000@163.com

Abstract. With the character of "playing while downloading", the streaming media technology becomes the most important transport technology of online multimedia information, especially video information. The application of streaming media technology in data processing changes the resource mode in online teaching.

Keywords: streaming media, online, video, resources library, design.

1 Introduction

Online teaching resource is an important part of the online teaching system as well as the prerequisite and foundation of online education. As an essential media styling in online teaching, video plays an irreplaceable role, for it can make teaching activities more vivid. With the development of network and multimedia technology, there will be some new application forms, so how to rationally and efficiently utilize the video resources has become a hot topic.

For a long time, due to the impact of network transmission speed and the video itself (such as large volume of data, complex procession, etc.), video teaching has been greatly limited in online teaching. In recent years, in order to solve the problem of download time consuming and thus adapt to the trend of online multi-media, the streaming media of "playing while downloading" emerges and gets rapid development. The application of streaming media technology in data processing changes the resource mode in online teaching.

2 The Teaching Role That Video Resources Play in Online Teaching

Compared with text, images and other teaching aids, video is more intuitive, vivid, information-intensive, etc. The combination of video and network has provided new opportunities for web-based instruction.

2.1 Video Can Arouse Students' Interest for Study

Sound, images, text and other comprehensive teaching environment provided by video can stimulate various senses of students to help them better grasp the course content.

H. Tan (Ed.): Knowledge Discovery and Data Mining, AISC 135, pp. 701–708.
springerlink.com

2.2 In the Classroom Teaching, There Exist Various Non-intellectual Factors in the Process of Teacher-Student Interaction

Such as teachers, emotion, attitude, action, etc. which may potentially affect the student's understanding and mastery of knowledge. However, for online learning based on text and still images, there is no face-to-face communication as well as real-time interaction; in fact, all those non-intellectual factors have been filtered [1]. Video utilized in online teaching can preserve those non-intellectual factors to some extent, thus optimizing the effect of online teaching.

2.3 Video Resources Can Improve Emotional Communication in Online Teaching to Some Extent

In the distance learning, learners lack emotional communication with both teachers and peers, thus always accompanying by strong aloneness. While the video resources are authentic and have continuous motion, which can give students an immersed sense and increase their intimacy with the teaching content and affect their intellectual and emotional investment.

3 Problems in the Resource Library of Current Online Videos

With the development of information technology in education, the hardware construction in colleges and universities continue to catch up. However, software development generally lags behind the hardware construction, in which the construction of teaching resource library has become the bottleneck of in-depth development of educational information. The main issues in the resource library of current videos are as follows:

3.1 Poor Transmissions in Network Teaching Videos

As the digitized audio and video information takes up a large amount of data, it needs huge storage space and high bandwidth requirements, which is difficult to transmit over the Internet without special encoding. Currently, the mode of "playing after downloading" is widely used, that is, first download audio and other multimedia information to a local hard disk, and then play them. Nevertheless, the download process often takes tens of minutes or hours.

3.2 Poor Interaction of Teaching Video Resources

Traditional video resources are the vanguard of traditional distance education and also play an important role in school education. But, on the other hand, there are also obvious shortcomings with the traditional video resources: the lack of two-way communication channels between teachers and students. They often cannot exchange with each other from a variety of aspects, and the learner is difficult to truly have deep-seated active option in aspects such as the degree of difficulty in the curriculum, learning time, place, etc.

3.3 Non-standardization in Resource Data Production

Non-standardization in data leads to data isolation and data tomb. On the one hand, this brings a great deal of inconvenience to data sharing, exchange and update, but also creates a repetitive construction of resource, wasting a lot of human, material and financial resources.

3.4 Undue Stress on "The Capacity "of Resources and Ignoring the Educational and Teaching Nature

Most of the video teaching resources currently emerge in the form of "library", ranging from tens of G to as many as several T. However, for measuring the quality of online teaching resource library, the amount of resources is not the absolute standard, but rather the efficiency that resources serve for practical teaching. From this point of view, the network teaching video resources requires carefully designed and appropriate treatment in order to achieve good teaching effect.

4 Overview of Streaming Media Technology

The so-called streaming media refers to the media format played in the Internet utilizing the streaming way. Its principle is to send a program as data packet by the video delivery server. Then the program will display after being unzipped by users.

4.1 The Advantages of Streaming Media

Compared with the simple download mode, streaming mode can avoid the shortcoming that users cannot view the video until the file has been downloaded from the internet.

4.1.1 Quick Start without Time Delay

The file of streaming media utilizes the mode of playing-transmitting. That is, in client, when users click the key of play, the file starts to play only after a short period of preset time (from several seconds to ten seconds), which has solved the problem of users' long time waiting.

4.1.2 Small Volume and Convenient Storage

Streaming media applies special technology of data compression and decompression, so compared with the sound file (.wav) as well as video file (. avi), the file of streaming media with the same content only accounts for 5% of them [2].

4.1.3 Less Demanding on the Bandwidth

Due to the fact that multimedia file has been dwindled after compression, so its demanding for bandwidth transmission is low, that is, ordinary dial-up Modem users can also watch the video.

4.1.4 Two-way Communication

The communication between streaming media server and streaming media player in client is the two-way communication. The server sends data and receives user's feedback simultaneously, so server and users keep in touch with each other in this process.

4.1.5 Effectively Ensure the Copyright of Programs

Streaming media server forbids users to download the file to a local directory but watch on the Internet, which can effectively protect the copyright of programs.

4.2 Popular File Formats of Streaming Media

Streaming media files which have been commonly used currently include Windows Media from Microsoft, RealMedia from RealNetworks, and Quicktime from Apple.

4.2.1 ASF format refers to the file compression format which can be used to watch online video. It applies MPEG-4 compression algorithm and has good compression ratio and image quality.

4.2.2 WMV format is the file compression format from Microsoft, which utilizes an independent coding and can be used for on-line real-time viewing of video. In addition, it is upgraded from the ASF format.

4.2.3 RM Format is the audio-and-video compression standard developed by Real Networks company, known as Real Media. It can develop different compression ratio in accordance with different network transmission rates, thus achieving the real-time transmission and play of image on the network with low rate.

4.2.4 RMVB format is a new video format upgraded from a RM video format. A dynamic bit rate is used in it to decrease the bit rate of a static picture and thus reach a subtle balance between the image qualities and file size.

4.2.5 MOV format is a video format developed by the U.S. Apple, which is commonly used in MAC system and PC platforms with the default player of Apple's QuickTime Player. It can maintain good image quality, simultaneously dwindle the file size.

5 Design of Online Video Resource Library Based on Streaming Media

5.1 Design Principles

To build the resource library of online video is a systematic project, which means we must regard "serve for teaching" as the fundamental point and ultimate goal, then make overall plans, and achieve it step by step. In the specific construction process, the following principles should be paid much attention to.

5.1.1 Educational Principles

The ultimate goal of building a resource library is serving for teaching, therefore, the needs of teaching must be taken into account both in content and function. During the construction of library, whether the resources are conducive to the teaching of discipline teachers and students' interest and motivation in learning should be considered.

5.1.2 Technical Principles

That system design should take full account of needs in future technology development. This includes not only the advanced nature of development content such as the database structure, data format and classification methods, but also the advancement in the development platform, operating system, programming model and other specific development technology.

5.1.3 The Principles of Integrity

Network video resources library system is a complicated system of complex media types, large data volume and dynamic change, which requires long-term construction and maintenance. Therefore, the whole structure of the system should be scientific, rational, and universal, both for construction but also facilitating exchanges between schools.

5.1.4 The Principle of Service

The ultimate goal of network teaching resource library building is to provide teaching services. If the library does not have convenient and fast service, the user is likely to give up using it due to the trouble or difficulty.

5.2 Analysis of Functional Goals

5.2.1 Digitalization of Audio and Video Data

This system can be used to transform analog signals of a lot of the old school teaching material, courseware, audio and video data into streaming media courseware which can be issued in the network for learners to play on demand.

5.2.2 Virtual Classroom Learning Environment

Students can study at any time in the dedicated electronic reading rooms and computer rooms opened up in the school, or student apartments with network connection and other places, and play teaching resources freely on demand according to their own learning progress and learning objectives.

5.2.3 Multimedia Assisted Instruction

Teachers play library resources in the classroom on demand for all students to watch, which supports multiple classrooms playing on demand at the same time.

5.2.4 Resource Sharing

Teachers can store some excellent resources in the library after editing and converting them, which can be played by staff and students at any time on demand. The user on demand can choose to watch different programs and have complete playback control (forward, back, stop, etc.).

5.3 Functional Design of Resource Library

5.3.1 Resources Management

Video on demand: directly play various videos on-demand

Download: provide available resources for download, which can be saved to a local directory.

Bookmark: some favorite resources can be added to the system's favorites for future use.

Publish: one or more resources can be selected for publication, and then the related information will be exhibited in the latest home board.

5.3.2 System Management

User register: new users can directly register in the system through users' client.

Information modification: users can modify personal information whenever necessary.

Permissions distribution: the system endows every user with different permission.

Board management: the administrator board in home page can be added, modified and deleted.

Plug-in management: some playback plug-ins can be added, modified and deleted.

6 Construction of Online Video Resource Based Streaming Media Technology

6.1 Platform Selection

Because there are great differences in the network environment of each school, so during the plan of building online resource platform based on streaming media technology, a further deep understanding from the angle of integrity, compatibility, and not enough third-party supporters is needed. The current Internet streaming video and audio technology solutions come from RealNetworks' RealMedia, Microsoft's Windows Media and Apple's Quick Time [3]. Here we choose the solutions from Microsoft Windows Media. Windows Media Services is a publishing platform of streaming multimedia with the ability to adapt to a variety of network bandwidth conditions, which can provide a set of solutions, including streaming media production, publishing, broadcasting and management. In addition, it also provides the development kit (SDK) for secondary development [4]. Library built by WindowsMedia Service has the characteristics of small investment (free full solution), short construction period, simple use, good scalability, easiness to maintain, so it is acclaimed.

6.2 System Composition

Windows Media Services system consists of three parts: Windows Media Tools, Windows Media Server and Windows Media Player.

6.2.1 Media Tools

Windows Media Tools is an important part of the whole program, which can provide a series of tools to help users generate multimedia streaming with ASF format, including real-time multimedia streaming, for online transmission and application.

Making flow:
(1)Collection of video resources
The acquisition of video resources can be live video recorded by camera, or TV recorded by VCR, or satellite signals, or recorded contents stored on video tape or disk.

(2) Acquisition and sorting of video material
Convert the analog signals of video into digital signals through the acquisition workstation, which is an important part of the resource library.

Acquisition workstation is the computer equipped with video capture card as well as encoder. It is the creation end of video content, that is, compress and convert different types of video sources into video files with streaming format. Windows Media Encoder 9.0 is selected as encoder, due to its convenience, good encoding technology as well as high quality of output. The VideumII Broadcaster video capture card from the company of Winnov is selected because it is compatible with Windows Media technology.

The property of each sample must be marked after digitization, which is convenient for index. In the system, different on-demand programs can be distinguished through unique identification information.

6.2.2 Media Server. After completing the production of Windows Media streaming media file, it can be put online for users to watch.

(1) Video Server
Video Server is the server for installing streaming media software, providing video program services, monitoring system operation, and storing video. The video server in the system is Windows Server 2003 operating system, and the Windows Media Service component is installed as the Windows Media server.

(2) WEB server and database server
Another server is used as the WEB server and database server. The Windows Advanced Server 2000 operating system is used on the server, and the Web service is provided by the built Windows2000 Server IIS components. SQL SERVER2000 is installed to provide database server for video data management.

(3) Media player
The clients of multimedia information resources include multi-media classrooms, electronic reading room, teaching and research room, computer room, library, and student dormitories. As a result of using the browser - server model, so clients only

need to install the browser and Windows Media Player, very convenient to use and basically without maintenance.

7 Conclusions

Streaming media technology is used to create video resource library. Teaching through the campus network platform is characterized by easiness to operate, wide application, less investment, simple maintenance and so on. Extensive video resource is the basis for achieving online teaching. On the other hand, the construction of teaching resources is a long-term and arduous task that requires teachers and professionals to develop and introduce resources as well as constantly exploring applications of streaming media technology on the campus network.

References

1. Bai, H.-Q.: The Probe on Video Used in Web-based Instruction. Journal of Yancheng Institute of Technology (Natural Science) 16(3), 9–11 (2003)
2. Chen, D.-W., Peng, Y.-X.: The Application of Streaming Media Technology in Campus Educational Resourse Delivering. E-education Research 3, 58–60 (2003)
3. Wang, P., Zhao, M.: The Key Technology based on Streaming Media. Journal of Anhui University 30(1), 28–31 (2006)
4. http://www.tele21.com.cn/64/2010_8_26/1_64_407_0_1282803826188.html

Scheduling Algorithm of Wireless Sensor Cluster Head Based on Multi-dimensional QoS

Wuqi Gao[1,*] and Fengju Kang[2]

[1] School of Computer Science and Engineering, Xi'an Technolical University, 710032, China
[2] College of Marine Engineering, Northwestern Polytechnical University, 710072, China
gaowuqi@126.com

Abstract. As for the tasks of how wireless sensor network selects cluster head and how cluster head node properly schedule a number of sensor nodes, cluster head scheduling algorithm based on multi-dimensional QoS is presented. Firstly, analytic hierarchy process in economic field is introduced into the resource scheduling algorithm to compute every dimensional parameters weight, then the tasks is allocated to appropriate cluster head according to customer satisfaction, QoS distance and loading equilibrium, etc. Finally, scheduling algorithm is analyzed by theory example and is simulated with CloudSim tool package. The experiment shows that the scheduling algorithm not only meets customer needs for multi-dimension QoS, but shortens the tasks of sensor nodes finishing time, and greatly improves sensor resource utility rate.

Keywords: wireless sensor network, resource scheduling algorithm, QoS. cluster heads, Multi-dimension QoS evaluation.

1 Introduction

In the application field of sensor network, in order to finish the task of monitor and measurement, a large number of low energy consumption sensor nodes, which have many functions such as perception, calculation and communication, are put into environments. These sensor nodes are used to perceive the happenings of events or gather data information like temperature, pressure, moisture, and etc. Then the sensors transfer the collected information to cluster head nodes or receiver nodes, particularly to those cluster head nodes with very limited resources, and try its best to finish the allotted tasks necessarily [1-3].

The traditional resource scheduling algorithms include FCFS (First Come First served) and Round-Robin algorithm. These methods can easily cause problems of task starvation and unbalanced resource, and etc[4]. Enlightening algorithm and QoS Guided Min-Min algorithm bring about problems of resource unused and task pressured crowded, and this two algorithms are not compatible with various QoS limitations. Thus, it is difficult for them to reflect dynamical heterogeneous[5-12]. A typical scheduling algorithm with the multi-dimension QoS guiding is QDDN[13].

[*] Micheal Johnson.

H. Tan (Ed.): Knowledge Discovery and Data Mining, AISC 135, pp. 709–716.
springerlink.com

Based only on the distance between tasks and resource to allot tasks, this algorithm does not take user's satisfaction, loading equilibrium, etc into account so as to lead to unsatisfactory scheduling effect. For the above problems to be solved, This paper proposes the cluster head scheduling algorithm based on multi-dimensional QoS S-CHSA. This algorithm can greatly reduce delay time of task transfer and prove its efficiency by theoretical example and simulation experiment.

2 QoS Parameter

This paper presents 4 QoS evaluation indexes: node residual energy, node throughput, stability and node packet loss rate. And they can be increased and decreased accordingly.

(1) Node Residual Energy

Its formula is $E_l = E_0 - E_{Total}$, E_l stands for node residual energy; E_0 initial energy; E_{Total} consumption energy. Node total consumption energy is: $E_{Total} = N \times \left[E_{CH} + \left(\frac{T}{N} - 1 \right) E_{non-CH} \right]$. Therefore, node residual energy is:

$$E_l = E_0 - N \times \left[E_{CH} + \left(\frac{T}{N} - 1 \right) E_{mon-CH} \right]$$

(2) Node Throughput

It is used to measure transmitting capability of the wireless sensor network and is related to protocol stack of the wireless sensor network and environment around. Throughput is divided into node throughput and network throughput: node throughput is defined as total data packet produced, forwarded and received by node in one measurement period; network throughput is defined as total throughput of all nodes in one measurement period.

(3) Stability

Its formula is $A = \frac{correctRunTime}{correctRunTime + mistakeRunTime}$, correctRunTime stands for normal running time of resource; mistakeRunTime stands for mistake time of resource

(4) Node Packet Loss Rate

It is defined as total data packet sent by the rest nodes of network to this node and difference value of total data packet actually received by this node and ratio of total data packet sent by the rest nodes of network to this node in one measurement period.

3 Multi-dimension QoS Evaluation

3.1 Qos Standardization

Due to the big difference of each dimension parameter values for resource and tasks, and for the purpose of a systemic evaluation to simulation QoS need and to resource service capability, standardizing the both matrixes is needed. This paper adopts efficiency coefficient method to finish the standardization, the following are the specific rules: the bigger and better index value is defined as maximum variables; the smaller and better index value as minimum variables; the better index value at a point as stability variables; the best index value at a range as range variables.

To design efficiency coefficient method for the above four variables respectively, table 1 is the efficiency coefficient method formula.

Table 1. Efficiency coefficient method Formula

Type	Formula	Condition	Type	Formula	Condition
Max	$(A-D)/(B-D)\times 40+60$	$A<B$	Min	$(A-C)/(B-C)\times 40+60$	$A>B$
	100	$A\geq B$		100	$A\leq B$
Stability	$[(C-A)/(C-B)]\times 40+60$	$A>B$	Range	$[(C-A)/(C-E)]\times 40+60$	$A>E$
				100	$F\leq A\leq$
	$[(A-D)/(B-D)]\times 40+60$	$A\leq B$		$[(A-D)/(F-D)]\times 40+60$	A<F

Notes: A---actual value B---satisfaction C---upper unallowable value D---lower unallowable value E—upper value F---lower value

3.2 Identifying User's QoS Synthetic Value with Synthetic Single Efficiency Coefficient Method

First of all, set up tier structural modeling. When using analytic hierarchy process to analyze the problems, we should categorize them at the very beginning to set up hierarchy structural modeling. The hierarchy can be divided into three classes:

(1) The highest tier, also called objective tier, is very often to analyze the expected objectives or ideal outcomes of the problems.

(2) The medium tier, namely principle one, includes intermediary involved in achieving the objectives, which consists of a lot of tiers.

(3) The lowest tier, measure or solution one, includes various available measures and solutions for achieving the objectives.

Then, through 9 classifications we try to set up judgment matrix Cij to get maximum character value and eigenvector. The analytical hierarchy process takes character vector of the judgment matrix as weight vector of every index, and uses summation to compute eigenvector of the said judgment matrix C. The followings are the specifics: Normalize every column of judgment matrix:

$$\overline{a_{ij}} = \frac{a_{ij}}{\sum_{k=1}^{n} a_{kj}} \quad (i,j=1,2,\cdots,n) \tag{1}$$

calculate the sums of all elements in every line:

$$W_i = \frac{\overline{W_i}}{\sum_{j=1}^{n} \overline{W_j}} \quad (i,j=1,2,\cdots,n) \tag{2}$$

normalize them:

$$\overline{W_i} = \sum_{j=1}^{n} \overline{a_{ij}} \quad (i = 1, 2, \cdots, n) \tag{3}$$

Wi means relative weight vector of every element in this tier to a element in previous tier. After identifying every QoS dimensional metrics, We can use formula (5) to compute synthetic value of QoS need.

$$\text{Uq(w)}=\sum s_{ij} w_j \tag{4}$$

3.3 User's Satisfaction

The sensor task scheduling algorithm should be used to try to meet user's QoS needs. On the condition that the bigger QoS parameter value in a dimension is, the higher the QoS is, formula (6) is adopted to compute user's obtained satisfaction in a dimension. Vice versa, on the condition that the bigger the QoS parameter value is, the lower the QoS is, formula (7) is adopted. In formula(7) j∈ [1,k]; rQi,j stands for service ability provisioned by QoS parameter at Dimension J in Resource Ri, tQi,j stands for need volume by QoS in Dimension J for Task I.

$$\text{Satisfyi,j}=\begin{cases} \dfrac{rQ_{i,j}}{tQ_{i,j}}, rQ_{i,j} < tQ_{i,j} \\ 1, \quad rQ_{i,j} \geq tQ_{i,j} \end{cases} \tag{5}$$

$$\text{Satisfyi,j}=\begin{cases} \dfrac{tQ_{i,j}}{rQ_{i,j}}, rQ_{i,j} < tQ_{i,j} \\ 1, \quad rQ_{i,j} \geq tQ_{i,j} \end{cases} \tag{6}$$

Thus, synthetic satisfaction in all operation resource dimensions for the user Task can be defined as:

$$\text{Satisfyi}=\frac{\sum_{j=1}^{k} satisfy_{i,j}}{k} \tag{7}$$

3.4 QoS Distance Measurement

QoS needs of various tasks are different, so are the QoS provisioned by various tasks. In order not to make higher QoS service capability resource occupied by lower QoS need tasks, which affects the operation of other tasks and leads to increasing total operating time, we try to have tasks allotted to its matched resource to be operated. So we use weight distance measurement method to compute the QoS distance between tasks and resources.

$$\text{Dis}=\sqrt{\sum_{j=1}^{k} w_j (ts_{i,j} - rs_{i,j)}^{\ 2}} \quad \text{w}_\text{j} \text{ ,j=1,2,\&,} \sum_{j=1}^{k} w_j^{\ 2} = 1 \tag{8}$$

4 S- CHSA Algorithm

Strategy of this S-CHSA is: to arrange the order and allot five percent of the order as optional cluster head according to synthetic efficiency coefficient method of tasks. At first, select cluster heads with higher satisfaction to allot. If there is only one cluster head that obtains the highest satisfaction, allot the task to it; if there are many of them with the highest satisfaction, select unused cluster head to achieve loading equilibrium; if there is no unused cluster heads, allot the task to the cluster head with shortest distance among the cluster heads and the task. The specific flowchart is showed in the following Fig. 2.

Table 2. Task Dimension QoS Need & the Available Index Value of Virtual Machine

Content	Node Residual Energy	Node Throughput	Stability	Node Packet Loss Rate
Cloudlet1	0.8	0.7	0.7	0.6
Cloudlet2	0.8	0.9	0.8	0.3
Cloudlet3	0.5	0.8	0.8	0.5
CH1	0.8	0.6	0.5	0.7
CH2	0.8	1.0	0.9	0.5
CH3	0.9	0.8	0.8	0.4
CH4	0.8	0.9	0.8	0.4
CH5	0.7	0.2	0.4	0.9

S-CHSA Algorithm Theory Test:Given that at present there are 3 cloud simulation tasks, t1,t2,t3, their QoS need dimensional value is 4, they are Node Residual Energy, Node Throughput, Stability, Node Packet Loss Rate respectively. Of them, the first three elements are positive vector, the last one negative vector. Task dimensional QoS need and the available QoS index value of cluster heads are indicated in the following Table2 For easy comparison, the index value is the relative value from 0 to 1 on average.

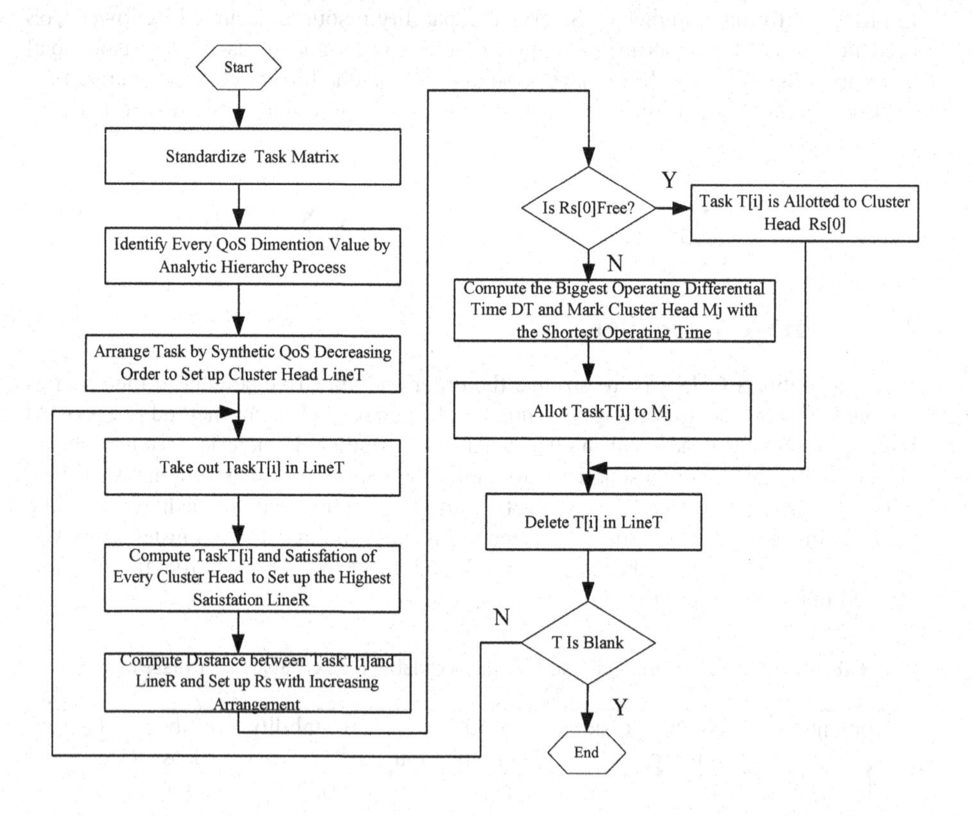

Fig. 2. S-CHSA Algorithm Flowchart

Get the standardized virtual machine Matrix vS3.4 and task Matrix cS3.4 by using Table 1 to standardize the positive and negative vectors respectively.

$$vS3,4=\begin{bmatrix} 0.75 & 0.50 & 0.20 & 0.33 \\ 0.75 & 1.00 & 1.00 & 0.67 \\ 1.00 & 0.75 & 0.80 & 0.83 \\ 0.75 & 0.87 & 0.80 & 0.83 \\ 0.50 & 0.00 & 0.00 & 0.00 \end{bmatrix}$$

$$cS3,4=\begin{bmatrix} 0.75 & 0.63 & 0.60 & 0.50 \\ 0.75 & 0.88 & 0.80 & 1.00 \\ 0.00 & 0.75 & 0.80 & 0.67 \end{bmatrix}$$

By the use of analytical hierarchy process, we can compute every dimensional QoS resource weight value, they are w1=0.123, w2=0.292, w3=0.319 and w4=0.266 respectively. By the use of formula (5) we can compute the total QoS need of every task, respectively they are u1=0.601, u2=0.870, u3=0.652. Therefore, the synthetic QoS of Task T2 is the biggest, Task T2 is firstly allotted to the operation line of the resource waiting for being operated, Task T1 is the last. By the use of formula (8) we can compute the satisfactions of cluster heads from v1to v5 in t2, respectively they are 0.68, 0.90, 0.91, 0.94, 0.48, and virtual machine v4 has task t2 get the highest satisfaction. Thus, task t2 is allotted to the operation line to be operated.

By formula (8) we can compute the satisfactions of cluster heads from v1 to v5 in t3, respectively they are 0.77, 1.0, 1.0, 1.0, 0.58, and the virtual machine v2, v3, v4 have task t3 get the highest satisfaction, and make themselves marked with the highest satisfaction for task t3. Thus, task t3 is allotted to the cluster heads with the shortest distance between it and the cluster heads v2, v3, v4 to be operated. By formula (9) we can compute the distances between Task t3 and v2, v3, v4, respectively they are 0.31、0.36 and 0.28. The distance between Task t3 and the virtual machine v4 is the smallest, however the virtual machine v4 has got Task t2 at this time, the distance between Task t3 and the virtual machine v2 is the second smallest, and v2 is free, so allotting Task t3 to v2 to be oprated achieves loading equilibrium among cluster heads.

By formula (8) we can compute the satisfactions of cluster heads from r1 to r5 in t1, respectively they are 0.86, 1.0, 1.0, 1.0, 0.60, the virtual machine v2, v3, v4 have task t1 get the highest satisfaction, and make themselves marked with the highest satisfaction for task t1. Thus, task t1 is allotted to the cluster heads with the shortest distance between it and the cluster heads v2, v3, v4 to be operated. By formula (9) we can compute the distances between Task t1 and v2, v3, v4, respectively they are0.31, 0.23, and 0.25. The distance between Task t1 and the virtual machine v1 is the smallest, and v3 is free at the time, so allotting Task t1 to v3 to be operated.

5 Conclusions

This paper proposes a cluster head scheduling model supporting QoS, and introduces S-CHSA algorithm in details on the bases of this model. The test shows S-CHSA algorithm guarantees satisfaction of user tasks and good performance in resource utility rate and loading equilibrium.

References

1. Consolini, L., Medagliani, P., Ferrari, G.: Adjacency Matrix-Based Transmit Power Allocation Strategies in Wireless Sensor Networks. Sensors 9, 5390–5422 (2009)

2. Wang, X., Ma, J.-J., Wang, S., Bi, D.-W.: Time Series Forecasting Energy-efficient Organization of Wireless Sensor Networks. Sensors 7, 1766–1792 (2007)
3. Qi, H., Kuruganti, P.T., Xu, Y.: The Development of Localized Algorithms in Wireless Sensor Networks. Sensors 2, 286–293 (2002)
4. Sun, Z.: Operating System tutorial. High Education Press (2003)
5. Li, J., Lu, X.-L.: Research on Resource Scheduling Strategies for Gird Based on GridSim. Computer Science 35(8), 95–97 (2008)
6. Shi, Y.-J., Shen, G.-J., Chen, W.: Solving Project Scheduling Problems using Estimation of Distribution Algorithm with Local Simplex Search. Information Technology Journal 10, 1374–1380 (2011)
7. Chen, Y.-P., Li, Z.-Z.: E-WsFrame: A Framework Support QoS Driven Web Services Composition. Information Technology Journal 6, 390–395 (2007)
8. Xu, Y., Wang, Z., Xu, F.: Optimized Grid Schedule based Qos Guided Min-Min. Computer Applications 27, 215–216 (2007)
9. Shen, H., Ding, Z., Chen, H.: Reliable Web Services Selection Based on Finite State Machine Model. Information Technology Journal 10, 1662–1672 (2011)
10. Yan, T., Li, W., Li, Y.: QoS Guided Task Scheduling Heuristic in Computational Grid Environments. Microelectronics and Computer 23(10), 107–110 (2006)
11. Yazdani, M., Gholami, M., Zandieh, M., Mousakhani, M.: A Simulated Annealing Algorithm for Flexible Job-Shop Scheduling Problem. Journal of Applied Sciences 9, 662–670 (2009)
12. Gong, H., Yu, J., Hou, Y., Liu, H.: User QoS and System Index Guided Task Scheduling in Computing Grid. Computer Engineering 35(7), 52–54 (2009)
13. Wu, Z., Luo, Z., Dong, F., et al.: QoS Deviation Distance Based Negotiation Algorithm in Grid Resource Advance Reservation. In: Shen, W., Yong, J., Yang, Y., Barthès, J.-P.A., Luo, J. (eds.) CSCWD 2007. LNCS, vol. 5236, pp. 316–330. Springer, Heidelberg (2008)
14. Calheiros, R.N., Ranjan, R., De Rose, C.A.F., Buyya, R.: CloudSim:A Novel Framework for Modeling and Simulation of Cloud Computing Infrastructures and Services, `http://www.cloudbus.org/reports/CloudSim-ICPP2009.pdf`
15. Calheiros, R.N., Ranjan, R., De Rose, C.A.F., Buyya, R.: CloudSim:A Toolkit for Modeling and Simulation of Cloud Computing Envirommentsand Evaluation of Resource Provisioning Algorithms, `http://www.buyya.com/papers/CloudSim2010.pdf`

An Improved Adaptive Genetic Algorithm

Tang Hongcheng

College of Automation and Electronic Engineering,
Qingdao University of Science and Technology, Qingdao 266042, China
Thch19@163.com

Abstract. Although the search process of GA may appear the global optimal solution, it can not guarantee that it is converged to the global optimal solution every time, but also the possibility of precocious defects occurs. For disadvantages of genetic algorithm, an improved adaptive GA is proposed with a real-coded, temporary memory set strategy, the improved cross-strategy and the improved mutation strategy. The results of demonstrate examples are proved that effectiveness of the improved GA is best.

Keywords: GA, improved algorithm, temporary memory set.

1 Introduction

Genetic algorithm(GA) is based by Darwin' evolution and Mendel' heredity[1].Although simple genetic algorithm (SGA) is effective on many problems, that conclusion can be made by the theoretical proof is: It cannot guarantee that SGA is converged to the global optimal solution every time, because the global optimal solution of SGA search process can't be retained, which SGA is directly related to the use of generation replacement technology. When parents are substitute by a perfect offspring, each generation of outstanding individuals may not produce offspring due to the probability characteristics of the search, which will result in slower searches and even than the most optimal solution[2].There may be shortcoming of SGA premature [3].

2 The Improved GA

Solving practical problems by GA, we hope that the optimal objective is converged in a wide range of space and the direction to the optimal solution as soon as possible, so the global optimal solution is found. In order to take into account both, an improved adaptive GA (IAGA) is proposed [4]. It is in the following areas:

2.1 Selection Strategy

Although the search process of SGA may appear the global optimal solution, it cannot guarantee that it is converged to the global optimal solution every time SGA [5]. In order to obtain the global optimal solution, the strategy of retaining the best individuals must be used. However, retaining the best individuals may lead to premature; the key

H. Tan (Ed.): Knowledge Discovery and Data Mining, AISC 135, pp. 717–723.
springerlink.com

reason is that the best retained individuals may be most or all of one super individual or very similar super individual. And when their fitness are very bigger than other individuals' fitness, the descendants of the best will be mostly selected in accordance with the fitness of individual offspring. Next the algorithm can't improve the fitness, so the algorithm can't go on iteration, which only leads to converge to local optimal individual. In order to prevent premature, this reference, a set of temporary memory strategies to improve the GA performance is proposed by mechanism of immune memory [3].

When antigens invade the body again, high-affinity antibody can be produced more than that of the initial immunization, known as immune response again or immune memory. To remember enough antigens, limited resources (memory cells) cannot increase indefinitely. The strategy of immune memory must be used by saving only a small amount of antigen with high affinity memory cells.

In this paper, the temporary memory strategies are proposed: Firstly, each fitness of individual is calculated. Then they are ranked according to fitness level. The difference between the former superior fitness of the two individuals is calculated. If the difference of fitness is less than a certain value, Euclidean distance is calculated. If Euclidean distance is less than a certain value, individual with lower fitness is eliminated and the difference between the fitness of eliminated individual and the fitness of next individual is calculated. If the difference of fitness is less than a certain value, Euclidean distance is calculated. If Euclidean distance is less than a certain value, individual with lower fitness is eliminated. ... Until the number of chosen individuals meet the requirement. These individuals for the temporary memory can be gone directly to the next generation. Then all individuals are selected according to roulette. It can effectively prevent that the optimal individuals be eliminated in first choice.

And considering the Euclidean distance prevent the super-individuals and their close relatives are all selected into the next generation, to prevent premature, and the better individuals also have evolution right to cross and mutate.

2.2 Crossover Strategy

Adaptive crossover methods: f_{max}—the biggest fitness of group, $\bar{f}$—the average fitness of group, f'—the bigger fitness between two crossed individuals, f—the fitness of mutated individual. The difference between f_{max} and $\bar{f}$ indicates the stability of the group. The smaller difference between f_{max} and $\bar{f}$ indicates the smaller difference between individual fitness in group, therefore, the likelihood of premature group is greater. In contrast the bigger difference between f_{max} and $\bar{f}$ indicates the bigger difference between individual fitness in group, which indicates divergence of individual characteristics. So P_c and P_m[6] is decided by f_{max}- $\bar{f}$.

To overcome premature, when f_{max}- $\bar{f}$ is small, Pc and Pm are increased. When f_{max}- $\bar{f}$ is big, P_c and P_m are reduced. It is that P_c and P_m are inversely proportional to f_{max}- $\bar{f}$ .When group has converge to the global optimal, for lack of distinguishing function, P_c and P_m are increased, which the probability of destroyed optimal individuals are increases. It overcomes premature, but it destroys optimal individual and makes the GA performance decline. Therefore, not only to overcome premature, but also to remain optimal individuals, different P_c and P_m are endowed in the same generation by different individuals:

Bigger fitness of the individual should reduce P_c and P_m to be protected, whereas smaller fitness of the individual should increase P_c and P_m.

So that not only P_c and P_m relate to $f_{max}-\bar{f}$, but also to f_{max}-f' and f_{max}-f. P_c must be adaptive adjusted:

$$P_c = \begin{cases} k_1(f_{\max} - f_c)/(f_{\max} - \bar{f}) & f_c \geq \bar{f} \\ k_2 & f_c < \bar{f} \end{cases} \quad (1)$$

Type of : k_1, k_2 are <1 normal number. The value of k1and k_2 have a great impact on performance of GA. Usually choose $0.8 \leq k_1 \leq 1.0$. As can be seen from the above equation, the best individual P_c is zero, which reflects the idea to retain an outstanding individual.

While such cross-adjust approach of adaptive algorithm helps to protect the best individual, easy to find the global optimum around local optimums, suitable for multi-peak solution, but this approach hinders the efficiency of evolution in the early evolution. For the best individual cross-rate being very small, almost not with the "outside contacting", so it's not an effective model with the spread to seriously affect the speed of evolution, and the best individual has not changed. It is possible soon occupy the entire population, resulting in "premature".

A improved crossover operator:

$$P_c = \begin{cases} k_2 & if \quad f_c ismemory \\ k_1(f_{\max} - f_c)/(f_{\max} - \bar{f}) & else \quad f_c \geq \bar{f} \\ k_2 & f_c < \bar{f} \end{cases} \quad (2)$$

High fitness individuals have been set into the temporary memory. Without crossing and mutating, they may have been set into the next generation; it can guarantee the best individual not to be eliminated.

High fitness individuals as part of a whole can be selected again because of the high fitness individuals is not necessarily a global optimal solution, which may be a local optimal solution, in order to get global optimal solution, so that high-fitness of individual is taken part in high degree crossing and mutating to search the global optimal solution.

2.3 Mutation Strategy

The adaptive Pm was proposed by Srinivas [7]:

$$P_m = \begin{cases} k_3(f_{\max} - f_c)/(f_{\max} - \bar{f}) & f_c \geq \bar{f} \\ k_4 & f_c < \bar{f} \end{cases} \quad (3)$$

While the mutation adjusting approach helps to protect the best individual, the approach hinders the evolution efficiency in the early evolution and the local best individual to develop and mutate in a right direction.

If search enforcement is strengthened around the second-best solution space, we can often get the global optimal solution as soon as possible to improve the convergence speed, to solve the slow rate of evolution at finish of evolution; or if the second-best solution cannot be evolved, it is difficult to search global optimal solution.

An improved mutation operation :

$$P_m = \begin{cases} k_4 & if \quad f_c ismemory \\ k_3(f_{\max} - f_c)/(f_{\max} - \bar{f}) & else \quad f_c \geq \bar{f} \\ k_4 & \quad f_c < \bar{f} \end{cases} \tag{4}$$

It is because the best individual has entered the temporary memory, so the best individual cannot be destroyed; the best individual, which can be used by a large mutation rate, may quickly search the global optimal solution. In order to speed up search, when the higher fitness of individuals in the temporary memory are mutated with small variation rate at late evolution, then small variations resulting individuals and temporary memory are got into the next generation. In order to search the global optimal solution, GA in the early stages of evolutionary computation must search within the given parameters. But we often get only second-best solution, rather than the true global optimal solution.

Once the algorithm is close to the global optimal solution, the GA is still used as a general global search strategy, making that the convergence rate decreased rapidly. Thus the performance of the current best individual does not change for a long time, resulting in slow convergence or even stagnation. This slow convergence in the late evolution o has been noticed by many researchers, and also some solutions are put forward.

For example, adaptive GA was proposed in document [7]: Mutation rate was increased, search capabilities were expand, and so on in late evolution. These methods still solve problems of later evolution by the global space. If search enforcement is strengthened around the second-best solution space, we can often get the global optimal solution as soon as possible to improve the convergence speed, to solve the slow rate of evolution at finish of evolution.

Strengthening search enforcement around the second-best solution space, that is, narrow the search to improve the local search capabilities.

Small variation is configured - each individual in the temporary memory is multiplied by the following factors:

$$ksm=((rand-0.5)*0.1)+1 \tag{5}$$

Rand ranges [0, 1], rand-0.5 ranges [-0.5, 0.5], so k_{sm} ranges [0.95, 1.05]. The good memorial individuals can be locally searched up and down to improve the ability of local search, which is very meaningful for algorithm around the global optimal solution.

The use timing of small variations - small variations are used in the later stages of the evolution, only after the search to sub-optimal solution, which is only very meaningful for algorithm close to the optimal value.

In this paper, the judgment method of the late evolution is: The difference between maximum fitness and average fitness is calculated; if which is less than a small value, small variations are implemented. Small variations offspring and individuals in temporary memory are selected together.

3 The Improved Genetic Algorithm Is Verified

To verify the proposed improved genetic algorithm, in this paper, a well-known test functions which is proposed by K.A De Jone [8] is used to verify improved genetic algorithm:

$$f_1(x)=100(x_1^2-x_2)^2+(1-x_1)^2 \quad (-2.048<x_i<2.048) \tag{6}$$

f_1 is a continuous unimodal function, and whose minimum point (1,1) is located in a narrow, parabolic-shaped, flat in the recess is more difficult to find, it is often used to test the premature convergence.

Three GA- SGA, AGA(which was proposed by[7]) and IAGA (which is proposed by this paper) are tested 100 by Matlab. Test results show in Table 1: 100 initial individuals are generated range of random (-2.048, 2.048). Their maximum evolutions are 200- generation. Specific GA parameters: SGA mutation rate P_m is 0.5 and crossover rate P_c is 0.5; AGA mutation rate coefficient k_3, k_4 are 0.5 and crossover rate coefficient k_1, k_2 are 0.6; IAGA mutation rate coefficient k_3,k_4 are 0.5 , crossover rate coefficient k_1,$k_2$are 0.6 and small variation threshold is 0.001.

Data in the table 1 were the results of 100 times evolution by Matlab. In which the obtained optimal value times is that the errors are ignored by Matlab; the obtained approximate optimal value times is that the errors of either the fitness or accuracy are beyond 0.0001; the obtained suboptimal value times is that the errors of either the fitness or accuracy are beyond 0.0001 and individuals are near (1,1) ; average generations close to optimal value is the average number of convergent evolution generation of sum of the optimal number and the approximate optimal number.

Table 1. Three GA100 test results

	The optimal value number	Approximate optimal value number	Suboptimal value number	Average generations close to optimal value
SGA	1	11	84	190
AGA	41	8	31	150
IAGA	50	6	36	132

It can be seen byTable1 that AGA and IAGA are super to SGA. An optimal value was searched by SGA. IAGA, which 50 optimal value and 132 generations are super to AGA. It proves that the proposed improved algorithm overcome a certain premature convergence, to avoid the occurrence of precocious.

Three GA changes of top 25 generations show as Fig. 1.

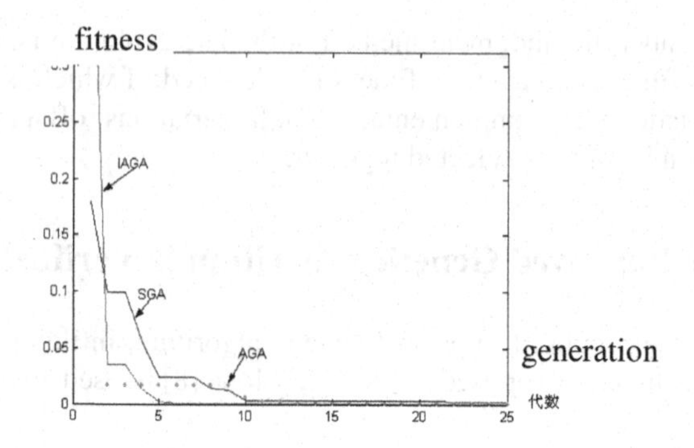

Fig. 1. Three GA changes of top 25 generations

Figure 1 shows that the proposed method of IAGA is the best, AGA is the better, SGA is the worst.

4 Conclusion

The improved genetic algorithm is better to retain the advantages of genetic algorithm, real-coded, the temporary memory strategy, the improved cross-strategy and improved the mutation strategy integrated into the algorithm, which effectively overcome the SGA shortcomings, such as prone to premature, poor local optimization, the evolution too many times by reasonable selecting mutation probability. Calculation results in the paper are proved the effectiveness of IAGA.

References

1. Huiren, Z., Wansheng, T., Ben, N.: Optimization of multiple traveling salesman problem based on hierarchical genetic algorithm. Application Research of Computers 26(10), 3754–3755 (2009)
2. Ding-li, L.: Summary of the genetic algorithms. Science and Technology of West China 8(25), 41 (2009) (in Chinese)
3. Yee, L., Gao, Y.: Degree of population diversity——A perspective on premature convergence in gas and it's Markov chain analysis. IEEE Trans. on NNs 8(5), 1132–1140 (1997)
4. Hongcheng, T.: Fault diagnosis technology and application based on artificial immune intelligent. Logistic Engineering University (2004) (in Chinese)
5. Ming, Z., Shulin, S.: Principle and application of Genetic Algorithms, Beijing, China (1999) (in Chinese)

6. Burnet, F.M.: The clonal selection theory of acquired immunity, London(1959)
7. Srinivas, M., Patnaik, L.M.: Adaptive probabilities of crossover and mutation in genetic algorithm. IEEE Transactions on System, Man and Cybernatics 24(4), 656–667 (1994)
8. Strassner, T., Busold, M., Herrmann, W.A.: MM3 parametrization of four-and five-coordinated rhenium complexes by a genetic algorithm. Journal of Computational Chemistry (23), 282–290 (2002)

Determinants of Financial Restatements in the Listed Companies in China

Guiling Zhang

Zhengzhou Institute of Aeronautical Industry Management, Zhengzhou, China
financepaperzhang@126.com

Abstract. This paper firstly introduces the latest Accounting Standards about financial restatement, and the current situation of financial restatement in listed company in China, then analyzes the determinants of financial restatements from company characters, ownership structure, internal governance structure and external governance, the results are as follows: the better performance of listed companies, the lower probability of financial restatement; ownership concentration is negative correlated with the probability of financial restatements; the number of independent directors is significantly positive correlated with the accounting information quality; the establishment of audit committee can reduce financial restatement; the supervision of the securities regulatory authorities can effectively reduce the financial restatement; audit by larger accounting firms can reduce financial restatement. And according to the above results, this paper gives the following suggestions for the improvement of listed company financial restatement: improve the governance structure; strengthen market supervision; perfect the relevant accounting standards, laws and regulations.

Keywords: Financial restatement, ownership structure, internal governance structure, external governance.

1 Introduction

In China, on February 2006, the Ministry of Finance issued Accounting Standards for Business Enterprises No.28–Changes in Accounting Policies and Estimates and Corrections of Errors, which included the definition, correction and disclosure of prior period errors. The prior period errors include calculation mistakes, mistakes in applying accounting policies, oversights or misinterpretation of facts, consequences of fraud, inventory overage, fixed asset overage, etc. An enterprise shall adopt the retrospective method to correct any important errors of prior period, however, unless it is impractical to recognize the amount of cumulative effects of the prior period error.

An enterprise shall, in its notes, disclose the following information related to the correction in prior period errors:

1. The nature of the prior period errors
2. The names of the affected items and the corrected amounts in the financial statements for all prior periods presented.
3. If it is unable to make a retrospective restatement, it shall state the facts, reasons, time point of beginning the correction of the prior period error, as well as the information about the concrete correction.

H. Tan (Ed.): Knowledge Discovery and Data Mining, AISC 135, pp. 725–730.
springerlink.com

The application of Accounting Standards for Business Enterprises No.28 formally marks the establishment of the financial restatement regulations in China.

2 The Current Situation of Financial Restatement in Listed Company in China

In recent years, financial restatements are becoming more and more frequent which has aroused wide concerns. At present, there are two main characters in financial restatement in listed company in China: firstly, large number, high ratio and growing trends. Chen Lingyun (2006) analyzed the financial restatement announcements from 2001 to 2004, and found that the rate of restatement companies reached to 20%. Secondly, wide range of restatement contents, part of the restatement even relates to the core accounting data, which would greatly influence the enterprise value. Lei Min (2006) analyzed the complementary and correcting notices issued by A-share listed companies, and found that more than half notices were concerned with profit, and this showed the disclosure of financial information was not normative.

3 The Determinants of Financial Restatements in Chinese Listed Companies

3.1 Characters of the Company

Company Size. For the relatively imperfect of the accounting system, small companies are more likely to make financial restatement; while, large companies use restatement as a tool to send company information to outside. Therefore, it is difficult to determine the impact of the company size on financial restatement in the listed companies. And the empirical research also support the above view, such as Ahmed and Goodwin (2007)

Company Performance. Due to wide concerns by investors and security analysts, the listed companies are under more operation pressure, so concealing its poor performance and financial condition become the main motivation of earnings management, therefore, earning management to avoid loss results in financial restatement. Li Changxin (2009) expressed that the probability of financial restatement in the listed companies was closely related to performance and financial condition, the better performance, the lower probability. That is earnings management is one of the main reasons of financial restatement.

3.2 Ownership Structure

The studies on ownership structure are mainly from ownership concentration and ownership nature. Ownership concentration reflects the influence on the managers by shareholders, and ownership nature has direct influence on shareholders' behavior and aims.

Ownership Concentration. In modern companies, ownership tends to be more and more decentralized, so owners' power is more and more in name, actually the companies are controlled by the managers. So the unsupervised managers may pursuit their own interests by sacrificing shareholders' interests. Therefore, shareholders should supervise the operating of the companies. If the company is success, more shares, more interests, and naturally major shareholders are trusted by all shareholders, so we can say that major shareholders supervision and control have positive effect on the improvements of company governance efficiency. That is to say ownership concentration will generate interest convergence effects, especially when the ownership is highly concentrated, the interests of the major shareholder are closely related to the company performance; the major shareholders are motivated to set effective accounting control procedures to find and prevent accounting errors. Therefore, ownership concentration and probability of financial restatements are negative correlation.

The Nature of Controlling Shareholders. In China, the most prominent feature of the ownership structure in listed companies is ownership relatively concentration, and ultimately controlled by the state. This structure leads to the 'owner vacancy' problem. As the agents of the state shares- the government officials don't have residual claim rights, they are lack of adequate economic interests to monitor and evaluate the managers, so it is difficult to supervise and restrain the managers. Because of the relatively low governance efficiency of the state-owned company, we can assume that the probability of financial restatement is significantly higher in state-owned listed companies. The study of Yu Peng (2007) confirmed the above assumption; he discussed the impacts of ownership structure governance efficiency on financial restatement. The results showed that, the sate-owned shares have relatively poor restrictions on the management of the listed company in China, which leads to higher possibility of financial restatement.

3.3 Internal Governance Structure

The Board Size and Structure. The board of directors is the core of corporate governance; it accepts the commission from shareholders, and supervises the management. So it is the agent, as well as the client, and plays a very important role in the internal corporate governance.

The size of the board is an important determinates of the accounting information quality. On one hand, organization theory states that large organizations need to spend more time on decision making, with large board, the communication in the directors will be hindered, and the efficiency will be reduced. On the other hand, small size board is lack of sufficient professional services, which leads to lower quality of accounting information. Therefore, neither too large nor too small board size would improve the accounting information quality. In China, the company law gives the upper and lower limits of director numbers in the board, so the size of the board isn't significantly different. But large size board makes "internal control" easier, which has negative effect on the financial information quality; that is the larger the board size, the higher probability of financial restatement.

Independent directors are often professionals or senior managers of other organizations or enterprises. Consideration their own reputation, they don't tend to be in collusion with the management. With rich experience and professional skills, the independent directors help the governance mechanism running better. Therefore it is believed that, the higher percentage of independent directors, the lower possibility of financial restatement. Xue Zuyun (2004) have the same research results with the above analysis, he stated that, the number of independent directors is significantly correlated with the accounting information quality.

Audit Committee. As internal supervision, compared with supervision from CPA and securities regulatory authorities, the audit committee can know the financial situation more timely and detailed. China Securities Regulatory Commission and Economic and Trade Commission jointly promulgated <Code of Corporate Governance for Listed Companies in China>. The following guidelines are provided: according to the relevant decision of general meeting, the board of directors can set up audit committee, the members of audit committee are all from the board, and more than half members should be independent directors, in addition, there should be at least one independent director of accounting professional in audit committee. This code is effective to improve the quality of accounting information and induce the financial restatement possibility. Therefore, the establishment of audit committee can reduce financial restatement.

Li Bin, Chen Lingyun (2006) discussed the effectiveness of the Audit Committee in Listed Companies in China. The results show that, the establishment of the audit committee has significantly reduced the possibility of financial restatement, meanwhile, with the improving of independence and authenticity, the possibility of financial restatement reduce further, and the quality of financial statement is higher. The results also showed that the audit committee in listed company had improved the quality of financial statements effectively in China.

3.4 External Governance

As the company's survival and development environment, the external governance influences the company's information disclosure significantly.

The Securities Regulatory Authorities. At present, in China the main securities regulatory authorities of Listed Companies information disclosure are China Securities Regulatory Commission, Shanghai securities exchange and Shenzhen securities exchange. China Securities Regulatory Commission is the management institution of securities supervision; also it is the most authoritative regulator. In the supervision of continuing information disclosure, the duty of the listed company supervision department in China Securities Regulatory Commission is to guide, supervise and check the supervision of the stock exchange on Listed Companies information disclosure, coordinate with stock exchange and the dispatched office on the supervision of listed companies' information disclosure. The Stock Exchange supervises the daily information disclosure of listed companies directly, its main responsibilities are to prior audit the temporary notice, and post audit the regular reports and continuous supervise the daily information disclosure of the listed companies. The supervision of the securities regulatory authorities can effectively reduce the financial restatement.

Public Accounting Firms. As an independent third party, the public accounting firms' high quality audit is important part of external supervision, and plays a special role in the improvement of financial reports reliability. If there is financial restatement in the listed companies, probably the audit of accounting firms did not find the related problems, that is professional incompetence, or the accounting firms found the problems but did not report, that is lack of independence. From the Enron event we can find that interests forced the CPA to compromise with the radical accounting policy of the listed companies, the CPA failed to correct the accounting errors in the financial statements in time, which leads to future financial restatement. Li Zexiong (2009) stated that larger accounting firms pay more attention to audit risk and their reputation, which leads to the improvement of professional performance and the reliability of financial report, and reduce financial restatement.

4 The Suggestions for Improvement of Listed Company Financial Restatement

4.1 Improve the Governance Structure in the Listed Companies

Firstly, improve the internal governance and internal controls to improve the quality of financial reports. Encourage the major shareholders especially the institutional investors to participate in corporate governance, strengthen the supervision of independent directors and the audit committee, all of the above will help to improve the quality of information in the listed companies. At the same time, regulation authorities should supervise the corporate governance of the listed companies. And warn or punish the listed companies which do not meet the legal requirements (such as the proportion of independent directors in the board is less than 1/3, no audit committee, etc). From the perspective of internal control, the listed companies should be absolutely responsible for the financial restatement. The management and primary financial accounting departments of listed companies should strictly control the process of making financial reports to ensure the information disclosed is true, accurate, complete and timely.

4.2 Continue to Strengthen the Market Supervision

For market supervision, we should improve the supervision efficiency of the stock exchanges and regulatory authorities, strengthen independent external audit of accounting firms. Regulatory authorities should put out appropriate operational guidelines such as "Corporate Disclosure Guidelines: Financial restatement"; at the same time, strengthen the supervision of information disclosure, punish the restatement concerning with intentional concealment, delay, false disclosure of information, and financial data manipulation. The stock exchanges should continue to focus on the abnormal changes in stock price trading volume, trading seats, accounts of the listed companies and should take mandatory suspension and inquiries on the companies which can not make a reasonable interpretation; Meanwhile, communicate with the listed companies, to find and settle the problems in information disclosure of listed companies. External audit institutions should strengthen the ranks of Certified Public

Accountants and the audit qualities, meanwhile, the External audit institutions should increase the CPA's professional and ethical standards.

4.3 Perfect the Relevant Accounting Standards, Laws and Regulations

For national legislation, we should further perfect the legal systems; improve information disclosure system, so as to provide legal support for information disclosure supervision. For example, enhance relevant laws and regulations on the illegal behavior of information disclosure; increase the intensity of punishment to raise illegal information disclosure cost, improve the securities civil lawsuit system and so on, in order to create a good environment for listed companies to provide high quality financial information.

References

1. Smith, T.F., Waterman, M.S.: Identification of Common Molecular Subsequences. J. Mol. Biol. 147, 195–197 (1981)
2. Agrawal, A., Chadha, S.: Corporate Governance and Accounting Scandals. Journal of Law and Economics (48), 371–406 (2005)
3. US General Accounting Office (GAO). Financial restatements: Update of public company trends, market impacts, and regulatory erdorcemerit activities. GAO-06—678, Washington D.C. (2006)
4. Ahmed, K., Goodwin, J.: An empirical investigation of earnings restatements by Australian firms. Accounting and Finance (47), 1–22 (2007)
5. Lei, M., Wu, W., Wu, C.: Financial Restatements of Listed Companies in China. Shanghai Management Science (4), 38–43 (2006)
6. Li, C.: Governance and financial restatements: evidence from listed companies. Shantou University, 31–32 (2009)
7. Yu, P.: Ownership structure and financial restatements: evidence from listed companies. Economic Research (9), 134–144 (2007)
8. Xue, Z., Huang, T.: Characteristics of board of directors and board of supervisors, and quality of accounting information-empirical analysis from china's capital market. The Theory and Practice of Finance and Economics (25), 84–89 (2006)

The Effects of Exercise on IL-2/sIL-2R

Liqiang Su

Department of Physical Education, Jiangxi University of Traditional Chinese Medicine, Nanchang, Jiangxi Province, China
s-2005100153@163.com

Abstract. The movement is a stress, it may cause the changes of organism immunity function which is influenced by many factors, in the immunity adjustment process, its interior has the very fine adjustment process. Through literature material, the purpose of this article is to summarize the influence of IL-2 /sIL-2R system.

Keywords: Training, INterleukin-2, The soluble intcrleukin-2 receptor.

1 Introduction

The results shows that training with big intensity or competition can influence organism immune function of athletes and increase the risk of infected organism, especially upper respiratory infection [1],and it is one of reasons of immune system lower that T-achroacyte activity of athletes is decreased caused by training with big intensity [2]. While stressing, free radical can influence immune function through attacking immunologic cells envelope to shape immunologic injury. [3] However, during immunological regulation, it has become a complex network due to interaction among cell factors that plays an important role in organism immune function. IL-2, mainly coming from activated CD4+T cells and a small quantity CD8+ T cells, plays an important role in immunological regulation, whose biological effects are T-cell growth factor, increment of mediate lymphocyte activation. The rise of soluble interleukin-2 receptor in serum can counteract IL-2 circulating, which makes immunity lower. Therefore, the disorder of IL-2 /SIL-2R system balance will make an important impact on body immunity.

2 The Proteinum Structure of Interleukin-2 and Physico-Chemical Property

The Interleukin 2 was discovered while researching long-time growth of T-cell. In 1976, Morgan etc stimulated T achroacyte to produce a factor through mitogen(PHA, canavaliae proteinum A(Con)A etc), it was named T cell growth factor(TCGF) at that time and later was formally named Interleukin 2 in lymphocyte factor meeting in 1979. [5]

The Interleukin in organism is mainly secreted from THI cells of I achroacyte ,The gene expression of Interleukin is made up of 153 residue antecedent N, signal peptide

H. Tan (Ed.): Knowledge Discovery and Data Mining, AISC 135, pp. 731–738.
springerlink.com

is made up of 20 amino acid residue, and they are cut while secreting. Human nature IL-2 is a polypeptide that is made up of 133 amino acid residue, 15kDa glycosylated protein intaked by molecule. isoelectric point (PI) 6.5-8.0, 3 aminothiopropionic acid residue(Cys), situates 58th,105th and 125th .Therefore, post-translational modification includes glycosylation sited Thr and Sulfur Bond sited 58th and 105th.glycosylation have no influence on proteinum activity ,while exact Sulfur Bond is necessary for activity, thus 125 free half sarcosine residue are usually mutated serin and alanine from recombined IL-2, which helps recombination protein to produce exact disulfide bond.[6]

3 The Resource and Biologic Activity of Interleukin-2

IL-2, mainly from actived CD4+T cells and a small quantity CD8+ T cells, is a lymphokine used for immunological regulation and hydrone proteinum with highly actived , multifunctional that is mostly produced achroacyte and mononuclear macrophage. The key biological functions of IL-2: regulating local immunity by autocrine and paracrine focus on target cells; enhancing local immunity by enhancing T cells activity generation (including CD4+T cells and CD8+T cells) and express cell factors (such asIL-4, IFN-γ and so on); encouraging B cell multiplication, differentiation and antibody production to induce Tc cells, achroacyte, lymphakine and activated K cells differentiation and improvement of their kill activity.

IL-2, whose biological effect is T cell growth, plays an important role in immunological regulation. The significant activity of immunological respond is activation of achroacyte which is sine quanon content in immunological respond cell mediated. There are three pacing factors for T cell activity: the first signal that is combineed MHC/ peptides that is superfic on the antigen presenting cell(or target cell)with TCR that is superfic on the T cells; the second signal that is combined surface molecule on antigen presenting cell with CD28 molecule of T cell; IL-1 and IL-2, etc. which are regarded as the third signal activated by T cell. Th cells have transcription of protooncogene c-fos and c-myc under co-stimulation of both IL-1 and other signals, further it causes transcription IL-2 and IL-2R genes. IL-2 is connected with IL-2R by autocrine and paracrine, and activate T cells by the pinocytosis of IL-2 and IL-2R compound of T cells. Therefore, the IL-2 and IL-2R on T cell surface are the important signals of the activation of T cells whose level reflects the function status of T cells, IL-2 is also regarded as an important regulatory factor of immunity, and the production ability of IL-2 can be as the important index of the immunity function in organism.

Unactivated CD4+T cells can immediately generate small amounts of cell factors such as IL-2, IL-4 and INF-r after antigenic stimulation , at this time, CD4+T cells is called Th0 cells which can be differentiated Th1 and Th2 at different conditions. Th1 excretes IL-2, TNF and INF-r, etc, encouraging immune response, promoting the function of B cells, NK cells, and activating macrophage. Th2 excretes IL-4, IL-5, IL-10 and IL-13, etc, inducing B cells multiplicated and differentiated antibody plasma cells, boosting composition of lgG and lgE. This is adjuvanticity T cell that is related to Humoral immune reaction.

In short, the biology of interleukin is mainly to excite T cellular, besides, it also includes other important functions such as to promote T cellular; to handle the whole NK activated, to differentiated and generated; to adjuste Nk cells to hold its function; to induce cytotoxicity T Lymphocyte production and generation; to derivn Lymphocyte production activated by lymphokine; to B cell multiplication and differentiation and synergistic effect with other interleukin.

4 The Relation between Interleukin 2 and Dissolubility Interleukin 2 Acceptor

The connection of IL-2 and IL-2R is actuating signal of T lymphocyte activation and proliferation. IL-2 is primarily produced by TH cells, with immunostimulation , causing various Inflammatory reaction, and its main function is to activate cytotoxicity T cell and to encourage composition of other cell factor, enhancing the effect of natural killer cell. IL-2 can boost cytotoxicity T cell and NK cytoactive to activate T cell multiplication. The educe of biological action of IL-2 have to rely on the mutual effect with IL- 2R on cell envelope, while the concentration of IL- 2, the express of IL-2R,and the affinity of IL- 2R on IL-2 have become important factors that decide T cell multiplication. IL-2R includes low, middle, high affinity, IL-2R with low affinity is constructed by a chain of 55KD, it is also called Tac; while IL-2R with middle affinity is constructed by β chain of 75KD; IL-2R with high affinity is constructed by a chain and β chain, usually only this kind of acceptor has biologic activity. [7] sIL-2R, is the cast of connection of coat and IL-2Ra chain, molecular mass is 45KD. We usually believe that sIL-2R released by T cell is the result of immunocell activation, so the advancement of sIL-2R is regarded as the signal of organism specific immunity.

Mutual effect among cell factors form a complicated network,which play important regulation role in organism immune system. Interleukin-2 is a T cell growth factor that activates T cell, mediates lymphocyte activation increment, a rise of solubility interleukin-2 in blood serum that make immunity lower can counteract IL-2 circulating, so the disorder of IL-2 /SIL-2R will make key influence on immunity. [8] Combination with low affinity from sIL-2R and IL-2 greatly blanket their immunity activity. obviously, sIL-2R is a negative growth factor of immune function, so some scholars regard sIL-2R as one of immune inhibitors whose abnormal rise shows that organism immune function is low. sIL-2R makes immune function mediated IL-2 lower whose mechanism: (1) sIL-2R membranes added with IL-2Ra is dropped, is also the main clearance way of membranes added with IL-2Ra ,and make activated T cell rest; (2) considerable sIL-2R has ablated on membrane of sIL-2R, and make activated lymphocyte in depletion, make immune function be in disorder through effecting IL-2/IL-2R immune system, the result is that IL- 2 is of lower level, anti-ambury is also degraded, while sIL- 2R has risen, competing with mIL-2R and connecting with IL- 2 have counteracted IL- 2 around T cell to further restrain immune function whose effect of immune monitor and immune clearance are attenuated. [9] however, sIL-2R, as Transport protein of IL-2, exists in T blood connecting with IL-2, this can extend IL-2 weak time in the body, and plays the role of norientation immune regulation through transporting IL-2 .[10] sIL-2R is expressed two-ways regulation of enhancement for low concentration and restraint for high concentr. [11]

The acceptor concentr of sIL-2R in serum is obviously increased, connecting with free IL-2, which interferes interaction between IL-2 and target cell and influences the effect of IL-2 in immune response in the body, therefore, we believe sIL-2R is high-strung amynology index. So,when sIL-2R is used as index of amynology, we should connect it closely with IL-2.

5 The Influence of Exercise on Interleukin -2

IL-2 findings are various because of different factors of training level of experiment objects, physiology and mental state, exercise load and nature etc. young bicycling athletes are regarded as research objects with strong training for 6 months, about 500 kilometres each week, and test immune function index before and after exercise, compared with common people without training. The results showed IL-2 is declined after training. Having observed immune reaction of rats acute exercise with long-time training , and wheel training for ten weeks, training group and control group are exercised for ten minutes with 70% maximal oxygen intake, as a result, the IL-2 of training group is descended after acute exercise. [12] The experiment of Jiaowei shows IL-2 secreted spleen cell in rats is obviously declined after exhaustion swimming. [13]. overtraining, defective nutrition and muscle damage will produce negative effect of IL-2 and leucocytic response.

Eight healthy amateur athletes are trained with medium intensity for 3 months, 1.8-2.5mmol/L blood lactate, three or four times every week, the aerobic level of training group is increased and IL-2 is not notably changed. There are other researches about eight healthy objects are trained for3-5 hours' running with medium intensity lasting 12 weeks, the results showed although Lactate Threshold increased, IL-2 didn't change notably. [14] some reports showed movement training can increase IL-2 through observing the activity changes of IL-2 secreted venous blood T leukomonocyte induced PHA for seven male athletes after an acute fatigue exercise, the results showed IL-2 activity secreted T leukomonocyte at immediate postexercise was slightly higher than that before exercising, and it notably increased for three hours after exercise, and it decreased to a little higher than that of the common. The author think it is the enhance response of immune function after acute exercise. other reports showed the IL-2 of athletes increased for sixteen hours after marathon. [15]. Sixty minutes group compared with one hundred and twenty minutes group: it had not burdened anything for sixty minutes every day for sixty minutes group and it also had not burdened anything for one hundred and twenty minutes for one hundred and twenty minutes group, the two groups were trained for eight weeks , compared with the control group , it showed the IL-2 in serum in one hundred and twenty minutes group was obviously decreased while the IL-2 in serum in Sixty minutes group was obviously higher than that in control group. [16]

Observing the effect of Taijiquan on IL-2 concentration of the old man, the 16 old women of 55- 65 aged in town are experiment objects, they are randomly divided into training group(10) and control group(6). The member of training group practice

Taijiquan every day while the control group member keep common activity and do not exercise on purpose. The results showed: the IL-2 concentration of the members of training group is obviously higher after Taijiquan for 6 months, one time boxing can notably enhance IL-2 concentration in serum. [17] The rats of training group of 30 minutes swim without any burden for 30 minutes each, and the rats of exercise group swim without burden for 60 minutes once, both groups are trained for 8 weeks, and rats are killed after 24 hours after the completion of exercise. The results showed that the IL-2 content in serum of the two group is obviously higher than that in control group. [18]. SD rats is adopted with an acute weight loading exercise, the weight loading is 3% of the body weight, rats are submerged Shuidi point for 5s before stopping exercise, and they are killed after 18 hours' exercise, the results showed that blood plasma IL-2 of the rats decreased but not notable differences.[19]

6 The Effect of Exercise on Dissolubility Interleukin-2 Acceptor

The consequences of report are various about the effect of movement training on sIL-2R because of different factors of objects such as training level, physiology, mental state, different loading and nature etc.

The research of movement training makes sIL-2R higher, the athletes are trained with medium intensity for 12 weeks(4-5 times a week, 30 minutes at a time, 65-70% maximal oxygen intake), then they are tested the immune reaction after cycle ergometer once(60% maximal oxygen intake, 60 minutes), the sIL-2R of training group for 12 weeks, increasing 33%, as compared with that of control group. 15 athletes are sampled with blood before 24 hours of exhaustion movement, 1 hour after training, 20 hours after exercise respectively, and the results showed the sIL-2R in serum of the athletes increased at 1 hour after training and it is basically recovered after 24 hours. [20] The research of training makes sIL-2R lower, observing amateur juvenile athletes who are trained with large intensity for a week, the results showed the sIL-2R of the athletes was obviously decreased after training every time and would be basically recovered after 24 hours after training. the researches into the effect of long-time exercise on immune system showed the sIL-2R concentration in serum of superior hand brigade athletes tendency to ascend. Compared with that before training, the sIL-2R concentration in serum was notably decreased after 36 hours after large loading training and a week before contest. [21] There are also researches reported that the effect of athletics aerobics training on sIL-2R that the sIL-2R is decreased after training, which showed large intensity can restraint the activity of T cells. [22] other reports showed that there was obvious effect of training on sIL-2R by observing 8 healthy amateurs who were trained with medium intensity for three months, 1.8-2.5mmol/L Blood Lactate Acid Concentration, 3-4 times every week, the oxygen level was increased of training group three months later, but sIL-2R was not obviously changed. Other researches although speed at Lactate Threshold was increased, the sIL-2R was not notably changed, by 8 healthy athletes with medium intensity running for 3-5 hours every week, lasting 12 weeks.

Most researches into training make IL-2 lower or fixed is long-time aerobic exercise that may enhance immune function and decrease the change extent of IL-2 compared with that of the control group, that is to say, the same loading will produce smaller immune reaction of aerobic exercise group. Some scholars believe that IL-2 is mainly produced CD4+ and CD16+, and IL-2 is not obviously changed due to CD4+ subpopulation decreasing and CD16+ subpopulation increasing while training. However, the IL-2 after exercise is higher than that before exercise, the reason is that the organism is in the state of large immune response induced by movement stress after large intensity exercise, and the immune response is not immediately ended. The explanation related to sIL-2R, some scholars think the production of sIL-2R is mainly from the dependence of T cells activation, so the rise of sIL-2R is regarded as the index of specific immune activation in organism, which is important immune suppression that has the function of restraining the activity of IL-2, its concentration is depended on the activity of T cells. Long-time aerobic exercise make organism in the state of balance of immune, sIL-2R is not significantly changed after exercise. The decrease of sIL-2R after exercise, the reason may be that the relief of immune restraint after exercise. In short, when evaluating the immune function, we should consider IL-2 level as well as other factors including the exact nature of training, stage, the actual status of athletes etc. because immune reaction is influenced by body conditions and free radical will produce certain effect on immune function during exercise, the effects of exercise on free radical are as follows.

7 In Summary

The change of exercise on immune system varies along with exercise form, exercise intensity and exercise time. High intensity and long-time exercise will cause temporary restraint of immune system function; [23] acute and short time exercise with medium intensity will activate immune system and improve immune function; [24, 25] long-time endurance exercise or Long-time reinforced exercise will restrain immune function. [26] in addition, different aged people will have different reaction of immune on the same intensity. a large amount of epidemiological datas show moderate exercise can enhance disease resistance of organism, including cell immunity and humoral immune function; overtraining will increase the affectability of organism on various diseases in particular kinds of bacteria and viral infection diseases. [27] The cells of different stages and different intracellular region will produce various reactions on free radical. The physiologic reaction of antigen-receptor will be different after activation, during the earlier period of accepting extra cellular signal, it needs free radical to promote the earlier transducer of extra cellular signal, while for the best function of certain nuclear factors will be expressed, persistent oxidative stress will produce the restraint. The organism will show the descend of immunity and affectability of organism, and this is bad for fitness, which is more important for the training of athletes who will get good grade by surpassing every aspect function, however, training with large intensity has negative effects on immunity of organism.

References

1. Nieman, D.C., Johansen, L.M., Lee, J.W., et al.: Infections episodes in runners befog and after the los Angels Marathon. Sports Med. Phys. Finmess. 30, 316–328 (1990)
2. Li, X., Wang, H., Zhao, A., et al.: The observation the effect of immune function in organism after large intensity by providing kiwi fruit. Chinese Sports Medicine 22(2), 187–188 (2003)
3. Hao, M.: The mechanism and regulation measures about decrease of immune functions of athletes. Sport Science Research 24(4), 47–49 (2003)
4. Tang, E.: Medical immunology, pp. 75–76. Sichuan University Press (2004)
5. Morgan, D.A., Ruscetti, F.W., Gallo, R.: Selectibe vitro growth of 1ymphoytes from normal human bonemarrows. Science 193, 1007 (1976)
6. Zheng, J.: The research status of interleukin-2. Channel Medicine 18(4), 1–3 (2006)
7. Zhang, X., et al.: The changes of sIL-2R in serum and clinical significance of Machine wounds patients. Huaxi Medicine 17(1), 24–25 (2002)
8. Zhou, X., et al.: The changes of IL-2和SIL-2R in serum during obstruction of biliary tract operation and the research of arginine. Chinese Modern Medicine 16(8), 1229–1234 (2006)
9. The effect of chronic myeloid leukemia on White Cells of CML and sIL-2R, IL-2 in serum. Study Journal of Traditional Chinese Medicine 20(4), 428–435 (2002)
10. Li, Y., et al.: The determination of the activity of sIL-2R of neonatal septicemia. Chinese Experiment Clinical Immunology Medicine 6(5), 28–30 (1994)
11. Wei, H.: The Research into the changes of immune function and mental status of female football player before and after contest. Magisterial Thesis of Beijing Sports University (2002)
12. Lin, Y.S., Jan, M.S., Tsai, T.J., Chen, H.: Immunomodulatory effects of acute exercise bout in sedentary and trained rats. Medicine and Science in Sports and Exercise 27(1), 73–79 (1995)
13. Jiao, W., et al.: The Preliminary Study of immune Profilin produced by violent exercise. Sports Science 18(3), 71–74 (1998)
14. Baum, M., Kloepping-Menke, K., Mueller-Steinhardt, M., Liesen, H., Kirchner, H.: Increased concentrations of interleukin 1-beta in whole blood cultures supernatants after 12 weeks of moderate endurance exercise. European Journal of Applied Physiology and Occupational Physiology 79(6), 500–503 (1999)
15. Huang, C.: The changes of T achroacyte activity of peripheral blood and blood plasm compound F, insuline level after acute fatigue exercise. Magisterial Thesis of Beijing Sports University (1994)
16. Yin, J.: The effect of different loading training on compound B of interleukin 2 in serum in rats and serum βendogenous opioid peptide and blood plasma Tachroacyte. Chinese Journal of Rehabilitation Medicine 21(7), 593–595 (2006)
17. Wang, X.: The effects of Taijiquan on interleukin 2 in the old man. Journal of Shandong Sports 19(58), 48–50 (2003)
18. Yin, J.: Study on the impact of medium and small exercise on corticosterone IL-2 in serum of mental stress rats and blood plasm T homeocyte subgroup. Magisterial Thesis of Yangzhou University (2003)
19. Chen, X.: Effect of an acute exercise on blood interleukin in rats. Beijing University Journal 23(3), 347–348 (2000)

20. Weinstock, C., Konig, D., Harnischmacher, R., Keul, J., Berg, A., Northoff, H.: Effect of exhaustive exercise stress on the cytokine response. Medicine and Science in Sports and Exercise 29(3), 345–354 (1997)
21. Li, X.: Study on the impact of training with large intensity swimming on the immune function of juveniles. Chinese Sports Medicine 20(3), 313–314 (2001)
22. Wang, Y., Li, X., Song, K.: Study on the impact of training with large intensity on the immune functions in aerobic athletes. Journal of Guangzhou Sports 26(2), 97–100 (2006)
23. Jin, H.: The progress status of sports immunology. Chinese Sports Medicine Magazine 19, 18–20 (2000)
24. Brahmi, Z., Thomas, J.E., Park, M., et al.: The effect of acute exercise on natural killer cell activity of trained and sedentary human subjects. Clin. Immunol. 5(5), 320–329 (1985)
25. Hall, T.J., Brostoff, J.: Inhabition of human natural killer cell activity by prostaglandin D2. Immunol. Lett. 7(3), 140–144 (1983)
26. Xu, Y., Lin, J.: The prospect of exercise on mechanism of immune system and research. Journal of PLA Institute of Physical Education 22(2), 77–79 (2003)
27. Wang, Y.: Sports, immune function and Viral infection. Zhejiang Sports Science 19(4), 39–41 (1997)

The Influence of Sports about Internet Addiction on Teenagers' Health in Different Dimensions

Wei Zhang

Department of Physical Education, Jiangxi University of Traditional Chinese Medicine, Nanchang, Jiangxi Province, China
jxzw666@163.com

Abstract. The paper explained hazard from internet addiction for teenagers in four different dimensions: physical health, mental health, social adaptation health and moral health respectively, and put forward the idea that sports played an positive role in perfecting health of teenagers with internet addiction. This provided a thread to treat "internet addiction" of teenagers.

Keywords: Sports, Internet addiction, Health.

1 Teenagers Addiction to Internet

Since the concept of "Internet addiction" has been proposed by Psychiatrist Goldberg of New York from 1990s,domestic and international educators have done a lot of theory and practice researchs, most of the objects are colledge students. However, teenagers who are in the stage of physical and psychological have become a high-risk groups. The China teenager Association for Network Announcement of the "China Youth Internet Addiction Data Report " in 2005,Up to 13.2% of teenagers have Internet Addiction, and Up to 13% of teenagers have Internet Addiction tendency [1]. In recent years, teenager Addiction to Internet has become a hot research in this world. Cheng Feng[2] reviewed the literature and research, they think Internet Addiction is a chronic, recurring fascination state because of excessive use of the Internet by teenagers, it can produce an irresistible desire to re-use In the psychological and behavioral, it also can produce the desired increase in use of time, increased tolerance, withdrawal symptoms appear and so on, so it can produce long-term dependency on the Internet .

At present, teenagers addiction to internet intervention on the research involves psychotherapy, medication and exercise intervention and other aspects. Investigation of exercise intervention studies focus on youth sports before and after intervention comparison analysis of Internet Addiction Disorder[2-4], while the research about sports IAD different dimensions of the health effects is low levels. however, this study on adolescent health problems of Internet Addiction, it is important for adolescents who are In the critical period of development.

H. Tan (Ed.): Knowledge Discovery and Data Mining, AISC 135, pp. 739–744.
springerlink.com

2 The Influence of Internet Addiction on Teenagers' Health in Different Dimensions

According to World Health Organization, the revised definition of health Should include physical health (physical health), mental health, social adaptation and moral health in 1947 and 1989[5].This article also start from the four aspects, discuss the effect of Internet Addiction on teenagers' health.

2.1 The Influence of Internet Addiction on Teenagers' Physical Health

Adolescence is a critical stage of human growth and development, the physiological function and anatomical structure is becoming more mature, but not yet reach adult levels. Compared with adults, mainly having the following features:(1)Motor system, teenagers' bone is more resilient and less rigid, the function of muscle contraction is weaker、poor endurance、easy fatigue;(2) Cardiovascular system, when teenagers' are in adolescence, their hearts have reached adult levels, but vascular development is still in a backward state;(3) Respiratory system, teenagers' thorax is smaller, respiratory muscle strength is weaker, so the lung capacity is smaller, respiratory frequency is quicker;(4) Nervous system, because teenagers' neural activity is instability, excitatory process is dominant, while excitement and inhibition spread easily in the cortex. So they are concentreless and restless.

However, teenagers engage in online activities is totally different from physiology and anatomy development requirements. They keep the same kind of action for a long time, belong to static activities (isometric contraction),lack of stimulation of the vertical movement and dynamic muscle activity, bone and muscle can not grow effectively, and long-term wrong posture can lead to deformation ,curved spine and muscle and joint disorders when teenagers' bone is forming. Long-term static activities can lead to venous blood decreasing and blood pressure increasing, it is not conducive to the development of cardiovascular system, and can cause various cardiovascular diseases. In the long process of accessing to the Internet for teenagers, they always breathe in the form of inspiratory muscle contraction and relaxation, lack of effective stimulation of expiratory muscle, it is not conducive to the balanced development of the respiratory muscles. In addition, without limiting smoking and poor air circulation environment can lead to pathologic changes in respiratory organs. In addition to above effects, teenagers' Circadian rhythm are disorder, dietary system is unreasonable, and overuse their eyes, these problems can lead to nervous system, digestive system, endocrine system, sensory system and the immune system disfunction, and affect the growth and development of these systems, even can evolve into organic disease.

2.2 The Influence of Internet Addiction on Teenagers' Mental Health

Compared with the physical health dimension, influence of Internet addiction in adolescent mental health dimension is more complex, many studies show that Internet addiction teenagers' psychological change characteristics is not identical. Foreign scholars found that[6], teenagers excessive use of network can increase their loneliness and depression,and reduce the psychological well-being. Comprehensived various research, emotionally, Internet addiction teenagers performance for easy impetuous, easy to enrage, often feel depressed, like solitude, and emotionally unstable; In cognitive, Many Internet addiction teenagers think themselves physically addicted to network, but at presentis difficult to control, and always like to delay the time. Generally speaking, personality characteristics of Internet addiction teenagers often present with self-abased, autism, cowardly, sensitive, alert, tend to abstract thinking and so on.

2.3 The Influence of Internet Addiction on Teenagers' Social Adaptation Health

Researching the social adaptation of the adolescents is mainly for their ability to adapt to the environment, the capability of changing their roles, the ability of cooperation and competition, the ability of self-regulation and facing frustration and the ability of cultural identity and so on. Observing and analyzing whether they can realize personal ideals and values and then make a creative contribution to society through maintaining a good interactive relationship with the society. For the Young people who have an Addiction to the Internet, they have a Long-term addiction to the world of the internet that is virtualization and procedures and many features are different even absolutely opposite form the real society. The better ability of mimicry in this age cause that they bring the wrong ways of exchanges and communions from the internet to the real life. Resulting that they can not properly deal with the relationship among their family, friends and classmates and they can not adapt to social life very well. And then they will separate themselves from groups gradually and seal up themselves and can not complete a good interaction with the community. Research has shown that[7] social alienation and dependence on the network had a significant positive correlation. The lack of composition of uniformity (the anonymous identity on the internet) and the lack of the experience of social interaction for face to face (they only have the social experience of communication on the internet) have become the two main aspects which have impacts on the health of social adaptation of young people.

2.4 The Influence of Internet Addiction on Teenagers' Moral Health

Moral health is considered to be the highest level in health dimensions. It is the development and sublimation of the mental health and social adaptation which is a psychology and behavior mode that is established on requirements of the ethics that the individual is based on social. For the Young people who have an Addiction to the Internet, Their lack of mental health and social adaptation may affect their moral health. The unhealthy state of psychology and social adaptation which are formed in the long online world will cause them to form an incorrect moral standard which may lead them to do some behavior and activities what they think are moral but are not in line with ethics actually. What is serious is that the assimilation and moral identity of part of the role in online games have become important factors that lead the youth to commit a crime.

3 Influence That Sports Have on Different Dimensions of Health Level of Internet Addiction Teenagers

Sports activities have an effect on internet addiction teenagers' health level mainly through two aspects.Firstly, physical activities can somehow cure internet addiction. As reported, for internet addiction, physical activities have become a therapy which had been widely used in many countries all over the world. Studies from home and abroad have convinced that physical activities had an obviously intervention effects on internet addiction teenagers.As a kind of alternative behaviour, physical activities are positive and non-material for Internet addiction,what's more, they brings the body peak experience by different forms, intensity and time, satisfy the internet addicts' psychological needs so as to control and reduce the influence they have on different dimensions of health level of internet addiction teenagers.Secondly, sports can promote internet addiction teenagers' health recovery because of its functions, such as improving psychological status, enhancing the social adaptability and promoting moral function.

3.1 Influences That Sports Have on Internet Addiction Teenagers' Physiological Health Dimensions

As functional treatment to internet addiction teenagers, physical activities on one hand reduce damages to health by means of decreasing teenagers' online frequency and time,cutting down the opportunities to stay statically which is easier tired for a long time.On the other hand, sports improve physiological health level by promoting the recovery and the enhancement of the body system's function of internet addiction teenagers. Mainly manifested in the following aspects: (1) Reasonable sports promote the growth of the adolescent's height (bone) and muscle, they can improve the bones' blood supply, make the tubular bones longer, the bone diameter and the muscle fiber thicker, muscle force more powerful. (2) By sustained physiological simulation to the circulation system of the teenagers, the physical exercise especially the long-time aerobic exercise can give rise to good adaptable changes such as making the contractive force of the myocardium strengthen, the cardiac output increase, heart capacity enhance, sinus cardiac eddy. (3) Physical exercise in adolescents' respiratory muscle strength and lung capacity. Additionally, reasonable exercises make a huge difference in recovering and promotong the nervous system, endocrine system, immune system and the digestive system of the internet addiction teenagers.

3.2 Influences That Sports Activities Have on the Mental Health Dimensionality of the Internet Addition Teenagers

In many studies, the change of the internet addition teenagers' health index is used as an important evaluation criterion to evaluate the intervention effect of the sports activities. Synthesize lots of research result, sports activities serve as a therapeutic tool for internet addition. Emotionally, reasonable sports activities intervention may make internet addition teenagers improve from easily irritable, infuriate, often feel being down in spirits, like to stay alone and emotional instability state to the emotion is more gentle and stable, the initiative is stronger, the sate that can procedurally communicate with parents. Cognitively, sports activities may promote internet addition teenagers hope themselves to take part in some term of sports activities, hope to gradually decrease their on-line time by sports activities. General speaking, after taking part in sports activities, the harmful personality characteristics peculiar to internet addition teenagers can be improved in a certain extent. Besides, according to the theory that physiology heath promotes mental health, reasonable sports may enhance internet addition teenagers' physiological function, boost their physical fitness. And that 'a health mentality in a health body', internet addition teenagers increasingly improve in health, gradually enhance in sports level, at the same time will also remit and eliminate mental health problem to a certain extent.

3.3 Influences That Sports Activities Have on the Social Adaptation Healthy Dimensionality of the Internet Addition Teenagers

Sports activities especially the team event that pay attention to group consciousness and spirit of cooperation, ask for the participants of the sports jointly achieve the match goal with the help of collective strength on the premise making the best of individual capacity. In addition, sports require fair competition, and the participants of the sports who can abide by the rules of sports activities frequently can consciously comply with social criterions. So when sports activities bring the internet addition teenagers from the cyber world, which is virtual, programmed and different from real society even that many features are completely opposite, into sports addition activities possessing social adaptive function, internet addition teenagers' environmental adaptation, social interaction, cooperation and competition ,self-control and withstanding frustration ability (by experiencing realization of fighting spirits in sports activities)are bound to greatly improved. However, the experiment research at this aspect is lack at present.

3.4 Influences That Sports Activities Have on the Moral Healthy Dimensionality of the Internet Addition Teenagers

Moral health is develop and sublimation of mental health and social adaptation, while moral health level is closely linked to mental health and social adaptation health. Sports activities, by means of improving the internet addition teenagers' mental health status and social adaptation condition, help them establish correct moral standards, understand correct code of ethics and know consciously abide by moral demand of the society. Nevertheless, that how to promote the internet addition teenagers' health level by reasonable sports activities is awaited for further study. To sum up, internet addition may affect four dimensionality including teenagers' physiology health, mental health, social adaption health and moral health level, while sports activities have a well improving effect on the internet addition teenagers' state of health. At the present state, to direct at individual health condition of the internet addition teenagers, that how to effectively design sports prescription including mode of sports, strength of sports, time and frequency of sports will be the emphasis that we will study next step.

References

1. China Youth Association for Network Development. Data Analysis for 2005 Statistical Report on Internet Addictions of Chinese Adolescents. China youth Daily, November 23 (2005)
2. Chen, F.: Discuss the sports missing of the Internet addiction to youth. Wuhan Institute of Physical Education, Wuhan (2009)
3. Liu, Y., Dan, J., Su, L.: Action Research on Sports Intervention of Internet Addiction Youth. Sports and Science 31(4), 9–13, 23 (2010)
4. Zhang, P.: Probe into the effect of physical exercising on teenagers' Guidance Mechanism for bad network behavior. The World of Sports. Academic Edition (8), 113–114 (2010)
5. Zeng, Z.: Interpretation and Evolution of the concept of health. Journal of Beijing Sport University 30(5), 618–619, 622 (2007)
6. Kraut, R.E., Patterson, M., Lundmark, V.: Internet paradox: A social technology that reduces social involvement and psychological well-being? American Psychologist 53(9), 1017–1032 (1998)
7. Cui, L., Liu, L.: The influence about Internet on college students' Social Development. Psychological Science 26(1), 64–66 (2003)

A Unified Frame of Swarm Intelligence Optimization Algorithm

Chen Jia-zhao, Zhang Yu-xiang, and Luo Yin-sheng

Xi'an Research Inst. of Hi-Tech,
Xi'an, Shaanxi Provice, P.R.C.
Jazhch@sina.com, yuxiangz@tom.com, jacobchn@hotmail.com

Abstract. Swarm intelligence optimization algorithm is a heuristic search algorithm based on the swarm intelligent behavior of biology, which shows excellent performance in deal with complex optimization problems. On the basis of analyzing particle swarm optimization algorithm, ant colony algorithm and artificial bee colony algorithm, the paper presents a unified frame of swarm intelligence optimization algorithm that is helpful for improving and perfecting swarm intelligence optimization algorithm.

Keywords: swarm intelligence optimization algorithm, particle swarm optimization algorithm, ant colony algorithm, artificial bee colony algorithm, unified frame.

1 Introduction

Swarm intelligence optimization algorithm stemmed from simulation of the evolution and swarm food-seeking living colony in nature. By studying living colony, people found that social living, such as ant, bird, fish and etc., shows unparalleled excellence in swarm than in single in food seeking or nest building, i.e., they shows great intelligence through some mechanism in swarm. Drawing inspiration from that, people design many optimization algorithm simulating colony living, such as ant colony algorithm[3], particle swarm optimization algorithm[1], shuffled frog-leaping algorithm[4], artificial bee colony algorithm[2] and etc., which are successfully applied in many fields.

Different kinds of swarm intelligence optimization algorithms have different steps and processes, but their substances are quite similar. To study their common regularities and unified expression is helpful in improving and optimizing the algorithms, and helps to design new algorithms.

2 Swarm Intelligence Optimization Algorithm

2.1 Particle Swarm Optimization Algorithm

Particle swarm optimization (PSO in short) algorithm, stemmed from simulation of the food-seeking behavior of bird flock, was presented by Eberhart and Kennedy in 1995 [1].

H. Tan (Ed.): Knowledge Discovery and Data Mining, AISC 135, pp. 745–751.
springerlink.com

In PSO algorithm, the solutions of a problem are supposed to a swarm of particles, and every particle has a fitness value determined by a function to be optimized and a speed to determine its flying direction and distance. The particles follow the current optimal particle to search in the solution space. PSO initializes a flock of random particles (i.e. random solution), and searches the optimal solution by iteration updating. In every iteration, a particle updates itself by tracking two extrema: one is the current optimal position found by the particle itself which is called individual extremum (pbest for short), and the other is the optimal position found by the whole flock which is called global extremum (gbest for short). At last, the optimal solution of the problem is found after many time's iteration. The velocity and position are updated by the following equations at iteration t:

$$v_{id}(t+1) = wv_{id}(t) + c_1 r_1 [p_{id} - x_{id}(t)] + c_2 r_2 [p_{gd} - x_{id}(t)] \tag{1}$$

$$x_{id}(t+1) = x_{id}(t) + v_{id}(t+1) \tag{2}$$

where, w is inertial factor; c_1 , c_2 are positive accelerating constants; r_1 , r_2 are uniformly distributed values from 0 to 1; $v_{id}(t)$, $x_{id}(t)$ are the components of dimension d of velocity and position of particle i when iterating the time t; p_{id} is the component of dimension d of pbest found by particle i ; p_{gd} is the component of dimension d of gbest found by the whole flock. Particle velocity components of every dimension are constrained in (- v_{max} , v_{max})(v_{max} >0).

The pseudo codes are:

Begin

Randomly initialize particle flock P;

For calculate the fitness $f(i)$ of each particle $i \in P$;

Compare fitness value of every particle with the individual extremum p_{id} , and replace p_{id} if the fitness value is better;

Compare fitness value of every particle with the global extremum p_{gd} , and replace p_{gd} if the fitness value is better;

Update the velocity and position of the particles by equation (1) and (2);

Terminate iteration if the terminating condition meets;

End for

End

Because PSO is simple in principle and easy to realize, it is widely applied in many fields such as production management, multi-object optimization, combinatorial optimization, and so on.

2.2 Ant Colony Algorithm

Ant Colony Algorithm, (ACA for short), presented by Dorigo.M in 1996[3], is a simulation of ant's swarm intelligent behavior in food seeking, where ants transmit messages to each other by releasing pheromone and can always find a shortest routing path form their nest to food source by swarm cooperation.

ACA was first presented combining with TSP problem, i.e., a salesman wants to find a shortest path starting from the present city, through all n cities to get, and going back the starting city at last, with each city passed through only once. ACA initializes m ants randomly locating in n cities, and ant k decides its next city to get (exclusive of reached cities) by the pheromone concentration on the traveling path and the transition probability determined by equation (3).

$$p_{ij}^{k}(t)=\begin{cases}\dfrac{\tau_{ij}^{\alpha}(t)\eta_{ij}^{\beta}(t)}{\sum\limits_{s\in allowed_k}\tau_{js}^{\alpha}(t)\eta_{js}^{\beta}(t)}, & j\in allowed_k \\ 0, & otherwise\end{cases} \tag{3}$$

Where, $p_{ij}^{k}(t)$ is ant k 's transition probability from city i to city j at moment t ; $\tau_{ij}(t)$ is the pheromone concentration on path (i , j) at moment t ; $\eta_{ij}(t)$ is heuristic function which may be a distance or fee between two cities, and generally $\eta_{ij}(t)=1/d_{ij}$, d_{ij} is the distance between city i and j ; α is message heuristic factor, and β is expectant heuristic factor; $allowed_k$ is the allowed next choice cities (exclusive of reached cities).

Pheromone concentration on the path is updated by equation (4).

$$\tau_{ij}(t+n)=(1-\rho)\tau_{ij}(t)+\sum_{k=1}^{m}\Delta\tau_{ij}^{k}(t) \tag{4}$$

Where, $\tau_{ij}(t+n)$ is the pheromone concentration on path (i , j) after $t+n$ moments; $\rho\in(0,1)$ is volatilizing factor of pheromone; $\Delta\tau_{ij}^{k}(t)$ is the pheromone left to path (i , j) by ant k in current cycle.

The pseudo codes of the algorithm are:

Begin

Initialize parameters with equal original pheromone on every path;

For

For choose next city for every ant by equation (3) and transfer to the next city;

If all cities are passed through;

End for

Update pheromone on every path by equation (4);

If all ants choose the same city or the maximum cycle index is got;

End for

End

ACA, a kind of algorithm with information positive feedback mechanism, is successfully applied in TSP, assignment problem, job shop scheduling, network route design.

2.3 Artificial Bee Colony Algorithm

Artificial bee colony algorithm (ABCA for short), presented by Karaboga.D in 2005 [2], is simulation of bee's intelligence showed in honey.

The algorithm designs three kinds of bees: employed bees, onlookers and scouts. Employed bees go out to honey, and a honey source corresponds only an employed bee; onlookers select honey source according to the honey source information brought back by the employed bees, and the quantity of onlookers is equal to that of employed bees, i.e., the two are equal to the quantity of honey sources; scouts search honey source randomly, but the quantity is very less, generally only one. In the algorithm, the position of a honey source represents a feasible solution of the problem to be optimized, and the honey amount of a honey source represents the fitness value of the problem. The bees look for honey sources with larger amount of honey by continuously searching, and at last find the source with largest amount of honey, i.e., the solution of the problem to be optimized.

For a d dimension problem to be optimized, the algorithm initializes randomly an original swarm with m solutions at first (m is equal to the quantity of employed bees and onlookers). Each solution x_i ($i = 1,2,\ldots,m$) is a vector with d dimensions. And then, employed bee creates a new honey source around the honey source in its memory:

$$v_{ij} = x_{ij} + \Phi_{ij}(x_{ij} - x_{kj}) \tag{5}$$

Where, k is selected randomly from $\{1,2,\ldots, m\}$, and $k \neq i$, $j \in \{1,2,\ldots,d\}$, $\Phi_{ij} \in [-1,1]$. If the amount of honey of new honey source is no less than that of the old source, the employed bee accepts the new one and gives up the old one. Otherwise, the employed bee keeps to honey in the old source. That is to make a choice between the new and old honey sources by greedy selection mechanism.

After every employed bee completes searching, they return nest to share honey source information with onlookers. According to the honey source information provided by the employed bees, onlookers select honey source by roulette:

$$p_i = \frac{fit_i}{\sum_{j=1}^{m} fit_j} \tag{6}$$

In the algorithm, if the honey amount of a honey source is found to keep at a low level after set iteration index "limit", the source is assumed having too less honey. Give up the source and replace it with the new honey source searched by the scout. If the given-up honey source is x_i and $j \in \{1,2,\ldots,d\}$, the replacement is as follows:

$$x_i^j = x_{\min}^j + rand()(x_{\max}^j - x_{\min}^j) \tag{7}$$

In this way, the algorithm realize a balance between global exploration and local exploitation by using employed bees and onlookers to carry out local search and onlookers and scouts to carry out global search.

The executive steps of the algorithm are:

Begin

Initialize a population randomly, and calculate its fitness value;

For

The employed bees create new positions by equation (5) and calculate the fitness values;

Select a position by greedy selection mechanism;

Calculate probability p_{ij} according to equation (6) so as to select x_{ij};

Create a new position according to the position selected by onlookers and calculate its fitness value;

Select a position by greedy selection mechanism;

Check if there is a honey source given up by bees. If it is, replace it with a new position created by scout according to equation (7);

Remember the best solution so far;

If the maximum iteration index meets;

End for

End

ABCA ensures its global optimization ability by introducing neighborhood source production mechanism with mutation operation function. It is applied to such fields as numerical optimization, artificial neural network training, optimization of sales network building, and so on.

3 A Unified Frame of Swarm Intelligent Optimization Algorithm

After a careful analysis, it can be found that the above swarm intelligent algorithms have certain of similarity in structure: initialize a population at first, and then update according to iteration equation, and at last terminate the algorithm when the termination condition meets. Although only PSO, ACA and ABCA are analyzed, the other swarm intelligent algorithms have the same conclusion.

In this paper, the swarm intelligent optimization algorithms are described as a unified frame:

Begin

Initialize population;

REPEAT

Update population information;

Exchange population information;

Population information compete;

UNTIL iteration termination

End

This paper thinks that all swarm intelligent algorithms are a course of information processing. The individuals are given some kind of information in population initialization, and the iteration is carried out by three operations, i.e., population information updating, population information exchange, and population information competition.

"Population information updating" means that individuals update their information to improve their environment adaptability (i.e. optimization ability) by sensing environment information. "Population information exchange" means that the whole population exchange and share inner information with each other during optimization.

"Population information competition" means that individuals select high quality information (near the optimal solution of the problem) and give up poor quality information (far away from the optimal solution of the problem) by comparing information. The order of three operations is not necessarily the case, but it is not difficult to find that almost all algorithms have the three operations. For example, PSO carries out information updating by updating velocity and position, and carries out information exchange by introducing individual extremum and global extremum when updating velocity and position, and carries out information competition by comparing every individual with the individual extremum and global extremum to ensure that the global extremum keeps the optimal information of the population from beginning to end.

4 Summary

Recent years, research workers continuously design new swarm intelligence algorithms to solve optimization problems. For example, according to fish shoals's food seeking behavior and living activity, Li Xiaolei and other people designed fish swarm algorithm[5]; according to rats's food seeking behavior, Liu Xuxun and other people designed rat swarm algorithm[6].There are so many new algorithms created, because there isn't a universal algorithm according to NFL theorem, and it needs new algorithms to solve a certain kind of problems. The unified frame can be a guide for new algorithm design.

References

1. Kennedy, J., Eberhart, R.: Particle Swarm Optimization. In: Proc. of International Conference on Neural Networks, pp. 1942–1948 (1995)
2. Karaboga, D.: An Idea Based On Honey Bee Swarm For Numerical optimization. Technical Report-TR06. Erciyes University (2005)
3. Dorigo, M., Maniezzo, V., Colorni, A.: The Ant System: Optimization by a colony of cooperating agents. IEEE Transactions on Systems, Man, and Cybernetics-Part B 26(1), 29–41 (1996)
4. Eusuff, M.M., Lansey, K.E.: Optimization of Water Distribution Network Design Using the Shuffled Frog Leaping Algorithm. Journal of Water Resources Planning and Management, ASCE 129, 3:210–3:225 (2003)
5. Li, X.-L., Shao, Z.-J., Qian, J.-X.: An Optimizing Method Based on Autonomous Animats: Fish swarm Algorithm. System Engineering Theory and Practice 11, 32–38 (2002)
6. Liu, X.-X., Cao, Y., Chen, X.-W.: Rat Swarm Algorithm Based on Moving Robot. Control and Decision 9, 1060–1064 (2008)

The Innate Problems and Suggestions on Inquiry Learning under Network Environment

Yangli Zhang[1] and Qiong Wu[2]

[1] College of Computer Science, Sichuan Normal University, Chengdu, Sichuan Province, China
[2] College of Electrical & Information Engineering, Southwest University for Nationalities, Chengdu, Sichuan Province, China
ylzhang@sicnu.edu.cn, wuqiong028@gmail.com

Abstract. Traditional teaching methods are faced with enormous challenges because of the rapid development of information technology. Compared with traditional teaching methods, inquiry learning under the network environment has big greater advantages. At the same time, it also has some innate problems, such as learning efficiency, learning conditions, the depth of learning and so on. Based on an analysis of these advantages and questions, some advices are given for inquiry learning.

Keywords: Network Environment, Inquiry Learning, Learning Environment, Learning Resources.

1 Challenges to Traditional Teaching Methods

Traditional teaching method, with its well-developed theoretical system, efficient dissemination and easy operation, is essentially a kind of reception learning that tends to be adopted by teachers. However, when facing the impact of information society and the challenges of knowledge economic, the interesting learning process is simplified by the traditional teaching style in which the students simply listen to the teachers' lectures bearing the hallmark of unidirectional and mechanical inculcating process [1]. It finally ends up as a single understanding activity lacking in cultivating students' creative mind and practical ability, renewing the students' ideas for learning, improving learning capacity, respecting students' diversity and personal development. The deficiencies are basically reflected in the following respects:

1.1 Teaching Objectives

The courses depend on cognitive approach most, while the presence of teaching objectives to achieve motor and mental skills is as weak as the policy section in cognitive approach. Thus, there exist deficiencies in cultivating students' long learning skills.

1.2 Course Contents

The teaching contents mostly come from teachers and textbooks relying on a single source of information, most of which are contents of cognitive approach. Thus, there exist deficiencies in strengthening students' information literacy.

H. Tan (Ed.): Knowledge Discovery and Data Mining, AISC 135, pp. 753–758.
springerlink.com

1.3 Patterns for Teaching and Learning

The classroom teaching limits the interaction among teachers and students. There exist deficiencies in cultivating students' team spirit and cooperative ability.

1.4 Curriculum Implementation

The students' initiatives are restricted by the teaching mode in which the students simply listen to teachers' lecture. There exist deficiencies in cultivating initiative spirit and innovation capacity.

1.5 Curriculum Evaluation

Teachers are usually the unique evaluators with the examinations acting as the main approach to evaluate. The evaluations is limited with importance attached to summative evaluations and relative evaluations, while at the same time it neglects the formative evaluations and absolute evaluations.

As the aspects mentioned above, even though the traditional teaching method has been contributing to human education, it is facing a great challenges with its deficiencies in cultivating students' practical and creative ability, integrative knowledge and ability, cooperative spirit and ability, information accomplishment and the ability to survive.

2 Advantages of the Inquiry Learning under Network Environment

Inquiry learning is a new learning style relative to reception learning. It aims at switching students' learning styles laying stress on probing the problems and the spirit of innovation, focuses on teaching students knowledge that will be useful for their whole life, and bears the hallmark of openness, comprehensiveness and practicality [2]. Inquiry learning is a rewarding attempt in order to make learners able to cope with changes quickly. In terms of teaching objectives, course contents, patterns for teaching and learning, curriculum implementation and curriculum evaluation, the inquiry learning, while acts as the complementation of traditional learning method, is quite different from the traditional one at the same time.

2.1 Teaching Objectives

The teaching objectives, apart from the requirement for the students to know well of the knowledge hierarchy of the course, lays stress on the cultivation of practical spirit, cooperative spirit and information literacy.

2.2 Course Contents

The source of the course content is plentiful as new knowledge can be generated through the students' exploring activities along with the inherent knowledge hierarchy. In other words, the process students engage in inquiry learning is a process of generating new knowledge in itself.

2.3 Patterns for Teaching and Learning

Learning in team collaboration, in which students can study, communicate and probe through interaction, is usually adopted to help to cultivate the students' cooperative spirits and ability.

2.4 Curriculum Implementation

Research project (problems solving) is the main line of the inquiry learning stressing that students should probe into their study initiatively under the supervision of teachers. Thus, students arouse and make full use of their initiative during the study, which help to cultivate the students' creativity and creative ability.

2.5 Curriculum Evaluation

Not only the teachers but also the students engage in the final evaluation. Besides examination, there are various approaches for the evaluation such as portfolio assessment and works exhibition. The evaluation concerns the mastery of the knowledge on text book as well as the students' learning attitudes and methods focusing on process evaluation and students' improvement of their own.

The development information technology such as computers, application of multimedia and communication network, has influences on human productivity, life and learning activity. Computer networking technology in particular has brought out a profound change of educational communications and technology. The computer network provides people with a huge amount of information resources, interactive intelligent resources, platform for communication crossing the space-time and open and free environment [3], presenting its incomparable advantages in the process of inquiry learning. It makes it possible for people to arrange their inquiry learning under network environment effectively with a new learning method and cognitive style.

3 Innate Problems from the Inquiry Learning under Network Environment

As previously described, there are deficiencies in curriculum objectives and contents, patterns for teaching and learning, curriculum implementation and evaluation existing in the traditional teaching method. It has been found that the inquiry learning under network environment covers the deficiencies of traditional teaching methods especially in arousing the initiatives, the desire for learning, and in cultivating team spirit, cooperative ability, creative spirit and innovation ability. However, there are innate problems from the inquiry learning under network environment. For instance, there are a series of problems concerning the setting of objectives, the implementation of the scheme, and evaluation strategies. Nevertheless, among those problems, the following innate ones call for criticism and profound consideration to inquiry learning under network environment.

3.1 The Learning Efficiency of Inquiry Learning under Network Environment

The inquiry learning under network environment covers the deficiencies of traditional teaching methods. Comparatively speaking, the inquiry learning is suitable for learning tacit knowledge and knowledge concerning the practical ability to solve problems which is not well-structured. Because they are ill-defined knowledge that can be gained through learners' personal experience and practice but can't be explained in words. The well-structured and explicit knowledge refers to those can be explained by concepts, propositions, formulas. The questions and knowledge concerning "what" and "how" are more suitable to be taught by the learning mode of transferring and receiving. So the learning of well-structured and explicit knowledge presents higher efficiency in the process of traditional learning method.

3.2 The Conditions Required for Inquiry Learning under Network Environment

Just like any other learning activities, the implementation of inquiry learning calls for certain learning conditions. Compared with the traditional learning methods, the inquiry learning under network environment has high requirements for learning conditions. It is basically reflected as the following aspects: The students studying under the network have greater motivation to acquire knowledge out of desires, curiosity, and interest. And that means the learners should have learning strategies and skills of higher level with stronger control abilities to their wills and more excellent IT skills. Besides, teachers are required to have wider knowledge and stronger organizing ability and coaching skills. Moreover, the study calls for more material resource.

3.3 The Depth of the Inquiry Learning under Network Environment

The inquiry learning under network environment is different from subject teaching whose exploring process is much more important than the result. Though personal experience, learners will have their access to the real society, understand the general process and approach of scientific research, manage to interact and cooperate with others, get to know the approach to get information apart from textbooks and make an attempt to solve the problems of the research project [4]. The main purpose of the inquiry learning is providing a platform for learners to have the experiences mentioned above. However, it may turn out that learners may find their research results small and one-sided with the superficial mastery of knowledge after a period of research. Emphasizing the process and neglecting may result in a mere formality of inquiry learning, and make the learner have no access to further study. How we can grasp the depth of the inquiry learning and deal with the relationship between the process and results is a problem to be solved when arranging inquiry learning under network environment.

4 Suggestions on Inquiry Learning under Network Environment

4.1 Choosing Appropriate Learning Content

As mentioned above, the inquiry learning is suitable for the learning of tacit knowledge and the knowledge which is not well-structured. The learning of well-structured and explicit knowledge presents higher efficiency in the process of traditional learning method. As it is, it's important to choose appropriate learning content for inquiry learning under network environment. The choosing of contents should adhere to the following guiding principles:

Firstly, there should be complexity in the learning contents emphasizing the overall mastery and multiple practical use of the knowledge.

Secondly, learning content should be a synthesis of cognitive learning, motor skill learning, and emotional learning.

Thirdly, learning content should be great in scope and depth related to the learners' knowledge background and existing cognitive structure.

4.2 Creating a Favourable Digital Environment for Inquiry Learning

Learning environment refers to the whole external conditions of inquiry learning under network environment which is the foundation and guarantee for learning activities. While the students start their initiative inquiry learning with the direction of teachers under network environment, the learning environment plays a more important role than it does in traditional teaching methods. The learning environment is of complexities and varieties with physical environment, immaterial conditions, network resource and resource from real world. Generally speaking, a favourable digital environment should be able to provide abundant information resource, great software acting as cognitive tools and platform for the communication crossing the space and time.

4.3 Combining the Inquiry Learning under Network Environment with the Traditional Teaching Method

Inquiry learning and traditional teaching method each has its merits and deficiencies. It is not proper and realistic to amplify the deficiencies of traditional teaching method. The reality and history tell us there is no perfect teaching method of almighty mode. Only when we choose the proper teaching style on account of practical experience, make full use of the advantages of both inquiry learning and traditional teaching, or combine the both perfectly adopting the good points with their weakness avoided, can we achieve a best teaching effect.

References

1. Zhong, Q., An, G.: Preliminary Study of Inquiry Learning Theory. Shanghai Education Publishing House, Shanghai (2003)
2. Zhang, L.: Inquiry Learning under Network Environment. Jilin Normal University Journal (Humanities & Social Science Edition) 4, 57–58 (2003)
3. Zhang, X., Yu, J.: Overview of Inquiry Learning under Network Environment. China Education Technology 3, 20–23 (2003)
4. Yan, Q.: Issue Guidance to Implement Strategies of Inquiry Learning, `http://www.chinaschool.org/sgzy/study/03-1118-05.htm`

Finding Maximum Noncrossing Subset of Nets Using Longest Increasing Subsequence

Xinguo Deng[1] and Rui Zhong[2]

[1] Center for Discrete Mathematics and Software College,
Fuzhou University CDM & SW, FZU Fuzhou, China
[2] National Engineering Research Center for Multimedia Software,
Wuhan University NERCMS, WHU Wuhan, China
xgdeng@fzu.edu.cn, zhongrui0824@126.com

Abstract. In the problem of maximum noncrossing subset of nets, the current algorithms of using either dynamic programming or the longest common subsequence have the complexity of O(n^2). In order to reduce the complexity of the existing algorithms, a more efficient algorithm of using longest increasing subsequence is introduced in this paper. The effectiveness of the algorithm with a time consuming complexity of O(nlogn) is illustrated through the theoretical analysis and by demonstrating the experiment results of corresponding C++ program.

Keywords: maximal noncrossing subset, longest increasing subsequence, binary searching, circuit wiring.

1 Introduction

A circuit consists of a set of modules and a set of nets. Each net specifies a subset of points, called terminals, on the boundary of the modules. The layout problem is to interconnect the modules as specified by the nets in terms of different technological design rules. Due to the complexity of the problem, VLSI layout design is typically decomposed into three phases: placement, global routing, and detailed routing. In the placement phase, circuit modules are geometrically positioned on a layout surface (chip). In the global routing phase, the routing region is partitioned into simple sub-regions, each called an elementary region, and global assignment of the wiring paths is determined for each net. In the detailed routing phase, detailed wirings of the individual routing regions are given.[1] The crossing distribution problem occurs before the detailed routing. It is observed that nets crossing each other are more difficult to route than those nets that do not cross. The layout of crossing nets must be realized in more than two layers, thus requiring a larger number of vias. [2]

Documents [3,4] studied the problem of maximum noncrossing subset of nets using dynamic programming [5,6,7] and longest common subsequence [8] respectively. However, the time complexity of the both algorithms is O(n^2). A more proficient algorithm using the longest increasing subsequence is put forward about the problem in this paper. The effectiveness of the algorithm whose time complexity is O(nlogn) is illuminated by the theoretical analysis and the experiment results of corresponding C++ program.

H. Tan (Ed.): Knowledge Discovery and Data Mining, AISC 135, pp. 759–767.
springerlink.com

The remainder of the paper is organized as follows. Section 2 introduces the problem of maximum noncrossing subset of nets in the circuit wiring. Section 3 illustrates the core thoughts of the algorithm using longest increasing subsequence and analyzes the time complexity of the algorithm. Section 4 is the C++ program implementation corresponding to the algorithm. Section 5 provides the experiment results of the C++ program corresponding to the algorithm. The last section concludes and gives the directions for future work in this field.

2 Maximum Noncrossing Subset of Nets

A routing channel with n pins on a side and a permutation C is given. Pin i on the top side of the channel is to be connected to pin C_i on the bottom side, $1 \leqslant i \leqslant n$. The pair (i, C_i) is called a **net**. In total, we have n nets that are to be connected or routed. Suppose that we have two or more routing layers, of which one is a *preferred layer*. For example, in the preferred layer it may be possible to use much thinner wires, or the resistance in the preferred layer may be considerably less than in other layers. Our task is to route as many nets as possible in the preferred layer. The remaining nets will be routed, at least partially, in the other layers. Since two nets can be routed in the same layer if they do not cross, our task is equivalent to finding a maximum noncrossing subset (MNS) of the nets. Such a subset has the property that no two nets of the subset cross. Since net (i, C_i) is completely specified by i, we may refer to this net as net i.[3]

Consider the example in Figure 1. The nets (1,8) and (2,7) (or equivalently, the nets 1 and 2) cross and so cannot be routed in the same layer. The nets (1,8), (7,9), and (9,10) do not cross and so can be routed in the same layer. These three nets do not constitute a MNS as there is a larger subset of noncrossing nets. The set of four nets {(4,2),(5,5),(7,9),(9,10)} is an MNS of the routing instance given in Figure 1.

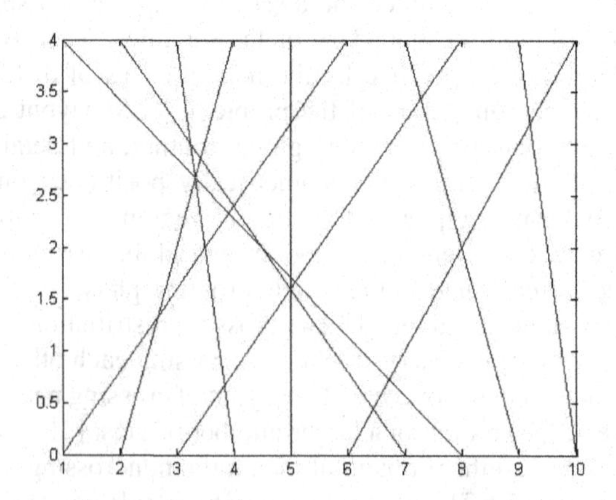

C=[8,7,4,2,5,1,9,3,10,6]

Fig 1. A wiring instance

Let $MNS(i,j)$ denote an MNS under the constraint that all pairs (u,C_u) in the MNS have $u \leq i$ and $C_u \leq j$. Let $size(i,j)$ be the size (i.e., number of nets) of $MNS(i,j)$. Note that $MNS(n,n)$ is an MNS for the input instance and that $size(n,n)$ is its size.

For the example in Figure 1, $MNS(10,10)$ is the answer we seek. From the result of running our C++ program, we can easily get the $size(10,10)$=4. Nets (1,8),(2,7),(7,9),(8,3),(9,10), and (10,6) cannot be the members of $MNS(7,6)$ because either their top pin number is greater than 7 or their bottom pin number is greater than 6 Therefore, we are left with four nets that are eligible for the membership in $MNS(7,6)$. The subset {(3,4),(5,5)} is a noncrossing subset of size two, and there is no noncrossing subset of size three. Hence, $size(7,6)$ =2.

3 Finding MNS Using LIS

3.1 Longest Increasing Subsequence Problem

The longest increasing subsequence problem is to find a subsequence of a given sequence in which the subsequence elements are in sorted order, lowest to highest, and in which the subsequence is as long as possible. This subsequence is not necessarily contiguous. Longest increasing subsequences are studied in the context of various disciplines related to mathematics, including algorithmics, random matrix theory, representation theory, and physics. The longest increasing subsequence problem is solvable in time O(n log n), where n denotes the length of the input sequence.[9]

For example, in the binary Van der Corput sequence 0, 8, 4, 12, 2, 10, 6, 14, 1, 9, 5, 13, 3, 11, 7, 15, a longest increasing subsequence is 0, 2, 6, 9, 13, 15. This subsequence has length six; the input sequence has no seven-member increasing subsequences. The longest increasing subsequence in this example is not unique: for instance, 0, 4, 6, 9, 11, 15 is another increasing subsequence of equal length in the same input sequence.

3.2 Finding MNS Using Longest Increasing Subsequence

Formally, the problem finding MNS is as follows:

Given nets $(1, C_1)$, $(2, C_2)$, … , (n, C_n), find the largest noncrossing subset for every $1 \leqslant i < j \leqslant n$, $C_i < C_j$.

From the above, the problem finding MNS is equivalent to finding the longest increasing subsequence of the corresponding pins on the bottom side as the positions of the pins on the top side increase.

The algorithm outlined below finds MNS with longest increasing subsequence efficiently, using only arrays and binary searching [10]. It processes the sequence elements in order, maintaining the longest increasing subsequence found so far. Denote the sequence values as C[1], C[2], etc. Then, after processing C[i], the algorithm will have stored values in two arrays:

lis[pos] — stores the position of the pin on the top side of the last value C[lis[pos]] such that pos ≤ i (note we have pos ≤ len ≤ i here) and there is an increasing subsequence of length len ending at C[lis[pos]].

pre[i] — stores the position of the predecessor of i in the longest increasing subsequence ending at C[lis[pos]].

In the algorithm, variables “len” and “pos” represent respectively the length and current position of the longest increasing subsequence found so far. Both of them are initialized with zeros.

Note that, at any point in the algorithm, the sequence C[lis[0]], C[lis[1]], ..., C[lis[pos]] is nondecreasing. For example, if there is an increasing subsequence of length “len” ending at C[lis[pos]], then there is also a subsequence of length “len-1” ending at a smaller value: namely the one ending at pre[lis[pos]]. Thus, we may do binary searches in this sequence in logarithmic time.

The algorithm, then, proceeds as follows.

```
input circuit wiring;
len = 0;
for i = 1, 2, ... n
{
binary search for the largest nonnegative pos ≤ len such that C[lis[pos]] < C[i] (or
pos = 0 if no such value exists);
    lis[pos]=i;
    if(pos==len) len=pos+1;
    pre[i]=lis[pos-1];
}
return len;
output MNS {(1, C1), (2, C2), … , (n, Cn)}.
```

The result of this is the length of the longest sequence in “len”. The actual longest sequence can be found by backtracking through the pre array: the last item of the longest sequence is in C[lis[len-1]], the second-to-last item is in C[pre[lis[len-1]]], and so on. Thus, the sequence has the form ..., C[pre[pre[lis[len-1]]]], C[pre[lis[len-1]]], C[lis[len-1]].

Because the algorithm performs a single binary search per sequence element, its total time is O(n log n).

4 C++ Implementation

4.1 Design

The methodology of top-down modular is adopted to design the program. Therefore, we considered the structure of the C++ program with three big modulars: 1. inputting the circuit wiring, 2. the central modular to find the MNS using longest increasing subsequence, 3. outputting the result.

The second modular computes the length of the MNS and finds the MNS using the longest increasing subsequence. This modular is further divided into three sub modules: LIS module, lower bound module and Compare module. The last modular outputs the result MNS processed in the second modular. A sixth module “Welcome” that displays the function of the program is also desirable. While this module is not directly related to the problem at hand, the use of such a modular enhances the user-friendliness of the program.

4.2 Program Plan

In last section, we have already pointed out the need for six program modules. A root (or main) module invokes four modules in the following sequence: Welcome module, InputCircuitWiring module, LIS module and Output module. LIS module invokes lower_bound module and the latter invokes Compare module.

A C++ program is designed by following the modular structure in Figure 2. Each program module is coded as a function. The root module is coded as the function "main"; "Welcome", "InputCircuitWiring", "LIS", "lower_bound", "Compare", and "OutputMNS" modules are implemented through different functions.

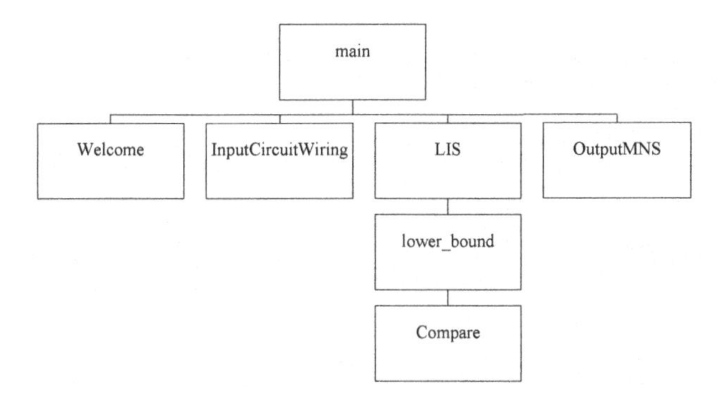

Fig. 2. Modular structure

4.3 Program Development

Function "Welcome" explains the function of the whole C++ program. Function "InputCircuitWiring" informs the user that the input is expected as a permutation of pins on the bottom of the channel corresponding to the pins on the top. The size of the group of the pins is determined first, so the total amount of pins is needed before an input begins. In our program, the input process is implemented by importing the input data from a text file called "input.txt". The result is outputted to the text file "output.txt".

The following Figure 3 details the struct of the circuit wire and core functions invoked by main function. Struct *Wire* consists of top side *top* and bottom side *bottom*. Bottom side is actually the wire permutation (i.e., C_i corresponding to i). The function "Compare" is invoked by binary search function "lower_bound". The function "lower_bound" helps to find the longest increasing subsequence, i.e., MNS.

The most important function "LIS()" finds the longest increasing subsequence using binary search. The function "lower_bound(lis, lis+len, wire[i].bottom, Compare)" is a type of "binary_search()". This function searches for the first place that *wire[i].bottom* can be inserted into and it is the ordered range defined by *lis* and *lis+len*. The return value of *lower_bound()* is an iterator that points to the location where *wire[i].bottom* can be safely inserted. Unless the comparison function *Compare* is specified, the < operator is used for ordering. Note that array *lis* records the position of the pin on the top side of current wire. The lenth of *lis[]* increases if *pos* is at the end. The predecessor of current wire is recorded. When the cycle terminates, the longest increasing subsequence (i.e., MNS) is stored in the array *lis,* and the length is computed.

Function "OutputMNS" records the inverse longest increasing subsequence via retracing from the last noncrossing wire to the first one, and then outputs MNS sequentially.

The details of other functions "Welcome", "InputCircuitWiring" and "main" are omitted here. The effects of these functions will be illustrated in the next section.

```
struct Wire {
    int top,bottom; //top side & bottom side
}wire[MAXN];

//compare function of lower_bound()
bool Compare(const int& x,const int& key)
{
    return wire[x].bottom<key;
}

//find a maximum noncrossing subset using longest increase subsequence
int LIS()
{
    int len=0,i;
    //wire[pos].botton increases monotonously
    for(i=1 ;i<=n ;i++)
    {
        //iterator lower_bound( iterator first, iterator last, const TYPE& val, CompFn f );
        int pos=lower_bound(lis,lis+len,wire[i].bottom,Compare)-lis;
        //current wire i
        lis[pos]=i;
        //lenth of lis[] increases if pos is at the end
        if(pos==len) len=pos+1;
        //predecessor of i
        pre[i]=lis[pos-1];
    }
    return len;
}

//output a maximum noncrossing subset
void OutputMNS(int len)
{
    printf("%d\n",len);
    int top=lis[len-1];//last noncrossing wire
    int cnt=0,i;
    while(len--)
    {
        //<top,bottom>
        answer[cnt++]=mp(top,wire[top].bottom);//retracing
        top=pre[top];
    }
    //output a maximum noncrossing subset
    for(i=cnt-1 ;i>=0 ;i--)
        printf("%d %d\n",answer[i].first,answer[i].second);//forward
}
```

Fig. 3. Struct and core functions

5 Experiment Results

5.1 Process of Progam Running

Table 1 shows the dynamic changes of variables and arrays in the process of program running for the example in Figure 1. Variable *i* represents repeating times. Variable *pos* returns the position of the current wire in the longest increasing subsequence using binary search algorithm. Array element *lis[pos]* records the position of the pin on the top side of the present wire. Variable *len* denotes the length of the longest increasing subsequence found so far. Array element *pre[i]* is the predecessor of *i* in the longest increasing subsequence ending at C[lis[pos]]. The longest increasing subsequence can be retraced via the array *pre[ ]*. Note that the algorithm can only find a longest increasing subsequence even if there are more than one solutions. The reason is that dynamical overwriting occurs in the process of binary search.

Table 1. Process of Program Running

i	1	2	3	**4**	**5**	6	**7**	8	**9**	10
pos	0	0	0	**0**	**1**	0	**2**	0	**3**	1
lis[pos]	1	2	3	**4**	**5**	6	**7**	8	**9**	10
len	1	1	1	**1**	**2**	2	**3**	3	**4**	4
pre[i]	0	0	0	**0**	**4**	0	**5**	0	**7**	8

5.2 Final Output

In Figure 4 our program outputs length and a maximum noncrossing subset of nets for the example in Figure 1. The length of MNS is 4. {(4,2),(5,5),(7,9),(9,10)} is a MNS of

```
//////////////////////////////////////////////////////////////////////////
 This program finds a maximum noncrossing subset
 of nets using longest increasing subsequence.
//////////////////////////////////////////////////////////////////////////
Input the size of wires:
Input the array of pins on the bottom side:
The length is 4.
A maximum nonccrossings subset of nets is :
top bottom
4  2
5  5
7  9
9  10
```

Fig. 4. Output of a maximum noncrossing subset

nets{(1,8),(2,7),(3,4),(4.2),(5,5),(6,1),(7,9),(8,3),(9,10),(10,6)}. Actually, {(3,4),(5,5),(7,9),(9,10)} is another MNS of nets in Figure 1. How to find all maximum noncrossing subsets of nets efficiently is our next research task.

6 Conclusion

In this paper we have studied the problem of maximum noncrossing subset of nets in the circuit wiring. An optimal algorithm to output a maximum noncrossing subset of nets and to compute its length was presented. In order to reduce the complexity of these problems, this algorithm adopts the longest increasing subsequence. Furthermore, predecessor of circuit wire is applied in the algorithm to improve the efficiency.

The complexity of this algorithm is analyzed theoretically. The overall complexity of this enhanced algorithm is O(nlogn). In contrast to this, the complexity of the algorithms of using either dynamic programming or longest common subsequence is $O(n^2)$. Further, a C++ program is developed to test this enhanced algorithm, and the satisfactory experiment results of the C++ program are presented.

The problem of finding all maximum noncrossing subsets of nets is still difficult to solve. Further study is directed towards the development of fast algorithms for solving the problem, thus minimizing wiring congestion.

Acknowledgment. The work was supported by the Natural Science Foundation of Fujian Province under Grant 2009J05142 and the Natural Science Foundation of Fuzhou University under Grant 0220826788.

References

1. Wolf, W.: Modern VLSI Design: System-on-Chip Design, 3rd edn., pp. 510–522. Pearson Education, Inc. (2003)
2. Song, X., Wang, Y.: On the crossing distribution problem. ACM Transaction on Design Automation of Electronic Systems 4(1), 39–51 (1999)
3. Sahni, S.: Data Structures, Algorithms, and Applications in C++, 2nd edn., pp. 735–740. McGraw-Hill, Inc. (2004)
4. Wang, X.: Algorithm Design and Analysis, 3rd edn., pp. 74–76. Publishing House of Electronics Industry (2007)
5. Skiena, S.S.: The Algorithm Design Manual, 2nd edn., pp. 273–315. Springer, Inc. (2008)
6. Han, Z., Ray Liu, K.J.: Resource allocation for wireless networks: basics, techniques, and applications, 1st edn., pp. 167–170. Cambridge University Press (2008)
7. Pandey, H.M.: Design Analysis and Algorithms, 1st edn., pp. 258–286. University Science Press (2008)

8. Iliopoulos, C.S., Sohel Rahman, M.: A New Efficient Algorithm for Computing the Longest Common Subsequence. Theory of Computing Systems 45(2), 355–371 (2009)
9. Krusche, P., Tiskin, A.: Parallel Longest Increasing Subsequences in Scalable Time and Memory. In: Wyrzykowski, R., Dongarra, J., Karczewski, K., Wasniewski, J. (eds.) PPAM 2009. LNCS, vol. 6067, pp. 176–185. Springer, Heidelberg (2010)
10. Neapolitan, R., Naimipour, K.: Foundations of Algorithms, 4th edn., pp. 91–144. Jones and Bartlett Publishers, LLC (2009)

An Optimization Algorithm for Minimizing Weight of the Composite Beam

Peng-yu Jiang, Zhu-ming, and Lin Jin Xu

Wuxi First Scientific Institute, Wuxi, 214035, China

Abstract. The particle swarm optimization was applied to the optimum design of the composite beam.The total weight of the structure was taken as the objective function.A technique of applying PSO integrated with general finite element code was developed for the optimization.Optimization was also conducted using zero-order method included in ANSYS and a comparison was made between zero-order method and PSO.Results demonstrate that PSO can find the global optimal design with higher efficiency regardless of the initial designs.for zero-order method the optimum solution is worst than the result of the PSO optimum.

Keywords: particle swam optimization, zero-order method, stacking sequence design.

1 Introduction

Fiber reinforced plastic (FRP) composites with superior stiffness-to-weight or strength-to-weight ratios are becoming the structural materials in aerospace, automobile, shipbuilding and other industries. The study of composite materials is of great importance. The optimum design of composite structures which can reduce the structure's weight without compromising its performance,or improve the performance without increasing its weight, provides the engineers with a tool that is essential in finding the best design among countless alternatives. In practical engineering problems, optimization may deal with functions which are discontinuous or un-differentiable, non-convex, multimodal, or contain noise. These make the optimization problem difficult or even impossible to be solved by traditional methods which require at least the first derivative of the objective function with respect to the design variables[1-2]. Moreover, for large solution spaces where extensive search is required, traditional methods are computationally expensive. In such cases, the intelligent optimization algorithms, which only use the evaluation of the function, should be taken as a promising tool to replace the traditional optimization techniques.

Particle Swarm Optimization (PSO) is an intelligent optimization algorithm developed recently[3-4]. Similar to Evolutionary algorithms (EAs), PSO is a population based optimization method. Distinct from the other EAs where knowledge is destroyed between generations, individuals in the population of PSO retain memory of known good solutions as the search for better solutions continues. The other advantage of PSO is that it's easy to implement and there are fewer parameters to adjust. However, as a new random search method, PSO encounters several problems such as premature

H. Tan (Ed.): Knowledge Discovery and Data Mining, AISC 135, pp. 769–775.
springerlink.com

convergence and / or slow search speed, too fast decrease of the variety of the particle swarm etc, resulting in a fail search in some cases[5-6]. References[7-10] indicated that the search capability of the algorithm can be enhanced by controlling the swarm variety.

This paper develops an improved PSO method for the beam structural optimization. Special mutation-interference operators are introduced to increase the swarm variety and improve the convergence. The novel strategy produces a variation of the speed, letting the algorithm explore the local and global minima effectively at the same time. Computational examples have shown that the proposed method can not only speed up the convergence significantly, but also prevent the premature convergence effectively. In the present paper, a commercial finite element code, ANSYS, is used to calculate interlaminar stresses at the free-edge of laminates. Optimization for minimizing interlaminar normal stresses is then conducted using the zero-order method (ZOM) incorporated in ANSYS. Results are presented for two different loading conditions: in-plane tensile load and uniform bending load. Numerical results demonstrate that the optimal solutions based on ZOM are sensitive to the initial designs. In order to overcome this disadvantage, a technique of applying the PSO algorithm integrated with FEM is developed. The ability to derive the global optimal and the potential of dealing with complicated problems in terms of practical time constraints are discussed and demonstrated. Examples of minimizing the weight of composite beam structure is modeled and solved.

2 Particle Swarm Optimization

Particle swarm optimization (PSO) is an evolutionary computation technique developed by Dr. Eberhart and Dr. Kennedy [3] in 1995. PSO was basically developed through simulation of bird flocking in two-dimension space. According to the research results for a flock of birds in finding food, one concludes that birds find food by flocking (not by each individual). The observation leads to the assumption that any information is shared inside flocking. Similar features exist in human groups, i.e., behavior of each individual (agent) is based on the behavior patterns authorized by the groups such as customs and other behavior patterns according to the experiences of each individual.

Searching procedures of PSO can be described as follows: The PSO initializes a group of random particles (individuals) firstly. Instead of using genetic operators as in genetic algorithm (GA), these individuals are “evolved” by cooperation and competition among themselves through generations. Each particle which represents a potential solution to a problem adjusts its flying according to its own flying experience and its companions’ flying experience. Let the ith particle is a point in the D dimensional space, $X_i = (x_{i1}, x_{i2}, \cdots, x_{iD})^T$ and the velocity (position change rate, or displacement increment) be $V_i = (v_{i1}, v_{i2}, \cdots, v_{iD})^T$. The best previous position (the position possessing the best fitness value) of any particle is recorded and denoted as $pbest$ and the best position of the whole group as $gbest$. Then the particles are manipulated according to the following equations:

$$v_{id}^{k+1} = \omega * v_{id}^{k} + c_1 * rand_1^k(\ ) * \left(pBest_{id}^{k} - x_{id}^{k}\right) + c_2 * rand_2^k(\ ) * \left(gBest_{id}^{k} - x_{id}^{k}\right) \tag{1}$$

$$x_{id}^{k+1} = x_{id}^{k} + v_{id}^{k+1} \tag{2}$$

In which v_{id}^{k} and x_{id}^{k} are respectively the current velocity and position of the particle i at iteration k, $rand_1^k()$ and $rand_2^k()$ are both random numbers uniformly distributed in the range [0,1]. Constants c_1, c_2 are called the cognitive and social parameter respectively, both taking the value of two in most cases. The inertia weight ω is restricted in [0.1,0.9], decreasing with the iteration as expressed below:

$$\omega = \omega_{\max} - iter \times \frac{\omega_{\max} - \omega_{\min}}{iter_{\max}} \tag{3}$$

Where $\omega_{\max}$ and $\omega_{\min}$ are the initial weight factor and the final one, $iter_{\max}$ is the maximum iteration number, and $iter$ the current iteration number. For a constrained optimization problem, the dynamic penalty function method is used to handle the constraints as follows.

$$\min F(X) = \min\left\{ f(x) + \sigma * \sum_{i=1}^{p} [\max(0,\ \ g_i(x))]^2 \right\} \tag{4}$$

$$g_i(x) \le 0 \qquad i = 1, 2, \cdots p$$

Here, σ is the penalty factor. To improve the performance of the algorithm, the penalty factor is taken to be correlated with the iteration number

$$\sigma(t) = e^{m \cdot t^n / T + \omega_0} \tag{5}$$

Where m and n are positive coefficients used to adjust the changing rate of σ; ω_0 is the initial penalty factor. In the present paper, m and n are assigned to be 5 and 1.2, respectively, and ω_0 is set to 10. The dynamic penalty factor is small at the beginning of iteration, which may help to search for the optimum solution in a larger design space. As the iteration number increases, the penalty factor gradually becomes larger, enforcing the constraint to be satisfied.

3 Optimization Procedure Combining Pso and Fem

The optimization procedure is essentially written in the software MATLAB. In the optimization process, the procedure first initializes the swarm. ANSYS is then called on the back stage to evaluate the objective and constraint functions of each particle through APDL commands written beforehand. ANSYS outputs the values of the functions to an external file. MATLAB reads them and computes the penalty function in Eq. (4) as the fitness of each particle. Every particle updates itself through Eqs. (1) and (2). The updated swarm is then returned to ANSYS for the next iteration. This process is repeated until the number of iterations reaches the pre-determined maximum iteration number. In the whole optimization process, data from ANSYS and MATLAB are exchanged back and forth with each other, as shown in Fig. 1. The number of times ANSYS is invoked for analysis equals the maximum iteration number multiplied by the number of particles. For PSO, the typical range for the number of particles is 20-40. For most problems, 10 particles are sufficient to obtain good results. The flow chart of optimization by PSO and FE code is showed in the Fig.1.

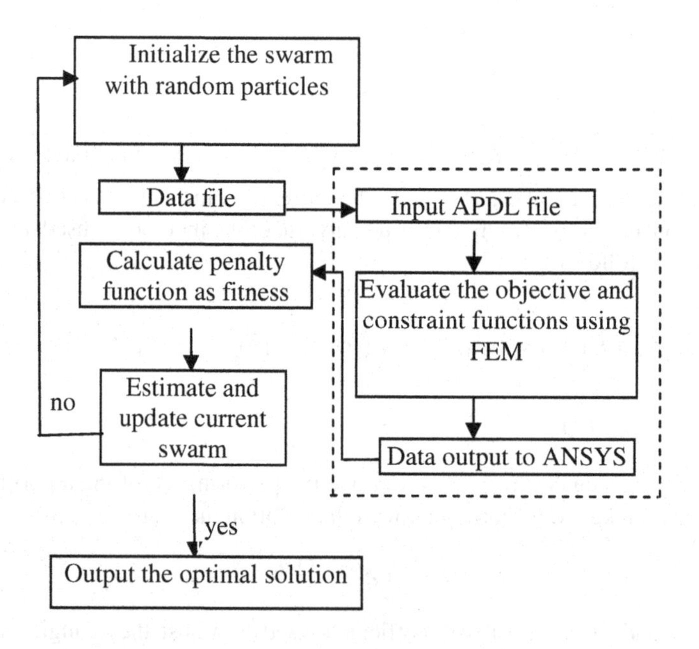

Fig. 1. Flow chart of optimization process by PSO and FE code

4 Optimization Example and Testing Results

4.1 Structure Analysis of a Box Beam

The optimization problems of the beam structure are solved. The beam is made of the composite laminate. The structure with twenty design variable has a stacking sequence of $[0/45/-45/90]_s$, where the number of is the design variable. The improved PSO algorithm is applied to the stacking sequence optimization of a symmetrically laminated composite plate. Only half of the plies of a laminate are coded in the seating bunch because of the symmetry.

The limit state function F based on the Tsai-Wu criterion is[12]

$$F = 1 - (F_{LL}\sigma_L^2 + F_{TT}\sigma_T^2 + F_{SS}\sigma_S^2 + 2F_{LT}\sigma_L\sigma_T + F_L\sigma_L + F_T\sigma_T) \quad (6)$$

In which

$$F_{LL} = 1/(X_T X_C) \quad F_{TT} = 1/(Y_T Y_C) \quad F_{SS} = 1/S^2$$

$$F_L = 1/X_T - 1/X_C \quad F_T = 1/Y_T - 1/Y_C \quad F_{LT} = (-1/2)\sqrt{F_{LL}F_{TT}}$$

In which X_T represents the tension strength in the longitudinal direction of a ply, Y_T the tension strength in the transverse direction, X_C and Y_C are the compression strengths in the longitudinal and transverse directions, and S the shear strength, respectively.

In applying the Tsai-Wu criterion, the beam is divided into a number of small elements in the plane, and the most dangerous one is depicted as the representative of the ply.

4.2 Minimum Weight Design of a Box Beam

The important aim of the optimum design for a composite beam is to minimize its cost or weight without compromising its performance. In the following, a beam is optimized to achieve the minimum weight while insuring certain system constrains. The total layers of per laminate are the variable and each layer's thickness is 0.4mm. The loading condition is 5 ton. The mechanical properties and strength parameters are listed in Table 1.

Table 1. Mechanical properties and strength parameters

property	E_1 (GPa)	E_2 (GPa)	G_{12} (GPa)	X_T (MPa)	X_C (MPa)	Y_T (MPa)	Y_C (MPa)	S (MPa)	ν_{12}
value	32.5	5.02	10.1	513	333	33.2	139	54.3	0.21

The design parameters and optimization results are listed in Table 2.

Table 2. Design parameters and optimization results

component	angle	ZOM numbers	PSO numbers
deck	$[0/45/-45/90]_s$	$[6/3/3/4]_s$	$[5/4/5/3]_s$
sidewall	$[0/45/-45/90]_s$	$[3/3/3/2]_s$	$[2/1/2/2]_s$
bottom	$[0/45/-45/90]_s$	$[2/3/3/4]_s$	$[2/3/4/3]_s$
bulkhead	$[0/45/-45/90]_s$	$[2/2/3/4]_s$	$[2/2/4/6]_s$
stiffener	$[0/45/-45/90]_s$	$[3/5/4/2]_s$	$[3/2/2/3]_s$

Based on the optimization results, the structure is made. The testing results is listed in Table 3.

Table 3. Comparison of the optimization results

	Weight (kg)	F	strain Computing analysis ($\mu\varepsilon$)	 Testing result ($\mu\varepsilon$)	displacement computing analysis (mm)	 Testing result (mm)
Initialization structure	11.29	0.267	5364	4921	2.82	2.13
PSO	5.156	0.961	6635	6173	6.79	5.42
ZOM	7.534	0.783	7142	6794	5.85	5.09

Fig. 2. The destroy shape of the structure. a)first picture for ZOM;b)second picture for PSO

The testing result is less than the computing analysis, the reason may be adding the strengthen components in making. But the optimization result can be seen. The structure in design load fulfils the request of strength and rigidity with less weight.

5 Conclusions

A technique involving the application of the exterior optimization algorithm PSO integrated with the FEM was developed for the optimization. ZOM incorporated in ANSYS was also used for comparison. It was found that PSO gives better results than ZOM. For a one design variable case, the results yielded by PSO are superior or equivalent to the best results among the different initial sets by ZOM with a comparable number of FE analyses. For the case of twenty design variables in the beam structure, PSO can yield results better than ZOM with fewer FE analyses and the result is also better than that available in the literature.

As the optimization results based on ZOM are sensitive to the initial design sets, when conducting optimization with ZOM, one should choose different initial design sets and compare the optimization solutions. The central problem is that it is not known which initial design can lead to a satisfactory optimum and whether another initial design exists from which a better result can be obtained. By contrast, with PSO it is not necessary to specify initial designs manually as it is a global stochastic method. Thus, it can be implemented in a straightforward manner and improvement of the objective function can be achieved simply by choosing a larger number of particles or iterations.

References

1. Reklaitis, G.V., Ravindran, A., Ragsdell, K.M.: Engineering optimization methods and applications. John Wiley and Sons Inc., New York (1983)
2. Arora, J.S.: Introduction to Optimum Design. McGraw-Hill Book Company, Singapore (1989)
3. Kennedy, J., Eberhart, R.: Particle Swarm Optimization. In: Proceeding of the IEEE International Conference on Neural Networks, Perth, Australia, pp. 1942–1948 (1995)
4. Kennedy, J., Eberhart, R.: Swarm Intelligence. Academic Press, San Diego (2001)
5. Clerc, M., Kennedy, J.: The particle swarm explosion, stability and convergence in a multidimensional complex space. IEEE Transactions on Evolutionary Computation 6, 58–73 (2002)
6. Trelea, I.C.: The particle swarm optimization algorithm: convergence analysis and parameter selection. Information Processing Letters 85(6), 317–325 (2003)
7. Angeline, P.: Using Selection to Improve Particle Swarm Optimization. In: Proceeding of the 1998 IEEE International Conference on Evolutionary Computation, Anchorage, USA, pp. 84–89 (1998)
8. Eberhart, R., Shi, Y.: Comparison between genetic algorithms and particle swarm optimization (1998)
9. Shi, X.H., Wang, L.M., Lee, H.P., Yang, X.W., Wang, L.M., Liang, Y.C.: An Improved Genetic Algorithm with Variable Population-Size and A PSO-GA Based Hybrid Evolutionary Algorithm. In: Proceedings of the Second International Conference on Machine Learning and Cybernetics, Xi'an, China, pp. 1735–1740 (2003)
10. Shi, X.H., Liang, Y.C., Lee, H.P., Lu, C., Wang, L.M.: An improved GA and a novel PSO-GA-based hybrid algorithm. Information Processing Letters 93(5), 255–261 (2005)
11. Runarsson, T., Yao, X.: Stochastic ranking for constrained evolutionary optimization. IEEE Trans. on Evolutionary Computation 4(3), 284–294 (2000)
12. Tsai, S.W., Wu, E.M.: A general theory of strength for anisotropic materials. Journal of Composite Materials 5, 58–80 (1971)

References

[illegible] A. R[illegible], K. M. [illegible] optimization methods [illegible] plications [illegible] New York (19[illegible])
[illegible] Optimization [illegible] McGraw-Hill Book Company, Singapore (19[illegible])
[illegible] Proceedings of the IEEE [illegible] Conference on Neural Networks [illegible] pp. 19[illegible]–1948 (19[illegible])
[illegible] Academic Press, San Diego [illegible]
[illegible]

Optimal Filtering for Stochastic Singular Systems with Uncertain Observations

Wang Huiying

Computer Department, School of Engineering, Northeast Agricultural University, Harbin, Heilongjiang

Abstract. A stochastic singular system with uncertain observations is converted to two reduced-order subsystems based on singular value decomposition. Using the projection method, the state filters are presented for two reduced-order subsystems with uncertain observations. State filter for the stochastic singular system is obtained. The estimation error covariance matrix is derived between two reduced-order filters. The filtering algorithm based on complete data in the previous literatures has lost the optimality. An example shows the effectiveness of the proposed algorithms.

Keywords: Stochastic singular system, uncertain observation, cross-covariance matrix, optimal filter.

1 Introduction

Descriptor systems have the extensive application backgrounds in electrical network, robotics, economics and complex chemical fields [1]. Compared with normal systems, descriptor systems is a class of dynamical systems and have more extensive forms. In 1974, Rosenbrock[2] firstly proposed the definition of the descriptor system in a complex electrical network system. Since then, many scholars find that the dynamic input and output model in economy and social system model in decision theory belong to descriptor systems.

In recent years, the state estimation problem for descriptor systems has attracted considerable attention due to the significant applications in system control and signal processing [3-5]. The state estimation for descriptor system has two forms, i.e., reduced-order filter [3-5] and full-order filter [6,7]. Currently, most results about state estimation for descriptor system are mainly focused on complete measurement data. However, in networked system, data loss often exists due to the unreliable communications network. Only the disturbance noise not the measurement signal can be received at some moments at data processing center. The missing measurements can be described by using the measurement equation with multiplicative noise. [8] designs the optimal linear filter in the linear minimum variance sense for discrete-time stochastic systems with multiplicative noise. [9] generalizes the results of [8] to the case of correlated multiplicative noise. The recursive filter and predictor with correlated noise at the neighboring moments are proposed for systems with missing measurements by applying the minimum mean-square error criterion. The robust filter[11] and ARMA signal filter[12]are also investigated for systems with missing measurements. Recently,

H. Tan (Ed.): Knowledge Discovery and Data Mining, AISC 135, pp. 777–784.
springerlink.com

most results about state estimation for systems with missing measurements are mainly focused on normal systems. However, the state estimation for descriptor systems with missing measurements is seldom reported.

In this paper, the filtering problem is investigated for descriptor system with missing measurements. The original descriptor system is equivalently transferred to two reduced-order subsystems based on singular value decomposition. The first reduced-order subsystem is a normal system with missing measurements and correlated noise. The state filter is easily obtained by the approaches in the existing references. The state filter of the second reduced-order subsystem is obtained by using the state filter of the first subsystem and the white noise filter. The white noise filter algorithm, the state filters of the two reduced-order subsystems and the filtering error variance matrix between the two subsystems are obtained by applying projection theory. Further, the state filter of the original descriptor system is obtained. The optimal filtering algorithm under complete measurement data in previous references has lost the optimality when there is the measurement data loss. The proposed algorithm is reduced to the optimal filtering algorithm under complete measurement data when there is not the measurement data loss.

2 Problem Formulation

Consider the following stochastic singular system with missing measurements:

$$Mx(t+1)=\bar{\Phi}x(t)+\bar{\Gamma}w(t) \tag{1}$$

$$y(t)=u(t)\bar{H}x(t)+v(t) \tag{2}$$

where $x(t)\in R^n$ is the state, $y(t)\in R^m$ is the measurement, $w(t)$ is the system noise, $v(t)$ is the measurement noise, $u(t)$ is a Bernoulli distributed random variable with the probabilities $\Pr ob\{u(t)=1\}=\alpha$, $\Pr ob\{u(t)=0\}=1-\alpha$, M, $\bar{\Phi},\bar{\Gamma},\bar{H}$ are the constant matrices with suitable dimensions.

Assumption 1. M is a singular square matrix, i.e., $\mathrm{rank}M=n_1<n$.

Assumption 2. $w(t)\in\mathbb{R}^r$ and $v(t)\in\mathbb{R}^m$ are white noises with zeros means and

$$\mathrm{E}\left\{\begin{bmatrix} w(t)\\ v(t)\end{bmatrix}[w^{\mathrm{T}}(k)\quad v^{\mathrm{T}}(k)]\right\}=\begin{bmatrix} Q_w & \bar{S}\\ \bar{S}^{\mathrm{T}} & Q_v\end{bmatrix}\delta_{tk} \tag{3}$$

where E is the mathematical expectation operator, δ_{tk} is the Kronecker delta function.

Assumption 3. The initial state $x(0)$ is uncorrelated with $w(t)$, $v(t)$ and $u(t)$, and satisfies that $\mathrm{E}x(0)=\mu_0$, $\mathrm{E}[(x(0)-\mu_0)(x(0)-\mu_0)^{\mathrm{T}}]=P_0$

Assumption 4. System (1) – (2) is regular.

Our aim is to find the linear minimum variance optimal state filter for the state $x(t)$ based on the received measurements $(y(t), y(t-1), \ldots, y(1))$, which will satisfy unbiasedness and optimality.

Firstly, we transform the original singular system (1)−(2) to the equivalent reduced-order subsystems using singular value decomposition. There are nonsingular matrices U and R such that [4]

$$UMR = \begin{bmatrix} I_{n1} & 0 \\ 0 & 0 \end{bmatrix}, U\bar{\Phi}R = \begin{bmatrix} A_{11} & A_{12} \\ A_{21} & A_{22} \end{bmatrix}, U\bar{\Gamma} = \begin{bmatrix} \Gamma_1 \\ \Gamma_2 \end{bmatrix},$$
$$\bar{H}R = \begin{bmatrix} H^{(1)} & H^{(2)} \end{bmatrix} \tag{4}$$

where $x(t) = R\begin{bmatrix} x^{(1)\mathrm{T}}(t) & x^{(2)\mathrm{T}}(t) \end{bmatrix}^{\mathrm{T}}, x^{(1)}(t) \in \mathbb{R}^{n_1}$, $x^{(2)}(t) \in \mathbb{R}^{n_2}$, $n_1 + n_2 = n$. A_{22} is an $n_2 \times n_2$ nonsingular matrix. Then, system (1)-(2) is transferred to the following two reduced-order subsystems:

$$\begin{cases} x^{(1)}(t+1) = \Phi x^{(1)}(t) + \Gamma w(t) \\ y(t) = u(t) H x^{(1)}(t) + \eta(t) \end{cases} \tag{5}$$

$$x^{(2)}(t) = B x^{(1)}(t) + C w(t) \tag{6}$$

where $\Phi = A_{11} - A_{12}A_{22}^{-1}A_{21}$, $\Gamma = \Gamma_1 - A_{12}A_{22}^{-1}\Gamma_2$, $H = H^{(1)} - H^{(2)}A_{22}^{-1}A_{21}$,
$\eta(t) = -u(t)H^{(2)}A_{22}^{-1}\Gamma_2 w(t) + v(t)$, $B = -A_{22}^{-1}A_{21}$, $C = A_{22}^{-1}\Gamma_2$.

Further, we have the following statistical information

$$\mathrm{E}\left\{ \begin{bmatrix} w(t) \\ \eta(t) \end{bmatrix} \begin{bmatrix} w^{\mathrm{T}}(k) & \eta^{\mathrm{T}}(k) \end{bmatrix} \right\} = \begin{bmatrix} Q_w & S \\ S^{\mathrm{T}} & Q_\eta \end{bmatrix} \delta_{tk} \tag{7}$$

where $S = \bar{S} - \alpha Q_w \Gamma_2^{\mathrm{T}} A_{22}^{-\mathrm{T}} H^{(2)\mathrm{T}}$, $Q_\eta = Q_v - \alpha H^{(2)} A_{22}^{-1} \Gamma_2 \bar{S}$
$-\alpha \bar{S}^{\mathrm{T}} \Gamma_2^{\mathrm{T}} A_{22}^{-\mathrm{T}} H^{(2)\mathrm{T}} + \alpha H^{(2)} A_{22}^{-1} \Gamma_2 Q_w \Gamma_2^{\mathrm{T}} A_{22}^{-\mathrm{T}} H^{(2)\mathrm{T}}$.

The reduced-order subsystem (5) is a normal system with missing measurements and correlated noise where data loss phenomenon is described by a Bernoulli distributed random variable which is the multiplicative noise. The state estimator of subsystem (5) can be obtained by using the existing results. The second subsystem is the linear combination of the state of the first subsystem and white noise. Hence, the state estimator of subsystem (6) can be obtained by using the state estimator of subsystem (5) and the white noise estimator. Next, we will respectively design the state estimator of the two subsystems.

3 Optimal Reduced-Order Filter

Reduced-order subsystem (5) is a normal system with multiplicative noise. Hence, the filter of reduced-order state $x^{(1)}(t)$ can be easily obtained via existing approach.

Lemma 1. [9] Under the Assumptions 1-4, the stochastic system (5) with missing measurements has the following optimal recursive Kalman filter

$$D(t) = \Phi D(t-1)\Phi^{\mathrm{T}} + \Gamma Q_w \Gamma^{\mathrm{T}} \tag{8}$$

$$\varepsilon(t) = y(t) - \alpha H\hat{x}^{(1)}(t \mid t-1) \tag{9}$$

$$Q_\varepsilon(t) = \alpha(1-\alpha)HD(t)H^{\mathrm{T}} + \alpha^2 HP^{(1)}(t \mid t-1)H^{\mathrm{T}} + Q_\eta \tag{10}$$

$$K_p(t) = [\alpha\Phi P^{(1)}(t \mid t-1)H^{\mathrm{T}} + \Gamma S]Q_\varepsilon^{-1}(t) \tag{11}$$

$$\hat{x}^{(1)}(t+1 \mid t) = \Phi\hat{x}^{(1)}(t \mid t-1) + K_p(t)\varepsilon(t) \tag{12}$$

$$K_f(t+1) = \alpha P^{(1)}(t+1 \mid t)H^{\mathrm{T}}Q_\varepsilon^{-1}(t+1) \tag{13}$$

$$\hat{x}^{(1)}(t+1 \mid t+1) = \hat{x}^{(1)}(t+1 \mid t) + K_f(t+1)\varepsilon(t+1) \tag{14}$$

$$P^{(1)}(t+1 \mid t+1) = [I_n - \alpha K_f(t+1)H]P^{(1)}(t+1 \mid t) \tag{15}$$

$$P^{(1)}(t+1 \mid t) = \Phi P^{(1)}(t \mid t-1)\Phi^{\mathrm{T}} + \Gamma Q_w \Gamma^{\mathrm{T}} - K_p(t)Q_\varepsilon(t)K_p^{\mathrm{T}} \tag{16}$$

where the initial values are $\hat{x}^{(1)}(0 \mid -1) = \mu_0^{(1)}$, $P^{(1)}(0 \mid -1) = P_0^{(1)}$, $D(0) = P_0^{(1)} + \mu_0^{(1)}\mu_0^{(1)\mathrm{T}}$ where $\mu_0^{(1)}$ is the first n_1 block components of $R^{-1}\mu_0$, $P_0^{(1)}$ is the top left corner $n_1 \times n_1$ sub-block of $R^{-1}P_0R^{-\mathrm{T}}$.

Lemma 2. [10] Under the Assumptions 1-4, the stochastic system (5) with missing measurements has the following optimal input white noise filter

$$\hat{w}(t \mid t) = M_w(t \mid t)\varepsilon(t) \tag{17}$$

where the gain matrix is computed by

$$M_w(t \mid t) = SQ_\varepsilon^{-1}(t) \tag{18}$$

Next, we will give the optimal filter for the reduced-order system (5).

Theorem 1. Under the Assumptions 1-4, the reduced-order subsystem (5) has the following linear minimum variance optimal filter

$$\hat{x}^{(2)}(t \mid t) = B\hat{x}^{(1)}(t \mid t) + C\hat{w}(t \mid t) \tag{19}$$

The filtering error variance matrix is computed by

$$P^{(2)}(t\mid t)=F(t)P^{(1)}(t\mid t-1)F^{\mathrm{T}}(t)+\alpha(1-\alpha)G(t)HD(t)H^{\mathrm{T}}G^{\mathrm{T}}(t)$$

$$+G(t)Q_{\eta}G^{\mathrm{T}}(t)+CQ_{w}C^{\mathrm{T}}-G(t)S^{\mathrm{T}}C^{\mathrm{T}}-CSG_{i}^{\mathrm{T}}(t) \tag{20}$$

$$F(t)=B[I_{n}-\alpha K_{f}(t)H]-\alpha CSQ_{\varepsilon}^{-1}(t)H \tag{21}$$

$$G(t)=BK_{f}(t)+CSQ_{\varepsilon}^{-1}(t) \tag{22}$$

Proof: Taking projection of both sides of (5) yields the reduced-order filter (19). From Lemma 1, we can respectively obtain the filtering error equation and innovation equation of reduced-order state $x^{(1)}(t)$ as follows

$$\tilde{x}^{(1)}(t\mid t)=[I_{n}-\alpha K_{fi}(t)H]\tilde{x}^{(1)}(t\mid t-1)-K_{f}(t)\eta(t)-$$

$$(u(t)-\alpha)K_{f}(t)Hx^{(1)}(t) \tag{23}$$

$$\varepsilon(t)=(u(t)-\alpha)Hx^{(1)}(t)+\eta(t)+\alpha H\tilde{x}^{(1)}(t\mid t-1) \tag{24}$$

From (5) and (17)-(19), we have the filtering error equation of reduced-order state $x^{(2)}(t)$ as follows

$$\tilde{x}^{(2)}(t\mid t)=B\tilde{x}^{(1)}(t\mid t)+Cw(t)-CSQ_{\varepsilon}^{-1}(t)\varepsilon(t) \tag{25}$$

Substituting (23) and (24) into (25) we have

$$\tilde{x}^{(2)}(t\mid t)=\{B[I_{n}-\alpha K_{f}(t)H]-\alpha CSQ_{\varepsilon}^{-1}(t)H\}\tilde{x}^{(1)}(t\mid t-1)$$

$$+Cw(t)-[BK_{f}(t)+CSQ_{\varepsilon}^{-1}(t)]\eta(t)$$

$$-[u(t)-\alpha][BK_{f}(t)+CSQ_{\varepsilon}^{-1}(t)]Hx^{(1)}(t) \tag{26}$$

where $F(t)$ and $G(t)$ are defined by (21) and (22). Then, (26) can be rewritten as

$$\tilde{x}^{(2)}(t\mid t)=F(t)\tilde{x}^{(1)}(t\mid t-1)+Cw(t)-G(t)\eta(t)$$

$$-(u(t)-\alpha)G(t)Hx^{(1)}(t) \tag{27}$$

From $\tilde{x}^{(1)}(t\mid t-1)\in\mathcal{L}(x^{(1)}(0),w(0),\ldots,w(t-1),\eta(1),\ldots,\eta(t-1))$, we have $w(t)\perp\tilde{x}^{(1)}(t\mid t-1)$, $\eta(t)\perp\tilde{x}^{(1)}(t\mid t-1)$ which together with $\mathrm{E}\{[u(t)-\alpha]^{2}\}=\alpha(1-\alpha)$ and $\mathrm{E}[u(t)-\alpha]=0$ leads to (20) by computing $P^{(2)}(t\mid t)=\mathrm{E}[\tilde{x}^{(2)}(t\mid t)\tilde{x}^{(2)\mathrm{T}}(t\mid t)]$, where $P^{(1)}(t\mid t-1)$ is computed by (16) of Lemma 1.

Theorem 2. Under the Assumptions 1-4, the filtering error cross-covariance matrix of the two states of the reduced-order subsystems (5) and (6) is computed by

$$P^{(12)}(t\mid t)=[I_{n}-\alpha K_{f}(t)H]P^{(1)}(t\mid t-1)F^{\mathrm{T}}(t)+$$

$$\alpha(1-\alpha)K_{f}(t)HD(t)H^{\mathrm{T}}G^{\mathrm{T}}(t)+K_{f}(t)Q_{\eta}G^{\mathrm{T}}(t)-K_{f}(t)S^{\mathrm{T}}C^{\mathrm{T}} \tag{28}$$

where the initial value $P^{(1)}(t\mid t-1)$ is computed by Lemma 1.

Proof: Substituting (23) and (27) into $P^{(12)}(t|t)=\mathrm{E}[x^{(1)}(t|t)x^{(2)\mathrm{T}}(t|t)]$ and using $w(t)\perp\tilde{x}^{(1)}(t|t-1)$, $\eta(t)\perp\tilde{x}^{(1)}(t|t-1)$, $\mathrm{E}\{[u(t)-\alpha]^2\}=\alpha(1-\alpha)$ and $\mathrm{E}[u(t)-\alpha]=0$, we can obtain (28).

Based on Lemmas 1-2 and Theorems 1-2, we have the state filter of the original descriptor.

Theorem 3. Under the Assumptions 1-4, the state filter of the original descriptor (1)-(2) is computed by

$$\hat{x}(t|t)=R[\hat{x}^{(1)\mathrm{T}}(t|t),\hat{x}^{(2)\mathrm{T}}(t|t)]^{\mathrm{T}} \tag{29}$$

The filtering error variance matrix is computed by

$$P(t|t)=R\begin{bmatrix} P^{(1)}(t|t) & P^{(12)}(t|t) \\ P^{(12)\mathrm{T}}(t|t) & P^{(2)}(t|t) \end{bmatrix}R^{\mathrm{T}} \tag{30}$$

where $P^{(1)}(t|t)$, $P^{(2)}(t|t)$ and $P^{(12)}(t|t)$ are computed by Lemma 1 and Theorems 1-2.

Proof: Taking projection on $x(t)=R[x^{(1)\mathrm{T}}(t)\ \ x^{(2)\mathrm{T}}(t)]^{\mathrm{T}}$, we have (29). Subtracting (29) from $x(t)$, we have

$$\tilde{x}(t|t)=R[\tilde{x}^{(1)\mathrm{T}}(t|t),\tilde{x}^{(2)\mathrm{T}}(t|t)]^{\mathrm{T}} \tag{31}$$

Substituting (31) into $P(t|t)=\mathrm{E}[\tilde{x}(t|t)\tilde{x}^{\mathrm{T}}(t|t)]$ yields (30).

Remark: Theorem 3 gives the optimal reduced-order filter algorithm for system (1)-(2) with missing measurements. When $\alpha=1$, the corresponding Lemmas 1-2 and Theorems 1-3 are reduced to the optimal filter algorithms under complete measurement data [5]. Hence, the filtering algorithms in previous references have lost the optimality under missing measurements.

4 Simulation Research

In this section, a simulation example is given to show the effectiveness of the proposed results.

Consider the descriptor system (1)-(2)

where $M = \begin{bmatrix} 1 & 0 & 0 & 0 \\ 0 & 0 & 1 & 0 \\ 0 & 0 & 0 & 0 \\ 0 & 0 & 0 & 0 \end{bmatrix}$, $\bar{\Phi} = \begin{bmatrix} 0.5 & 0 & -0.1 & 0 \\ -2 & 0 & 0.2 & 0 \\ 2.55 & -0.5 & -0.93 & 0 \\ -2.4 & -1 & -0.76 & 2 \end{bmatrix}$, $\bar{\Gamma} = \begin{bmatrix} -0.1 & -0.4 \\ 0.3 & 2.1 \\ 0.4 & -1.1 \\ 1.7 & 2.01 \end{bmatrix}$,

$\bar{H} = \begin{bmatrix} 1 & 1.5 & 0.7 & 0 \\ 0.2 & 0 & 1.2 & 1 \end{bmatrix}$. The Bernoulli distributed random variable $u(t)$ satisfies the probability $\text{Prob}(u(t)=1)=\alpha, \text{Prob}(u(t)=0)=1-\alpha$ and is uncorrelated with measurement noise and system noise. Measurement noise $v(t)$ and system noise $w(t)$ satisfy the relation $v(t)=\beta w(t)+\gamma(t)$ where β is the correlated coefficient. The Gauss noise $\gamma(t)$ with zero mean and variance Q_γ is independent of $w(t)$. Our aim is to find the filter $\hat{x}(t \mid t)$.

In simulation, we set $\alpha = 0.7$, $\beta_1 = 0.2$, $Q_w = \begin{bmatrix} 1 & 0 \\ 0 & 1 \end{bmatrix}$, $Q_\gamma = \begin{bmatrix} 0.36 & 0 \\ 0 & 0.36 \end{bmatrix}$. The initial values are $x(0) = \begin{bmatrix} 0 & 0 & 0 & 0 \end{bmatrix}^{\mathrm{T}}$, $P_0 = 0.2I_4$, $D(0) = 0.2I_2$. From Lemma 1, we can obtain the filter $\hat{x}^{(1)}(t \mid t)$ of the first reduced-order state $x^{(1)}(t)$. From Theorem 1, we have the filter $\hat{x}^{(2)}(t \mid t)$ of the second reduced-order state $x^{(2)}(t)$. Further, we can obtain the state filter $\hat{x}(t \mid t)$ of the original descriptor system.

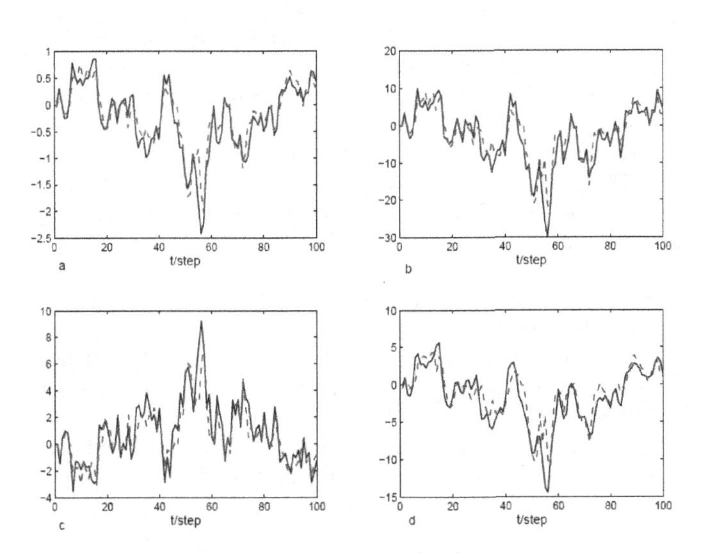

Fig. 1. State filter of descriptor system with missing measurements

a~d of Fig. 1 denote the true values and optimal filters of the four components, respectively, where the solid curves are the true values and the dashed ones are the filters. From the simulation results, we see that the effectiveness of the proposed estimation algorithms.

5 Conclusion

In this paper, we investigate the linear minimum variance optimal state filter for the stochastic singular system with missing measurements. The original system is equivalently converted to two reduced-order subsystems by using the singular value decomposition. Then, the corresponding reduced-order state filter and input white noise filter are obtained for the two subsystems. The filtering error cross-covariance matrix between the two reduced-order subsystems is derived. Further, the state filter of the original descriptor system is obtained by using nonsingular inverse transform. The optimal filtering algorithm under complete measurement data in previous references has lost the optimality when there is the measurement data loss.

References

1. Yang, D.M., Zhang, Q.L., Yao, B.: Descriptor systems. Science Press, Beijing (2004)
2. Rosenbrock, H.H.: Structural properties of linear dynamic systems. Int. J. Contr. 20(2), 191–202 (1974)
3. Wang, E.P., Wang, C.Z.: Optimum recurrence filtering method for singular discrete stochastic system (1). Acta Automatica Sinica 14(6), 409–415 (1988)
4. Qu, D.M., Ma, J., Sun, S.L.: Distributed fusion filter for stochastic singular systems with unknown disturbance. In: WCICA, Beijing, pp. 1431–1435 (2010)
5. Sun, S.L., Ma, J.: Distributed reduced-order optimal fusion Kalman filters for stochastic singular systems. Acta Automatica Sinica 32(2), 286–290 (2006)
6. Nikoukhah, R., Willsky, A.S., Levy, B.C.: Kalman filtering and Riccati equations for descriptor systems. IEEE Trans. Automat. Contr. 37(9), 1325–1341 (1992)
7. Sun, S.L., Ma, J.: Optimal filtering and smoothing for discrete-time stochastic singular systems. Signal Processing 87(1), 189–201 (2007)
8. Nahi, N.E.: Optimal recursive estimation with uncertain observation. IEEE Trans. Inform. Theory 13, 457–462 (1969)
9. Hadidi, M.T., Schwarts, S.C.: Linear recursive state estimators under uncertain observations. IEEE Trans. Automat. Contr. 24(6), 944–947 (1979)
10. Hermoso, A., Linares, J.: Linear estimation for discrete-time systems in the presence of time-correlated disturbances and uncertain observations. IEEE Trans. Automat. Contr. 39(8), 1636–1638 (1994)
11. Wang, Z., Yang, F., Ho, D.W.C., Liu, X.: Robust H-infinity filtering for stochastic time-delay systems with missing measurements. IEEE Trans. Signal Process. 54, 2579–2587 (2006)
12. Yan, S.W., Sun, S.L., Qu, D.M.: Filter design for ARMA signals with uncertain observations. Science Technology and Engineering 10(22), 5396–5400 (2010)

Centralized Fusion White Noise Estimator for Multi-sensor Systems with Missing Measurements

Wang Huiying

Computer Department, School of Engineering, Northeast Agricultural University,
Harbin, Heilongjiang

Abstract. This paper is concerned with the information fusion estimation problem about the white noise. White noise estimator has an important application in oil seismic exploration. The centralized fusion white noise estimator in least mean square sense is presented for multi-sensor discrete-time stochastic linear systems with missing measurements based on the innovation analysis method, where different sensors have different missing measurement rates. Compared to local estimators, the proposed centralized fusion estimator has better estimation accuracy. An example shows the effectiveness.

Keywords: Centralized fusion, white noise estimator, missing measurement, multi-sensor system, innovation analysis method.

1 Introduction

Mendel[1] has proposed the input white noise estimator for the dynamic systems by taking the oil seismic exploration as the application background. The principle of exploration is that the reflection coefficient sequence of the seismic wave in the oil layer, which is generated by the explosives buried under the ground, is used to judge whether there is oil field and the geometry size of the oil field. The reflection coefficient sequence can be represented by Bernoulli-Gaussian white noise. So the white noise estimation problem is the key of seismic exploration. The input of sensor is the white noise reflection sequence and the output is the received signal. The estimator of input white noise is from the output signal, which can be called as white noise deconvolution filter. At present, the research for the multi-sensor information fusion estimation of white noise has gained the extensive attention. However, the white noise estimation problems with uncertain observation are seldom studied.

Uncertain observations exist in many practical applications. For example, in the wireless sensor networks, the observation data loss may exist in the target tracking such that the observation uncertainty will be in the measurement equation. As for the systems with uncertain observations, many scholars have been undertaking study. Nahi[2] has derived the linear minimum variance filtering algorithm where the noise are mutually independent. The steady state estimation error variance is further obtained by applying the unbiased estimation and the prescription methods in [3]. In [4], the adjacent time recursive filter and predictor are presented for systems with uncertain observation by using the linear minimum mean square error criterion. At present, most of

H. Tan (Ed.): Knowledge Discovery and Data Mining, AISC 135, pp. 785–791.
springerlink.com

the results are only for the single sensor systems. While the results for multi-sensor systems with uncertain observations are few.

Since the multiple sensors can provide the abundant information in time and space, so the research for the multi-sensor information fusion estimation problem has an important theoretical significance. There are two main methods presently for the multi-sensor fusion estimation problems: the centralized fusion and the distributed fusion. The centralized estimation has the global optimality, and the distributed estimation is often sub-optimal. Thus the centralized estimation usually has better accuracy than the distributed estimation.

In this paper, the white noise estimation problem is investigated for the stochastic discrete-time linear system with multiple sensors having uncertain observations where each sensor may have different observation data loss rate. In the case of correlated noises, the multi-sensor centralized optimal input white noise estimators in the linear minimum variance sense are given by applying Kalman filtering method, which has the global optimality.

2 Problem Formulation

Consider the multi-sensor stochastic system with uncertain observations:

$$x(t+1) = \Phi x(t) + \Gamma w(t) \tag{1}$$

$$y_i(t) = u_i(t) H_i x(t) + v_i(t) \text{ , } i = 1, 2, \cdots, L \tag{2}$$

where $x(t) \in R^n$ is the state, $y_i(t) \in R^{m_i}$ are the observations, $\Phi \in R^{n \times n}$, $\Gamma \in R^{n \times r}$ and $H_i \in R^{m_i \times n}$ are constant matrices, the subscript i denotes the ith sensor, L denotes the number of sensors. $u_i(t)$ is the Bernoulli distributed random variables which is uncorrelated with other random variables and satisfying $\Pr ob\{u_i(t) = 1\} = \alpha_i$, $\Pr ob\{u_i(t) = 0\} = 1 - \alpha_i$. $w(t)$ can be described by the following Bernoulli-Gaussian white noise as $w(t) = b(t) g(t)$, where $b(t)$ is the white Bernoulli distribution noise whose value is 1 or 0, and the probability is $\Pr ob\{b(t) = 1\} = \lambda$, $\Pr ob\{b(t) = 0\} = 1 - \lambda$, $0 < \lambda < 1$. $g(t)$ is the zero-mean Gaussian white noise with covariance σ_g^2 which is independent of $b(t)$. Obviously, the mean of $w(t)$ is zero and the covariance is $\sigma_w^2 = \lambda \sigma_g^2$.

Assumption 1. The input noise $w(t)$ and the measurement noise $v_i(t)$ satisfy the statistical characteristics

$$\mathrm{E}\left\{ \begin{bmatrix} w(t) \\ v_i(t) \end{bmatrix} [w^{\mathrm{T}}(k) \quad v_j^{\mathrm{T}}(k)] = \begin{bmatrix} Q_w & S_j \\ S_i^{\mathrm{T}} & Q_{vij} \end{bmatrix} \delta_{tk} \right., \begin{bmatrix} Q_w & S_i \\ S_i^T & Q_{vi} \end{bmatrix} > 0 \tag{3}$$

where $Q_{vii} = Q_{vi}$, E is the symbol of mean, T is the symbol of transpose, and δ_{tk} is the Kronecker delta function.

Assumption 2. The initial value $x(0)$ is uncorrelated with $u_i(t)$, $w(t)$, $v_i(t)$, and

$$\mathrm{E}[x(0)] = \mu_0,\quad \mathrm{E}\{[x(0)-\mu_0][x(0)-\mu_0]^T\} = P_0 .$$

Our aim is to find the centralized multi-sensor information fusion estimators $\hat{w}(t \mid t+N)$ in the linear minimum variance sense for input white noise $w(t)$ based on the measurement $(y_i(t)\cdots y_i(1))$, $i=1,2,\cdots,L$ which are called as filter, smoother and predictor as for N=0, N>0 and N<0, respectively.

3 Centralized White Noise Estimators

The augmented measurement equation can be obtained from each sensor observation equation. Thus the measurement equation of the original system can be transformed into the equivalent model

$$Y_c(t) = U_c(t) H_c x(t) + V_c(t) \tag{4}$$

where

$$Y_c(t) = \begin{bmatrix} y_1(t) \\ y_2(t) \\ \vdots \\ y_L(t) \end{bmatrix},\quad H_c = \begin{bmatrix} H_1 \\ H_2 \\ \vdots \\ H_L \end{bmatrix},\quad V_c(t) = \begin{bmatrix} v_1(t) \\ v_2(t) \\ \vdots \\ v_L(t) \end{bmatrix},$$

$$U_c(t) = \begin{bmatrix} u_1(t) I_{m_1} & & & \\ & u_2(t) I_{m_2} & & \\ & & \ddots & \\ & & & u_L(t) I_{m_L} \end{bmatrix} \tag{5}$$

I_{mi} represent the identify matrices with suitable dimensions. The covariance matrix of the augmented measurement white noise $V_c(t)$ is

$$Q_{v_c} = \begin{bmatrix} Q_{v1} & Q_{v12} & \cdots & Q_{v1L} \\ Q_{v21} & Q_{v2} & \cdots & \vdots \\ \vdots & \vdots & \ddots & \vdots \\ Q_{vL1} & \cdots & \cdots & Q_{vL} \end{bmatrix} \tag{6}$$

The correlation matrix between $V_c(t)$ and $w(t)$ is

$$S_c = \mathrm{E}[w(t) V_c^{\mathrm{T}}(t)] = \begin{bmatrix} S_1 & S_2 & \cdots & S_L \end{bmatrix} \tag{7}$$

The mean of augmented matrix $U_c(t)$ is

$$\alpha_c = \mathrm{E}[U_c(t)] = \begin{bmatrix} \alpha_1 I_{m1} & & & \\ & \alpha_2 I_{m2} & & \\ & & \ddots & \\ & & & \alpha_L I_{mL} \end{bmatrix} \tag{8}$$

Theorem 1: For stochastic systems (1) and (4) with uncertain observations satisfying Assumptions 1 and 2, the centralized optimal input white noise estimators are given as follows

$$\hat{w}_c(t \mid t+N) = 0, (N < 0) \tag{9}$$

$$\hat{w}_c(t \mid t+N) = \sum_{j=0}^{N} M_c(t \mid t+j)\varepsilon_c(t+j), (N \geq 0) \tag{10}$$

where the gain matrices can be calculated by the following equations

$$M_c(t \mid t) = S_c Q_{\varepsilon_c}^{-1}(t) \tag{11}$$

$$M_c(t \mid t+j) = D_{w_c}(t)\Psi_{pc}^{\mathrm{T}}(t+j, t+1)H_c^{\mathrm{T}}\alpha_c^{\mathrm{T}}Q_{\varepsilon_c}^{-1}(t+j) \tag{12}$$

The error covariance matrix of the input white noise is

$$P_c^w(t \mid t+N) = Q_w - \sum_{j=0}^{N} M_c(t \mid t+j)Q_{\varepsilon_c}(t+j)M_c^{\mathrm{T}}(t \mid t+j), (N \geq 0) \tag{13}$$

where

$$D_{w_c}(t) = Q_w \Gamma^{\mathrm{T}} - S_c K_{pc}^{\mathrm{T}}(t), \quad \Psi_{pc}^{\mathrm{T}}(t+j, t+k) = \prod_{k=1}^{j}[\Phi - K_{p_c}(t+j-1)\alpha_c H_c]^{\mathrm{T}}, \quad \Psi_{pc}(t,t) = I_n \tag{14}$$

Proof: According to the projection theory [6], it can be known that (9) and (11) are true, where the smoothing gain $M_{w_c}(t \mid t+N)$, $(N \geq 0)$ is given by

$$M_{w_c}(t \mid t+N) = \mathrm{E}[w(t)\varepsilon_c^{\mathrm{T}}(t+N)]Q_{\varepsilon_c}^{-1}(t+N) \tag{15}$$

From the definition of innovation, we have

$$\varepsilon_c(t+N) = [U_c(t+N) - \alpha_c]H_c x(t+N) + V_c(t+N) + \alpha_c H_c \tilde{x}_c(t+N \mid t+N-1) \tag{16}$$

$$\tilde{x}_c(t+1 \mid t) = x(t+1) - \hat{x}_c(t+1 \mid t) = \Psi_{pc}(t)x_i(t \mid t-1) + \Gamma w(t) - K_{pc}(t)V_c(t) - (U_c(t) - \alpha_c)K_{pc}(t)H_c x(t) \tag{17}$$

The N-step prediction error covariance matrix can be obtained by iteration

$$\tilde{x}_c(t+N\mid t+N-1)=\Psi_{p_c}(t+N,t)\tilde{x}_c(t\mid t-1)+$$
$$\sum_{k=1}^{N}\Psi_{p_c}(t+N,t+k)[\Gamma w(t+k-1)-K_{pc}(t+k-1)V_c(t+k-1)$$
$$-(U_c(t+k-1)-\alpha_c)K_{p_c}(t+k-1)H_c x(t+k-1)] \tag{18}$$

Where $\Psi_{p_c}(t+j,t+k)$ is defined by (14). Substituting (18) into (16), we have

$$\varepsilon_c(t+N)=(U_c(t+N)-\alpha_c)H_c x(t+N)+V_c(t+N)+$$
$$\alpha_c H_c\{\Psi_{pc}(t+N,t)\tilde{x}_c(t\mid t-1)+$$
$$\sum_{k=1}^{N}\Psi_{pc}(t+N,t+k)[\Gamma w(t+k-1)-K_{pc}(t+k-1)V_c(t+k-1)$$
$$-(U_c(t+k-1)-\alpha_c)K_{pc}(t+k-1)H_c x(t+k-1)] \tag{19}$$

From (10), we can obtain the centralized estimation error equation of the white noise as

$$\tilde{w}_c(t\mid t+N)=w(t)-\sum_{j=0}^{N}M_c(t\mid t+j)\varepsilon_c(t+j) \tag{20}$$

The error covariance matrix can be obtained from (20) as

$$P_c^w(t\mid t+N)=\mathrm{E}\{[w(t)-\hat{w}_c(t\mid t+N)][w(t)-\hat{w}_c(t\mid t+N)]^T\}$$
$$=Q_w(t)+\sum_{j=0}^{N}M_c(t\mid t+j)\varepsilon_c(t+j)M_c^T(t\mid t+j)-$$
$$\sum_{j=0}^{N}\mathrm{E}[w(t)\varepsilon_c^T(t+j)]M_c^T(t\mid t+j)-$$
$$\sum_{j=0}^{N}M_c(t\mid t+j)\mathrm{E}[\varepsilon_c(t+j)w^T(t)] \tag{21}$$

By the definition of (15), we have

$$\mathrm{E}[w(t)\varepsilon_c^T(t+N)]=M_{wc}(t\mid t+N)Q_{\varepsilon c}(t+N) \tag{22}$$

Substituting (22) into (21) yields (13).

4 Simulation Research

Consider the discrete-time linear time-varying stochastic system with three sensors of uncertain observations

$$x(t+1)=\begin{bmatrix}1 & 0\\ 0.1+\sin(2\pi t/N) & 0\end{bmatrix}x(t)+\begin{bmatrix}\cos(2\pi t/N) & 0.5\\ -0.4 & 1\end{bmatrix}w(t)$$

$$y_i(t)=u_i(t)H_i(t)x(t)+v_i(t)$$

$$v_i(t)=c_iw(t)+\xi_i(t),\quad i=1,2,3$$

where $x(t)$ is the state, $y_i(t)$, $i=1,2,3$ are the observation signals of three sensors, time-varying measurement matrices are $H_1(t)=\begin{bmatrix}1 & 0.8\cos(2\pi t/N)\\ -0.8 & 0.5\end{bmatrix}$, $H_2(t)=\begin{bmatrix}1 & 0.5\cos(2\pi t/N)\\ 0 & 1\end{bmatrix}$, $H_3(t)=\begin{bmatrix}0.5 & 0\\ 0.2 & 0.5\sin(2\pi t/N)\end{bmatrix}$. $u_i(t)$, $i=1,2,3$ are Bernoulli distributed white noises which satisfying $\Pr ob(u_i(t)=1)=\alpha_i$, $\Pr ob(u_i(t)=0)=1-\alpha_i$, $i=1,2,3$. $v_i(t)$, $i=1,2,3$ are measurement noises of three sensors, respectively, which are correlated process noises $w(t)=[\beta(t)\quad \delta(t)]^T$ satisfying Bernoulli-Gaussian distribution and $w(t)=b(t)g(t)$, where $b(t)$ is Bernoulli distributed white noise with $\Pr ob(b(t)=1)=\lambda$, $\Pr ob(b(t)=0)=1-\lambda$. $g(t)$ is the zero-mean Gaussian white noise with the variance matrix Q_g. Correlation coefficients c_i, $i=1,2,3$ are constants. $\xi_i(t)$, $i=1,2,3$ are Gaussian white noise with mean zero and covariance $Q_{\xi i}$ which are independent of $b(t)$ and $g(t)$. Our aim is to find the optimal information fusion one-step smoother $\hat{w}_o(t\,|\,t+1)$ of the input white noise $w(t)$.

In the simulation, we choose $Q_g=4I_2$, $Q_{\xi 1}=0.01I_2$, $Q_{\xi 2}=0.02I_2$, $Q_{\xi 3}=0.03I_2$, $c_1=0.9$, $c_2=0.7$, $c_3=0.6$, $\lambda=0.3$, $\alpha_1=0.3$, $\alpha_2=0.5$, $\alpha_3=0.6$. The initial values are $x(0)=0$, $P_0=0.1I_2$, $D(0)=0.1I_2$, and the number of the sampled data is 100.

Applying Theorem 1, we can obtain the multi-sensor centralized white noise estimators and the error covariance matrices. The centralized fusion one-step smoother of the white noise is shown in figure 1, where the ends of the line are the true value of white noise, and the asterisk are the estimator. It can be seen that the centralized smoother has the better estimation performance. Figure 2 is the error accuracy comparison of the one-step smoother for the white noise. We can see that the centralized fusion smoother has better accuracy than any local smoother based on single sensor.

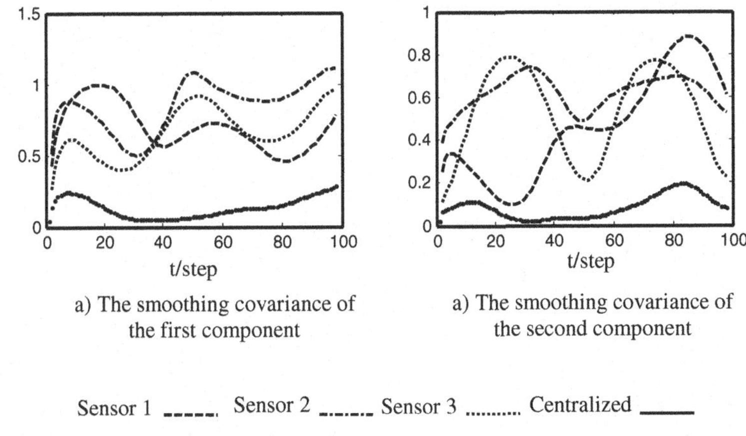

a) The smoothing covariance of the first component

a) The smoothing covariance of the second component

Sensor 1 -----. Sensor 2 _._._._ Sensor 3 .... Centralized ____

Fig. 2. The error precision comparison of one-step smoother for white noise

5 Conclusion

Due to the influence of environment and the different performance of different sensors, the observation data loss rate may be different for each sensor. In this paper, we study the centralized fusion white noise estimation problem for discrete-time linear systems with multiple sensors of uncertain observations. The centralized fusion white noise estimation algorithms in the linear minimum variance sense are derived based on the projection property. Compared to the local estimator based on single sensor, it has better estimation accuracy. Because the measurement augmentation method is adopted, it has the global optimality.

References

1. Mendel, J.M.: White-noise estimators for seismic data processing in oil exploration. IEEE Trans. Automatic Control 25(5), 694–706 (1977)
2. Nahi, N.E.: Optimal recursive estimation with uncertain observation. IEEE Trans. Inform. Theory 15, 457–462 (1969)
3. Ediwin, W.N., Yaz, E.: Recursive estimator for linear and nonlinear systems with uncertain observations. Signal Prosessing 62, 215–228 (1997)
4. Hermoso, A., Linares, J.: Linear estimation for discrete-time systems in the presence of time-correlated disturbances and uncertain observations. IEEE Trans. Automat. Contr. 39(8), 1636–1638 (1994)
5. Willner, D., Chang, C.B., Dumm, K.P.: Kalman filter algorithm for a multisensory system. In: Proceedings of IEEE Conference on Decision and Control, Clearwater, Florida, pp. 570–574 (December 1976)
6. Sun, S.L., Deng, Z.L.: Multi-sensor optimal information fusion Kalman filter. Automatica 40(6), 1017–1023 (2004)

Correction to: Knowledge Discovery and Data Mining

Honghua Tan

Correction to:
H. Tan (Ed.), *Knowledge Discovery and Data Mining*, Advances in Intelligent and Soft Computing 135, https://doi.org/10.1007/978-3-642-27708-5

In the original version of the book, the Preface content has been completely removed. The erratum book has been updated with the changes.

The updated version of the book can be found at
https://doi.org/10.1007/978-3-642-27708-5

H. Tan (Ed.): Knowledge Discovery, and Data Mining, AISC 135, p. C1,
https://doi.org/10.1007/978-3-642-27708-5_109

Author Index

Printed in the United States
By Bookmasters